Volumes already published

Volume 1: Trees I (1986)
Volume 2: Crops I (1986)
Volume 3: Potato (1987)
Volume 4: Medicinal and Aromatic Plants I (1988)
Volume 5: Trees II (1989)
Volume 6: Crops II (1988)
Volume 7: Medicinal and Aromatic Plants II (1989)
Volume 8: Plant Protoplasts and Genetic Engineering I (1989)
Volume 9: Plant Protoplasts and Genetic Engineering II (1989)
Volume 10: Legumes and Oilseed Crops I (1990)
Volume 11: Somaclonal Variation in Crop Improvement I (1990)
Volume 12: Haploids in Crop Improvement I (1990)
Volume 13: Wheat (1990)
Volume 14: Rice (1991)
Volume 15: Medicinal and Aromatic Plants III (1991)
Volume 16: Trees III (1991)
Volume 17: High-Tech and Micropropagation I (1991)
Volume 18: High-Tech and Micropropagation II (1992)
Volume 19: High-Tech and Micropropagation III (1992)

Volumes in preparation

Volume 20: High-Tech and Micropropagation IV (1992)
Volume 21: Medicinal and Aromatic Plants IV (1993)
Volume 22: Plant Protoplasts and Genetic Engineering III
Volume 23: Plant Protoplasts and Genetic Engineering IV
Volume 24: Medicinal and Aromatic Plants V
Volume 25: Maize

Biotechnology in Agriculture and Forestry 19

High-Tech and Micropropagation III

Edited by Y.P.S. Bajaj

With 231 Figures

Springer-Verlag Berlin Heidelberg GmbH

Professor Dr. Y.P.S. BAJAJ
A-137
New Friends Colony
New Delhi 110065, India

ISBN 978-3-642-08104-0

Library of Congress Cataloging-in-Publication Data. (Revised for volume 3). High-tech and micropropagation. (Biotechnology in agriculture and forestry ; 17) Includes bibliographical references and index. 1. Plant propagation—In vitro. I. Bajaj, Y.P.S., 1936– . SB123.6.H54 1991 631.5′3 91-22304
ISBN 978-3-642-08104-0 ISBN 978-3-662-07770-2 (eBook)
DOI 10.1007/978-3-662-07770-2

Originally published by Springer-Verlag Berlin Heidelberg New York in 1992
Softcover reprint of the hardcover 1st edition 1992

Production Editor: Ingrid Samide, Heidelberg
Typesetting: Macmillan India Limited, Bangalore
31/3145-5 4 3 2 1 0 – Printed on acid-free paper

Dedicated to my Father
Late S. Harnam Singh Bajaj

Preface

Micropropagation of plants is a multibillion dollar industry being practiced in hundreds of small and large nurseries and commercial laboratories throughout the world. At the present, it is the only component of plant biotechnology which has been commercially exploited on such a large scale, especially for the production of ornamentals. Now micropropagation of trees and medicinal plants is assuming great importance. With recent progress made in the propagation of fruit and forest trees, and the immediate need for afforestation and planting of orchards, propagules and plantlets are required quickly and in large numbers. Although the basic methodology of in vitro culture of plant tissues has been fairly mechanized over the last 25 years, conventional micropropagation methods are relatively slow, time-consuming, expensive, and labor-intensive; up to 50–80% of the cost of plant production is attributed to labor. In order to reduce the cost and to increase the efficiency and scale up of plant production, automation is being incorporated. Robots, microcomputers, and other automated systems of artificial intelligence have been employed to surface-sterilize tissues, dispense media, and transfer plants to pots and field. The automatic control of environments in greenhouses to protect cultivation is another area of tremendous commercial importance.

The use of artificial intelligence, image analysis, video cameras, bioreactors, mistifiers, automatic production lines, synthetic seed, storage and cryopreservation of elite genetic stocks, and the computerization of information management will facilitate micropropagation and protected cultivation. Judging from the enthusiasm and interest of biotechnology enterprises, it is envisaged that it will not be long before we see the routine jobs being done by robots for industrial micropropagation.

Taking the above-mentioned points into consideration, it appeared necessary to review the literature and to present the state of the art on micropropagation of plants. Thus, 118 chapters contributed by experts from 30 countries are compiled in a series composed of the following four parts:

High-Tech and Micropropagation I comprises 29 chapters on various aspects of micropropagation and automation, i.e., setting up a commercial laboratory, meristem culture, somatic embryogenesis, graft incompatibility, vitrification, acclimatization, nature and origin of explants, problems with explant exudates, disposable culture vessels, induction of rooting, artificial substrates, rockwool, autotrophic micropropagation, storage and cryopreservation of cultures, quiscent somatic embryos, synthetic seed, automated systems, artificial intelligence, use of robots in transplantation, information management, bioreactors, and computerized greenhouses.

High-Tech and Micropropagation II comprises 26 chapters on American sweetgum, black locust, birches, cedar, cocoa, larches, Persian walnut, poplars, *Quercus*, sandalwood, *Sorbus*, apple, pear, peach, plum, cherry, *Citrus*, raspberry, strawberry, grapevine, kiwi, Japanese persimmon, papaya, pineapple, date palm, and olive tree.

High-Tech and Micropropagation III comprises 34 chapters on vegetable and tuber crops, medicinal and aromatic plants, fiber and oilseed crops, and grasses. The specific plants dealt with include: asparagus, lettuce, horseradish, cucumber, potato, sweet potato, artichoke, white yam, cardamom, fennel, celery, thyme, leek, mentha, turmeric, lavender, agave, yucca, cotton, jute, sunflower, ryegrass, zoysiagrass, and various species of *Aconitum, Artemisia, Camelia, Centaurium, Digitalis, Dioscorea, Glehnia, Parthenium,* and *Pinella.*

High-Tech and Micropropagation IV comprises 29 chapters on ornamental species of *Alstroemeria, Antirrhinum, Begonia, Chrysanthmum, Cornus, Euphorbia, Gardenia, Gladiolus, Hyacinthus, Impatiens, Iris, Lycoris, Nematanthus, Phalaenopsis, Paeonia, Pelargonium, Rhododendron, Ruscus, Saintpaulia, Senecio, Syringa,* orchids, cacti, roses, and Boston ferns.

These books will be of use to advanced students, research workers, and teachers in horticulture, forestry, tissue culture, and plant biotechnology in general, and also to individuals interested in commercial micropropagation.

New Delhi, June 1992

Professor Y. P. S. Bajaj
Series Editor

Contents

Section I Vegetables, Tubers, and Root Crops

I.5 Micropropagation of *Cucumis* spp.
M. Ziv (With 2 Figures)

I.6 Micropropagation of Potato (*Solanum tuberosum* L.)
J. H. Dodds, D. Silva-Rodriguez, and P. Tovar
(With 6 Figures)

I.7 Micropropagation of Sweet Potato (*Ipomoea batatas* L.)
R. P. Chée, J. R. Schultheis, and D. J. Cantliffe
(With 1 Figure)

I.8 Micropropagation of Artichoke (*Cynara scolymus*)
V. Rossi and G. de Paoli (With 8 Figures)

I.9 Micropropagation of White Yam
(*Dioscorea rotundata* Poir.)
S. Y. C. Ng (With 11 Figures)

I.10 Micropropagation of Cubio (*Tropaeolum tuberosum* R & P)
O. TORRES, M. PEREA-DALLOS, and T. J. FANDIÑO
(With 7 Figures)

Section II Medicinal and Aromatic Plants

II.1 Micropropagation of Cardamom
(*Elettaria cardamomum* Maton)
B. R. REGHUNATH and Y. P. S. BAJAJ (With 11 Figures)

II.2 Micropropagation of Fennel (*Foeniculum vulgare* Miller)
G. HUNAULT and J. DU-MANOIR (With 12 Figures)

II.3 Micropropagation of Celery (*Apium graveolens* var. *dulce*)
K. F. TOTH and M. L. LACY (With 3 Figures)

II.4 Micropropagation of Thyme (*Thymus vulgaris* L.)
M. FURMANOWA and O. OLSZOWSKA (With 9 Figures)

II.9 Micropropagation of *Agave* spp.
M. L. Robert, J. L. Herrera, J. L. Chan, and F. Contreras (With 8 Figures)

II.10 Micropropagation of Yucca (*Yucca glauca* Nutt.)
S. E. Bentz and H.-J. Talbott (With 3 Figures)

II.11 Micropropagation of Monkshood (*Aconitum* spp.)
R. L. Cervelli (With 7 Figures)

II.12 Micropropagation of French Tarragon (*Artemisia dracunculus* L. var. *sativa*)
S. L. Kitto and W. A. Mackay (With 7 Figures)

II.13 Micropropagation of *Camellia* spp.
A. M. Vieitez, M. L. Vieitez, A. Ballester, and E. Vieitez (With 8 Figures)

Creemers-Molenaar, J., Barenbrug Holland BV, Stationsstraat 40, 6678 AC Oosterhout, The Netherlands. (Correspondence address: Foundation for Agricultural Plant Breeding, P.O. Box 117, 6700 AC Wageningen, The Netherlands)

Desjardins, Y., Centre de Recherche en Horticulture, Département de Phytologie, Faculté des Sciences de l'Agriculture et de l'Alimentation, Université Laval, Québec, QC, Canada G1K 7P4

Diettrich, B., Institute of Pharmaceutical Biology, Martin-Luther-University Halle-Wittenberg, Weinbergweg 15, O-4050 Halle, FRG

Dodds, J. H., International Potato Center, P.O. Box 5969, Lima, Peru

Fandiño, T. J., Lab. Cult. Tej. Veg., Depto. Biología, Facultad de Ciencias, Universidad Nacional de Colombia, Bogotá, A.A. 23227, Colombia

Fowlds, D. L., UN/FRD Research Unit for Plant Growth and Development, Department of Botany, University of Natal, P.O. Box 375, Pietermaritzburg 3200, Republic of South Africa

Furmanowa, M., Department of Biology and Pharmaceutical Botany, Medical Academy, Banacha 1, 02-097 Warsaw, Poland

Gill, M. S., Punjab Agricultural University, Ludhiana, India

Górecka, K., Research Institute of Vegetable Crops, 96-100 Skierniewice, Poland

Haque, M. M., Department of Botany, University of Dhaka, Dhaka-1000, Bangladesh

Hatano, K., Faculty of Pharmaceutical Sciences, Kyushu University, Higashiku, Fokuoka 812, Japan. (Present address: Cooperate R and D Plant Technology Institute. Ube Industries Ltd., 1978-5 Ogushi, Ubeshi, Yamaguchi 755, Japan)

Herrera, J.L., División de Biología Vegetal, Centro de Investigación Científica de Yucatán, A. C. Apdo. Postal 87 Cordemex, 97310 Mérida, Yucatán, México

Hiraoka, N., Niigata College of Pharmacy, 5-13-2 Kamishinei-cho, Niigata 950-21, Japan

Hoque, M. I., Department of Botany, University of Dhaka, Dhaka-1000, Bangladesh

Huang, F. H., Department of Horticulture and Forestry, University of Arkansas, Fayetteville, AR 72701, USA

Huizhu, Xu, Wuhan Institute of Botany, Academia Sinica, Wuhan 430074, China

Hunault, G., Université P. et M. Curie, Station de Biologie Végétale A. de Richelieu, Cherré, 72400 La Ferté-Bernard, France

ISLAM, A. S., Department of Botany, University of Dhaka, Dhaka-1000, Bangladesh. (Present address: Department of Botany, The University of Texas at Austin, Austin, Texas 78713-7640, USA)

KING, J. W., Department of Agronomy, University of Arkansas, Fayetteville, AR 72701, USA

KITTO, S. L., Delaware Agricultural Experiment Station, Department of Plant Science, College of Agricultural Sciences, University of Delaware, Newark, DE 19717-1303, USA

LACY, M. L., Department of Botany and Plant Pathology, Michigan State University, Plant Biology Building, East Lansing, MI 48824, USA

LEIKE, H., Institute of Breeding Research, Neuer Weg 22, O-4300 Quedlinburg, FRG. (Present address: Am Hackelbach 13, O-4300 Quedlinburg, FRG)

LUCKNER, M., Institute of Pharmaceutical Biology, Martin-Luther-University Halle-Wittenberg, Weinbergweg 15, O-4050 Halle, FRG

LUPI, M. C., Department of Agricultural Plant Biology, Genetics Section, Via Matteotti 1/B, 56100 Pisa, Italy

MACKAY, W. A., Department of Horticulture, University of Maryland, College Park, MD 20742, USA

MANOIR DU, J., Centre de Recherches Pernod-Ricard, Laboratoire d'Agronomie, 120, Av. du Maréchal Foch, 94015 Créteil Cédex, France

MINZHI QIAN, Wuhan Institute of Botany, Academia Sinica, Wuhan 430074, China

NG, S. Y. C., The International Institute of Tropical Agriculture (IITA), Oyo Road, P.M.B. 5320, Ibadan, Nigeria

NISHIOKA, I., Faculty of Pharmaceutical Sciences, Kyushu University, Higashiku, Fukuoka 812, Japan

OLSZOWSKA, O., Department of Biology and Pharmaceutical Botany, Medical Academy, Banacha 1, 02-097 Warsaw, Poland

OYANAGI, M., Niigata College of Pharmacy, 5-13-2 Kamishinei-cho, Niigata 950-21, Japan

PANIZZA, M., Department of Biologia delle Piante Agrarie, University of Pisa, Viale delle Piagge, 23, 56124 Pisa, Italy

PAOLI DE, G., Centrale Ortofrutticola-Vitrocoop, Via Masiera I, 1191, 47023 Cesena, Italy

PEFFLEY, E. B., Department of Agronomy, Horticulture and Entomology, Texas Tech University, Lubbock, TX 79409-2134, USA

Perea-Dallos, M., Lab. Cult. Tej. Veg., Depto. Biología, Facultad de Ciencias, Universidad Nacional de Colombia, Bogotá, A.A. 23227, Colombia

Pink, D. A. C., Horticulture Research International, Wellesbourne, Warwick, CV35 9EF, UK

Reghunath, B. R., USDA Project on Tissue Culture, College of Agriculture, Vellayani P.O., Trivandrum-695 522, Kerala, India

Robert, M. L., División de Biología Vegetal, Centro de Investigación Científica de Yucatán, A. C. Apdo. Postal 87 Cordemex, 97310 Mérida Yucatán, México

Rossi, V., Centrale Ortofrutticola-Vitrocoop, Via Masiera I, 1191, 47023 Cesena, Italy

Schultheis, J. R., Vegetable Crops Department, University of Florida, Gainesville, FL 32611, USA

Seraj, Z. I., Department of Botany, University of Dhaka, Dhaka-1000, Bangladesh

Shoyama, Y., Faculty of Pharmaceutical Sciences, Kyushu University, Higashiku, Fukuoka 812, Japan

Silva-Rodriguez, D., International Potato Center, P.O. Box 5969, Lima, Peru

Staden van, J., UN/FRD Research Unit for Plant Growth and Development, Department of Botany, University of Natal, P.O. Box 375, Pietermaritzburg 3200, Republic of South Africa

Sugaya, A., Department of Pharmacognosy, Faculty of Pharmaceutical Sciences, Josai University, Sakado-shi, Saitama-ken, 350-02, Japan

Talbott, H.-J., USDA-ARS, National Plant Germplasm Quarantine Laboratory, 11601 Old Pond Drive, Glenn Dale, MD 20769, USA

Thompson, L. F., Department of Agronomy, University of Arkansas, Fayetteville, AR 72701, USA

Tognoni, F., Department of Biologia delle Piante Agrarie, University of Pisa, Viale delle Piagge, 23, 56124 Pisa, Italy

Torres, O., Lab. Cult. Tej. Veg., Depto, Biología, Facultad de Ciencias, Universidad Nacional de Colombia, Bogotá, A.A. 23227, Colombia

Toth, K. F., Department of Botany and Plant Pathology, Michigan State University, Plant Biology Building, East Lansing, MI 48824, USA

Tovar, P., International Potato Center, P.O. Box 5969, Lima, Peru

Vágnerová, H., Institute of Experimental Botany, Czechoslovak Academy of Science, Ke dvoru 15, 166 30 Praha 6, Czechoslovakia

VIEITEZ, A. M., Plant Physiology, CSIC, Apartado 122, 15080 Santiago de Compostela, Spain

VIEITEZ, E., Department of Biology, Faculty of Biology, University Santiago de Compostela, Spain

VIEITEZ, M. L., Department of Biology, Faculty of Biology, University Santiago de Compostela, Spain

ZIV, M., Department of Agricultural Botany and the Warburg Center for Biotechnology in Agriculture, The Hebrew University of Jerusalem, P.O. Box 12, Rehovot 76100, Israel

Section I Vegetables, Tubers, and Root Crops

I.1 Micropropagation of Hybrid Lines in Vegetable Breeding

H. Leike[1,2] and W. Bauch[1,3]

1 Introduction

Vegetables include more than 100 species belonging to many botanical families. Most of them are of local importance and used only in tropical or subtropical zones, but about 25–30 vegetables, mostly originating in the warmer temperate zones, are of greater economic importance and are found worldwide (Table 1). They represent nine botanical families. Most of them are seed-propagated annual or biennial plants. Some species are herbaceous perennial plants, and in this case generally propagated vegetatively. The special use of biotechnological methods for breeding depends on the different species and is strongly influenced by the life cycle of the plant.

2 Micropropagation of Vegetables

2.1 Development of Methods

In vitro techniques in vegetables are less developed at present but the situation has changed in the past few years (Doré 1987). Many usable procedures are known for the Brassicaceae, while in the families Asteraceae, Apiaceae, some Liliaceae (asparagus, garlic), some Solanaceae (tomato), and some Cucurbitaceae successful propagation methods were carried out, problems remain for most grain legumes (except *Pisum*), red and green pepper (Fari 1986), and the Chenopodiaceae. No usable propagation method has yet been found for spinach (Nesković and Culafić 1988).

The aim of clonal vegetative propagation in breeding is to produce copies of an original plant selected for its desirable traits. In vitro techniques are clearly capable of bringing about enormous improvements in the efficiency of propagation. In general, the use of in vitro techniques leads to the rationalization of the entire breeding cycle, especially in hybrid lines in F_1 breeding. The propagules must maintain strict genetic conformity and stability over thousands of individuals. To fulfill this requirement the most reliable micropropagation method is to use existing lateral buds or shoot tips as initial explants. In Brussels sprout, the

[1] Institute of Breeding Research, Neuer Weg 22, 0-4300 Quedlinburg, FRG
[2] Present address: Hackelbach 13, 0-4300 Quedlinburg, FRG
[3] Gemüsezuchtstation, Nienhäger Chaussee, 0-2560 Bad Doberan, FRG

Biotechnology in Agriculture and Forestry, Vol. 19
High-Tech and Micropropagation III (ed. by Y.P.S. Bajaj)

Table 1. Selected vegetable crops of worldwide importance

Botanical family	Parts of plants used for human consumption Leaves, stems, buds	Roots	Bulbs, tubers	Inflorescences	Fruits, seeds
Brassicaceae (Cruciferae)	Cabbage Brussels sprouts Kale (*Brassica oleracea* ssp.) Chinese cabbage (*Brassica chinensis*, *B. pekinensis*)	Turnip (*Brassica rapa*) Radish (*Raphanus sativus*)		Cauliflower Broccoli (*Brassica oleracea* ssp.)	
Liliaceae	Onions (*Allium* spp.) Leek (*Allium porrum*) Asparagus (*Asparagus officinalis*)		Onion (*Allium* spp.) Garlic (*Allium sativum*) Leek (*Allium porrum*)		
Apiaceae (Umbelliferae)	Celery (*Apium graveolens*)	Carrot (*Daucus carota*) Celery (*Apium graveolens*) Parsnip (*Pastinaca sativa*)			
Solanaceae					Tomato (*Lycopersicon lycopersicum*) Red, green pepper (*Capsicum annuum*) Aubergine (*Solanum melongena*)

Asteraceae (Compositae)	Lettuce (*Lactuca sativa*) Chicory (*Cichorium intybus*) Endive (*Cichorium endivia*)		Artichoke (*Cynara scolymus*)	
Fabaceae (Leguminosae)				Pea (*Pisum sativum*) Chick pea (*Cicer arietinum*) Different beans (*Phaseolus* spp.) Soybean (*Glycine max*) Broad bean (*Vicia faba*)
Chenopodiaceae	Spinach (*Spinacia oleracea*) Chard (*Beta vulgaris*)	Red beet Beetroot (*Beta vulgaris*)		
Cucurbitaceae				Cucumber (*Cucumis sativus*) Melon (*Cucumis melo*) Pumpkin, squash (*Cucurbita pepo*)
Polygonaceae	Rhubarb (*Rheum rhaponticum*)			

numerous bud meristems of the sprout heads permit quick and simple cloning of the selected plants (Clare and Collin 1974). A similar technique, using lateral shoot tips, is found to be effective in asparagus (Murashige et al. 1972). Different approaches have to be adopted in celery, cabbage, cauliflower, broccoli, onion, witloof chicory, horseradish, and other vegetables. The use of small segments of young inflorescences, inflorescent axes, shoots, roots, and other plant organs, the induction of short callus phases, and the quick formation of several adventitious shoots in the primary callus have also proved to be useful methods leading within a few months to many thousands of plants usable in breeding work. By using inflorescence explants, also preexisting meristems may be included.

There are many review papers summarizing the recent development in biotechnology in vegetables, including micropropagation methods on cauliflower (Grout 1988), cabbage (Leike 1988), Brussels sprout (Walkey and Pink 1988), cucumber (Malepszy 1988), chicory (Schoofs and De Langhe 1988), and lettuce (Alconero 1988).

2.2 Practical Use

The direct use of in vitro-cloned plants in the field or greenhouse for production of vegetables is limited for economic reasons. It is restricted to those vegetatively propagated vegetable crops in which micropropagation results in high yields and improved homogeneity, for example, globe artichoke (*Cynara scolymus* L.) in Italy (Ancora 1985), garlic (*Allium sativum* L.) in Czechoslovakia (Novák 1983), triploid genotype of watermelon [*Citrullus lanatus* (Thurnb.) Mansf.] in Romania (Anghel and Rosu 1985), rhubarb (*Rheum rhaponticum* L.) in Belgium (Roggemans and Boxus 1988), and horseradish (*Armoracia rusticana* G. M. Sch.) in Germany.

Any vegetable crops reproducible by seeds can be propagated in this way less expensively than by in vitro cloning. While the price per in vitro-propagated plantlet in most cases may exceed 25 US cent generally in agriculture crops the single seed can be produced for less than 0.01 US cent.

Micropropagation has been increasingly introduced in vegetable breeding and seed production. In general, during breeding in vitro cloning is not used exclusively but combined with generative reproduction. As a rule, the micropropagated plantlets show high uniformity. By repeated micropropagation the breeder is able to rapidly create uniform plant stocks. However, after repeated change between vegetative and generative reproduction, the usual procedure during the course of breeding, homogeneity again decreases, although some authors have stated the contrary.

In vegetable breeding and seed production, micropropagated plants are at present used as follows (Table 2):

— In vegetatively propagated vegetable crops, for instance horseradish, garlic, globe artichoke, and rhubarb, highly productive plants with desirable traits were selected and micropropagated on a large scale. In this way morphologically very uniform plant stocks originate with high yields in comparison to the original population, if we assume that the higher productivity of the selected plants has been genotypically conditioned and not only modified.

Table 2. Use of in vitro-cloned plants in vegetable breeding and seed production

Micropropagated objects		Kind of vegetable	Aim of micropropagation
1	Vegetatively propagated vegetable crops	Horseradish Garlic Globe artichoke Rhubarb	— Uniformly productive plant stocks — Elimination of virus infection during breeding
2	Generatively propagated vegetables, cross-pollinators	Cabbage Brussels sprout Cauliflower Broccoli Asparagus Celery Chicory Carrot Onion, etc.	— Propagation and maintenance of selected genotypes with desirable traits — Subsequent reproduction by seeds again
2.1	Dioecious hybrid lines	Asparagus	— Large-scale micropropagation of a selected single female and a male plant with checked high combination effectivity — F_1 hybrid seed production by using the female and male clones
2.2	Self-incompatible (si) hybrid lines	White cabbage Red cabbage Brussels sprout Chicory	— Vegetative propagation of desirable sister lines, which cannot be generatively propagated or only with little effectivity — Subsequent production of combined sister lines generatively
2.3	Self-compatible (sc) hybrid lines and maintainers	White cabbage Red cabbage Brussels sprout Cauliflower Chicory Celery Carrot Onion	— Vegetative propagation of selected single plants — Subsequent generative reproduction of the clones (kind of inbreeding, which causes a great number of progenies)
2.4	Male-sterile hybrid lines	Celery	— Permanent vegetative propagation of male sterile hybrid lines. Search of maintainers will be avoided — Subsequent production of F_1 seed production with the male-sterile plants by using them as seed plants
3.1	Late varieties of annual vegetable crops	Cauliflower Iceberg lettuce	— Return to the juvenile stage, late varieties do not reach the stage of formation of mature seeds till autumn. Therefore, selection during stage of marketable crops, cutting of usable explants, micropropagation — In the following year generative reproduction of the clones
3.2	Seed plants of cross-pollinator of vegetables	White cabbage Red cabbage Brussels sprout Onion Carrot	— Aim of the breeders: improvement of seed plant morphology — Return to juvenile stage. Selection of seed plants during the stage of seed formation, cutting of usable explants, micropropagation of the selected plants — Subsequent generative reproduction

Micropropagation is also used in clonal hybrids of asparagus to obtain pathogen- and virus-free plant stocks. After transferring the clones to the greenhouse or field, reinfection cannot be prevented. Nevertheless, during the first years perennial vegetables grow better and attain a higher biomass by using virus-free plantlets.

- In any generatively reproducible vegetable crop, especially in cross-pollinators, it is possible to multiply selected single plants with desirable traits by in vitro techniques and maintain them for breeding purposes in subsequent years, possibly in combination with some method of long-term storage. Later, variants can be checked in generative reproduction.
- In vitro cloning is of great importance in F_1 hybrid breeding of vegetables. In the dioecious species asparagus, in general, it is impossible to produce hybrid lines without using in vitro propagation (Sect. 2.3.1). F_1 hybrid breeding in cabbage usually works on the principle of self-incompatibility. After bud pollination the self-incompatible inbred lines often lose their reproducibility (Bauch 1986). If this occurs in isogenic sister lines they can be effectively micropropagated (Sect. 2.3.2). The self-incompatible inbred lines in chicory are also cloned in vitro for breeding purposes and F_1 seed production (Theiler-Hedtrich and Badoux 1986).

Self-compatible lines are used in vegetable breeding in various ways, for instance: in white and red cabbage (*Brassica oleracea* L. var. *capitata* L.), in Brussels sprout (*B. oleraceae* var. *gemmifera* DC.), cauliflower (*B. oleracea* var. *botrytis* L.) as pollinator for modified single crosses (Bauch 1982), in chicory (*Cichorium intybus*) for inbreeding self-incompatible lines (Schoofs and De Langhe 1988), in onion (*Allium cepa* L.), and carrot (*Daucus carota* L.) to maintain the genetic-plasmatic male-sterile hybrid lines and as pollinator (Stein et al. 1985).

Micropropagation is also helpful during inbreeding to improve the homogeneity of the lines and through it the hybrid varieties in different ways:

- After selfing, the seed yield of the self-compatible single plant is usually very small and the number of progeny minimal. After cloning the selected plant and planting in a flowering group, many seeds can be obtained. The quantity of progeny ensures that testing can be carried out. The basis of selection is enlarged and the selected plants can be used for hybrid seed production immediately.
- In vegetable crops with male sterility and with a high reproduction rate, for example, celery (*Apium graveolens* L.), a simplification of the hybrid breeding system seems to be possible (Browers and Orton 1986). After a period of introgressive inbreeding, necessary to achieve homogeneous hybrids, the male-sterile lines were completely cloned in vitro. In this case the development of maintainers is unnecessary, and hybrid breeding thus requires less time. Additionally, the possibility that male-fertile plants may segregate in the female (male-sterile) line can be avoided. This idea is at present only theoretical. Commercial F_1 hybrid varieties developed in this way are unknown.
- Another effective use of micropropagation is to revert from a later stage of ontogenesis to a younger one, from senescence to the juvenile phase in the life cycle. This is particularly important for long-lived plants, e.g., trees and shrubs

(Jensen 1983), but this possibility also helps in special vegetable breeding programs, in both conventional and F_1 hybrid breeding.

In northern and southern climatic zones with a short vegetation period, some vegetables, for instance, late varieties of cauliflower and iceberg lettuce (*Lactuca sativa* L. var *capitata nidus jaggeri* Helm.), reach the marketable stage only in the late summer, but cannot form seeds able to germinate. In this case the breeder can select plants with the desired traits. Suitable explants cut from these plants can be transferred to the juvenile phase, micropropagated, and in the following year used as seed plants (Hall 1983).

If the breeder intends to change the morphology of seed plants in cross-pollinators, he can start the selection work during flowering and seed formation. At that time a separate generative reproduction of the selected positive genotypes has become impossible, but also in this case suitable explants of these senescent plants can be used to obtain a multiple repetition of this procedure, thus an improvement in seed plant growth can be expected.

2.3 Examples of the Use of Micropropagation in Breeding Special Vegetable Crops

2.3.1 Asparagus

1. Advantages of F_1 Hybrids. Internationally, the leading position is set by F_1 varieties, which in the last few years have dominated the market to an increasing extent. In combination with the locally sold cultivars, the special advantages of F_1 hybrid varieties are:

— considerably higher yields owing to the great number of spears per plant,
— better quality of spears due to the higher uniformity.

2. Micropropagation and Production of Rooted Plantlets. Asparagus has a very low rate of vegetative multiplication under classical horticultural conditions. On the other hand, breeders urgently need multiplication techniques for selected single plants to force the breeding process. Micropropagation methods in asparagus were published as long as 20 years ago, generally different modifications of shoot tip culture (Murashige et al. 1972; Hasegawa et al. 1973) were used.

Reuther (1978) used rapidly proliferating callus lines for shoot organogenesis and micropropagation techniques. High productive single male plants were selected and propagated, each plant in the field coming from the same origin. In this way, Reuther and Becker (1985) created new high-yielding male asparagus varieties.

Doré (1980) developed two techniques, shoot tip culture (meristem culture) and in vitro cutting. To establish sterile in vitro culture, generally meristem culture is used. The original explants are cut from the apex of young spears. Numerous meristems are located under the scales of the upper part of an asparagus spear. After sterilization the explants were cultivated on the medium of Murashige et al. (1972), and many of the preexisting lateral meristems grew and formed shoots.

The second technique, in vitro cutting, consists of cultivating basal nodes of young stems. When grown in vitro, successive sections of stems obtained from these nodes will lead to shoots of normal morphology and later to rooted plantlets.

With any technique used, the multiplication rate of asparagus shoots is high, but rooting and transplantation to the soil remain critical steps. We induced rooting by cultivating the young shoot tips on the medium of Linsmaier and Skoog (1965) with 0.2 mg/l kinetin and 5 mg/l IBA added. If we used explants with a length of 1 cm in test tubes, the rooting rate did not exceed 30%. After cutting only the apices with a length of 1–2 mm and cultivating them first in Petri dishes, the rooting rate increased to 50–60% on average. Within 4–6 weeks the apices grew to young rooted plants very similar to an asparagus seedling.

The plantlets were then transferred to a larger culture flask and developed into plants with many spears (5–10 cm long), ready for transfer to the soil. The best conditions were as follows:

— during the daily light period 6000 lx,
— temperature during the day +28 °C,
— temperature during the night +12 °C,
— humidity for the first 3 weeks 80–100%,
— humidity subsequently to 8 weeks 60%,
— substrate: sterilized sandy soil.

The rate of loss of plantlets is high, even under optimal conditions not less than 15%. Multiplication rate, rooting, and general losses of plantlets depend also on the genotype. Differences observed between male and female plants are generally not significant.

3. Integration of Micropropagation in Breeding Schemes. In asparagus, micropropagation is of fundamental importance for F_1 hybrid breeding. We must distinguish between three different breeding methods: (a) double hybrids, (b) clonal hybrids (= simple hybrids), and (c) all-male F_1 hybrids (Reuther 1983). Clonal hybrids and all-male F_1 hybrids can only be realized by using micropropagation for multiplication of the hybrid lines.

a) Clonal hybrids. At the beginning of breeding, female and male plants with high yields and other desirable traits were selected from the same or different dioecious populations. Under controlled conditions the female and male plants were crossed and their progeny tested for combining ability. Those female and male plants which after crossing gave rise to the most productive hybrids were then selected, micropropagated on a large scale, and transferred to the greenhouse or field. F_1 seed production resulted from open pollination between one clone originating from a male and another originating from a female plant. The F_1 hybrid variety is dioecious and segregates to 50% female and 50% male plants (Fig. 1). The homogeneity of the hybrids is generally high, but the degree of uniformity depends on the homogeneity of the parents.

b) By micropropagation in clonal hybrids it is possible to fix the high heterosis observed in hybrid plants in every plant of the clonal hybrid variety.

c) All-male F_1 hybrids. In the dioecious population of asparagus the male plants produce a greater number of spears and have a longer life than the female ones, altogether the males are more productive than the female plants. Therefore, for yield, a population with 100% male plants is superior to the general dioecious population. Breeders are thus interested in all male hybrids.

In asparagus, the inheritance of sex is determined by sex chromosomes; the female plants are characterized by xx, the male by xy. The creation of an all-male hybrid is possible by using supermales, plants characterized by yy. The supermales can be produced in two ways:

— Conventionally by selfing of andromonoecious plants. The progenies consist of xy and yy males, which cannot be checked by cytological methods, because the differences between the sex chromosomes are too small, but only by test crosses with xx female plants and afterwards checking the progeny (for further details,

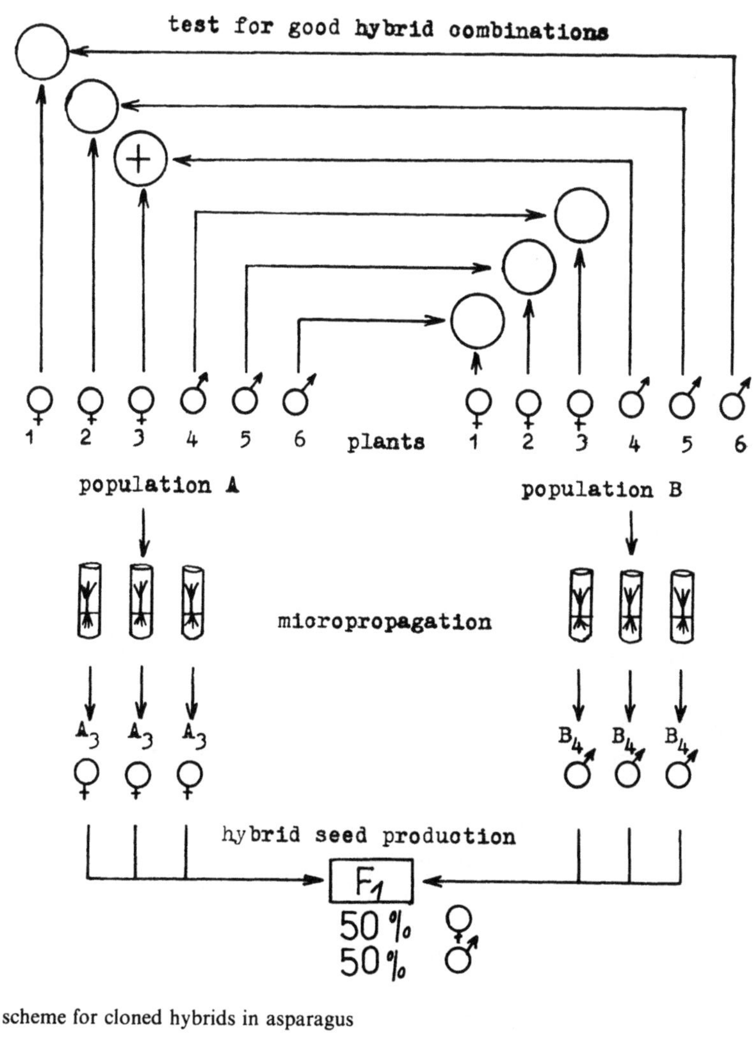

Fig. 1. Breeding scheme for cloned hybrids in asparagus

see Reuther 1983). Approximately 2% of the male plants are yy supermales (Weit and Stein 1981).

— The second, more effective way to obtain supermale plants is by using pollen-derived haploid plants. After diploidization, about 50% of them must be yy supermales. As soon as there are flowering yy plants, the breeder must check their combining ability by crossing with several heterozygous or homozygous female plants of different origin. Afterwards the best supermale and female plants must be micropropagated on a large scale to start F_1 seed production (Fig. 2).

The homogeneity of the all-male F_1 hybrid varieties depends on that of the supermale and female parent plants. The most consistent hybrid varieties can be

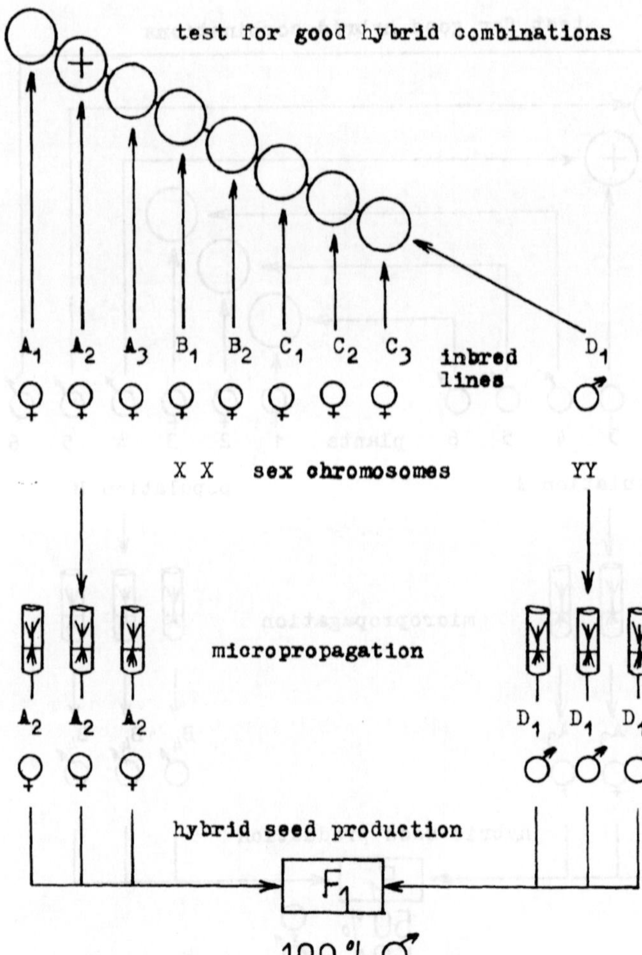

Fig. 2. Breeding scheme for all-male F_1 hybrids in asparagus

produced by using female and male dihaploids, e.g., pollen-derived regenerated chromosome-doubled plants, although in this case very extensive controls concerning combining ability are necessary to find those hybrids in which the inbreeding depression of the parent plants is completely overcome.

4. Use of in Vitro-Cloned Plants for F_1 Seed Production.[2] As micropropagation starts with either one male or one female plant, the number of plants propagated per year is limited. In order to obtain maximum seed production under temperate zone conditions, the seed plants should be planted in nonheated greenhouses (Fig. 3). In comparison with seed production outdoors, five- or sixfold quantities of seeds are harvested.

The in vitro-cloned plants were first cultivated in flowerpots 10–12 cm in diameter for 6 months. They can be planted if the root system is well developed and the spears reach a length of about 50–60 cm. During the entire vegetation period from April to August, the plants can be put in the greenhouse. Later, developed in vitro plants spend the winter in cool but frost-free rooms, without being covered

Fig. 3. F_1 hybrid seed production in asparagus in the greenhouse

[2] From experiments at the Seed Breeding Station Möhringen of the VEB Saat- und Pflanzgut Magdeburg, Germany (Gottwald and Wolterstorff 1989)

with foliage or other protection against low temperatures. The planting distance in the greenhouse is 0.5 m, row spacing, depending on the type of greenhouse, is 1.5–2 m. The ratio between female and male hybrid parents is 4:1. The best pollination effect is obtained if the male plants are regularly distributed over the whole area, e.g., in the planting row each fifth plant is male. During flowering after the middle of May, bees are needed for pollination; one well-developed bee colony is sufficient to pollinate plants over 500 m^2. The seed harvest starts after the 2nd year with considerable yields, the highest yield of hybrid seeds being obtained 3–6 years after planting in the greenhouse, about 50–60 g/m^2. Then 2000 m^2 area and 2000 cloned plants (1600 females and 400 males) are needed to produce 100 kg F_1 hybrid seed annually. Depending on the organization of micropropagation, this number of plants can be reached within 2 years (Fig. 4). This requires micropropagation of the selected female plant to be started with at least 80 shoot apices as explants. The cycle from cutting the explant to the in vitro-regenerated plant takes 3 months. By repeating the micropropagation cycles, it is possible to obtain about 5000 female and 1250 male in vitro plants within 1 year. In the course of transfer to soil at present the losses are very high.

2.3.2 Brassica Vegetables (*Brassica oleracea* L.) White Cabbage. Red Cabbage (*B.o. capitata* L.), Brussels Sprout (*B.o. gemmifera DC.*), and Cauliflower (*B.o. botrytis* L.)

1. Advantages of F_1 Hybrids. For many years, the share of F_1 varieties in the international assortment of *Brassica* vegetables has been increasing. The essential advantages of the F_1 hybrids in comparison to seed varieties are the following:

— Improved homogeneity. For instance, the heads of white and red cabbage of one batch are more uniform in size and thus better commercially; in cauliflower and early cabbage the crops become marketable more regularly, so that the number of harvesting operations can be reduced; in Brussels sprout, the lateral sprouts are more uniform and of similar size.

— The possibility of combining the desirable traits in one hybrid variety more rapidly, for instance, earliness and resistance to splitting, or good storing ability with high yield in white cabbage, the dark red color and earliness in red cabbage, loose distribution of the sprouts and high yield in Brussels sprout, low tendency to discoloration of the curds and high yield in cauliflower, and many other traits.

2. Micropropagation and Production of Rooted Plantlets. In all *Brassica* vegetables there are high regeneration potentials. Many review papers concerning in vitro techniques, including many micropropagation methods have been published (see Sect. 2). In cabbage, first successful results in micropropagation were obtained by using stem and root segments, cotyledons, and leaves from young seedlings (Bajaj and Nietsch 1975). In a breeding program it is desirable to have a rapid multiplication technique for producing large numbers of parental lines in which the explants for propagation can be taken from cabbage heads selected by the breeder. Since there are only 15–35 axillary buds available for culture from the head, this means that the direct use of axillary meristems is not applicable in breeding systems.

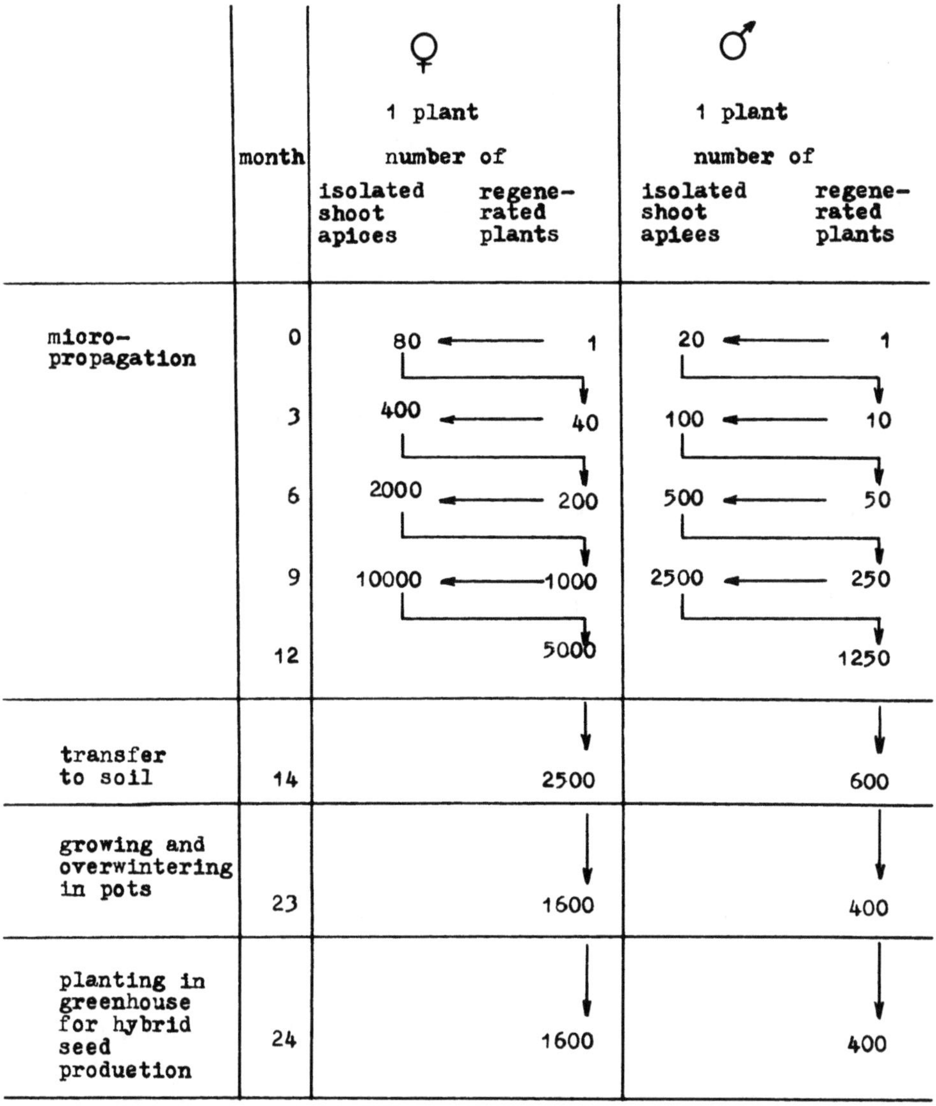

Fig. 4. Micropropagation of female and male plants in asparagus for 100 kg F_1 hybrid seed production

Inducing rapid proliferation and the formation of multiple shoots on the base of isolated axillar buds (Walkey et al. 1980; Leike 1981) proved to be successful. After 5–6 weeks the adventitious shoots can be isolated and rooted. To further improve effectivity, numerous other explants have been tested. The best results have been achieved with pieces of inflorescences with a length of 5 mm (Leike and Wieczorrek 1982). Up to 30 adventitious shoots were formed in some explants and one single flowering plant of head cabbage may be a source of more than 10000 explants. This method can also be used successfully in flowering plants of cauliflower and Brussels sprout. By means of this technique, propagation rates increased to 1/10000 and more within 10 weeks, 95–98% of rooted shoots developing into true type-to-type plants.

In cauliflower, floral meristem culture is the most commonly applied micropropagation technique (Walkey and Woolfitt 1970; Crisp and Walkey 1974). Small pieces of mature curd are cultured as explants, and produce leafy shoots at culture temperature, i.e., approximately 25 °C. It is estimated that a mature curd contains some 10 000 floral primordia which will revert to a vegetative pattern of development. The excision of even a very small portion of the total tissue mass can give a relatively large number of clonal regenerants. Transplantation of in vitro cauliflower plantlets is described in detail by Grout and Aston (1977, 1978).

In Brussels sprout, Clare and Collin (1974) were the first to show that meristem tips taken from axillary buds could be regenerated into plantlets. Further studies in micropropagation were made by Dunwell and Davies (1975).

Rooting in cabbage, cauliflower, and brussels sprout shows great similarities, although different methods have been developed. In some cases the rooting of the isolated adventitious shoots occurs without difficulties, taking place generally without auxin, but it can also be promoted by 3–5 mg/l IBA.

Establishing in vitro plantlets in soil is more problematic. It has been shown in studies with cauliflower that these difficulties mainly concern maintaining the plantlet during the morphological and physiological adaptation to a dry environment (Grout and Aston 1977). These studies, which are probably equally applicable to brussels sprout and cabbage, demonstrated that in culture the plantlet's vascular system, connecting roots, and stem were frequently poorly developed. The epiculicular wax that helps to control transpiration was present only in reduced quantities. There are some differences between *Brassica* crops, cauliflower appearing to be most sensitive, and cabbage considerably more resistant. Consequently, control of humidity in the environment of the plantlets following transfer is critical for survival. The plantlets must be kept in the greenhouse under a plastic cover for 7 to 10 days to maintain high humidity. During the next 2 weeks, the plastic cover is removed for a longer period each day.

Various sterilized soil and compost mixtures have been used. During the day, the plants need a temperature of 20–24 °C.

3. Integration of Micropropagation in Breeding Schedules. In *Brassica* hybrid breeding, homogeneous F_1 hybrid varieties can be produced in three ways: (a) single crosses, (b) sister line single crosses, and (c) modified single crosses.

a) Single crosses. Two self-incompatible inbred lines, reproduced by bud pollination, are planted together in the greenhouse and flower with open pollination (Fig. 5). As a rule, in single crosses it is ineffective to propagate self-incompatible hybrid lines by in vitro cloning, because the reproduction rate is tenfold higher when using generative bud pollination (Table 3) and because in this step a great number of in vitro plantlets are needed (cf. following Subsect. 4). Nevertheless, micropropagation can be helpful for the breeder to quickly produce a greater amount of seeds necessary to release the new F_1 hybrid in a shorter time. In Germany, self-incompatible lines of F_1 hybrids of cabbage Lena, and Brussels sprout Karat, already developed by the breeders, together with generative reproduction, were cloned one or two times and the in vitro plantlets used as seed plants for F_1 seed production.

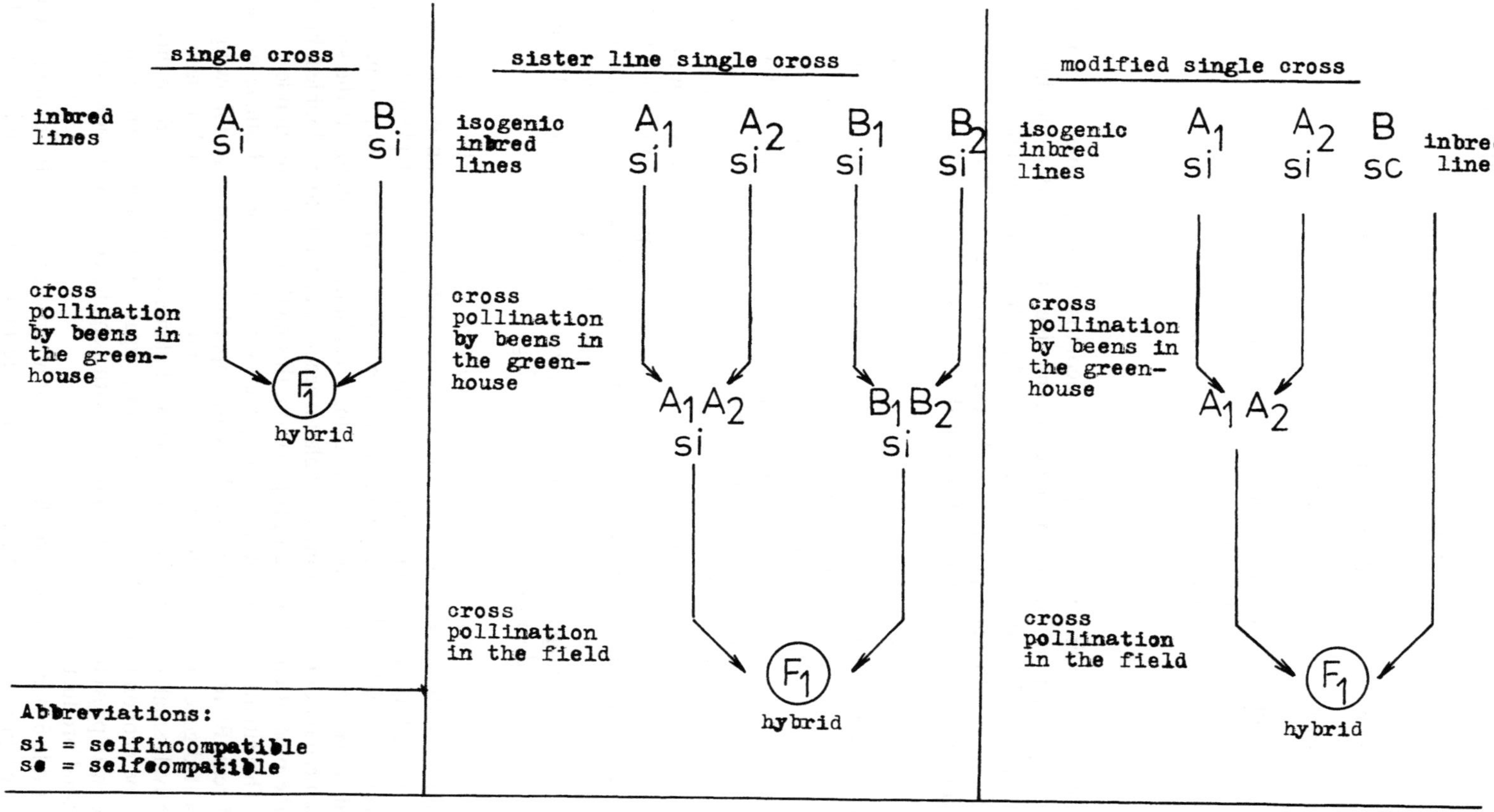

Fig. 5. Breeding schemes of F_1 hybrids in *Brassica oleracea* L.

Table 3. Reproduction rates of *Brassica* vegetable by using different propagation methods

Breeding plant	Reproduction method	Reproduction rate
Open-pollinated varieties	Open pollination	1:25 000
Self-incompatible inbred line	Selfing by bud pollination	1:2500
Self-incompatible inbred line	Micropropagation	1:250

b) Sister line single crosses. Four self-incompatible inbred lines are propagated by bud pollination. Two of them are isogenic lines, the only distinguishing feature being the S-allele. In the first step of breeding, both isogenic lines flower in the greenhouse with open pollination. In the second step, hybrid seeds have been produced by open pollination between the two sister-line combinations outdoors (Fig. 5).

In breeding practice, sometimes one or two of the needed four lines, as a result of inbreeding, cannot generatively propagate sufficiently. In this case, hybrid seed production by micropropagation of these self-compatible isogenic lines is reliable, because the number of plants needed is not too great (cf. following Subsect. 4).

c) Modified single crosses. Hybrids originate by open pollination of one sister-line combination and one self-compatible inbred line (Fig. 5). In this case, micropropagation can also be used very effectively (Fig. 6). The situation in the self-compatible lines is similar to that in the sister-line single crosses. In the course of inbreeding the self-compatible pollinators, micropropagation is advantageous: through combination of vegetative and generative propagation a great amount of selfing seeds can be produced, to widen the basis of selection work (Table 4). The selection of homogeneous self-compatible inbred lines is accelerated. In this way uniform modified single crosses originate (Fig. 7).

4. Use of in Vitro-Clones Plantlets for F_1 Seed Production. Whether or not in vitro cloning can be effectively used for F_1 hybrid seed production depends on the chosen method of breeding. In single crosses in vitro-cloned plantlets must be directly used for seed production. The number of plants needed to produce 100 kg seeds is very high (Fig. 8), calculated for greenhouse conditions. If the seeds are produced outdoors, the number of plants needed must at least be doubled. Because of the present costs of micropropagation, this method is not profitable.

The situation in sister lines, single crosses, and modified single crosses is different. By in vitro cloning of the self-compatible lines and the following twice-repeated generative cross-pollination, to produce 100 kg of F_1 hybrid seeds only a relatively small number of micropropagated plants is needed (Fig. 8). In this case it can be economically effective to micropropagate permanently self-incompatible lines with proven combining ability but with a small generative production rate for F_1 seed production.

Table 5 gives examples of the temporal integration of in vitro cloning used for F_1 seed production in different *Brassica* vegetables. In the biennial types, from

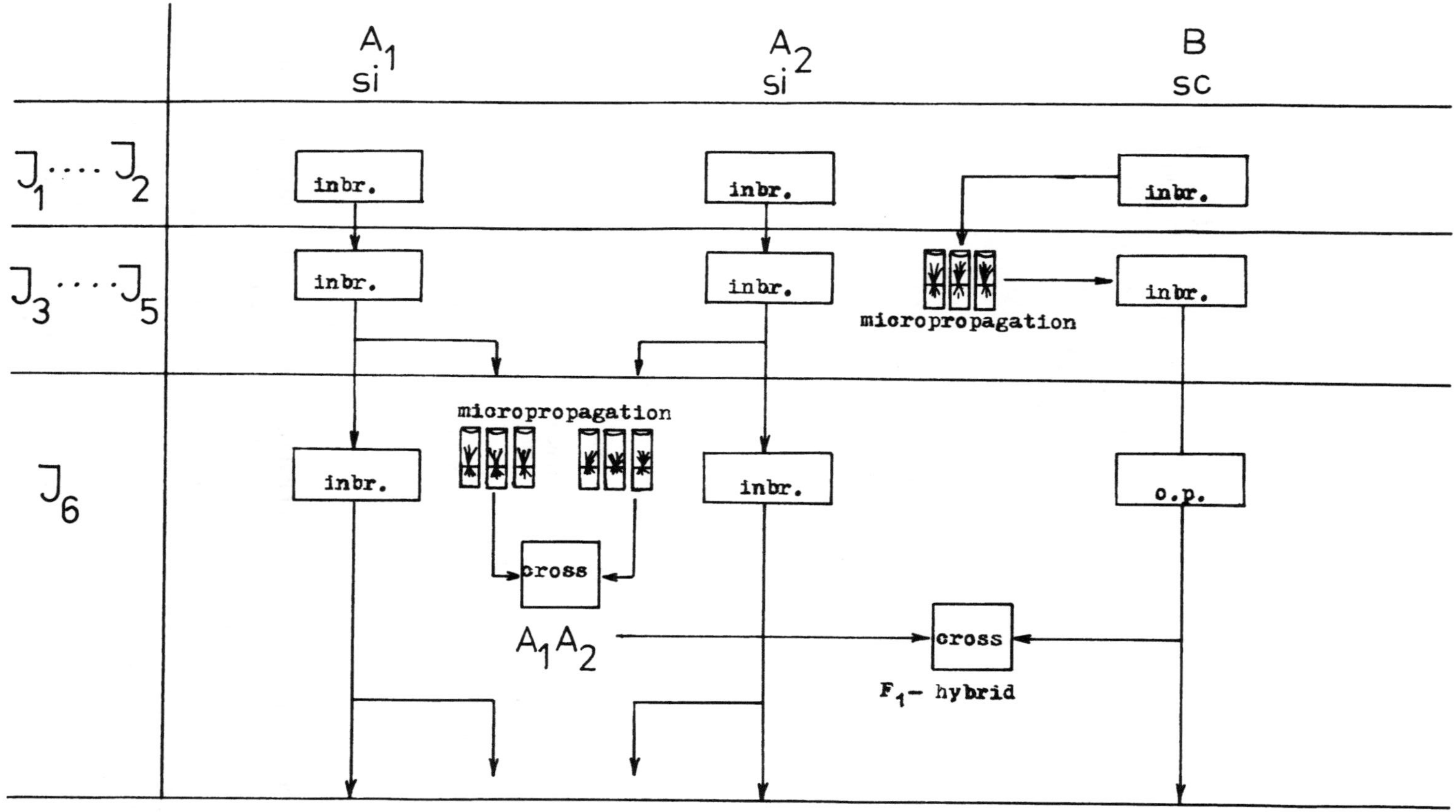

Abbreviations: si = selfincompatible ; sc = selfcompatible : inbr. = inbreeding
cross = cross pollination : o.p. = open pollination

Fig. 6. Breeding scheme of modified single cross, including micropropagation, in cabbage

Fig. 7. Experimental modified single hybrid in white cabbage in the breeding plot

Table 4. Yield of seeds during the inbreeding of one self-compatible plant of white cabbage dependent on different propagation procedures

Propagation	No. of original plants	Vegetative	Generative	
		No. of cloned plants	Kind of pollination	Harvested seeds (g)
Generative	1	–	Manual	5
Vegetative and generative	1	250	Open pollination	1000

micropropagation to the beginning of hybrid seed production, it takes 15 months, in the annual cauliflower, 12 months. In F_1 seed production the following time schedule was tried and tested:

a) White cabbage, red cabbage, late types. The flower shoots used as source for explants are grown in the greenhouse in March/April, in the case of storing the heads, in May. Until the end of the vegetation period in October, the micropropagated plantlets reach a stage with 12–15 leaves and cannot be planted outdoors. The clones winter in flowerpots in the greenhouse. Because the cloned plants are smaller in spring than the better-developed seed-propagated lines, seed production will need five times as long. Vernalization takes at least 12 weeks. It is also necessary to obtain synchronous flowering of the lines.

hybrid types	structure of hybrids	number of micropropagated si - plants
single cross	$A_{si} \times B_{si}$	25000
sister line single cross	$(A_{1\,si} \times A_{2\,si}) \times (B_{1\,si} \times B_{2\,si})$	250
modified single cross	$(A_{1\,si} \times A_{2\,si}) \times B_{sc}$	500

Fig. 8. Micropropagation of self-incompatible plants in cabbage for 100 kg F_1 hybrid seed production by different hybrid types

Table 5. Temporal integration of the in vitro cloning in the F_1 hybrid seed production in different *Brassica* vegetables

	White cabbage, red cabbage – late		Brussels sprout		Cauliflower	
	Ontogenic stage	Month	Ontogenic stage	Month	Ontogenic stage	Month
In vitro cloning explant source	Second vegetation year, flower shoots	April–July	Second vegetation year, flower shoots	March–June	First vegetation year, curd	October–January
Transplantation in soil	3–5 leaves	August	3–5 leaves	June	3–5 leaves	February
Cultivation in flowerpots (8 cm)	Until 8th leaf developed	September	Until 8th leaf developed	July	Until 8th leaf developed	March
Continued cultivation						
— in flowerpots (10 cm) in the greenhouse	Until 12th–15th leaf developed	October	—	—	—	—
— planted outdoors	—	—	About 50% of final size	August–November	Until flowers come into bloom	April–June
Vernalization in the greenhouse						
— in flowerpots (10 cm)	12th–15th leaf	November–February	—	—	—	—
— planted into plot	—	—	About 50% of final size	November–January	—	—
Flowering	Flower formation	April–May	Flower formation	February–March	Flower formation	July
Harvest of seeds	Formation of mature seeds	August	Formation of mature seeds	June	Formation of mature seeds	October

b) Brussels sprout. The flower shoot grows earlier than in cabbage, therefore, micropropagation starts already in March. After transplantation, plants are rooted in soil, then in August are transferred to the open field, and by November develop to 50% of their final size. Afterwards the plants are transferred to the greenhouse and vernalized there. Flowering begins in February.

c) Cauliflower. Lines and populations which are intended for micropropagation are cultivated in the summer, so that the curds can be selected in October. Regenerated plants are planted outdoors in April, seed plants flower in July. In cauliflower there are no differences in size and seed yield between seed-propagated and in vitro-cloned seed plants.

3 Conclusions

Micropropagation has not yet become a standard practice for most vegetables. The different stages of development of the in vitro techniques depend on the differing economic importance and on the different botanical families to which the vegetables belong.

Nevertheless, the starting costs are high for initiating a program, labor costs can remain high even after the establishment of the laboratory (Ng 1986). The new breeding technology facilitates the task of plant breeding without removing its characteristic element, but it would be wrong to suppose that there will be a simple step from the culture tube to the ideal crop plant.

The decisive factor in economic effectiveness is the integration of biotechnological methods and conventional breeding procedures. Consequently, the integration of micropropagation depends on a continuous dialogue between breeders and the scientists involved in basic research. Only first steps have been realized in Quedlinburg; until now seven new vegetable varieties have been created using micropropagation methods, and have been confirmed by government testing

Table 6. Varieties from the Institute of Breeding Research Quedlinburg bred by biotechnological methods (1990)

			Breeder
F_1 hybrids	Brussels sprout	Karat	R. Neubert
	Cabbage	Baltikol	W. Bauch
	Cabbage	Lena	W. Bauch
	Cabbage	Juna	W. Bauch
	Red cabbage	Maturus	W. Bauch
Clon F_1 hybrids	asparagus	Schneekop	M. Stein
	asparagus	Apollo	J. Gottwald
Seed varieties	Cauliflower	Jako	W. Jahr
	Celery	Anita	E. Weit

procedures (Table 6). It is not possible in every case to produce top-quality varieties with outstanding traits immediately. However, in the case of cabbage, the F_1 hybrids are clearly superior to the comparable varieties: Baltikol in terms of storage life, Lena and Juna in terms of early maturity and total yield.

References

Alconero R (1988) Lettuce (*Lactuca sativa* L.). In: Bajaj YPS (ed) Biotechnology in agriculture and forestry, vol 6. Crops II. Springer, Berlin Heidelberg New York, pp 351–369

Ancora G (1985) Plant improvement through tissue culture at ENEA. In: Biotechnology in international agricultural research. IRRI Manila, pp 158–167

Anghel I, Rosu A (1985) In vitro morphogenesis in diploid, triploid, and tetraploid genotypes of watermelon — *Citrullus lanatus* (Thunb.). Mansf Rev Roumaine Biol 30: 43–55

Bajaj YPS, Nietsch P (1975) In vitro propagation of red cabbage (*Brassica oleracea* L. var. *capitata*). J Exp Bot 26: 883–890

Bauch W (1982) Bewertung verschiedener Hybridzuchtverfahren bei Kopfkohl (*Brassica oleracea* var. *capitata* L.). Arch Züchtungsforsch 12: 135–145

Bauch W (1986) Ergebnisse bei der Selektion auf Inzuchtverträglichkeit bei Weißkohllinien. Tag Ber Akad Landwirtsch Wiss DDR 239: 3–10

Browers MA, Orton TJ (1986) Celery (*Apium graveolens* L.). In: Bajaj YPS (ed) Biotechnology in agriculture and forestry, vol 2. Crops I. Springer, Berlin Heidelberg New York, pp 405–420

Clare MV, Collin HA (1974) Meristem culture of Brussels sprouts. Hortic Res 13: 111–118

Crisp P, Walkey DGA (1974) The use of aseptic meristem culture in cauliflower breeding. Euphytica 23: 305–313

Doré C (1980) Different utilizations of in vitro cloning techniques. In: Reuther G (ed) Proc 5th Int Asparagus Symp, Sec Veg, Aug 28 — Sept 1 1979, Geisenheim, pp 146–149

Doré C (1987) Application of tissue culture to vegetable crop improvement. In: Greer CE, Somers DA, Hackett WP, Biesboer DD (eds) Plant tissue and cell culture. Proc Int Congr Plant tissue and cell culture, Aug 3–8 1986, Univ Minnesota. Alan R Liss, New York, pp 419–432

Dunwell JM, Davies DR (1975) Techniques for propagating inbred brussels sprout strains by means of tissue culture. Grower, pp 105–106

Fari M (1986) Pepper (*Capsicum annuum* L.). In: Bajaj YPS (ed) Biotechnology in agriculture and forestry, vol 2, Crops I. Springer, Berlin Heidelberg New York, pp 345–362

Grout BWW (1988) Cauliflower (*Brassica oleracea* var. *botrytis* L.). In: Bajaj YPS (ed) Biotechnology in agriculture and forestry, vol 6, Crops II. Springer, Berlin Heidelberg New York, pp 211–225

Grout BWW, Aston MJ (1977) Transplanting of cauliflower plants regenerated from meristem culture. I. Water loss and water transfer related to changes in leaf wax and to xylem regeneration. Hortic Res 17: 1–7

Grout BWW, Aston MJ (1978) Transplanting of cauliflower plants regenerated from meristem culture. II. Carbon dioxide fixation and the development of photosynthetic ability. Hortic Res 17: 65–71

Hall O (1983) Användning av vävnadskulturteknik i blomkals – och sallatsförädling. Sver Utsädesfören Tidskr 93: 31–34

Hasegawa PM, Murashige T, Takatori PH (1973) Propagation of *Asparagus* through shoot apex culture. II. Light and temperature requirements, transplantability of plants, and cytohistological characteristics. J Am Soc Hortic Sci 98: 143–148

Jensen CJ (1983) Cell and tissue culture in plant breeding and research. Acta Agric Scand 23

Keith A, Walker PD (1986) Automation of in vitro plant propagation: has its time finally arrived? Genet Eng News 6/7: 52

Leike H (1981) Probleme bei der Linienerhaltung und -vermehrung in der Hybridzüchtung mit Hilfe der Gewebekultur. Tag Ber Akad Landwirtsch Wiss DDR 191: 19–27

Leike H (1988) Cabbage (*Brassica oleracea* var. *capitata* L.). In: Bajaj YPS (ed) Biotechnology in agriculture and forestry, vol 6. Crops II. Springer, Berlin Heidelberg New York, pp 226–251

Leike H, Wieczorrek G (1982) In-vitro-Vermehrung bei Kopfkohl (*Brassica oleracea* L. var. *capitata*). Arch Züchtungsforsch 12: 375–383

Linsmaier EM, Skoog F (1965) Organic growth factor requirements of tobacco tissue cultures. Physiol Plant 18: 100–127

Malepszy S (1988) Cucumber (*Cucumber sativus* L.). In: Bajaj YPS (ed) Biotechnology in agriculture and forestry, vol 6. Crops II. Springer, Berlin Heidelberg New York, pp 277–293

Murashige T, Shabde PH, Hasegawa FH, Takatori FH, Jones JB (1972) Propagation of *Asparagus* through shoot apex culture. I. Nutrient medium for formation of plantlets. J Am Soc Hortic Sci 97: 158–160

Nešković M, Ćulafić L (1988) Spinach (*Spinacia oleracea* L.). In: Bajaj YPS (ed) Biotechnology in agriculture and forestry, vol 6. Crops II. Springer, Berlin Heidelberg New York, pp 370–385

Ng TJ (1986) Use of tissue culture for micropropagation of vegetable crops. In: Zimmermann RH et al. (eds) Tissue culture as a plant production system for horticultural crops. Martinus Nijhoff, Dordrecht, pp 259–270

Novák FJ (1983) Production of garlic (*Allium sativum* L.) tetraploids in shoot-tip in vitro culture. Z Pflanzenzücht 91: 329–333

Reuther G (1978) Verklonung von weiblichen und männlichen Spargelstämmen durch Gewebekultur. Gartenbauwissenschaft 43: 1–10

Reuther G (1983) Application of tissue culture methods for improvement of *Asparagus officinalis* L. Asparagus Res Newslett 1/2: 42–95

Reuther G, Becker U (1985) Gewinnung von F_1-Klonhybriden bei *Asparagus officinalis* durch in vitro-Vermehrung von Zuchtstämmen. In: Pflanzliche Zellkulturen. BMFT-Statusseminar, May 9, Jülich, pp 57–79

Roggemans J, Boxus P (1988) Rhubarb (*Rheum rhaponticum* L.). In: Bajaj YPS (ed) Biotechnology in agriculture and forestry, vol 6. Crops II. Springer, Berlin Heidelberg New York, pp 339–350

Schoofs J, De Langhe E (1988) Chicory (*Cichorium intybus* L.). In: Bajaj YPS (ed) Biotechnology in agriculture and forestry, vol 6. Crops II. Springer, Berlin Heidelberg New York, pp 294–321

Stein M, Weit E, Wolfram H (1985) Zuchtmethodische Aspekte der Nutzung der Pollensterilität bei Zwiebeln und bei Möhren. Arch Gartenbau 33: 345–353

Theiler-Hedtrich R, Badoux S (1986) Utilisation de la multiplication in vitro dans un programme de la selection de la chicorée amére rouge (*Cichorium intybus* L.). Rev Suisse Vitic Arboric Hortic 18: 271–275

Walkey DGA, Neely HA, Crisp P (1980) Rapid propagation of white cabbage by tissue culture. Sci Hortic 12: 99–107

Walkey DGA, Pink DAC (1988) Brussels sprout (*Brassica oleracea* var. *gemmifera*) and broccoli (*B. oleracea* var. *italica*). In: Bajaj YPS (ed) Biotechnology in agriculture and forestry, vol 6. Crops II. Springer, Berlin Heidelberg New York, pp 252–276

Walkey DGA, Woolfitt JMG (1970) Rapid clonal multiplication of cauliflower by shake culture. J Hortic Sci 45: 205–206

Weit E, Stein M (1981) Fragen des Spargelsortiments und der Neuzüchtung von rein männlichen Spargelsorten. Arch Gartenbau 29: 177–186

Wolterstorff B, Gottwald J (1988) Gesunde und leistungsfähige Spargelanlagen durch gezielte Maßnahmen des Pflanzenschutzes. Gartenbau 35: 140–142

I.2 Micropropagation of Asparagus (*Asparagus officinalis* L.)

Y. Desjardins[1]

1 Introduction

Asparagus, a member of the family Liliaceae, is a perennial of which the spears or growing stems are eaten. The perennial structure of the plant is a crown constituting a ramified rhizome formed by the consecutive initiation and development of buds and storage roots. Vegetative propagation by fractionation of the crown is possible on a small scale, but is unpractical on a commercial scale. The plant is thus propagated from seeds sown in nurseries at high densities. In the spring of each year, spears develop from buds on the crown and give rise to ramified stems. These stems bear cladophylls which are leaf-like structures morphologically related to stems. The actual leaf is a small scale located along the stem.

The plant is dioecious with 2n = 20 (Jessop 1966). There is still controversy as to the origin of sex determination in asparagus. Sex dimorphism is controlled by a single genetic factor (Lazarte and Garrison 1980). Thuesen (1960) reported chromosomes pair # V to be heteromorphic. However, Löptien (1979) localized the sex determinants on this chromosome pair but described it as a homomorphic pair. Female plants are typically homogametic XX, and male plants are heterogametic XY or homogametic YY. Normal open-pollinated populations thus have an equal number of male and female plants.

On some very rare occasions (less than 1%) a functional pistil can form on a staminate plant, giving rise to andromonoecious plants (Thévenin 1966). In this case, hermaphroditic flowers, upon pollination, will produce two genotypes: XY male and YY super-male. Crossing supermales with females produces all-male plants. Many breeding programs around the world are focusing on the production of such all-male hybrids since staminate plants are most desirable for commercial production because of higher yield, vigor, and longevity (Ellison et al. 1960). Moreover, increased disease resistance and drought tolerance are associated with maleness (Ellison 1986).

Asparagus is now produced worldwide. The European Economic Community and the United States used to be the sole producers, but now, Taiwan, New Zealand, Chile, Canada, and Australia are producing important quantities (Table 1). The harvested product from asparagus is the growing spear which can be green, or white if previously blanched. European countries and Taiwan mostly produce white asparagus, while the United States and other countries grow green

[1] Centre de Recherche en Horticulture, Département de Phytologie, Faculté des sciences de l'Agriculture et de l'Alimentation, Université Laval, Québec, QC, Canada G1K 7P4

Biotechnology in Agriculture and Forestry, Vol. 19
High-Tech and Micropropagation III (ed. by Y.P.S. Bajaj)

Table 1. Asparagus production in the most important producing countries (1990)

Country	Approximate area planted (ha)
Argentina	2840[a]
Australia	3510[b]
Canada	2000[c]
Chile	4300[b]
France	20000[a]
Germany (East)	3100[a]
Germany (West)	5200[a]
Italy	6000[d]
Japan	9970[b]
Mexico	3100[e]
The Netherlands	2840[f]
New Zealand	3000[b]
Spain	22000[f]
Taiwan	10000[b]
USA	45870[g]
California	17180
Michigan	9320
Washington	13640
New Jersey	870

[a] Asp. News Letter (1990) 8(1): 3–8.
[b] Acta Hortic (1990) 271: 25–31.
[c] Canadian Asparagus Research Report 1989.
[d] Eurostat (1988).
[e] SARH-DGEA, Asp. News Letter (1986) 3(2): 1–5.
[f] Asp. News Letter (1989) 6(2): 3–11.
[g] USDA Agricultural Statistics (1989).

asparagus. Asparagus is grown in temperate climates in rapidly warming soil. It is one of the earliest crops in the spring, allowing growers to obtain a rapid cash flow at the beginning of the summer. The crop can also adapt well to subtropical climates where special growing techniques have been developed (mother-stalk technique, Taiwan) (Wang 1973). The plant grows best on well-drained, neutral sandy soil. Most cultivated varieties of asparagus are lines of the species *A. officinalis*. Until 1980, most varieties were actually synthetic varieties achieved by cross-pollination of superior parents and by subsequent mass selection among the progeny. However, since then, clonal hybrids have replaced the original varieties. The new varieties are the result of strict parental selection on the basis of the largest possible general combining ability, and rely on the in vitro propagation of genitors for the production of the hybrid seeds. Advances in anther culture (Doré 1975, 1977, 1990; Falavigna et al. 1985) open the way for the production of real F_1 hybrids originating from homozygous parental lines.

Healthy asparagus fields can be profitable for many years. However, important root diseases (*Fusarium oxysporum* Schlecht f. sp. *asparagi* Cohen and *F. moniliforme* Sheldon) have caused the premature removal of numerous fields throughout the world. Important breeding programs have been initiated in New Jersey, California, France, Canada, Germany, Holland, and Italy to tackle this problem but also to improve yield. Because of its perennial nature, asparagus requires heavy financial input at the time of establishment, and because profits will not be obtained for 5 or 6 years, high-yielding varieties are required for the production to be worthwhile. In this context, tissue culture and micropropagation have evolved as essential requirements of breeding programs.

2 Micropropagation

2.1 Need for Micropropagation

Vegetative propagation of asparagus, although possible, has proven slow and practically impossible on a large scale (Yang 1977). Micropropagation was thus developed to provide a means of supplying breeding programs with the large numbers of identical clones necessary for hybrid production purposes. Tissue culture is now a definite requirement for many aspects of breeding programs. If anther culture, parasexual hybridization, and somaclonal variation are used in the breeding scheme at one stage or the other, micropropagation will be necessary in order to rapidly accumulate the number of individual plants necessary to produce a hybrid (Clonal hybrids, Doré 1975).

With the recent advances in micropropagation, great expectations have been placed in the possibility of large-scale cloning of superior individual selections. One of the factors which has prevented the wide commercial application of asparagus clones is certainly the lack of evidence for advantage in yield. Even though Ellison (1985) reported a yield of 14000 kg/ha, ten times the average yield obtained in Québec (Canada), these results were preliminary and have not been confirmed since on a larger scale. Very few variety trials have incorporated material of in vitro origin for comparison with conventional varieties. In variety trials established in Canada, tissue-cultured plants, selected from outstanding plants at the University of Guelph, have never performed to the level of varieties originating from seeds, and have not outyielded the best cultivar, Jersey Giant (Table 2). In these trials, an important proportion of the total yield was composed of small nonmarketable spears. Personal communication with breeders in the USA, Italy, and France confirms the lower yield obtained from tissue-cultured clones. In a similar manner, despite the 237% higher productivity of tissue-cultured material as compared to the cultivar Schwetzinger Meisterchuß reported by Reuther (1977), a large proportion of the yield was also found to be of lower grade categories. Lower yield and grade of spears produced by tissue-cultured plants may be aftereffects of tissue culture, or can most likely be attributed to the selection criteria applied for the choice of outstanding plants, based on a large number of spears after harvest rather than on individual plant productivity. In general, in vitro-cultured plants do not demonstrate the same type of vigor and uniformity as transplants or crowns (Fig. 1d). These results are deceptive, and do not correspond to the yield which was expected from clones originating from outstanding plants identified by segregating open-pollinated varieties or from very old plantations. This is quite paradoxical considering the fact that tissue-cultured plants are mericlones and should thus have

Table 2. Marketable yield of conventional varieties and outstanding selections cloned by tissue culture at three locations in Canada in 1989[a]

		Year of establishment and locations		
		1985 Guelph, Ontario	1986 Kemptville, Ontario	1985 Ste-Anne de Bellevue, Québec
Cultivar			Marketable yield (kg/ha)	
Jersey Centennial		2551 (100)[b]	889 (71)	4827 (97)
Jersey Giant		2188 (85)	1237 (100)	4975 (100)
Viking KB_3		1973 (77)	896 (72)	4067 (81)
Clone				
	G188	1582 (62)	—	4355 (87)
	G127	1011 (39)	—	4450 (89)
	G136	939 (36)	1031 (83)	3617 (72)
	G124	668 (26)	514 (41)	2303 (46)
	G182	598 (23)	567 (45)	—
	G273	88 (3)	—	—

[a] Canadian Asparagus Research Report 1989.
[b] Numbers in parenthesis represent the yield percentage compared to the best cultivar in the specific cultivar trial.

an identical genetic background and demonstrate a high degree of uniformity. As will be discussed in greater depth later, tissue-cultured plants suffer severe stress during transfer to micropropagation stage IV or to the field. This stress may contribute to their slow recovery and establishment during the first year, and explain their poor overall performance. It thus appears that before large-scale cloning by tissue culture can become a common practice, a number of problems need to be addressed.

2.2 Review of in Vitro Studies

The first report of asparagus tissue culture was that of Loo (1945, 1946), who presented evidence that a stem tip can grow indefinitely on a basic medium supplemented with sucrose and some organic constituents; however, there was no root growth. Galston (1947, 1948) also cultured asparagus in vitro, but in his experiments, asparagus was used as a simple biological model plant for studying growth and differentiation in stems and roots.

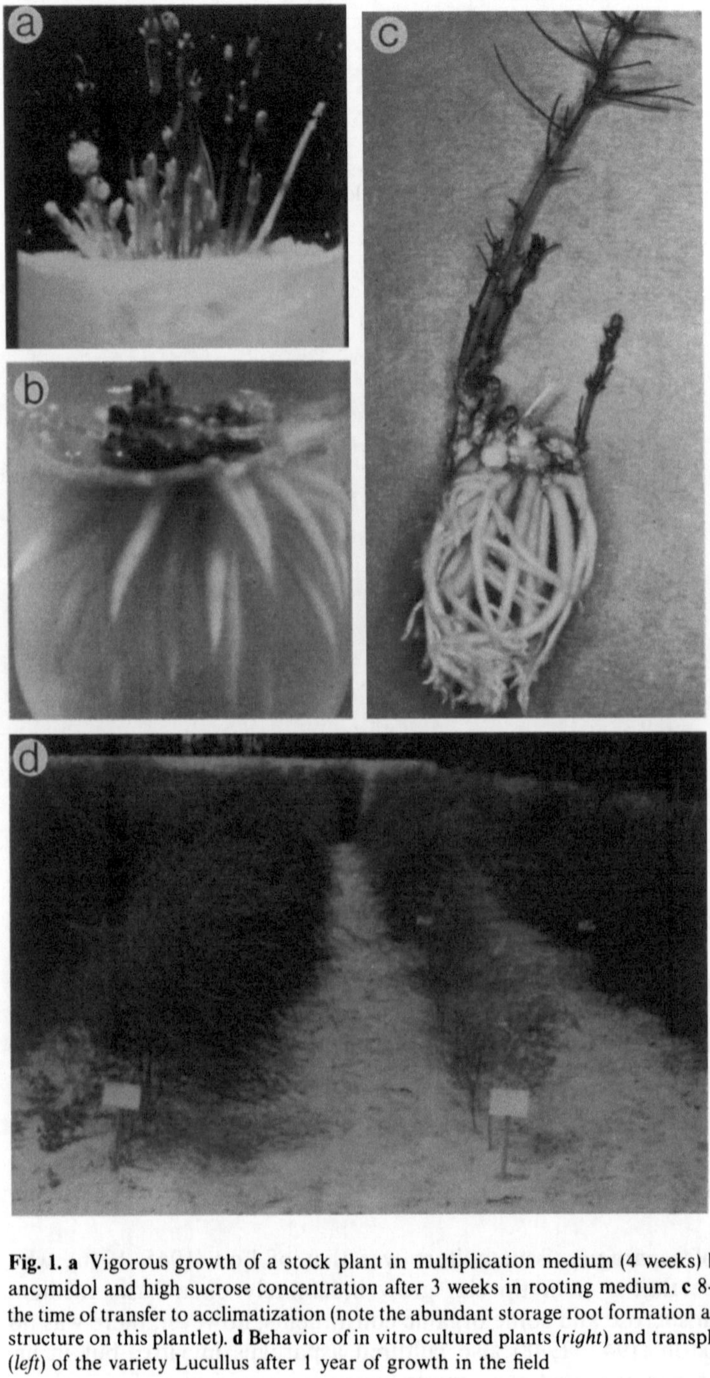

Fig. 1. a Vigorous growth of a stock plant in multiplication medium (4 weeks) **b** Root promotion by ancymidol and high sucrose concentration after 3 weeks in rooting medium. **c** 8-week-old plantlets at the time of transfer to acclimatization (note the abundant storage root formation and the massive crown structure on this plantlet). **d** Behavior of in vitro cultured plants (*right*) and transplants grown from seed (*left*) of the variety Lucullus after 1 year of growth in the field

Gorter (1965) transferred in vitro plantlets to soil, and developed a technique by which roots were induced to grow from callus in the dark, followed by transfer to liquid medium augmented with boric acid to allow for shoot growth. Wilmar and Hellendorn (1968) induced embryo formation on vigorously growing callus in agar medium containing a low concentration of 2,4-D (0.1 mg/l). Moreover, Takatori et al. (1968) produced callus from spear slices and were able to induce plantlet formation on MS medium (Murashige and Skoog 1962) containing 0.5 mg/l NAA, 15% coconut milk, and 50 mg/l adenine sulfate. An extensive description of a callus regeneration system which maintained plant integrity and conformity was provided by Reuther (1977, 1984). In this system, callus proliferation was obtained by 1 mg/l NAA and kinetin, and organogenesis was stimulated by 1 mg/l IAA and BAP. Hunault (1973, 1974, 1975, 1981) also intensively studied the proliferation of callus. His results suggest that auxin and cytokinin are necessary for optimal development of callus, with 2,4-D and BA being the most efficient. Illumination of callus prevents organogenesis; this inhibition can be overcome partly by a high concentration of cytokinins. Growth of callus was optimized on medium containing high thiamine levels and the presence of both NH_4^+ and NO_3^-.

Although callus culture is feasible for the vegetative propagation of asparagus, some researchers have observed genetic instability in the cultures, and it is thus not recommended as a standard method of propagation. Indeed, Wilmar and Hellendorn (1968), Malnassy and Ellison (1970), and Yakuwa et al. (1971a, b, 1972, 1982) observed polyploidy as a result of callus regeneration. The inadequacy of callus culture for the propagation of asparagus has been overcome by using meristem cultures to bypass the callus stage and therefore prevent a shift in ploidy levels. Two methods have been developed to obtain true-to-type cultures: meristem culture and micropropagation by axillary branching. Murashige et al. (1972) and Hasegawa et al. (1973) used shoot apices on defined medium to regenerate plantlets without the prior induction of callus. Their medium was an MS medium supplemented with 0.3 mg/l NAA, 0.1 mg/l kinetin, 40 mg/l adenine sulfate, 500 mg/l Difco Bacto malt extract, and 170 mg/l $NaH_2PO_4 \cdot H_2O$. The optimal pH was found to be 5.8, and the optimal sucrose level for shoot induction was 2.5%. Although a large number of plants could be produced, the technique is not suited to commercial production (Yang and Clore 1973). It is extremely difficult and time-consuming to excise the minute apical meristem, and new apices have to be gathered continuously from spears growing in the field, which represents a serious problem (Doré 1975).

A major breakthrough in in vitro propagation of asparagus was a technique developed by Yang and Clore (1973), who successfully induced axillary branching on nodal sections of asparagus. The first step of the technique is the establishment of stock plants from which nodal sections are taken continuously. Initially, spears are sterilized in a 1% sodium hypochlorite solution for 20 min followed by two rinses in sterilized distilled water. A lateral bud, located under a scale, is excised and transferred onto basal medium to allow shoot development. The medium used is a modified MS medium supplemented with 0.1 mg/l kinetin and NAA (Yang and Clore 1973). The shoots are grown for 4 to 6 weeks before being cut into single bud segments (nodal sections) and cultured on basal medium for 10 to 12 weeks or until multiple shoots and roots develop. The rooting of these plantlets is achieved by successive transfer to basal medium. It takes from 6 to 8 months to establish vigorous growth of stock plants and to obtain a sufficient number of plants for the multiplication step (Fig. 1a). The clones are multiplied by taking lateral buds and culturing them on a basal medium. According to Doré (1988), the gradual subculturing of the stock plant allows for the rejuvenation of the plant material and improves its ability to produce shoots and roots. The length of time required for this stage is dependent on the physiological stage of the original explant. Lateral bud production appears to be under the control of a strict apical dominance. This suggestion is justified by the fact that buds which are distant from the apex have a greater tendency to form roots and branches as was shown by Yang and Clore (1973, 1974a, b). Nodal sections with branches also root more readily than nodal sections without branches (Yang and Clore 1975).

It has been found that one important factor necessary for the successful multiplication and rooting of asparagus nodal sections is the initiation of a structure known as a crown. In fact, Doré (1988) mentions that the most important factor of the micropropagation system was precisely the induction of a crown in vitro, formed by the successive development of lateral buds at the base of the shoots. She observed that only plantlets which have initiated a crown have the capacity to withstand transfer to acclimatization and achieve proper survival to field conditions. Rivière and Muller (1974) have also shown that repeated transfer of nodal sections to fresh medium induced the formation of a functional crown. In a comprehensive histological study, they showed that vascular connections between shoots and roots were weak through callus and took a long time to establish. The presence of the crown ensured proper vascular connections. This may explain why the presence of a crown was shown to improve the transfer of plantlets to soil (Doré 1974, 1975; Thévenin and Doré 1976). In general, the formation of the crown appears to be under the influence of apical dominance and the juvenile status of the explant. In this respect, the presence of certain amino acids and of polyamines at the beginning of the multiplication stage, which are correlated to the juvenile status of explants (Doré 1988), could represent useful biochemical markers for the determination of the propagation and rooting potential of in vitro cultured plantlets.

Until now, despite the fact that micropropagation techniques have been widely used for the production of clonal hybrids in different breeding programs, many factors have prevented the large-scale multiplication of superior clones by this technique. Of these, two factors have been more restrictive: (1) the low and erratic rooting behavior of some clonal selections and (2) the difficulty to transfer the in vitro clones to field conditions and their slow recovery afterwards. In addition, as was mentioned previously, very few data exist on the performance of selected superior clones, and the high price of material originating from tissue culture has limited the utilization of the technique.

2.3 Rooting of Nodal Sections

Yang and Clore (1974a) reported low rooting success of asparagus nodal sections of 10 to 50%, depending on the source and the age of the explants. Rooting behavior of nodal sections was considerably improved by the addition of ancymidol, a growth retardant, to the culture medium (Chin 1982). In addition, this growth retardant promoted vigorous root and shoot growth and suppressed callus formation in some instances (Fig. 1b). Retardation of growth by ancymidol appears to be mediated by its antigibberellin activity. Indeed, ancymidol blocks the oxidation of ent-kaurene, ent-kaurenol, and ent-kaurenal by inhibiting kaurene oxidase (Coolbaugh and Hamilton 1976). Other growth retardants have demonstrated the same growth inhibition properties in vitro (Fig. 2; Khunachack et al. 1987). In our experiment, however, the antigibberellin compounds did not improve the rooting percentage (Desjardins 1984). Both ancymidol and paclobutrazol demonstrated more potent growth inhibition properties than CCC, SADH, and phosphon-D (Desjardins 1984) or B-995 (Khunachack et al. 1987). According to Khunachack et al. (1987), all growth retardants except CCC increased the vigor of nodal sections in vitro. Chin (1982) suggested that asparagus possesses a high titer of endogenous gibberellic acid that might be inhibitory to root formation. It is well documented in the literature that exogenous applications of gibberellic acid on cuttings inhibit adventitious root formation in many species (Batten and Goodwin 1978). There is also evidence that gibberellins block auxin activity during root primordium development (Brian et al. 1955; Haissig 1972). Root morphology was comparable with ancymidol and paclobutrazol, with thick storage-type roots and branching and thickening of lateral buds (Fig. 1b, c; Khunachack et al. 1987). The time at which the ancymidol treatment is applied can also play a very important role in the induction of rooting. In previous experiments, rooting was not always promoted by ancymidol, as clonal variations occurred (Desjardins et al. 1987). Doré (1988) has rightly shown that in a preliminary stage, following initiation of explants on initiation medium, ancymidol had no effect on rooting. After a certain time, which was dependent on the number of subcultures and the physiological age of the donor plants, ancymidol induced prolific rooting and contributed to the induction of a crown on explants.

Other factors have also been found to have an influence on rooting of nodal sections in vitro. Indeed, the absolute sucrose concentration in the culture medium has an important effect on rooting of explants (Fig. 3). Increasing the sucrose concentrations to about 7% resulted in the most efficient rooting. This improved

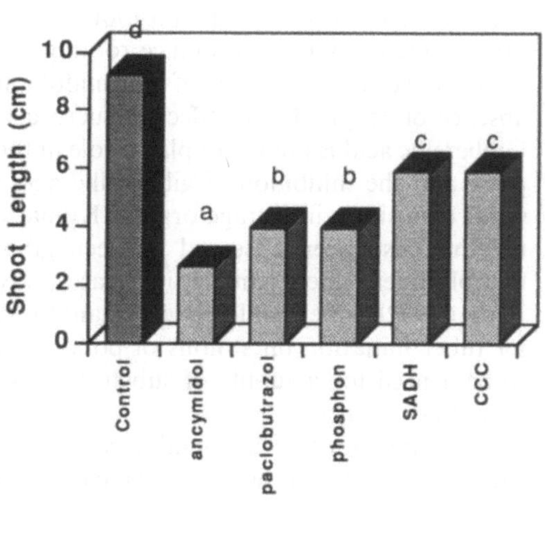

Fig. 2. The effect of growth retardants on shoot elongation (cm) of nodal sections of asparagus (clone G-97) after 9 weeks of culture in a modified MS medium. (Mean separation by protected LSD at the 5% level). (Desjardins 1984)

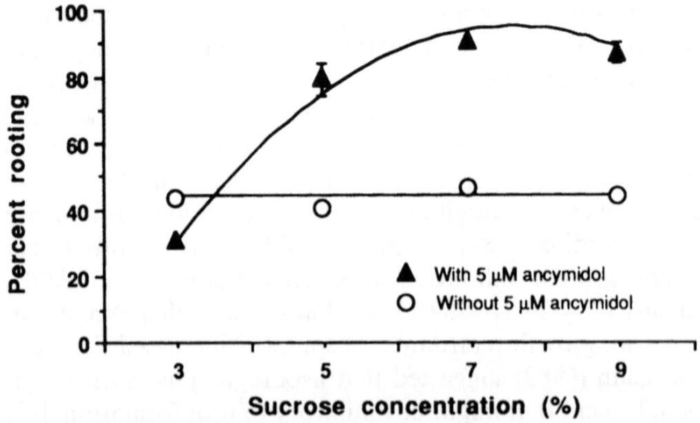

Fig. 3. Effect of sucrose concentration on percent rooting of nodal sections of Jersey Centennial asparagus cultured in medium with or without 5 μM ancymidol

Table 3. The effect of sucrose and mannitol, alone and in combination, on percent rooting of nodal sections of Jersey Centennial asparagus 9 weeks after subculture in a modified MS medium supplemented with 5 μM ancymidol. (Desjardins et al. 1987)

Treatment	Mean % rooting
3% Sucrose	42 b[a]
7% Sucrose	90 a
3% Sucrose + iso-osmotic mannitol concentration equivalent to 4% sucrose	39 b
Mannitol equivalent to 3% sucrose	0 c
Mannitol equivalent to 7% sucrose	0 c

[a] Mean separation in columns by Duncan's multiple range test, at 1% level of significance.

rooting did not result from an osmotic effect of sucrose, as iso-osmotic concentrations of mannitol did not induce rooting (Table 3). It appears that an interaction exists between the presence of ancymidol and sucrose in the medium, since in the absence of ancymidol no effect of sucrose was observed (Desjardins et al. 1987). Gibberellic acid is known to play a role in the carbohydrate metabolism within the plant and the inhibition of gibberellic acid synthesis has been shown to lead to sugar deposition in storage organs (Kumar and Wareing 1974). A strong correlation has also been observed between gibberrellic acid content and activity of hydrolytic enzymes, namely, invertase (Gayler and Glasziou 1969; Palmer and Barker 1972), which could be important for the breakdown of sugars. As is the case for tuber initiation on stolons of potatoes (Lovell and Booth 1967), there seems to be a need for a supply of substrate to sink regions to form storage roots on asparagus in vitro.

Two functions of ancymidol can thus be suggested for the stimulation of rooting. First, it may disrupt the apical dominance and lead to the initiation of side shoots and to the formation of a crown. As we showed earlier, the induction of the crown in vitro appears to be a factor severely limiting the initiation of roots in culture. Second, antigibberellin compounds may have an effect on carbohydrate metabolism which would lead to the deposition of the polyfructan inulin in storage roots.

2.4 Acclimatization

Another problem which has been observed with clones is their poor recovery and slow growth upon transfer to acclimatization and field conditions. It has often been suggested that this response was a persisting effect of tissue culture and a poor adaptation of clones to external conditions. Plantlet mortality during acclimatization ex vitro has mostly been attributed to poor control over water loss (Grout and Aston 1977; Brainerd and Fuchigami 1982). Indeed, anatomical and physiological modifications induced by tissue culture have been shown to hinder the transfer to acclimatization. In asparagus, water loss of in vitro-cultured material is important when compared with control plants (De et al. 1992; Fig. 4). Since the anatomy and physiology of leaves initiated in vivo change slowly towards that of completely acclimatized greenhouse-grown plants (Donnelly and Vidaver 1984; Donnelly et al. 1985; Grout and Millam 1985), it is essential to provide conditions of high relative humidity during acclimatization to prevent desiccation.

Tissue-cultured plants have also been shown to possess insufficient photosynthetic activity to provide a positive carbon balance upon transfer to soil conditions (Donnelly and Vidaver 1984). However, according to the classification of Grout and Price (1987), asparagus can be considered as photosynthetically competent in vitro. Indeed, in vitro-rooted plantlets were shown to possess a photosynthetic activity comparable to young greenhouse seedlings (De et al. 1992). Upon transfer to acclimatization, photosynthesis declined markedly, as a result of

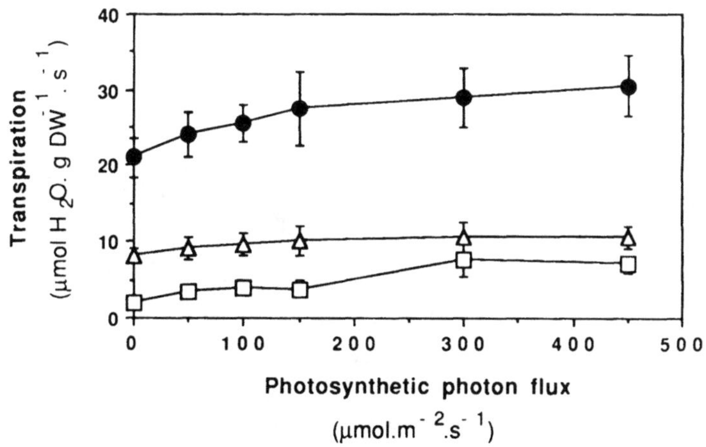

Fig. 4. Transpiration rates of in vitro-cultured plantlets (*closed circles*), acclimatized plants (*open squares*) and seedlings (*open triangles*) at different PPF. *Vertical bars* represent SE of the mean of five replications. (De et al. 1992)

the severe water stress. As for anatomical characteristics, physiology and more precisely photosynthetic capacity is gradually regained as new leaves develop in vivo. Experiments with asparagus (Desjardins et al. 1990) have shown that it was possible to influence the development and adaptation of new leaves during acclimatization by controlling the environmental growth conditions during this period. Most noticeably, CO_2 enrichment at 1500 ppm and supplementary lighting supplied during 4 weeks of acclimatization ex vitro resulted in significant improvement of dry weight of clones after this period (Fig. 5). These treatments were also very beneficial for the establishment and growth of clones in the field. For example, supplementary lighting and CO_2 enrichment provided during the acclimatization period enabled clones to reach a size comparable to that of transplants after two seasons in the field (Table 4). It appears that increased size obtained in acclimatization led to improved growth in the field. Strong correlation between dry weight accumulation in asparagus crowns and yield was observed by Ellison et al. (1960). Similarly, the yield of tissue-cultured material in the field can be improved by supplying proper growth conditions during acclimatization, as newly adapted leaves are forming, thus providing more reserves. If one is to properly evaluate the yield potential of the material originating from in vitro culture, it seems thus essential to provide proper growing conditions during acclimatization ex vitro. This may have been one of the pitfalls of previous clone trials.

The possibility of interacting with adaptation of asparagus plantlets ex vitro and stimulating growth during acclimatization and after field transfer suggests that it might also be possible to induce similar adaptations in test tubes. Moreover, recent report by Fujiwara et al. (1987) shows that restricted gas exchange and low CO_2 concentration within the tissue culture environment could increase growth of tissue cultures (Mousseau 1986; Desjardins et al. 1988). In experiments conducted in especially designed growth chambers (Laforge et al. 1990, 1991), nodal sections of asparagus, in the rooting stage, were also submitted to high CO_2 concentrations and photosynthetic photon flux treatments. On average, CO_2 enrichment at 3000 ppm caused a 72% increase in fresh weight accumulation (Table 4). Under uncontrolled CO_2 concentrations (330 ppm treatment), light had

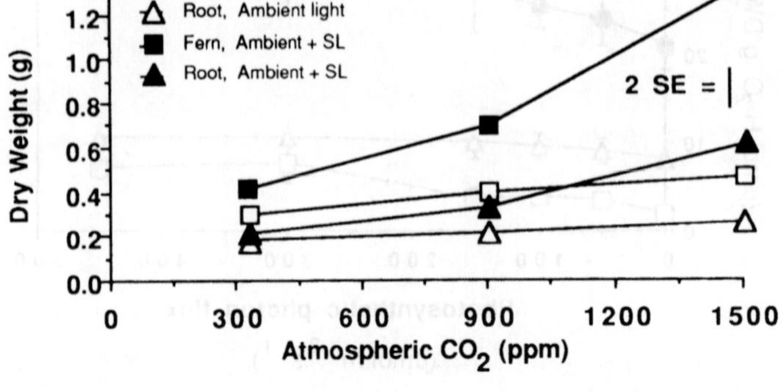

Fig. 5. Fern and root dry weight accumulation of asparagus clones after 42 days of growth in stage IV acclimatization, under CO_2 enrichment and supplementary lighting (ambient, and ambient + 80 μmol/m^2/s^1). (Desjardins et al. 1990)

Table 4. Effect of carbon dioxide enrichment and photosynthetic photon fluxes (PPF) provided during the in vitro rooting stage on fresh weight of asparagus plantlets in vitro and after 4 weeks of acclimatization

Carbon dioxide (μl/l[1])	PPF (μmol/m[2]/s[1])	Fresh weight (g) Rooting stage (4 weeks)	Acclimatization (4 weeks)
330	250	0.79 ± 0.18[a]	0.99[b]
	125	1.03 ± 0.19	1.09
	80	0.88 ± 0.13	1.06
Mean		0.90 ± 0.13	
1650	250	2.03 ± 0.39	1.40
	125	0.97 ± 0.16	0.62
	80	0.82 ± 0.17	0.83
Mean		1.27 ± 0.23	
3000	250	1.46 ± 0.21	1.90
	125	1.81 ± —	1.06
	80	1.38 ± 0.24	0.83
Mean		1.55 ± 0.22	
Treatment significance			
CO_2 Linear			0.009
PPF Linear			0.008

[a] Mean ± SE, n = 20 (Laforge et al. 1990).
[b] Significance level according to a T-test (Laforge et al. 1991).

no effect on fresh weight accumulation. However, in the presence of high photosynthetic photon fluxes, higher CO_2 concentrations, i.e., 1650 and 3000 ppm, significantly increased fresh weight. Very interestingly, providing CO_2 in the in vitro environment during rooting of nodal sections resulted in better development during acclimatization (Table 5). The improved growth in acclimatization observed following CO_2 enrichment treatment in vitro appears to result from greater reserve accumulation in vitro. In a manner similar to that observed in field-grown asparagus (Ellison et al. 1960), a strong correlation probably exists between weight accumulation in vitro and vigor of plantlets during acclimatization. Research shows that low photosynthesis of in vitro-cultured asparagus plantlets is not the cause of poor plant recovery upon acclimatization. Indeed, the photosynthetic capacity of in vitro plantlets was comparable to that of greenhouse seedlings after 5 weeks of culture (De and Desjardins, data not given). Poor adaptation of plantlets to low relative humidity thus appears to cause severe stress, which might cause poor recovery. As was mentioned previously, maintenance of conditions of high humidity upon transfer to ex vitro acclimatization is therefore required to prevent undesirable loss of plants.

3 Protocol

The best explants used for initiating in vitro cultures are the buds located under the scale of an actively growing spear (10–15 cm). The tip (≈ 4 cm) and the base (≈ 3 cm) sections are discarded. Only the scales remaining on the stem are used. They are carefully excised and sterilized in a 15% commercial solution of sodium hydroxide for 15 min followed by three rinses in sterile distilled water.

They are then placed on a modified MS medium as described by Yang and Clore (1973). Axillary shoots developing from the bud scale are excised (cut back) to about 1 cm of the base and the whole scale is subcultured onto new medium. This procedure is repeated four or five times until a strong stock plant is obtained. The basic multiplication medium consists of MS modified medium with 0.1 mg/l NAA and kinetin and 5 μM ancymidol (Elanco, Eli Lilly). Under the action of these growth regulators, plantlets will develop vigorous, strong shoots, often a crown structure, and usually storage roots. Propagation is obtained by excising the roots and separating the crown in multiple sections. When strong axillary branches (two or three branches) are located at the base of in vitro spears, they are also subcultured on the same medium. Weak branches or apices can be subcultured but they will usually take a very long time to form a crown structure and roots.

Plantlets are rooted in the above medium, supplemented with 7 to 8% sucrose, for a period of 5 to 7 weeks. In vitro plantlets are then transferred to a Pro-Mix Bx artificial substrate (Premier Peat Moss) for acclimatization and greenhouse culture. Upon transfer to acclimatization, they are placed under a humidity tent, where relative humidity is maintained at saturation for a minimum of 2 weeks. After gradual reduction of the humidity, they are placed under supplementary lighting (20 W/m^2) and supplied with carbon dioxide enrichment (+ 900 ppm). At this stage, mycorrhizal fungi (*Glomus vesiculiferum* and *G. versiforme*) can be very beneficial. A fertilization schedule for asparagus clones in the greenhouse has been presented in Desjardins et al. (1990).

4 Conclusion

Some attempts have been made by private laboratories to multiply exceptional asparagus selections by tissue culture in France, Holland, Israel, Italy, and the United States. As described in a previous section, the low yield of clonal material, high production costs associated with the intensive labor requirements, poor survival of clones which reduces the efficiency of the process, and slow recovery and growth in the field have prevented the large-scale commercial propagation of asparagus for the purpose of cloning exceptional plants. However, private laboratories are involved in the commercial propagation of genitors intended for clonal hybrid seed production. For example, the breeding program of Rutger University, New Jersey, is contracting out and licencing commercial laboratories for the propagation of female and super-male parent plants. This has allowed the rapid release of a number of new cultivars from this program.

Needless to say, micropropagation of asparagus still remains a highly desirable technique, which justifies research to optimize its procedures. If commercial laboratories are to propagate certain outstanding selections, they will have to present strong evidence as to the long-term yield and economic return for growers. Even if in vitro material were more expensive than either crowns or seedlings, proof of its higher yield potential should overcome the growers' apprehension.

Acknowledgements. The author thanks Dr. Yelle for constructive comments and critical review of the manuscript. He would also like to express his gratitude to the Canadian Horticultural Council, The Canadian Food Processors Association, Agricultùre Canada, and Québec's Ministry of Agriculture,

Fisheries, and Food for their financial support. Paper no. CRH-30 of the Centre de recherche en horticulture, Université Laval.

References

Batten DJ, Goodwin PB (1978) Phytohormones and the induction of adventitious roots. In: Letham DH, Goodwin PB, Higgins TJV (eds) Phytohormones and related compounds. A comprehensive treatise, vol II. Phytohormones and the development of higher plants. Elsevier/North Holland, New York

Brainerd KE, Fuchigami LH (1982) Stomatal functioning of in vitro and greenhouse apple leaves in darkness, mannitol, ABA and CO_2. J Exp Bot 33: 388–392

Brian PW, Hemming HG, Radley M (1955) A physiological comparison of gibberellic acid and some auxins. Physiol Plant 8: 669–681

Chin CK (1982) Promotion of shoot and root formation in asparagus in vitro by ancymidol. HortScience 17: 590–591

Coolbaugh RC, Hamilton R (1976) Inhibition of ent-kaurene oxidation and growth by (p-methoxyphenyl)-5-α-cyclopropyl-α-(p-methoxyphenyl)-5-pyrimidine methyl alcohol. Plant Physiol 57: 245–248

De Y, Desjardins Y, Lamarre M, Gosselin A (1992) Photosynthesis and transpiration of in vitro-cultured asparagus plantlets. Sci Hortic 49: 9–16

Desjardins Y (1984) In vitro rooting of nodal sections of *Asparagus officinalis* L. MSc Thesis, University of Guelph, pp 96

Desjardins Y, Tiessen H, Harney PM (1987) The effect of sucrose and ancymidol on the in vitro rooting of nodal sections of asparagus. HortScience 22: 131–133

Desjardins Y, Laforge F, Lussier C, Gosselin A (1988) Effect of CO_2 enrichment and high photosynthetic photon flux on the development of autotrophy and growth of tissue-cultured strawberry, raspberry and asparagus plants. Acta Hortic 230: 45–53

Desjardins Y, Gosselin A, Lamarre M (1990) Growth of transplants and in vitro-cultured clones of asparagus in response to CO_2 enrichment and supplemental lighting. J Am Soc Hortic Sci 115: 364–368

Donnelly DJ, Vidaver WE (1984) Leaf anatomy of red raspberry transferred from culture to soil. J Am Soc Hortic Sci 109: 172–176

Donnelly DJ, Vidaver WE, Lee KY (1985) The anatomy of tissue cultured red raspberry prior to and after transfer to soil. Plant Cell Tissue Organ Cult 4: 43–50

Doré C (1974) Production de plantes homozygotes mâles et femelles à partir d'anthères d'asperge. C R Acad Sci Paris Ser D 278: 2135–2138

Doré C (1975) La multiplication clonale de l'asperge (*Asparagus officinalis* L.) par culture in vitro: son utilisation en sélection. Ann Amelior Plantes 25: 201–224.

Doré C (1977) In vitro technique as an efficient tool in asparagus breeding, Acta Hortic 78: 89–93

Doré C (1988) Nouveau regard sur le comportement de l'asperge (*Asparagus officinalis* L.) multipliée in vitro. Agronomie 8: 843–850

Doré C (1990) *Asparagus* anther culture and field trials of dihaploids and F_1 hybrids. In: Bajaj YPS (ed) Biotechnology in agriculture and forestry, vol 12. Haploids in crop improvement I. Springer, Berlin Heidelberg New York, pp 322–345

Ellison JH (1985) Breeding true varieties of asparagus. In: Lougheed EC, Tiessen H (eds) Eucarpia Proc 6th Int Asparagus Symp, Guelph, pp 7

Ellison JH (1986) Asparagus breeding. In: Basset MJ (ed) Breeding vegetable crops. AVI, Westport, Conn, 584 pp

Ellison JH, Scheer DF, Wagner JJ (1960) Asparagus yield as related to plant vigor, earliness and sex. Proc Am Soc Hortic Sci 75: 411–415

Falavigna A, Tacconi MG, Casali PE (1985) Asparagus breeding through anther in vitro culture: eight years of activity in Italy. In: Lougheed EC, Tiessen H (eds) Eucarpia Proc 6th Int Asparagus Symp, Guelph, pp 31–45

Fujiwara K, Kozai T, Oki H, Watanabe I (1987) Fundamental studies on environments in plant tissue culture vessels. (3) Measurements of carbon dioxide gas concentration in closed vessels containing tissue cultured plantlets and estimates of net photosynthetic rates of plantlets. J Agric Meteorol 43: 21–30

Galston AW (1947) Nutritional requirements for root initiation in excised asparagus stem tips. Am J Bot 34: 599–600
Galston AW (1948) On the physiology of root initiation in excised asparagus stem tips. Am J Bot 35: 281–287
Gayler KR, Glasziou KT (1969) Plant enzyme synthesis: hormonal regulation of invertase and peroxidase synthesis in sugar cane. Planta 84: 185–195
Gorter CJ (1965) Vegetative propagation of *Asparagus officinalis* by cuttings. J Hort Sci 40: 177–179
Grout BWW, Aston MJ (1977) Transplanting of cauliflower plants from meristem culture. I. Water loss and water transfer related to changes in leaf wax and to xylem regeneration. Hortic Res 17: 1–7
Grout BWW, Millam S (1985) Photosynthetic development of micropropagated strawberry plantlets following transplanting. Ann Bot 55: 129–131
Grout BWW, Price (1987) The establishment of photosynthetic independance in strawberry cultures prior to transplanting. In: Ducate G, Jacob M, Simeon A (eds) Proc Symp Florizel 87: Plant micropropagation in horticultural industries. Arlon, Belgium, pp 55–60
Haissig BE (1972) Meristematic activity during adventitious root primordium development. Influence of endogenous auxin and applied gibberellic acid. Plant Physiol 49: 886–892
Hasegawa PM, Murashige T, Takatori FH (1973) Propagation of asparagus through shoot apex culture. II. Light and temperature requirements, transplantability of plants, and cyto-histological characteristics. J Am Soc Hortic Sci 98: 143–148
Hunault G (1973) Influence de diverses concentrations d'acide 2,4-dichlorophénoxyacétique et de benzyladénine sur la croissance des tissus d'asperge (*Asparagus officinalis* L.) cultivés in vitro. C R Acad Sci Paris 276 Ser D: 3135–3138
Hunault G (1974) Influence de l'éclairement et de la teneur du milieu de culture en cytokinines sur la formation, la croissance et l'organogénèse du cal de tige d'asperge (*Asparagus officinalis* L.). C R Acad Sci Paris 278 Ser D: 1849–1852
Hunault G (1975) Influence de différent milieux de culture sur la croissance de tissus d'asperge (*Asparagus officinalis* L.) cultivés in vitro. C R Acad Sci Paris 280 Ser D: 2661–2664
Hunault G (1981) La nutrition azotée des tissus d'asperge (*Asparagus officinalis* L.) cultivés in vitro. Influence de quelque acides aminés sur la croissance. Ann Sci Nat Bot Paris 13(2–3): 137–146
Jessop JP (1966) The genus *Asparagus* in southern Africa. Bothalia 9: 31–96
Khunachak A, Chin CK, Le T, Gianfagna T (1987) Promotion of asparagus shoot and root growth by growth retardants. Plant Cell Tissue Organ Cult 11: 97–111
Kumar D, Wareing PF (1974) Studies on tuberization of *Solanum andigena*. II. Growth hormones and tuberization. New Phytol 73: 883–840
Laforge F, Desjardins Y, Graham MED, Gosselin A (1990) Miniature growth chambers for the study of environmental conditions in vitro. Can J Plant Sci 70: 825–836
Laforge F, Lussier C, Desjardins Y, Gosselin A (1991) Effect of light intensity and CO_2 enrichment during in vitro rooting stage on subsequent growth of plantlets of strawberry, raspberry and asparagus. Scientia Hortic 47: 259–269
Lazarte JE, Garrison SA (1980) Sex modification in *Asparagus officinalis* L. J Am Soc Hortic Sci 105: 691–694
Loo SW (1945) Cultivation of excised stem of asparagus in vitro. Am J Bot 32: 13–17
Loo SW (1946) Further experiments on the culture of excised asparagus stem tips in vitro. Am J Bot 33: 156–159
Löptien H (1979) Identification of the sex chromosome pair in asparagus (*Asparagus officinalis* L.). Z Pflanzenzücht 82: 162–173
Lovell PH, Booth A (1967) Effects of gibberellic acid on growth, tuber formation and carbohydrate distribution in *Solanum tuberosum*. New Phytol 66: 525–537
Malnassy P, Ellison JH (1970) Asparagus tetraploids from callus tissue. HortScience 5: 444–445
Mousseau M (1986) CO_2 enrichment in vitro. Effect on autotrophic and heterotrophic cultures of *Nicotiana tabacum* (var. Samsun). Photosynth Res 8: 187–191
Murashige T, Skoog F (1962) A revised medium for rapid growth and bio-assays with tobacco tissue cultures. Physiol Plant 15: 473–497
Murashige T, Shabe MN, Hasegawa PM, Takatori FH, Jones JB (1972) Propagation of asparagus through shoot apex culture. I. Nutrient medium for formation of plantlets. J Am Soc Hortic Sci 97: 158–161
Palmer CE, Barker WG (1972) Changes in enzymes activity during elongation and tuberization of stolons of *Solanum tuberosum* L. cultured in vitro. Plant Cell Physiol 13: 681–688
Reuther G (1977) Adventitious organ formation and somatic embryogenesis in callus of asparagus and iris and its possible application. Acta Hortic 78: 217–224

Reuther G (1984) Asparagus. In: Sharp WR, Evans DA, Ammirato PV, Yamada Y (eds) Handbook of plant cell culture, vol 2. Macmillan, New York, pp 211–242
Rivière S, Muller JF (!974) Résultats obtenus par la multiplication végétative in vitro des bourgeons souterrains de *l'Asparagus officinalis* L. C R Acad Sci Paris Ser D 279: 889–902
Takatori FH, Murashige T, Stillman JI (1968) Vegetative propagation of asparagus through tissue culture. HortScience 3: 20–22
Thévenin L (1966) Les problèmes d'amélioration chez *Asparagus officinalis* L. 1. Biologie et amélioration. Ann Amelior Plantes 17: 33–66
Thévenin L, Doré C (1976) L'amélioration de l'asperge (*Asparagus officinalis* L.) et son atout majeur, la culture in vitro. Ann Amelior Plantes 26: 655–674
Thuesen A (1960) Cytogenetical studies in *Asparagus officinalis* L. R. Vet Agric Univ, Yearb, Copenhagen, 47: 47–71
Wang CS (1973) The mother-stalk culture of asparagus in Taiwan. In: 4e Réunion sur la sélection de l'asperge. CRNA, Versaille, pp 27–31
Wilmar C, Hellendorn M (1968) Growth and morphogenesis of asparagus cells cultured in vitro. Nature 217: 369–370
Yakuwa T, Harada T, Saga K, Shiga Y (1971a) Studies on the morphogenesis of asparagus. I. Callus formation originating in the pith tissue of asparagus spears in tissue culture. J Jpn Soc Hortic Sci 40: 230–236
Yakuwa T, Harada T, Saga K, Shiga Y (1971b) Studies on the morphogenesis of asparagus. II. Effect of auxins and 6- benzyladenine on callus formation and organ formation of stem pieces cultured in vitro. J Jpn Soc Hortic Sci 41: 347–353
Yakuwa T, Harada T, Inagaki N, Shiga Y (1972) Studies on the anther culture of horticultural crops. I. Callus and organ formation in anther culture of asparagus. J Jpn Soc Hortic Sci 42: 272–280
Yakuwa T, Harada T, Tsuji H (1982) Studies on the morphogenesis of asparagus. IV. The effect of transplanting on callus and organ formation of stem segments cultured in vitro. J Fac Agric Hokkaido Univ 61: 151–159
Yang HJ (1977) Tissue culture technique developed for asparagus propagation. HortScience 12: 140
Yang HJ, Clore WJ (1973) Rapid propagation of asparagus through lateral bud culture. HortScience 8: 141–143
Yang HJ, Clore WJ (1974a) Development of complete plantlets from moderately vigorous shoot of stock plants of asparagus in vitro. HortScience 9: 138–139
Yang HJ, Clore WJ (1974b) Improving the survival of aseptically cultured asparagus plants in transplanting. HortScience 9: 235–236
Yang HJ, Clore WJ (1975) In vitro reproductiveness of asparagus stem segments with branched-shoots at a node. HortScience 10: 411–412

I.3 Micropropagation of Lettuce (*Lactuca sativa* L.)

D.A.C. PINK[1]

1 Introduction

1.1 Origin

Lettuce (*Lactuca sativa*) is a cool-season leafy vegetable belonging to the Cicoreae tribe of the family Compositae. Evidence from Egyptian tomb paintings indicates that the crop has been cultivated since before 4500 B.C. It is thought to have originated from a progenitor similar to *L. serriola* (Durst 1930; Ryder and Whitaker 1976), cultivated possibly as a forage or oil crop; primitive forms of *L. sativa* are reportedly cultivated for such purposes in Egypt today (Lindqvist 1960). Selection during domestication resulted in a loss of spines, reduction in latex content, and an increase in the time taken to bolt. Seed size increased and the development of nonreflexed involucres prevented seed shattering. Changes in leaf shape and development of the heading character have resulted in six broad morphological types of lettuce: crisphead, butterhead or cabbage, cos, leaf, stem and Latin. Full descriptions of these are given by Ryder (1979). The crisphead is the most extensively grown type in the US and has become increasingly popular in Europe. Stem lettuce is a Chinese vegetable and is thought to have evolved separately from a primitive lanceolate leaf type which migrated eastwards (Helm 1954).

1.2 Distribution, Importance, and Use

Lettuce is grown on all continents, but the greatest producers and consumers are Europe and the United States (Table 1). Large areas of lettuce are also grown in southeastern Australia, Japan, China, northern Mexico, Chile, Argentina, Brazil, and Peru. In the USA, lettuce is grown principally as an outdoor crop with seasonal rotation of production. Thus in California, which accounts for 70% of US production, summer crops are grown in the cool coastal valleys while the warmer desert valleys are used to produce winter crops. In northern Europe outdoor production is limited to the summer months; during the winter, lettuce is either produced as a protected crop or imported from Spain, Israel, or the USA. The value of the US crop is nearly $800 million (USDA 1990) and in the UK lettuce is the single most valuable vegetable crop, with a value of £75 million (MAFF 1986).

[1] Horticulture Research International, Wellesbourne, Warwick, CV35 9EF, United Kingdom

Biotechnology in Agriculture and Forestry, Vol. 19
High-Tech and Micropropagation III (ed. by Y.P.S. Bajaj)

Table 1. Area of lettuce cultivation and production in EEC countries (Eurostat 1989) and the United States (USDA 1990)

	Area (1000 ha)	Production (1000 t)
EEC	75.9	1866
Spain	25.0	565
Italy	19.9	383
France	13.0	338
United Kingdom	7.0	222
Fed. Rep. Germany	3.8	82
Netherlands	3.5	110
Greece	3.0	65
Belgium	2.4	85
Portugal	1.3	29
Eire	0.3	5
Denmark	0.2	8
Luxembourg	0.0	0
USA	98.2	3819
California	68.2	2909

Although there are recorded instances of it being used as a tobacco substitute (Ryder 1979) lettuce is almost exclusively consumed as a fresh raw salad vegetable, the exception being stem lettuce, which is cooked before being eaten. In terms of nutritional value, lettuce makes a minimal contribution to diet since on a fresh weight basis, 95% of the plant is water. However, lettuce is the single most popular vegetable in the UK, and when overall bulk is considered, lettuce is rated fourth behind tomato, citrus, and potato in terms of overall contribution to nutrition (Rick 1978). In the USA it ranks only behind potato (Anon 1983). Lettuce provides significant amounts of phosphorus, iron, sodium, and potassium (Watt and Merrill 1963). Other nutrients are present in varying amounts in the different lettuce types (Ryder 1979).

1.3 Conventional Breeding

Cultivated lettuce has a narrow genetic base and its close wild relatives offer breeders a potential source of useful variation especially for resistance to diseases such as downy mildew (*Bremia lactucae* Regel). *L. sativa*, together with *L. serriola*, *L. saligna*, and *L. virosa* constitutes a group with 2n = 18, which is reproductively isolated from the remainder of the genus. *L. sativa* and *L. serriola* can be crossed easily to produce fertile hybrids and are often considered as the same species. Crossing *L. sativa* with *L. saligna* is more difficult, while crosses between *L. sativa* and *L. virosa* are rarely successful and produce sterile hybrids. Nevertheless, at least one lettuce cultivar has *L. virosa* germplasm in its pedigree (Thompson and Ryder 1961).

Lettuce is a seed-propagated inbreeding species in which a low level of outcrossing (ca. 5%) occurs naturally. Controlled cross-pollinations can be achieved using a variety of methods (Ryder 1985) but none of these reliably avoid the production of selfed seed. The genetics of lettuce has, until recently, not been studied extensively. However, Michelmore and coworkers (Landry et al. 1987) are attempting to produce a genetic map of the lettuce genome using conventional genetic analysis and analysis of restriction fragment length polymorphisms (RFLPs).

Intensive breeding effort has been expended only on the crisp and butterhead types. In some cases single plant selections from commercial crops have resulted in improved plant lines. Conventional breeding programmes have, however, generally used backcrossing and pedigree selection methods depending upon the diversity of the parental material. Single-seed descent (SSD) has been shown to be potentially more efficient than pedigree selection (Maxon-Smith 1984).

1.4 The Need for in Vitro Culture of Lettuce

Somaclonal variation, induced by regeneration from callus (Sibi 1976, 1979; Brown et al. 1986) and protoplasts (Engler and Grogan 1983) has been demonstrated for lettuce.

Although many of the variants produced were not novel (Brown et al. 1986), the frequency with which they were recovered suggests this may be an efficient method of inducing variation and widening the genetic base. Similarly, protoplast fusion techniques may also widen the gene pool available to breeders by allowing the introduction of useful genes from wild species which are difficult or impossible to cross with *L. sativa*.

Some characteristics can only be assessed after the plant has been harvested, e.g., russet spotting during cold storage. In such cases the ability to recover selected plants by micropropagation is of great value to the breeder. Similarly, being able to induce rapid flowering by maintaining regenerated plants at or above 18 °C (Sibi 1976; Pink and Carter 1987, Sect. 3.4) would be useful for breeding purposes. This is especially so for a single-seed descent programme in crisp lettuce, which requires a lot of effort to keep plants healthy during their vegetative stage.

2 Review of Tissue Culture Work

A large number of reports of in vitro culture and regeneration of lettuce are found in the literature. Doerschug and Miller (1967) were the first to regenerate lettuce plants successfully; since then regeneration has been reported from various explants including: meristems (Koevary et al. 1978; Bloksberg and Saltveit 1986; Pink and Carter 1987), protoplasts (Berry et al. 1982; Engler and Grogan 1983; Brown 1986), callus (Doerschug and Miller 1967; Sibi 1976, 1979; Brown et al. 1986), suspension cultures (Alconero 1983), and hypocotyls and cotyledons (Doerschug and Miller 1967; Webb et al. 1984). Studies have also been carried out on tissue culture of related wild *Lactuca* species (Alconero 1983; Brown et al. 1987).

Because lettuce is amenable to tissue culture, in vitro systems have also been used as tools in physiological studies with no attempt at regeneration of plants. Thus lettuce callus was one of several types used to investigate the effect of 2,4-dichlorophenoxyacetic acid (2,4-D) on the production of volatiles (Thomas and Murashige 1979). Similarly, callus cultures derived from lettuce stem and petiole were among the systems used to study chlorophyll production in vitro (Vasil and Hildebrandt 1966; Fukami and Hildebrandt 1967). Suspension cultures of lettuce have been used to study growth energetics in vitro (Lamport 1964), lignification (Ganapathy and Hildebrandt 1967), and the production of antimicrobial substances (Veliky and Latta 1974). Roberts and co-workers used in vitro culture of lettuce pith explants to study tracheary differentiation (Dalessandro and Roberts 1971; Johnson and Roberts 1978; Tucker et al. 1986). Eenink (1983) investigated the usefulness of storing pollen as an aid to lettuce breeding. He assessed the viability of stored pollen by in vitro germination. Lettuce was also among 77 species used by Binding et al. (1981) in a comparative study of protoplast regeneration; however, they failed to regenerate plants from lettuce protoplasts.

3 Micropropagation

As stated above, regeneration of lettuce plants has been achieved via the in vitro culture of various explants. However, many of these techniques do not fall within the general description of micropropagation, i.e., they are not suitable for the mass production of clonal material. For the purposes of the present chapter, attention is restricted to micropropagation using meristem cultures, callus and suspension cultures, or organogenesis from cotyledons or hypocotyls. For a general account of lettuce tissue culture see Alconero (1988).

3.1 Sterilization of Explants

The procedures that have been used for surface sterilization of lettuce explants are similar, and generally involve immersion in a chlorine solution followed by several washes in sterile distilled water or in some cases (Berry et al. 1982; Brown et al. 1986) sterile tap water. Some workers have also included a pre-sterilization treatment. These include washing in water and Tween 80 (Bloksberg and Saltveit 1986) and dipping in 40% ethanol for a few seconds (Alconero 1983).

The majority of workers have used diluted domestic bleach (e.g., NaOCl) as their surface sterilant, but a 0.1% $HgCl_2$ solution was used by Doerschug and Miller (1967) to sterilize seeds and Pink and Carter (1987) used a solution of sodium dichloroisocyanurate ($C_3N_3O_3Cl_2Na$) to sterilize axillary buds. The latter gives a more controlled amount of available chlorine than bleach. The strength of chlorine solution used has been as low as 0.1% (Koevary et al. 1978) but a 1% solution is more usual. The time of immersion in the surface sterilant varies between workers. Bloksberg and Saltveit (1986) carried out a factorial experiment in which the concentration of hypochlorite and period of immersion was varied.

Table 2. Summary of in vitro culture of lettuce

Explant	Growth response	Medium	Plant growth regulators (mg/l)	Reference
Apical segments and expanded axillary buds	Shoots	Miller's	0.5 IAA + 0.5 K	Koevary et al. (1978)
	Roots			
Axillary buds	Shoots	MS salts +	1 IAA + 0.5 BA	Bloksberg and Saltveit (1986)
	Roots	B5 vitamins	None	
Axillary buds	Shoots	MS	1–2 K + 6.4 IAA	Pink and Carter (1987)
	Roots		4.8–6.4 IAA	
Cotyledons	Callus	MS macroelements + Heller's microelements + Morel's vitamins	0.02 NAA + 0.2 K	Sibi (1976)
	Shoots	As above	0.02 NAA + 0.02 K	
	Roots	As above	0.003 IBA	
Cotyledons	Callus and shoots	MS	0.5 BA + 0.05 NAA	Brown et al. (1986)
	Roots	MS	None	
Leaf segments	Callus and shoots	Miller's	5 IAA + 0.5 K	Koevary et al. (1978)
	Roots	Miller's	1 IBA	
Seedling roots	Callus, roots, and shoots	Miller's	0.5 K + 5 IAA	Doerschug and Miller (1967)
Hypocotyl	Callus	MS	0.1 2,4-D + 1 K + 80 adenine sulphate	Thomas and Murashige (1979)

Petiole and stem	Callus	T	15% coconut milk + 0.1 NAA 2.5 Ca pantothenate	Fukami and Hildebrandt (1967)
Petiole and stem	Callus	MS	10 IAA + 10 K	Vasil and Hildebrandt (1966)
Leaf and root segments	Callus	Venketeswaran	Not given	Bolton and Bozarth (1974)
Leaf segments	Callus	B5 or MS	0.1 Picloram or 1–2 NAA + 0.5–1 K or BA	Alconero (1983)
	Suspension culture	B5	1–2 NAA, 0.5 BA	
	Shoots	B5 or MS	1–3 K or BA	
	Roots	B5	1.2 NAA	
Petiole	Callus	C	0.1 NAA + 15% coconut milk	Ganapathy and Hildebrandt (1967)
	Suspension culture	C	As above	
		D	0.1 NAA, + 0.006 2,4-D + 15% coconut milk	
Seed	Suspension culture	White (1943)	20% deproteinized coconut water	Lamport (1964)
Cotyledons and hypocotyls	Organogenesis	Millers	0.5 K + 5 IAA	Doerschug and Miller (1967)
Cotyledons	Organogenesis	MS	0.5 K or 0.5 K + 5 IAA	Webb et al. (1984)
Hypocotyls	Organogenesis	MS or White's	Various	Sasaki (1975, 1979 a, b, c; 1982)
Cotyledons	Organogenesis	MS	1 IAA + 1 K	Kadkade and Seibert (1977)

They found that the best protocol was to shake the explants (axillary buds) for 20 min in four times their volume of 1% NaOCl solution, with two drops of Tween 80 added to each 100 ml of solution. In contrast, Engler and Grogan (1983) immersed leaves for only 30 s in a 0.5% NaOCl solution. However, they were dealing with leaves from plants grown in a growth chamber, whereas Bloksberg and Saltveit were working with axillary buds taken from field-grown plants. Significantly higher levels of contamination have been found in cultures initiated from axillary buds taken from field-grown plants compared with cultures initiated from glasshouse-grown plants (Pink and Carter 1987). The majority of contaminants were bacteria, which it was presumed were present in the intercellular spaces. These might be eliminated by initiating onto a medium containing an antibiotic. However, some caution should be exercised in using antibiotics, since gentamicin at least can have an inhibitory effect on lettuce tissue grown in vitro (Dodds and Roberts 1981). Bolton and Bozarth (1974) eliminated the problem of contamination in callus cultures by growing their "donor" plants axenically from surface sterilized seeds.

3.2 Media

The majority of workers have used Murashige and Skoog's (1962) medium (MS) in their studies. However, various other media have been used to culture lettuce in vitro. These are summarized in Table 2.

3.3 Regeneration

3.3.1 Meristem Culture

Koevary et al. (1978) successfully regenerated lettuce plants from apical segments and expanded axillary buds. Regeneration was not possible from nonexpanded axillary buds. In contrast, Pink and Carter (1987) successfully regenerated plants of the butterhead cv. Capitan and butterhead and crisp breeding lines from both expanded and nonexpanded buds (Fig. 1). They did, however, find that nonexpanded buds took longer to develop. Plants have also been regenerated from axillary buds of the cv. Winterhaven (Bloksberg and Saltveit 1986).

Shoot growth has been promoted on media containing indole-acetic acid (IAA) and benzyladenine (BA) (Bloksberg and Saltveit 1986) or IAA and kinetin (Koevary et al. 1978; Pink and Carter 1987), and root formation has been induced using IAA (Pink and Carter 1987), indole-butyric acid (IBA) (Koevary et al. 1978), or on medium containing no growth hormones (Bloksberg and Saltveit 1986) (Table 2). Pink and Carter found that the optimum concentration of IAA was dependent upon the concentration of kinetin in the shoot growth medium. Thus cultures initially grown on medium containing 1 mg/l kinetin rooted best on MS + 6.4 mg/l IAA, while the best root formation in cultures initially grown on medium containing 2 mg/l kinetin occurred on MS + 4.8 mg/l IAA (Table 3).

Fig. 1. Lettuce plantlet regenerated from axillary bud, 18 days after initiation on MS and 1 mg/l kinetin

Table 3. Effect[a] of concentration of growth substances on the root (R) and shoot(s) production of meristem cultures of lettuce cv. Capitan. (Pink and Carter 1987)

Concentration of IAA in rooting medium (mg/l)	Concentration of kinetin in shoot growth medium (mg/l) 0.5	1	2
3.2	R/S	R/S	R/S
4.8	60/40	30/90	70/60
6.4	40/50	100/60	90/20
8.0	70.50	10/30	10/10
Significance of interaction between kinetin and IAA concentrations for:		Shoots $p < 0.001$ Roots $p < 0.001$	

[a] Expressed as % cultures showing good shoot or root growth.

In some studies root initiation was enhanced by reduced light intensity (Koevary et al. 1978; Bloksberg and Saltveit 1986), while in others the opposite was found (Pink and Carter 1987).

3.3.2 Callus and Suspension Cultures

Regeneration has been successful from callus initiated from cotyledons (Sibi 1976, 1979; Brown et al. 1986; Brown 1986), leaf segments (Koevary et al. 1978), and from seedling roots (Doerschug and Miller 1967). However, regeneration was not very efficient from the latter. Regeneration was more successful and quicker from callus initiated from segments taken from the periphery of the leaf than from callus derived from segments taken from the center (Koevary et al. 1978). The regenerative capacity of callus was genotype-dependent (Brown 1986) with relatively high frequencies of regeneration for cvs. Salad Bowl, Red Salad Bowl, Lobjoits Cos Green, and Pennlake. Some cultivars, particularly butterhead types, produced large numbers of regenerants but these were difficult to root. Callus produced from cv. Sabine completely lacked regenerative capacity.

Sibi (1976) used a medium based upon MS macroelements, microelements of Heller (Gautheret 1959), and vitamins of Morel (Gautheret 1959), with NAA and kinetin added for shoot initiation and IBA for root formation (Table 2). Sasaki (1975) found that a combination of a low concentration of 2,4-D (0.01 mg/l) and high concentration of kinetin (0.5 mg/l) enhanced the induction of adventitious buds in callus derived from lettuce hypocotyls. Brown (1986) used MS with NAA and BA to initiate and maintain callus from cotyledons. Regeneration of shoots occurred on this medium and rooting was achieved on MS with no plant growth regulators. Doerschug and Miller (1967) and Koevary et al. (1978) regenerated plants using Miller's (1963) medium.

Light intensity had a differential effect upon shoot and root formation (Koevary et al. 1978). High light intensity promoted earlier shoot development, whereas a low intensity promoted root initiation without shoot development. Temperature also markedly influenced rate of regeneration. The optimum regime found by Koevary et al. (1978) was to culture leaf sections at 25 °C until callus had developed and buds formed, followed by transfer to 22.5 °C for plantlet development.

Sibi (1979) noted variation among regenerants for leaf morphology, color, and development of axillary buds. Brown et al. (1986) found that approximately 6% of the progeny of regenerants exhibited variation at the seedling stage. Most of the variants were deleterious recessive mutations. However one line, derived from cv. Salad Bowl, had increased vigor, flowered earlier, and had a higher chlorophyll content. There was also variation among regenerant lines for reaction to lettuce mosaic virus and downy mildew (*Bremia lactucae*).

Regeneration has also been achieved from cell suspension cultures (Alconero 1983). These were derived from callus produced from young expanded leaves cultured in the dark on Gamborg's B5 (Gamborg et al. 1968) or MS + picloram or NAA and kinetin or BA, for one month at 24–27 °C. Suspension cultures were maintained in B5 medium + NAA and BA. Regeneration of shoots occurred after plating onto solid B5 or MS medium containing kinetin or BA and root formation was induced on B5 + NAA (Table 2).

3.3.3 Organogenesis

Doerschug and Miller (1967) regenerated plants from both cotyledons and hypocotyls of several lettuce cultivars, but found cotyledons to be more responsive. Webb et al. (1984) found that cotyledons from 4-day-old seedlings gave the best regeneration of shoots and roots; root formation decreased with increasing age. Sasaki (1979a, b, c; 1982) has carried out a detailed investigation into the factors influencing adventitious bud formation in hypocotyls. His protocol involved preculturing hypocotyls on an "initial" medium followed by subculture on a "transfer" medium. No differences in responsiveness were detected between MS or White's basal medium. The addition of casamino acids to both media had a beneficial effect. However, these were beneficial across a wide range in the initial medium (0.2–0.8%), whereas the optimum in the transfer medium was restricted to 0.2%. Similarly, kinetin and 2,4-D were beneficial over a wide range of concentrations in the initial medium (Sasaki 1979b). The presence of 2,4-D in the transfer medium was inhibitory and bud initiation occurred on media containing kinetin only. Webb et al. (1984) induced bud formation on MS + IAA and kinetin or MS + kinetin alone. In contrast, Doerschug and Miller (1967) could not initiate buds on media containing only kinetin. Adenine could be substituted for kinetin, and NH_4NO_3 also induced bud formation (Sasaki 1979a). IAA, IBA, and NAA did not stimulate bud formation whereas zeatin, 2-isopentenyl adenine (2-ip), BA, and kinetin were all active (Sasaki 1979c). The ribosides of these cytokinins were much less active. The addition of sucrose and maltose to the initial medium enhanced organogenesis but fructose was detrimental. However, addition of fructose together with glucose and sucrose to the transfer medium was beneficial (Sasaki 1979a).

Bud formation was stimulated by low temperature followed by high temperature (Sasaki 1982) and by light (Sasaki 1979c; Webb et al. 1984). Exposure to red light (660 nm) resulted in a 2- to 2.5-fold increase in number of shoots (Kadkade and Seibert 1977). This effect could be reversed by subsequent exposure to far red light (744 nm).

3.4 Establishment of Plants

One of the major difficulties in establishing plants is desiccation. It is therefore necessary to maintain conditions of high humidity when transplanting plantlets to the glasshouse. Generally, this is done by enclosing the plantlets under some form of plastic covering. They are then "weaned" by slowly reducing the humidity over a 1–3-week period. Koevary et al. (1978) reduced transplanting shock by transferring plantlets together with the intact agar medium. However, this can lead to problems of infection by fungal and bacterial pathogens, and other workers have removed the media by washing in distilled water (Bloksberg and Saltveit 1986).

Infection can be reduced by transferring plantlets to sterilized vermiculite (Koevary et al. 1978) or soil-less compost (Bloksberg and Saltveit 1986). Plantlets

Fig. 2. Regenerated plant established in Jiffy 7 peat block, 7 days after transfer from culture

can be established in Jiffy 7 peat blocks (Fig. 2) and fungal infection can be reduced if these are re-hydrated in a solution of benomyl (Pink, unpubl.). Using these techniques, workers have reported success rates varying from 50% (Bloksberg and Saltveit 1986) to 95% (Koevary et al. 1978; Pink and Carter 1987).

The temperature at which established plants are grown has a marked effect upon their growth. Plants grown at 18 °C or above failed to heart, and bolted to produce flower stalks approximately 6 weeks after establishment (Fig. 3) (Pink and Carter 1987). Photoperiod had no effect upon bolting (Sibi 1976) and control plants grown from seed under identical conditions grew normally.

Variation has been reported among regenerants by some workers. Often this variability decreases with age (e.g., Koevary et al. 1978). Sibi (1976, 1979) and Brown et al. (1986) carried out detailed studies of somaclonal variation in the progeny of regenerated plants. Their plants, however, were regenerated from callus. In general, plants regenerated from meristematic tissue displayed little or no somaclonal variation (e.g., Pink and Carter 1987).

4 Commercial Aspects

The various in vitro tissue culture techniques now available can be used to assist or replace conventional methods of plant propagation and breeding (Walkey 1986). In the case of lettuce, it is unlikely that micropropagated plants will replace conventional seed-raised plants for the establishment of a commercial field crop. At the

Fig. 3. Flowering regenerated plant, 18 weeks after initiation of in vitro cultures

time of writing, the approximate cost of a plant produced via meristem culture is £(UK) 0.25. It is unlikely that the unit cost will ever compete with that of seedlings raised using modern module systems, which is slightly over 1 penny. However, work is being carried out at HRI-Littlehampton (formerly the Glasshouse Crops Research Institute) into methods of "capturing" hybrid lettuce selections for glasshouse crops. F_1 hybrid lettuce seed is both difficult and expensive to produce, but by micropropagating selected F_1 plants it is possible to assess their performance against traditional inbred cultivars. Great problems have been encountered with premature flowering of the micropropagated plants. In some instances, plants flower while still in the culture tubes; however, in general the plants only bolt once established in the glasshouse (Fig. 4). Although this may be an advantage for the breeder for shortening his generation times, it is obviously disadvantageous when using micropropagated plants to establish a crop. Although there appear to be genotypic differences for the time taken to flower, all genotypes so far tested have flowered prematurely after establishment from culture.

The economic feasibility of using an in vitro system for establishing a glasshouse crop depends upon whether F_1 hybrid crisp lettuce has sufficient advantage

Fig. 4. F_1 hybrid between two winter crisphead cultivars. *Left* multiplied by micropropagation via axillary buds and showing a high proportion of bolting plants; *right* grown from seed and showing no bolting. (Photograph courtesy of Dr. F.A. Langton, Horticulture Research International, Littlehampton)

over inbreds to allow glasshouse production in the UK in midwinter. On present costings it would seem unlikely (Dr. A Langton, pers. commun). Similarly, workers in Japan are investigating the possibility of producing glasshouse crops of F_1 hybrid lettuce using artificial seeds. These are produced by encapsulation of lettuce shoots regenerated from callus derived from young leaf material on MS + BA and IAA. (Dr. M Sanada, pers. commun). Although it is possible to obtain regeneration rates of 10^5 shoots per gram of callus in a 6-month period, the cost of artificial seed is still three to four times that of true seed. It would appear, therefore, that at present it is only when fewer plants of selected valuable breeding lines are involved that the costs of micropropagation are acceptable (Koevary et al. 1978; Bloksberg and Saltveit 1986; Pink and Carter 1987). The routine establishment of commercial crops using in vitro techniques will not occur until methods to reduce costs (e.g., mechanization of micropropagation) are developed, or unless the resultant crop displays sufficient advantage over conventional crops (e.g., in disease resistance or vigor).

5 Protocol

Excise axillary buds, remove outer leaves of larger buds to give an explant 2–3 mm long. Surface sterilize in a 1% chlorine solution for 5 min followed by two washes in sterile distilled water. Place explant on surface of 8 ml MS medium at pH 5.7 containing: 30 g/l sucrose; 0.1 mg/l thiamine HCl; 100 mg/l inositol; 8 g/l agar; 1 mg/l kinetin and 6.4 mg/l IAA contained in a 75 × 25 mm glass tube sealed with sterile unglazed polypropylene film (GS Packaging Ltd.[2]). Transfer regenerated shoots (Fig. 1) to 8 ml MS medium at pH 5.7, containing sucrose, inositol, and thiamine HCl at above concentrations plus 6 g/l agar and 6.4 mg/l IAA to induce rooting. Carefully remove rooted plants from tubes and wash off agar medium. Transplant plantlets to Jiffy 7 peat blocks pre-soaked in a 0.1% benomyl (Benlate, DuPont) solution (Fig. 2) and place plants in a plastic propagator (Ward) with vents closed. Wean plants by opening vents a small amount each day over a 1–2-week period. Once weaned, transplant to pots of Levington's soil-less compost (Fisons Plc) and grow on in glasshouse. Plants grown at 18 °C or above will not heart and will bolt early (Fig. 3).

References

Alconero R (1983) Regeneration of plants from cell suspensions of *Lactuca saligna*, *Lactuca sativa* and *Lactuca serriola*. Hortic Sci 18: 305–307

Alconero R (1988) Lettuce (*Lactuca sativa* L.). In: Bajaj YPS (ed) Biotechnology in agriculture and forestry, vol 6. Crops II. Springer, Berlin Heidelberg New York Tokyo p351–369

Anon (1983) Agricultural statistics. United States Department of agriculture. US Govt Print Office, Washington, DC

Berry SF, Lu DY, Pental D, Cocking EC (1982) Regeneration of plants from protoplasts of *Lactuca sativa* L. Z Pflanzenphysiol 108: 31–38

Binding H, Nehls R, Kock R, Finger J, Mordhurst G (1981) Comparative studies on protoplast regeneration in herbaceous species of the Dicotyledonae class. Z Pflanzenphysiol 101: 119–130

Bolton WE, Bozarth GA (1974) Establishment of tissue cultures from mature axenic lettuce plants. Phyton 32: 77–80

Bloksberg LN, Saltveit ME (1986) Regeneration of plants from axillary buds of harvested and stored heads of field-grown Iceberg lettuce. Hortic Sci 21: 1201–1203

Brown C (1986) The generation of variability in lettuce (*Lactuca sativa L.*) using tissue culture methods. Ph.D. Thesis, Univ Nottingham

Brown C, Lucas JA, Crute IR, Walkey DGA, Power JB (1986) An assessment of genetic variability in somaclonal lettuce plants (*Lactuca sativa* L.) and their offspring. Ann Appl Biol 109: 391–407

Brown C, Lucas JA, Power JB (1987) Plant regeneration from protoplasts of a wild lettuce species (*Lactuca saligna* L.) Plant Cell Rep 6: 180–182

Dalessandro G, Roberts LW (1971) Induction of xylogenesis in pith parenchyma explants of *Lactuca*. Am J Bot 58: 378–385

Dodds JH, Roberts LW (1981) Some inhibitory effects of gentomycin on plant tissue cultures. In Vitro 17: 467–470

Doerschug MR, Miller CO (1967) Chemical control of adventitious organ formation in *Lactuca sativa* explants. Am J Bot 54: 410–413

Durst CE (1930) Inheritance in lettuce. Ill Agric Exp Stn Bull 356: 237–341

Eenink AH (1983) Preliminary results of research on storage and in vitro germination of lettuce pollen as an aid in lettuce breeding. Euphytica 32: 521–526

Engler DE, Grogan RG (1983) Isolation, culture and regeneration of lettuce leaf mesophyll protoplasts. Plant Sci Lett 28: 223–229

[2] GS Packaging Ltd, Aber Works, Aber Road, Flint, Clwyd, UK.

Eurostat (1989) Agriculture Statistical yearbook. Statistical office of the European communities, Luxembourg, Brussels

Fukami T, Hildebrandt AC (1967) Growth and chlorophyll formation in edible green plant callus tissues in vitro on media with limited sugar supplements. Bot Mag Tokyo 80: 199–212

Gamborg OL, Miller RA, Ojima K (1968) Nutrient requirements of suspension cultures of soybean root cells. Exp Cell Res 50: 151–158

Ganapathy PS, Hildebrandt AC (1967) Suspension cultures and lignification in *Lactuca sativa* L. Indian J Exp Biol 5: 196–197

Gautheret RJ (1959) La culture des tissus végétaux. Masson, Paris

Helm J (1954) *Lactuca sativa* in morphologisch-systematischer Sicht. Kulturpflanze II

Johnson DC, Roberts LW (1978) Waterstress influences xylogenesis in cultured explants of *Lactuca*. Phytomorphology 28: 297–209

Kadkade P, Seibert M (1977) Phytochrome-regulated organogenesis in lettuce tissue culture. Nature (London) 270: 49–50

Koevary K, Rappaport L, Morris LL (1978) Tissue culture propagation of Head lettuce. Hortic Sci 13: 39–41

Lamport DTA (1964) Cell suspension cultures of higher plants: isolation and growth energetics. Exp Cell Res 33: 195–206

Landry BS, Vesseli RV, Farrara B, Michelmore RW (1987) A genetic map of lettuce (*Lactuca sativa* L.) with restriction fragment length polymorphism, isozyme, disease resistance and morphological markers. Genetics 116: 331–337

Lindqvist K (1960) On the origin of cultivated lettuce. Hereditas 46: 319–350

MAFF (1986) Basic horticultural statistics for the United Kingdom, calendar and crop years 1976–1985. Ministry of Agriculture, Fisheries and Food, London

Maxon-Smith JW (1984) Glasshouse lettuce. Glasshouse Crops Res Inst, Annu Rep 1984: 24–29

Miller CO (1963) Kinetin and kinetin-like compounds. In: Linskens HF, Tracey MV (eds) Modern methods of plant analysis, vol 6. Springer, Berlin Heidelberg pp 194–202

Murashige T, Skoog F (1962) A revised medium for rapid growth and bioassays with tobacco cultures. Physiol Plant 15: 473–497

Pink DAC, Carter, PJ (1987) Propagation of lettuce (*Lactuca sativa*) breeding material by tissue culture. Ann Appl Biol 110: 611–616

Rick CM (1978) The tomato. Sci Am 239: 67–76

Ryder EJ (1979) Leafy salad vegetables. AVI, Westport, Connecticut, pp 13–94

Ryder EJ (1985) Lettuce breeding. In: Bassett M (ed) Breeding vegetable crops. AVI, Westport, Connecticut

Ryder EJ, Whitaker TW (1976) Lettuce *Lactuca sativa* (Compositae) In: Simmonds Evolution of crop plants. NW (ed) Longman, London, pp 39–41

Sasaki H (1975) Physiological and morphological studies on development of vegetable crops. III Adventitious bud formation of callus tissues derived from lettuce hypocotyl. J Jpn Soc Hortic Sci 44: 138–143

Sasaki H (1979a) Physiological and morphological studies on development of vegetable crops. IV Effect of various media on the adventitious bud formation of lettuce hypocotyl tissue cultured in vitro. J Jpn Soc Hortic Sci 47: 479–484

Sasaki H (1979b) Physiological and morphological studies on development of vegetable crops. V. Action of 2,4-D and kinetin on the adventitious bud formation of lettuce hypocotyl tissue cultured in vitro. J Jpn Soc Hortic Sci 48: 61–66

Sasaki H (1979c) Physiological and morphological studies on development of vegetable crops. VI. Effect of several auxins, cytokinins and cytokinin-ribosides on the adventitious bud formation of lettuce hypocotyl tissue cultured in vitro. J Jpn Soc Hortic Sci 48: 67–72

Sasaki H (1982) Effect of temperature and light on adventitious bud formation of lettuce hypocotyl tissue cultured in vitro. J Jpn Soc Hortic Sci 51: 187–194

Sibi M (1976) La notion de programme genetique chez les végétaux superieurs. Ann Amelior Plant 26: 523–547

Sibi M (1979) Expression of cryptic factors in vivo and in vitro. Proc Conf Broadening Genet Base Crops, Wageningen 1978. Pudoc, Wageningen, pp 339–340

Thomas DS, Murashige T (1979) Volatile emissions of plant tissue cultures. II Effects of the auxin 2, 4-D on production of volatiles in callus cultures. In Vitro 15: 659–663

Thompson RC, Ryder EJ (1961) Description and pedigrees of nine varieties of lettuce. USDA Tech Bull 1244

Tucker WQJ, Wilson JW, Greshoff PM (1986) Determination of tracheary element differentiation in lettuce pith explants. Ann Bot 57: 675–679
USDA (1990) Agricultural statistics 1990. United States Government Printing Office, Washington
Vasil IK, Hildebrandt AC (1966) Growth and chlorophyll production in plant callus tissues grown in vitro. Planta (Berlin) 68
Veliky IA, Latta RK (1974) Antimicrobial activity of cultured cells and tissues. Lloydia 37: 611–620
Walkey DGA (1986) Micropropagation of vegetables. In: Micropropagation in Horticulture. Univ Nottingham Symp Handbook, March 1986
Watt BK, Merrill AL (1963) USDA Agric. Handbook 8. Composition of Food
Webb DT, Torres LD, Fobert P (1984) Interactions of growth regulators, explant age, and culture environment controlling organogenesis from lettuce cotyledons in vitro. Can J Bot 62: 586–590
White PR (1943) A handbook of tissue culture. Cattell, Lancaster

I.4 Micropropagation of Horseradish (*Cochlearia armoracia* L.)

K. GÓRECKA[1]

1 General Account

1.1 Botany

Horseradish [*Cochlearia armoracia* L., *Armoracia rusticana* (Gaerth)] belongs to the *Cruciferae* family, and is the most commonly cultivated species of this genus. It was originally grown in southeastern Europe and western Asia. In the Asian part of the USSR, in Siberia, and the Far East, a wild species of horseradish, *Armoracia sisimbrioides* D.S., is found. Horseradish was first mentioned in a historical commentary from the 12th century.

According to some data, horseradish was primarily used by Slavic folks, and from these it was spread to almost all parts of Europe. Aleksieieva (1987) stated that horseradish was one of the oldest cultivated vegetable species in the European part of the continent. Jankowski (after Dąbrowska 1977) wrote that in Poland it was a commonly planted species in the 15th century.

Horseradish is a perennial species, the edible part consisting of true roots, very massive and well developed (Fig. 1). Under cultivation, the main root is long, cylindrical, covered with whitish, rough skin (cortex). The inner cylinder is white and fibrous. Although in central Europe horseradish produces plenty of flowers from May to July, it does not produce seeds without special treatment (Fig. 2). Brzeziński (after Dabrowska 1977) obtained fertile seeds after girdling a main horseradish root. Moravec and Kvasnička (1963) obtained horseradish seeds by pollination of geographically distant lines. All the researchers who succeeded in forcing horseradish to set seeds emphasize that the plants thus obtained represented a wide variety of phenotypes.

1.2 Importance of the Plant

Horseradish is one of the most important spicy plants in moderate zones. As already stated, the useful part is the root, which possesses a typically hot taste, due to the compound isothiocyan allyl, which is released only on grinding the root tissue. The roots are also used by the food industry as an ingredient for sauces and mustard dressings. Its addition improves the taste, quality, and storage of

[1] Research Institute of Vegetable Crops, 96-100 Skierniewice, Poland

Biotechnology in Agriculture and Forestry, Vol. 19
High-Tech and Micropropagation III (ed. by Y.P.S. Bajaj)

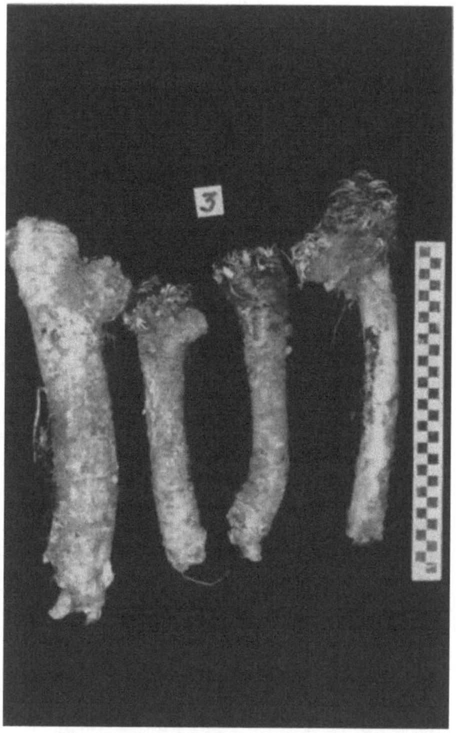

Fig. 1. Edible parts of horseradish roots as they appear in fall. The thinner roots at the base, used as root cuttings, are already removed. *Ruler length* 30 cm. (Górecka, unpubl.)

preserved cucumbers, tomatoes, and mushrooms. Horseradish helps in the digestion of animal protein, so it often goes well with meats. A Polish speciality is red beets and horseradish with bacon.

Presently, interest in horseradish is growing, due to its disinfectant and fungistatic properties. In some countries, horseradish is used medicinally, e.g., for some liver diseases. The root is a source of peroxidase for enzyme production.

1.3 Conventional Propagation

Commercially, horseradish is propagated exclusively by means of root cuttings (Fig. 3). The cutting is prepared from lateral roots, 0.8–1.4 cm in diameter and 30–35 cm long. The cutting cannot be shorter than 30 cm, because later it becomes the edible part of the plant. First-quality edible roots should be at least 25 cm long, straight, without lateral stems or roots. During the growing season the cutting does not grow longer, but forms lateral roots on the basal end and a rosette of leaves on the apical end. The cutting thickens very markedly up to 3–5 cm in diameter. Roots should be as long and thick as possible. The lateral roots are uniformly thick over the entire length, so the bases of the cuttings are marked by a slanted cut, which enables the cuttings to be planted out correctly. The bunches of prepared cuttings are stored in cold frames or cellars over the winter. In spring the cuttings are inspected for visible disease or injuries, and planted in the field.

Fig. 2. Horseradish plant in the field at the flowering phase, June 9, 1989. Sprouting of the aboveground part started in spring (March). (Górecka, unpubl.)

There are two ways of planting:

1. Cuttings are laid flat one by one on the ground and covered with shallow soil; spacing ca. 50 cm between the rows.
2. Cuttings are planted at an angle of a 15 or 45° in furrows.

1.4 Need for in Vitro Culture

One of the limiting factors in horseradish production is the viral diseases, of which more than ten exist, as listed in Table 1 in order of importance. The most common is turnip mosaic virus (TuMV). According to results of Macias (pers. commun.), TuMV infected almost 100% of the inspected horseradish cuttings. Paludan (1973) reported that TuMV caused a root yield lower by 37%. There have also been attempts to produce TuMV-free plants by means of thermotherapy. These trials, carried out in the Virusology Lab., Research Institute of Vegetable Crops, Poland, met with failure. The plants were not TuMV-free. Holmes (1965) reported that TuMV-free plants were obtained in only one of the experiments conducted. Paludan (1973), using thermotherapy, procured root cuttings free of TuMV. He obtained the best results at 38 °C for 60 days – 83% virus-free plants. However,

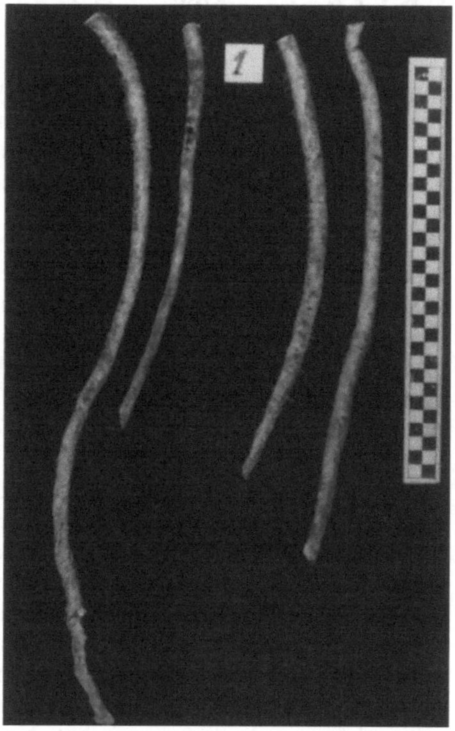

Fig. 3. Root cuttings of horseradish ready for planting out in spring. *Ruler length* 30 cm. (Górecka, unpubl.)

Table 1. Virus species found in horseradish plants. (Górecka and Macias, in press)

Virus	Abbreviation
Turnip mosaic virus	TuMV
Arabis mosaic virus	A MV
Cauliflower mosaic virus	Ca MV
Tomato black ring virus	TBRV
Tobacco mosaic virus	TMV
Beet wester yellows virus	BWYV
Malva mosaic virus	MMV

Błaszczak et al. (1977) stated, that during a 30- or 35-day incubation period at 37 °C, many cuttings died or deteriorated, and the cuttings which survived the treatment were not TuMV-free. For these reasons, attempts to produce TuMV-free plants by means of in vitro propagation were undertaken. Additionally, a rapid micropropagation method was sought by horseradish breeders to propagate and evaluate breeding material from generative crosses of different lines (clones) or varieties. Traditional vegetative propagation methods are neither rapid nor effective enough.

2 Brief Review of Tissue Culture

Horseradish, as a species, is not a widely researched plant, and the literature on tissue culture of horseradish is rather sparse, one of the reasons being its relatively low importance in comparison with other vegetable crops. It is also difficult to establish the initial shoot meristem culture, since the meristems are located on heavily infected underground roots. It is difficult to disinfect these meristems effectively without also killing the tissues (Górecka et al. 1978). This is true also with underground buds of other species. To avoid this problem in raspberry, Sobczykiewicz (1980) first sprouted the shoots and took meristems from them, well above ground level.

Hickman and Varna (1968) obtained horseradish plants from tissue cultures of meristems taken from 2-cm-long buds. These buds were washed in 96% ethanol and soaked in 5% calcium hypochlorite for 15 min. The meristems were grown on Morel and Müller medium. Mori (1971) quoted horseradish as one of 12 species for which methods of virus-free plant production had been elaborated in Japan. Horseradish was made free of TuMV virus, using 0.5 mm meristems.

Paludan (1971) propagated horseradish from meristems along with other species. Later (Paludan 1973) he applied the in vitro propagation method to produce TuMV-free plants. Liquid Murashige-Skoog's (1962) (MS) medium with paper bridges or agar media were applied. Ninety seven percent of the resulting plants were TuMV-free. The concentrations of hormones used, sterilization methods, meristem size, and culture conditions were, however, not given, and the results were reported from only one experiment. Novák and Havránek (1973) took explants from root pith, and grew the cultures in a three-stage procedure. Firstly, a green, firm callus tissue was obtained in MS medium with the addition of 2 mg/l IAA, 2 mg/l 2,4-D and 1 mg/l kinetin. Next, the callus was transferred onto a MS medium without 2,4-D, but kinetin was enriched to 10 mg/l. On this medium shoots were formed. For root formation shoots plus callus were cultured on the medium containing 1 mg NAA/l, and roots were formed in 30–40 days. Up to eight shoots were obtained from one explant. In 1977, Meyer and Milbrath reported that the highest propagation rate (eight to ten plants from one leaf fragment in 6 weeks) was obtained using LS medium (Linsmaier and Skoog 1965) with 1 mg/l NAA and 0.25 mg/l kinetin. They did not mention the range of concentrations or other growth regulators tested.

Because of limited literature on horseradish propagation in vitro, it was necessary to broaden the search, and include data for other species propagated from leaf fragments. According to different authors, there are many factors which play a role in the process of organogenesis. In particular, the role of temperature in the growth chamber was emphasized (Fonnesbeck 1974; Anderson and Carstens 1977). However, the majority of the researchers paid more attention to the hormone composition of the growth media. Data on the optimal media are very variable and inconsistent.

The present author (Górecka 1987) tried to find an effective medium for rapid propagation. The best results, 41.5 of shoots from one leaf fragment in 28 days, were obtained on LS medium containing 0.8 mg/l IAA and 4 mg/l kinetin. The resulting shoots were subsequently easily rooted for 14 days on the same medium without kinetin, with almost 100% success. The shoots were also rooted with almost 100% success directly in a peat-sand substrate, without hormones. Yamada et al. (1987) discussed the possibility of commercial production of peroxidase, utilizing horseradish tissue and cell culture techniques.

3 Micropropagation

In the first experiment, the meristems used as a source of initial explants were obtained from adventitious buds formed on roots. The roots were placed in Ca-neutralized peat substrate, in a warm greenhouse. The outgrowing buds were taken off the roots, cleaned, outer scales removed, washed with soap, and rinsed with running tap water for a 24-h period, and then sterilized. Horseradish buds are difficult to sterilize, because the roots, which grow in soil, are usually contaminated by a large spectrum of microorganisms. Therefore, strong chemicals need to be used in high concentrations and a relatively long period of treatment is often necessary. Some methods reported by different authors were also tried (Górecka et al. 1978).

The best method of sterilizing meristems in our experiments was as follows (Table 2): (1) ethanol 80% for 6 min, then chloramine T 5%, 30 min; (2) as above, but chloramine T 7.5%; (3) as above, but chloramine T 10% for 15 min. The percentage of contamination after 20 days was very low (0.4) with these methods of sterilization, the percent of plantlets transferred into pots was rather high after 40 days (52–64). In some experiments the effectiveness of sterilization seemed to be more dependent on the origin of roots than on the sterilization method. The best means of sterilization were also the most universal. They were successfully applied for meristems of roots from different regions of horseradish cultivation. It was noted that it was more beneficial to apply lower chloramine concentrations for a longer period of time.

Another problem in horseradish propagation from meristems was their size. Mostly, meristems of about 1 mm were used. A general rule was observed that the larger the meristem was, the better its growth on the medium, and the higher the probability of virus transmission to the resulting plantlets. With smaller meristems, there exists a higher probability of obtaining virus-free plants, but the meristems are likely to die more easily, and have more difficulty in starting their development on the medium. Cultures were maintained under 16-h fluorescent light (3000 lx), at 20 °C.

As already mentioned, new plants were obtained from meristems maintained on the medium for 6–7 weeks. The resulting plants were planted out into pots containing peat substrate and then were kept under plastic film, with beds mulched with moist peat to maintain higher air humidity. The inner surfaces of the plastic tunnel were also frequently sprayed with water. The plants obtained from in vitro culture adapted very well to external conditions. After 5–7 days, the plants were gradually exposed to the open air, and after a few days were ready for planting in the field. The loss of plants during this stage did not exceed a few percent.

These plants resulting from meristems were subjected to tests for the presence of TuMV. These tests were conducted by the traditional method of mechanical

Table 2. Influence of different methods of sterilization on the development and contamination of meristems (%). Experiment begun May 3–11, 1977. (Górecka et al. 1978)

Method of sterilization	Results (%) after: 20 days					40 days				
	Contaminated	Necrotic	Healthy nondeveloped	Developing	Transplanted into pots	Contaminated	Necrotic	Healthy nondeveloped	Developing	Transplanted into pots
Ethanol 80% for 6 min, chloramine T 5% for 30 min	4.1	0	0	62.6	33.3	18.7	0	0	29.3	52.0
Ethanol 80% for 6 min, chloramine T 7.5% for 30 min	0	0	0	66.0	34.0	7.1	0	0	28.6	64.7
Ethanol 80% for 6 min, chloramine T 10% for 15 min	0	0	0	74.5	25.5	9.8	0	0	27.5	62.7

inoculation of test plants: *Chenopodium quinoa*, *Nicotiana glutinosa*, *Nicotiana tabacum*. It was found with this test that about 40% of the plants obtained in vitro from TuMV-infected meristems were TuMV-free.

In our first experiments in the 1970s, we obtained one plant from one meristem on a MS medium with 1 mg/l NAA and 0.2 mg/l kinetin, solidified with 0.65% agar. Further experiments aimed at obtaining higher propagation. This was accomplished by modification of the medium. On these media several rooted plantlets and unrooted "rosettes" were obtained from one meristem. These rosettes, transferred onto an identical, fresh medium, were subsequently propagated. It was found that among the progeny obtained from one meristem TuMV-free plants were present as well as TuMV-infected ones. It was therefore decided to propagate only a small quantity of virus-free plants from meristems and in turn to propagate these healthy plants, using a more efficient method. In the paper by Meyer and Milbrath (1977), an interesting method of multiplication from leaf fragments of horseradish was described. It was decided to try this method, which offered the following advantages:

1. access to leaves is easier than to meristems at the proper stage of maturity;
2. it is more convenient to cut leaves than to isolate meristems under a microscope;
3. sterilization of leaves is not very difficult because leaves are not in contact with the soil.

The leaf fragments were triangles with sides about 1 cm long. The cuts were made in such a manner that the crossing of the veins was about 1 mm from the cut side. This intersection was planted basipetally in a culture medium. The leaf fragments were sterilized by shaking them in 2.75% calcium hypochlorite with 0.5% Tween 20 for 20 min. We repeated the experiment according to Meyer and Milbrath (1977), but we did not obtain such good results (they reported eight to ten plants from one leaf fragment). Their paper, however, did not contain all the detailed information, so an explanation for our results was not possible. The majority of authors emphasized the importance of the hormone contents in the media used. The recommended concentrations for propagation media differ over a wide range. Therefore the present author has done many experiments in order to find a well-balanced medium for rapid horseradish propagation.

Firstly, experiments on the effect of a medium composition, particularly of its hormonal compounds, were carried out with leaf fragments. (1) Different hormones from the same group were tried; (2) hormone concentrations were changed to increase the amount of cytokinin to auxin ratio; (3) other compounds such as vitamins and adenine sulfate were included.

The first results gave us information on the general directions in which explants responded to different media. For further experiments only media were chosen on which explants formed shoot bunches without roots ("rosettes"). It was necessary to subculture these rosettes. For each medium tested, several variations were prepared:

1. propagation medium, identical to the initial one;
2. rooting medium, with lower – as compared to the initial medium – concentration of cytokinin, for small rosettes;
3. rooting medium, with cytokinin omitted, for larger rosettes;

4. medium to increase organogenesis, with higher cytokinin content as compared to the initial medium.

Following the above scheme, several experiments were carried out. The best media from these experiments were collected in a final experiment, the results of which are given in Table 3. After 28 days of culturing, the highest number of rosettes obtained from one leaf piece was 41.5 on a medium containing IAA 0.8 mg + kinetin 4 mg/l (Fig. 4A, B). On the medium recommended by Meyer and Milbrath (1977) for horseradish fragments, only 1.2 rosettes and 1.3 plantlets were noted (treatment 1). They reported eight to ten plantlets from one piece in 6 weeks of culture. On the medium developed for broccoli (Johnson and Mitchel 1978) 8.7 shoots on an average were observed from one explant (treatment 2). In general, media containing IAA produced more shoots, whereas those with NAA tended to produce a lesser number of rooted plantlets ready for transplantation. Some beneficial effect of vitamin C was noted with medium containing NAA.

Table 3. Effect of hormones on the propagation rate of horseradish initiated from leaf fragments. The explants were cultured for 28 days. (Górecka 1987)

Treatment	Hormone mg/l	Av. no. from one explant	
		Plantlets	Shoots
1	NAA 1 K 0.25	1.3	1.2
2	IAA 8 K 4	0.3	8.7
3	NAA 8 K 4	0.0	0.8
4	IAA 1 K 0.25	0.4	20.5
5	NAA 1 K 0.25 Vit. C 50	0.6	6.5
6	IAA 8 K 4 Vit. C 50	0.3	10.2
7	NAA 0.1 K 0.25	1.8	3.5
8	IAA 0.8 K 4	0.0	41.5
9	NAA 0.1 BA 0.2	0.25	4.2
10	IAA 0.8 BA 4	0.0	7.5

NAA – naphthaleneacetic acid
IAA – indoleacetic acid
K – kinetin
BA – benzyladenine

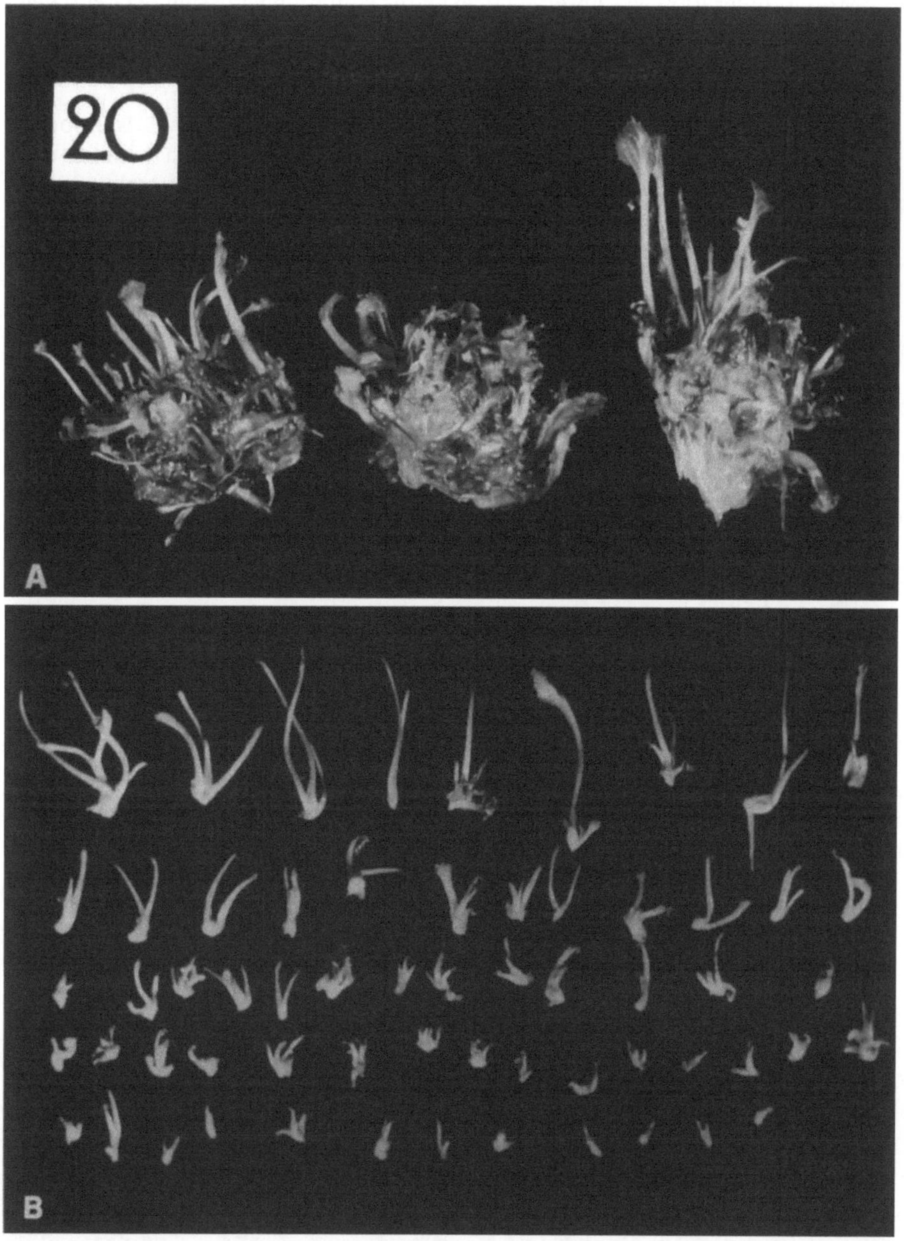

Fig. 4. A, B. Horseradish rosettes resulting from one leaf fragment during 4-week culture on LS medium (IAA 0.8 mg/l + kinetin 4 mg/l). **A** Before dividing. (Górecka, unpubl.). **B** After dividing into separate rosettes. (Górecka 1987)

More proliferation was observed in explants containing crossing of lateral veins in contrast to crossing of veins at the main rib of leaf.

Transferring individual shoots onto the same medium initiated rapid multiplication, as observed in four subsequent subcultures.

These results were confirmed in the next experiments. Large, well-formed rosettes, separated during passages and placed on rooting media without cytokinins, rooted very quickly and easily to almost 100% within 2 weeks. Smaller rosettes, intended for rooting on media with lower cytokinin concentrations, also rooted very well. In this case a simultaneous process of shoot multiplication was observed, but not as intensive as on the propagating medium. The adaptation of these plants occurred in the same manner as described for plants obtained from meristems (Figs. 5, 6). Such an easy rooting of rosettes, and data from Sobczykiewicz (1980) on rooting ex vitro, gave an indication for similar trials with horseradish. Two different substrates were tried: peat-sand 1 : 1 (v/v), and pure sand. Under a rooting layer of the substrate, a thick peat pad was placed to ensure adequate humidity. The bases of the rosettes were treated with powdered auxin preparations, containing 50 or 200 mg IBA per 1000 g and untreated control. These rosettes were inserted in the rooting substrate in boxes. Regardless of treatment and substrate used, all rosettes rooted in 15 days (Table 4). This markedly simplified the procedure. The flat boxes were placed under a plastic tent in a greenhouse.

Fig. 5. TuMV-free horseradish transplants obtained by in vitro culture, after removal from a plastic tent. The multiplication stage was 28, rooting 14, and adaptation 14 days, respectively. (Górecka, unpubl.)

Fig. 6. TuMV-free horseradish plants obtained by in vitro culture, growing in a field. Transplants were planted out on April 15, 1987. The picture was taken on August 20, 1987. (Górecka, unpubl.)

Table 4. The effect of IBA treatment and substrates on rosettes rooting in 15 days (%). (Górecka 1987)

Treatment	Rosettes rooted (%)	
	Peat:sand 1:1 (v/v)	Sand
1. Control, untreated	100	100
2. IBA 50 mg/l	96.3	100
3. IBA 200 mg/l	100	100

4 Summary and Conclusions

The method developed for obtaining and rapid multiplication of TuMV-free horseradish plants in vitro is based on many years' experiments (Górecka et al. 1989). There were two stages of healthy plant production:

1. obtaining plant material free of TuMV;
2. multiplication of TuMV-free plants.

Stage 1. Roots infected with TuMV were grown in a greenhouse so that buds could be excised. The buds were sterilized, the best results being obtained with 80%

ethanol for 6 min followed by chloramine T 5–15% for 15–30 min. After sterilization, meristems of about 1 mm long were isolated, and cultured on MS medium with 1 mg/l NAA and 0.2 mg/l kinetin. The plants obtained after 6–7-week culture were planted into pots (peat substrate was used), under plastic film cover. Special small plastic tunnels were constructed for this purpose and located on greenhouse benches. After acclimation and hardening, the plants were tested for the presence of TuMV.

Stage 2. TuMV-free plants were used as a source of explants for propagation of healthy plants. Young leaves were cut from the mother plants, and triangular pieces of leaves, about 1 cm in size, were taken. Each leaf piece contained a crossing of veins about 1 mm from one of the sides. The leaf fragments were sterilized with 2.75% calcium hypochlorite and 0.5% Tween for 20 min. The explants were cultured on MS medium containing 0.8 mg IAA and 4 mg kinetin 1. About 40 rosettes were obtained in 4 weeks from one explant. The rosettes were multiplied on the same medium. The larger rosettes were rooted on the same medium, but with cytokinin omitted. They rooted very well in 2 weeks and were potted and hardened. The rosettes were also rooted directly in a peat/sand substrate 1 : 1 (v/v) under a double plastic tunnel in a greenhouse for 2–3 weeks. In this case, growth regulators were not necessary, and over 90% of the plants survived.

5 Protocol

For production of virus-free horseradish plants, the best explants were meristems isolated from the buds on the roots. For rapid propagation of TuMV-free plants, the best explants were fragments of young leaves. The best growth of the meristems was observed on the MS medium solidified with 6.5 g/l agar and supplemented with NAA 1 mg/l and kinetin 0.2 mg/l. For leaf fragments the best was LS medium with IAA 0.8 mg/l and kinetin 4 mg/l.

Optimal meristem growth was obtained under 16-h fluorescent light (300 lx) at 20 °C and leaf fragment cultures grew under the same conditions but with somewhat more intensive light.

References

Aleksieieva MW (1987) Perennial vegetable plants. Rossielhozizdat, Moscow, pp 93 (in Russian)
Anderson WC, Carstens JB (1977) Tissue culture propagation of broccoli, *Brassica oleracea* (Italica group), for use in F_1 hybrid seed production. J Am Soc Hortic Sci 102(1): 69–73
Błaszczak W, Fiedorow Z, Gotębniak B (1977) Virus diseases and blackening of horse-radish roots. Biul Warzywniczy 20: 259–274 (in Polish)
Dąbrowska B (1977) Horse radish. In: Detailed cultivation of vegetable crops PWRiL, Warszawa, pp 602–614 (in Polish)
Fonnesbeck M (1974) Temperature effects on shoot and root development from *Begonia × cheimantha* petiole segments grown in vitro. Physiol Plant 32: 282–286
Górecka K (1987) In vitro propagation of horse radish. Acta Hortic 212(2): 671–674
Górecka K, Macias W (in press) Obtaining and multiplication of TuMV-free horse radish plants in vitro. Zesz Probl Postępów Nauk Roln
Górecka K, Srzednicka W, Jankiewicz LS (1978) Tissue culture of horse radish (*Cochlearia armoracia* L.) meristems: sterilization of buds and comparison of media. Acta Agrobot 31(1/2): 195–204

Górecka K, Macias, Jankiewicz LS (1989) A tissue culture method for TuMV-free horseradish plants production. Biul Warzywniczy (Suppl) Part 2 (in Polish)

Hickman AJ, Varma A (1968) Viruses in horse radish. Plant Pathol 17: 26–30

Holmes FO (1965) Elimination of Turnip Mosaic Virus from stock of horse radish. Phytopathology 55(5): 530–532

Johnson BB, Mitchel D Jr (1978) In vitro propagation of broccoli from stem, leaf and leaf rib explants. Hort Sci 13(3): 246–247

Linsmaier EM, Skoog F (1965) Organic growth factor requirements of tobacco tissue cultures. Physiol Plant 18: 100–127

Meyer MM Jr, Milbrath GM (1977) In vitro propagation of horse radish with leaf pieces. Hort Sci 12(6): 544–545

Moravec J, Kvasnička S (1963) Possibilities of increasing yield and quality of horse radish. Olomouc Bull 7 (in Czech)

Mori K (1971) Production of virus-free plants by means of meristem culture. Jarq 6(1)

Murashige T, Skoog F (1962) A revised medium for rapid growth and bioassays with tobacco tissue cultures. Physiol Plant 15: 473–497

Novák FJ, Havránek P (1973) Organogenesis in callus cultures of horse radish (*Armoracia lapathifolia* Gilib.) In: Plant tissue cultures in genetics and breeding. 1st Colloq Olomouc, pp 209–215

Paludan N (1971) The establishment of virus-free meristem-tip culture of horticultural plants. Tidsskr Plant 75(3): 387–410 (in Danish)

Paludan N (1973) Turnip mosaic virus in horse radish (*Armoracia lapathifolia* Gilib.). Tidsskr Plant 77(2): 161–169 (in Danish)

Sobczykiewicz D (1980) Preliminary note on mass production of raspberry plants through placing unrooted plantlets obtained from meristem cultures directly in the soil. Fruit Sci Rep 7(1)

Yamada Y, Kobayashi S, Watanabe K, Hayashi U, Yajima Y, Inouse H (1987) Production of horse radish peroxidase by plant cell culture. J Chem Tech Biotechnol 38: 31–39

I.5 Micropropagation of *Cucumis* spp.

M. ZIV[1]

1 General Account

The genus *Cucumis*, belonging to the Cucurbitaceae, grows throughout the world, from the tropics to the temperate zones. *Cucumis sativus* L. is believed to have been cultivated in India some 3000 years ago and has been known in the Mediterranean region since the ancient Greek and Roman civilizations. *Cucumis* comprises nearly 40 species, of which several are important market crops, *C. sativus* L. and *C. melo* L. being the main ones (Whitaker and Davis 1962; Esquinaz-Alcazar and Gulick 1983).

The cucumber (*C. sativus*) originated in south Asia and includes cultivars having 2n = 14. Some of its wild relatives with the same chromosome number are still found in India. The muskmelon (*C. melo*) is thought to have originated in Africa, where wild species with the same 2n = 24 (or higher) are found. Cultivars of both species originated through successive natural and artificial hybridization and are presently grown in open fields or glasshouses (Whitaker and Davis 1962).

Most *Cucumis* species are herbaceous plants, characterized by long trailing or climbing vines which can grow to several feet in length. The genus *Cucumis* includes monoecious or andromonoecious cultivars, it is open-pollinated and self-compatible. Genetic and hormonal control of sex expression in *C. sativus* and *C. melo* has been studied extensively (Frankel and Galun 1977).

The fruit, which is an inferior berry (pepo), is the economic organ of the *Cucumis* species. The cucumber fruit is used in its immature stage as a fresh or pickled vegetable. *C. melo* is mostly used for its fruity characteristics as a mature fleshy fruit.

Both cucumber and muskmelons are propagated by seeds and commercial production of F_1 hybrid cucumber or muskmelon seeds is dependent on gyno-ecious lines. An extensive hybridization program has introduced several high-quality hybrids. Hybridization, where barriers do not exist, depends on hand pollination and long breeding programs.

Conventional crosses and transfer of desirable traits, especially disease resistance, from wild species, have not been accessible in *Cucumis* (Esquinas-Alcazar and Gulick 1983). Tissue culture techniques can contribute to improve breeding programs, which are at present slow and tedious.

[1] Department of Agricultural Botany and the Warburg Center for Biotechnology in Agriculture, The Hebrew University of Jerusalem, P.O. Box 12, Rehovot 76100, Israel

Biotechnology in Agriculture and Forestry, Vol. 19
High-Tech and Micropropagation III (ed. by Y.P.S. Bajaj)

The need for genetic variation, the knowledge of biosystematic relation, and possibilities to overcome crossability barriers can be greatly improved by in vitro techniques. Production of gynoecious populations as well as exploitation of somaclonal variation for pathogen, herbicide, and pesticide resistance can be advanced by tissue culture. In addition, plant transformation through genetic engineering and gene transfer technology will contribute to improved genetic variation (Kroon et al. 1979).

Tables 1 and 2 present the area, yield, and production of cucumber and muskmelon for the year 1987 (FAO 1987). The two species are cultivated the world over, the major producing countries being China, Japan, Turkey, the USA, The Netherlands, and Romania for cucumber, and China, Spain, Iran, Romania, Iraq, and Mexico for muskmelon. Several countries, particularly in Europe, Japan, and the Middle East, cultivate some of the *C. sativus* cultivars totally or partially under glass, which is reflected in the large yields per hectar, as can be seen in Table 1 (FAO 1987).

Studies on *Cucumis* diseases as well as pesticide and herbicide resistance have been carried out extensively. Interspecific hybrid production for the incorporation of disease and insect resistance or stress tolerance has been limited due to the existence of incompatibility barriers. The most compatible crosses produced only slightly fertile F_1 hybrids and self-sterile F_2 progeny. The success of a cross between *C. melo* and *C. sativus* was limited. Often the production of hybrids through the use of a third species helped to overcome incompatibility (Esquinas-Alaczar and Gulick 1983).

Table 1. Area, yield, and production of *C. sativus* in different parts of the world for 1987. (FAO 1987)

	Area 1000 ha	Yield kg/ha	Production 1000 mt[a]
World	845	15006	12674
Africa	22	16348	360
North, Central America	74	13373	987
USA	44	13048	571
South America	3	16602	46
Asia	439	15640	6868
China	240	15516	3716
Japan	24	43333	1040
Turkey	44	17241	750
Europe	130	21227	2768
Netherlands	2	221100	442
Romania	30	15000	450
Spain	9	31667	285
UK		357894	85*
Yugoslavia	16	7889	128
USSR	175	9286	1625
Total developed countries	382	16234	6196
Total developing countries	463	13994	6478

[a] mt = 10^6 tons.

Table 2. The area, yield and production of *C. melo* in different parts of the world for 1987. (FAO 1987)

	Area 1000 ha	Yield kg/ha	Production 1000 mt[a]
World	603	14869	8967
North, Central America	86	15327	1325
Mexico	29	12176	352
USA	46	19365	885
South America	28	10889	305
Asia	311	14152	4402
China	118	19695	2324
Iran	69	6599	452
Iraq	32	10938	350
Europe	130	16523	2145
Romania	17	23529	400
Spain	67	13313	892
Total developed countries	198	17552	3478
Total developing countries	405	13556	5490

[a] mt = 10^6 tons.

2 In Vitro Studies

2.1 Embryo Culture

Cross-combination between various *Cucumis* species from African and Asian origin often results in the development of proembryos which abort even when rescued and cultured in vitro (Custers and Kruit 1982). One species, *C. zeheri* Son, which is used for hybridization, develops embryoids in vitro if rescued from the beginning of the intermediate stage onward. The medium which supported growth included 0.1 g/l casein hydrolysate, kinetin 0.46 or 4.6 μM, IAA 0.06 μM, and 3.5% sucrose (Custers and Kruit 1982). Development of embryos from *C. metulifernis* × *C. africanus* isolated at the heart stage was better than in those isolated at earlier stages. Complete plants developed on a medium with 0.4 μM kinetin and 3.5% sucrose (Custer 1981; Custer and Kruit 1982). For further details on *Cucumis* and other cucurbit embryo culture see Table 3 in Jelaska (1986).

2.2 Callus and Cell Suspension

Both cucumber and muskmelon explants responded well to combinations of auxins and cytokinins, and produced callus (Table 3). Most of the workers reported on the use of hypocotyl and cotyledons from axenic seedlings, as well as axenic leaves, as the best explant source for callus. Expanding seedling leaves produced callus when coconut milk was added to 2,4-D and kinetin or BA (Fadia and Mehta 1976;

Table 3. Morphogenetic response of various cultivars of *Cucumis sativus* (cucumber) and *Cucumis melo* (melon) explants *in vitro*

Cultivar	Explant	Medium and growth Regulators (μM)	Morphogenetic response	Reference
A. *Cucumis sativus*				
Ashley, China	Leaves and cotyledonary protoplasts	Liquid and solid 1/2 Harada	Callus	
		MS, 2,4-D5,4 K 4.6	Roots	Coutts and Wood (1977)
Carolina	Axillary buds	MS, NAA 0.5, K 4.6	Shoots, roots	Handley and Chambliss (1979)
Hokus	Hypocotyl	MS, 1 g/l CH, K 45.6, IAA 0.57	Callus, buds	
	Cotyledons	MS, 1 g/l CH, K 45.6, IAA 0.57	Buds	
	Immature embryos and their cotyledons	MS, 1 g/l CH, K 45.6, IAA 0.57	Occasional buds	Custers and Bergervoet (1980)
85 cv. and lines	Cotyledons, hypocotyls	MS, NAA 5.4 BA 4.4	Shoots, roots (in 28/85 cv.)	Wehner and Locy (1981)
SMR58	Anthers	Nitsch and Nitsch Raffinose 2%	Callus	
		MS, 50 mg/l ascorbate, 160 mg/l phloroglucinol, BA 4.6, GA 0.3, IBA 0.2, sucrose 2%	Embryoids, plants	Lazarate and Sasser (1982)
Superpickle		MS liquid, NAA 1.5 or 5.0, BA 0.5 or 1.0	Callus	
		2,4-D 5.0, BA 0.5 + 1 g/l CH	Proembryoids, embryoids adventitious buds	
		MS	Shoots, plants	Rajasekaran et al. (1983)

Table 3. (Continued)

Cultivar	Explant	Medium and growth Regulators (μM)	Morphogenetic response	Reference
Skierniewicki, Boroszczagowki	In vitro leaves	MS, 2, 4, 5T 0.47 BA 3.5, or 2,4-D 1.8, 2-ip 1.8 or 2,4-D 3.9	Callus	
		2,4-5-T 0.23 or	Plants	Malepszy and Nadolska-Orczyk (1983)
		BA 3.5 or 2,4-D 0.09	Plants	
GY-3	In vitro leaves; protoplasts	MS NAA 5.4 2-ip 4.9	Callus	
		2,4-D 1.8 2-ip 3.9	Plants	Orczyk and Malepszy (1985)
Straight Eight and various breeding lines	Cotyledons, roots, cotyledonary protoplasts	MS, 2,4-D 4.5 NAA 5.4, BA 2.2	Callus & somatic embryoids	
		NAA 5.4, BA 2.2	Mature embryoids	
		MS	Plants	Trulson and Shahin (1986)
Delilah, Alma	Hypocotyls, cotyledons	MS 2,4-D 5.0 BA 1.0 + 0.1 g 1 CH	Callus	
		MS liquid	Proembryogenic, callus, embryoids	
		$\frac{1}{2}$ MS solid and liquid in a double layer, zeatin 1.0, ABA 0.4, 0.5% AC	Plants	Ziv and Gadasi (1986)
Jinyan-2, 512, Mai, 192, 395, Changchunmici, gynoecious and F_1 hybrids	Cotyledonary protoplasts	Modified DPD liquid 2,4-D 2.2, K 4.6	Cell division	

Chiyonichirohukou-2		Modified MS solid 2,4-D 0.02, K 4.6	Callus	
F_1 hybrid 37-GX78-50		MS, IAA 0.35, K 23.2, BA 8.8	Embryoids	
		$\frac{1}{2}$ MS	Plantlets, shoots, roots	Jia et al. (1986)
Poinsett 76	Leaves	MS solid, liquid 2,4,5-T 5.0 BA 4.0	Callus, cell suspension	
	MS liquid, 0.5% AC	Embryoids		Chee and Tricoll (1988)
Hokus, Salty GY-3 Boroszczgowski	Cotyledons leaves, protoplasts	$\frac{1}{2}$ MS macro, MS micro 2,4-D 4.0, BA 4.0	Callus	
		$\frac{1}{2}$ MS macro, MS micro	Embryoids, plants	Colijn-Hooymans et al. (1988)
Delilah	Hypocotyl	$\frac{1}{2}$ MS solid full $CaCl_2$, 2, 4-D 5 + NAA 5	Callus	Ziv et al. (1989)
		$\frac{1}{2}$ MS liquid full $CaCl_2$, 2,4-D 5 + NAA5	Embryos	
Delilah	Cotyledons, hypocotyl, cell suspension	Liquid cultures in bioreactors, $\frac{1}{2}$ MS full $CaCl_2$, 0.5 ABA	Proembryogenic callus	Ziv (1989)
		Solid $\frac{1}{2}$ MS, full $CaCl_2$, 0.5 ABA Polymer "K"	Plants	Ziv et al. (1989)
B. *Cucumis melo*				
	Hypocotyl, peduncle	MS, NAA 1.1, K 9.2	Callus	
		MS, NAA 1.1, K 0.09–0.9	Shoots	Abak and Dumas de Vaulx (1980)

Table 3. (Continued)

Cultivar	Explant	Medium and growth Regulators (μM)	Morphogenetic response	Reference
Cucumis melo utilissimes, Dultrie and Fuller	Hypocotyl, cotyledons	MS, NAA 5.4, 10.8 CW 15% v/w or IBA 25.2, BA 4.4	Callus	
		MS, IBA 4.9, BA 4.4r or NAA 5.4 adenine 33.7 mg/l	Shoots	Halder and Gadgil (1982)
Cantaloup charentais, Valenciniano	Apical buds	MS, NAA 13.5 BA 4.4, 0.1% YE		
pinyonet, Pinyonet Piel, deSapo	Protoplasts from axenic leaves	B5, NAA 2.7, 2,4-D 4.4, BA 4.4	Shoots, callus	Moreno et al. (1984, 1985a, b)
Amarillo Oro	Hypocotyl, cotyledons	MS 17.1–34.2 IAA 3.0 K 7.6–21.0	Callus Shoots Plants	Moreno et al. (1985a, b)
Sunday Akigata	Mature seed cotyledons	MS, 2,4-D 4.5, BA 0.44	Callus, cell suspension	
	Cell suspension	MS basal	Embryoids, plantlets	Oridate and Oosawa (1986)
Cantaloup Charentais T, Doublon, DMIA187, Piboula, Ogon 9	Hypocotyl cotyledons	MS, IAA 11.4, K 9.3	Callus	
			Buds and plantlets	
(All except Ogon 9)	Cotyledons			
		MS basal	Rooted plants	Bouabdallah and Branchard (1986)

Tokyo Early, Valenciano Pinyonet, Cantaloup Charentais, Tendral verde Tarido, Melonica (and) 10 others cvs. with less than 5% morphogenic response)	Axenic cotyledons	MS solid, IAA 34.2, K 27	Callus, buds, shoots	
	Basal MS	Rooted plant		Orts et al. (1987)
Pusa Sharbati	Axenic leaves	MS, BA 1, 2-ip 1 BA 5, 2-ip 10 IAA 0.5	Callus Shoots Rooted plants	
	Hypocotyl	MS IAA 5.7 K 2.3	Callus	
		ABA 0.4, GA 0.11 2-ip 2,5, BA 2.2	Shoots	Kathal et al. (1986)
	Hypocotyl	1-5 BA or 3 BA + 2-ip	Buds	
Accent, Galia 4215, Preco, Viva	Axenic leaves imbibed cotyledons	MS, BA 4.6	Callus, shoots	Dirks and van Buggenum (1989)

Coutts and Wood 1975, 1977). In some cucumber cultivars, a better response was obtained when 2,4,5-T and 2-ip were used instead of BA and 2,4-D alone or in combination (Malpszy and Nadolska-Orczyk 1983). The addition of 0.1 to 1 g/l casein hydrolysate (CH) was reported to improve cucumber as well as muskmelon callus development (Rajasekaran et al. 1983; Ziv and Gadasi 1986; Moreno et al. 1984, 1985a).

Cell suspension cultures from callus which developed on various explants were established in both *C. sativus* and *C. melo* (Vajranabhaijah and Mehta 1977; Rajasekaran et al. 1983; Oridate and Oowsawa, 1986; Ziv and Gadasi, 1986; Chee and Tricoli 1988; Bergervoet et al. 1989; Ziv et al. 1989).

Trulson and Shahin (1986) reported that when a combination of both auxins, 2,4-D and NAA, was used together with BA, a better pro-embryonic callus resulted from cucumber and muskmelon explants. This was true for another cultivar, Dalilah (Ziv et al., unpubl.). In *C. melo* IAA ranging from 0.5 or 1.6 μM gave a better morphogenic callus with an improved regeneration potential (Moreno et al. 1985a). An induction treatment of 6 or 10 days with 2,4-D-induced callus in cucumber hypocotyl explant resulted in improved embryogenesis and subsequent plant development as compared to continuous exposure to 2,4-D for 24–28 days (Ziv and Gadasi, unpubl.).

2.3 Protoplast Culture

Successful protoplast isolation, cell division, and microcalli and callus development were reported in several cultivars (Table 3). However, the number of cultivars which regenerated embryoids or plants from protoplast calli is rather limited.

Jia et al. (1986) found that liquid medium induced morphogenic callus from cotyledonary protoplasts in the presence of 2,4-D and kinetin. Transfer of the callus to growth regulator-free half-strength MS medium enhanced shoot, root, and plant regeneration.

When protoplasts isolated from axenic cotyledons and leaves of *C. sativus* were embedded in agarose beads, they produced callus and shoots. The callus sustained its regenerative capacity for a year (Dons and Bouwer 1986; Colijn-Hooymans et al. 1988). Another improvement in protoplast calli regeneration was by the addition of glycine (Orczyk and Malepszy 1985). Organized callus from cotyledonary protoplast of two cultivars of *C. melo* on IAA and kinetin combination with 1 g/l CH regenerated when subcultured through a series of media with increased IAA and kinetin level. A final subculture to IAA-free (or low) medium with 0.44 μM BA induced plant development (Moreno et al. 1984, 1985b). A 5–12% frequency for protoplast fusion was achieved in *C. melo metuliferous* and in *C. melo* × *Cucurbita martinezi* (Roig et al. 1986).

Protoplasts isolated from cotyledons and leaves of two pickling inbred lines of *C. sativus* produced callus on various 2,4-D/BA or NAA/BA hormone combinations. A plating density of 3.5×10^4 protoplasts was required for sustained cell division. Calli subcultured on low hormone or hormone-free medium developed somatic embryos, shoots, or plants. The highest frequency of callus and organized development was achieved when the protoplasts were embedded in 0.4% agarose (Punja et al. 1990a). Protoplasts from the pickling cultivars were transformed with

A. tumefaciens strain LBA4404 carrying the NPTII gene. Transformation was confirmed in regenerated plants by assays for NPTII activities. Progeny from transformed fertile plants pollinated with nontransformed plants carried the NPTII gene and segregated at a 1:1 ratio (Punja et al., pers. commun.).

2.4 Anther Culture

Anther culture was reported for one cultivar on Nitsch and Nitsch (1969) medium, which induced callus, embryoids and plants in the presence of ascorbate, phloroglucinol, BA, and GA (Lazarate and Sasser 1982).

2.5 Micropropagation

Micropropagation in *Cucumis* species was established through organogenesis and/or somatic embryogenesis. Bud development and proliferation from apical and axillary buds was mainly through existing meristems. In hypocotyl or cotyledonary explants direct organogenesis was observed. Callus initiation was necessary for adventive organogenesis or embryogenesis or for morphogenic response from anthers, cell suspension, protoplasts, or in transformation studies.

2.5.1 Somatic Embryogenesis

Somatic embryogenesis has been reported for several cucumber and melon cultivars (Table 3). In general, the same explants which were induced to form callus responded in embryoid formation. Both liquid and agar-solidified media were inducive to embryoid development. Various growth regulators used for callus induction had to be either lowered or removed. Complete removal was better for embryogenesis in most cultivars (Table 3). Gelrite instead of agar in the multiplication stage and culture under a 12-h photoperiod improved somatic embryogenesis in Clinton, a pickling cultivar (Ladyman and Girard 1990). Gelrite was used by Kim and Janick (1989) to induce callus on zygotic embryos in the presence of 9 μM 2,4-D. Callus transferred to 2,4-D-free liquid medium produced embryos in the presence of only sucrose as a carbohydrate source. A major problem in embryoid development was the formation of neomorphs and secondary embryogenesis. Ziv and Gadasi used 0.1% activated charcoal in a double-layer culture to which 0.4 μM filter sterilized ABA was added, and were able to decrease both abnormal phenomena. Washing embryoids with 0.5% charcoal improved their germination into plants (Chee and Tricoli 1988). Glycerol added to the medium also eliminated the phenomenon of secondary embryogenesis in three *Cucumis sativus* cultivars (Ziv et al. unpubl.).

Punja et al. (1990b) reported regeneration through somatic embryogenesis from both leaf and petiole explants when 2,4-D/BA were added to the medium. Pretreatment in the dark for 2–3 weeks improved regeneration frequencies.

2.5.2 *Organogenesis*

Bud and shoot regeneration was achieved in *Cucumis* either directly on various explants or via callus which developed on the explant. Gambley and Dodd (1990) reported direct adventitious bud formation from the base of *C. sativus* cotyledonary explants in the presence of 4 mg/l (or less) of BA, kinetin, or 2-ip. Dirks and van Buggenum (1989) reported on massive shoot proliferation from *C. melo* cotyledon sections, dissected from mature imbibed nongerminated seeds, on MS with 4.4 µM BA. A similar response was observed in explants from fully expanded axenic leaves on the same medium.

Shoot regeneration in *C. melo* (cv. Pusa sharbati) was influenced by the size and the age of leaves which were isolated from axenic seedlings. The decline in regeneration potential was accentuated by higher levels (5–10 µM) of cytokinins (Kathal et al. 1988).

Axillary buds from *C. sativus* cultured on MS with 0.54 µM NAA and 0.46 µM kinetin proliferated into shoots and roots with little callus. The plants thus produced survived transplanting to soil (Handley and Chambliss 1979).

Shoot and root development from cotyledons and embryonal axis explants were induced in *C. melo* by NAA, kinetin, and coconut milk (CM) (Halder and Gadgil 1982). When the explants were isolated from the chalazal zone of the ovary's nucellus CM was not necessary for organogenesis (Kumari 1978).

2.6 Plant Regeneration and Establishment

In several cultivars, plant regeneration frequencies were high, and normal plants were established ex vitro (Table 3). Cotyledon, leaf, and petiole explants from inbred lines of F_1 hybrids of *C. sativus* produced callus, and regenerated plants. Survival of the transplanted plants was approximately 80%. In *C. melo* only petiole explants regenerated plants (Punja et al. 1990b). In some cases, culture conditions induced early flowering in plants both in vitro and ex vitro. Flowers were either staminate or pistilate (Rajasekaran et al. 1983; Ziv and Gadasi 1986). Ziv (unpubl.) was able to delay flowering to the 7th or the 8th node when the transplanted plants were cultured under short days (8/16 h) at 20/15 °C day/night temperature. Following transfer to long days, the pistillate flowers produced fruits (Fig. 2). Kim et al. (1988) reported that plant regeneration from diverse cvs. of *C. sativus* was genotype-dependent, with only three out of ten plants regenerated. One of the best-responding cultivars, Manchoonchoungjang, produced plants which were found to be tetraploid.

2.7 Transformation in Cucumber and Melon

Recently, *Agrobacterium*-mediated transformation and regeneration via embryogenesis or organogenesis was reported as a successful system to introduce genetic material into plant species belonging to the Cucurbitaceae. Cotyledons from 3–5-day-old seedlings were used as explants to transfer genetic material into *C. sativus* cv. Poinsett 76 by using an avirulent strain of *Agrobacterium tumefaciens*. Kanamycin-resistant embryonic callus differentiated embryos which produced transformed plants (Chee 1990). Using *Agrobacterium rhizogenes*, transformed callus was obtained from the hairy roots induced on hypocotyl sections, as well as from *A. tumefaciens*-transformed cotyledonary disc. The transformed callus from the cotyledon discs regenerated shoots resistant to kanamycin (Srivastava et al. 1990).

Transgenic *C. melo* plants were produced by inoculating cotyledon explants with *A. tumefaciens* with the NPTII gene for kanamycin resistance. The regenerated shoots developed into morphologically normal plants, of which 85% contained the NPT gene (Fang and Grumet 1990). Dong et al. (1990) reported a regeneration frequency of about 50% of *A. tumefaciens*-transformed *C. melo* cotyledonary explants. The plants appeared morphologically normal and were tested for the presence of the 35S GUS gene.

Cucumber plants transformed with *A. tumefaciens* containing CMV strain C coat protein were field-tested, and showed a lower rate of infection by CMV than control plants (Slighton et al. 1990).

2.8 The *C. sativus* System

In *C. sativus* (cv. Delilah), callus was initiated from cotyledons and hypocotyl explants isolated from axenic 8–10-day-old seedlings. MS minerals, CH, and a combination of 5 µM 2,4-D and 1 µM BA induced callus on the explants in liquid shake cultures. Two types of calli developed on the explants: friable white callus, which remained disorganized; and a yellow nodular callus, which developed into proembryogenic clumps upon subculture to hormone-free medium. The addition of activated charcoal and ABA decreased the production of friable callus (Table 4). Activated charcoal probably adsorbed some of the growth regulators and reduced their effect. In the basal medium, embryoids as well as "neomorphs" (abnormal embryoids) developed (Table 4, Fig. 1a). "Neomorphs" tended to develop a disorganized shoot system (Fig. 1a, arrow). Normal embryoid development was increased by the addition of ABA to the basal liquid medium; in agar medium free of auxin but with 0.4 µM ABA organogenesis was observed (Fig. 1b). Plantlets

Table 4. The effect of various media on callus, embryogenesis and regeneration. (After Ziv and Gadasi 1986)

Medium	Nodular Callus (% FW)	Neomorphs	Embryoids	Plants
		No./culture		
Basal	22[a]	28 ± 3[b]	7 ± 0.3	2 ± 0.2
+ CH (0.1%)	22	20 ± 2	12 ± 1	5 ± 0.6
+ AC (0.5%)	45	15 ± 2	8 ± 0.5	8 ± 0.9
+ Zeatin 1 µM	25	25 ± 2	11 ± 2	3 ± 0.4
+ ABA 0.4 µM	29	18 ± 1	20 ± 2	5 ± 0.6
+ AC + ABA	58	14 ± 1	10 ± 1	9 ± 1

[a] Ten cultures/treatment.
[b] Mean ± SE of eight cultures/treatment.

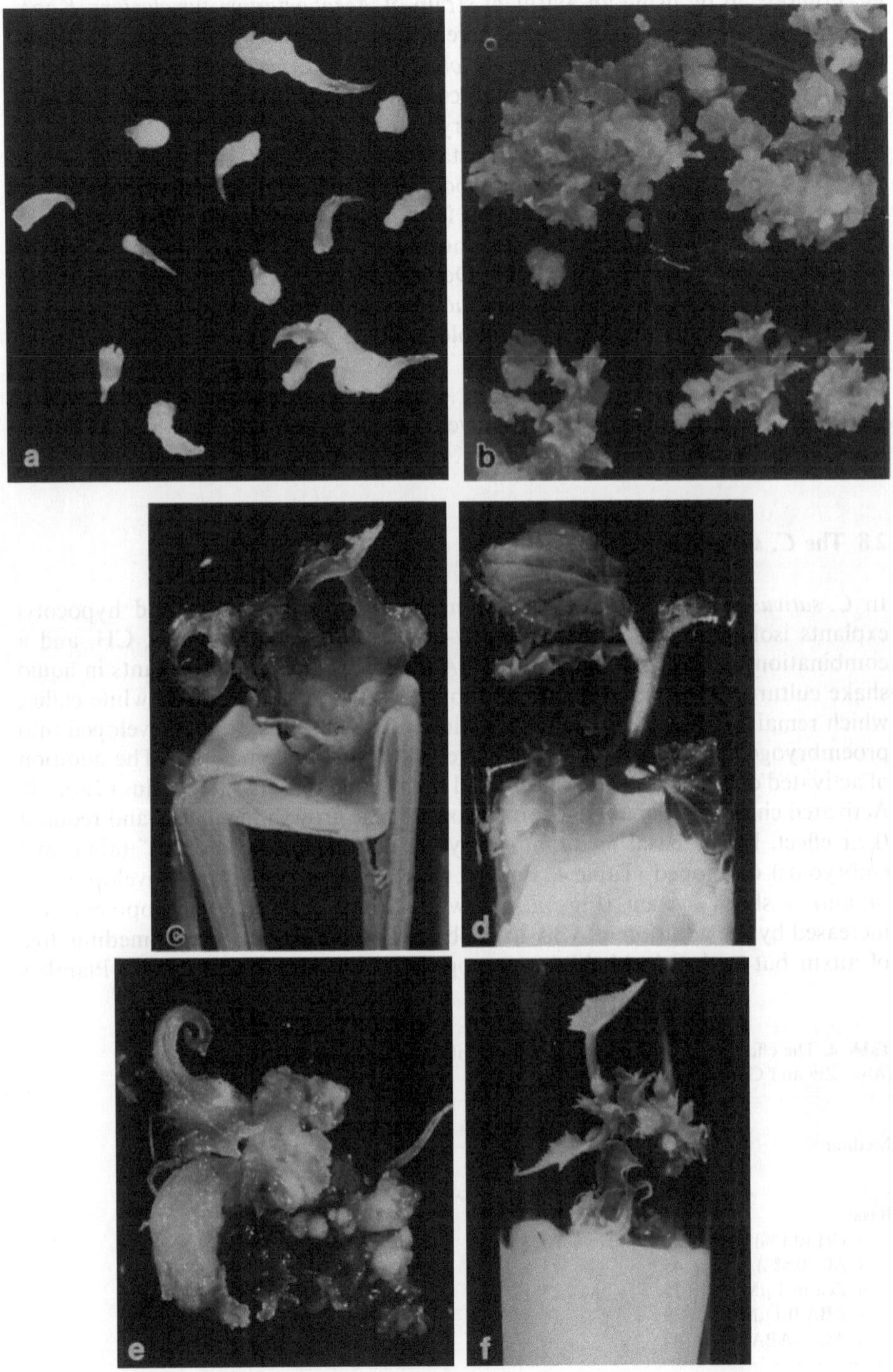

developed best in liquid medium which contained both activated charcoal and ABA. However, all the plantlets had vitrified leaves (Fig. 1c) and could not be transplanted ex vitro due to leaf desiccation as compared to normal plants (Fig. 1d). Furthermore, when subcultured to liquid medium on filter paper bridges or to agar media, they developed secondary callus, which underwent secondary embryogenesis (Fig. 1e). A different approach to obtain normal plants was by the use of a double layer culture. The lower layer consisted of an agar medium with 0.5% activated charcoal on top of which 2 ml half-strength MS mineral ($CaCl_2$ at full strength) with 0.4 μM ABA was added. Proembryogenic callus subcultured from liquid medium to double layer cultures regenerated the highest frequency of normal plants upon subculture to liquid medium on filter paper bridges under SD (8/16 h) photoperiod (Fig. 1c).

Plantlets subcultured for further hardening on half-strength MS, 0.4 μM ABA activated charcoal, and 2% sucrose tended to flower in vitro (Fig. 1f). Plants hardened in the same medium and transferred to a desiccator cylinder with anhydrous $CaSO_4$ for 8 days before transplanting ex vitro flowered on the eight to ten nodes (Fig. 2A). Flowers on the lower nodes were staminate, while those on higher nodes were pistilate. The extent of flowering was reduced and their appearance started at the eight to ten nodes by subjecting the plants to a short-day photoperiod (8/16 h) and lower day/night temperature (20/15 °C) (Fig. 2B) (Ziv and Gadasi, unpubl.). Flowering plants which were established ex vitro (Fig. 2C) developed fruits on the upper nodes (Fig. 2D).

Scaled-up micropropagation was carried out in bioreactors (Ziv 1989). A bubble bioreactor was constructed by welding a sintered glass funnel to an 8-cm glass cylinder 60 cm long. Air was bubbled through the sintered glass from an air pump through filters, and sterile water to prevent the medium from evaporation.

The growth value $[(FW_{FINAL} \cdot FW_{INITIAL})/FW_{INITIAL}]$ was compared in liquid cultures in Erlenmeyer flasks on a gyratory shaker to that in the bioreactors. As can be seen in Table 5, the growth value in the bioreactor was three fold higher than in the flasks, which could be due to the improved aeration in the bioreactor. A better morphogenic response was also observed in the bioreactor. The number of organized green centers and embryoids developing from proembryonic clusters was higher in bioreactor cultures (Table 5). They developed into shoots upon subculture to double layer cultures or to stationary liquid medium (half-strength MS, full $CaCl_2$, 0.4 μM ABA) on paper bridges or plugs (as shown in Fig. 2B). The medium contained a polymer mixture, "K" (supplied by Plant Biotech Industries, Israel). The shoots rooted within 12–15 days, were transplanted to pots, kept for additional 15 days under 8/16-h 20/15 °C photo- and thermoperiods. Plants with five or six leaves were established in the greenhouse and flowered at the 12th node stage (Ziv 1989).

Fig. 1. a Embryoids and "neomorphs" from hormone-free suspension cultures of *Cucumis sativus*. × 2.2. **b** Organogenesis in proembryonic aggregates subcultured to hormone-free $\frac{1}{2}$ MS (full $CaCl_2$) agar medium with 0.4 μM ABA. × 1.6. **c** Vitreous plantlets developing on filter paper bridges in MS liquid medium with 0.5 μM 2,4-D. × 1.8. **d** Normal plant on $\frac{1}{2}$ MS (full $CaCl_2$) medium with 0.4 μM ABA under SD, 8/16 h. × 1.8. **e** Secondary embryogenesis on MS agar medium with 0.5 μM 2,4-D. **f** Young plant with vegetative and flower buds on filter paper bridge in liquid MS with 0.5 μM 2,4-D. × 1.4. (After Ziv and Gadasi 1986; Ziv 1989)

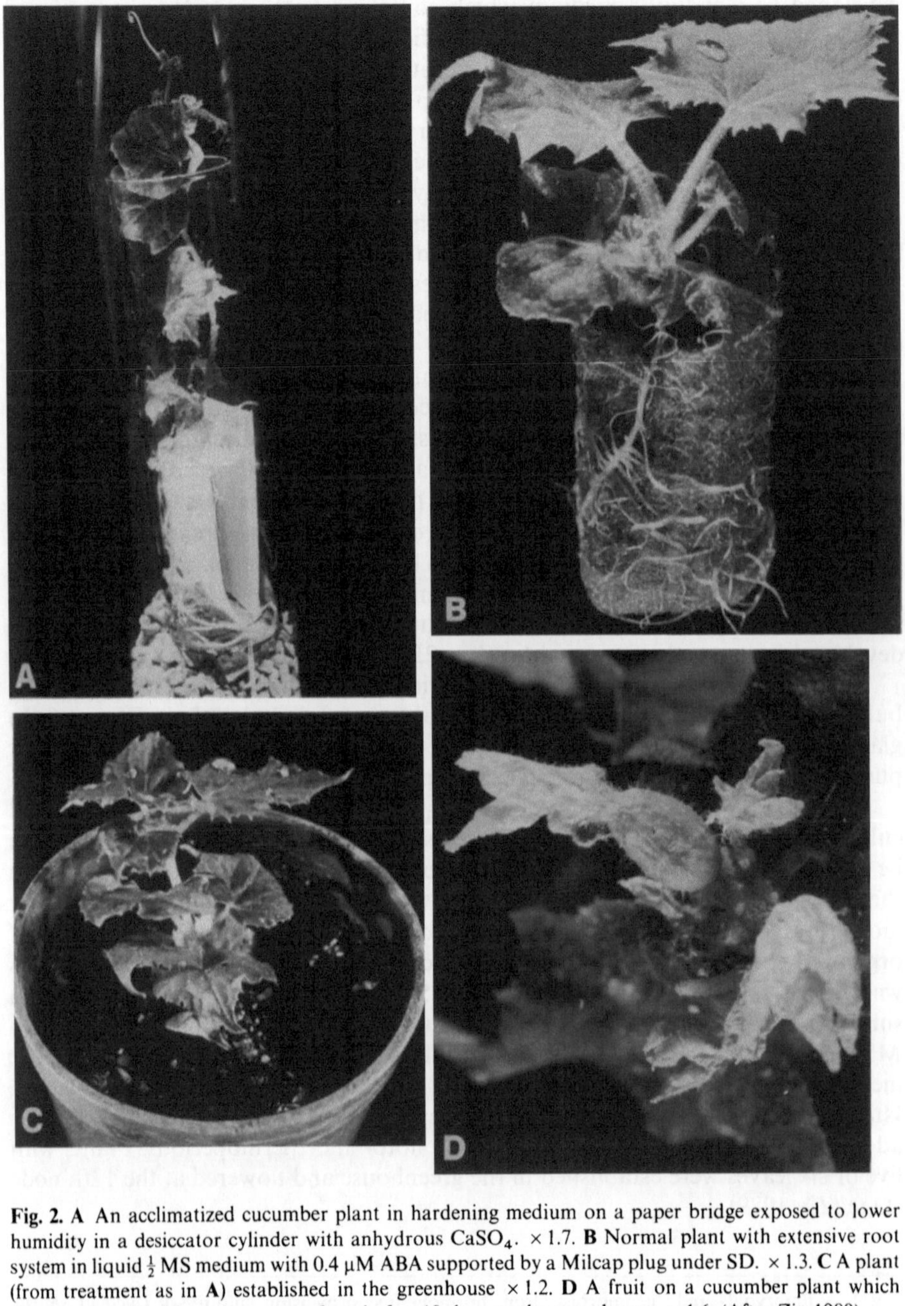

Fig. 2. A An acclimatized cucumber plant in hardening medium on a paper bridge exposed to lower humidity in a desiccator cylinder with anhydrous $CaSO_4$. × 1.7. **B** Normal plant with extensive root system in liquid $\frac{1}{2}$ MS medium with 0.4 μM ABA supported by a Milcap plug under SD. × 1.3. **C** A plant (from treatment as in **A**) established in the greenhouse × 1.2. **D** A fruit on a cucumber plant which developed under SD and was transferred after 18 days to the greenhouse. × 1.6. (After Ziv 1989)

Table 5. Growth and differentiation in *C. sativus* embryogenic callus in Erlenmeyer flasks and bioreactors after 24 days in culture. (Ziv et al. 1989)

Culture condition[a]	FW_I (g)	FW_F	Growth value[b]	No. embryoids/ 100 mg callus (FW)[c]
Flask − ABA	3.5	9.8	1.8	10
Flask + ABA	3.6	10.1	1.8	12
Bioreactor − ABA	15.0	54.0	2.6	24
Bioreactor + ABA	15.2	66.1	3.4	36

[a] Eight flasks and three bioreactors/treatment.
[b] $(FW_F - FW_I)/FW_I$.
[c] Three samples of 100 mg callus were taken from the replicates of each treatment.

Embryogenic callus subcultured every 21–25 days for over 3 years in shake liquid cultures or bioreactors retained the morphogenetic potential and plants regenerated upon subculture to stationary cultures (Ziv and Gadasi, unpubl.).

3 Summary and Conclusions

The application of tissue culture methods for the improvement of *Cucumis* species and other cucurbits has advanced rapidly in the last few years. This is especially true as far as plant regeneration and establishment from explants, callus, or cell suspension is concerned. The improved techniques can now be utilized more readily in crop improvement of *Cucumis*. There are, however, several problems which have to be solved in the field of embryo culture, protoplast culture, and fusion and plant regeneration from somatic hybrids, from transformed protoplasts, or transgenic plant development. Genetic investigation at the molecular level in establishing phylogenetic relationships can enable the genetic resources of cultivated *Cucumis* to be increased, as was recently established for *Cucumis melo* (Perl-Treves and Galun 1985).

The failure to rescue globular embryos from interspecific hybrids could perhaps be overcome by the use of nurse cultures of cotyledonary stage embryos supporting young (pro- or globular stage) embryos. Induced somaclonal variation can be achieved in callus cultures by mutagenesis in vitro (Semel 1986). Transformation of cucumber with *Agrobacterum rhizogenes* plasmid reported by Trulson et al. (1986) and by *A. tumafeciens* (Chee 1990; Srivastava et al. 1990; Fang and Grumet 1990; Dong et al. 1990) is a promising method for the insertion of genes and the creation of desired genetic variation. The low frequencies in protoplast fusion reported by Roig et al. (1986) for *Cucumis* could perhaps be improved by medium manipulation and callus conditioning, as reported by Hamilton et al. (1985), who used a membrane raft for protoplast culture. Cloning in vitro of selected or transformed genotypes which cannot propagate conventionally can be carried out in large-scale liquid cultures in bioreactors with a higher and more efficient regenerative frequency (Ziv 1989).

4 Protocol

1. Surface sterilize seeds in 9% calcium hypochlorite for 30 min, wash two or three times with sterile water.
2. Germinate seeds on moist filter paper axenically in the light for 8–10 days.
3. Remove cotyledons and/or hypocotyl section 1 cm below the cotyledons.
4. Prepare explants 0.5–1 cm from the cotyledons and 1 cm long from the hypocotyls.
5. Transfer to agar-solidified MS medium for callus initiation. Supplement the medium with 2.2–4.56 μM BA and 1–5 μM 2,4-D with 0.1–1 g/l CH.
6. Initiate suspension cultures by transferring callus to liquid MS medium with 0.5–1 μM 2,4-D and 0.1 g/l CH in Erlenmeyer flasks.
7. Subculture the morphogenic aggregates after sieving through a 800–1000 μM sieve, to remove the small nonmorphogenic aggregates, to MS with low (0.2–0.5 μM 2,4-D) or to hormone-free agar medium (for organogenesis) or to liquid cultures (for embryogenesis).
8. Regenerate shoots (from buds) or plants (from embryoids) on half-strength MS medium. To prevent secondary embryogenesis or de-differentiation in the callus, add 0.5% activated charcoal (AC) with 0.4 μM ABA in a double layer culture, the lower part – agar with AC, the upper – liquid with ABA.
9. Harden plants on filter paper bridges or synthetic plugs (Milcap) with half-strength MS minerals and 1–2% sucrose liquid medium for 12–18 days. If flowering is observed, transfer to 8/16-h 20/15 °C photo-thermoperiod.
10. Uncover the culture tubes and maintain in a desiccator with anhydrous $CaSO_4$ for 6–8 days.
11. Wash excess medium from the roots which developed on the filter paper or in the plugs and transplant to the greenhouse.
12. To maintain nonmorphogenic callus subculture every 18–21 days to a fresh MS liquid medium with 1 μM each of 2,4-D and BA with 0.1 g/l CH.

Acknowledgements. The author would like to thank Plant Biotech Industries for financial support and the supply of Polymer "K", and Ms Nili Ben Yehezkel for typing and editing the manuscript.

References

Abak K, Dumas de Vaulx R (1980) in vitro callus and shoot induction from hypocotyl and peduncle of muskmelon *Cucumis melo*. Cucurbit Gen Coop Rep 3: 27–29

Bergervoet JHV, van der Mark F, Custers JBM (1989) Organogenesis versus embryogenesis from long-term suspension cultures of cucumber (*Cucumis sativus* L.). Plant Cell Rep 8: 116–119

Bouabdallah L, Branchard M (1986) Regeneration of plants from callus cultures of *Cucumis melo* L. Z Pflanzenphysiol 96: 82–85

Chee PP (1990) Transformation of *Cucumis sativus* via *Agrobacterium tumefaciens*. In: Nijkamp HJJ, Van der Plaas LHW, Van Aatrijk J (eds) Progress in plant cellular and molecular biology. Kluwer, Dordrecht, pp 85–90

Chee PP, Tricoli DM (1988) Somatic embryogenesis and plant regeneration from cell suspension cultures of *Cucumis sativus* L. Plant Cell Rep 7: 274–277

Colijn-Hooyamans CM, Bouwer R, Dons JJM (1988) Plant regeneration from cucumber protoplasts. Plant Cell Tissue Organ Cult 12: 147–150

Colijn-Hooymans CM, Bouwer R, Orczyk M, Dons JJM (1988) Plant regeneration from cucumber (*Cucumis sativus*) protoplasts. Plant Sci 57: 63–71

Coutts RHA, Wood KR (1975) The isolation and culture of cucumber mesophyll protoplasts. Plant Sci Lett 4: 189–193

Coutts RHA, Wood KR (1977) Improved isolation and culture method for cucumber mesophyll protoplasts. Plant Sci Lett 9: 45–51

Custers JBM (1981) Heart-shape stage embryos of *Cucumis* species more successful in embryo culture than advanced stage embryo. Cucurbit Gen Coop Rep 4: 48–49

Custers JBM, Bergervoet JHW (1980) In vitro adventitious bud formation on seedlings and embryo explant of *Cucumis sativus* L. Cucurbit Gen Coop Rep 3: 2–4

Custers JBM, Ee van G (1980) Reciprocal crosses between *Cucumis africanum* L.f. and *C. metuliferus* Naud. II. Embryo development in vivo and in vitro. Cucurbit Gen Coop Rep 3: 50–51

Custers JBM, Kruit S (1982) Embryo culture with *Cucumis* species. In: Fujiwara A (ed) Plant tissue culture 1982. Maruzen, Tokyo, pp 777–778

Custers JBM, Ee van G, Deb Nijs APM (1980) Tissue culture methods and interspecific hybridization in the breeding of cucumber (*Cucumis sativus* L.). In: Doré C (ed) Application de la culture in vitro à l'amélioration des plantes potagères, Sect Legumes. Inra, Versailles, pp 132–137

Dirks R, van Buggenum M (1989) In vitro plant regeneration from leaf and cotyledon explants of *Cucumis melo* L. Plant Cell Rep 7: 626–627

Dong JZ, Yang MZ, Jia SR, Chua NH (1990) Transgenic melon plants (*Cucumis melo* L.) obtained by *Agrobacterium tumefaciens*-mediated gene transfer. Abstr IAPTC Congr, Amsterdam, p 53

Dons JJM, Bouwer R (1986) Improving the culture of cucumber protoplasts by using agarose-disc procedure. In: Nuclear techniques and in vitro culture for plant improvement. IAEA, Vienna, pp 497–504

Esquinas-Alcazar JT, Gulick PJ (1983) Genetic resources of Cucurbitaceae. IBPGR, Rome

Fadia VP, Mehta A (1976) Tissue culture studies on cucurbits: chlorophyll development in *Cucumis* callus cultures. Phytomorphology 26: 170–175

Fang G, Grumet R (1990) *Agrobacterium tumefaciens*-mediated transformation and regeneration of muskmelon plants. Plant Cell Rep 9: 160–164

Food and Agriculture Organization (FAO) (1987) Production yearbook, vol 41. Rome, pp 185–186; 201

Frankel R, Galun E (1977) Pollination mechanisms, reproduction and plant breeding. Springer, Berlin Heidelberg New York, 281 pp

Gambley RL, Dodd WA (1990) An *in vitro* technique for the production de novo of multiple shoots in cotyledon explants of cucumber (*Cucumis sativus* L.). Plant Cell Tissue Organ Cult 20: 177–183

Halder T, Gadgil VM (1982) Morphogenesis in some plant species of the family Cucurbitaceae. In: Rao A (ed) Tissue culture of economically important plants. Natl Univ Singapore, COSTED, pp 98–103

Hamilton R, Petresen H, Chin C (1985) Plant tissue culture on membrane rafts. Biotechniques 3: 96

Handley LW, Chambliss OL (1979) In vitro propagation of *Cucumis sativus* L. Hortic Sci 14: 22–23

Jelaska S (1986) Cucurbits. In: Bajaj YPS (ed) Biotechnology in agriculture and forestry, vol 2. Crops I. Springer, Berlin Heidelberg New York, pp 372–386

Jia SR, Fu YY, Lin Y (1986) Embryogenesis and regeneration from cotyledon protoplast culture of cucumber (*Cucumis sativus* L.). J Plant Physiol 124: 393–398

Kathal R, Bhatnagar SP, Bhojwani SS (1986) Regeneration of shoots from hypocotyl of *Cucumis melo* cv. Pusa Sharbati. J Plant Physiol 126: 59–62

Kathal R, Bhatnagar SP, Bhojwani SS (1988) Regeneration of plants from leaf explants of *Cucumis melo* cv. Pusa Sharbati. Plant Cell Rep 7: 449–451

Kim M, Janick J (1989) Somatic embryogenesis and organogenesis in cucumber. Hortic Sci 24: 702

Kim SG, Chang JR, Cha HC, Lee KW (1988) Callus growth and plant regeneration in diverse cultivars of cucumber (*Cucumis sativus* L.). Plant Cell Tissue Organ Cult 12: 67–74

Kroon GH, Custers JBM, Kho YO, Den Nijs APM, Varekamp HQ (1979) Interspecific hybridization in *Cucumis* (L.) I. Need for genetic variation, biosystematic relations and possibilities to overcome crossability barriers. Euphytica 28: 723–728

Kumari A (1978) Part I: Effects of cucurbitacins on SE in vitro. Part II: Cucurbitaceae nucellus cultures as a correlative study on the probable role of cucurbitacins in somatic embryogeny. Ph.D. Thesis, Univ Delhi

Ladyman JAR, Girard B (1990) Non-hormonal factors that improve the development of somatic embryos in cucumber. Abstr XXIII Int Hortic Cong, Firenze, Italy, 148 pp

Lazarte JE, Sasser CC (1982) Asexual embryogenesis and plantlet development in anther culture of *Cucumis sativus* L. Hortic Sci 17: 88

Malepszy S, Nadolska-Orczyk A (1983) In vitro culture of *Cucumis sativus*. I. Regeneration of plantlets from callus formed by leaf explants. Z Pflanzenphysiol 111: 273–276

Moreno V, Zubeldia L, Roig LA (1984) A method for obtaining callus cultures from mesophyll protoplasts of melon (*Cucumis melo* L.). Plant Sci Lett 34: 195–201

Moreno V, Garcia-Sogo M, Granell I, Garcia-Sogo B, Roig LA (1985a) Plant regeneration from calli of melon (*Cucumis melo* L., cv. Amarillo Oro). Plant Cell Tissue Organ Cult 5: 139–146

Moreno V, Zubeldia L, Garcia-Sogo B, Nuez F, Roig LA (1985b) Somatic embryogenesis in protoplast-derived cells of *Cucumis melo* L. In: Horn W, Jensen CJ, Odenbach W, Schieder O (edn) Genetic Manipulation in Plant Breeding. De Gruyter, Berlin, pp 491–493

Nitsch JP, Nitsch C (1969) Haploid plants from pollen grains. Science 163: 85–87

Orczyk W, Malepszy S (1985) In vitro culture of *Cucumis sativus* L. V. Stabilizing effect of glycine on leaf protoplasts. Plant Cell Rep 4: 269–273

Oridate T, Oosawa K (1986) Somatic embryogenesis and plant regeneration from suspension callus culture in melon (*Cucumis melo* L.). Jn J Breed 36: 424–428

Orts MC, Garcia-Sogo B, Roche MV, Roig LA, Moreno V (1987) Morphogenetic response of calli derivated from primary explants of diverse cultivars of melon. Hortic Sci 22: 666

Perl-Treves R, Galun E (1985) The *Cucumis* plastome: physical map, intragenetic variation and phylogenetic relationships. Theor Appl Gen 71: 417–429

Punja ZK, Tang FA, Sarmento GG (1990a) Isolation, culture and plant regeneration from cotyledon and mesophyll protoplasts of two pickling cucumber (*Cucumis sativus* L.) genotypes. Plant Cell Rep 9: 61–64

Punja ZK, Abbas N, Sarmento GG, Tang FA (1990b) Regeneration of *Cucumis sativus*, vars. *sativus* and *hardwickii*, *C. melo* and *C. metuliferus* from explants through somatic embryogenesis and organogenesis: Influence of explant source, growth regulator regime and genotype. Plant Cell Tissue Organ Cult 21: 93–102

Rajasekaran K, Mullins MG, Nair Y (1983) Flower formation in vitro by hypocotyl explants of cucumber (*Cucumis sativus* L.) Ann Bot 52: 417–420

Roig LA, Zubeldia L, Orts MC, Roche MV, Moreno V (1986) Plant regeneration from cotyledon protoplast of *Cucumis melo* L. cv. Cantaloup Charentais. Cucurbit Gen Coop Rep 9: 74–77

Semel J (1986) Concluding remarks. In: Semel J (ed) Somaclonal variations and crop improvement. Martinus Nijhoff, Dordrecht, pp 276–277

Slighton JL, Chee PP, Gonsalves D (1990) Field testing of cucumber plants which express the CMV coat protein gene: field plot design to test natural infection pressures. In: Nijkamp HJJ, Vanderplas LHW and Van AArtrijk J (eds) Progress in plant cellular and molecular biology. Kluwer, Dordrecht, pp 201–206

Srivastava DK, Andrianov VM, Piruzian S (1990) Genetic transformation and foreign gene expression in cucumber tissues (*Cucumis sativus* L. cv. Kustovoi). Abstr IAPTC Cong Amsterdam, p 77

Trulson AJ, Shahin EA (1986) In vitro plant regeneration in the genus *Cucumis*. Plant Sci 47: 35–43

Trulson AJ, Simpson RB, Shahin EA (1986) Transformation of cucumber (*Cucumis sativus* L.) plants with *Agrobacterium rhizogenes*. Theor Appl Gen 73: 11–15

Vajranabhaiah SN, Mehta AR (1977) Effect of kinetin on growth and nucleic acid metabolism in suspension cultures of *Cucumis melo* L. Ann Bot 41: 483–491

Wehner TC, Locy RD (1981) In vitro adventitious shoot and root formation of cultivars and lines of *Cucumis sativus* L. Hortic Sci 16: 759–760

Withaker TW, Davis GN (1962) Cucurbits: botany, cultivation and utilization. Interscience, New York, 250 p

Ziv M (1989) Use of bioreactors for scaling-up somatic embryogenesis in cucumber. Plant Biotechnol Ind (Israel) (submitted)

Ziv M, Gadasi G (1986) Enhanced embryogenesis and plant regeneration from cucumber (*Cucumis sativus* L.) callus by activated charcoal in solid/liquid double-layer cultures. Plant Sci 47: 115–122

Ziv M, Gadasi G, Nadel B, Altman A (1989) Somatic embryogenesis and plant regeneration in *C. sativus* propagated in large-scale liquid cultures. Plant Biotechnol Ind (Israel) (submitted)

I.6 Micropropagation of Potato (*Solanum tuberosum* L.)

J. H. Dodds, D. Silva-Rodriguez, and P. Tovar[1]

1 Introduction

The potato is one of the world's most economically important crop plants. Subjecting the plant to a period of thermotherapy followed by meristem culture now routinely allows the production of pathogen-tested (virus-free) stock cultures, and in vitro cultures free from bacteria, fungi, and viruses are used for micropropagation of large quantities of disease-free plants (Quak 1977; Bajaj 1981; Lizarraga et al. 1986; Estrella 1987). These disease-free plants form the nuclear stock of material in a potato seed tuber program. Disease-free plants are high-yielding and produce tubers of better marketable quality and resulting higher price.

International distribution of potato germplasm is routinely carried out for two principle reasons: (1) commercial export of seed tubers to those countries that lack the technical or physical facilities to produce their own high-quality seed, and (2) the International Potato Center distributes germplasm for evaluation by national potato programs. There is an increasing tendency to export germplasm in vitro because of the ease of meeting plant quarantine regulations. The Potato Center exported over 7500 in vitro plants in 1987 to over 40 different countries worldwide (Dodds et al. 1986) (Table 1).

Table 1. Countries where germplasm was distributed during 1987/1988

Argentina	Ethiopia	Pakistan
Australia	France	Paraguay
Austria	Germany	Peru
Bangladesh	Guatemala	Philippines
Bhutan	Greece	Rwanda
Bolivia	Holland	Spain
Brazil	Hungary	Sri Lanka
Canada	India	Sweden
Chile	Iran	South Africa
China	Italy	Taiwan
Colombia	Jamaica	Thailand
Costa Rica	Japan	Turkey
Cuba	Kenya	UK
Dominican Republic	Madagascar	USA
Ecuador	Mexico	Venezuela
El Salvador	Nepal	

[1] International Potato Center P. O. Box 5969, Lima, Peru

Biotechnology in Agriculture and Forestry, Vol. 19
High-Tech and Micropropagation III (ed. by Y.P.S. Bajaj)

2 Micropropagation

Many potato seed programs use in vitro pathogen-tested plantlets as starting material. The initial stages of the potato seed program can make use of varying amounts of in vitro micropropagation depending on the size and location of the program. The basic methods used, however, are very similar in most institutions and are based on the rapid growth on solid or liquid culture media of single node cuttings or stems with multiple nodes (Espinoza et al. 1984). The basic micropropagation methods used at CIP and many other institutions are as described below.

Growth of Single Node Cuttings. Single nodes with leaves are excised from small in vitro plantlets; in some genotypes the large leaves are carefully removed. The removal of the leaves ensures uniform growth of the newly developing shoots. If large leaves are left on the single node pieces, it is believed that hormones from the senescing leaf can inhibit growth of the newly developing shoot. Each node is then inoculated onto the surface of agar-solidified medium. The axillary bud quickly grows out (Fig. 1), and in 3–4 weeks a plantlet with six or seven more nodes becomes available for subculture.

Liquid Shaken Cultures. In vitro plantlets are cut into stem cuttings each with three to four nodes; the large leaves are removed. Each stem piece is placed in 15 ml of liquid medium and the flasks are shaken (80 rpm). After 2–3 weeks of rapid growth each flask contains 60 to 70 nodes (Fig. 1).

The strength/vigor of these in vitro plantlets grown in large plastic boxes (Magenta Corp.) has sometimes been weak. It is possible to greatly invigorate the plantlets by the addition of silver salts to the medium. Silver thiosulfate at approx. 1000 ppm is used as a stock solution and this is then added to the liquid culture medium at 1.5 ml/l (Perl and Galun pers. commun.).

Once a suitable number of small plantlets has been produced, they can be transferred to nonsterile conditions, i.e., transplanted into beds or pots for tuber production.

2.1 Transplantation

In vitro plantlets grown in either of the two containers (test tubes or magentas) can be transplanted when they are 3–4 cm long to either pots, jiffy pots, or seed beds. The planting mix used is a mixture of peat moss and sand (1:2 by volume). The pots are filled with the peat moss/sand mix, covered with aluminum foil, and sterilized for 1 h, if an autoclave is available. If an autoclave is not available, the pots must be washed with detergent, rinsed with running water, and the planting mix and some extra sand are sterilized by any other means (heat, steam, fungicides, etc.). Jiffy pots are already sterile.

The pots are then irrigated with a small amount of water, and they are prepared to receive the plantlet by making a hole in the center of the peat moss/sand mix with a clean stick or pencil. With the aid of sterilized forceps (flamed to red hot

Fig. 1. A Growth of single node cuttings on solid media. **B** Shaken cultures with stem segments in liquid medium containing GA_3. **C** Plastic box with 25 individual rooted plantlets ready for transfer to pots or seed beds. **D** In vitro plantlets a few days after transplant to jiffy pot. (Courtesy CIP)

and cooled), the plantlets are pulled out of the test tubes or magenta. If the plantlets come from test tubes, the agar from the roots must be washed off by gently immersing them several times in sterilized water, trying not to wet the rest of the plantlet. Plantlets are separated from each other and are planted individually in the holes in the potting mix with the roots plus one or two nodes below the surface. Sterilized sand is placed around the plantlet and pressed lightly to keep the plantlets straight in the pot.

The entire plantlet is covered with an inverted beaker to prevent dehydration until the roots are established (approximately 10 days). The pots are kept in a clean area, at 18–25 °C with 14–16-h illumination. The plantlets are irrigated lightly with tap water (if it has a low salt content) until they are well rooted. They must not be overwatered. When roots are established, supplementary nutrients may be dissolved in the irrigation water. Gradually the plants are exposed to the normal atmosphere by removing the beakers for short periods each day. When the plants are well rooted, normal fertilizers can be dissolved in the irrigation water (5–50–17) or (12–12–12). The same procedure is followed in the case of transplanting directly to seed beds.

2.2 Mechanization of Micropropagation

One of the largest cost inputs to any commercial (private or public sector) micropropagation program is cost of labor. In developed countries labor cost can easily be 60–70% of production cost. Various methods are being developed that will allow mechanization of some aspects of this labor-intensive activity.

There has long been interest in the idea of using some type of robotic mechanism for micropropagation. Recently various commercial companies have developed such mechanized micropropagation systems. For reason of patent law, the details of this procedure cannot be explained in this chapter; however, it is expected that once the patents are approved and in use more details will become available (also see Bajaj 1991).

At the International Potato Center we have been studying more simplistic ways of mechanizing the micropropagation process. For example, the use of a modified Wareing blender that will cut nodal cuttings, also the use of a devise comprising a block of scalpel blades or razor blades. Preliminary results for these simple methods are most encouraging.

The use of mechanization both sophisticated and simple should lead to increased success and economic competitiveness of micropropagation systems.

2.3 In Vitro Tubers as Part of a Micropropagation System

In recent years, interest has developed in many countries on the induction of potato tubers under in vitro conditions. Several different methods are available to bring about the induction process (Stallknecht and Farnsworth 1982; Koda and Okasawa 1983; Hussey and Stacey 1984; Bajaj 1987). At CIP, we have developed a rapid, cost-effective method that involves the addition of BAP, CCC (chlorocholine chloride), and sucrose to the liquid medium used for propagation. We have shown that this technique is not only rapid and efficient but is applicable to a wide range of potato genotypes (Estrada et al. 1986).

Figure 2a shows diagramatically the basic methods used for the induction process, while Fig. 2b shows tubers induced under these conditions. Programs in many developed and developing countries have started in vitro tuberization programs. The tubers are normally utilized in two different ways. Firstly, for distribution of germplasm either nationally or internationally; secondly, these small tubers can be used as an additional component to the standard methods of rapid propagation used in a seed tuber production program.

At CIP, the use of in vitro tubers offers some advantages for distribution of germplasm. The tubers are produced under in vitro conditions from pathogen-tested plantlets and thus comply easily with international quarantine requirements. However, unlike in vitro plantlets, if the package is delayed in transit there are fewer problems of losing the material. Figure 3a shows freshly packed in vitro tubers for export while Fig. 3b shows the same tubers after 8 weeks at 22 °C packed in the same way as distributed tubers; even after this period of time material can still be utilized. However, there are disadvantages. The unit cost production of an in vitro tuber is significantly more than that of an in vitro plantlet. For that reason CIP is using this type of distribution only to certain locations where normal transit times are prolonged.

A number of methods have been used for the induction of in vitro tubers; Table 2 shows a comparison of four published methods (Wang and Hu 1982; Kim 1982; Hussey and Stacey 1984). Table 3 shows the media used at CIP for the standard method we have applied to over 200 genotypes.

2.4 Effect of Various Carbon Sources and Growth Regulators on In Vitro Tuberization

Since the first report on in vitro tubers, published in 1953, several authors have reported the induction of tubers on in vitro potato plants through the use of growth regulators or special environmental conditions with inductive characteristics (Baker 1953; Claver 1967; Schilde-Rentschler et al. 1984).

Among the growth regulators that affect in vitro tuberization, cytokinins have been found to be essential for tuber initiation. Some authors suggest that a specific tuber-forming substance is responsible for the tuberization process and that this substance is cytokinin-like in nature. BAP at a concentration of 0.25 to 10 mg/l can easily induce in vitro tubers in different genotypes, as revealed by various authors, whilst kinetin has the same effect at concentrations from 2.5 to 5 mg/l (Palmer and Smith 1970; Minglo-Castel et al. 1976).

Gibberellic acid has been reported to inhibit the tuberization process (Tizio 1971). This inhibition can be overcome by the addition of CCC to the medium. When included in MS medium at a concentration of 500 mg/l together with 5 mg/l of BAP and 8% sucrose, CCC can induce tubers in a broad range of potato genotypes within a period of 4 weeks (Schilde-Rentschler et al. 1984).

Whilst ABA is generally considered a tuberization promotor, contradictory reports have been written on the role of ethylene in in vitro tuberization. Some authors report that ethrel at a concentration of 50 mg/l increased the number of

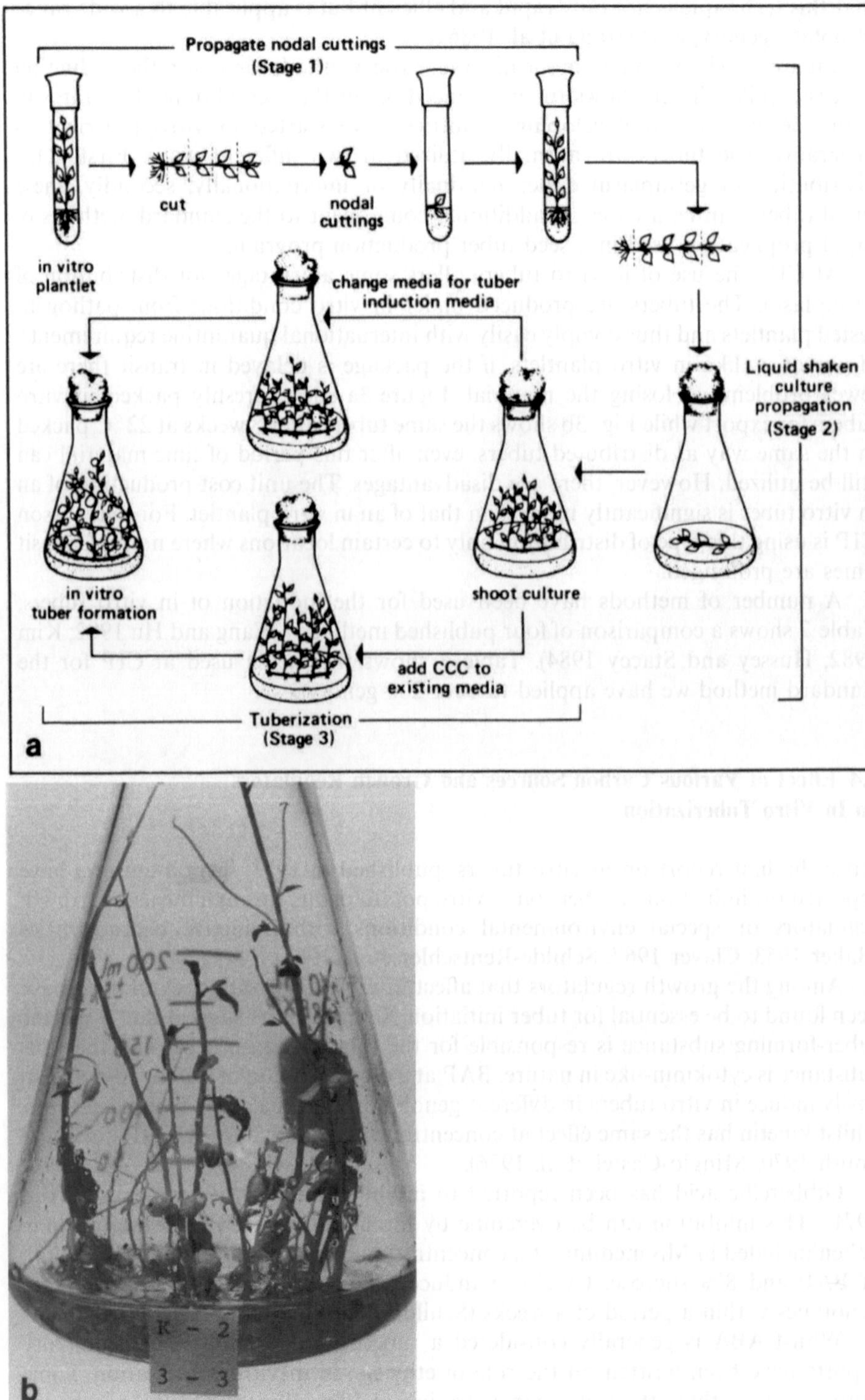

Propagate nodal cuttings
(Stage 1)
in vitro
plantlet
cut
nodal
cuttings
change media for tuber
induction media
Liquid shaken
culture
propagation
(Stage 2)
in vitro
tuber formation
shoot culture
add CCC to
existing media
Tuberization
(Stage 3)
a
200 ml
150
K – 2
3 – 3
b

Fig. 3. a Sterile Petri dishes containing in vitro tubers; these dishes are packed with cotton wool or vermiculite and shipped air freight to their destination. **b** A similar group of in vitro tubers after 4 months of storage at room temperature. Although sprouts are long, plantlets can be rescued from this material. (Courtesy CIP)

tubers; other scientists found that it influences stolon growth but is not directly related to tuber initiation (Koda and Okasawa 1983; Claver 1970).

Sucrose as a carbon source plays an important role in in vitro tuberization, as reported by several authors. Different concentrations of sucrose, alone or in combination with CCC, cytokinins, or coumarin, have been tried. Sucrose concentrations ranging from 2 to 12% can induce in vitro tubers in potato plants (Jung et al. 1972; Tovar et al. 1987).

The process of tuberization is also controlled by environmental conditions: short days and low night temperatures stimulate the process whereas long days and high temperatures are noninductive conditions. For in vitro tuberization the optimum incubation temperature is between 18 to 20 °C and total darkness or

Fig. 2. a Schematic representation of the basic methodology used for in vitro tuber induction at CIP. **b** A typical flask containing in vitro tubers after 1 month on induction medium. (Courtesy CIP)

Table 2. A comparison of four methods used for in vitro induction

Process	CIP	Wang and Hu (1982)	Kim (1982)	Hussey and Stacey (1981)
Plant material	Stems of in vitro plantlets	Stems of in vitro plantlets	Single nodes from in vitro plantlets	Single nodes from in vitro plantlets
Propagation medium	MS[a] + 0.5 BAP + 0.4 GA + 0.01 NAA	MS + 0.005 NAA	MS + 0.1 GA + 0.5 K	
Tuberization medium	MS + 5 BAP + 500 CCC + 8% sucrose	MS + 10 BAP + 8% sucrose	MS + 5 BAP + 6% sucrose	MS + 2 BAP + 6% sucrose
Temperature (°C)	18	20	20	20
Lighting (h)	0	8 (100 lx)	0	8
Time for tuber production (weeks)	4	16	4–8	6–8
Tuber yield (no.)	10 tubers/ flask	50 tubers/ flask	1–2 tubers/ node	1/node

[a] All media preparation were based on the medium of Murashige and Skoog (1962) salts mixture (MS).

NOTE: The addition of all growth regulators is given in mg/l. Add BAP, GA, NAA, K, CCC.
BAP = benzylaminopurine; GA = gibberellic acid; NAA = naphthalene acetic acid;
K = kinetin; CCC = chlorocholine chloride.

short days of low light intensity (less than 500 lx) are optimal photoperiods for this process (Koda and Okasawa 1983; Palmer and Smith 1969; Laurence and Barker 1963; Hussey and Stancey 1981).

The standard method for in vitro tuber induction used at the International Potato Center has been described above. However, some aspects of this methodology have recently been improved to render larger tubers and an increase in number.

The growth and induction phases are divided into three stages:

Stage I. Initial Propagation Through the Use of Single Node Cuttings. In vitro potato plants from the pathogen tested collection stock are propagated by single node cuttings and inoculated into test tubes containing solid Murashige and Skoog medium enriched with the addition of several vitamins (Medium MSA, Table 4). Vigorous plantlets with thick stems and small leaflets, suitable for the induction of in vitro tubers, are obtained.

Stage II. Preparation of Plants for Further Induction of Tuberization. Several stem segments containing six nodes each are inoculated into 250-ml Erlenmeyer flasks containing 20 cm^3 of Murashige and Skoog medium supplemented with growth regulators (Medium 9, Table 4). Flasks are kept under noninductive conditions at

Table 3. Composition of culture media for conservation, propagation, and tuberization in vitro

Compound (mg/l)	Single node propagation	Shaken liquid propagation	In vitro conservation	In vitro tuber induction
NH_4NO_3	1650	1650	1650	1650
KNO_3	1900	1900	1900	1900
$CaCl_2 \cdot 2H_2O$	440	440	440	440
$MgSO_4$	370	370	370	370
KH_2PO_4	170	170	170	170
H_3BO_3	6.2	6.2	6.2	6.2
$MnSO_4 \cdot 4H_2O$	16.9	16.9	16.9	16.9
$ZnSO_4 \cdot 7H_2O$	8.6	8.6	8.6	8.6
K	0.83	0.83	0.83	0.83
$Na_2MoO_4 \cdot 2H_2O$	0.25	0.25	0.25	0.25
$CuSO_4 \cdot 5H_2O$	0.025	0.025	0.025	0.025
$CaCl_2 \cdot 6H_2O$	0.025	0.025	0.025	0.025
Na_2EDTA	37.3	37.3	37.3	37.3
$FeSO_4 \cdot 7H_2O$	27.8	27.8	27.8	27.8
Thiamine HCl	0.4	0.4	0.4	0.4
Ca pantothenic acid	2.0	2.0		
Gibberellic acid	0.25	0.4		
Benzylaminopurine				5.0
Naphthalene acetic acid		0.01		
Chlorocholine chloride				500
Inositol	100	100	100	100
Sucrose	3%	2%	3%	8%
Mannitol			4%	
Agar	0.8%		0.8%	

Table 4. Media composition

MSA medium		Medium 9	
MS		MS	
+ Gibberellic acid	0.1 mg/l	+ Gibberellic acid	0.4 mg/l
Glycine	2.0 mg/l	Benzylaminopurine	0.5 mg/l
Nicotinic acid	0.5 mg/l	Naphtalene acetic acid	0.01 mg/l
Pyridoxine	0.5 mg/l		
Thiamine	0.4 mg/l		
Gelrite	1.9 mg/l		
Sucrose	25.0 mg/l		

16-h light photoperiod and a temperature of 22 °C. Orbital horizontal shakers are used to give the plants a better distribution of nutrients inside the flasks. By using single node cuttings instead of stem segments with several nodes we can improve the quality of the plants that will be induced to tuberize in vitro. Thirty single nodes per flask and no agitation gave 26 to 30 vigorous plantlets per flask in four different clones (Table 5). In this way almost all the nodes grow into plantlets and the distribution of plantlets inside the flask is uniform in space and oxygen supply.

On the other hand, the nutrient medium was enriched with several vitamins (MSA medium) and studies were made to see the effect of the use of growth

Table 5. Mean number of plantlets from 30 single nodes inoculated in Erlenmeyer flasks. The data correspond to 20 replicates

Clone	No. of plants
Seseni	27.6
LT-8	26.8
Montsama	31.2
I-1035	26.3

Table 6. Mean number and weight of tubers grown in Erlenmeyer flasks in two different media during propagation stage. The data correspond to ten replicates

Clone	Medium 1 without growth regulators		Medium 2 with growth regulators	
	Mean no. of tubers per flask	Mean weight of tubers per flask	Mean no. of tubers per flask	Mean weight of tubers per flask
Seseni	18	0.69	25	0.99
Montsama	13	0.55	26	0.99
LT-7	6	0.28	16	0.65
I-1035	18	1.17	19	0.89
LT-8	18	1.06	6	0.32

regulators in this stage. The results in Table 6 show that there is a great influence of the genotype on the final weight and number of tubers.

Stage III. Induction of Tuberization. Plantlets that have been growing in Erlenmeyer flasks for 4 weeks are induced to tuberize by changing the propagation medium to a tuberization medium consisting of Murashige and Skoog salts supplemented with 5 mg/l BAP, 500 mg/l CCC, and 8% sucrose. This medium induces in vitro tubers in 4 to 6 weeks.

Some changes in the formulation medium have improved the number and size of the tubers that can be obtained from 30 initial single node cuttings. By reducing the nitrogen concentration (Table 7), using B5 medium instead of Murashige and Skoog medium, we obtain a better tuber yield per flask.

Table 7. Tuber yield in Erlenmeyer flasks with high and low nitrogen, using single nodes as inocula. The data correspond to six replicates

Clone	Low nitrogen		High nitrogen	
	Mean no. of tubers per flask	Mean weight of tubers per flask	Mean no. of tubers per flask	Mean weight of tubers per flask
DTO-33	16.1	734.5	7.5	394.1
LT-2	17.0	1105.8	15.6	403.2

In a further experiment, the CCC in the induction medium was replaced by different concentrations of ABA, coumarin, and BAP to study their effects as tuber inducers. The results shown in Table 8 indicate that coumarin at a concentration of 50 mg/l increases tuber number and total tuber weight. Different carbon sources: sucrose, glucose, and fructose at concentrations ranging from 4 to 12% were also studied in relation to the tuberization process. Eight percent sucrose included in the media composition was found to be the best for the induction of in vitro tubers (Table 9).

We have also tried direct in vitro tuberization including growth and induction of the plantlets in one single step. Stem segments containing two nodes each were inoculated both in test tubes and in large plastic vessels (magentas). In the case of test tubes, one single stem segment was included per test tube and the inoculation was made in such a way that one of the nodes was included inside the agar whilst the other was above it. In the plastic vessels 12 explants were inoculated, forming three rows of four stem segments each. In both cases the vessels contained B5 medium supplemented with 5 mg/l BAP, 8% sucrose, and 1.9% Gelrite. Once inoculated, they were directly placed under dark conditions at a temperature or approximately 18 °C for a period of 8 weeks. Table 10 shows the results of test tubes where tuberization was induced after 8 weeks.

Table 8. Effect of different inducers on in vitro tuberization clone BR-63.65

Medium	(mg/l)	Mean no. of tubers/flask	Mean weight tubers/flask (mg)	Mean weight/tuber (mg)
ABA	1.5	4.2	464.5	110.6
	15.0	1.4	125.3	89.5
	50.0	0.0	0.0	0.0
Coumarin	2.5	2.8	176.9	63.2
	25	7.6	570.1	75.0
	50	16.0	682.6	42.6
BAP	5	3.6	478.6	132.9
	10	3.4	217.8	78.8
	50	1.8	125.1	69.5
Standard control (ccc)		2.8	436.7	155.9
Noninduction control		1.6	157.0	98.1

Table 9. Effect of different carbon sources on in vitro tuberization of clone Ticahuasi

Medium		Mean no. of tubers/flask	Mean weight tubers/flask (mg)	Mean weight/tuber (mg)
Sucrose	4%	2.0	61.7	30.8
	6%	8.0	606.2	75.8
	8%	9.2	731.8	140.7
	10%	6.8	552.1	81.2
	12%	9.2	671.0	72.9
Glucose	4%	3.2	64.7	20.2
Fructose	4%	0.0	0.0	0.0
	12%	0.8	87.1	108.8

Table 10. Results of direct in vitro tuberization in test tubes containing MS medium supplemented with 5 mg/l BAP and 8% sucrose, and grown in darkness

Clone	No. of test tubes inoculated	No. of test tubes with tubers	% Tuberization
LT-8	96	78	81%
Montsama	96	42	44%
I-1035	96	67	70%
Atlantic	48	23	45%
Seseni	96	78	81%

2.5 In Vitro Tuber Characteristics

Morphological Characteristics. Potato clones can be characterized on the basis of a number of morphological and biochemical characteristics such as tuber shape and color, leaf shape, and an electrophoretic pattern of tuber storage proteins. In a sample of clones, we made a comparative analysis between in vitro and field-produced tubers for both shape and color. In all cases, both shape and color characteristics were markedly similar (Fig. 4).

Storage of in Vitro Tubers. Tuber dormancy is related to the hormonal balance within the tuber. Thus, the different methods of in vitro tuber induction play a regulatory role in controlling in vitro tuber dormancy (Coleman 1987). We have found that the environmental conditions during induction play a critical role in dormancy. When tubers were induced in total darkness, harvested, and stored at 4 °C, the average natural dormancy period was 210 days; however, if the tubers were induced under 8 h of light (1000 lx), the natural dormancy was 60 days.

Attempts to break dormancy chemically, for example the use of gibberellic acid, have so far given highly erratic results; this is probably related to physiological size and state of the tuber, since similar problems are encountered with small tubers

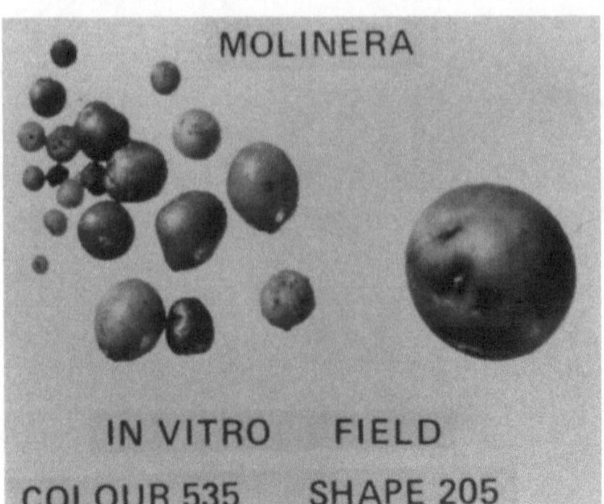

Fig. 4. Comparison of tuber characteristics for in vitro tubers and field-produced tubers of the same clones. (Courtesy CIP)

induced in nursery beds. The most rapid and reliable way to break dormancy was to harvest the in vitro tubers and store them in the diffuse light in sterile Petri dishes sealed with parafilm at 4 °C for 4 weeks followed by storage at 18–22 °C for 4–8 months.

Incorporation into a Seed Multiplication Program. In vitro tubers can be used as the initial pathogen-tested material in a seed multiplication program and may offer some advantages over in vitro plantlets. In vitro tubers can be planted in pots in greenhouses to produce pathogen-tested mother plants from which cuttings can be taken as part of a conventional rapid multiplication program, or they can be planted at high density (up to 100 tubers/m^2) in nursery beds similar to those described by Wiersema et al. (1987). The growth of these in vitro, tuber-derived plants in nursery beds offers an excellent method for producing high-quality seed tubers (pre-basic seed).

Four clones were used in an experiment to determine seed tuber yield from in vitro tubers planted in beds (Table 11).

In a 90-day period we produced in some genotypes more than 900 small, high-quality tubers in an area of 1 m^2. In another experiment, in vitro tubers were graded into three different size categories and planted in nursery beds at a density of 100 in vitro tubers/m^2. After 90 days tubers were harvested and graded into six weight categories. Table 12 shows the distribution of weight categories obtained in the harvest of the variety Mariva, the majority of the tubers being between 1–20 g. These high-quality tubers can then be further propagated within a seed mutliplication program. It should also be noted that cuttings can also be taken from the in vitro, tuber-derived plants growing in nursery beds as a method of increasing overall propagation rate.

Table 11. Number of seed tubers from in vitro tubers (all sizes) planted in nursery beds at a density of 100 tubers/m^2. (Wiersema et al. 1987)

Clone	Tubers harvested/ 1 m^2 bed
DTO-2	330
Mariva	648
Atacama	492
Piñaza	195

Table 12. Distribution of tuber size (yield) in a 1-m^2 bed planted with different sizes of in vitro tubers var. Mariva. Planting density 100 in vitro tubers/m^2. Huancayo, Peru. (Wiersema et al. 1987)

Initial in-vitro tuber size (diameter)	No. of tubers						
	1 g	1.5 g	5.10 g	10.20 g	20.40 g	40 g	Total
3 mm	123	150	76	61	14	3	427
3–7 mm	184	214	171	148	56	3	776
7 mm	207	330	191	151	29	4	912

2.6 Field Performance of Micropropagated Plants

Rarely are micropropagated plants transplanted directly to the field; the normal stages are for large-scale micropropagation in tubes or boxes (Fig. 5) to be taken and transplanted at high density in seed beds (Fig. 6). The small high-quality tubers harvested from these seed beds are then taken to the field for further multiplication.

To estimate field performance of pathogen-tested material is normally a somewhat complex affair since it involves interactions of genotype, pathogen, and environment. However, in the most drastic cases, Potato Leaf Roll Virus can, for example, lead to a 60% loss in full yield and 80% loss of total marketable value.

The failure to plant in vitro material directly in the field is often related to propagule cost, should mechanized systems of micropropagation reduce the unit cost sufficiently, then it is feasible that direct planting could take place.

3 Summary

Micropropagation of potato is now a standard basic for seed multiplication in almost all potato programs worldwide. The process is based on the rapid multiplication of pathogen-tested materials under careful conditions thus avoiding yield/quality loss due to pathogens.

Potato has often been described as a model crop plant for tissue culture studies. Methods are well established for meristem culture, and various forms of micropropagation using plantlets, in vitro tubers, etc. The multi-

Fig. 5. Massive in vitro plantlet production in test tubes

Fig. 6. Transplant of in vitro plants to seed beds at high density to produce mini-tubers

plication of potato using these methods is common in both private and state run schemes. In this chapter details are given of the basic methods used for multiplication at the International Potato Center (CIP). CIP has responsibility for maintainance and distribution of potato germplasm to developing country national programs worldwide and to achieve this objective relies heavily on tissue culture methods.

References

Bajaj YPS (1981) Regeneration of plants from potato meristems freeze-preserved for 24 months. Euphytica 30(1): 141–145

Bajaj YPS (ed) (1987) Biotechnology in agriculture and forestry, vol 3. Potato. Springer, Berlin Heidelberg New York

Bajaj YPS (ed) (1991) Biotechnology in agriculture and forestry, vol 17. High-Tech and Micropropagation I. Springer, Berlin Heidelberg New York Tokyo

Baker WG (1953) A method for the in vitro culturing of potato tubers. Science 118: 384

Claver FK (1967) Influence of long day and temperature on tuberization of shoots and plantlets of potato in vitro. Revista de Investigaciones Agropecuarias INTA, Buenos Aires, Rep Argentina Ser 2, Biol Prod Veg IV (12): 223–230

Claver FK (1970) The effects of abscisic acid on tuberization of potato sprouts in vitro. Phyton 27: 25–29

Coleman WK (1987) Dormancy release in potato tubers: a review. Am Potato J 64: 57–68

Dodds JH, Silva-Rodriguez D, Bryan JE (1986) Transport, receipt and propagation of in vitro potato plantlets. Int Potato Cent, Lima, Peru, 14 pp
Espinoza N, Estrada R, Tovar P, Bryan J, Dodds JH (1984) Tissue culture micropropagation, conservation and export of potato germplasm. Specialized Technology Document No. 1. Int Potato Cent, Lima, Peru, 20 pp
Estrada R, Tovar P, Dodds JH (1986) Induction of in vitro tubers in a broad range of genotypes. Plant Cell Tissue Organ Cult 7: 3–10
Estrella D, Muñoz F, Naranjo H (1987) Production of certified seed from tissue culture. An innovative approach in Ecuador. Am Potato J 64(8): 435
Hussey G, Stacey NJ (1981) In vitro propagation of potato (*Solanum tuberosum*) Ann Bot 48: 787–796
Hussey G, Stacey NJ (1984) Factors affecting the formation of in vitro tubers of potato (*Solanum tuberosum* L.) Ann 53: 565–578
Jung ML, Irvine BR, Baker WG (1972) In vitro tuberization of the common potato (*Solanum tuberosum*) is not a response to the osmotic concentration of the medium. Can J Bot 50: 603–605
Kim YC (1982) In vitro tuber formation from proliferated shoots of potato (*Solanum tuberosum*) as a method of aseptical maintenance. Ph.D. Thesis, South Korea
Koda Y, Okasawa Y (1983) Influences of environmental, hormonal and nutritional factors in potato tuberization in vitro. Jpn J Crop Sci 52(4): 582–591
Laurence GH, Barker WG (1963) A study of tuberization in the potato. *Solanum tuberosum* 40(10): 349–356
Lizarraga R, Tovar P, Jayasinghe U, Dodds JH (1986) Tissue culture for elimination of pathogens. Specialized technology document No. 3. Int Potato Cent, Lima, Peru, 21 pp
Minglo-Castel AM, Young RE, Smith AE (1976) Kinetin-induced tuberization of potato in vitro: on the mode of action of kinetin. Plant Cell Physiol 17: 557–570
Murashige T, Skoog F (1962) A revised medium for rapid growth and bioassays with tobacco tissue cultures. Physiol Plant 15: 473–497
Palmer CE, Smith OE (1969) Cytokinin and tuber initiation in the potato *Solanum tuberosum* L. Nature (London) 221: 279–280
Palmer CE, Smith OE (1970) Effect of kinetin on tuber formation on isolated stolons of *Solanum tuberosum* L. cultured in vitro. Plant Cell Physiol 11: 303–314
Quak F (1977) Meristem culture and virus-free plants. In: Reinert J, Bajaj YPS (eds) Applied and fundamental aspects of plant cell, tissue, and organ culture. Springer, Berlin Heidelberg New York, pp 598–615
Schilde-Rentschler L, Espinoza N, Estrada R (1984) Induction of tubers in vitro and their utilization for storage and distribution of potato germplasm. In: Abstr Conf Papers 9th Triennial Conf Europ Assoc Potato Res, Interlaken, Schweiz, 1–6 July, Switzerland
Stallknecht GF, Farnsworth S (1982) General characteristics of coumarin-induced tuberization of axillary shoots of *Solanum tuberosum* L. cultured in vitro. Am Potato J 59: 69–75
Tizio R (1971) The action and probable role of certain gibberellins A-1, A-3, A-4, A-7, A-9, A-13 on stolon growth and tuberization of potato *Solanum tuberosum*. Potato Res 14: 193–204
Tovar P, Chandra R, Ruiz de Arcaute R, Dodds JH (1987) Effect of medium composition on in vitro tuberization. Am Potato J 64: 462
Wang PJ, Hu CY (1982) In vitro mass tuberization and virus-free seed potato production in Taiwan. Am Potato J 59: 33–37
Wiersema SG, Cabello R, Tovar P, Dodds JH (1987) Rapid seed multiplication by planting into beds microtubers and in vitro plants. Potato Res 30: 117–120

I.7 Micropropagation of Sweet Potato (*Ipomoea batatas* L.)

R. P. Chée, J. R. Schultheis, and D. J. Cantliffe[1]

Introduction

Sweet potato (*Ipomoea batatas* (L.) Lam.) ranks seventh among crops grown for food, and tenth among all crops grown worldwide (FAO 1986). Besides food production, sweet potato has good potential as a biomass crop for ethanol and methane (Smith and Frank 1984). The accumulated carbohydrate in the roots can be efficiently digested and converted into ethanol or methane gas and used for energy (Smith et al. 1987).

Sweet potato is a member of the Convolvulaceae (morning glory) family and is usually grown for its edible roots. It is often confused with yam since both storage organs look and taste similar. Yams, however, belong to the genus *Dioscorea* (family Dioscoreaceae) and are monocots, diploid (2n = 20), and their storage organs are tubers, whereas sweet potatoes are dicots, hexaploid (2n = 90), and the storage organs are botanically classified as roots (Wilson and Collins 1988). The literature often confuses the issue by incorrectly calling the sweet potato's storage organ a tuber (Kays 1985). Sweet potato originated in tropical America (Henderson et al. 1984). Since then it has been distributed worldwide.

Sweet potatoes require a growing season that has at least 4 to 5 frost-free months to produce ample-sized storage roots. Warm conditions are required since temperatures at or below 15 °C result in little or no plant growth (Pierce 1987). Above 15 °C, Harter and Whitney (1927) determined plant growth rates continued to increase up to a temperature of 35 °C, while Pierce (1987) reported optimum growth occurred at air temperatures of 29 °C.

The crop grows best in well-aerated, sandy-type soils, with a pH ranging between 5 and 7.8 (O'Hair 1984). However, good yields of 17.9 and 17.0 MT/ha have been obtained at soil pH values of 4.3 and 4.6 (Abruna et al. 1978). Soil aeration, when limiting, can inhibit storage root formation and development (Wilson 1970).

2 Need for Micropropagation

Commercial plantings of sweet potatoes are primarily for food production, but attention in recent years has focused on this crop as a biomass crop for energy

Florida Agricultural Experiment Station Journal Series No. N-00418
[1] Vegetable Crops Department, University of Florida, Gainesville, FL 32611, USA

Biotechnology in Agriculture and Forestry, Vol. 19
High-Tech and Micropropagation III (ed. by Y.P.S. Bajaj)

production. Southern Japan, in some instances, has used sweet potato for energy production (Yen 1974). Mass production, however, is limited by the high cost of vegetative propagation (Cantliffe et al. 1987). Problems with this planting method include the following: (1) length of time needed for plants to survive stressful conditions, (2) maintenance and production of disease-free plants, (3) spatial requirements for large-scale production, and (4) labor costs to maintain the plant propagules (Cantliffe et al. 1987).

Sweet potato plants generally do not produce seeds since most cultivars are self-incompatible (Hernandez and Miller 1964; Wang 1964; Onwueme 1978). Those plants which are produced from seeds are highly heterozygous and differ genetically from the mother plant. Seeds mature approximately 40–60 days after anthesis (Matsuo et al. 1984), and are about 3 mm in length with one surface flattened and the other convex (Hayward 1938; Onwueme 1978). They also have hard, nearly water-impervious seed coats. Scarification either with sulfuric acid for about 45 min or by mechanically clipping part of the seed coat is necessary for optimum germination, which may reach approximately 90% (Jones and Dukes 1982). Germination usually occurs within 1 to 2 days, while plant emergence takes an additional day or two under favorable growing conditions (Hayward 1938). The use of tissue culture has been suggested as a means to economically produce crops normally propagated vegetatively, such as sweet potato. The rapid multiplication of disease-free plant material would be of great value to plant breeders and might open the door to hybridization and commercial reproduction of sweet potato hybrids.

3 In Vitro Studies/Micropropagation

Relatively few studies have dealt with cell and tissue culture of sweet potato. Root and adventitious shoot formation was obtained from callus derived from root plugs (Gunkel et al. 1972; Yamaguchi and Nakajima 1973). Responses to various hormone levels and their effects on the rate of callus proliferation and shoot regeneration were tested in two cultivars (Litz and Conover 1978). Murashige and Skoog (1962) medium (MS) supplemented with indolacetic acid (IAA) stimulated callus production in White Star explants. The addition of activated charcoal inhibited callus growth, and 1 mg/l 6-benzylaminopurine (BAP) promoted shoot proliferation averaging 8.5 shoots 5 weeks after culture establishment. Cultivar PI 315343 averaged 3.4 less shoots, while rooting and subsequent plant formation was enhanced with 1 mg/l α-naphthaleneacetic acid (NAA).

Root and shoot formation were obtained from callus derived from leaf explants on MS media supplemented with 10–20 mg/l adenine or 0.1–0.5 mg/l kinetin (K) (Sehgal 1975). Sehgal (1978) also reported nodulation followed by plant formation from anther callus obtained on MS media containing adenine (10 or 20 ppm) plus 2,4-dichlorophenoxyacetic acid (2,4-D) (1 ppm). Adventitious shoots were generated from root discs in one of three species tested, while shoots readily formed in all three species from petiole sections (Hwang et al. 1983). Callus formed best from stem, leaf, and storage root explants when placed on MS media supplemented with

1 mg/l NAA and 10 mg/l indole-butyric acid (IBA). Roots and shoots also formed when roots were subcultured on medium containing 1 mg/l NAA and 0.1 mg/l IBA. Plant formation was improved by light. Scaramuzzi (1986) induced roots and shoots in sweet potato by meristem tip culture using MS medium supplemented with kinetin and IAA or NAA. The use of BAP did not induce shoot production.

Micropropagation is not much different from conventional vegetative propagation since similar vegetative loci are also stimulated to produce new organs. For example, in shoot tip micropropagation, new shoots are produced from axillary buds of shoots placed in culture. This cycle is repeated, leading to the production of large numbers of shoots in a short time. Rapid proliferation is obtained through the recycling of very small organs. Proliferation is further promoted by optimizing nutrient media and incubation conditions. Plant growth regulators are also used to enhance organ growth. Differences in yield between conventional propagation and micropropagation can be illustrated with a woody perennial (Chée et al. 1984). In grapes, for example, a nursery might obtain ten new vines a year from a stock plant using hardwood cuttings, while 2000 vines can be produced in a month through a shoot micropropagation system.

Our initial studies with *I. batatas* consisted of selecting an explant containing meristematic cells that were either competent for or already determined for embryogenesis, and from which embryogenic callus could be repeatedly produced. Concomitantly, there was the necessity to define a culture medium capable of inducing the determination for embryogenesis in competent cells and/or to promote the proliferation of cells determined for embryogenesis. This dual task of finding an adequate explant and a matching culture medium was investigated by screening root, stem, leaf, petiole, and apical shoot explants on culture media containing relevant plant growth hormones (Liu and Cantliffe 1984, 1985). Embryogenic callus was obtained from leaf, shoot tip, stem, and root explants. The formation of embryogenic callus from shoot apices was most efficient and was investigated in detail (Cantliffe et al. 1987; Liu et al. 1989). Histological studies demonstrated embryogenic callus initiation from the periphery of the apical dome, and from the youngest pair of leaf primordia. The frequency of explants producing embryogenic callus declined from 90 to 30% as the number of leaf primordia pairs was increased from one to three (Cantliffe et al. 1987; Liu et al. 1989). The 2,4-D concentration of the media was important. While explants on media with 0 and 4.0 mg/l 2,4-D either produced nonembryogenic callus or failed to survive, at 0.5 to 3.0 mg/l 2,4-D over 80% of explants consisting of the apical dome formed embryogenic callus (Cantliffe et al. 1987).

The application of somatic embryogenesis to synthetic seed production requires an abundance of embryogenic callus. In sweet potato, embryogenic callus is yellow, opaque, and firm with a nodular surface (Liu and Cantliffe 1984; Jarret et al. 1984; Chée and Cantliffe 1988b). Nonembryogenic callus is white, translucent, and friable. Embryogenic callus grows slowly and produces fast-growing nonembryogenic callus which soon dominates the culture. The selective proliferation of embryogenic callus was obtained on a modified MS agar solidified medium with 10 μM 2,4-D and 1 μM BAP and in liquid medium with 5 μM 2,4-D (Chée and Cantliffe 1988b).

Because growth and development is also influenced by inorganic nutrients, proliferation of embryogenic callus in response to ammonium, potassium, chloride, and sodium was investigated (Chée et al. 1991). Growth was optimum at 15 mM when the ammonium concentration was varied from 2.5 to 20 mM. Embryogenic callus growth increased with increased potassium concentrations from 6.25 to 61.25 mM while nonembryogenic callus growth decreased. Additions of up to 40 mM NaCl did not affect embryogenic callus growth.

Long-term maintenance of embryogenic cultures, without embryo formation, was achieved on agar solidified medium and in suspension culture. Eight embryogenic lines have been successfully maintained for 34 months on solidified medium by subculturing callus every 6–8 weeks (Chée and Cantliffe 1988b). The average production in the 15th passage (30 months) was, at 8 weeks, 46 mg of embryogenic callus per mg of embryogenic callus subcultured. Long-term maintenance of the callus lines was also achieved in suspension culture by subculturing every 14 days, 6 mg of drained material (fresh weight) per ml of fresh medium.

Sweet potato embryogenic suspension cultures consist of heterogenous populations of embryogenic cell aggregates, nonembryogenic cells and their aggregates, and calli composed of all three (Chée and Cantliffe 1988a). Embryogenic cell aggregates are yellow aggregates of cytoplasmic cells 10–30 μm in diameter. Nonembryogenic cell aggregates are white, translucent, and composed of vacuolated cells 50–200 μm in diameter. Proliferation of embryogenic material in *I. batatas* suspension cultures occurs primarily through cycles, including detachment of discrete cell aggregates from calli and formation of new calli from those subunits. To study their composition, sweet potato embryogenic suspension cultures were fractionated with sieves having mesh openings ranging from 63 to 1000 μm and the embryogenic fraction was identified (Chée and Cantliffe 1989a). The material larger than 355 μm was composed primarily of calli and made up 75% of the total mass of cultures. These calli were composed of embryogenic and nonembryogenic subunits, and 98% of the embryogenic subunits measured 355–1000-μm (Table 1). The

Table 1. Composition of calli from 14-day old suspension cultures of *Ipomoea batatas* Lam[a] (see Chée and Cantliffe 1989)

Fraction (μm)	Cultures with embryogenic subunits[b] (%)	Embryogenic subunits/ fraction[c] (% of total)	Embryogenic subunits/ embryogenic culture	Embryogenic subunits forming embryos (%)
710–1000	100	76.4 ± 5.7[d]	180.8 ± 22.5[d]	70.0 ± 7.7[d]
500–710	100	19.0 ± 3.9	46.7 ± 9.3	19.5 ± 13.1
355–500	100	6.2 ± 1.3	15.8 ± 4.2	5.5 ± 5.5
250–355	50	1.3 ± 0.8	3.5 ± 1.9	trace
180–250	17	0.1 ± 0.1	0.2 ± 0.2	trace
125–180	17	0.1 ± 0.1	0.2 ± 0.2	trace
90–125	0	0.0	0.0	0.0
63–90	0	0.0	0.0	0.0

[a] Six replicates; for each replicate, 710–3000 μm calli collected from a suspension culture were separated into their component subunits and the resulting suspension was fractionated with sieves; each fraction was cultured on basal medium and embryo formation was evaluated after 12 days. NA = not available.
[b] Embryogenic subunits = yellow cytoplasmic cell aggregates; six cultures per fraction.
[c] Average total number of embryogenic subunits per replicate = 240.
[d] Means ± standard errors.

125–355 μm fraction of suspension cultures was composed of cell aggregates of which 20% were embryogenic. The 63–125-μm fraction contained only 2% of embryogenic cell aggregates.

The application of somatic embryogenesis to synthetic seed production will require both high embryo numbers per amount of embryogenic callus cultured, and the concomitant formation of individualized embryos. Since embryogenic and nonembryogenic cell aggregates of various sizes are released when calli undergo fragmentation (Chée and Cantliffe 1989a), embryo yields could be optimized if the callus or embryo production potentials of cell aggregates of different sizes were known. Then the appropriate suspension culture fractions could either be used to produce embryos or be recycled for further embryogenic callus production, and methods for obtaining the desired cell aggregates sizes could be developed. The embryo production from cell aggregates of sizes ranging from 180 to 1000 μm was studied (Chée and Cantliffe 1991). The percentage of aggregates forming embryos decreased with decreasing aggregate size; 20% of the 710–1000-μm aggregates and only 2% of the 180–250-μm aggregates formed embryos (Table 2). Embryo production decreased from 8.8 embryos per 710–1000 μm aggregate to 1.3 embryos per 180–250-μm aggregate. Conversely, embryo production per mg of cultured embryogenic callus increased quadratically with decreasing aggregate size. Individual embryos were produced from cell aggregates 180–250 μm in size. Thus, embryo production per unit mass of cultured callus could be optimized using 180–250-μm cell aggregates. However, under our present suspension culture conditions, embryos initiated on large cell aggregates develop to the torpedo and cotyledonary stages, while they remain arrested at the globular stage on small cell aggregates (Chée and Cantliffe 1989b, c).

In order to demonstrate that embryos could develop to advanced stages on small aggregates, we compared embryo formation from cell aggregates 250–355 μm in diameter cultured either in suspension in liquid medium, placed on agar solidified medium, or immobilized on alginate beads floated in liquid medium (Chée and Cantliffe 1989b). Embryos were initiated but remained arrested in their globular stage on cell aggregates cultured in suspension. Embryos developed to the torpedo stage from cell aggregates cultured on solidified medium and from cell aggregates anchored on alginate beads. Thus, embryos continued to develop beyond the globular stage when a structural polarity was created. This may have

Table 2. Effect of cell aggregate size on embryo production at 21 days in *Ipomoea batatas* (L.) Lam.[a]

	Cell aggregate diameter (μm)					
	710–1000	500–710	355–500	250–355	180–250	Trend
Aggregates per replicate	10	44	52	54	60	
Aggregates with embryos (%)	20	10	4	3	2	L**
Embryo production:						
per aggregate	8.8	4.9	3.8	2.0	1.3	L**
per mg of aggregate	23	37	83	121	223	L**Q**

[a] Statistical analysis was by the component of variance method; F test for significance was used; * = significant at P = 0.05; ** = significant at P = 0.05; L = linear trend; Q = quadratic trend; 10 replicates per treatment.

led to the establishment of a physiological polarity. A structural polarity exists also in calli which are composed of cell aggregates, and from which embryos develop beyond the globular stage. In that case we provided indirect evidence that in *I. batatas* embryo development is subjected to polar transport of endogenous IAA from embryogenic loci (Chée and Cantliffe 1989c). In those experiments, embryo development was increasingly restricted to lesser advanced stages with increased concentrations in the agar solidified media of either 2,4-D or triiodobenzoic acid (TIBA), a specific inhibitor of polar auxin transport. Consequently, inhibition of IAA transport might be the physiological mechanism through which exogenous auxins, widely used in protocols for somatic embryogenesis, inhibit embryogenesis and promote the proliferation of embryogenic callus.

I. batatas somatic embryos are found arrested at all stages of development, and show great variability in developmental patterns and morphologies (Chée and Cantliffe 1988a). A conventional development pathway through globular and heart stages leads to embryos at the torpedo and cotyledonary stages. Assignment of an embryo to a certain category must not rely strictly on morphological resemblance but must also account for the presence of similar histogens. For example, many variants of the torpedo stage were found but all had in common functional shoot and root apices, a cotyledon histogen, and a nonexpanded hypocotyl (Chée and Cantliffe 1988a).

Three types of somatic embryos capable of growing into plants were selected (Chée and Cantliffe 1988a). They included the torpedo and cotyledonary embryos, and a torpedo embryo with an elongated hypocotyl. They differed fundamentally from mature embryos obtained from seed in their morphology, lack of large cotyledons, and did not undergo any apparent phase of dormancy before producing plants. All three types of embryo converted rapidly into plantlets when growth hormonal additives especially auxin as NAA (Table 3) were added to the germination media (Chée et al. 1990).

Table 3. Response of sweet potato [*Ipomoea batatas* (L.) Lam.] somatic embryos to NAA[a] (see Chée et al. 1990)

	% Embryos responding								
	NAA (μM)								
Criterion	0	0.1	0.2	1	2	5	10	F	R^2
Shoot growth	17	41	38	33	25	26	13	L*	0.23
Root growth	47	76	78	80	74	51	20	L**	0.49
Shoot and root growth	3	38	33	28	23	14	5	L**	0.39
Basal callus growth	3	8	8	5	9	3	53	Q*	0.48

[a] Results at 21 days. Statistical analysis by the component of variance method; regression analysis for NAA treatments only, F = F test of significance, L = linear, Q = quadratic, ** = significant at $P = 0.01$, * = significant at $P = 0.05$, R^2 = determination coefficient; the contrasts "0 vs 0.1 μM NAA" were significant at $P = 0.01$ for root growth and callus growth, and at $P = 0.05$ for embryos with both shoot and root growth; data are means of four blocks of ten embryos.

4 Summary and Conclusions

Micropropagation of sweet potato has been demonstrated by several workers (Henderson et al. 1984; Scaramuzzi 1986). Our research has demonstrated the use of somatic embryogenesis as an extremely efficient method to propagate sweet potato (Fig. 1). Callus is obtained from apical domes of disease-free stock material. Embryogenic callus is initiated in solid culture with 2,4-D and embryos are produced without 2,4-D (Tables 4 and 5). The use of somatic embryos offers a

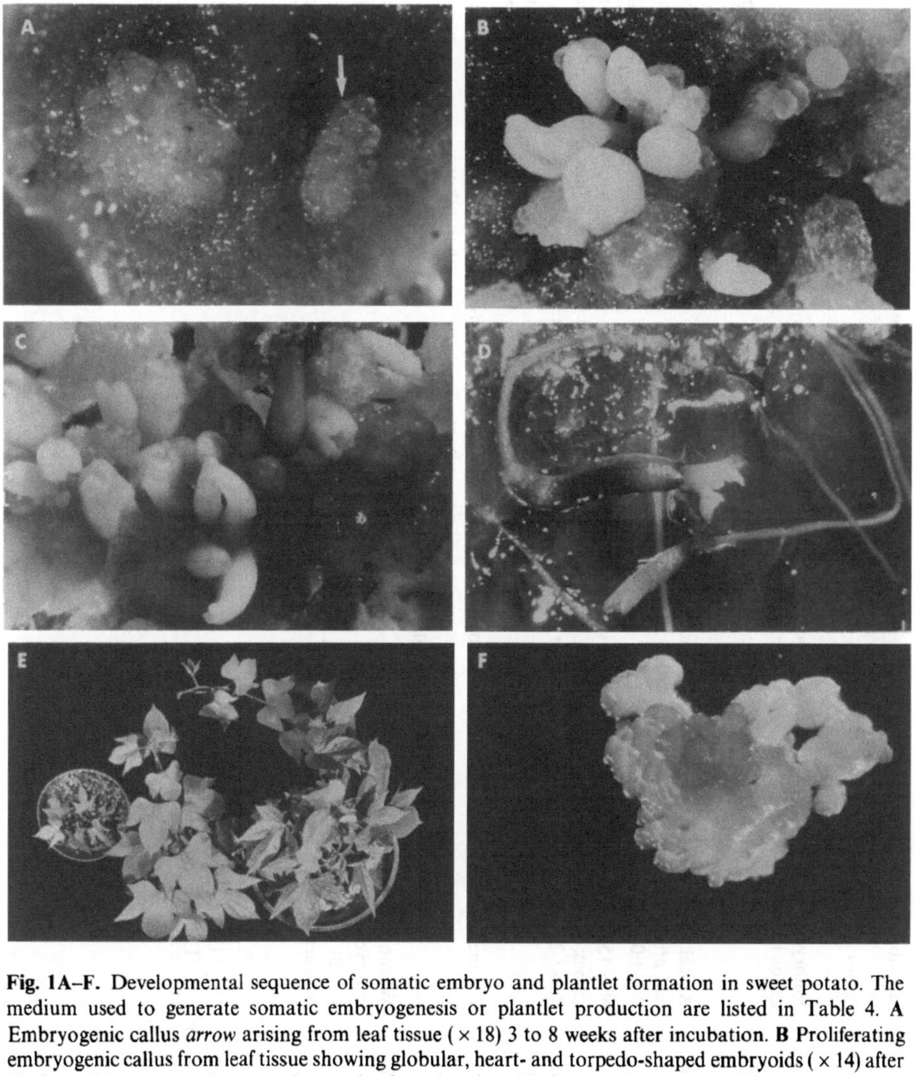

Fig. 1A–F. Developmental sequence of somatic embryo and plantlet formation in sweet potato. The medium used to generate somatic embryogenesis or plantlet production are listed in Table 4. **A** Embryogenic callus *arrow* arising from leaf tissue (× 18) 3 to 8 weeks after incubation. **B** Proliferating embryogenic callus from leaf tissue showing globular, heart- and torpedo-shaped embryoids (× 14) after 10 days. **C** Numerous heart- and torpedo-shaped embryoids from embryogenic callus (× 7.5) after 14 days. **D** Germination of embryoids into plantlets (× 7) 14 days after placing on conversion medium. **E** Maturing plants regenerated from embryogenic callus after 5 weeks. **F** Subcultured embryogenic callus with a low rate of embryoid production (× 15) subcultured every 8 weeks (see Liu and Cantliffe 1984)

Table 4. Protocol for somatic embryogenesis and plantlet production in sweet potato [*Ipomoea batatas* (L.) Lam.]

Procedure	Step 1	Step 2		Step 3		Step 4
	Embryogenic callus induction	Embryogenic callus proliferation		Embryo production		Conversion to plantlets
		Agar cultures	Suspensions	Agar cultures	Suspensions	
Explant, inoculum or propagule	Apical dome plus 1–2 leaf primordia (ca. 0.2–1.0 mm)	Embryogenic callus from stage 1, 2; 1 mg/cm^2	Embryogenic callus from stage 1, 2; 6–40 mg/ml	Embryogenic callus from stage 2; 1 mg/cm^2	Embryogenic callus from stage 2; 6 mg/ml	Somatic embryo from stage 3; 1 embryo/cm^2
Incubation condition	60 × 15 mm plastic Petri dishes; 12.5 ml medium; dark; 27 °C	100 × 15 plastic Petri dishes; 25 ml medium; dark; 27 °C	125 ml, 250-ml Erlenmeyer flasks; 20, 40 ml medium; rotary shaker 100 rpm; dark; 27 °C	100 × 15 mm plastic Petri dishes; 25 ml medium; 1st week in dark; 10-h photoperiod; 27 °C	125-ml or 250-ml Erlenmeyer flasks; 20 or 40 ml medium; rotary shaker 100 rpm; dark; 27 °C	100 × 25 mm disposable plastic Petri dishes; 25 ml medium; 10 h photoperiod 25 °C
Medium	CI[a] medium	CP[b] medium for agar cultures	CP medium for suspension cultures	EP[c] medium for agar cultures	EP medium for suspension cultures	EC[d] medium
Recultures	—	Every 8 weeks	Every 2 weeks	—	—	—
Formation of embryogenic callus, somatic embryo and plantlet	Within 4 weeks, nonembryogenic callus, within 8 weeks, embryogenic callus	—	—	Globular embryos within 8 days, first mature embryos in 12 days, mature embryos arrested at 21 days	Clusters of embryos in 21 days	Root growth 8–14 days, visible shoot growth 14–21 days

[a] CI = callus induction.
[b] CP = callus proliferation.
[c] EP = embryo production.
[d] EC = embryo conversion.

Table 5. Composition of media used for each culture stage in somatic embryogenesis and plant production in sweet potato [*Ipomoea batatas* (L.) Lam.]

	Culture stage			
	Embryogenic callus induction (CI)	Embryogenic callus proliferation (CP)	Embryo production (EP)	Embryo conversion to plantlets (EC)
Inorganic salt	MS	Modified MS 30 mM KCl	Modified MS 10 mM NH_4NO_3	MS
Organic substances (μM)				
Thiamine · HCl	5	5	5	5
Nicotinic acid	10	10	10	10
Pyridoxin · HCl	5	5	5	5
Myo-inositol	500	10	10	500
Growth regulators (μM)				
2,4-D	10	10/5[a]	0/0, 0.5	—
BAP	—	1/0	—	—
ABA	—	—	0, 1, 10/0, 10	—
NAA	—	—	—	0, 0.1
Sucrose (%)	3	3	3	1.6
Agar (%)	0.6	0.6	0.8	0.8
pH before autoclaving	5.8	5.8	5.8	5.8

[a] Concentration in agar solidified media/concentration in liquid media.

unique opportunity for the integration and automation of rapid clonal multiplication and seeding of sweet potato.

References

Abruna F, Vicente-Chandler J, Rodriguez J, Badillo J, Silva S (1978) Crop response to soil acidity factors in utisols and ozisols in Puerto Rico. V. Sweet potato. J Agric Univ P R 63: 250–267

Cantliffe DJ, Liu JR, Schultheis JR (1987) Development of artificial seeds of sweet potato for clonal propagation through somatic embryogenesis. In: Smith WH, Frank JR (eds) Methane from biomass – a systems approach. Elsevier, New York, pp 183–195

Chée RP, Cantliffe DJ (1988a) Somatic embryony patterns and plant regeneration in *Ipomoea batatas* Poir. In Vitro 24: 955–958

Chée RP, Cantliffe DJ (1988b) Selective enhancement of *Ipomoea batatas* Poir. embryogenic and non-embryogenic callus growth and production of embryos in liquid culture. Plant Cell Tissue Organ Cult 15: 149–159

Chée RP, Cantliffe DJ (1989a) Composition of embryogenic liquid cultures of *Ipomoea batatas* (L.) Lam. and production of individualized embryos from cell aggregate suspensions. Plant Cell Tissue Organ Cult 17: 39–52

Chée RP, Cantliffe DJ (1989b) Embryo development from discrete cell aggregates in *Ipomoea batatas* (L.) Lam. cultured in vitro in response to structural polarity. In Vitro Cell Dev Biol 25: 757–760

Chée RP, Cantliffe DJ (1989c) Inhibition of somatic embryony in response to 2,3-5-triiodobenzoic acid and 2,4-dichlorophenoxyacetic acid in *Ipomoea batatas* (L.) Lam. J Plant Physiol 135: 398–403

Chée RP, Cantliffe DJ (1992) Embryogenic callus and somatic embryo production from *Ipomoea batatas* (L.) Lam. cell aggregates of various sizes. Sci Hortic 27: in press.
Chée RP, Pool RM, Bucher D (1984) A method for large scale in vitro propagation of *Vitis*. New York's Food and Life Sciences Bulletin 109. New York State Agric Exp Stn, Geneva, New York
Chée RP, Schultheis JR, Cantliffe DJ (1990) Plant recovery from sweet potato somatic embryos. Hortic Sci 25: 792–797
Chée RP, Leskovar DI, Cantliffe DJ (1992) Embryogenic callus and embryo growth in *Ipomoea batatas* (L.) Lam. in response to NH_4^+, NO_3^-, K^+, Cl^-, and Na^+. J Amer Soc Hort Sci 117: in press
FAO (1986) Production yearbook. Food and Agricultural Organization, Rome, Italy, vol 40
Gunkel JE, Sharp WR, Williams BH, West WC, Drinkwater WO (1972) Root and shoot initiation in sweet potato explants as related to polarity and nutrient media variations. Bot Gaz 133: 254–262
Harter LL, Whitney WA (1927) Influence of soil temperatures and soil moisture on the infection of sweet potato by the stem rot organism. Agric Res 34: 435–441
Hayward HE (1938) Convolvulaceae (*Ipomoea batatas*). In: The structure of economic plants. Macmillan, New York
Henderson JHM, Phills BR, Whatley BT (1984) Sweet potato. In: Sharp WR, Evans DA, Ammirato PV, Yamada Y (eds) Handbook of plant cell culture, vol 2. Macmillan, New York, pp 302–326
Hernandez TP, Miller JC (1964) Further studies on the incompatibility in the sweet potato. Proc Am Soc Hortic Sci 85: 426–429
Hwang LS, Skirvin RM, Casao J, Bouwkamp J (1983) Adventitious shoot formation from sections of sweet potato grown in vitro. Sci Hortic 20: 119–129
Jarret RL, Salazar S, Fernandez R (1984) Somatic embryogenesis in sweet potato. Hortic Sci 19: 397–398
Jones A, Dukes PD (1982) Longevity of stored seed of sweet potato. Hortic Sci 17: 756–757
Kays SJ (1985) The physiology of yield in sweet potato. In: Bouwkamp JC (ed) Sweet potato products: a natural resource for the tropics. CRC, Boca Raton, pp 79–132
Komaki K, Chée RP, Cantliffe DJ (1989) Development of a synthetic seed system of sweet potato at the University of Florida. Nogyo Gijutsu Kyoki 44: 204–207
Liu JR, Cantliffe DJ (1984) Somatic embryogenesis and plant regeneration in tissue cultures of sweet potato (*Ipomoea batatas* Poir.) Plant Cell Rep 3: 112–115
Liu JR, Cantliffe DJ (1985) Tissue culture propagation development and its application to energy crops. In: 1984 Int Gas Res Conf Proc, Govern Inst, pp 622–629
Liu JR, Cantliffe DJ, Simonds SC, Yuan JF (1989) High frequency somatic embryogenesis from cultured shoot apical meristem domes of sweet potato (*Ipomoea batatas*) SABRAO J 21: 93–101
Litz RE, Conover RA (1978) In vitro propagation of sweet potato. Hortic Sci 13: 659–660
Matsuo TS, Itoo S, Murofushi N, Takahashi N, Kobayashi M, Chishiki T (1984) Identification of gibberellins in seeds of sweet potato [*Ipomoea batatas* (L.) Lam.] and several other Convolvulaceae plants. Agric Biol Chem 48: 2935–2941
Murashige T, Skoog F (1962) A revised medium for rapid growth and bioassays with tobacco tissue culture. Physiol Plant 15: 473–497
O'Hair SK (1984) Farinaceous crops, pp. 109–136. In: CRC Handbook of tropical food crops. CRC, Boca Raton
Onwueme IC (1978) The tropical tuber crops. Wiley, New York, pp 167–195
Pierce LC (1987) Vegetables: characteristics, production, and marketing. Wiley, New York
Scaramuzzi F (1986) Sweet potato (*Ipomoea batatas* Poir.) In: Bajaj YPS (ed) Biotechnology in agriculture and forestry, vol 2: Crops I. Springer, Berlin Heidelberg New York, pp 455–461
Sehgal CB (1975) Hormonal control of differentiation in leaf cultures of *Ipomoea batatas* Poir. Beitr Biol Pflanz 51: 47–52
Sehgal CB (1978) Regeneration of plants from anther cultures of sweet potato (*Ipomoea batatas* Poir.). Z Pflanzenphysiol 88: 349–352
Smith WH, Frank JR (1984) Comparative biomass yields of energy crops. In: Palz W, Coombs J, Hall DO (eds) Energy from biomass. Elsevier, New York, pp 323–329
Smith PH, Bordeaux FM, Goto M, Shiralipour A, Wilkie A (1987) Biological production of methane from biomass. In: Smith WH, Frank JR (eds) Methane from biomass – a systems approach. Elsevier, New York, pp 291–334
Wang H (1964) A study on self- and cross-incompatibilities in the sweet potato in Taiwan (Formosa). Proc Am Soc Hortic Sci 84: 424–430

Wilson LA (1970) The process of tuberization in sweet potato (*Ipomoea batatas* (L.) Lam.). Proc Int Symp Trop Root Crops 2: 24–26

Wilson LG, Collins WW (1988) What is the difference between a sweet potato and a yam? NC Agric Ext Ser Hortic Inf Feb 1988, Leaflet 23-A

Yamaguchi T, Nakajima T (1973) Hormonal regulation of organ formation in cultured tissue derived from root tubers of sweet potato. In: Proc 8th Int Conf Plant Growth Substances. Tokyo, Jpn, pp 1121–1127

Yen DE (1974) The sweet potato and Oceania. Bishop Museum Press, Honolulu, Hawaii

I.8 Micropropagation of Artichoke (*Cynara scolymus*)

V. Rossi and G. De Paoli[1]

1 Introduction

1.1 Botanical Description

The genus *Cynara* belongs to the Compositae (subfamily Tubuliflorae). The globe artichoke (*Cynara scolymus*) is a unique vegetable, having fibrous, fleshy rhizomes, with buds that develop into several tomentose and branched stems. The most vigorous varieties may reach 1.20–1.30 m in height. The bearing stem is erect, ribbed, and round in cross-section, ending with a floral head (capitulum). Axillary buds develop into several branches terminating in a capitulum which is smaller than the main one and which also matures later.

The capitulum is composed of a receptacle surrounded by a series of many bracts. The receptacle is discoidal, pulpy, and hairy in its upper face where the florets are inserted, while the bracts are pulpy at their bases and their shape and colors change depending on the varieties. The more or less closed bracts are generally oblong and triangular with violet coloring, sometimes terminating into a spine. The leaves are broad, lobed, or deeply divided; their color is dark green on the upper face and lighter on the lower, tomentose one.. The fruit is an oblong achene crowned by a characteristic pappus with plumose bristles (Marzi 1981). The artichoke is a monoecious self-sterile plant and its capitulum is composed of several tubular and bluish-violet fertile florets opening from the outside inwards. When the flowers open, the dehisced anthers shed the pollen, which germinates at once, but as the stigmata is not yet mature, no self-pollination occurs. The pollen, which remains alive for 4–5 days, is sticky, and can pollinate other capitula of the same plant only if carried by insects. Thus, protandry and entomophilous pollination ensure a high degree of heterozygosity (Ryder et al. 1987).

1.2 Distribution and Importance

In recent years, artichoke cultivation has been expanding in some Mediterranean regions where it has been valuable crop since ancient times, while in other countries of the world it is seldom found, and sometimes even unknown, though it could find optimum growing conditions. According to the data of the 1987 FAO yearbook (Fig. 1) Europe, of all the continents, has the largest surface (74.5%) used for

[1] Centrale Ortofrutticola–Vitrocoop, Via Masieta I, 1191, 47023 Cesena, Italy

Biotechnology in Agriculture and Forestry, Vol. 19
High-Tech and Micropropagation III (ed. by Y.P.S. Bajaj)

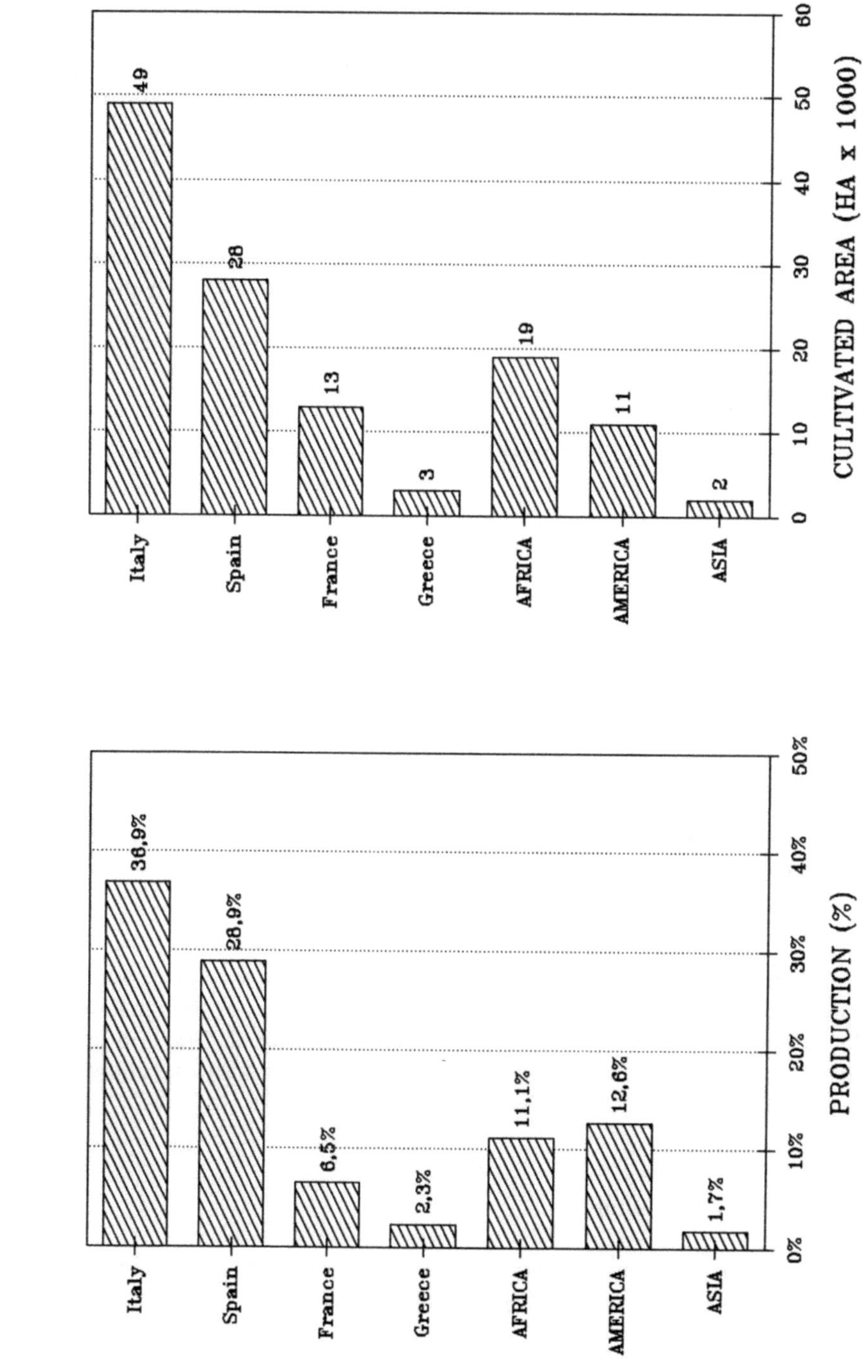

Fig. 1. Production and cultivated areas of artichoke (FAO 1987)

artichoke cultivation, and is the main producer, followed by Africa, America, and Asia. Argentina (6.5%) and the USA (California 4.48%) are the main artichoke producers on the American continent. About 95% of world production is limited to the countries lining the Mediterranean, particularly Italy, Spain, and France, which together produce 80% of the world total. Some states have lately been expanding this cultivation, for example, Spain and Greece. However, Italy is still the main producer on a world scale, with an annual production representing 36% of the total, and a cultivation area of 50 000 ha, which has been diminishing over the last few years. The cultivation areas are mainly located in southern Italy and in the islands, where the climate allows good winter production, which is valued for its quality and for being one of the few crops produced in this season. Artichoke cultivation is a good source of income if the grower can count on early production.

1.3 Conventional Propagation and Genetic Improvement

The plant parts employed for vegetative propagation of artichoke are: (1) "crown pieces"; (2) "ovoli"; and (3) offshoots. The choice depends on the tradition of the cultivation area. While crown pieces are more commonly employed in the USA, Italian growers prefer ovoli, which are enlarged buds along the stem in a dormant state. In Italy this propagation method allows earlier production compared to offshoots (Dellacecca and Pace 1976; La Malfa and Foury 1976), while in France the contrary is true. Offshoots are axillary buds on the plant base that develop into roots during the spring and are thus good propagation material, mainly employed in France and Spain. Today's artichoke cultivars are rather faulty, since they are generally local populations having high phenotypic variability and unsatisfactory yield, even though they are valued for other characteristics (Noto 1981). In order to obtain more rewarding varieties, their genetic improvement is aimed at:

— increase in yield per plant;
— early and simultaneous production;
— disease tolerance or resistance;
— greater resistance to salinity, cold, drought, etc.

Artichoke breeding may be carried out by both vegetative, clonal, and sexual selection (Ancora 1986). The latter could be more advantageous though it does not produce cheap varieties (Bigot and Foury 1984; Pecaut et al. 1985). In contrast, improvement via vegetative selection has been giving good results, allowing the quick cultivation of better cultivars (Tesi 1976; Noto 1981).

1.4 Need for in Vitro Culture

Traditional propagation favors spread of disease and develops a phythopathogenic situation hard to control, since field-grown plants can be easily attacked by several pathogen agents such as viruses, bacteria, fungi, nematodes, and insects. However, in vitro culture allows the production of disease-free plants, which are of great value to artichoke growers. In fact, Pecaut et al. (1985) found that virus-free artichokes of the cultivar Camus de la Bretagne propagated through meristem cultures gave a yield increase of 30–50% compared to controls, for 3 years at least. Afterwards there was reinfection by the most common viruses. Finally, artichoke micropropagation is a powerful tool for the breeder, who can without delay offer the results of his research to artichoke growers. In contrast, with traditional methods, several years are needed to obtain a large quantity of plants for mass production.

2 Review of Micropropagation

Several researchers have been experimenting with new methods for in vitro artichoke cultivation; in 1973, De Leo and Greco were already studying this technique with the aim of propagation. In 1979, during the Third International Artichoke Congress at Bari, several authors presented their results on in vitro culture of this plant and pointed out that the greatest problems concern the rooting phase. Better results were obtained by Harbaoui and Debergh (1980), who succeeded in obtaining more than 55% rooted shoots, but in some cases, vitrescence caused degeneration of the in vitro tissues (Debergh et al. 1981). In 1981, Ancora et al. reached 58% rooting and pointed out the genetic stability of the cv. Romanesco artichoke; while in 1982, Pecaut et al. obtained eight virus-free micropropagated varieties with 70% rooting.

Further physiological studies carried out by Moncousin and Gaspar (1983) found that in vitro artichoke peroxidase activity and rhizogenesis are very similar to those in plants obtained by seed germination.

3 Micropropagation (Scheme 1, Figs. 2–8)

3.1 Mass Production by in Vitro Propagation

As far as vegetables are concerned, micropropagation is employed in order to produce disease-free plants and/or to propagate a limited number of species supporting genetic improvement (e.g., rapid propagation of hybrid parental lines). General interest in artichoke micropropagation is growing fast, and there is abundant literature dealing with the difficulties met in producing this plant. Remarkable results were obtained for some cultivars and they can be applied for mass propagation, although with some difficulties. Most of the authors have reported very interesting scientific methods; however, practical approaches to them are not at all easy. Today, just a few private laboratories are able to put tens of thousands of artichoke plants onto the market.

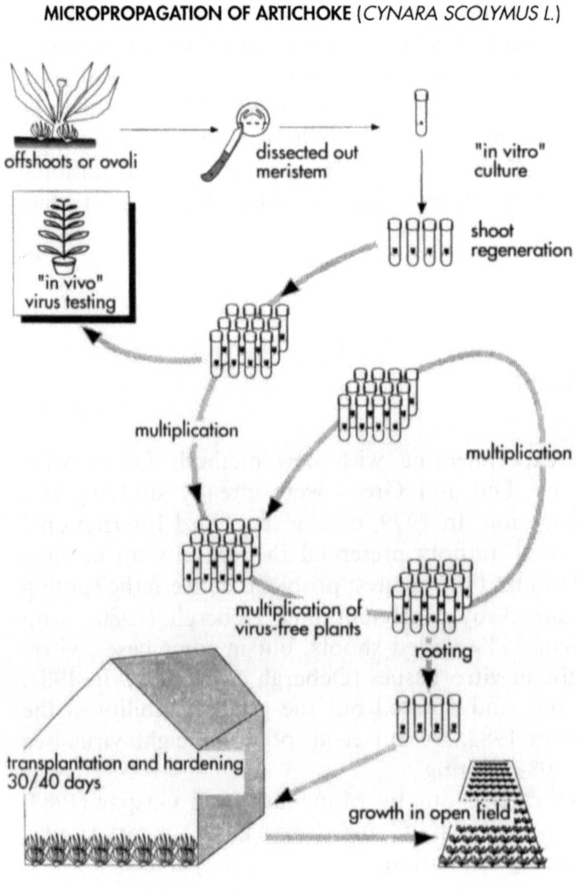

Scheme 1.

3.2 Sterilization of Explants

The sterilization of artichoke explants can be difficult if they are obtained from field-grown plants, highly contaminated by external microorganisms. Large numbers of bacteria and viruses found within mother plant tissues can be reduced by growing the explants under controlled hygienic conditions for a few months (Zuccherelli et al., 1978; Scortichini 1988). Buds are generally explanted from offshoots: during spring they react positively to in vitro growth because they are at the beginning of their vegetative season. For in vitro artichoke cloning, explants are collected from shoot apices of 0.1–1 cm in length, from mother plants previously selected for their agronomic behavior. Explanting meristematic tips of 0.6–0.8 mm in length, preceded (or not) by thermotherapy, allows rapid production of virus-free stock. For explant sterilization several chemicals such as 96% ethanol, mercury chloride, and sodium or calcium hypochlorite can be utilized. All of them are generally quite effective. The protocol we adopted for achieving sterile explants is as follows:

Fig. 2. Early selection of the cv. Romanesco originating from micropropagation during the production of floral heads in the open field

Fig. 3. In vitro artichoke shoots grown in the culture room

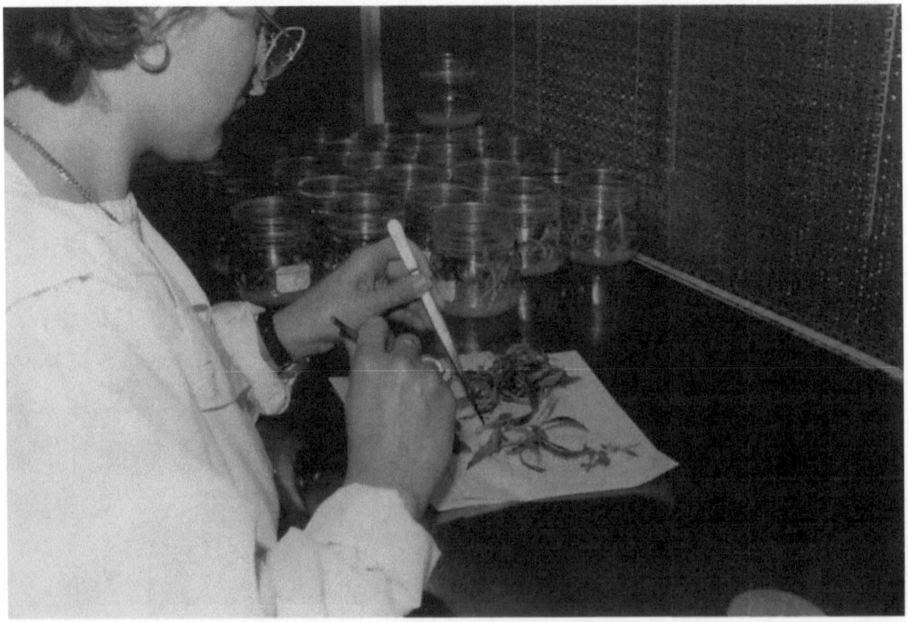

Fig. 4. Worker operating under an aseptic flow, during artichoke mass propagation

Fig. 5. Proliferated shoots of the cv. Verde Brindisino

Fig. 6. Vegetative aspects of three different micropropagation stages: multiplication on the *left*; before rooting phase in the *middle*; rooted plants on the *right*

Fig. 7. In vitro rooted plants

Fig. 8. Plants of the cv. Romanesco after 20 days of acclimatization in the greenhouse

1. removal of external leaves in order to obtain buds of 2–3 cm in size;
2. 20 min immersion in 2–3% sodium hypochlorite, followed by drying under aseptic flow;
3. excision of apical buds (or meristems) using stereo-microscope and sterile scalpels;
4. further washing of explants in a sterile solution containing citric acid (100 mg/l) and ascorbic acid (100 mg/l) to avoid tissue photooxidation;
5. in vitro culture of the explants by transferring them to tubes containing the appropriate medium solidified with agar.

This procedure allowed us to obtain 40–50% sterile explants.

If bacterial contaminations appear, antibiotics, such as penicillin, gentamycin, streptomycin, or chloramphenicol (10–100 mg/l), can be added to the medium (Harbaoui and Debergh 1980) but is highly recommended to start with absolute sterile explants.

3.3 Media Composition

Tables 1 and 2 show some media for in vitro artichoke propagation. The MS macronutrient composition (Murashige and Skoog 1962) is widely used both for proliferation and rooting, though sometimes it is modified or the salt concentration is even reduced to half. As far as micronutrients and organic supplements are concerned, the Nitsch and Nitsch (1969) or Gamborg et al. (1968) media have been

Table 1. Shoot proliferation media

	Ancora et al. (1981)	Bigot and Foury (1984)	Harbaoui and Debergh (1980)	Moncousin (1981)	Pecaut et al. (1983)	Rossi and De Paoli (1986)
Macroelements	MS	MS Mod	MS	MS	MS	MS Mod
Microelements	MS	MS	Nitsch and Nitsch	Gamborg	MS	MS
Fe EDTA	MS	MS	MS	N.S.	MS	MS
Vitamins and amino acids (mg/l)	MS	MS	—	N.S.	MS	MS
Myo-inositol	100	—	100	—	—	—
Thiamine	—	—	0.4	—	—	—
Growth regulators (mg/l)						
2-iP	—		—	10	—	—
K	5	1	2–5	—	1	—
BAP	—	—		—	—	0.05
GA_3	—	—	0.01	0.5	—	—
IAA	0.5	—	1	0.1		0.5
NAA	—	0.1	—	—	0.1	—
Carbohydrates (g/l)						
Sucrose	40	30	20	20	30	20
Other components (mg/l)						
Adenine	—	40	—	100	40	—
NaH_2PO_4	—	—	—	—	85	—
Agar (%)	0.7	0.7	0.6	0.7	N.S.	0.7
pH	5.5	5.5	5.6	5.7	N.S.	5.8

N.S. = Not specified

Table 2. Rooting media

	Ancora et al. (1981)	Bigot and Foury (1984)	Harbaoui and Debergh (1980)	Moncousin (1981)	Pecaut et al. (1983)	Rossi and De Paoli (1986)
Macroelements	MS/2	MS/2	MS	MS/2	MS/2	MS Mod
Microelements	MS	MS	Nitsch and Nitsch	MS	N.S.	MS
Fe EDTA	MS	MS	MS			MS
Vitamins and amino acids (mg/l)		MS	—	N.S.	N.S.	MS
Myo-inositol	100	—	100	N.S.	N.S.	—
Thiamine	—	—	0.4			—
Ascorbic acid	10	—	—			
Growth regulators (mg/l)						
K	—	—	—	—	0.05	—
IBA	—	—	2	—	—	—
IAA	—	—	—			2
NAA	2	0.1	—	0.1	0.5	2
Carbohydrates (g/l)						
Sucrose	20	20	20	20	N.S.	20
Other components						
Activated charcoal (gr/l)		2	—	—	—	—
L-tyrosine (mg/l)		—	—	50	—	—
Agar (%)	0.7	0.7	0.6	0.8	N.S.	0.7
pH	5.5	5.5	5.6	N.S.	N.S.	5.8
Rooting %	58%	77%	55%	75%	70%	48.5%

N.S. = Not specified

successfully adopted besides the MS. Several growth regulators are generally added. For inducing proliferation of shoots, the most frequently used cytokinins are: kinetin (1–5 mg/l) or BAP, at very low concentrations (0.05 mg/l) or 2-iP (10 mg/l).

GA_3 is effective for growing dormant buds during the initial stage of culturing, and later for proliferation, though at lower concentrations (0.01–0.5 mg/l). The most effective auxin for rooting is NAA (0.1–2 mg/l) also combined with IAA (2 mg/l). Only few researchers have added other constituents such as sulfate-adenine (40–100 mg/l) during multiplication, and activated charcoal (2 g/l) or L-tyrosine (50 mg/l) at the rooting stage.

The sugar employed as carbon source is sucrose (20–40 g/l). All the media are solidified with agar (6–8 g/l) and the required pH is 5.5 – 5.8 (see Tables 1 and 2).

3.4 Culture Conditions

The initial explants, both meristems and apical shoots, are first put into culture tubes and then transferred to larger pots (200–500-ml capacity) for quicker and easier mass propagation (Druart and Gruselle 1986).

The optimum environmental conditions of the growth room are:

— temperature kept at 20 °C (with few fluctuations);
— 16:8-h photoperiod, or continuous light;
— constant light intensity (3000–4000 lx) provided by cool fluorescent lamps.

After 15–20 days the explants are subcultured to a fresh medium, in order to avoid tissue degeneration, a quite frequent phenomenon. Some bacterial contaminations may occur and they are probably due to the presence of endogenous bacteria growing after a few subcultures. Transferring these shoots to a basic medium supplemented with Bacto Difco Peptone (265 mg/l) and Bacto Yeast Extract (88 mg/l) induces rapid bacterial growth so that it will be possible to select sterile material (Boxus and Terzi 1985). Another obstacle met during in vitro propagation of artichoke is vitrescence (Debergh et al. 1981). Several factors are implicated in this phenomenon, such as the kind of agar and its dose, temperature fluctuations in the growth room, chemical toxic substances accumulating in the media, etc. (Paques and Boxus 1985).

According to our experiments, it is possible to decrease vitrification by keeping shoots in a juvenile condition through continuous cutting of the oldest and vitrified portions. Our experience taught us that artichoke in vitro cannot be kept in a cold storage room (at 2–4 °C) longer than a few months.

3.5 In Vitro Propagation

The initial stage of in vitro artichoke growth is really slow, the explants are adjusting to artificial conditions. If the starting explants are from meristems, this phase is obviously longer, and 20–25% tissue degeneration may occur. Harbaoui and Debergh (1980), Ancora et al. (1981), and Pecaut et al. (1983) suggest utilizing a medium containing lower concentrations of salts and phythormones compared to proliferation media. Browning and/or yellowing of the media are often observed during the initial phase of artichoke culture as in some fruit trees, such as apple, olive, almond, and pistachio (Rugini 1984). This phenomenon can be reduced by adding antioxidants such as GSH (500 mg/l), citric acid (10–100 mg/l), etc. to the media and lowering the culture room light intensity for 1 week. In vitro artichoke proliferation is quite fast: the average multiplication rate is 2.5 axillary buds every 3 weeks for the most active clones i.e., 9×10^5 shoots per year in theory (Bigot and Foury 1984).

However, the most crucial phase for artichoke propagation is rooting. Generally, shoots develop into roots more easily if, for 10 days at least, they are transferred to an elongation medium containing lower cytokinin concentrations compared to proliferation media. This proved to be effective for several fruit trees (Zuccherelli 1979), and we found for artichoke, too. Several media have been suggested by the authors quoted (Table 2), but unfortunately none of them is really excellent for rooting induction. The rooting percentages obtained vary from 48.5% (Rossi and De Paoli 1986) to 77% (Bigot and Foury 1984), depending on the protocol followed and the varieties studied. In 1981, Ancora et al. obtained more than 58% rooting shoots of the cultivar Romanesco cultured on half-strength MS with addition of thiamine (1 mg/l), myo-inositol (100 mg/l), ascorbic acid (10 mg/l), and NAA (2 mg/l).

Pecaut et al. (1983) achieved good results by culturing eight French varieties coming from shoots grown on half-strength MS with addition of NAA (0.5 mg/l) and kinetin (0.05 mg/l) and kept in the dark, at 25 °C, for 21 days. Once transferred to the greenhouse, 70% shoots developed into roots after 40 days. Some physiological studies carried out on the cv. Vert de Laon by Moncousin and Gaspar (1983) show that by adding 10 mg/l D2 vitamin and 1 mg/l NAA to the culture medium, it is possible to obtain 75% rooting.

We know by experience that only those shoots that have grown on root-inducing media, but still have no roots, can be successfully transferred for acclimatization. However, their leaves must be well opened, tomentose, and dark green in color.

This experiment show that for artichokes in vivo rooting is possible, as has already been attempted with ornamental plants (Debergh and Maene 1981) and fruit trees (Zimmermann et al. 1985). This method would be very advantageous to lower the production costs, and to make in vitro propagation less expensive.

3.6 Acclimatization and Development in the Field

The final step of micropropagation is acclimatization of the in vitro artichoke plants. Its success mainly depends on the previous phases. If the plants reach 3–4 cm in height and have a rooting system composed of three or four roots, their transfer to the greenhouse will be satisfactory. In contrast, small weak plants and only one root will have little chance of surviving.

The technique employed for transferring to the greenhouse has not been widely studied by the authors quoted above; however, some of them (Bigot and Foury 1984) report difficulty in obtaining a good survival percentage, which never reaches 60%. Plants are generally transferred to plastic pots (diameter 7.5 cm) containing

80% peat and 20% perlite (Ancora et al. 1981; Rossi and De Paoli 1986). Irrigation must be moderate, since artichokes are very sensitive to high humidity. In contrast, in the hardening-off area relative humidity should be about 100% (Montalti 1979) during the first 15–20 days. The entire acclimatization period lasts about 4–5 weeks and consists of two phases: greenhouse followed by transfer to a shaded environment. Greenhouse temperature must be below 30 °C, since higher values will endanger plant survival (Bigot and Foury 1984).

The optimum times for acclimatization are:

— end of the winter season and throughout the spring;
— end of summer and beginning of fall.

However, fall production is not always recommended in the northern cultivation areas of Italy and France, because the young plants do not have enough time to grow and yield good spring production.

The plants, once hardened off according to the previously described procedure, will reach 10–15 cm in height and have a well-developed rooting system. These physical characteristics will give them a 100% chance of survival once transferred to the open field, where they will grow fast and vigorously. As confirmed by several authors, the yield increase of field-grown micropropagated artichokes is really interesting. The most studied varieties are Camus de la Bretagne and Blanc Hyerois (Pecaut et al. 1985), which have undoubtedly been giving better results than plants from traditional propagation methods.

The agronomic behavior of micropropagated plants is excellent especially if they are virus-free (e.g., cv. Romanesco studied by Ancora in 1986). Nevertheless, positive results can be obtained even from plants originating in vitro that are infected by viruses.

Early varieties with a fall-winter production and particularly the cv. Violet de Provence, have been giving some trouble: Pecaut (1985–1986) observed genetic instability shown by plants that were not true to type, with indented leaves and later production.

Finally, the results of our recent experiments show that some micropropagated plants of the cultivars Spinoso Sardo and Violetto di Foggia show juvenile behavior once transferred to the open field. They are called wild and heads are produced later.

3.7 Commercial Aspects of Micropropagation

Micropropagation of vegetables is a well-known technique, often employed to produce virus-free mother plants (potato, garlic, etc.) and/or to clone valuable genotypes supporting genetic improvement (asparagus, chicory, cauliflower, etc). As far as the artichoke is concerned, micropropagation is successfully employed for renewal and cloning selection. It is also used for nursery production. In fact, this technique is slowly but steadily expanding in Italy and France. The Italian market alone needs 10 million artichoke plants per year. In 1988, Italy and France put some 500 000 plants on the market by means of direct floral head production. The expansion of this technique is still slow because it is expensive; in fact, the cost of each micropropagated plant is almost 75 cents (USA). Private laboratories are interested in this technique and are trying to improve it and to lower production costs to a price more accessible for artichoke growers. Finally, although in vitro propagated plants cost more than those produced by traditional methods, artichoke growers well know their value and are willing to face higher prices for micropropagated plants.

4 Conclusions

The micropropagation technique described here is certainly new and revolutionary for artichoke cultivation. Nevertheless, some cultivars have been used successfully while others need more research, both in vitro and in the field. In vitro cultivation as a standard method for artichoke reproduction could lead to rapid dispersal of this crop in areas where it is still limited.

In fact, vegetative propagation is the main factor at present limiting its expansion, producing varieties that are not good from a genetic and agronomic point of view.

In such a context, micropropagation of artichoke, besides favoring restoration and genetic improvement of the cultivated varieties, could lead to a worldwide increase in cultivation. Recently, improvements have been made in finding a good mass propagation procedure. The advantages offered by this technique are remarkable and make us hope for its application also in those varieties which are as yet not easily micropropagated.

5 Protocol

1. Explants

Explants obtained from selected mother plants and sterilized by a 20-min immersion in 2% sodium hypochlorite solution.

2. Media

a) Proliferation media. MS (1962) modified in macroelement compositions (MS*)[2] with growth regulators added (BAP 0.01 mg/l and IAA 0.5 mg/l), sucrose 20 g/l and agar 0.7%; pH adjusted to 5.8.

b) Rootong media: see Table 2.

3. Optimal Conditions

Culture rooms kept at 20 °C, constant light intensity (3000 lx).
Subculture: once every 3 weeks.

[2] (MS*) = NH_4NO_3 410; KNO_3 475; $CaCl_2$ 440; $MgSO_4$ 370; KH_2PO_4 170 (mg/).

References

Ancora G (1986) Globe artichoke (*Cynara scolymus* L.). In: Bajaj YPS (ed) Biotechnology in agriculture and forestry vol 2. Crops I. Springer, Berlin Heidelberg New York, pp. 471–483
Ancora G, Belli Donini ML, Cuozzo L (1981) Globe artichoke plants obtained from shoot apices through rapid in vitro micropropagation. Sci Hortic 14: 207–213
Benoit H, Ducreux G (1981) Etude de quelques aspects de la multiplication végétative in vitro de l'artichaut (*Cynara scolymus* L.). Agronomie 1 (3): 225–230
Bigot C, Foury C (1984) Multiplication in vitro d'artichaut (*Cynara scolymus* L.) à partir de semences: comparaison au champ de quelques clones à la lignée dont ils sont issus. Agronomie 4(8): 699–710
Boxus P, Terzi FM (1985) Big losses due to bacterial contaminations can be avoided in mass propagation scheme. In: Int Symp In vitro problems related to mass propagation of horticultural crops. Gembloux, Acta Hortic 212: 91–93
Debergh P, Maene L (1981) A scheme for commercial propagation of ornamental plants by tissue culture. Sci Hortic 14: 335–345
Debergh P, Harbaoui Y, Lemeur R (1981) Mass propagation of globe artichoke (*Cynara scolymus* L.): evaluation of different hypotheses to overcome vitrification with special reference to water potential. Physiol Plant 53: 181–187
De Leo P, Greco B (1976) A new technique for propagation of artichoke: in vitro culture of apical meristems. 2nd Int Congr Artichoke, Bari, Italy, pp 657–665
Dellacecca V, Pace M (1976) Influence of propagation material on yield and earliness of artichoke. 2nd Int Congr Artichoke, Bari, Italy, pp 419–425
De Paoli G, Rossi V (1987) Carciofo in vitro. Terra Vita 12: 55–61
Druart P, Gruselle R (1986) Plum (*Prunus domestica*). In: Bajaj YPS (ed) Biotechnology in agriculture and forestry, vol 1. Trees I. Springer, Berlin Heidelberg New York, pp 130–154
Gamborg OL, Miller RA, Ojima K (1968) Plant cell cultures. I. Nutrient requirements of suspension cultures of soybean root cells. Exp Cell Res 50: 151–158
Harbaoui Y, Debergh P (1980) Multiplication in vitro de clones sélectionnées d'artichaut (*Cynara scolymus* L.). Eucarpia, Versailles
La Malfa G, Foury C (1976) Relationships among propagation material, biology and yield of artichoke. 2nd Int Congr Artichoke, Bari, Italy, pp 407–418
Magnifico V (1987) Alcuni aspetti agronomici della coltivazione del carciofo. Informatore Agrar 4: 57–64
Marras F, Foddai A, Fiori (1982) La selezione sanitaria del carciofo "Spinoso sardo" Inf Fitop 9: 47–50
Marzi V (1969) Current status of world artichoke production *Cynara scolymus* L. 1st Int Congr Artichoke Bari, Italy, pp 319–342
Moncousin C (1981) Multiplicatiòn végétative accéléré. Rev Hortic Suisse 54: 105–116
Moncousin C, Gaspar Th (1983) Peroxidase as a marker for rooting improvement of *Cynara scolymus* L. cultured in vitro. Biochem Physiol Pflanz 178: 263–271
Montalti G (1979) Influenza del substrato nell'ambientamento di alcune piante da frutto derivate da coltura in vitro. In: Tecniche di colture in vitro per la propagazione su vasta scala delle specie ortofrutticole. S.O.I Pistoia, Italy
Morone Fortunato I, Vanadia S, Lattanzio V (1979) First results on micropropagation of artichoke (*Cynara scolymus* L.). 3rd Int Congr Artichoke, Bari, Italy, pp 247–252
Murashige T, Skoog F (1962) A revised medium for rapid growth and bioassays with tobacco tissue cultures. Physiol Plant 15: 473–497
Nitsch JP, Nitsch C (1969) Haploid plants from pollen grains. Science 163: 85–87
Noto G (1981) First results in a breeding programme of *Cynara scolymus* L. 3rd Int Congr Artichoke, Bari, Italy, pp 831–842
Paques M, Boxus Ph (1985) Vitrification: review of literature. In: Boxus P, Read PE, O'Riordain F (eds). In vitro problems related to mass propagation of horticultural plants. Acta Hortic 212: 155–166
Pecaut P, Dumas de Vaulx R, Lot H (1983) Virus-free clones of globe artichoke (*Cynara scolymus*) obtained after in vitro propagation. Acta Hortic 131: 303–309
Pecaut P, Corre J, Lot H, Migliori A (1985) Intéret des plants sains d'artichaut régénérés par la culture in vitro. Rev Hortic 256
Pecaut P (1985–1986) Researches sur l'artichaut. INRA, Montfavet

Rana GL, Russo M, Gallitelli D, Martelli GP (1982) Artichoke latent virus: characterization, ultrastructure and geographical distribution. Ann Appl Biol 101: 279–289

Rosati P, Tognoni F (1979) Propagazione su vasta scala delle specie ortofrutticole: tecnica di coltura ed organizzazione del laboratorio. In: Tecniche di colture in vitro per la propagazione su vasta scala delle specie ortofrutticole. S.O.I. Pistoia, Italy

Rossi V, De Paoli G (1986) Micropropagation of artichoke (*Cynara scolymus* L.). In: 6th Int Cong Plant tissue and cell culture, Minnesota, p 105

Rugini E (1984) In vitro propagation of some olive (*Olea europea* L.) cultivars with different root ability, and medium development using analytical data from developing shoots and embryos. Sci Hortic 24: 123–134

Ryder EJ, De Vos NE, Bari MA (1987) The globe artichoke (*Cynara scolymus* L.). Hortic Sci 18(5): 646–653

Scaramella Petri R, Ricci A (1981) Micropropagation of *Cynara scolymus* L. var. sativa, by tissue culture. 3rd Int Congr Artichoke, Bari, Italy, pp 231–238

Scortichini M (1988) L'inquinamento batterico nelle colture dei tessuti vegetali in vitro. Inf. Fitopatol 12/88. Italy. pp 231–238

Tesi R (1976) First results on breeding programme of Tuscan varieties of *Cynara cardunculus* var. *scolymus* L. 2nd Int Congr Artichoke, Bari, Italy, pp 747–763

Uncini L (1987) Speciale carciofo. Agric Ricerca 75–76: 65–111

Zimmermann RH, Yae BW, Fordham I (1985) Comparison of rooting methods for apple cultivars in vitro. In: Boxus P, Read PE, O'Riordan F. (eds) In vitro problems related to mass propagation of horticultural plants. Acta Hortic 212: 303–310

Zuccherelli G (1979) Moltiplicazione in vitro dei portinnesti clonali del pesco. Frutticoltura (Italy) 41: 15–20

Zuccherelli G, Venturi V, Damiano C (1978) Rapid propagation on large scale of Damasco 1969 peach rootstock by in vitro culture. In: Round-table conference in vitro multiplication of woody species, Gembloux, 6–8 June

I.9 Micropropagation of White Yam (*Dioscorea rotundata* Poir.)

S.Y.C. NG[1]

1 Introduction

White yam (*Dioscorea rotundata* Poir.) belongs to the family Dioscoreaceae, genus *Dioscorea*. It is also known as white Guinea yam, Guinea yam, and 8-months' yam. The center of origin of white yam is in West Africa. *D. rotundata*, a cultigen, is so close to *D. cayenensis* that it has been regarded as a subspecies. It belongs to section Enantiophyllum, which comprises the most economically useful species of *Dioscorea*. *D. rotundata* appears to be of hybrid origin. It was speculated by Coursey (1976) that one parent is *D. cayenensis*, a forest species, adapted to a short dry season and having limited tuber dormancy. The other must be a savanna species, *D. praehensilis*, which bears the greatest resemblance to some *D. rotundata* forms, although *D. abyssinica* and *D. togoensis* may also have contributed. It was taken early in post-Columbian times to the West Indies, but it is not grown in Asia. The chromosome number is probably similar to *D. cayenensis*, as suggested by Purseglove (1976), and it was reported by Coursey (1976) as $2n = 2x = 40$.

It is an annual plant, a crop which is produced in 8 to 10 months' growing period with 3 to 4 months' dormancy period. It is a monocotyledon with cylindrical stem, 10–12 m long, twining to the right, usually spiny; leaves usually opposite, simple, $10–12 \times 6–8$ cm, broadly cordate, accuminate. It is dioecious, both male and female flowers are produced in the leaf axils. The male flowers are borne in panicles while the females are in spikes. The male flowers are small and inconspicuous, having green or creamish white color. The sepals and petals are three in number, with three or six stamens. The female flowers are slightly larger with the same number of sepals and petals, and have an inferior ovary with three locules, each of which contains two ovules. There are three stigmas, and pollination is by insects. The yam fruit is dry and dehiscent with three locules, the junctions of which extend into flat wings. At dehiscence the locules separate and release the seed. The seed is flattened and with wings, which are adapted for wind dispersal (Coursey 1972; Lawton 1967; Onwueme 1978).

The underground tubers, the edible part of the plant, is believed to be initiated from the organ intermediate between stem and root – the hypocotyl (Lawton and Lawton 1969). The tubers are normally cylindrical with rounded or pointed ends, smooth brown skins, and white mealy flesh, but sometimes they may be irregular. Normal tuber weight is 2–5 kg, but in some areas 10 kg is not uncommon, and 20–25 kg has also been recorded. The tubers are also used as planting material.

[1] The International Institute of Tropical Agriculture (IITA), Oyo Road, P.M.B. 5320, Ibadan, Nigeria

Biotechnology in Agriculture and Forestry, Vol. 19
High-Tech and Micropropagation III (ed. by Y.P.S. Bajaj)

1.1 Importance of the Plant

White yam is the most widely grown and eaten yam species in West Africa, especially in Nigeria, and it is the most important in the whole world. Yam tubers are a source of carbohydrate. They are good energy sources, derived mainly from carbohydrates, since fats are low in yam tubers. According to Diehl (1982), yam contributes ca. 20% to the total daily calorific intake in the rural areas of Nigeria. The total carbohydrate in white yam tubers ranged from 27 to 33.2 mg/100 g edible portion (Eka 1985; Watson 1971). Yam tubers also contain vitamins such as carotene, thiamine, riboflavin, niacin, and ascorbic acid. White yam tubers, or yams in general are fairly high in most of the essential amino acids. They are rich in arginine, leucine, isoleucine, and valine but low in histidine, methionine, cystine, and tryptophan. The crude protein in the part which is consumed, is 11.21% (Eka 1985). The quantity of yam consumed as staple in West Africa is sufficient to supply almost one-third of the basic protein requirement of an adult male (Ayensu and Coursey 1972). However, the tannin content in the tubers tends to reduce the value of yam as a suitable source of protein. The tannin may combine with the protein and thus render it nutritionally unavailable. White yam tubers are also rich in minerals, such as calcium, phosphorus, and iron (Eka 1985). The crude fibre content in white yam peels is 9.47% of dry matter (Oyenuga 1968).

According to a saying of the Igbo, a Nigerian tribe, yam was given to mankind by God, and is thus closely linked to the origin of mankind. Yam is also part of the religious heritage of several Nigerian tribes and to date often plays a key role in religious ceremonies and social events (Arinze 1970). The new yam festival, which marks the onset of yam harvesting, is an outstanding social event almost everywhere in the yam-growing belt of West Africa (Coursey and Coursey 1971). Therefore white yam is truly an indigenous crop in both the cultural and the biological sense.

1.2 Distribution and Area Under Cultivation

Yams are essentially tropical plants; except for *D. japonica* and *D. opposita*, they cannot tolerate frosty conditions (Coursey 1972; Purseglove 1976). Temperatures between 25 and 30 °C are required for normal growth and development. Yams perform best when supplied with ample moisture throughout their growing period. Yams require soils of high fertility and perform best in loamy soils. Sandy soils with a low water- and nutrient-holding capacity are not suitable for yam cultivation. Yam also requires well-drained soil. Poorly drained soils, and the resultant waterlogging, cause the roots and tubers to rot.

Light is also an important factor in yam production, daylength playing an important role in tuber formation. Short days favor tuber formation, while long days favor vegetative growth (Njoku 1963; Okezie 1981). Therefore yam cultivation and production have been adapted to coastal forest areas and flood plains with high rainfall as well as to the Northern Guinea Savannah condition.

White yam is the dominant yam species cultivated in West Africa; the yam belt which stretches from the Côte D'lvoire, Benin, Ghana, Nigeria, Cameroon, and to some East African countries, represents the major yam-producing countries in

Table 1. Production and cultivation of yams. (FAO 1988)

Region/country	Production (1000 mt)	Area (1000 ha)
World	26968	2566
Africa	25780	2429
North and Central America	387	62
South America	365	42
Asia	186	15
Ocenia	248	18

Africa. White yam is also cultivated in Caribbean areas. Separate data on cultivation and production of different yam species are not available. A breakdown on the production and cultivation of yams throughout the world over the past years is presented in Table 1. According to FAO (1988), world yam production in 1987 was estimated at 26968 thousand metric tons (mt) with 25780 thousand mt produced from Africa accounting for more than 95% of world production. The major producing countries are Nigeria (19000 thousand mt), Côte D'lvoire (2900 thousand mt) and Ghana (1000 thousand mt). In fact, Nigeria accounts for 70% of the world yam production. The areas under yam cultivation throughout the world are estimated at 2475 thousand ha, of which 2340 thousand ha. are in Africa. FAO (1988) statistics showed that over the 3 years from 1985 to 1987, more than 95% of the total world yam production was produced in Africa. The yam species that are important as staples in the tropics are: white yam, water yam, yellow yam, trifoliate yam, aerial yam, and Chinese yam (Hahn et al. 1987).

1.3 Diseases and Pests

The diseases and pests discussed in the following refer to white yam in particular, but in most cases apply to yams in general. A detailed account of common African pests and diseases of yam was given by Theberge (1985).

1.3.1 Diseases

Yam mosaic virus. A virus disease on white yam was described with the typical foliage characteristics of green vein-banding and mottling and quite often accompanied by shoe-stringing and bushiness. This type of reaction to the virus is believed to be due to host/pathogen interaction. The virus is transmitted mechanically and by *Aphid gossypii* (Terry 1976). It occurs in the major yam-growing areas of West and Central Africa.

Apollo Disease. This is a foliage disease caused by *Colletotrichum* spp. The first symptom is the appearance of small necrotic spots on the tips of the lamina on the abaxial side of the mature leaves, lower down the veins. The spots are subcircular in shape and later turn dark brown in color. Eventually the whole leaf yellows, shrivels, and finally the brownish color of the spot changes to a dirty brown or metallic (silvery) color. Petioles and veins can also be affected. (Nwankiti 1982).

Yam Bacterial Rot. The causal agent is *Erwinia carotovora* (Jones) Holland. It is spread worldwide and occurs in *Dioscorea* spp., occuring during storage, and causing watery soft rot of the tubers. Losses can be dramatic if proper sanitary and insect control measures are not undertaken.

Yam Storage Rot. Several pathogens are involved. These include *Fusarium* spp., *Rhizopus nodosus, Rosellinia* spp., *Botryodiplodia theobromae* Pat., *Aspergillus* spp., *Penicillium* spp., and numerous other fungi and bacteria. The disease results mainly from wounding and bruising during harvesting and transporting of the tubers.

1.3.2 Pests

Nematodes. Nematodes are the most important pests that attack yams, the three major ones being *Scutellonema bradys* (yam nematode), *Pratylenchus coffeae* (lesion nematode), and *Meloidogyne* spp. (root-knot nematode). The yam and lesion nematodes are of primary importance in the damage they cause to tubers during storage either directly or indirectly; root-knot nematode affects plant growth and causes yield losses in the field and in tuber quality when heavily infested.

The yam nematode has been reported to occur in West Africa, the Caribbean, Brazil, and India. Tubers damaged by yam nematode are found to be attacked by pathogenic and saprophytic organisms, particularly fungi and bacteria, which gradually bring about the destruction of the tubers in the ground and in storage (Baudin 1956; Nadakal and Thomas 1967; Smit 1967; Bridge 1972; Adesiyan et al. 1975).

Lesion nematode is the major cause of yam tuber dry rot known as "burn" in Jamaica and Puerto Rico (Bridge 1982). It has been recorded as a pest of yams in Jamaica, Puerto Rico, and the British Solomon Islands. Infected tubers showed severe necrosis, dry rot, and deep cracks in the outer tissues, causing deterioration of tubers and marked reduction in quality.

Another species of *Pratylenchus* that occurs in yams is *P. brachyurus*, which is found in root and tubers in the Côte d'Ivoire (Luc and de Guiran 1961; Miege 1957; Smit 1967) and Nigeria (Bridge 1973), and in yam roots in Guatemala (Jenkins and Bird 1962).

Root-knot nematodes are found in all yam-growing areas of the world, causing severe tuber yield loss. Infestation by root-knot nematodes causes light to heavy galling of the yam, rots, and knobby or bumpy tuber surfaces. *M. incognita* and *M. javanica* are widespread on *D. rotundata* and *D. alata*.

Control of nematodes in yams is by chemical (treating tubers to prevent dissemination of nematodes in propagative tuber materials) and cultural means (crop rotation and fallow) in the field to reduce infestation. Treatment of propagative tuber material prior to planting by immersion in nematicides or hot water treatment is more successful than field treatment in controlling nematodes. Hot water treatment at 50 and 55 °C for 40 min killed over 98% of yam nematodes without damage to white yam tubers (Bridge 1975).

Yam Tuber Beetles (*Heteroligus spp.*). Yam tuber beetles make feeding holes on the yam tuber, often before harvest, resulting in low marketability. Yam setts are also attacked after planting and the vegetative development of the plants can be limited.

Stored Yam Beetles. The pests *Araecerus fasciculatus* (De Geer), *Carpophilus dimidiatus* (Fabricius), *C. freemani*, and *Palorus subdepressus* occur throughout the world. They attack not only yam but also other stored plant products such as cassava, maize, coffee, and others. They are storage pests, making feeding holes on the yam tuber, the adults and the larvae tunneling through the tubers. Those tubers that are damaged during harvesting are more vulnerable to attack.

Yam Leaf Beetles. Crioceris livida and *Lema armata* occur especially in West Africa. Both adults and larvae feed on the yam leaves and tender growing vein, causing vein dieback and defoliation. Yam leaf beetles are minor pests and, the damage is generally localized.

1.4 Conventional Practices for Propagation, Production, and Improvement

Yam production requires high input and is labor-intensive. The crop is also difficult to grow because of the expense and unavailability of planting materials (seed yam or planting setts). In fact, one third of the harvest is often reserved by farmers for the next planting. Seedbed preparation is laborious and tedious, staking is required, weeding is necessary, and harvesting needs special care. The labor requirements for cropping 1 ha of yam were estimated at 1815 man h/ha. (Hahn 1984). According to FAO (1988) statistics, there was a decline in yam production (from 1985 to 1987), though the area devoted to yam cultivation had increased slightly.

Propagation of planting setts for research has been improved by using yam vein cuttings (Akoroda and Okonmah 1982) and for both research and farm levels by using minisetts (Otoo et al. 1987). Leafy vein cuttings with four to six nodes were rooted in soil or sand. An average of 47.2% of the vein cuttings of four varieties of white yams were able to produce roots and became established (Akoroda and Okonmah 1982). Conventional seed yam production either uses the whole small tubers or small pieces of tuber weight between 80–100 g as planting materials. A new seed yam production technology, the minisett technology, using tuber pieces of 25 g as planting material, described by Otoo et al. (1987), has been adopted by seed yam growers in the yam-producing areas. Tubers weighing about 1000 g are ready for harvesting at 5 to 6 months after planting of the minisetts.

Hybridization in white yam, seed set, and growing large populations of seedlings has been developed (Doku 1973; Sadik 1976; Sadik and Okereke 1975a,b; Wilson 1982). The seeds were planted directly into the seed nursery bed for evaluation (Wilson 1982).

2 In Vitro Approaches

2.1 Need to Incorporate Unconventional Methods

Despite the importance of yams in the tropics, comparatively little research has been devoted to them. Thus, technological development has been slow and only limited impact has been made.

Although there was success in hybridization and genetic recombination in white yam, the breeding process is slow as a result of dormancy and the long growth period. It takes a long time (6 to 8 years) to select new varieties. As mentioned earlier, yams are normally dioecious, with male and female flowers produced on different plants, usually with more male plants than female plants; the synchronization of male and female flowers is a problem. However, many yam cultivars do not produce flowers at all. These are the handicaps in yam improvement. Due to the flowering behavior and nonflowering in yam, the choice for parents in a hybridization program is limited to those that produce flowers.

2.2 Summary of Tissue Culture Work

In vitro culture of yam dates back as far as 1958, with the experiment by Sawada et al. on the effects of sucrose and peptones on organogenesis of *D. opposita*. However, over the years, attention has been given to the disogen-producing yam species only. Very little work has been devoted to edible yams.

It appears from the literature that tissue culture technology has developed well in disease elimination and rapid multiplication, using meristem/shoot tip cultures for the former, and meristem, shoot tip, and nodal cultures for the latter (Table 2). It is possible to regenerate plants from petiole, stem, leaf, and tuber tissue of yams through direct plant regeneration or via callus and suspension culture. Other tissue culture methods such as embryo and protoplast culture have also been reported.

2.3 Micropropagation

One of the most frequent applications in plant tissue culture is that of disease elimination and rapid multiplication of plant materials, especially among vegetatively propagated crops. It has been commercialized in many ornamentals. The

Table 2. Summary of tissue culture work on white yam.

Species	Explant	Regeneration/ study	Reference
D. cayenesis	Node	Plantlet	Arnolin (1985)
D. rotundata	Node	Plantlet	Mantell et al. (1978) Cortés-Monllor and Lin (1983) Ng and Hahn (1985) Ng (1986d)
D. rotundata	Meristem	Plantlet	Ng (1986c)
D. rotundata	Tuber	Callus and plantlet regeneration	IITA (1974)
D. rotundata	Tuber	Callus	Ng (1984)
D. rotundata	Tuber	Plantlet	Ng (1984)
D. rotundata	Embryo	Plantlet	Okezie et al. (1983)
D. cayenensis	Embryo	Callus and plant regeneration	Viana and Mantell (1989)
D. rotundata	Protoplast from tuber	Isolation and purification	Onyia et al. (1984)
D. rotundata	Buds	Irradiation	Mbanaso and Ene (1988)

induction of single or multiple shoot formation from meristem tips, axillary buds, and nodal culture is an effective means for clonal propagation.

Control of bacterial and fungal diseases may be obtained through the application of chemicals. However, viral diseases have not been successfully controlled through chemical means. Chemicals which can affect virus multiplication usually exert high phytotoxicity on the host plant.

Viruses and nematodes are spread in vegetatively propagated crops such as yam not only by their natural vectors but also through the successive generation with infested vegetative propagules. Eventually, the entire crop may become infected. Therefore, viruses have seriously hampered the international exchange of vegetative materials as well as reducing the yield.

The most successful tissue culture technique applied to yam is probably in the area of disease elimination. Both meristem tip culture or the combination of heat treatment and meristem culture are being used. Working on *D. alata*, Mantell et al. (1980) were able to regenerate apical meristems to plantlets on MS medium with 1 or 0.5 mg/1 NAA and 0.2 or 0.1 mg/1 BAP. Plantlets were obtained in 15 to 20 weeks. Meristems obtained from shoot apex or axillary buds of *D. rotundata* and *D. alata* were also successfully regenerated into plantlets on MS medium with 3% sucrose, 0.12 mg/l BAP, 0.2 mg/l NAA, 0.08 mg/l GA_3, 20 mg/l cysteine, and 0.6% agar (Ng 1984, 1986c).

Ribavirin (1-B-D-ribofuranosyl-1,2,4-triazole-3-carboxamide) has been incorporated in the culture media to increase the efficacy of disease elimination in other crops, such as potato (Wambugu et al. 1985). However, higher concentrations of ribavirin is toxic to the explant tissues. The attempt to utilize ribavirin to replace heat treatment was not very successful. Ng (1988b) carried out studies on the effects of ribavirin on plantlet regeneration in white yam and to evaluate the efficacy of virus elimination. The concentrations of ribavirin were 10 and 20 mg/l. Media without ribavirin were used as control. Liquid media with filter paper bridges were used.

Four months after culturing, 40% of the meristems in all treatments showed green growth. The meristems were then transferred to solid media without ribavirin for further growth and development. Two months after the transfer, 30% of the meristems in ribavirin-free media developed into plantlets, 20% at 10 mg/l ribavirin and none at 20 mg/l ribavirin.

Using another white yam variety, TDr 179, Ng (unpubl.) found that explant materials (meristems) obtained from heat-treated mother plants (the mother plants were subjected to a constant temperature of 37 °C for 1 month) greatly increased the regeneration rate. Three months after culturing of the meristems, plantlet formation was obtained from the heat-treated meristems, whereas the meristems that were cultured on different ribavirin treatments, as well as the control, had only swelled and become green. Two months after transfer to fresh culture media, the heat-treated meristems developed into plantlets that were ready for transplanting whereas in the control and ribavirin treatments, only shoots or green leaves were obtained. This showed that heat treatment facilitates the growth and development of the meristems. Thus plantlets can be obtained in a shorter period of time and a higher regeneration rate can be achieved.

The plantlets were transplanted to soil and virus indexing was performed. Virus-free yam plants were obtained from heat-treated plant materials.

The elimination of flexous rod virus infections in *D. alata* was possible when meristems were obtained from heat-treated mother plants grown at 36 ± 1 °C for at least 14 days. The meristems obtained from heat-treated plants had a higher regeneration rate and faster growth. Virus-free plants of *D. trifida* were also obtained by culturing newly formed node after the mother plants had been grown at 42 °C for 22 days (Balagne 1985). Therefore, it is possible to obtain virus-free yams by culturing heat-treated meristems or even node cuttings. Several selected white yam varieties were free from yam mosaic virus using the heat-treated meristems as the explant (Ng, unpubl.).

Plantlets were obtained within 3 to 4 months after culturing, depending on the varieties. Figure 1 shows the development of meristem into plantlet. The meristems turned black after excision and inoculation to the culture media. However, 5 to 7 days after inoculation, they turned green and developed into a plantlet suitable for transplanting/subculturing within 3 to 4 months. The culture room was maintained at a temperature ranging from 25 to 30 °C with 16-h daylength. The regenerated plantlets were transplanted to insect-free isolation for virus indexing and negatively indexed plants multiplied rapidly.

The detailed procedure for yam (*D. rotundata* and *D. alata*) multiplication using node cuttings as explants was described by Mantell et al. (1979) and Ng (1984, 1986d). Node cuttings were regenerated to plantlets on MS medium supplemented with 2% sucrose, 0.5 mg/l kinetin, 20 mg/l cysteine and 0.6% agar (Ng, 1984, 1986d). At IITA, the certified yam plants were rapidly multiplied in vitro using single node cuttings from the in vitro plantlets which were maintained by sub-culturing meristem-regenerated plantlets before they were transplanted for virus

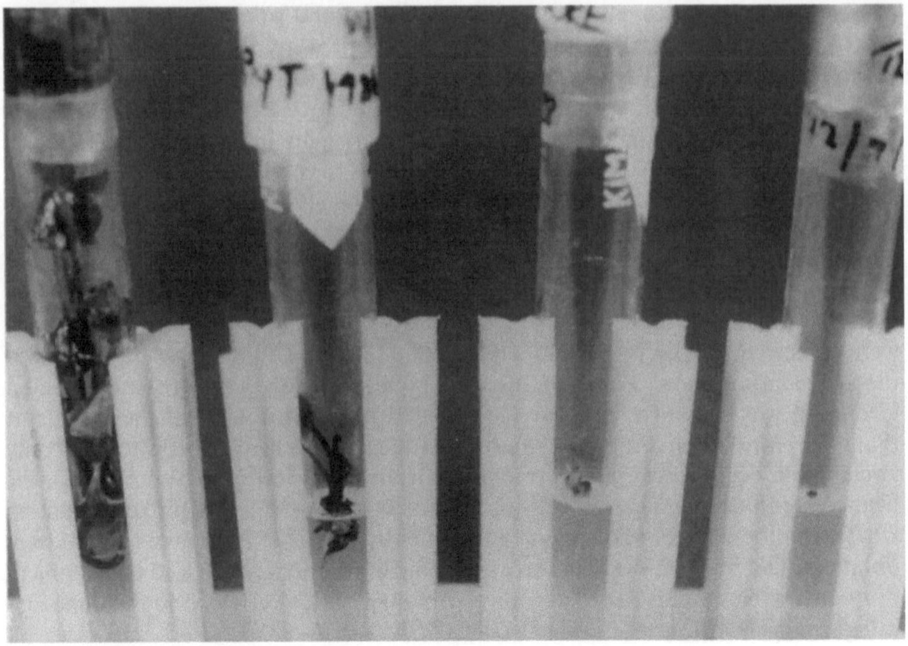

Fig. 1. Development of white yam meristem to plantlet, *Right* swelled meristem; *left* complete plantlet

indexing (Fig. 2). Plantlets grown in solid culture media in sterile plastic culture tubes are used for international distribution (Fig. 3). The multiplication rate in such culture medium was five times in 2 to 3 months of culturing.

Using the virus-indexed yam materials from meristem culture, virus-free white yams were distributed to more than 15 countries in Africa (Ng, unpubl.).

Cortès-Monllor and Lin (1983) achieved fourfold multiplication in *D. rotundata* within 2 to 3 months using a nodal segment of 0.5 to 1 cm grown on MS medium with 2 mg/l IAA and 2 mg/l kinetin followed by transferring the shoot to MS medium with 1 mg/l NAA to obtain roots. They found that the age of nodal segment influenced growth of the explants. Node cuttings that were 2 to 6 months old had the greatest production of plantlets, whereas node cuttings of 7 months old grew very slowly and thus the production of number of plants was reduced. The axillary buds of single node cuttings of *D. floribunda* were induced to proliferate to shoots on MS medium with 2 mg/l BAP and on MS medium with 15 mg/l adenine sulfate and 0.1 mg/l IAA (Chaturvedi 1975). Plantlets were also obtained from node cuttings cultured on MS medium with 2 mg/l kinetin (Lakshmi Sita et al. 1976). However, higher rates of sprouting and faster shoot growth were obtained when kinetin at 5 mg/l was added to the medium. Node cuttings of *D. floribunda* (Forsyth and van Staden 1982) and *D. alata, D. cayenensis*, and *D. trifida* (Arnolin 1985) were also regenerated to plantlets and successfully established in the field.

Ng (1986a) studied the effects of different cytokinins and physical status of the culture media on plantlet development in white yam using node cuttings. Three

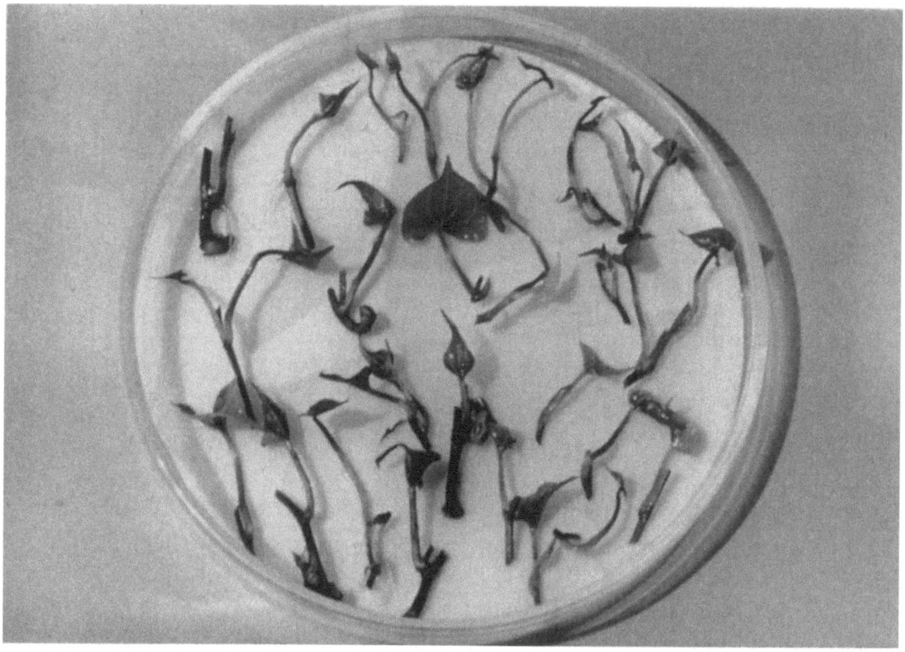

Fig. 2. Single node cuttings obtained from in vitro white yam plantlet used for multiplication. (Ng and Ng, 1991)

Fig. 3. White yam plantlet in sterile culture tube ready for international distribution

kinds of cytokinins were used, kinetin, 6-(r,r-dimethylallylamino) purine (2-ip) and 6-benzyl amino purine (BAP) at two levels, 5 mg/l and 10 mg/l both on solid and liquid media. Among the three cytokinins, it was found that kinetin gave the highest number of nodes that can be used for subculturing, followed by 2-ip. In all treatments, the highest number of nodes produced per explant were obtained from nodes grown on liquid medium. When grown in liquid media, the node cutting developed faster and grew vigorously. Within 6 weeks a plantlet with 5 to eight nodes with multiple shoots were obtained (Fig. 4).

The effects of daylength on white yam and water yam growth and development were also studied (Ng 1986a). Cultures of node cuttings were exposed to different daylengths: 24, 16, 12, and 8 h. In white yam, the number of nodes produced increased with the increase of daylength. However, water yam behaved completely differently. Shorter daylength or day-neutral tended to enhance shoot growth and thus increase the number of nodes.

The importance of pre-treatment of mother plants under long daylength has been realized (Mantell et al. 1978, 1979). Nodal cuttings from long day-treated plants developed into plantlets on basal MS medium without growth hormone. Nodal cuttings obtained from 12 h daylength mother plants had a very poor regeneration rate. Mantell et al. (1979) have estimated that it is possible to obtain 65000 plantlets from a single node segment culture within a period of 6 months.

Fig. 4. White yam node cuttings grown in liquid culture medium

Ng (1984) was able to obtain direct plant regeneration from tuber discs of *D. rotundata* on MS medium supplemented with 0.5 or 1 mg/l BAP (Fig. 5). Tuber discs of 1 cm^3 were obtained from newly harvested yam tubers after the removal of the corky periderm. The surface sterilized tuber discs were then inoculated to culture media containing BAP. With BAP (0.5 or 1 mg/l) in the culture medium, the upper surface of the tuber discs burst 1 month after culturing. Bud formation was observed in the 2nd month, followed by root initiation and elongation. Complete plantlets were obtained within 3 months after culturing. The formation of plantlets from tuber discs would also provide an excellent opportunity for rapid multiplication of yam for the immediately following growing season, especially as breeding material.

2.4 Microtuber Formation

The phenomenon of in vitro tuberization was observed in *D. bulbifera* (Uduebo 1971; Ammirato 1976, 1984; Forsyth and Van Staden 1984), *D. alata* (Ammirato 1976, 1984), *D. rotundata* and *D. alata* (Ng 1986a), and *D. rotundata* (Ng 1988a,b), using node cuttings as explants. Larger bulbils were obtained from *D. bulbifera* when culture media contained 5 to 10 mg/l IAA and in the presence of 0.5 mg/l kinetin; the bulbil dry weight was also increased to a maximum under such treatment (Uduebo 1971). Increase in sucrose concentration from 2 to 8% and/or a decrease in cytokinin concentration in the basal MS medium also induced tuberization in *D. bulbifera* (Forsyth and van Staden 1984). Study on the effects of

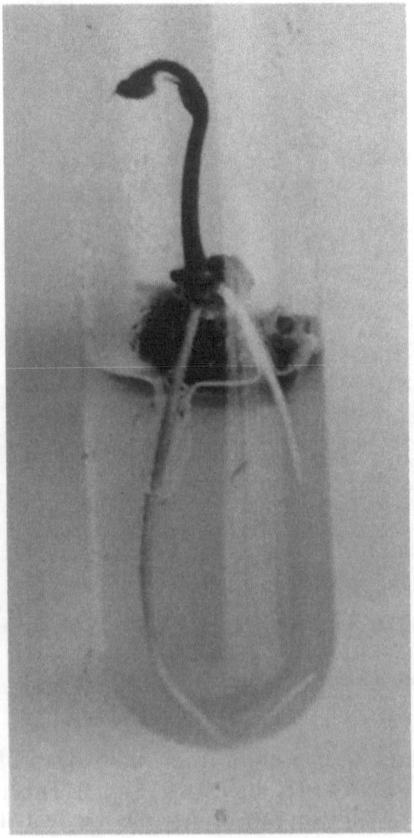

Fig. 5. Yam plantlet developing from the tuber disc

hormone on tuberization of both *D. alata* and *D. bulbifera*, Armmirato (1976) observed that NAA or ABA stimulated tuber formation, GA_3 promoted shoot growth, and a zeatin and NAA combination stimulated large abnormal tubers. Tuberization in *D. rotundata* was also influenced by the sucrose concentration in the media (Ng 1988a).

Ng (1988a) reported that microtubers of different sizes were obtained from more than 30 varieties of white yam. Tuberization occurred mainly at the base of the stem/node (Fig. 6). Occasionally, microtubers were obtained from the nodes which were just above the surface of the culture media. The frequency of tuberization varies depending on the variety. The shape of the microtubers was either round or oblong with diameter ranging from 5 to 10 mm and length from 10 to 20 mm (Fig. 7).

Shoot growth was strongly suppressed at higher concentrations of sucrose in the culture medium. At 8 and 10% sucrose, shoot growth was retarded and there was massive growth of roots at the base. The effect of sucrose and daylength on tuberization and microtuber formation is shown in Table 3. Under 12 and 16 h light, the frequency of tuberization was highest at 5% sucrose and declined at a higher or lower sucrose concentration. The same trend was obtained in the number

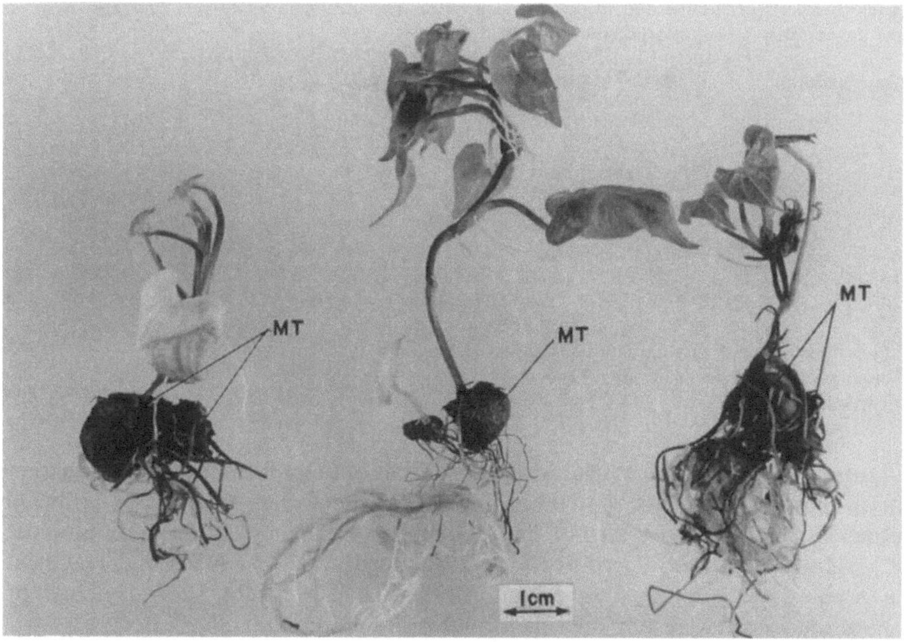

Fig. 6. White yam plantlets with one or two microtubers (*MT*) developed at the base of the stem. (Ng 1988a)

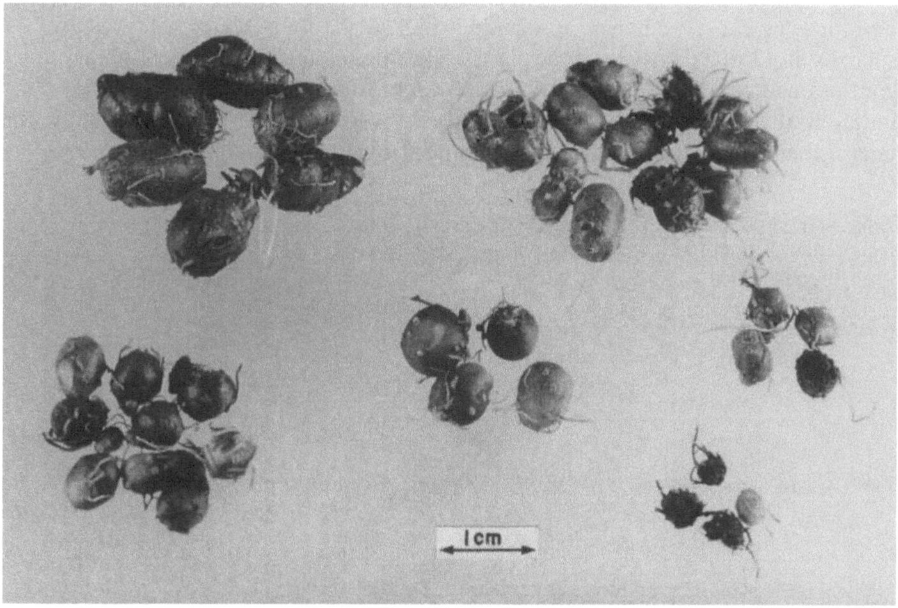

Fig. 7. Microtubers of different sizes and shapes harvested from in vitro white yam plantlets. (Ng 1988a)

Table 3. Effects of sucrose and daylength on in vitro tuberization in white yam (in transparent culture vessels with 60-ml capacity). (Ng 1988a)

Daylength (h)	% Sucrose							
	3		5		8		10	
	%T[a]	NT[b]	%T	NT	%T	NT	%T	NT
8	16.7	0.3[c]	33.0	1.8[c]	25.0	1.7[c]	37.5	0.8[c]
12	41.7	2.0	62.5	2.3	37.5	2.0	25.0	2.0
16	37.5	2.5	50.0	2.8	37.5	1.5	12.5	1.0
24	25.0	1.5	25.0	3.0[c]	87.5	3.3[c]	50.0	2.5[c]

[a] Percentage of plants that initiate tuberization at 2 months.
[b] Number of tubers per plant after 5 months.
[c] Tuber-like structure.

of tubers per plant. At 8 h light, while some signs of tuber formation were apparent at all sucrose treatments, only tuber-like structures (swollen shoot bases with calli attached) were observed. At 24 h light the highest frequency of tuberization was at 8% sucrose; however, microtubers were obtained at 3% sucrose only. The most favorable daylengths for tuber induction were 12 and 16 h light. Short daylength or continuous light were not suitable for tuber induction.

On the other hand, using semi-transparent plastic containers with a greater capacity (500 ml), aerial tubers were obtained. Under such conditions, the effects of sucrose concentration (3 and 5%) and daylength (8, 12, 16, and 24 h) were studied (Ng 1988b). The results are presented in Table 4. The percentage of tuberization and number of tubers formed per plant were higher at 5% sucrose than at 3% sucrose in all daylength treatments. The greatest number of tubers per plant obtained was at 5% sucrose under 24 h light (7.0 tubers/plant) followed by 5% sucrose at 12 h light (6.5 tubers/plant). The high number of tubers/plant under continuous light was attributed to aerial tuber formation. The aerial tubers were initiated at a later stage of growth, after 3 to 4 months of culturing. Thus, the microtuber production rate can be enhanced by such a method.

Table 4. Effects of sucrose concentration and daylength on yam tuberization in semi-transparent culture vessels (with 500 ml capacity). (Ng 1988b)

Photoperiod (h)	% Sucrose			
	3		5	
	%T[a]	NT[b]	%T	NT
8	18.8	2.0[c]	58.3	4.2
12	37.5	3.0	50.0	6.5
16	62.5	2.5	75.0	4.3
24	43.8	4.5	52.1	7.0

[a] Percentage of plant initiated tubers at 2 months.
[b] Number of tubers produced per plant after 5 months.
[c] Tuber-like structure.

About 80% of the harvested microtubers sprouted after 2 to 3 months of storage under aseptic conditions with moistened filter paper. The sprouted microtubers were transplanted to jiffy peat pellet (Fig. 8) for further growth. One month after transplanting to jiffy peat pellet, the plants were transplanted to a seed bed in the screenhouse (Fig. 9). Tubers were harvested after 7 months of growth. The tubers obtained ranged from over 40 g to 400 g per tuber. About 59% of the harvested tuber weighed more than 100 g; these tubers could then be used as planting setts for commercial yam production. One to two tubers were usually obtained from each plant.

The important role of microtubers in the international distribution of virus-tested plant materials cannot be overemphasized (Ng 1988a). They can be packed in sterile containers with moistened cotton wool or filter paper for international distribution (Fig. 10). The advantages of using microtubers for international germplasm exchange are:

- they can be produced from virus-tested plants, thus the danger of introducing diseases is eliminated;
- they can be kept under dark conditions for a long period of time (3 months) without loss of viability, thus facilitating despatch by mail;
- they can be easily transplanted and established in soil with minimum care.

The in vitro microtuber production system not only provides microtubers for storage, planting, or distribution, but at the same time the shoots can also be used for further multiplication. This increases the overall multiplication rate.

Fig. 8. The germination and growth of microtuber. *Left* dormant microtuber; *right* plant developed from microtuber

Fig. 9. White yam plants from microtubers grown in a seedbed

2.5 Germplasm Conservation

The maintenance of yam germplasm as living collections is very difficult. The loss of germplasm materials during growth in the field and in storage is tremendous. Tissue culture is/can be used to maintain germplasm of vegetatively propagated crops.

The use of axillary buds and nodal cuttings grown on reduced growth media and or incubated at reduced temperature are the commonly used technique for in vitro germplasm conservation of vegetatively propagated crops (Ng and Ng 1991).

Contaminations such as bacteria and fungi are most common in tissue culture. Bacterial contamination sometimes becomes evident only after several subcultures. This sometimes creates problems, especially for germplasm conservation.

In an experiment, antibiotics were incorporated in the culture media to study their effects on plantlet development and control of bacteria contaminations. The

Fig. 10. Microtubers packed in Petri dishes. (Ng 1988a)

results showed that the antibiotics tetracycline, penicillin, and cyclohexamine are all equally effective in controlling bacteria contamination both at 10 and 20 mg/l. Lower concentrations of penicillin and cyclohaxamine had no effect on plantlet regeneration in yam. However, higher concentration of these two antibiotics, as well as tetracycline at both concentrations, drastically reduced plantlet regeneration in yam. These results suggested that antibiotics can effectively control bacterial contaminations, while at the same time plant growth and development are not affected.

Fungus contamination in yam cultures using nodal cuttings obtained directly from field-grown plants was also a serious problem. A double disinfection method proved to be very effective in eliminating fungus contaminations using calcium hypochlorite (Zok, pers. commun.) and sodium hypochlorite (Ng, unpubl.) were developed.

Ng and Hahn (1985) and Ng (1986b) reported the use of lower incubation temperature to maintain germplasm of *D. alata* and *D. rotundata*. The plantlets can easily be kept for 1 year under lower incubation temperature, 18–20 °C (Fig. 11). Studies on the effects of culture media, especially in the presence of mannitol and higher sucrose level, on plant regeneration ability and growth rate were carried out (Ng 1988b). The cultures were kept under normal incubation conditions. The results of this study are presented in Table 5. The highest percentage plantlet formation was with media III, IV, and V. The lowest plant height was with medium IV, followed by medium VI. Media I, II, III, and V had almost the same plant height. This indicated that MS media supplemented with 5% sucrose, 3% mannitol, and 0.5 mg/l kinetin (medium IV), where percentage plantlet formation was maximum and at the same time plant height was the lowest, was the best culture

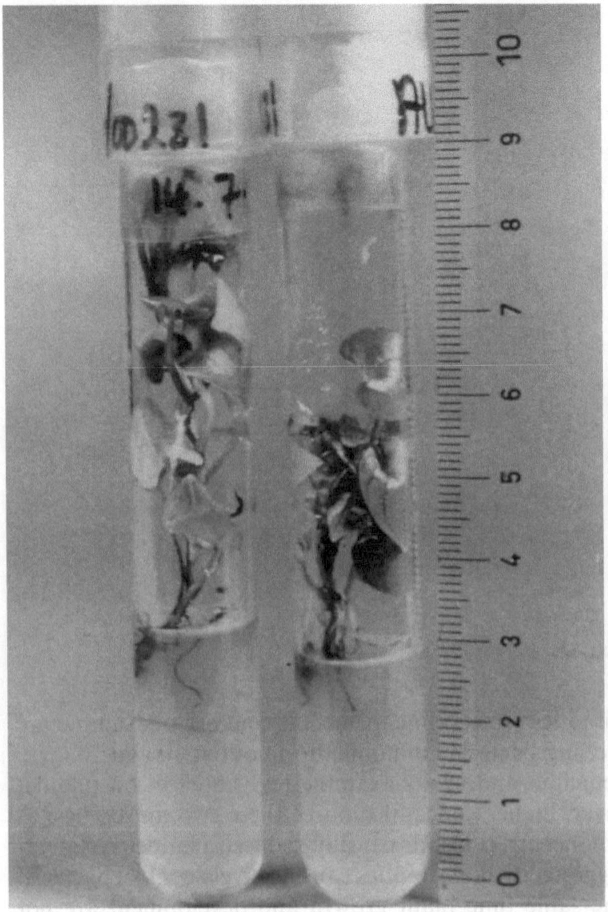

Fig. 11. White yam plantlets maintained at a lower incubation temperature (18–20 °C) for 6 months (*right*) and 1 year (*left*). (Ng and Ng, 1991)

Table 5. Effects of sucrose and mannitol on in vitro yam growth. (Ng 1988b)

Treatment (MS media +)	Plantlet (%)	Plant height (mm)
I 3% sucrose + 0.5 mg/l kinetin (K) (control)	80	16.9
II 5% sucrose + K	80	17.1
III 3% sucrose + 3% mannitol + K	100	17.3
IV 5% sucrose + 3% mannitol + K	100	11.3
V 3% sucrose + 1 mg/l BAP	100	17.7
VI 5% sucrose + 1 mg/l BAP	60	15.0

medium for yam germplasm conservation. Media VI had a relatively slow growth rate, but the percentage of plantlet formation rate was the lowest. It is therefore not suitable for conservation purposes.

IITA is using both the field cultivation and the in vitro reduced growth storage methods to maintain its yam germplasm collections. Currently, there are more than

600 accessions of yams in its in vitro collection and it is targeted to transfer the entire collection (more than 1500 accessions) to be maintained in vitro (Ng, unpubl.). The yam plantlets can be kept at 18–20 °C for 1–2 years.

2.6 Callus/Cell Suspension Culture and Plant Regeneration

The ability to establish a callus/cell suspension culture and regenerate plants from these cultures is very important in the application of advanced biotechnological techniques for plant improvement. It is possible to regenerate plants from yams, though the regeneration rate is very low. Direct plant regeneration from explant materials, somatic embryogenesis, and plant regeneration from callus/cell suspension culture have been very useful for rapid clonal propagation and for obtaining somaclonal variants.

Yam tuber blocks of 50 mg per piece were induced to form callus on MS medium with 2 mg/l 2, 4-D or NAA (IITA 1974). On transfer to MS medium with 4 mg/l kinetin, the callus regenerated shoots. Plantlets were also obtained by transferring shoots to medium containing 1 mg/l NAA. Callus formation was also obtained from tuber discs and nodal segment of *D. alata* (Mantell et al. 1978) as well as tuber discs of *D. rotundata* (Ng 1984).

More recently, embryogenic suspension cultures were established and full somatic embryos were obtained from a white yam variety. Shoot tip differentiation was obtained from these embryos (Novak, pers. commun.). Somatic embryos were also obtained from *D. composita* and *D. cayenensis* (Viana and Mantell 1989) and *D. opposita*, (Finer, pers. commun.) and plantlets were obtained. Studies on the formation of callus using different types of explant of *D. alata* and *D. trifida* (Fautret et al. 1985) concluded that young leaf petiole was best for such a type of culture. Calli from the petiole of *D. alata* were regenerated to shoots but not from *D. trifida*. The highest regeneration rate was 6% in a medium that contained 10^{-5} M 2,4,5-T and 5×10^{-6} M kinetin. They also studied the effect of total nitrogen supply and the ratio of the reduced and oxidized form in the medium on the shoot regeneration rate in *D. alata*. The results showed that 36 mM total amount of nitrogen supplied with equal proportion of both forms of nitrogen gave the highest percentage of callus regenerated to shoots.

In *D. floribunda*, calli were obtained from node or internode segments. Shoot primordia developed when the callus was cultured on MS or White's medium with 2,4-D or NAA combined with BAP or kinetin. Upon transfer to half-strength MS basal medium, shoots developed. Excised leaves of *D. floribunda* and *D. composite* obtained from in vitro plants were induced to form multiple shoots from the pulvinus in a medium containing 0.5 mg/l BAP, 50 mg/l adenine sulfate, and 0.1 mg/l NAA (Sinha and Chaturvedi 1979). The pulvinus first swelled after an incubation of 5 to 10 days followed by slight callus formation. After another 5 days, shoot buds emerged from the callused base. Fully developed shoots were obtained 50 days later. An average of six shoots per explant were obtained. Stem segments of *D. composite* were also induced to form callus on MS medium with 1 mg/l 2,4-D. Calli were regenerated to plants on a medium containing 0.5 mg/l NAA and 1.5 or 1 mg/l BAP.

2.7 Embryo Culture

Embryo or seed culture is very useful for those seeds that have a dormancy period or cannot germinate easily under normal situations. Plantlets were obtained from embryo cultures of *D. rotundata*. Okezie et al. (1983) compared germination of the excised embryo grown on artificial culture media with that of the normal intact seeds at increasing time intervals. They found that embryos of *D. rotundata* germinated faster when cultured on MS or LS medium than if left in intact seeds. The ability of the embryos to germinate increased with time from seed harvest. This indicated that the seeds had gradually recovered from dormancy. They also found that seeds at harvest contained immature embryos and that embryos increased in size and transformed from globular to fan-shaped appearance during the 3–4-month period of dormancy. Plantlets were also obtained from mature embryos of *D. opposita* seeds which did not germinate using conventional germination methods (Yakuwa et al. 1981).

Degras (1989) mentioned that Arnolin and Gamiette were able to obtain plantlets from embryos of *D. alata* from less definite origin. This provides opportunity for rescuing embryos at an early developmental stages.

2.8 Protoplast Culture, Transformation and Mutation Studies

The successful isolation of protoplast from tuber meristem tissue of *D. rotundata* reported by Onyia et al. (1984) is very encouraging. They obtained a very good yield of viable protoplasts using a two-step enzyme digestion method. However, they did not attempt to culture the isolated protoplasts. Degras, in his recent paper (1989), mentioned that protoplasts were also isolated from cotyledon and callus of white yam. The protoplasts were successfully cultured and cell division occurred; nevertheless, plant regeneration or embryo formation were not obtained. Schafer et al. (1987) demonstrated the ability of *Agrobacterium* to infect yams, and provided direct evidence for the integration of T-DNA into the nuclear genome of *D. bulbifera*.

Mutation techniques, brought about either by chemical mutagens or irradiation, are a powerful tool for obtaining variability in crop plants. Preliminary work on this aspect was carried out on yam node cuttings (Mbanaso and Ene 1988; Ng and Novak, unpubl.). IITA, in collaboration with the International Atomic Energy Agency at Vienna, has evaluated some gamma-irradiated *D. rotundata* material (Ng and Novak, unpubl.). Results obtained showed that there were no apparent differences in terms of tuber shape, leaf shape, vein morphology, and pigmentation. However, the harvested tubers are now being planted for further observations. Arnolin and Degras (1984) have reported on the exposure of plantlet of *D. alata* and *D. trifida* to gamma irradiation in view of obtaining mutants.

3 Protocol

Protocols for micropropagation of white yam via meristems and node cuttings are described here.

3.1 Meristems

a) Sprouted yam tubers are planted in pots in the greenhouse.
b) After 2 months, the plants are transferred to a growth chamber for heat treatment at 37 °C for 1 month.
c) Both apical and lateral buds are collected in moistened containers and brought into the laboratory for disinfection.
d) The buds are disinfected with 70% ethanol for 5 min, then by 10% sodium hypochlorite solution with Tween 20 for 20 min. The explants are then rinsed with three changes of sterile distilled water and dissected under a stereoscope in a microflow transfer hood.
e) The meristems are innoculated to a culture medium with the following composition: MS medium supplemented with 3% sucrose, 20 mg/l cysteine, 0.2 mg/l NAA, 0.12 mg/l BAP, 0.08 mg/l GA_3, 80 mg/l adenine sulfate, and 0.6% agar.
f) The cultures are incubated in a culture room with temperature range from 25 to 30 °C, with 16-h photoperiod and 5000 lx light intensity.
g) Plantlets develop in 2 to 3 months after incubation.
h) Regenerated plantlets can be used for subculture using node cuttings (for micropropagation) or transplanting to soil for virus indexing.

3.2 Node Cuttings

a) Node cuttings are collected from greenhouse-grown plants, field-grown plants, and virus-indexed plants during the 2nd to 4th months of vegetative growth.
b) They are disinfected with 70% ethanol, followed by 10% sodium hypochlorite solution with Tween 20 for 20 min. After rinsing with three changes of sterile distilled water, the explants are further treated with 5% sodium hypochlorite solution for 10 min and rinsed with three changes of sterile distilled water.
c) In a microflow transfer hood, the node cuttings are then placed in a sterile Petri dish with sterile filter paper to remove excess water.
d) The node cuttings are then placed on MS medium supplemented with 2% sucrose, 20 mg/l cysteine, 0.5 mg/l kinetin, and 0.6% agar.
e) Cultures are incubated at 25 to 30 °C, with 16-h photoperiod and 5000 lx light intensity.
f) Plantlets are obtained 2 months after culturing.
g) Plantlets are then divided into single node cuttings and subcultured onto liquid culture media with the same composition as in procedure (d).
h) Node cuttings in liquid media produced multiple shoot within 6 weeks of culturing. They can be further multiplied in liquid medium or subcultured on solid medium for root development and transplanting to soil.

4 Summary and Conclusion

Meristem/shoot tip and nodal cutting cultures are well-established methodologies in yam. These methodologies have made a tremendous impact on agriculture in terms of production of virus-free plants which are then able to cross the national quarantine boundaries with a high degree of confidence. Rapid multiplication is another area which can have a great impact in promoting productions of these crops, particularly in the area of the production of good-quality planting setts. The other area is the use for germplasm storage, which can allow the maintenance of clonal material under reduced growth conditions. However, the problem of genetic instability needs to be monitored. Cryopreservation in this crop should be explored for longer-term storage. In vitro microtuber formation in yams can be used for the selection of characteristics such as tuber shape, besides its usefulness in the international distribution of germplasm. The ability to regenerate plants from callus/suspension cultures of yams is encouraging. Studies on the ability of *Agrobacterium* to infect certain yam species are underway aiming at promoting the protein content and quality in yam tubers (Osuji 1988 and pers. commun.). Somaclonal variation can be explored for selecting new plant variety. Chemical mutagens and irradiation research provide excellent possibilities. Symbiosis between bacteria and the West African wild yam species, *D. macroura* (Orr 1926) and *D. sansibarensis* (Miller and Reporter, in press) has been reported. Bacteria were also found in leaf tips of over 600 accessions of different yam species, this includes the majority of white yam accessions (Hahn et al., unpubl.). The number of bacteria accommodated in the leaf tips differs between different accessions, which showed that there are varietal differences in the degree of symbiosis. Studies on the function of such bacteria on the plant and possible genetic manipulation are areas that need to be explored.

Endophytic bacteria were also found in aseptically micropropagated plant cultures of *D. alata* and *D. cayenesis* (Mantell, et. al. pers. commun.).

References

Adesiyan SO, Odihirin RA, Adeniji MO (1975) Histopathology studies of the yam tuber (*Dioscorea rotundata* Poir.) infected with *Scutellonema bradys* (Steiner & LeHew). Int Biodetn Bull 11: 48–55

Akoroda MO, Okonmah LU (1982) Sett production and germplasm maintenance through vine cuttings in yams. Trop Agric (Trinidad) 59(4): 311–314

Ammirato PV (1976) Hormonal control of tuber formation in cultured axillary buds of *D. bulbifera* and *D. alata*. Plant Physiol (Suppl) 57: 66

Ammirato PV (1984) Yams. In: Ammirato PV, Evans DA, Sharp WR, Yamada Y (eds) Handbook of plant cell culture, vol 3. Crop species. Macmillan, New York, pp 327–354

Arinze FA (1970) Sacrifice in Ibo religion. Ibadan Univ Press, Ibadan, Nigeria

Arnolin R (1985) Different reaction of yam (*Dioscorea* L.) to in vitro node cutting, in reference to the cultivars. Abstr 7th Symp Int Soc Tropic Root Crops. INRA, Guadeloupe, July 1–6, p 51

Arnolin R, Degras L (1984) In vitro culture, results and prospects for food yams. In: Proc 6th Symp Int Soc Tropic Root Crops. CIP, Lima, Peru, pp 641–648

Ayensu ES, Coursey DG (1972) Guinea yams. The botany, ethnobotany, use and possible future of yams in West Africa. Econ Bot 26(4): 301–318

Balagne M (1985) In vitro thermotherapy of *Dioscorea trifida*. In: Abstr 7th Symp Int Soc Tropic Root Crops. INRA, Guadeloupe, July 1–6, p 52

Baudin P (1956) Les maladies parasitaires des ignames en Côte d'Ivoire. Rev Mycol, Paris, Suppl 21(2): 87–111

Bridge J (1972) Nematode problems with yams (*Dioscorea* spp.) in Nigeria. Pest Artic News Summ 18: 89–91

Bridge J (1973) Nematodes as pests of yams in Nigeria. Meded. Fac. Landbouwwet. Rijksuniv Gent 38: 841–852

Bridge J (1975) Hot water treatment to control plant parasitic nematodes of tropical crops. Meded Fac. Landbouwwet Rijksuniv Gent 40: 249–259
Bridge J (1982) Nematodes of yams. In: Miege J, Lyonga SN (eds) Yams, ignames. Clarendon, Oxford, pp 253–264
Chaturvedi HC (1975) Propagation of *Dioscorea floribunda* from in vitro culture of single-node stem segments. Curr Sci 44(23): 839–841
Cortés-Monllor R, Lin LJ (1983) Tissue culture propagation of yam in Puerto Rico. J Agric Univ Puerto Rico 67(4): 419–428
Coursey DG (1972) The Origin and domestication of yams in Africa. In: Harlen JR (ed) The origin of the African plant; domestication. Mouton Publishers, The Hague, pp 383–408
Coursey DG (1976) Yams. In: Simmonds NW (ed) Evolution in crop plants. Longman, London, pp 70–74
Coursey DG, Coursey CK (1971) The New yam festivals of West Africa. Anthropos Stud 66: 444–454
Degras L (1989) Yam and biotechnology. Paper presented at CTA/FAO Symp Plant Biotechnology for Developing Countries. Luxembourg, June 26–30
Diehl L (1982) Smallholder farming systems with yam in the Southern Guinea Savannah of Nigeria. GTZ, Germany
Doku EV (1973) Sexuality and reproductive biology in Ghanaian yam (*Dioscorea* species) cultivars. 1. Preliminary studies. Proc 3rd Int Symp Tropic Root Crops. Ibadan, Nigeria
Eka OU (1985) The chemical composition of yam tuber. In: Osuji GO (ed) Advances in yam research – the biochemistry and technology of the yam tubers. Biochem Soc Nigeria collabor Anambra State Univ Technol, pp 51–74
FAO (1988) FAO Production Yearbook 1987. FAO, Rome
Fautret A, Dublin P, Chagvardieff P (1985) Callus formation and neoformation with two edible *Dioscorea* species: *D. alata* and *D. trifida*. Paper presented on 7th Symp Int Soc Tropic Root Crops. INRA, Guadeloupe, July 1–6
Forsyth C, van Staden J (1982) An improved method of in vitro propagation of *Dioscorea bulbifera*. Plant Cell Tissue Organ Cult 1: 275–281
Forsyth C, van Staden J (1984) Tuberization of *Dioscorea bulbifera* stem nodes in culture. J Plant Physiol 115: 79–83
Hahn SK (1984) Tropical root crops – their improvement and utilization. Agriculture Conference paper No.2. Int Inst Tropic Agric, Ibadan, Nigeria, 28 pp
Hahn SK, Osiru DSO, Akoroda MO, Otoo JA (1987) Yam production and its future prospects. Outlook Agric 16(3): 105–110
International Institute of Tropical Agriculture (IITA) (1974) Yam tissue culture. In: IITA 1973 Annu Rep Root and Tuber Improvement Program. IITA, Ibadan, Nigeria, 35 pp
Jenkins WR, Bird GW (1962) Nematodes associated with wild yam *Dioscorea* sp. with special reference to the pathogenicity of *Meloidogyne incognita*. Plant Dis Rep 46: 659–860
Lakshmi Sita G, Bammi RK, Randhawa GS (1976) Clonal propagation of *Dioscorea floribunda* by tissue culture, Indian. J Hortic Sci. 51: 551–554
Lawton JRS (1967) A key to the *Dioscorea* species in Nigeria. J West Afr Sci Assoc 12: 3–9
Lawton JR, Lawton JRS (1969) The development of the tuber in seedlings of five species of *Dioscorea* from Nigeria. Bot J Linn Soc 62: 223–232
Luc M, de Guiran G (1961) Les nematodes associes aux plantes de l'Ouest Africain. Liste preliminaire. Agron Trop Nogent 15: 434–449
Mantell SH, Haque SQ, Whitehall AP (1978) Clonal multiplication of *Dioscorea alata* L. and *Dioscorea rotundata* Poir. yams by tissue culture. J Hortic Sci 53: 95–98
Mantell SH, Haque SQ, Whitehall AP (1979) A rapid propagation system for yams. Yam Virus Project Bull 1, CARDI
Mantell SH, Haque SQ, Whitehall AP (1980) Apical meristem tip culture for eradication of flexous rod viruses in yam (*Dioscorea alata*). Tropic Pest Manage 26(2): 170–179
Mbanaso ENA, Ene LSO (1988) Aspects of biotechnology for the improvement of cassava (*Manihot esculenta*) and yams (*Dioscorea* spp.) at the National Root Crops Research Institute, Umudike. In: The use of biotechnology for the improvement of cassava, yams and plantain in Africa. IITA Meet Rep Ser 1988/2, pp 24–28
Miége J (1957) Influence de quelques caractères des tubercules semences sur la leveé et le rendement des ignames cultivées. J Agric Trop Bot Appl 4: 315–342

Miller IM, Reporter M (in Press) Bacterial leaf symbiosis in *Dioscorea sansibarensis*. Morphology and ultrastructure of the accuminate leaf gland.
Nadakal AM, Thomas N (1967) Observations of nematodes associated with "dry rot" of *Dioscorea alata* L. Sci Cult 33: 142–143
Ng SYC (1984) Meristem culture and multiplication. In: IITA Annu Rep 1983, pp 133–134
Ng SYC (1986a) Tissue culture. In: IITA root and tuber improvement program Annu Rep 1985. IITA, Ibadan, Nigeria
Ng SYC (1986b) Application of tissue culture in root crops conservation. Paper presented at the 13th Annu Conf Genet Soc of Nigeria held on 10–14 February, 1986. Forest Res Inst Nigeria, Ibadan,. Nigeria
Ng SYC (1986c) Virus elimination in sweet potato, yam and cocoyam. In: Proc Regional Root Crops Workshop held at Cali. Colombia, Sep 13–16, pp 97–102
Ng SYC (1986d) Rapid propagation techniques for sweet potato, yam and cocoyam. In: Proc Regional Root Crops Workshop held at Cali, Colombia, Sep 13–16, pp 117–118
Ng SYC (1988a) In vitro tuberization in white yam (*Dioscorea rotundata* Poir.) Plant Cell Tissue Organ Cult 14: 121–128
Ng SYC (1988b) Tissue culture. In: IITA Root and Tuber Improvement Program Annu Rep 1986. Int Inst Tropic Agric, Ibadan, Nigeria
Ng SYC, Hahn SK (1985) Application of tissue culture to tuber crops at IITA. Proc Inter-Centre Seminar on International Agricultural Research Centres and Biotechnology. JRRJ, Los Banos, Philippines, April 23–27, pp 27–40
Ng SYC, Ng NQ (1991) Reduced growth storage of germplasm. In: Dodds, JH (ed), In vitro methods for conservation of plant genetic resources. Chapman and Hall, London, pp. 11–39
Njoku, E (1963) The propagation of yams (*Dioscorea* spp.) by vine cuttings. J W Afr Sci Ass 8: 29–32
Nwankiti AO (1982) Symptomatology, etiology, and incidence of a leaf disease of yam (*Dioscorea* spp.) originally called "Apollo" disease. In: Miege J, Lyonga SN (eds) Yams, Ignames. Clarendon, Oxford, pp 274–279
Okezie CEA (1981) Morphogenetic and growth analysis studies on the white Guinea yam (*Dioscorea rotundata* Poir) raised from seed. Ph.D. Thesis, Univ Nigeria, Nsukka, Nigeria
Okezie CEA, Nwoke FIO, Okonkwo SNC (1983) In vitro culture of *Dioscorea rotundata* embryo. In: Proc 2nd Triennial Symp Int Soc Tropic Root Crops – Africa Branch held in Douala, Cameroon. Aug 14–19, pp 121–124
Onwueme IC (1978) The tropical tuber crops. Wiley Sons, Chichester, 234 pp
Onyia GOC, Gahan PB, Norman H (1984) The use of new probes for protoplast integrity following isolation and purification of protoplasts from tubers of white yam (*Dioscorea rotundata* Poir). Plant Sci Lett 33: 231–238
Orr MY (1926) On the secretory organs of the Dioscoreaceae. Notes R Bot Gard (Edinb), 3: 133–147
Osuji GO (1988) The prospects of improving yams genetically through recombinant DNA technology. In: The use of biotechnology for the improvement of cassava, yams and plantain in Africa. IITA Meet Rep Ser 2: 34–40
Otoo JA, Osiru DSO, Ng SYC, Hahn SK (1987) Improved technology for seed yam production. Int Inst Tropic Agric, Ibadan, Nigeria, 56 pp
Oyenuga VA (1968) Nigeria foods and feeding stuffs, their chemistry and nutritive value, 3rd edn. Univ Press Ibadan, Nigeria
Purseglove JW (1976) Tropical crops. Monocotyledons. Longman, London, pp 97–117
Sadik S (1976) A review of sexual propagation for yam improvement. Proc 4th Int Symp Tropic Root Crops. Cali Columbia, pp 40–44
Sadik S, Okereke OU (1975a) A new approach to improvement of yam *Dioscorea rotundata*. Nature (Lond) 254: 134–135
Sadik S, Okereke OU (1975b) Flowering, pollen grain germination, fruiting, germination and seedling development of white yam, *Dioscorea rotundata* Poir. Ann Bot 39: 597–604
Sawada E, Yakuma T, Imakawa S (1958) Studies on the formation of aerial tubers in Chinese yam. II. On the aerial tuber formation in sterile culture vine segments. J Hortic Assoc Jpn 27: 241–244
Schafer W, Gorz A, Gahl G (1987) T-DNA integration and expression in a monocot crop plant after induction of *Agrobacterium*. Nature (Lond) 327: 529–532
Sinha M, Chaturvedi HC (1979) Rapid clonal propagation of *Dioscorea floribunda* by in vitro culture of excised leaves. Curr Sci 48(4): 176–178
Smit JJ (1967) Nematodes. In: Coursey DG (ed) Yams. Longmans, London, pp 115–120

Terry ER (1976) Incidence, symptomatology, and transmission of a yam virus in Nigeria. Proc 4th Symp Int Tropic Root Crops, Cali, Columbia, pp 170–173

Theberge RL (1985) Common African pests and diseases of cassava, yam, sweet potato and cocoyam. Int Inst Tropic Agric, Ibadan, Nigeria, 108 pp

Uduebo AE (1971) Effect of external supply of growth substances on axillary proliferation and development in *Dioscorea bulbifera* Ann Bot 35: 159–163

Viana AM, Mantell SH (1989) Callus induction and plant regeneration from excised zygotic embryos of the seed-propagated yam *Dioscorea composita* Hemsl. and *D. cayenensis* Lam. Plant Cell Tissue Organ Cult 16: 113–122

Wambugu FM, Secor GA, Gudmestad NC (1985) Eradication of potato virus Y and S from potato by chemotherapy of cultured axillary bud tips. Am Potato J 62: 667–672

Watson JD (1971) Investigation on the nutritive value of some Ghanaian foodstuffs. Ghana J Agric Sci 4: 95–111

Wilson JE (1982) Progress in the breeding of yam, *Dioscorea* spp. In: Miege J, Lyonga SN (eds) Yams, ignames. Clarendon, Oxford, pp 17–22

Yakuwa T, Harada T, Kasai N, Araki H (1981) Studies on the botanical characteristics of genus *Dioscorea*. II. On the formation and germination of seed in Chinese yam (cv. Nagaimo). J Fac Agric Hokkaido Univ. 60: 220–229

I.10 Micropropagation of Cubio (*Tropaeolum tuberosum* R & P)

O. TORRES, M. PEREA-DALLOS, and T.J. FANDIÑO[1]

1 Introduction

Tropaeolum belongs to the family Tropaeolaceae, that includes about 100 species. It takes its name from the Greek word τρωπαLoν = trophy, in allusion to the armor of ancient knights-errant. Leaves from some species look like shields and the longstalked, solitary and occasionally double flowers are orange to scarlated in color. Smaller than those of the garden nasturtium (*Tropaeolum majus*). They are distributed throughout the neotropical region from Mexico to Brazil and Argentina. Cubio (*Tropaeolum tuberosum* Ruíz and Pavon 1802) is the most important species because it produces edible tubers (Fig. 1), which are consumed by inhabitants of the Andean regions.

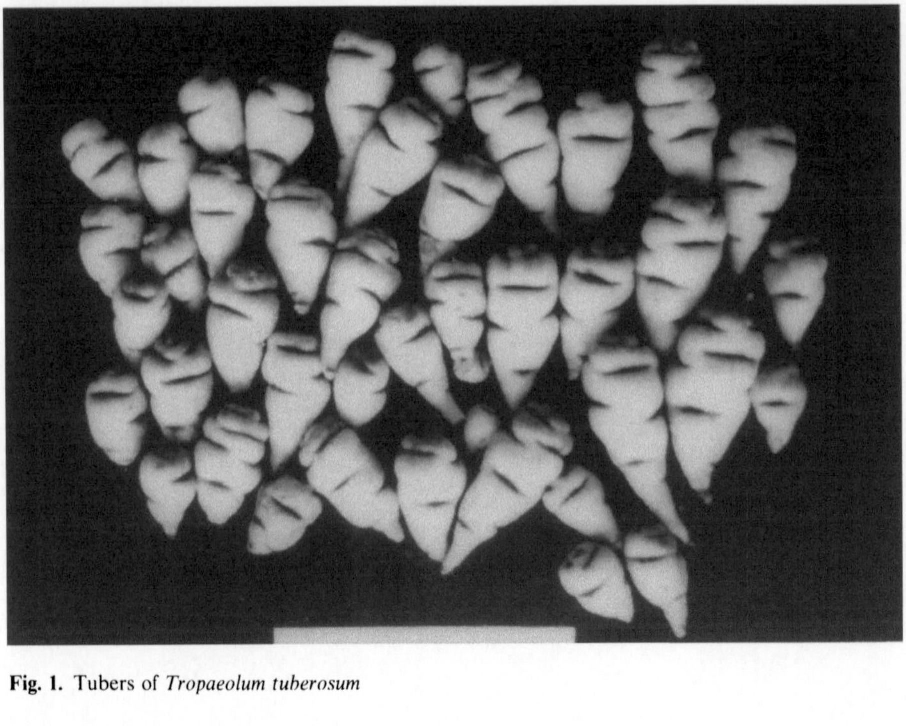

Fig. 1. Tubers of *Tropaeolum tuberosum*

[1] Laboratorio Cultivo Tejidos Vegetales, Departamento de Biología, Facultad de Ciencias, Universidad Nacional de Colombia, A.A. 23227, Bogotá, Colombia

Biotechnology in Agriculture and Forestry, Vol. 19
High-Tech and Micropropagation III (ed. by Y.P.S. Bajaj)

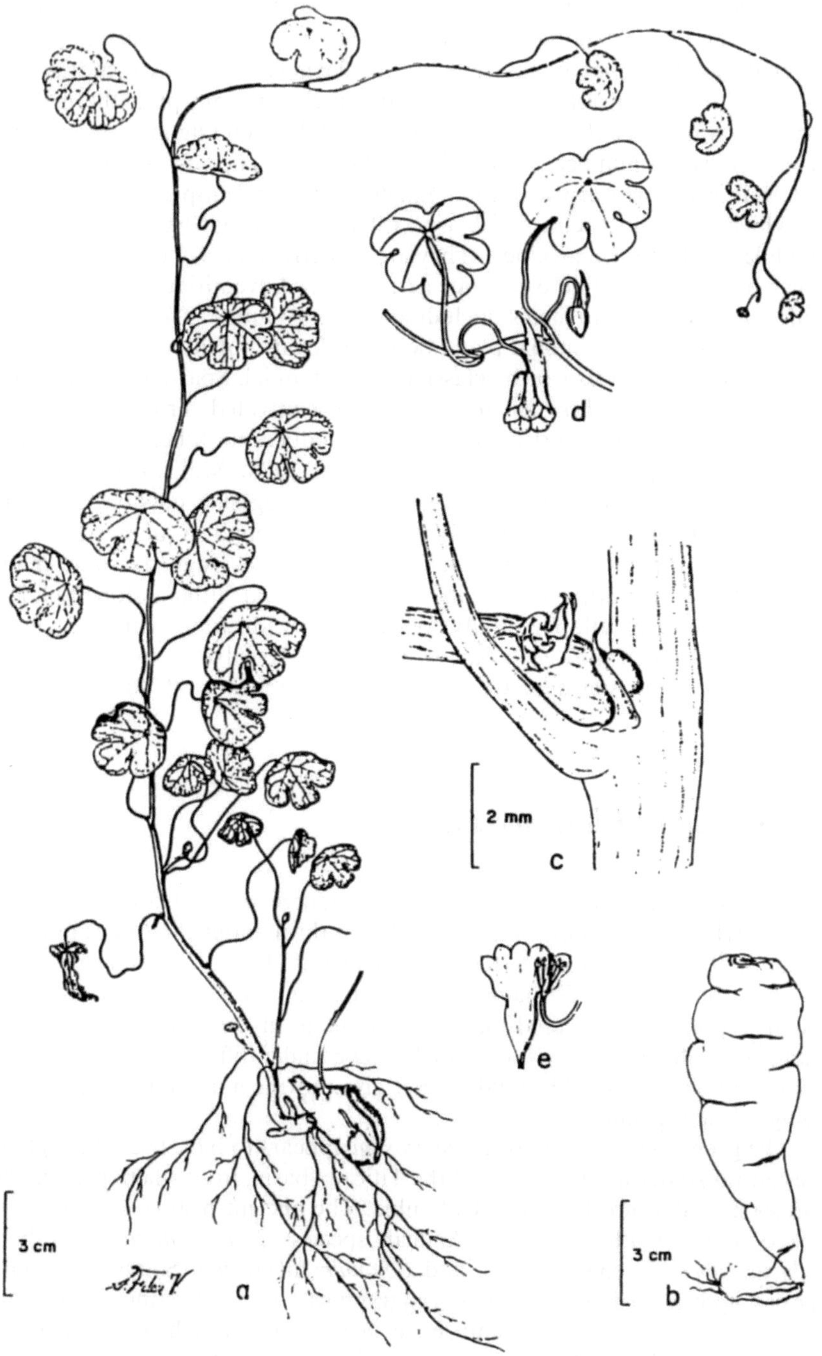

Fig. 2a–e. *Tropaeolum tuberosum*: **a** Habit. **b** Tuber. **c** Detail of leaf axil. **d** and **e** Flower. (**d** and **e** from Ruíz and Pavon 1802)

T. tuberosum (Fig. 2) is a prostrate or climbing herb growing to over 1–1.5 m in diameter and 0.5–0.8 m high. Aerial stems are cylindrical and slender. Leaves are alternate, their petioles are long (10–15 cm), flexible, and serve as tendrils, the blade is peltate three-to-five-lobed. Flowers are single, zygomorphic; the peduncles as long as the petioles; the calyx is reddish, five-lobed, ending in an elongate calcar; petals are five in number, usually dark yellow or orange; there are eight stamens. The fruit is a schizocarp with three indehiscent carpels (Sparre 1973). The tubers, for which the plant is cultivated, are axillary shoots that have become thickened and filled with storage products. They are subterranean. Tubers bear axillary buds which have the potential to generate aerial stems. Adventitious roots arise from the nodes on underground stems, including tubers (White 1975).

T. tuberosum is an annual plant; however, for its potential for continual tubers production this species can be classified as perennial. Sparre (1973) distinguishes two subspecies: *T. tuberosum tuberosum*, the cultivated form, and *T. tuberosum silvestre*, a naturally occurring form which produces no tubers. The chromosome number of *T. tuberosum* is $2n = 52$ (Gibbs et al. 1978). Multivalent formation and the occurrence of other meiotic abnormalities suggest a possible autopolyploid origin. *T. tuberosum tuberosum* is a probable tetraploid in an $x = 13$ series. The polyploid nature of this subspecies in such a series suggests some interesting possibilities for the origin of the tuber-bearing cultivar: either autopolyploidy from subsp. *silvestre*, or allopolyploidy following hybridization with another closely related species such as *T. cochabambae* ($2n = 26$). To solve the problem, cytological studies of *T. tuberosum silvestre* are required (Gibbs et al. 1978).

2 Distribution and Importance

Tubers of *T. tuberosum* are an important source of food for about 9 million people living at elevations of 2500 to 4000 m throughout the Andes in Colombia, Ecuador, Peru, and Bolivia (King and Gershoff 1987). *T. tuberosum* was called cubio by the ancient Chibcha people of the Colombian High Andes, and this is still the most popular name used today in Colombia. In the other countries, Ecuador, Peru, and Bolivia, *T. tuberosum* is well known under different names, mashua, añu, isaño, and piña-mama (Montaldo 1967), given by the numerous indigenous people living in the highlands.

The geographical origin of *T. tuberosum* appears to have been the highlands of the central Andes in the vicinity of the Titicaca basin, from where presumably this species migrated northward to Colombia. The ancient history of *T. tuberosum* is linked to three other tuberous Andean species: *Oxalis tuberosa* (Oxalidaceae), *Ullucus tuberosus* (Basellaceae), and *Solanum tuberosum* (Solanaceae). The four species appear to have the same geographical origin. Pre-Inca pictograms representing these tuberous plants have been found, which are evidence of the importance of these tubers even in those days (Hodge 1946).

Cubio (English name: Mashua) is cultivated in highlands from 2600 to 4000 m. The best yields are obtained near 3000 m altitude, and can reach 30 000 kg/ha per year. As the life cycle is short (5–6 months), it is possible to obtain two harvests per year. Around the Bogotá high plateau, tuberization is simultaneous with flowering, which is a useful sign to the farmer (Villamizar 1985).

The nutritional value of cubio is similar to that of other tubers. Gonzalez and Obando (1980) estimated that cubio is better food than cassava and arracacha, and similar to potato. Tubers of cubio contain all the essential amino acids (Gonzalez and Obando 1980; King and Gershoff 1987), and possess a high level of ascorbic acid (Gonzalez and Obando 1980). Table 1 shows the composition of cubio, potato, cassava, oca (*O. tuberosa*), and ulluco (*U. tuberosus*). Since the tubers of cubio, oca, and ulluco are cooked and eaten whole, including the husks, they present only a minor loss in nutritional content. These neglected Andean tuberous plants might be good alternatives for the farmers of the highlands, who depend on potatoes for their subsistence. Potato monoculture often creates great losses for farmers because of pests and diseases, or overproduction, that can lead to low prices.

Other uses have been attributed to *T. tuberosum*. Gonzalez and Obando (1980) estimate that the starch type of the tubers indicates their potential use in the wafer industry. Tubers and foliage of *T. tuberosum* have been used as food for domestic animals (Villamizar 1985). Indigenous people attribute medicinal properties to the tubers and the Indians believe that the tubers are an aphrodisiac when eaten (Hodge 1946; Villamizar 1985); however, research is necessary to study the truth of these assertions. Finally, the beautiful flowers and foliage of *T. tuberosum* are very attractive, so that this plant is often found in home gardens.

Table 1. Nutritional composition of some tubers[a]. Table de Composición de Alimentos Colombianos. ICBF. Ministerio de Salud, Bogotá (1978)

Component	Tuber				
	Potato	Cassava	Cubio	Oca	Ulluco
Edible portion (%)	80	80	95	95	95
Calories	91	146	58	55	58
Water (g)	75	61	83	84	84
Protein (g)	1.9	0.8	1.6	1.1	1.5
Fat (g)	0.1	0.1	0.1	0.1	0.0
Carbohydrate (g)	21	35	13.1	12	13
Fiber (g)	0.5	0.9	0.9	0.7	0.3
Ash (g)	1.0	0.9	0.9	0.9	0.8
Calcium (mg)	2	27	7	8	15
Phosphorus (mg)	28	35	49	42	57
Iron (mg)	1.0	0.4	0.7	0.5	0.4
Thiamine (mg)	0.08	0.04	0.05	0.04	0.04
Riboflavin (mg)	0.08	0.03	0.04	0.03	0.03
Niacin (mg)	0.9	0.5	0.6	0.4	0.4
Ascorbic ac. (mg)	16.0	30.0	120.0	25.0	20.0
Vitamin A I.U.	0	10	0	0	

[a] Based on 100 g of the edible portion.

3 Conventional Propagation and Breeding

T. tuberosum is vegetatively propagated by tubers. Multiplication by scions is possible because adventitious roots are easily induced from the aerial stem. The polyploid nature of *T. tuberosum* seems to be related to floral abnormalities and marked levels of pollen sterility, so that this species has been protected from sterility problems by asexual tuber propagation (Gibbs et al. 1978). Seed tubers require special conditions of storage at least 2 months before sowing. They must be protected from excessive moisture, dryness, or sunlight. Seed tubers are sown when buds have begun to emerge. *T. tuberosum* is able to produce botanical seeds (Cárdenas 1969); however, the literature has no reports of seeds being used for cultivation. Pollen sterility, a common occurrence in polyploid vegetatively propagated crops such as *T. tuberosum*, is a serious problem for breeding. Besides, several authors have reported great variability in many features of *T. tuberosum* (Bukasov 1930; Cárdenas 1969; White 1975; King and Gershoff 1987); therefore, research into the genetic nature of those variations is necessary. White (1975) observed fasciation from aerial stems and tubers and suggested a relation between tuberization, fasciation, and polyploidy.

4 Need for in Vitro Culture

Despite its good nutritional value, many people do not consume this food because they consider the tubers of *T. tuberosum* to have a disagreeable odor, which is produced by the volatile mustard oils (White 1975). The tuber flavor could also be improved. We believe that improvement of commercial properties is also necessary; better flavor and perhaps more attractive features in the tubers, e.g., color, aroma, and texture, are required to increase consumer acceptance. Together with these characteristics, protein quality could be improved. However, the high level of pollen sterility reported by Gibbs et al. (1978) for cultivars of various regions from Peru and from a long-established horticultural cultivars in Scotland, presents a serious problem for breeding, as in other polyploid crops such as potato (Miller and Lipschutz 1984). Plant cell and tissue culture methods offer alternatives for crop improvement by somaclonal variation and somatic hybridization. Besides, other *Tropaeolum* species, e.g., *T. cochabambae*, could be a source of viable pollen for interspecific hybridizations. In this case, embryo rescue and/or in vitro fertilization would be necessary.

Diseases of *T. tuberosum* have been little studied. In the literature we encountered only two reports of diseases of *T. tuberosum*. Delhey and Monasterios (1977) reported a mosaic disease of *T. tuberosum* from Bolivia. *Spongospora subterranea*, a well-known pathogenic fungus of potato, also affects *T. tuberosum* (White 1975). Since cubio is cultivated together with potato, other potato diseases could also affect it.

Propagation by tubers is very slow and inadequate for large-scale cultivation. The experience acquired with in vitro tuberization of potato could be applied in *T. tuberosum* to produce disease-free seed tubers for propagation and germplasm conservation. King and Gershoff (1987) claim the need to conserve the variability of Andean tubers because they have been gradually replaced by other foodstuffs with lower nutritional value.

5 In Vitro Approaches

Before beginning our research on *T. tuberosum*, we found no reports of in vitro studies for this species. In the literature surveyed there were only three reports of in vitro tissue culture of the nasturtium *Tropaeolum majus*. Ball (1946) obtained complete plants by aseptic culture of shoot tips consisting of apical meristem with three leaf primordia and some subjacent stem tissue. Smith and Murashige (1970) achieved development of complete plants by in vitro culture of isolated meristematic dome tissue of *T. majus* and other species. Kirkland et al. (1971) found glucosinolates in cell suspension cultures of *T. majus*.

Perea-Dallos et al. (1986) reported callus induction from leaves on semi-solid MS medium containing 1 mg/l kinetin and 1–10 mg/l 2,4-D. After subculture, morphogenesis was observed. Shoot proliferation from isolated buds on semi-solid MS medium with 5 mg/l BAP and 1 mg/l kinetin was also obtained. Shoots were separated and rooted on the same basal medium containing 1–2 mg/l NAA. Estrada et al. (1986) reported micropropagation from *T. tuberosum*, *O. tuberosa*, and *U. tuberosus*. These authors mentioned a single method for micropropagation of the three tuber crops from buds using three different media: semi-solid MS with 0.25 mg/l GA_3 (A medium); semi-solid MS without growth regulators (B medium) and liquid MS containing 0.4 mg/l GA_3, 0.5 mg/l BAP and 0.1 mg/l IAA (C medium). This last medium was used for rapid propagation, while A and B media were used for following subcultures during 6 and 3 weeks respectively before transfer to soil.

Starting from our previous reports (Perea-Dallos et al. 1986; Fandiño et al. 1987; Torres et al. 1988) and more recent unpublished observations, we have established the following protocols for tissue culture of *T. tuberosum*. The basal medium for all experiments consists of semi-solid standard MS (Murashige and Skoog 1962) with an initial pH of 5.8. Table 2 summarizes our in vitro studies with *T. tuberosum*.

5.1 Protocol for Callus Culture

1. Prepare basal medium and add 2 mg/l 2,4-D and 0.5 mg/l kinetin.
2. Dispense 8 ml of medium into 40-ml flasks and sterilize.
3. The best explants are aerial stem, petiole, and leaf blades, including main veins.
4. Take pieces of shoots from field-grown plants including stem and leaves. Rinse with tap water and dilute detergent solution.
5. Cut pieces from stem (1 cm long), petiole (1 cm long), and blade (1 cm^2).
6. Surface sterilize the excised pieces in 70% ethanol for 1 min.
7. Rinse the pieces three times in sterile distilled water.
8. Transfer the pieces to a 1% sodium hypoclorite solution containing 0.1% Tween 20 for 2 min.
9. Rinse the pieces four times in sterile distilled water.

10. Place one explant in each culture flask.
11. Incubate the cultures at 25 °C under light (1000 lx, 16-h photoperiod) or dark conditions. Callus induction is detected after 4 days in petiole or stem explants because they become swollen. From leaf explants cell clusters are observed after 10 days. Callus from leaf blade grows more quickly than callus from petiole or stem explants after 20 days.
12. To obtain friable calli subculture the callus on basal medium with 5 mg/l BAP for 20 days. After 30 days nonsubcultured callus becomes yellowish and compact.

Callus induction from stem can be obtained by culture on basal medium with 4–5 mg/l Dicamba or 5 mg/l Picloram. The same concentrations of these growth regulators inhibit callus formation from leaf blade explants. Often, systemic bacterial contaminants present on plants from the field make in vitro establishment difficult. Therefore, we have obtained plants from meristem culture on basal medium without hormones in order to obtain bacteria-free explants.

Table 2. Summary of our in vitro studies with *T. tuberosum*

Explant	Growth regulators ($mg\,l^{-1}$)			Response
	Auxin	Cytokinin	Other	
Leaf, stem	2,4-D (2)	K (0.5)		Callus
Leaf blade	2,4-D (1)	K (8)		Tuber-like structures
Stem	DIC (5)	ADE (1)		Callus
Stem	PIC (5)	ADE (1)		Callus
Callus		BAP (5)		Friable calli
Callus	NAA (2)	BAP (3)		Tuber-like structures
Stem bud		BAP (5) + K (1)		Multiple shoot formation, 5–50 shoots from each bud
Stem bud		(1) BAP + K (5)		Single shoot from each bud
Stem bud		(3)		Single shoot from each bud
Stem node	NAA (2–5)	BAP + K (3)	Sucrose 6%	Complete plantlets bearing tuber-like structures
Stem node	NAA (2)		GA_3(5)	Complete plantlets bearing tuber-like structures
Stem node		BAP (5)	Sucrose 6%	Shoots without tuber-like structures
Stem node Aseptic shoots	NAA (1–2)			Shoot lengthening, adventitious roots arise from nodes after 4 days. Tuber-like structures sometimes appear after 20 days

Note: basal medium was MS for all experiments.

Abbreviations: 2,4-D = 2,4-dichlorophenoxyacetic acid; K = kinetin; DIC = Dicamba; PIC = Picloram; ADE = Adenine; BAP = 6-benzylaminopurine; NAA = naphthalene acetic acid; GA_3 = gibberellic acid.

5.2 Protocol for Micropropagation by Multiple Shoot Formation

1. Prepare basal medium and add 5 mg/l BAP and 1 mg/l kinetin.
2. Dispense 25 ml of medium into 100-ml flasks and sterilize.

3. Take aerial stems from field-grown plants. Rinse the stems and surface sterilize, separating axillary buds following the same procedure as used for callus explants.
4. Place one bud in each flask.
5. Incubate the cultures at room temperature (18–22 °C) and daylight (12-h photoperiod). Multiple shoot induction (Fig. 3) is obtained after 20 days. Under these conditions shoot clusters can be maintained for 10–12 months without subculture for medium-term conservation of germplasm. After 10–12 months shoots can be separated and rooted. They develop into normal plants.
6. After 30 days, subculture shoot clusters on the same medium (MS + 5 mg/l BAP and 1 mg/1 kinetin) at 22–24 °C and 1000 lx illumination (16-h photoperiod) for rapid shoot proliferation. At higher temperatures (up 26 °C) shoots become chlorotic and die.
7. Separate each shoot from shoot clusters and transfer to culture tubes containing basal medium with 1 mg/l NAA. Incubate the cultures at 20–22 °C in daylight (12-h, photoperiod). Adventitious roots arise from nodes after 4–6 days. After 15–20 days plantlets (Fig. 4) can be transferred to soil.
8. Transfer plantlets to soil into glass boxes at high humidity. After 1 week remove the plastic cover and allow the growth of plantlets for 1–2 weeks before transfer to pots (Fig. 5) or field.

5.3 Propagation by Nodal Cuttings

1. Surface sterilize shoots from young field-grown plants using the same procedure as indicated for callus explants.
2. Transfer surface-disinfected (or in vitro) nodal cuttings (1 cm long) to test tubes containing basal medium with 1–2 mg/l NAA.
3. Incubate the cultures at 20–22 °C in daylight (12-h photoperiod). Axillary buds develop to complete plantlets, including roots, within 3–4 weeks.
4. Transfer plantlets to soil following the same procedure as indicated for plantlets obtained from multiple shoot formation.

Fig. 3. Multiple shoot formation from cubio axillary bud cultivated on MS medium with 5 mg/l BAP and 1 mg/l kinetin after 30 days

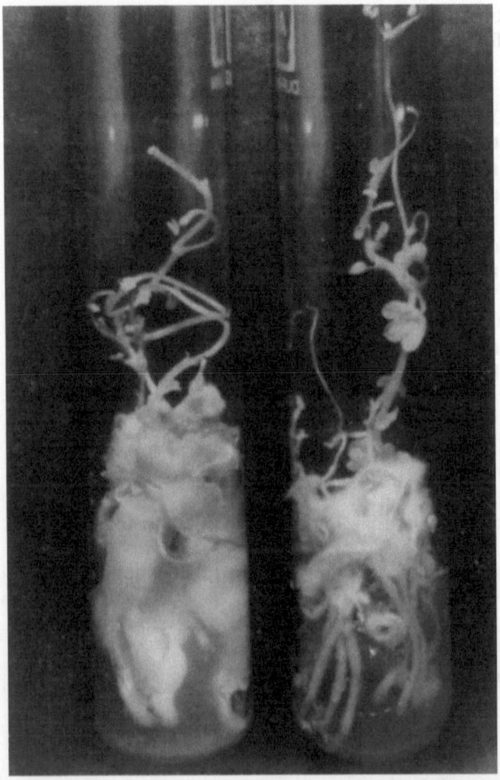

Fig. 4. Rooting of *T. tuberosum* plantlets on MS medium with 1 mg/l NAA after 20 days

Fig. 5. Micropropagated plant of *T. tuberosum* 30 days after transfer to soil

Fig. 6. Organogenesis from callus of *T. tuberosum* cultivated on MS medium with 2 mg/l NAA and 3 mg/l BAP after 30 days

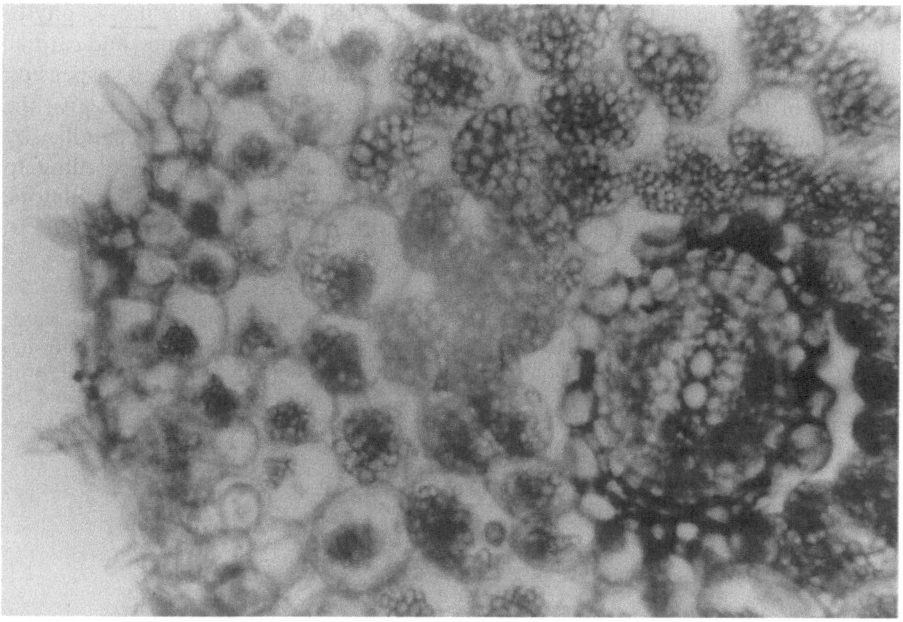

Fig. 7. Section of the organoid shown in Fig. 6

5.4 Organogenesis

We have obtained formation of adventitious tuber-like structures from leaf explants, calli (Fig. 6), shoot clusters (Fig. 3), and shoots transferred onto rooting medium (Fig. 4), under various environmental conditions, as well as different hormonal and sucrose supplements. Sections of some of these structures were examined (Fig. 7) and showed anatomy similar to sections of young adventitious roots from *T. tuberosum* described by White (1975, Fig. 38a). Anatomy of these thick structures differs significantly from the normal tuber anatomy of *T. tuberosum*, which shows a circle of collateral vascular bundles surrounding ground tissue compound by uniform parenchyma according to White (1975) and our observations. Plantlets bearing these thick tuber-like structures were transferred to soil. They develop into normal plants but tuber-like structures were early disintegrated.

6 Conclusions and Prospects

Tubers of *Tropaeolum tuberosum* and two other neglected species oca (*Oxalis tuberosa*) and ulluco (*Ullucus tuberosus*), possess a high nutritional level. However, these tuberous crops have not acquired the commercial importance of their companion in origin, *Solanum tuberosum*. Potato has become one of the most important crops of the world, while cubio, oca, and ulluco have remained in their homeland as species once widely known but being gradually abandonned (Hodge 1946; King and Gershoff 1987). Their flavor and other properties have been judged as not attractive for massive consumer acceptance. Therefore, research for improvement of *T. tuberosum*, *O. tuberosa*, and *U. tuberosus* must be directed to obtaining more commercially attractive tubers and thus biotechnology holds great promise.

Secondary product formation on in vitro systems is another field for investigation in *T. tuberosum*. González and Obando (1980) found by qualitative phytochemical analysis flavonoids, alkaloids, steroids, sesquiterpen-lactones, and cardiac heterosides. Kirkland et al. (1971) found mustard oil glycosides in cell suspension cultures of the garden nasturtium *Tropaeolum majus*. These compounds give the nasturtium-like odor to species of the Tropaeolaceae and Brassicaceae families by release of volatile mustard oils from myrosin cells (White 1975). According to Steward and Krikorian (1971), mustard oil glycosides can act as growth regulators, e.g., as germination and plant growth inhibitors, or making plants resistant to infection by microorganisms.

Acknowledgments. The authors dedicate this paper to our Alma Mater Universidad Nacional de Colombia. We are indebted to Dr. Hernan Cardozo, who originally suggested working with this neglected species, and for his collaboration. We express our thanks to Dr. Edward G. Kirby for useful comments concerning the manuscript. We are also grateful to Jorge Restrepo, Olga Ruíz, and the students Amparo Salamanca, Adelaida López, and José Mikán. Finally, we thank Mr. Silvio Fernández for the drawings and Mr. Julio Cabra for his photography work.

References

Ball E (1946) Development in sterile culture of stem tips and subjacent regions of *Tropaeolum majus* L. and of *Lupinus albus* L. Am J Bot 33: 301–318

Bukasov S (1930) The cultivated plants of Mexico, Guatemala and Colombia. Bull Appl Bot Genet Plant Breed (Leningrad) Suppl 47, 553 p

Cárdenas M (1969) Manual de plantas económicas de Bolivia. Icthus, Cochabamba, 421 p

Delhey R, Monasterios T (1977) A mosaic disease of isaño (*Tropaeolum tuberosum*) from Bolivia. Pflanzenkr Pflanzensch 84 (4): 224–231

Estrada R, Manya W, Pulache C, Sánchez H, Yonamine T (1986) Maintenance, micropropagation and seed production of the Andean tuber crops: oca, olluco and mashua. In: Abstr 6th Int Congr Plant tissue and cell culture, Univ Minneapolis, Aug 3–8, p 103

Fandiño TJ, Torres O, Perea-Dallos M (1987) Morfogénesis y Micropropagación de *Tropaeolum tuberosum*. Bol Cient Asoc Colombiana Estud Veg In Vitro (ACEVIV) 1 (2): 29–35

Gibbs P, Marshall D, Brunton D (1978) Studies on the cytology of *Oxalis tuberosa* and *Tropaeolum tuberosum*. Notes R Bot Gard Edinburgh 37: 215–220

Gonzalez G, Obando R (1980) Estudio preliminar de la composición química del Cubio. Tésis, Dep Química, Fac Cienc, Univ Nacional de Colombia, Bogotá, 111 p

Hodge W (1946) Three neglected Andean tubers. J New York Bot Gard 47: 214–224

King S, Gershoff S (1987) Nutritional evaluation of three underexploited Andean tubers: *Oxalis tuberosa* (Oxalidaceae), *Ullucus tuberosus* (Basellaceae) and *Tropaeolum tuberosum* (Tropaeolaceae). Econ Bot 41 (4): 503–511

Kirkland D, Matsuo M, Underhill E (1971) Glucosinolates and myrosinase in plant tissue cultures. Lloydia 34: 195–198

Miller S, Lipschutz L (1984) Potato. In: Ammirato PV, Evans DA, Sharp WR, Yamada Y (eds) Handbook of plant cell culture, vol 3. MacMillan, New York, pp 291–326

Montaldo A (1967) Bibliografía de raíces y tubérculos tropicales. Rev Fac Agron Univ Cent Venez (Alcance) 13: 492–493

Murashige T, Skoog F (1962) A revised medium for rapid growth and bioassays with tobacco tissue cultures. Physiol Plant 15: 473–497

Perea-Dallos M, Fandiño TJ, Torres O (1986) Morphogenesis and micropropagation of *Tropaeolum tuberosum*. In: Abstr 6th Int Congr Plant Tissue and Cell Culture, Univ Minneapolis, Aug 3–8, p 104

Ruiz H, Pavon J (1802) Flora Peruviana et Chilensis, vol 3. Madrid, p 77

Smith R, Murashige T (1970) In vitro development of the isolated shoot apical meristem of angiosperms. Am J Bot 57 (5): 562–568

Sparre B (1973) Tropaeolaceae. Flora of Ecuador. Oper Bot Ser 2, 30 p

Steward F, Krikorian A (1971) Plants, chemicals and growth. Academic Press, New York, 232 p

Torres O, Fandiño TJ, Perea-Dallos M (1988) Aspectos anatómicos y fisiológicos en cultivos in vitro de *Tropaeolum tuberosum*. Bol Dep Biol, Fac Cienc, Univ Nacional de Colombia 2 (8)

Villamizar ME (1985) Aporte al conocimiento de *Ullucus tuberosus*, *Oxalis tuberosa* y *Tropaeolum tuberosum*. Tésis, Dep Biol, Fac Cienc, Univ Nacional de Colombia, Bogotá, 148 p

White J (1975) Notes on the biology of *Oxalis tuberosa* and *Tropaeolum tuberosum*. Senior Honors Thesis Biol, Harvard College, 96 p

Section II Medicinal and Aromatic Plants

II.1 Micropropagation of Cardamom (*Elettaria cardamomum* Maton)

B.R. Reghunath[1] and Y.P.S. Bajaj[2]

1 Introduction

1.1 Morphology and Cytology

Cardamom (*Elettaria cardamomum* Maton) is a tall perennial shrub (Fig. 1), belonging to the family Zingiberaceae. The aerial part of the plant consists of 10–20 erect leafy shoots (pseudo-stem) 2–5.5 m tall and made of leaf sheaths. In mature plants (bearing stage) several erect or decumbent panicles of length 0.6 to 1.2 m are also present. Both the shoots and panicles emerge from a horizontal subterranean woody rhizome. Each panicle bears numerous small, white or pale green flowers characterized by a conspicuous labellum with violet streaks radiating from the center. The flowers are hermaphrodite, zygomorphic, and have an average size of 4 × 1.7 cm. The ripe fruit is an ovoid trilocular capsule containing 15 to 20 aromatic seeds embedded in mucilage and enclosed in a semi-hard fleshy pale green to yellow pericarp. On drying, the pericarp becomes fibrous or leathery and the mucilage disappears. The dried capsules with the aromatic seeds inside are the commercial cardamoms.

The basic chromosome number of *Elettaria* is x = 12 (Darlington and Wylie 1955) and the somatic number of *E. cardamomum* is 2n = 48 (Gregory 1936; Sharma and Bhattacharyya 1959; Ramachandran 1969). Chakravorti (1948), however, reported 2n = 52.

1.2 Distribution

E. cardamomum is a tropical plant native to the western ghats in southern India. It grows well on high ranges with damp soil rich in humus and under partial shade of forest trees. It prefers altitudes between 760 and 1400 m above mean sea level, with an annual rainfall of 1500–5000 mm, and a temperature range of 10–35 °C (Purseglove et al. 1981).

Cultivation of *E. cardamomum* in India is concentrated mainly in the south in the states of Kerala, Karnataka, and Tamil Nadu. Guatemala, Tanzania, and Sri

[1] USDA Project on Tissue Culture, College of Agriculture, Vellayani P.O., Trivandrum-695 522, Kerala, India

[2] Former Professor of Tissue Culture, Punjab Agricultural University Ludhiana, India. (Present address: A-137 New Friends Colony, New Delhi 110065, India)

Biotechnology in Agriculture and Forestry, Vol. 19
High-Tech and Micropropagation III (ed. by Y.P.S. Bajaj)

Fig. 1. *Elettaria cardamomum* plant at the bearing stage (age: 5 years)

Lanka are the other major cardamom-producing countries. New Guinea, Costa Rica, El Salvador, and the Honduras have also emerged in recent years as producers and suppliers of cardamom on the world market. The average annual world production of cardamom is presently estimated at around 10 000 t, of which India provides 40%, Guatemala 45%, and other countries 15% (Cherian 1987).

1.3 Varieties, Types, or Races

The "true cardamom" in trade is the dried fruit (capsule) of *E. cardamomum* var. *cardamomum*. Spice is also obtained by drying the fruits of various *Amomum* and *Aframomum* species and also that of wild *Elettaria* varieties, often referred as "false cardamoms", but these are regarded as much inferior in aroma and flavor quality (Purseglove et al. 1981). Two botanical varieties of true cardamom are recognized, based on the size of the fruit, namely var. *major* Thwaites and var. *cardamomum* (syn. var. *minor* Watt; var. *minuscula* Burkill). The fruit of the var. *major* is larger, longer (2.5–5 cm), with more numerous, larger, and less aromatic seeds than that of the var. *cardamomum*. It is known as Long Wild Cardamom and is less common in trade than the var. *cardamomum*. True cardamoms are further distinguished into two main types, namely Malabar and Mysore, based on the appearance of the plant and fruit and the chemical characteristics of the seeds. These types are indigenous to India, but both have been introduced into other countries. Mysore and Malabar types, having grown together for so long in India, a naturally bred hybrid type with mixed plant characters of these types originated there, and is known as the Vazhukka type (John 1988).

1.4 Uses and Importance of the Plant

Cardamom is used directly as a flavoring material in three forms: whole, the decorticated seeds, and ground. In international trade the whole form predominates over the other forms. It is from the decorticated seeds that the essential oil (steam-distillation) and oleoresin (solvent-extraction) are extracted.

Cardamom in the whole or ground form is used mainly for domestic culinary purposes. In Asia, it is a masticatory as well. It is used as a flavorant in a variety of spiced rice, meat and vegetable preparations, puddings, sweets, and certain beverages. In the Middle East, cardamom has been traditionally used for flavoring coffee. In the European countries the spice is used in curry powders, sausages, soups, canned fish, and also to flavor tobacco. A range of baked goods including cakes, buns, pastries, and bread are largely flavored with cardamom.

Cardamom oil finds its main application in the flavoring of processed foods, cordials, bitters, and liqueurs, and occasionally in perfumery. The oleoresin also has similar applications, but is less extensively used.

Cardamom also has distinctive medicinal properties, and is found in most pharmacopeias. In India, the tincture of cardamom seed is used in the indigenous systems of medicine to strengthen the digestive, respiratory, and urino-genital systems (Pruthi 1976; Warrier 1989). Kirtikar and Basu (1935) reported tonic, stimulant, stomachic, and emmineagogic properties in the spice. According to them, it is administered internally in diseases of the liver and the uterus, and externally to tumors of the uterus. Cardamom is reported to have curative properties in rheumatism, bronchitis, poisoning, itching, and furuncles (Warrier 1989). The aqueous extract of the rhizome of the cardamom plant is a good tonic and is claimed to have laxative properties. Cardamom oil has anti-bacterial activity (Sreenivasamurthy and Krishnamurthy 1959). Nayak and Dutta (1961) reported the bactericidal property of cardamom oil against *Vibrio comma* in vitro at dilutions of 1:1000–1:15000.

1.5 Propagation Methods

E. cardamomum can be propagated from seed or by division of the subterranean rhizomes. In India, seed propagation is common, whereas in Guatemala and Sri Lanka, the vegetative method is popular (Ridley 1912; Anon 1977). Seeds do not carry the mosaic or Katte disease causing systemic virus (Varma and Capoor 1958; Rao and Naidu 1973); the method is hence prevalent in Katte-infested areas. The major disadvantage of using seedlings is the high variability that occurs in the productivity of the crop (Purseglove et al. 1981).

Vegetative propagation using division of rhizomes (each division consisting of at least one old and one young shoot) is simple, reliable, cheap, and permits the use of selected clonal material (Purseglove et al. 1981). This method enables earlier and higher yield when compared to seedlings. Parameswar et al. (1979) recorded nearly tenfold increase in the productivity of cardamom plantations raised by vegetative methods compared to seedling plantations. However, when this method is used, there is always a shortage of planting material for planting large areas. Vegetative propagation is dangerous when diseases and pests are present, particularly mosaic virus and thrips (Purseglove et al. 1981).

Recently, micropropagation methods have been standardized in *E. cardamomum*, and offer tremendous scope for rapid vegetative propagation of elite clones eliminating systemic pathogens including viruses (Nadgauda et al. 1983; Kumar et al. 1985; Reghunath 1989). Replanting of old seedling plantations with selected clones propagated by this method is expected to give a five to six fold increase in the current average productivity of *E. cardamomum* in India (Anon 1989).

2 Micropropagation

It has been reported (Purseglove 1975) that cardamom seed germination is often poor and very irregular, and some seeds may even take more than 1 year to germinate. Under natural conditions the seeds may be dormant for a considerable period of time. The germination of seeds under ordinary conditions is about 30% within a period of 70 days. Kologi et al. (1973) reported 22% seed germination outdoors and 35% germination in a plastic house maintained at 23–26 °C and 58–95% relative humidity.

Elettaria cardamomum is often infested by several pests and diseases. Mosaic or Katte disease of cardamom, with an intensity ranging from 0.01 to 99%, is the major constraint in cardamom production. A crop loss of up to 68% during the 3rd year of infection has been noted in certain Indian plantations. Nonavailability of sufficient planting materials in high yielding clones, crop loss caused by cardamom thrips, and increasing drought in the cardamom-growing tracts of India are the other constraints.

E. cardamomum plants can be effectively protected from infestation by mosaic or Katte disease, by controlling the insect vectors and by resorting only to seed propagation. However, the predominance of the seed propagation method in a cross-pollinated crop like cardamom has resulted in nonconsistent yield and drastic reduction in productivity. Thus micropropagation is a logical alternative. Following applications of micropropagation techniques for medicinal and aromatic plants have been outlined by Bajaj et al. (1988):

1. increase in the propagation rate of plants;
2. availability of plants throughout the year, i.e., in all seasons;
3. protection of plants against pests and pathogens under controlled conditions;
4. uniform clones;
5. uniform production of secondary metabolites.

Three approaches have been followed for the micropropagation of cardamom, (1) through callusing, (2) through adventitious bud formation, and (3) through enhanced axillary branching.

The regeneration of plants from callus cultures was first reported in 1982 (Srinivasa Rao et al.); however, no successful reports of direct formation of adventitious buds in vitro other than from the primordia of shoot and inflorescence are available in the literature. Nadgauda et al. (1983) released an average number of three precocious shoots per shoot primordium (10–20 mm long); cultured on MS medium supplemented with BAP, kinetin, and coconut milk, the plantlets could be rooted. Later, Kumar et al. (1985) reported direct shoot formation by culturing inflorescence (panicle) primordium (15–25 mm long) on MS medium containing NAA, BAP, and kinetin. The plantlets were rooted by incubating them in darkness. Our work on the micropropagation of cardamom adopting various in vitro methods and different explant is summarized here.

2.1 Micropropagation Through Callus Culture

2.1.1 The Culture Method

Explant. In our culture experiment, shoot primordium tips (3 mm) (Fig. 2) excised from both greenhouse-grown and in vitro-raised plants of the type Mysore were utilized for initiating callus cultures. Explants from greenhouse-grown plants were given a series of surface sterilization treatments (Table 1). A quick dip in absolute alcohol (95%) followed by treatment in mercuric chloride (0.15%) solution for about 12 min gave reasonably good sterilization. Habituation of the explant to in vitro conditions and the reduced chance of microbial contamination in in vitro explants had resulted in initiation of callus growth in 100% of the cultures compared to cultures from greenhouse-grown plants.

Culture Medium. Two media, MS and Schenk-Hildebrandt (SH) (1972), were tested. Callusing was more or less similar in these media; however, the dry weight of callus obtained from MS medium was comparatively greater than that from SH medium (Table 2).

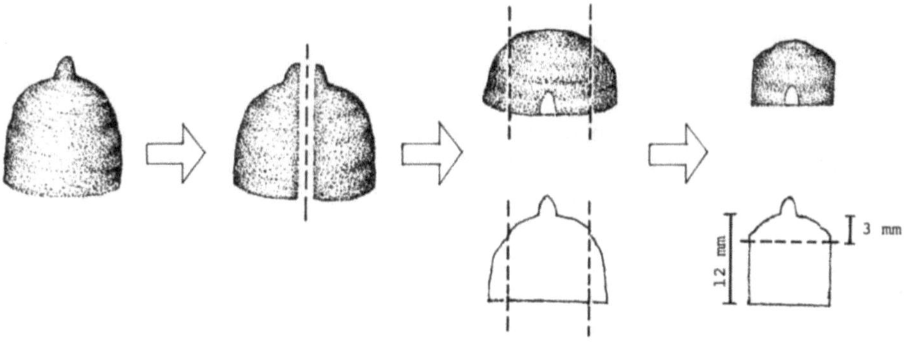

Fig. 2. Sequential steps of excising meristem from shoot primordium

Table 1. Trials on the surface sterilization of *E. cardamomum* explants

Explant	Sterilant	Concentration (%)	Duration
Shoot primordium	Absolute alcohol	95	5–15 s
	Sodium hypochlorite	2.00–4.00	5–15 min
	Mercuric chloride	0.05–0.20	5–15 min
	Combination of the best treatment of the above (serial treatment)	–	–
Inflorescence primordium	Absolute alcohol	95	5–15 s
	Sodium hypochlorite	2.00–4.00	5–15 min
	Mercuric chloride	0.05–0.20	5–15 min
	Combination of the best treatment of the above (serial treatment)	–	–
Immature inflorescence segments	Absolute alcohol	95	5–15 s
	Mercuric chloride	0.05–0.20	5–15 min
Immature capsule	Absolute alcohol	95	5–15 s
	Mercuric chloride	0.05–0.20	5–15 min

Table 2. Callus production potential of *E. cardamomum* shoot primordium in two different culture media (Reghunath 1989)

BAP (mg/l)	MS medium		SH medium	
	Callus fresh weight: dry weight (mg)[a]			
	0.5	1.0	0.5	1.0
Auxin (mg/l)				
NAA 0.5 + 2,4-D 0.1	308:6.17	453:13.83	406:4.83	420:9.33
NAA 2 + IAA 1	348:8.67	483:9.17	393:7.67	453:9.50
2, 4-D 0.5 + IAA 1	320:6.33	527:15.50	350:4.17	428:10.67
2, 4-D 1	403:6.67	411:10.50	291:5.67	410:10.17
NAA 4	495:9.57	615:16.33	375:8.00	446:10.83

[a] Mean of three replications; culture period 28 days.

Culture conditions. Each explant was planted on a semi-solid medium, gelled with 0.6% Difco-Bactor agar, in glass flasks and cultured at 23 ± 2 °C temperature under white fluorescent light of 1000 lx intensity and 16-h photoperiod.

2.1.2 The Cultures Raised

Effect of Growth Regulators. The cultures raised by supplementing 4 mg/l of NAA and 1 mg/l of BAP in the culture medium registered maximum callus production

(Table 2). Treatments involving 2, 4-D, however, produced profuse, white, friable callus, and did not initiate shoot organogenesis (caulogenesis) in any of the explants when they were subjected to caulogenesis induction treatments. However, rhizogenesis was observed in certain cultures.

The calli obtained from the treatment combination involving NAA and BAP were compact and semi-hard (Fig. 3B), and when subcultured on an auxin-free medium containing 3 mg/l of BAP and 0.5 mg/l of kinetin or 2 mg/l of BAP and 1 mg/l of kinetin, they started caulogenesis. Each culture produced six to nine meristemoids, which on subculture to same medium produced shoots (9 ± 2 mm) within 28 days (Table 3). The different stages of development of a shoot tip explant to a whole plant through callusing is depicted in Fig. 3C. In a comparative study with Mysore type on the potentiality of callusing and subsequent organogenesis, showed that two other major types of cardamom, Mysore and Vazhukka, were comparable, whereas Malabar was slightly inferior (Table 4).

Effect of Coconut Milk. Caulogenesis was remarkably favored by the addition of coconut milk to the culture medium. The mean number of green meristemoids produced in the callus mass in a flask was enhanced to 12 by the supplementation of 5% coconut milk.

2.2 Micropropagation Through the Enhanced Axillary Branching Method

2.2.1 The Culture Method

Explant. Shoot primordium explants (Fig. 4A) and inflorescence primordium explants (Fig. 4B) excised from greenhouse-grown plants of the type Mysore were utilized in the initial experiments. The optimum length of an explant was 20–30 mm (Fig. 5A). In further trials, the precocious shoots (20–30 mm) obtained from initial shoot primordium cultures were used. Shoot primordia used in the initial cultures were subjected to certain chemical and/or physical pre-treatments for enhancing axillary bud break (Table 5). Among the treatments, cytokinin infusion (BAP/kinetin 200 mg/l for 6-24 h) was found favorable (Fig. 5B). Inflorescence primordium explants, although they produced vegetative shoots (Fig. 5C, D), were found to be less efficient in multiple shoot production than shoot primordium.

Culture Medium. Two media, MS and SH, were tried. In MS medium, the regenerated shoots showed symptoms of leaf tip drying and chlorosis. Reduction of the inorganic salts of MS medium to one-half, or use of SH medium, produced normal shoots (Fig. 6A, B). The SH medium was superior to half-strength MS medium, since it registered 31% greater shoot dry weight (Table 6).

Culture Conditions. Both semi-solid (gelled with 0.6% Difco-Bactor agar) and liquid cultures were tried. Liquid cultures were kept either static or on a gyratory shaker revolving at 100 rpm. Precocious shoot explants cultured in liquid medium in conical flasks under gyratory shaking produced 111.57% more axillary branches than those cultured in semi-solid medium (Table 7). The cultures were maintained

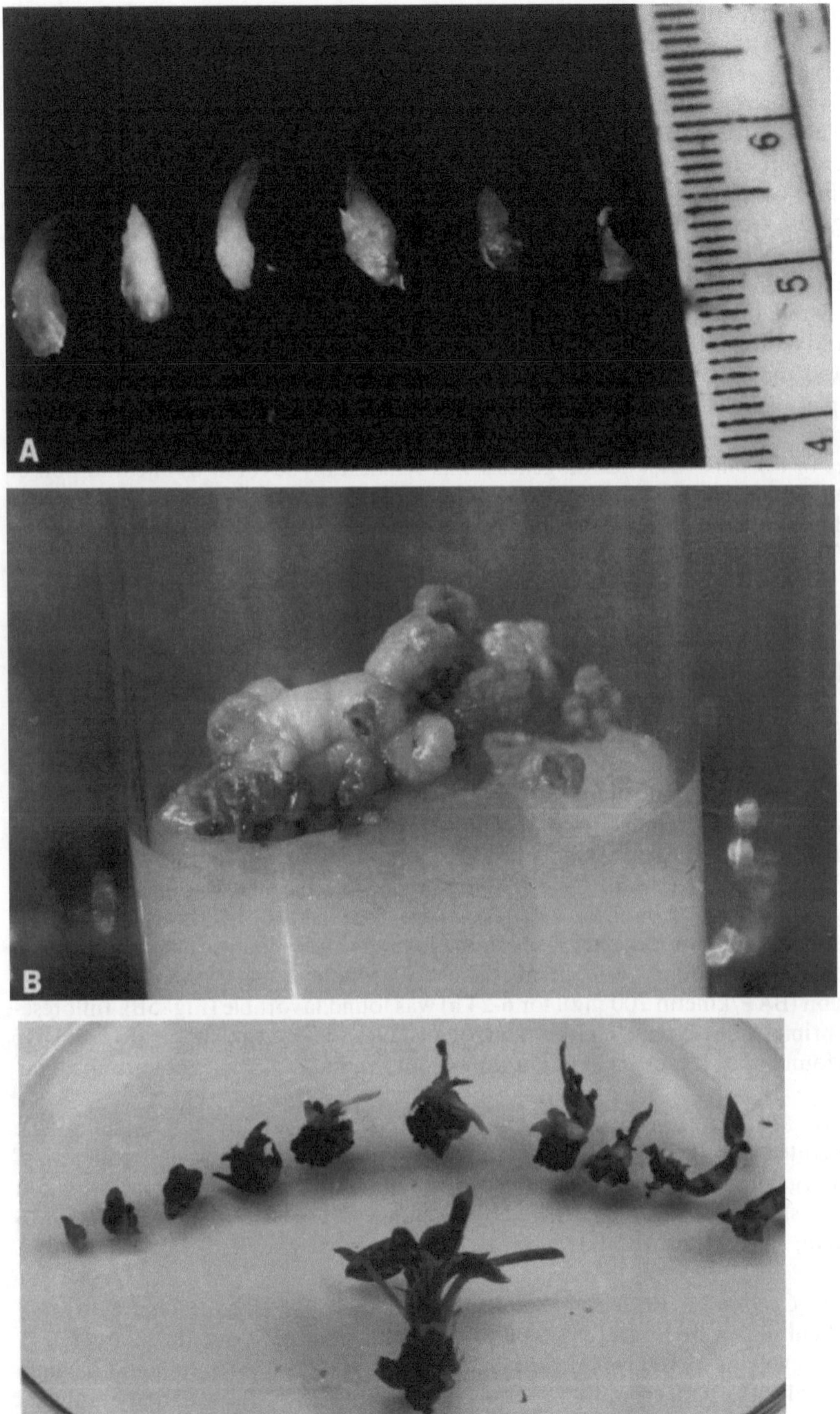
A
B
C

Table 3. Caulogenesis in callus cultures of *E. cardamomum*. (Reghunath 1989)

BAP[b] (mg/1)	MS medium				
	Caulogenesis[a]				
	0	0.5	1	2	3
Kinetin (mg/l)					
0	$+^{0}$	$+^{0}$	$+^{2}$	$+^{2}$	$+^{3}$
0.5	$+^{0}$	$+^{1}$	$+^{2}$	$+^{3}$	$+^{4}$
1	$+^{1}$	$+^{2}$	$+^{2}$	$+^{4}$	$+^{3}$
2	$+^{1}$	$+^{2}$	$+^{2}$	$+^{3}$	$+^{0}$
3	$+^{2}$	$+^{3}$	$+^{2}$	$+^{0}$	$+^{0}$

[a] Based on three visual observations (similarity $83.2 \pm 16.6\%$); culture period 28 days.

[b] 6-Benzylaminopurine

$+^{0}$ Not tested.

$+^{1}$ Callus turns to compact pale yellow masses without any further development.

$+^{2}$ Callus turns to compact masses and certain white tubular protuberances appear from beneath.

$+^{3}$ As in $+^{2}$ above; produces one or two greenish meristemoids from above.

$+^{4}$ As in $+^{3}$ above; produces six to nine greenish meristemoids which on subculture to same medium produces shoots (9 ± 2 mm) within 28 days.

Table 4. Comparative performance of major cardamom types on the induction and growth of callus and subsequent organogenesis in shoot tip cultures. (Reghunath 1989)

	Basal medium: MS + NAA 4 mg/l + BA 1 mg/l + CM 5% + biotin 1 mg/l				
Variety	Cultures[a] initiating callus (%)	Growth score[a] (gr)	Callus index	Organogenesis[c]	
				Rating[b]	Remarks[d]
Mysore	100	1.80	180.0	Very good	Callus grows rapidly and forms compact calloid masses with 10–12 green buds
Vazhukka	100	2.00	200.0	Very good	Callus grows rapidly and forms compact calloid masses with 10–12 green buds
Malabar	100	1.50	150.0	Good	Callus grows rapidly and forms compact calloid masses with 6–8 green buds

[a] Mean of four replications; culture period 28 days.

[b] Based on visual observation which showed 90% similarity and 5–10% variation.

[c] Medium: MA + BA 2 mg/l + kinetin 1 mg/l + CM 5% + biotin 1 mg/l.

[d] Observations taken after 4–6 weeks of callus subculturing.

◄

Fig. 3A–C. Establishment of callus and the regeneration of plants. **A** Sequential steps of preparing a shoot-tip explant (3 mm) from a precocious shoot. **B** Compact callus induced from precocious shoot-tip cultured for 28 days on modified MS medium containing NAA and BAP. **C** Whole plant regeneration from a shoot tip explant via callus-mediated process: different stages

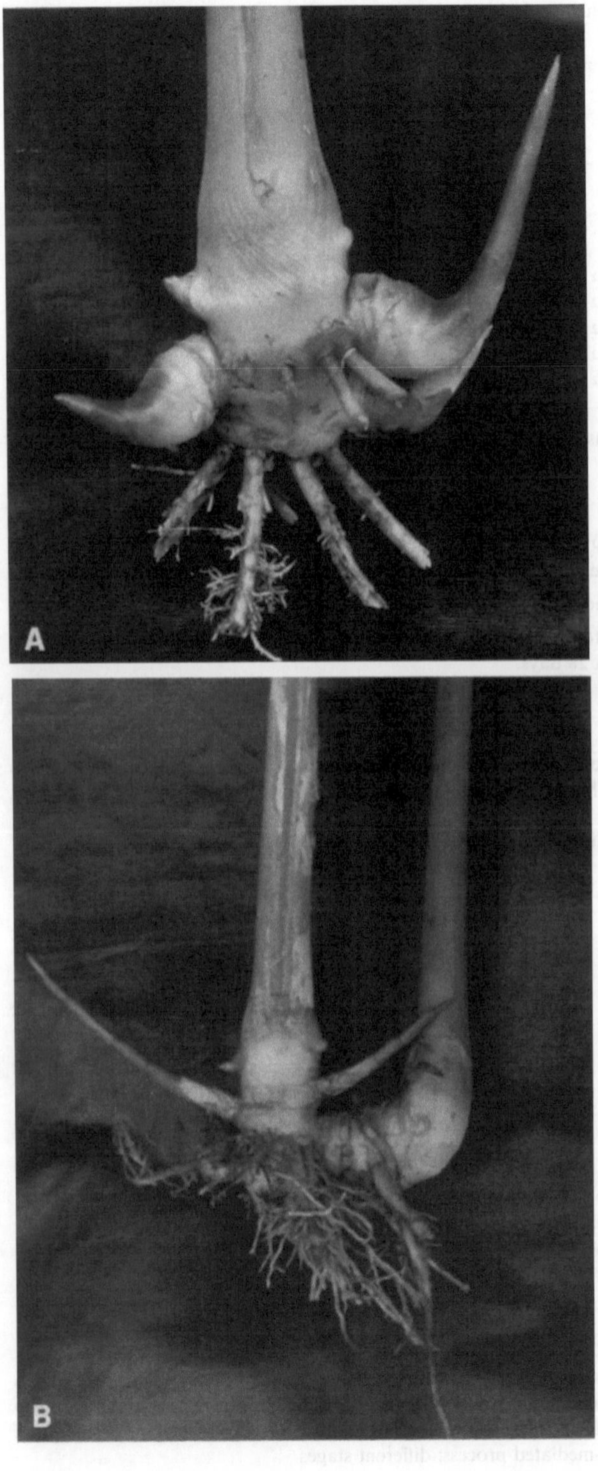

Fig. 4A,B. Meristems of cardamom rhizome. **A** Shoot primordium. **B** Inflorescence (panicle) primordium

Table 5. Pretreatment of shoot and inflorescence primordia in *E. cardamomum*

Method of treatment	Treatments
Chemical	1. Kinetin (0, 200, 600, 1000 mg/l) infusion under in vivo conditions for 6, 12, and 24 h
	2. BA (0, 200, 600, 1000 mg/l) infusion under in vivo conditions for 6, 12, and 24 h
Physical	1. Killing bud apex using sterile red-hot needle, just before cutting
	2. Scooping out bud apex using a sterile surgical blade just before culturing
	3. Scoring bud apex with 2-4 vertical incisions of 3–4 mm depth

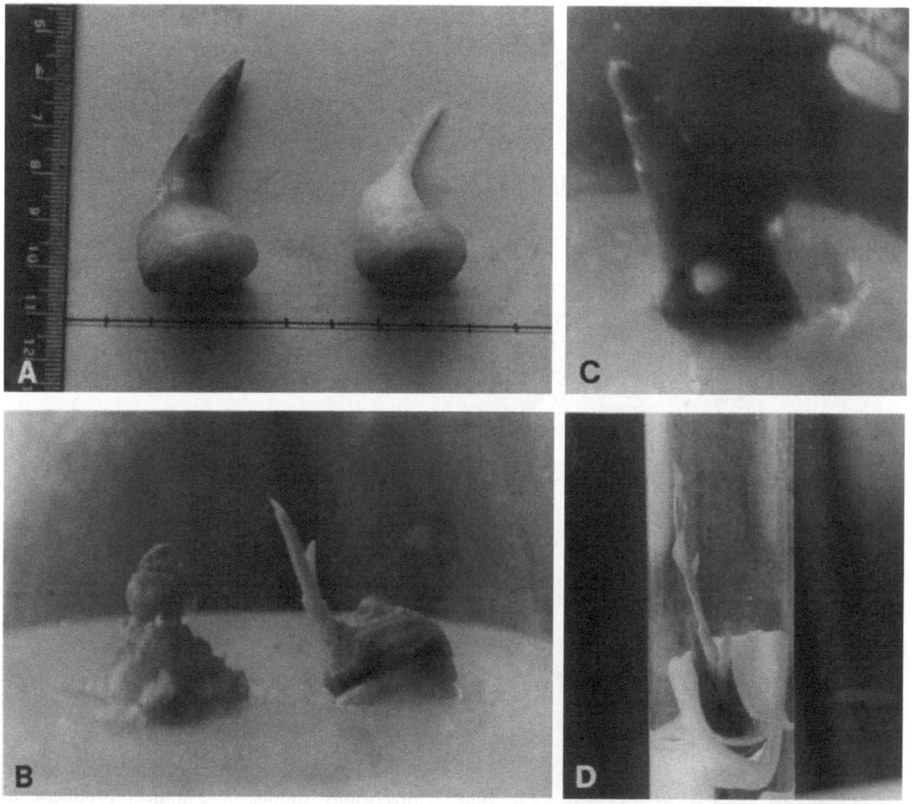

Fig. 5A–D. Axillary branching in shoot and inflorescence primordium. **A** A shoot primordium of optimum size before and after the removal of outer scale leaves. **B** pretreated (cytokinin infusion) shoot primordium showing emergence of axillary branches, 21 days after culture (medium: SH + 4 mg/l BAP + 0.5 mg/l NAA). C,D. Production of vegetative shoot from an inflorescence primordium (medium: SH + 4 mg/l BAP + 0.5 mg/l NAA). **C** Induction of vegetative bud (21 days after culture). **D** Emergence of vegetative shoot (49 days after culture)

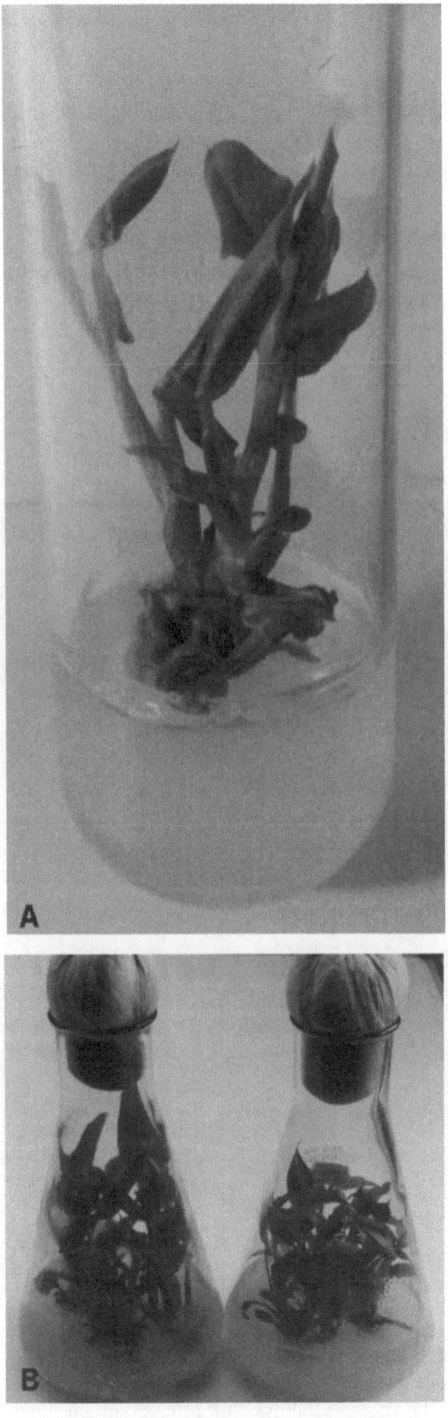

Fig. 6A,B. Induction of axillary branches in precocious shoots. **A** Plantlets produced in modified SH medium after 35 days of culture. **B** Cultures (63 days old) showing healthy plants growing in SH basal medium (*left*) and in half-strength MS mineral salt medium (*right*)

Table 6. Axillary branching in precocious shoots of *E. cardamomum* in two different culture media. (Reghunath 1989)

Observations[b]	MS medium[a]			SH medium[b]		
	Inorganic salt strength (%)					
	25	50	100	25	50	100
No. of Axillary branches/shoot	3.25	5.50	5.00	3.00	4.75	5.75
Fresh weight of a branch (mg)	1370	1430	1500	1310	1390	1460
Dry weight of a branch (mg)	72.60	100.10	109.50	65.50	82.00	131.40

[a] Medium contained 4 mg/l 6-BAP and 0.5 mg/l α-NAA.
[b] Mean of four replications: culture period 35 days.

Table 7. Axillary branching in precocious shoots of *E. cardamomum* in different physical conditions of culture medium. (Reghunath 1989)

Physical conditions	SH medium[a]
	Axillary branches/shoot explant[b]
Semi-solid–agar 0.6%	7.17
Liquid – with shaking for 35 days	15.17
Liquid – without shaking for 35 days	7.83
Liquid – with shaking for 14 days; then without shaking for 21 days	14.33

[a] Medium contained 3 mg/l 6-BAP; 0.5 mg/l kinetin; 1 mg/l α-NAA; 12% coconut milk.
[b] Mean of six replications.

at 23 ± 2 °C temperature under white fluorescent light of 3000 lx intensity and 16-h photoperiod.

2.2.2 The Cultures Raised

Effect of Growth Regulators. The number of axillary branches produced was maximum when the medium contained 4 mg/l BAP and 0.5 mg/l NAA. Use of 1.5 mg/l kinetin along with 3 mg/l BAP and 0.5 or 1 mg/l NAA also gave similar results (Table 8). With higher levels of BAP and/or kinetin than the above,

Table 8. Axillary branching in shoot primordia and precocious shoots of *E. cardamomum* in media containing BAP and various combinations of NAA and BAP. (Reghunath 1989)

Basal medium: MS

Treatment (mg/l)	Shoot primordium		Precocious shoot	
	Shoots[a]/ initial explant	Fairly elongated shoots[a]/ explant	Shoots[a]/ initial explant	Fairly elongated shoots[a]/ explant
BAP1	0.46 (1.21[b])	0.00 (1.00[b])	1.25	0.21 (1.10[b])
BAP2	0.72 (1.31)	0.00 (1.00)	1.75	0.21 (1.10)
BAP4	1.79 (1.67)	0.21 (1.10)	2.00	0.88 (1.37)
BAP10	1.43 (1.56)	0.00 (1.00)	3.50	0.99 (1.41)
NAA 0.1 + BAP1	0.46 (1.21)	0.21 (1.10)	2.25	0.66 (1.29)
NAA 0.1 + BAP2	1.43 (1.56)	0.72 (1.31)	3.25	1.82 (1.68)
NAA 0.1 + BAP4	3.49 (2.12)	1.72 (1.65)	3.50	1.96 (1.72)
NAA 0.1 + BAP10	2.65 (1.91)	0.93 (1.39)	3.50	1.37 (1.54)
NAA 0.5 + BAP1	1.69 (1.64)	0.99 (1.41)	2.50	1.46 (1.57)
NAA 0.5 + BAP2	2.96 (1.99)	0.88 (1.37)	3.75	1.40 (1.55)
NAA 0.5 + BAP4	4.52 (2.35)	1.99 (1.73)	5.00	2.24 (1.80)
NAA 0.5 + BAP10	3.45 (2.11)	0.93 (1.39)	4.25	1.92 (1.71)
NAA 1.0 + BAP1	1.46 (1.57)	0.99 (1.41)	3.00	1.13 (1.46)
NAA 1.0 + BAP2	3.45 (2.11)	1.22 (1.49)	3.25	1.37 (1.54)
NAA 1.0 + BAP4	3.97 (2.23)	1.96 (1.72)	4.75	2.20 (1.79)
NAA 1.0 + BAP10	3.49 (2.12)	1.46 (1.57)	3.50	1.76 (1.66)
CD (5%)	0.32	0.28	1.70	NS
SEM ±	0.11	0.10	0.60	0.19

[a] Mean of four replications; culture period 35 days.
[b] Data in parentheses indicate transformed values ($\sqrt{x+1}$).

although they produced more branches, a large proportion of them were highly compressed and small.

Effect of Coconut Milk. Addition of coconut milk to the culture medium significantly enhanced the production of axillary branches. The percentage improvement in axillary branch induction compared to that of the control was 55 when 12% coconut milk was added to the medium (Table 10). The influence of coconut milk (CM) was also superior to that of casein hydrolysate (CH) (Table 11).

Effect of Light. Axillary branching in shoot primordia and precocious shoots was favored when the light was maintained at 2000-3000 lx intensity. The percentage increase in the number of branches produced at 2000 and 3000 lx light intensities compared to 500 lx light intensity was 26.92 (Table 12).

Table 9. Axillary branching in shoot primordia and precocious shoots of *E. cardamomum* in media containing various combinations of IAA and BAP. (Reghunath 1989)

Basal medium: MS

Treatment (mg/l)	Shoot primordium		Precocious shoot	
	Shoots[a]/ initial explant	Fairly elongated shoots[a]/ explant	Shoots[a]/ initial explant	Fairly elongated shoots[a]/ explant
IAA 0.5 + BAP1	1.00	0.72 (1.31)	2.00	1.00
IAA 0.5 + BAP2	1.50	0.72 (1.31)	3.50	1.00
IAA 0.5 + BAP4	3.50	1.62 (1.62)	4.25	2.00
IAA 0.5 + BAP10	3.00	0.88 (1.37)	4.00	2.00
IAA 1.0 + BAP1	1.75	0.93 (1.39)	2.50	1.00
IAA 1.0 + BAP2	3.25	1.22 (1.49)	3.50	1.00
IAA 1.0 + BAP4	4.50	1.69 (1.64)	4.75	2.00
IAA 1.0 + BAP10	3.75	1.04 (1.43)	4.00	2.00
IAA 2.0 + BAP1	1.50	0.99 (1.41)	3.00	1.25
IAA 2.0 + BAP2	2.00	0.93 (1.39)	3.00	1.25
IAA 2.0 + BAP4	4.00	2.61 (1.90)	4.75	2.50
IAA 2.0 + BAP10	3.50	1.16 (1.47)	4.25	2.25
CD (5%)	0.93	NS	1.22	0.65
SEM ±	0.32	0.19	0.42	0.23

[a] Mean of four replications; culture period 35 days.

[b] Data in parentheses indicate transformed values ($\sqrt{x + 1}$).

Table 10. Axillary branching in precocious shoots of *E. cardamomum* as influenced by growth regulators and coconut milk. (Reghunath 1989)

Treatment	MS medium
	Axillary branches/shoot explant[a]
4 mg/l BAP + 0.5 mg/l NAA	5
1.5 mg/l K + 3 mg/l BAP + 0.5 mg/l NAA	5
12%CM + 4 mg/l BAP + 0.5 mg/l NAA	7.75

[a] Mean of four replications; culture period 35 days.

Performance of E. cardamomum Types. Of the three types compared, Mysore, Vazhukka and Malabar, the types Mysore and Vazhukka produced a greater number of axillary branches from shoot primordia as well as from precocious shoots when compared to the type Malabar. The growth of induced branches was also superior in the Mysore and Vazhukka types (Table 13).

Table 11. Axillary branching in shoot primordia and precocious shoots of *E. cardamomum* in media containing casein hydrolysate (CH) and coconut milk (CM) : (Reghunath 1989)

Basal medium: MS + BAP + 4 mg/l + NAA 0.5 mg/l

Treatment	Shoot primordium			Precocious shoot		
	Shoots[a]/ initial explant	Length (cm)[a] of		Shoots[a]/ initial explant	Length (cm)[a] of	
		Longest shoot	Longest leaf		Longest shoot	Longest leaf
CH 50 mg/l	5.00	3.10	2.15	5.00	4.35	4.15
CH 100 mg/l	5.25	3.25	2.05	5.25	4.00	4.10
CH 500 mg/l	4.75	2.80	1.60	6.25	3.80	3.75
CH 1000 mg/l	4.25	2.65	1.45	6.50	3.55	3.45
CH 2000 mg/l	4.25	2.60	1.35	6.50	3.55	2.65
CM 5%	6.25	3.80	2.45	6.00	6.00	4.00
CM 10%	6.50	4.00	2.65	7.50	6.45	4.80
CM 12%	6.50	4.40	3.10	7.75	6.40	5.10
CM 15%	6.75	4.40	2.80	7.50	6.00	5.00
CM 20%	6.00	4.15	2.50	7.00	5.65	4.50
Control (Basal medium)	4.25	3.80	2.00	5.50	5.85	4.50
CD (5%)	1.09	1.00	0.54	1.15	0.83	0.89
SEM ±	0.38	0.35	0.19	0.40	0.29	0.31

[a] Mean of four replications; culture period 35 days.

Table 12. Axillary branching in shoot primordia and precocious shoots of *E. cardamomum* cultured under different light intensities. (Reghunath 1989)

Basal medium: SH

Treatment	Shoot primordium[a]			Precocious shoot[b]		
	Shoots[c]/ initial explant	Length (cm)[c] of		Shoots[c]/ initial explant	Length (cm)[c] of	
		Longest shoot	Longest leaf		Longest shoot	Longest leaf
500 lx	5.20	4.46	3.60	6.00	6.48	4.56
1000 lx	6.00	4.28	3.50	7.00	6.00	4.00
2000 lx	6.60	4.20	3.00	7.80	6.00	4.86
3000 lx	6.60	4.00	3.20	7.00	5.20	5.00
No light	1.20	4.00	0	–	–	–
CD (5%)	1.02	NS	NS	NS	NS	NS
SEM ±	0.35	0.24	0.27	0.51	0.33	0.31

[a] Medium contained BA 4 mg/l + NAA 0.5 mg/l + CW 12% + biotin 1 mg/l.
[b] Medium contained BA 3 mg/l + kinetin 0.5 mg/l + NAA 1 mg/l + CM 12% + biotin. 1 mg/l
[c] Mean of five replications; culture period 35 days.
0 = No response.
– = Not tried.

Table 13. Comparative performance of shoot primordia and precocious shoots of major *E. cardamomum* types in axillary branching and growth. (Reghunath 1989)

Basal medium: SH + BA 4 mg/l + NAA 0.5 mg/l + CM 12% + biotin 1 mg/l

Types	Shoot primordium			Precocious shoot		
	Shoots[a]/ initial explant	Length (cm)[a] of		Shoots[a]/ initial explant	Length (cm)[a] of	
		Longest shoot	Longest leaf		Longest shoot	Longest leaf
Mysore	6.50	4.40	3.10	7.67	6.45	5.10
Vazhukka	7.00	5.10	3.55	8.17	7.15	6.00
Malabar	4.50	3.65	2.85	6.50	5.25	4.50
CD (5%)	1.05	0.93	NS	1.17	0.68	0.90
SEM ±	0.35	0.31	0.21	0.39	0.23	0.30

[a] Mean of six replications; culture period 35 days.

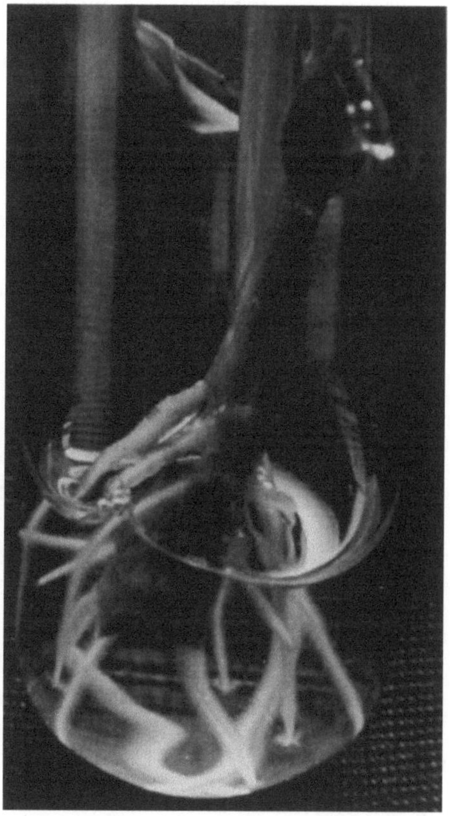

Fig. 7. Rooted plantlet: 28 days after transfer to half strength MS liquid medium

2.3 Transplantation to Soil

2.3.1 *In Vitro Rooting and Hardening*

Rooting of excised axillary shoots (30–40 mm) was achieved by culturing them initially in a semi-solid medium (agar 0.6%) containing half-strength MS salts ($MS\frac{1}{2}$) and 0.5% activated charcoal for 1 week. They were then subcultured onto a liquid $MS\frac{1}{2}$ medium containing 1.5 mg/l IBA for another week under high light intensity (3500 lx). Rooted shoots (Fig. 7) were then transferred to nonsterile $MS\frac{1}{2}$ liquid medium containing only mineral salts and were shifted to a greenhouse where partial sunlight was available. After 1–2 weeks, planting out was done in polythene bags filled with different types of potting mixture. Of the seven mixtures tested, vermiculite + fine sand mixture (1:1 v/v) was found to be the best, giving 91.7% survival of plantlets. The plantlets were kept in a mist chamber (90–100% RH) for 2–3 weeks. By this method the rooting was 100% (Fig. 8), and the survival rate was 83.3% (Table 14).

2.3.2 *Field Planting*

A comparative performance evaluation of micropropagated plants of major *E. cardamomum* types at different stages of growth after field planting indicated that the Mysore and Vazhukka type were significantly superior in field survival and growth, except in plant height than the Malabar type, at 24th week of planting

Fig. 8. Rooted plantlets after planting out in vermiculite-sand mixture under mist for 14 days

Table 14. Influence of potting mixtures, age of plantlets and hardening treatments on the survival of planted out *E. cardamomum* plantlets. (Reghunath 1989)

		Survival of plantlets (%)[a]
A. Potting mixture		
1. Perlite (Soilrite)		33.3
2. Perlite + sand (1:1 v/v mixture)		41.7
3. Green moss		25.0
4. Vermiculite		66.7
5. Vermiculite + sand (1:1 v/v mixture)		91.7
6. Sand + soil + dried and powdered cattle manure (1:1:1 v/v mixture)		41.7
7. Coir dust		8.3
B. Planting out and hardening[b]		
Age of the plantlets	Hardening	
1. Direct planting without rooting	3 weeks under partial sunlight	0
2. Direct planting after dipping the shoot ends in Seradix-B (No. 3) rooting hormone powder	3 weeks under partial sunlight	0
3. Planting after rooting in $\frac{1}{2}$ MS + IBA 1.5 ppm[c] liquid medium for 3 weeks	Whole rooting period under 3000 lx light	41.7
4. Planting after rooting in $\frac{1}{2}$ MS + IBA 1.5 ppm[c] liquid medium for 3 weeks	1 week under 1000 lx light and 2 weeks under partial sun light	58.3
5. Planting after rooting in $\frac{1}{2}$ MS + IBA 1.5 ppm[c] liquid medium for 3 weeks	1 week under 3000 lx light and 2 weeks under partial sun light	83.3
6. Planting after rooting in $\frac{1}{2}$ MS + IBA 1.5 ppm semi-solid medium for 3 weeks	1 week under 3000 lx light and 2 weeks under partial sun light	66.7

[a] Mean of 12 replications.
[b] Potting mixture: vermiculite + sand (1:1 v/v).
[c] IBA and sucrose were eliminated after 1 week.

out (Table 15). At the 12th week after field planting, a comparative performance study of micropropagated plantlets of the Vazhukka type with Vazhukka seedlings of similar size and vegetative characters indicated that, in spite of the initial set back, the growth and establishment of micropropagated plants were on par with or even better than the seedlings. Maximum plant height (97.4 cm), greater number of leaves (18.8), and higher rate of tillering (50% more) were recorded by micropropagated plants (Fig. 9, Table 16). Two years after planting, micropropagated plants produced inflorescences (Fig. 10) whereas seedlings took one more year to

Table 15. Comparative performance of micropropagated plantlets of major *E. cardamomum* types on the survival and growth under mist and under planted out conditions[a]. (Reghunath 1989)

Parameter	Varieties[b]			CD
	Mysore	Vazhukka	Malabar	(5%)
1. After 3 weeks under mist				
Survival (%)	91.7	100.0	83.3	–
Plant height (cm ± SE)	9.8 ± 0.4	10.6 ± 0.4	8.7 ± 0.4	1.0
Leaves/plant (No. ± SE)	3.0 ± 0.2	3.3 ± 0.2	2.8 ± 0.2	NS
Length of longest leaf (cm ± SE)	8.0 ± 0.3	8.4 ± 0.3	5.5 ± 0.3	0.7
2. 12 weeks after planting out				
Survival (%)	83.3	91.7	75.0	–
Plant height (cm ± SE)	19.0 ± 0.3	19.8 ± 0.3	17.2 ± 0.3	0.8
Leaves/plant (No. ± SE)	7.3 ± 0.4	8.0 ± 0.4	6.3 ± 0.4	1.0
Length of longest leaf (cm ± SE)	14.3 ± 0.5	16.6 ± 0.5	15.1 ± 0.5	1.4
3. 24 weeks after planting out				
Survival (%)	83.3	83.3	66.7	–
Plant height (cm ± SE)	61.2 ± 5.8	64.8 ± 5.8	56.9 ± 5.8	NS
Leaves/plant (No. ± SE)	13.3 ± 0.5	14.2 ± 0.5	11.3 ± 0.5	1.3
Length of longest leaf (cm ± SE)	30.5 ± 1.3	33.3 ± 1.3	27.2 ± 1.3	3.7

[a] Planting medium: ordinary potting mixture containing sand, soil, and dry-powdered cattle manure (1:1:1 v/v).
[b] Mean of 12 replications.

start flowering. A diagramatic representation of various routes adopted in the present study of in vitro propagation in cardamom is presented in Fig. 11.

3 Summary and Conclusions

For mass multiplication of virus-free clonal plants, in vitro shoot tip culture seems to be most suitable. The enhanced axillary branching technique described in this chapter may be effectively utilized for rapid clonal propagation of pathogen-free stock of *E. cardamomum* because of the high rate of shoot multiplication and excellent shoot growth. The in vitro method of shoot tip callus culture and subsequent shoot organogenesis offers great promise in the evolution of stress-tolerant varieties in *E. cardamomum*. Intensive research is required to evolve drought-tolerant varieties, since drought is a recurring problem in the cardamom-growing tracts of India.

Fig. 9. In vitro-obtained plantlets transferred from test tube to soil in polythene bags, 16 weeks after transfer from test tube

Table 16. Comparative performance of micropropagated plantlets of *E. cardamomum* and seedlings of type Vazhukka on the survival and growth under field conditions[a] (Reghunath 1989)

Parameter	Tissue culture plants[b]	Seedlings[b]
1. At field planting (i.e., 24 weeks after planting out)		
Establishment (%)	100.0	100.0
Plant height (cm ± SE)	64.8 ± 4.3	64.5 ± 4.3
Tillers/plant (No. ± SE)	3.6 ± 0.3	3.0 ± 0.3
Leaves/plant (No. ± SE)	13.7 ± 0.6	14.3 ± 0.6
2. 12 weeks after field planting		
Establishment (%)	80.0	100.0
Plant height (cm ± SE)	97.4 ± 5.4[c]	94.0 ± 4.8
Tillers/plant (No. ± SE)	5.9 ± 0.4[c]	4.1 ± 0.3
Leaves/plant (No. ± SE)	18.8 ± 0.6[c]	18.4 ± 0.5

[a] At Pampadumpara (Idukki Dist. India), conditions.
[b] Mean of ten replications.
[c] Mean of eight replications.

Fig. 10. Micropropagated plants transferred to field showing precocious flowering at 2nd year of planting

4 Protocol

1. Best Explant

   Shoot-bud primordium (20–50 mm long), given pretreatment with 200 mg/l kinetin or BA in vivo for 6–24 h followed by standard sterilization treatment in 95% ethanol and 0.15% mercuric chloride. The donor stock plants should be pathogen-indexed and should be in bearing stage in order to ensure high productivity.

2. Best Multiplication Medium

   SH basal medium + 0.5 to 1 mg/l NAA, 0.5 to 3 mg/l kinetin or 3 to 4 mg/l BA, 12 to 15% coconut milk.

3. Optimal Conditions

   21–25° C, 16-h photoperiod (2000–3000 lx). Culture the explants on liquid-shake culture (100 rpm) for 2–3 weeks and then transfer to agar-gelled medium.

4. Best Rooting Medium

   MS salts (half-strength) gelled with agar and 0.5–1% activated charcoal. After 1 week transfer to MS salts (half-strength) in liquid state containing 1–1.5% IBA.

5. Best Acclimation

   3–4 weeks at high humidity (90-100% RH) at 20 °C in vermiculite-sand (1 : 1 v/v) mixture.

Acknowledgment. The first author wishes to acknowledge the Indian Council of Agricultural Research for the Senior Research Fellowship awarded to him for conducting the present study. He also wishes to express his gratitude to Dr. P.K. Gopalakrishnan, Associate Dean (Retd.), College of Horticulture, Trichur (Kerala) for his guidance and for providing laboratory facilities.

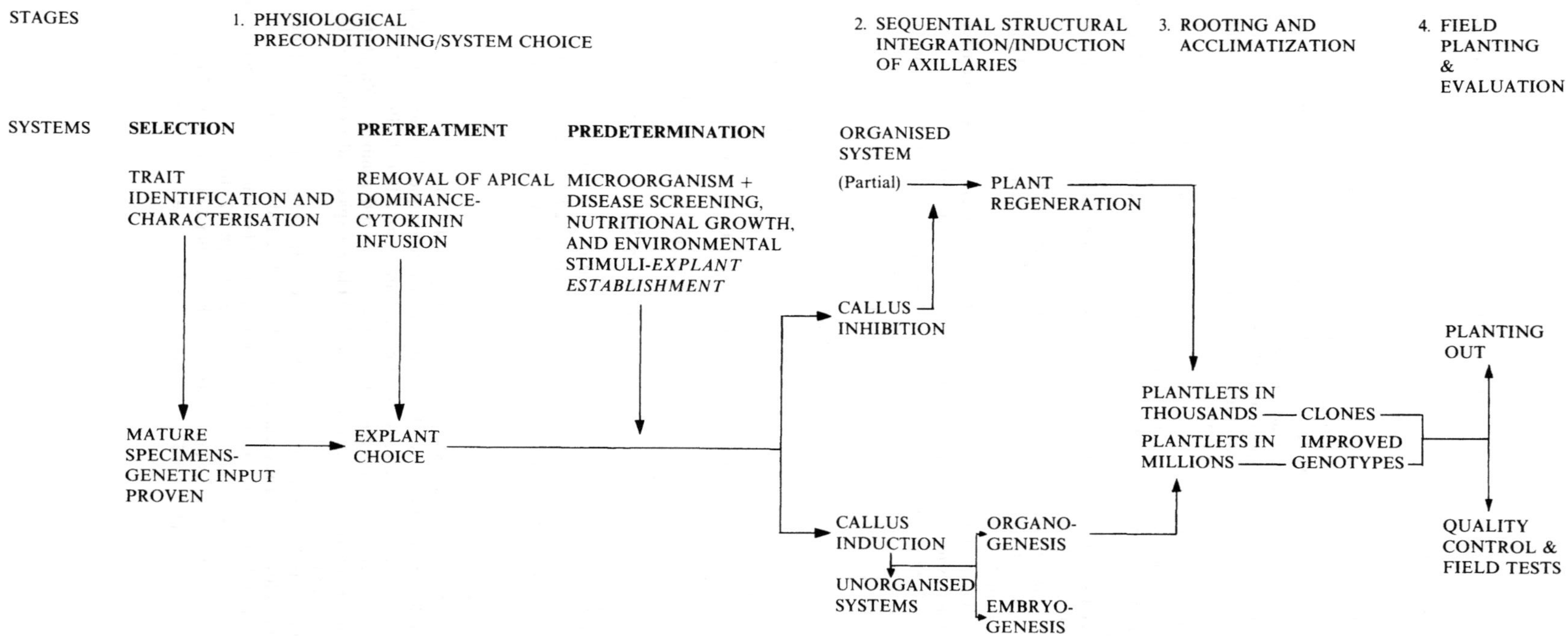

Fig. 11. Diagramatic representation of various routes of in vitro propagation in cardamom

References

Anon (1977) Reports of the Indian cardamom delegation to Guatemala. Cardamom 9: 3–10
Anon (1989) Plan to boost cardamom output. Spice India 2: 26
Bajaj YPS, Furmanowa M, Olszowska O (1988) Biotechnology of the micropropagation of medicinal and aromatic plants. In: Bajaj YPS (ed) Biotechnology in agriculture and forestry, vol 4: Medicinal and aromatic plants I. Springer, Berlin Heidelberg New York Tokyo, pp 60–103
Chakravorti AK (1948) Multiplication of chromosome numbers in relation to speciation in Zingiberaceae. Sci Cult 14: 137–140
Cherian A (1987) Recent trends on increasing productivity and reducing cost of production of cardamom (small) in India. Cardamom 20: 5–11
Darlington CD, Wylie CP (1955) Chromosome atlas of flowering plants, 2nd edn. Allen and Unwin, London
Gregory PJ (1936) The floral morphology and cytology of *Elettaria cardamomum* Maton. J Linn Soc London 50: 363–393
John K (1988) Choice of cardamom crop. Spices News Lett 22: 59
Kirtikar KR, Basu ED (1935) *Elettaria cardamomum*. In: Indian medicinal plants, vol 4 1975, reprint edn. Periodical excerpts. Delhi, pp: 2442–2444
Kologi SD, Pattanshetti HV, Prasad ABN (1973). Germination of cardamom seeds can be improved by treatment with nitric acid and sowing in plastic house. Curr Res 2: 3
Kumar KB, Prakashkumar P, Balachandran SM, Iyer RD (1985) Development of clonal plantlets from immature panicles of cardamom. J Plant Crops 13: 31–34
Murashige T, Skoog F (1962) A revised medium for rapid growth and bioassays with tobacco tissue cultures. Physiol Plant 15: 473–497
Nadgauda RS, Mascarenhas AF, Madhusoodanan KJ (1983) Clonal multiplication of cardamom (*Elettaria cardamomum* Maton) by tissue culture. J Plant Crops 11: 60–64
Nayak KP, Dutta NK (1961) Role of essential oils and allied drugs in experimental cholera of the rabbit. Indian J Med Res 49: 51–54
Parameswar NS, Haralappa HS, Maheshagowda HP (1979) Cardamom yields can be increased by clonal propagation. Curr Res 8: 150–151
Purseglove JW (1975) Tropical crops: Monocotyledons, vol 1, 2. The English Language Book Society and Longman, London, 529 pp.
Purseglove JW, Brown EG, Green CL, Robbins SRJ (1981) Spices, vol 2. Longman, London, pp 581–643.
Pruthi JS (1976) Spices and condiments. National Book Trust, New Delhi, India, pp 53–54
Ramachandran K (1969) Chromosome numbers in Zingiberaceae. Cytologia 34: 213–221
Rao DG, Naidu R (1973) Studies on "Katte" disease of small cardamom. J Plant Crops 1 (Suppl): 129–136
Reghunath BR (1989) In vitro studies on the propagation of cardamom (*Elettaria cardamomum* Maton). Ph.D. Thesis, Kerala Agric Univ, Trichur, India
Ridley HN (1912) Spices. Macmillan, London, pp 324–359
Schenk RU, Hildebrandt AC (1972) Medium and techniques for induction and growth of monocotyledonous and dicotyledonous plant cell cultures. Can J Bot 50: 199–204
Sharma AK, Bhattacharyya NK (1959) Cytology of several members of Zingiberaceae. Cell 59: 297–346
Sreenivasamurthy V, Krishnamurthy K (1959) Place of spices and aromatics in Indian dietary. Food Sci 8: 284–288
Srinivasa Rao NK, Narayanaswamy S, Chacko EK, Doreswamy R (1982). Regeneration of plantlets from callus of *Elettaria cardamomum* Maton. Proc Indian Acad Sci 91: 37–41
Varma PM, Capoor SP (1958) Mosaic disease of cardamom and its transmission by the banana aphid *Pentalonia nigronervosa* Coq. Indian J Agric Sci 28: 97–108
Warrier PK (1989) Spices and ayurveda – the combination that heals. In: Nayar KG (ed) Spice Fair 1989 Cochin, India: April 9–11, 1989 Commemorative volume. Spices Board, Cochin, India p 105

II.2 Micropropagation of Fennel *(Foeniculum vulgare* Miller*)*

G. Hunault[1] and J. Du Manoir[2]

1 Introduction

Fennel (*Foeniculum vulgare* Miller) has been used since the remotest antiquity for its aromatic (anise fragrance), medicinal, and nutritive properties (Hegi 1966; Badoc 1988). Originating in the Mediterranean basin, this plant is now more or less extensively cultivated in all the countries where climate is compatible with its growth, notably in Europe and Asia. As a species, *Foeniculum vulgare* actually comprises numerous types of fennel that differ in the aspect and height of their vegetative system, shape and size of their fruit, duration of their life span, and composition of their essential oil. These varied types have usually been classified into two subspecies: ssp. *piperitum* (Ucria) Coutinho and ssp. *capillaceum* (Gilib) Holmboe (Hegi 1966). The ssp. *piperitum* consists of uncultivated perennial plants with short, rigid lobed leaves and narrow umbels that produce small fruit devoid of anise fragrance. In contrast, ssp. *capillaceum* is characterized by linear long, supple lobed leaves, wider umbels, and more or less anise-tasting fruit. The various types now under cultivation belong to the latter subspecies and are grouped into three varieties. The variety *azoricum* (Miller) Thellung (bulb fennel, Italian fennel, Florence fennel) is an annual in which the hypertrophied sheaves of the basal leaves constitute a "bulb" similar to that of celery; it is also used as a vegetable (mainly in the north mediterranean countries). The variety *dulce* (Miller) Thellung (sweet fennel, French fennel) comprises generally undersized annuals or biannuals producing sweet anise-tasting fruit mainly used as a condiment (biscuits, bread, etc.). Their essential oil is also used for its therapeutical properties (aromatic stimulant, eupeptic, carminative). The variety *vulgare* (Miller) Thellung (bitter fennel) is composed of heterogeneous, often elevated perennials whose fruit leaves a bitter after-taste. Richer in essential oil than the variety *dulce*, this variety is cultivated, particularly in Europe, for the production of *trans*-anethole, an aromatic component utilized as a flavoring (anise fragrance) in the food industry (drinks, sweets, cakes, etc.). Many wild types are also included in this variety despite the fact that their umbels and their fruits are generally much smaller than those of the cultivated ones.

Most of the essential oil of the aerial vegetative system is to be found in the fruit (2 to 6% of their fresh weight). The main essential oil constituent generally is

[1] Université P. et M. Curie, Station de Biologie Végétale A. de Richelieu, Cherré, 72400 La Ferté-Bernard, France

[2] Centre de Recherches Pernod-Ricard, Laboratoire d' Agronomie, 120, Av. du Maréchal Foch, 94015 Créteil Cédex, France

Biotechnology in Agriculture and Forestry, Vol. 19
High-Tech and Micropropagation III (ed. by Y.P.S. Bajaj)

trans-anethole (50%–80%), which gives the plant its anise-like smell. The essential oil also contains several other aromatic components, notably estragole (2%–10%), and various terpenes, mainly fenchone (5%–20%), limonene (1%–15%), α-pinene (0.5%–8%), α-phellandrene (0.5%–5%), and camphene (0.1%–0.5%). The bitter taste of the var. *vulgare* fruits is due to their high amount of fenchone. The heterogeneity of *Foeniculum vulgare* is also to be found in its essential oil, whose composition notably varies according to accessions. The amount of estragole can therefore exceed that of anethole (Betts 1976; Hethelyi and Tetenyi 1979; Badoc 1988). Besides, some chemotypes produce no or hardly any estragole (Toth 1967; Ishibashi et al. 1967) or anethole (Betts 1968, 1976; Shah et al. 1969, 1970; Badoc 1988). According to Badoc (1988), the essential oil from ssp. *piperitum* shows neither anethole nor estragole, but it contains about 40% limonene. The most typical chemotypes in the variety *dulce* contain only a very small quantity of fenchone (Ishibashi et al. 1967; Osisiogu 1967; Toth 1967; Karlsen et al. 1969; Badoc 1988), this being generally compensated by a higher amount of anethole. The root-derived essential oil is used mainly for its therapeutical properties, as a diuretic. It differs totally from that of the aerial parts and contains mainly dillapiole (> 50%) and apiole (10%–15%) (Toth 1967; Trenkle 1972; Fujita et al. 1980).

The natural heterogeneity of this species has made it possible to select numerous cultivars fit for local use. However, though such cultivated types often differ markedly from wild plants, fennel has so far been subject to only limited improvement. The research now being carried out with *dulce* and *vulgare* varieties aims at raising fruit and anethole yields. Like many other Apiaceae, fennel is mainly allogamous under natural conditions of growth, which maintains a high level of heterozygosity among its populations. Such allogamy can eventually lead to a drop in cultivar performance after accidental crossing with wild plants. Maintaining the quality of these cultivars is only possible by controlling this heterozygosity. Likewise, the realization of cultivar-improving hybridizations requires the capacity to maintain every interesting genotype for a long time and to dispose them of in sufficient quantity when needed. Yet the absence of any kind of vegetative multiplication in the species makes both conservation and in vivo multiplication of genotypes difficult, hence the interest in micropropagation techniques. In vitro regeneration of fennel plants is possible either from multiple shoots or from somatic embryos.

2 Micropropagation Through Axillary Budding

Various attempts at micropropagation from in vitro-cultivated buds have been made to overcome the difficulties in cloning this species in vivo (Table 1). As early as 1976, Paupardin, working on the variety *dulce*, showed that it was possible to induce new shoot formation by growing seedling apical meristems on MS medium with addition of either IAA and BA (Paupardin 1976; Garcia-Rodriguez et al. 1978) or with NAA and BA (Paupardin et al. 1980). Neoformed buds can root and regenerate whole plants when they are cultivated with 1 mg/l IBA or without any growth regulator. The same protocol enabled Badoc (1982) to obtain nearly

Table 1. In vitro bud culture studies on fennel

Fennel type	Explant	Medium			Reference
		Salts	Growth regulators (mg/l)	Other additives	
S	Buds from seedlings	MS	IAA (0.01) + BA (1)	Glucose (3%), thiamine (?)	Garcia-Rodriguez et al. (1978)
S	Buds from seedlings	MS	NAA (0.1) + BA (0.1)	Glucose (3%), thiamine (1)	Paupardin et al. (1980)
B	Buds from seedlings	MS/2	NAA (0.1) + BA (0.1)	Glucose (2%), thiamine (1) Ca pantothenate (0.5), Biotin (0.1), meso-inosit (100) Riboflavine (0.1), nicotinic acid (1), folic acid (0.01)	Badoc (1982)
B, S	Apical buds from mature plants	SD	IAA (0.1) + BA (0.1)	Sucrose (2%), MS vitamins, Biotin (0.1), folic acid (0.1)	Du Manoir et al. (1985)

Abbreviations: S = Sweet, B = Bitter; MS = Murashige and Skoog (1962); SD = Shah and Dalal (1980); IAA = indole-acetic acid; NAA = naphthalene-acetic acid; BA = benzyladenine.

Fig. 1A–F. Axillary budding micropropagation. **A** Primary explant (apical bud from young shoot). **B** Multiple shoots obtained in the presence of 0.1 mg/l IAA and BA. **C** Plantlets derived from the rooting of microcuttings (3 weeks on 0.1 mg/l IBA medium). **D** Greenhouse stage. Plantlets have been transferred to small pots (Paperpot) filled with a sand, mold, and peat mixture (1 : 1 : 1). **E** Young plants about to be taken to the field (1 month after transfer from the test tube). **F** Open-ground plants, about 10 weeks after transplantation. (**A**, **C**, **F** Du Manoir, unpubl.; **B**, **D**, **E** Hunault, unpubl.)

identical results with seedling apices of the variety *vulgare*. According to Badoc, induction of multiple budding does not require the addition of an auxin. From a practical point of view, agronomical interest in fennel genotypes can only be estimated on mature plants. The use of seedling shoot apices as primary explants thus arouses limited interest. This led one of us (Du Manoir et al. 1985) to establish and standardize a protocol using apical buds from adult plants (Fig. 1). The macroelement solution of Shah and Dalal (1980), initially used for liquorice, gives better shoot multiplication than MS or MS 1/2. Though not indispensable, addition of folic acid and biotin also induces increased ramification. During the multiplication stage IAA should be preferred to IBA, for the latter tends to induce hyperhydric transformation of some buds. Dissection of shoot clumps makes it possible to isolate microcuttings, which root rapidly when transferred to the same medium without cytokinins. A few weeks later most of the plantlets thus obtained can be acclimatized to soil in the greenhouse and later planted out in the field.

Our team now makes regular use of this protocol to clone bitter fennel (var. *vulgare*) genotypes that exhibit interesting agronomical characteristics. The plants thus obtained are then used in the improvement program, notably to create clone hybrids. Some genotypes, however, proved relatively refractory to this protocol and produced only a few axillary shoots.

3 Micropropagation Through Somatic Embryogenesis

3.1 Review of Somatic Embryogenesis Studies on Fennel (Table 2)

Attempts at in vitro culture of fennel explants were first made by Sehgal (1964), who then observed somatic embryo formation. By cultivating immature fruit on a modified White's medium with added yeast extract or casein hydrolysate, this author could induce limited proliferation of the zygotic embryo, at the periphery of which later appeared many supernumerary embryos able to develop directly into plantlets (Sehgal 1964; Johri and Sehgal 1966). An interesting feature of these assays is that embryogenesis was triggered by nitrogenous additives only and not by growth regulators. A year later Maheshwari and Gupta (1965) obtained fennel tissue cultures by placing stem fragments on White's medium. These tissues produced embryos and plantlets after being kept on Nitsch medium enriched with coconut milk (15%). Their embryogenic aptitudes were preserved after being turned to suspension culture. Becker (1970) could also isolate tissue strains of this species (among them var. *azoricum*) so as to study their essential oil composition. He pointed out that the tissues were able to generate "shoots", but neither their exact nature nor their origin were studied. One of us (Hunault 1981) confirmed that fennel stem tissues are apt to produce somatic embryos. Similarly petiole, hypocotyl, and cotyledon pieces can generate embryogenic colonies. A multistage protocol was perfected (Hunault 1984), making it possible to take plants from embryogenic cell suspensions to the field.

The isolation of protoplasts from embryogenic suspensions was realized by Miura and Tabata (1986). In the presence of 2,4-D and kinetin, a small percentage

Table 2. In vitro tissue culture studies on fennel

Fennel type	Explant	Medium			Morphogenesis	Reference
		Salts	Growth regulator (mg/l)	Other additives		
?	Young fruits	WM	None	CH (0.1%) or YE (0.1%)	Polyembryony	Sehgal (1964)
?	Stem pieces	N	2,4-D (5)	Sucrose (2%)	Roots, embryoids,	Maheshwari and Gupta (1965)
?	Stem pieces	N	None	CM (15%) + sucrose (2%)	Plantlets	Maheshwari and Gupta (1965)
?A	Stem pieces	MS	NAA (2) + K (0.5)	Sucrose (3%)	Buds (?)	Becker (1970)
S	Stem pieces	H	NAA (0.5) + K (0.5)	Glucose (3%), thiamine (1)	Roots	Paupardin et al. (1980)
B, S	Stem, hypocotyl, petiole, cotyledon pieces	MS	IBA (1) or NAA (1)	Glucose (3%), LS vitamins	Roots	Hunault (1981, 1984)
B, S	Stem, hypocotyl, petiole, cotyledon pieces	MS	2,4-D (1) + K (0.3)	Glucose (3%), LS vitamins	Embryoids	Hunault (1981, 1984)
B	Petiole pieces	SD	2,4-D (0.1) + BA (0.3)	Sucrose (3%), MS vitamins, Folic acid (0.1), Biotin (0.1)	Embryoids	Simard (1985)
?	Protoplasts from embryogenic suspensions	MS	None	Glucose (9%), LS vitamins	Embryoids	Miura and Tabata (1986)
?	Petiole pieces	MS	2,4-D (0.2) + K (0.2)	Sucrose (3%), LS vitamins	Embryoids	Miura et al. (1987a, b)

Abbreviations: A = Bulb fennel; WM = White modified (Ranga Swamy 1961); N = Nitsch (1951); H = Heller (1953); 2,4-D = 2,4-dichlorophenoxyacetic acid; IBA = indolebutyric acid; K = kinetin; CH = casein hydrolysate; YE = yeast extract; CM = coconut milk; LS = Linsmaier and Skoog (1965). For other abbreviations see Table 1.

of these protoplasts (0.08%) proliferated to produce new embryogenic colonies. In contrast, in the absence of any growth regulator, every reactive protoplast (0.16%), after dividing in an assymetrical way, directly generated an embryo.

3.2 Development of Embryogenic Tissues

When placed on an embryogenesis-inducing medium, fennel explant tissues first proliferate to give rise to a hardy structured callus essentially composed of meristematic parenchyma (primary callus). Primary calli differ according to the nature of the explants or the regulator used for induction. They are often formed of irregular, soft, light-colored, hyperhydric-looking nodules. When such a callus is maintained on the inducing medium, it produces on its surface some granulous, compact, white, yellowish or grayish aggregates, whose morphology differs completely from that of the primary callus. Their frequently rapid proliferation leads to the appearance of friable colonies (secondary calli). When transferred to a medium devoid of auxin (development medium), these tissues give embryos that evolve directly into plantlets (Hunault 1981, 1984) (Fig. 2). Under some culture conditions (for example, in the case of induction on a medium with low auxin content or if previously induced explants are prematurely transferred onto a medium devoid of growth regulators) primary calli can directly produce embryos and plantlets without it being possible to detect any secondary callus stage. According to our observations, embryogenic secondary calli actually result from a change, under the influence of auxinic regulators, of the ontogenesis of the first embryonic structures produced by the primary callus. These structures, instead of evolving into embryos, proliferate in a disorganized way to give the granules that constitute the secondary callus (Hunault 1981 and unpubl. observ.). The way in which this callus develops is quite in keeping with Kohlenbach's conceptions (1977, 1985) about the origins of embryogenic tissues. The first phase of the embryogenic process (primary embryogenesis) corresponds to the dedifferentiation of the primary callus parenchymatous elements and leads to the appearance of embryo mother cells (zygote-like cells). Since the inducing medium generally disturbs the evolution of these cells into embryos, instead, they generate tissue clumps (secondary calli) whose elements still preserve the initial embryogenic potentialities and may evolve into embryos as soon as the level of auxin is reduced (adventive secondary embryogenesis). As emphasized by Kohlenbach (1985), the key phenomena of somatic embryogenesis do not occur during secondary embryogenesis but during primary embryogenesis, during the reprograming of the explant parenchymatous cells, via the primary callus, towards the "zygote-like" state.

3.3 Parameters Affecting Somatic Embryogenesis in Fennel

Whichever type of explant is used, the appearance of embryogenic colonies never occurs on all of them. Most often it only affects less than one third of the explants cultured. Moreover, for any given explant, the appearance of the secondary callus or of the embryos generally occur in a very restricted and localized area of the primary callus, which shows that only a few cells of the latter take part in the

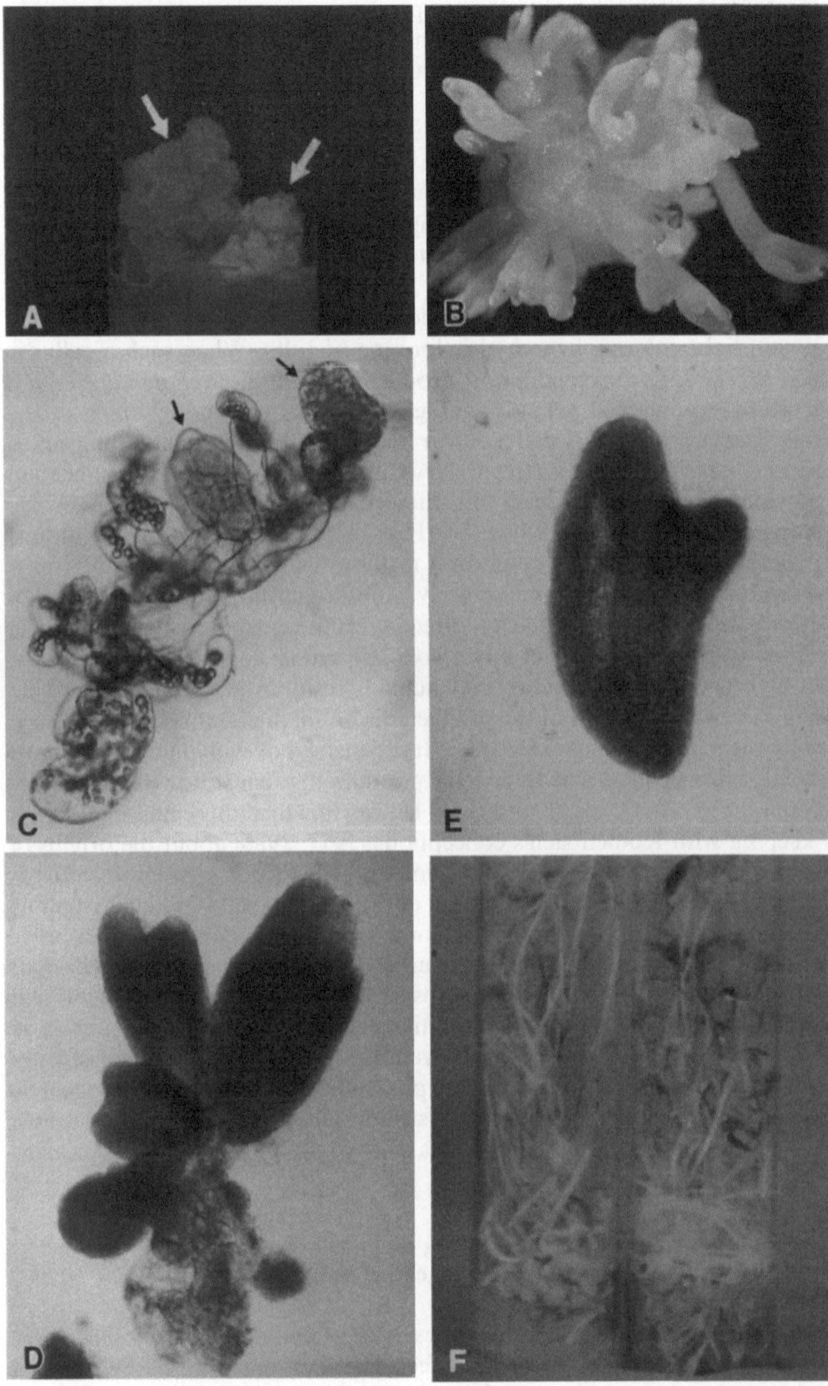
A
B
C
D
E
F

embryogenic process. Several assays aiming at increasing the efficiency of induction have been conducted by our team (unpubl.). The first explants that we used were hypocotyl pieces from aseptically cultivated seedlings. These pieces were permanently kept on the inducing medium and embryogenesis was assessed by counting the fragments that produced secondary calli. Because this hypocotyl system had various drawbacks (rather low reactivity, very slow productivity of secondary calli, heterogeneity of the mother seedlings due to the high heterozygosity of fennel) we later preferred to use petiole fragments originating from vitroplants obtained by micropropagation (see Sect. 2). These fragments are usually cultivated on an inducing medium for 4 or 5 weeks, and then transferred in such conditions as to allow the evolution of embryos and plantlets (development medium). The effect of treatments is then measured by counting the embryogenic explants 2 and 3 months after transfer.

As with most other embryogenic systems, the medium on which hypocotyl or petiole fragments are cultivated must contain an auxin. Figure 3 shows the effect of

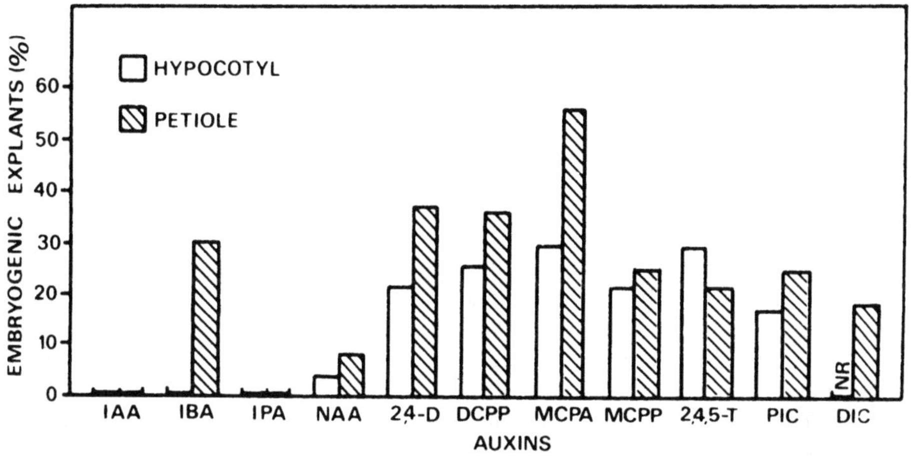

Fig. 3. The influence of different auxins on the embryogenic reaction of hypocotyl and petiole fragments. Hypocotyl fragments were continuously maintained (24 weeks) on the inducing medium (1 mg/l auxin + 0.5 mg/l kinetin, in light) while petiole fragments were submitted for 5 weeks to the inducing treatment (1 mg/l auxin + 0.1 mg/l kinetin, in darkness) and then transferred to development medium (13 weeks). *IAA* Indoleacetic acid; *IBA* indolebutyric acid; *IPA* indolepropionic acid; *NAA* naphthaleneacetic acid; 2,4-D: 2,4-dichlorophenoxyacetic acid; *DCPP* 2-(2,4-dichlorophenoxy)-propionic acid; *MCPA* 4-chloro-2-methylphenoxyacetic acid; *MCPP* 2-(2-methyl-4-chlorophenoxy)-propionic acid; *2,4,5-T* 2,4,5-trichlorophenoxyacetic acid; *Pic* 4-amino-3,5,6-trichloropicolinic acid (Picloram); *Dic* 2-methoxy-3,6-dichlorobenzoic acid (Dicamba). (Hunault, unpubl.)

Fig. 2A–F. Plantlet production by somatic embryogenesis. **A** Tissues obtained by growing stem fragments in the presence of 2,4-D (1 mg/l) and kinetin (0.3 mg/l) (1st subculture). Primary callus (*left*) gave rise to embryogenic secondary callus (*right*). **B** Secondary callus granule with somatic embryos on its surface. **C** Primary stage of cell suspension-derived embryos. These are linked at their base by long cells not unlike those suspensor cells (*arrows*). **D** Aggregate of somatic embryos (more advanced stage). **E** Embryo detail. **F** Embryo-derived plantlets. (**A**, **B**, **E** Hunault 1981: **C**, **D** Hunault 1984; **F** Hunault, unpubl.)

various auxins on both types of explants. Weak auxins (IAA, IBA, IPA, NAA) are usually less efficient, even when their concentration is higher than 1 mg/l. IBA can, however, trigger noticeable embryogenesis on petiole fragments. Among the strong auxins we have examined, the best results were obtained with MCPA, which also induced strong proliferation of the primary callus tissues. Plantlets regenerated with MCPA are robust but they often display root malformations. With Picloram and Dicamba (two compounds used to trigger embryogenesis in other systems) the effect is generally no better than with 2,4-D.

Cytokinins have a less clearcut influence. In contrast to what happens with auxins, it is possible to obtain the formation of secondary callus or embryos without any exogenous cytokinin, even if the growth of the primary callus is restricted under such conditions. There actually is some interaction between the effects of cytokinins and light. Explant proliferation, particularly that of hypocotyl fragments, is very limited in continuous light if the medium contains only an auxin. Under such conditions, adding a cytokinin greatly enhances the development of the primary callus (see Hunault et al. 1989, Fig. 2B). It also triggers intensification of embryogenesis, though in a more limited way (Fig. 4). In fact, embryogenesis appears to be reduced when cytokinin concentrations are optimal for callus growth. On the other hand, noticeable proliferation occurs in darkness without cytokinin. Although embryogenesis seemed to be stimulated in some experiments (see Fig. 7, for example), we could not repeatedly demonstrate any positive effect of cytokinins on the embryogenic induction.

Maintaining explants for several months in the presence of regulators is by no means imperative except when a strain of embryogenic tissue has to be isolated. Figure 5 shows that the strongest embryogenic reaction is actually obtained with fragments grown on the inducing medium for 2 to 4 weeks. When this treatment is prolonged further, the ability of the explant to produce embryos notably decreases, reaching a minimum within 8 to 10 weeks. It then slightly increases again as if a new wave of induction were arising. The origin of these variations remains unexplained.

The influence of the basal medium composition has not been examined in detail. Most of the tests were carried out on the mineral solution of Murashige and Skoog (1962) (MS). Nevertheless, we were able to obtain results almost equivalent

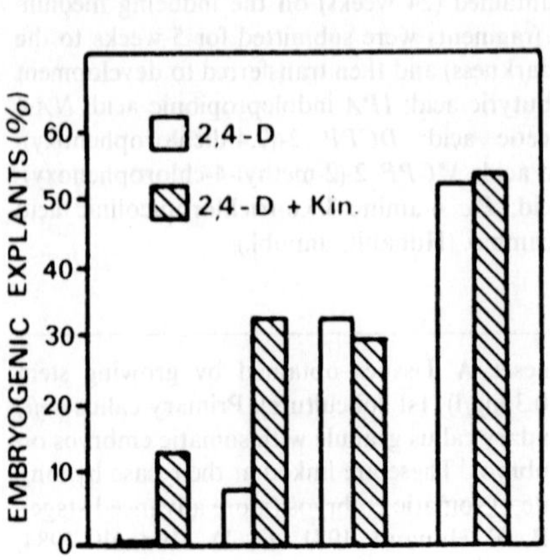

Fig. 4. Variations in embryogenic reaction according to light and kinetin. *a* Hypocotyl fragments derived from light-grown plantlets, also exposed to light during induction, *b* hypocotyl fragments excised from dark-grown plantlets, cultivated in light. *c* hypocotyl fragments excised from dark-grown plantlets and cultivated in darkness; *d* vitroplant petiole fragments cultured in darkness during the induction stage. Adding a cytokinin is mainly necessary when induction takes place in the light. (Hunault, unpubl.)

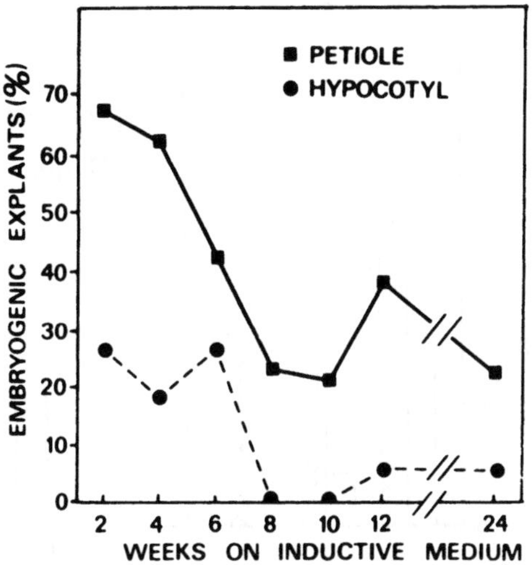

Fig. 5. Influence of induction duration on embryogenic reaction. Explants have been submitted to the inducing treatment (0.5 mg/l 2,4-D + 0.1 mg/l kinetin, in darkness) for various lengths of time, and then transferred to development medium. Evaluation of embryogenic reaction was made 24 weeks after induction. The embryogenic reaction of the explants decreases when inducing treatment is prolonged further than 4 to 6 weeks. (Hunault, unpubl.)

to the latter with the macroelement solutions of Shah and Dalal (1980) and Chu et al. (1975), whose concentrations are notably inferior, particularly in $NH4^+$, to that of MS. B5 solution (Gamborg and Eveleigh 1968) seems to be a little less favorable. That of Heller (1953) permits only very limited embryogenesis. Adding amino acid mixtures (mixture of Nickell and Maretzki 1969; yeast extract; Biosoyase) to the MS solution does not trigger stimulation of the explant reactivity. Likewise incorporating various amino acids (glutamine, asparagine, arginine, proline, tryptophan, phenylalanine, tyrosine) was utterly inefficient, probably because the MS medium already contains a considerable quantity of reduced nitrogen.

Among glucosides, glucose, and sucrose constitute excellent carbohydrate sources. For petiole fragments, the optimal dose of glucose is around 2–3%, and from 1 to 3% for sucrose (Fig. 6). Fructose and galactose also enable good embryogenesis.

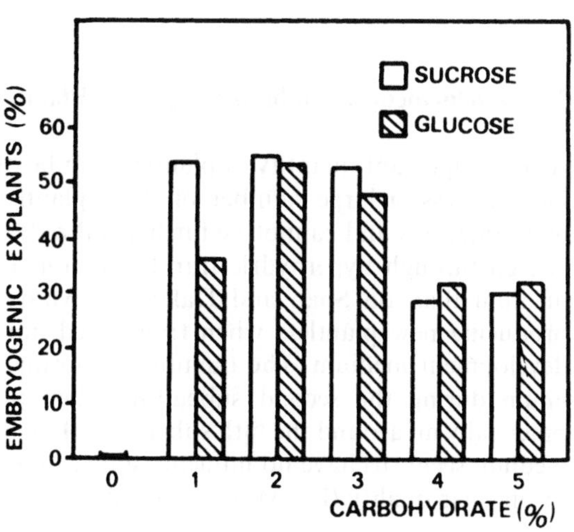

Fig. 6. Influence of glucose and sucrose on petiole fragment embryogenic reaction. (Hunault, unpubl.)

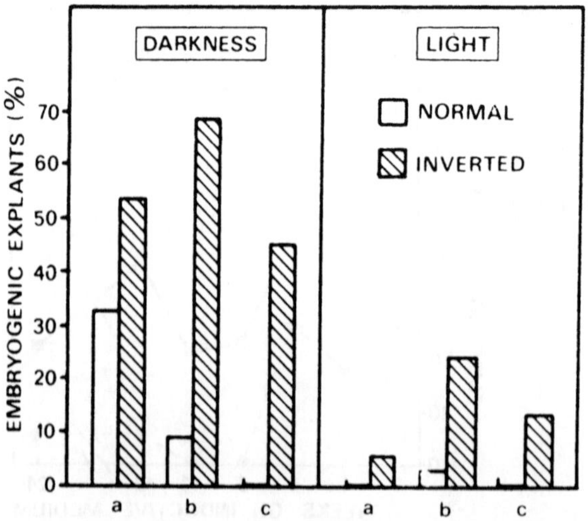

Fig. 7. Variations in somatic embryogenesis according to explant lay-out and light. Petiole fragments were sown in normal position (proximal end inside medium) or in inverted position (distal end inside medium), either in continuous light or in darkness. *a* Induction in the presence of 2,4-D only (1 mg/l), development without growth regulators; *b* induction on 2,4-D + kin (0.3 mg/l), development without growth regulators; *c* induction on 2,4-D + kinetin, development on kinetin. Inverted position and darkness favor embryogenic induction. (Hunault, unpubl.)

The intensity of embryogenesis also depends on exposure to light and on how the explants are set on the culture medium. Continuous light, when applied during the induction phase, has a clearly inhibiting effect both on primary callus formation and on embryogenic reaction (Figs. 4, 7). On the other hand, it does not seem to have any negative effect when applied during the development phase. Polarity also definitely influences the behavior of petiole pieces, which generally react more intensively when laid in an inverted position, that is when their distal extremity is set inside the medium. Under such conditions, embryos most often appear next to the proximal extremity, away from the culture medium.

Taking these different observations into account makes it possible to obtain a noticeable increase of the embryogenic reaction. Yet this phenomenon affects only a very limited fraction of explant cells, and there is no doubt that the physiological state of the explants just before induction plays an essential part in their reaction. The embryogenic ability also differs largely according to the clone used for the experiment. A genotypic dependence of embryogenesis is thus very likely.

3.4 Development of Embryo-Originated Plantlets

A very important quantity of plantlets can be obtained from embryogenic tissues. Nevertheless, a large number of these plantlets display more or less extensive malformations and cannot be further utilized (Hunault 1981, 1984). The plantlets altered through hyperhydric transformation seem to be more numerous on MS medium than on Shah and Dalal (Du Manoir, unpubl.). The callus keeps on producing new plantlets when transferred during successive subcultures on the development medium. The quantity of normal-looking plantlets reaches a maximum during the second subculture, then progressively decreases to become practically nil around the fifth subculture (Fig. 8). Adding activated charcoal to the medium seems to have no influence either on the number of regenerated plantlets or on their quality (Du Manoir, unpubl.).

Fig. 8. Evolution of normal plantlet production during successive subcultures of petiole fragment-derived callus (8-week subcultures) on development medium. Best plantlet production occurred during second subculture. (Du Manoir, unpubl.)

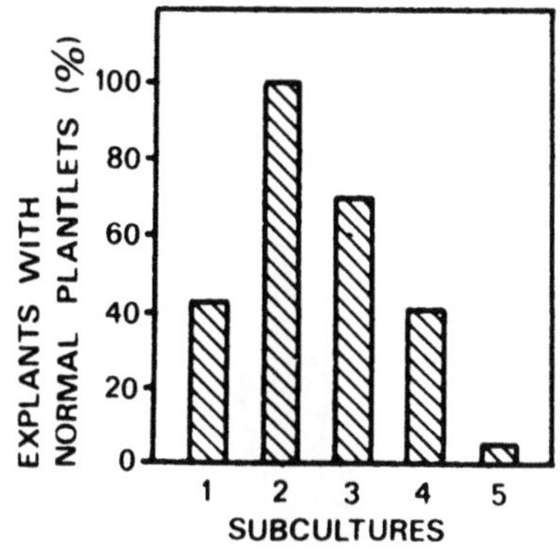

3.5 Somaclonal Variation During Somatic Embryogenesis

To evaluate the conformity and homogeneity of embryo-originated plants, Miura et al. (1987a) conducted a field trial and compared the morphological and chemical characteristics of a population of 41 plants originating from somatic embryos to an equivalent number of plants obtained from seeds. The population raised from somatic embryos proved to be more homogenous than that coming from seeds. Given the important heterozygosity of fennel, and the high variability generally existing among cultivars, these findings are not at all surprising. This does not necessarily mean that no variation occurs during somatic embryogenesis in fennel.

For the same genotype Fig. 9 shows both the distribution of a population of about 150 bitter fennel plants obtained via axillary budding and that of an equivalent number of plants derived from somatic embryos. The statistical analysis of these two populations showed that they were significantly different (0.05% level) in fruit weight and anethole content (Du Manoir, unpubl.). The embryo-derived populations are therefore less homogenous than those produced by axillary buds.

Recently we examined the conformity of embryo-derived plants in a randomized field trial with three replications (Hunault and Desmarest 1989). The plants used for this trial were all obtained through somatic embryogenesis from the petiole fragments of a single bud clone kept in vitro since 1984. Each of these plants was first multiplied via axillary budding so that each initial embryo could be represented in the field not by a single individual only but by a clonal population able to be assessed statistically. As in the work of Miura et al. (1987a), these plants were subjected to morphological observations, biometrical measurements, and analysis of their essential oil composition. Twenty-eight somatic embryo-derived clones were thus compared to the original clone, chosen as a control. For each studied character of the vegetative system, several clones noticeably differed from the control (Fig. 10). Of the 28 clones, four proved to be morphologically very different from the original. Notable variations were also observed in essential oil composition (Fig. 11). About two thirds of the clones fruited less than the control, some of them even producing an insignificant quantity of fruit (Fig. 12). On the other hand, several yielded better than the control. All these observations show that, in fennel, variations from the mother plant can be generated during the formation of somatic embryos. Even if the extent of these variations cannot

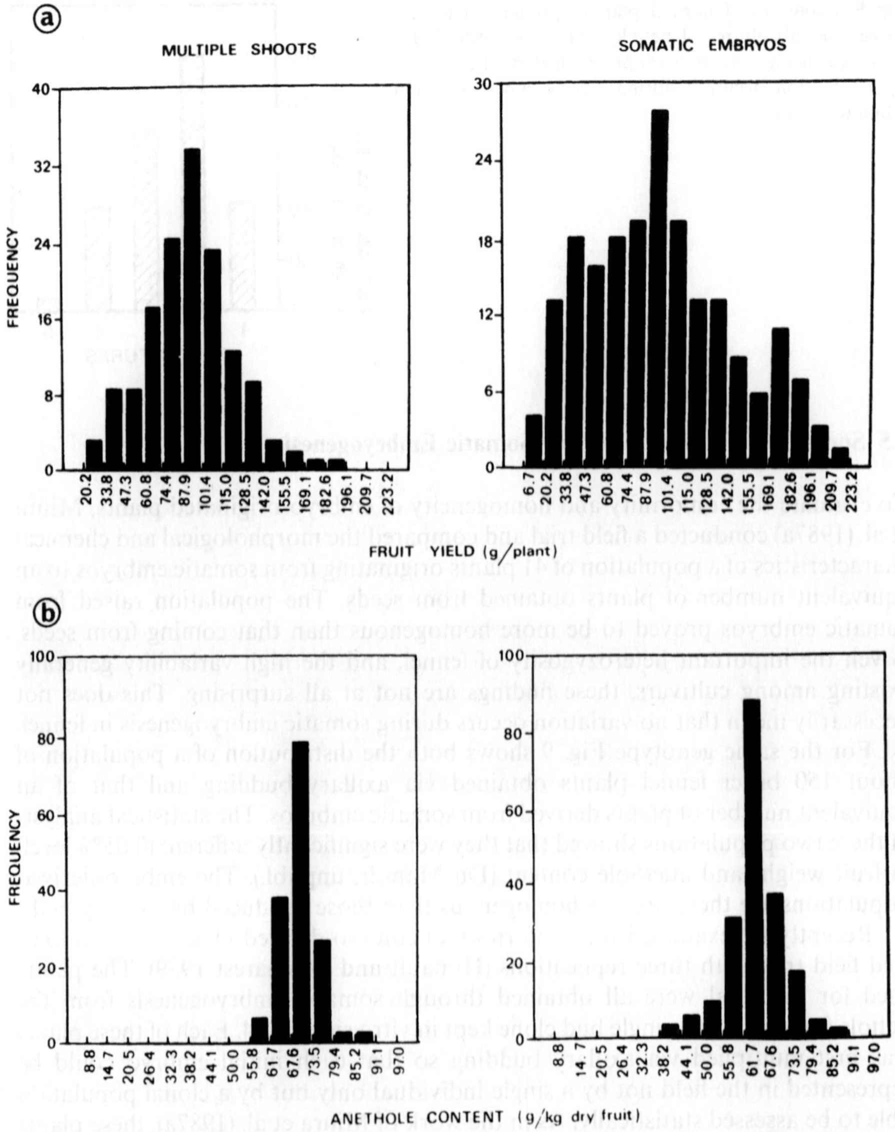

Fig. 9a,b. Distribution patterns in terms of: **a** produced fruit weight (g/plant) and **b** anethole yield (g/kg dry fruit) in a population of multiple shoot-derived plants (148 individuals) and in a somatic embryo-derived population (133 individuals). Values obviously differ more widely within the embryo-derived population. (Du Manoir, unpubl.)

be compared with that observed among wild or even cultivated populations, somatic embryogenesis cannot be considered as a true-to-type method of micropropagation, especially as fruiting can be greatly disturbed. Despite the fact that this variation often leads to regression in agronomically useful characteristics, a

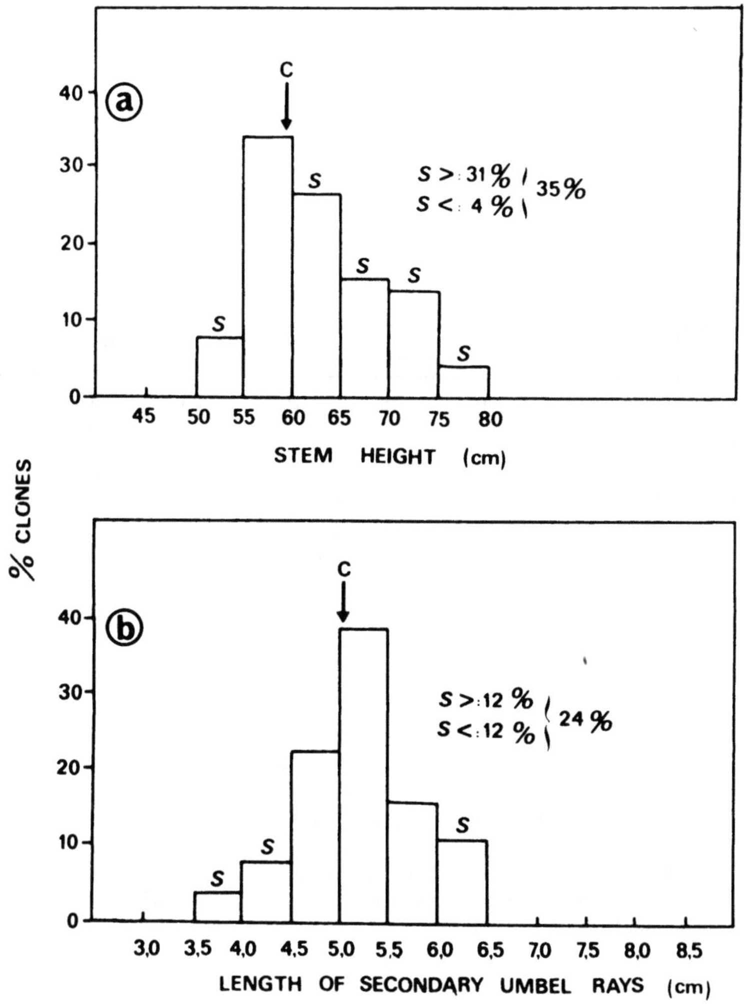

Fig. 10a,b. Distribution patterns in terms of: **a** main stem average height (measured at primordial umbel level) and **b** secondary umbel average diameter (outer ray length) in a population of 28 clones (each clone obtained through axillary budding from the apex of a somatic embryo-derived plantlet.) *C* control position; *S* indicates the presence of clones significantly different from control (0.05% level) in the corresponding column. *S* >, *S* < percentage of clones significantly superior or inferior to control. (Hunault and Desmarest 1990)

low percentage of the regenerated plants still present interesting features (higher fruit or anethole yield, earlier flowering), which could be used in an improvement program. Nevertheless, as our observations have so far dealt only with plants of the first generation, we do not know to what extent these interesting characteristics are maintained in the progeny.

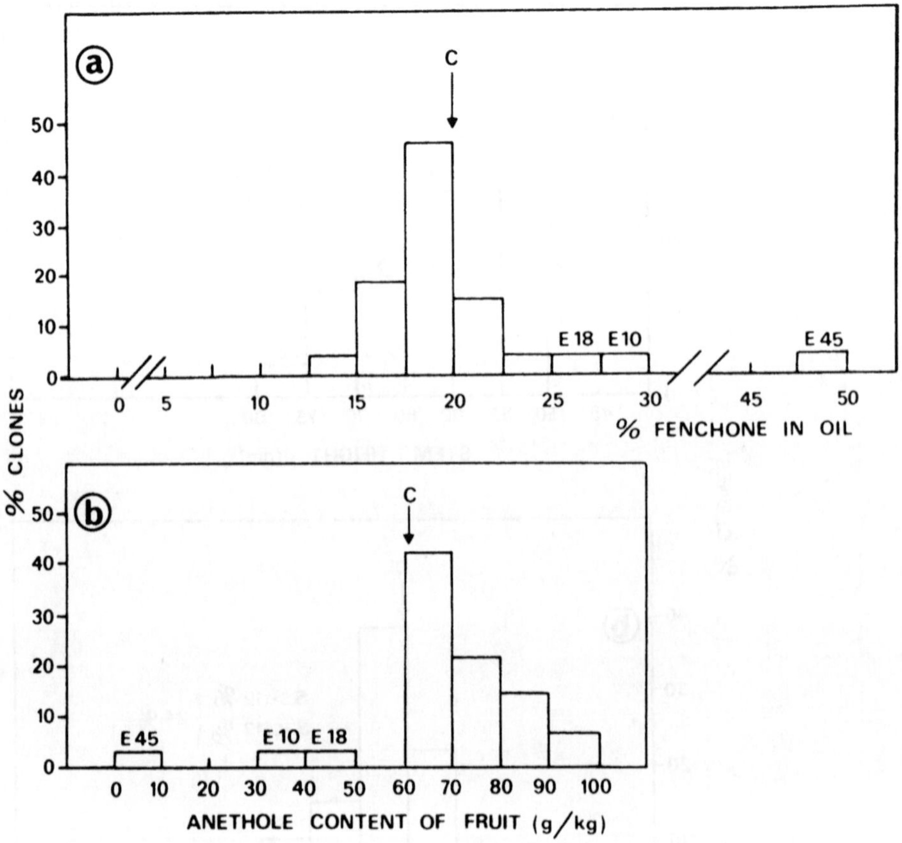

Fig. 11a,b. Distribution patterns in terms of: **a** fenchone amount in essential oil and **b** anethole content in fruit (see Fig. 10). *E10, E18, E45* refer to clones with a high fenchone and a low anethole content. (Hunault and Desmarest 1990)

4 Summary and Conclusions

In vitro regeneration of whole fennel plants was obtained either via axillary budding (multiple shoots) or by somatic embryogenesis. Somatic embryo-derived plants may differ to a greater or lesser degree from the original plant, both on the morphological level and in the composition of their essential oil. Even if the differences observed seem limited in comparison with the natural variability of fennel populations, somatic embryogenesis cannot be considered as a completely reliable micropropagation technique. This variation commonly results in reduced agronomical interest of the regenerated plants, especially owing to a considerable drop in their fruiting capacity. Yet some of these plants (about 10% of the population in our assays) can display interesting particularities within an improvement program. The stability of these variations in progeny and their real interest in this naturally very heterogenous species remain to be assessed. Up to now the

Fig. 12a,b. Frequency diagrams showing: **a** number of fruit per secondary umbel and **b** fruit weight per plant (see Fig. 10). (Hunault and Desmarest 1990)

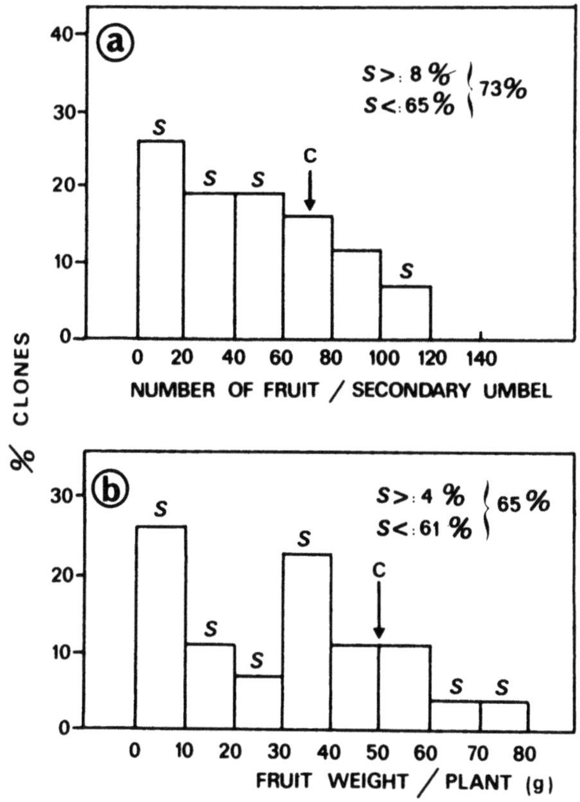

multiple shoot technique is the only one that makes it possible to obtain plants strictly identical to the mother plant. During the last 5 years, this technique has been widely used by our team both to create clones from adult fennel plants selected for their agronomical interest (better seed yield, high amount of anethole, male sterility, etc.) and to multiply plants originating from somatic embryos. During field trials carried out within the improvement program of bitter fennel, notably during the creation of clone hybrids or of synthetic varieties of clones, each genotype could thus be represented by a perfectly homogenous population, and not by a single individual, which obviously is very advantageous.

5 Protocol

1. Micropropagation Through Multiple Shoots

a) Sterilize young shoots for a few seconds in 95% alcohol, and then 15 min in a filtered solution of 70 g/l calcium hypochlorite. Rinse three times in sterile distilled water.
b) Grow apical buds on following medium: Shah and Dalal (1980) macroelements, MS (1962) microelements, MS vitamins, without glycine, but with 0.1 mg/l biotin and folic acid, 2% sucrose, and 0.7% agar.
c) After 4 weeks, transfer to same medium added with 0.1 mg/l IAA and 0.1 mg/l BA to trigger axillary bud development. Subculture buds about every month under the same conditions.

d) To obtain whole plants, transfer buds onto the same medium devoid of BA. The best results are obtained in the presence of 0.1 mg/l IAA or IBA.
e) When roots are 2 cm long (about 3 weeks) move plantlets to small pots containing a horticultural substrate. Transfer to the greenhouse and maintain a high level of atmospheric humidity during the first week. Four weeks later plantlets can be transferred to the field.

2. Somatic Embryogenesis

Isolation of embryogenic tissue strains
a) Cut off explants about 1 cm long (petiole, stem, hypocotyl fragments) from vitroplants or aseptically grown seedlings.
b) Lay out these explants in an inverted position or flat onto an inducing medium containing MS mineral salts, LS vitamins, 2% glucose or sucrose, 1 mg/l 2,4-D, 0.1 mg/l kinetin, and 0.7% agar at 25 °C in the light. Embryogenic tissues can be recognized by their white or grayish color and by their finely granulous texture. Maintain them on the same medium.
c) To obtain embryos and plantlets, transfer embryogenic tissues to a medium lacking growth regulator and containing Shah and Dalal (1980) macroelements (or half-strength MS macroelements), 0.4 mg/l thiamine HCl, and 2% glucose. Better plantlet development can be obtained by first growing the embryogenic tissues in a liquid medium and then laying them out onto a solidified development medium (Hunault 1984).

Rapid plantlet production from primary explants
d) When embryogenic tissue strains are not required, grow explants for 1 month only on the inducing medium, in darkness, then transfer to development medium. After 1 to 2 months, embryos and plantlets develop from the small primary callus formed on the explants.

Cloning of embryo-derived plantlets
e) Cut off the buds under the cotyledonary node as soon as the first true leaf is fully spread and grow them on the same medium as for adult plant buds.

References

Badoc A (1982) Contribution à l'étude des phénomènes d'organogenèse et de callogenèse de tissus de Fenouil vulgaire (*Foeniculum vulgare* subsp. *capillaceum* var. *vulgare* (Mill.) Thellung) cultivés in vitro; analyse des constituants de l'huile essentielle des explants. D.E.A., Lille Univ, France, 74 pp

Badoc A (1988) Contribution à l'étude du genre *Foeniculum*. Doctoral Thesis, Lille Univ, France, 246 pp

Becker H (1970) Untersuchungen zur Frage der Bildung flüchtiger Stoffwechselprodukte in Calluskulturen. Biochem Physiol Pflanz 161: 425–441

Betts TJ (1968) Examination of fennel fruits by gas chromatography without preliminary distillation. J Pharm Pharmac (Suppl) 20: 61S–64S

Betts TJ (1976) Possible new Western Australian fennel chemovar. Aust J Pharmacol Sci 5: 78–79

Chu CC, Wang CC, Sun CS, Hsu C, Yin KC, Chu CC, Bi FY (1975) Establishment of an efficient medium for anther culture of rice through comparative experiments on the nitrogen source. Sci Sin 28: 659–668

Du Manoir J, Desmarest P, Saussay R (1985) In vitro propagation of fennel (*Foeniculum vulgare* Miller). Sci Hortic 27: 15–19

Fujita SI, Asami Y, Nozaki K (1980) The constituents of the essential oils from *Foeniculum vulgare* Miller. J Agric Chem Soc Jpn 54: 765–767

Gamborg OL, Eveleigh DE (1968) Culture methods and detection of glucanases in suspension cultures of wheat and barley. Can J Biochem 46: 417–421

Garcia-Rodriguez MJ, Paupardin C, Saussay R (1978) Sur la formation d'un tissu sécréteur et la synthèse d'anéthole par des tissus de Fenouil (*Foeniculum vulgare* Mill. var. *dulce*) cultivés in vitro. CR Acad Sci Paris 287: 693–696

Hegi G (1966) Illustrierte Flora von Mitteleuropa, Bd 5, 2. Hauser, München

Heller R (1953) Recherches sur la nutrition minérale des tissus végétaux cultivés in vitro. Ann Sci Nat, Bot Biol Végét 14: 1–223

Hethelyi I, Tetenyi P (1979) Pirolizises gaskromatografia alkalmazasa a gyogynovenykutatasban. Herba Hung 18: 87–96

Hunault G (1981) La culture in vitro des tissus de Fenouil (*Foeniculum vulgare* Miller). Premières observations sur le comportement des explantats primitifs et des cals. CR Acad Sci Paris 293: 553–558

Hunault G (1984) In vitro culture of fennel tissues (*Foeniculum vulgare* Miller) from cell suspension to mature plant. Sci Hortic 22: 55–65

Hunault G, Desmarest P (1990) Field assessment of somaclonal variation during somatic embryogenesis in bitter fennel. Bull Soc bot Fr, Actualités Botanques 137: 45–49

Hunault G, Desmarest P, Du Manoir J (1989) *Foeniculum vulgare* Miller: Cell culture, regeneration, and the production of anethole. In: Bajaj YPS (ed) Biotechnology in agriculture and forestry, vol 7: Medicinal and aromatic plants II. Springer, Berlin Heidelberg New York Tokyo, pp 185–212

Ishibashi K, Katsuhara J, Hashimoto K, Kobayashi M (1967) Oil constituents of star anise and fennel seed. J Chem Soc Jpn, Ind Chem Sect 70: 1195–1197

Johri BM, Sehgal CB (1966) Growth responses of ovaries of *Anethum*, *Foeniculum* and *Trachyspermum*. Phytomorphlogy 16: 364–378

Karlsen J, Baerheim Svendsen A, Chingova B, Zolovitch G (1969) Studies on the fruits of *Foeniculum* species and their essential oil. Planta Med 17: 281–293

Kohlenbach HW (1977) Basic aspects of differentiation and plant regeneration from cell and tissue cultures. In: Barz W, Reinhard E, Zenk MH (ed) Plant tissue culture and its biotechnological applications. Springer, Berlin Heidelberg New York, pp 355–366

Kohlenbach HW (1985) Fundamental and applied aspects of in vitro plant regeneration by somatic embryogenesis. In: Schäfer-Menuhr A (ed) In vitro techniques, propagation and long term storage. Martinus Nijhoff/Dr W Junk, Dordrecht, pp 101–109

Linsmaier EM, Skoog F (1965) Organic growth factor requirements of tobacco tissue cultures. Physiol Plant 18: 100–127

Maheshwari SC, Gupta GRP (1965) Production of adventitious embryoids in vitro from stem callus of *Foeniculum vulgare*. Planta 67: 384–386

Miura Y, Tabata M (1986) Direct somatic embryogenesis from protoplasts of *Foeniculum vulgare*. Plant Cell Rep 5: 310–313

Miura Y, Fukui H, Tabata M (1987a) Clonal propagation of chemically uniform fennel plants through somatic embryoids. Planta Med 1: 92–94

Miura Y, Ogawa K, Fukui H, Tabata M (1987b) Changes in the essential oil components during the development of fennel plants from somatic embryoids. Planta Med 1: 95–96

Murashige T, Skoog F (1962) A revised medium for rapid growth and bioassays with tobacco tissue cultures. Physiol Plant 15: 473–497

Nickell LG, Maretzki A (1969) Growth of suspension cultures of sugarcane cells in chemically defined media. Physiol Plant 22: 117–125

Nitsch JP (1951) Growth and development in vitro of excised ovaries. Am J Bot 38: 566–577

Osisiogu IUW (1967) Essential oils of Nigeria. Part. II. A study of the oil of fennel produced at Nsukka. Planta Med 15: 30–31

Paupardin C (1976) Sur la différenciation d'un tissu sécréteur et la formation d'huile essentielle par des tissus végétaux cultivés in vitro. Actes 101e Congr Soc Savantes, Lille, France, pp 629–641

Paupardin C, Garcia-Rodriguez MJ, Bricout J (1980) Multiplication végétative de quelques plantes aromatiques: problèmes posés par la production d'essence. CR Acad Agr Fr 66: 658–666

Ranga Swamy NS (1961) Experimental studies on female reproductive structures of *Citrus microcarpa* Bunge. Phytomorphology 11: 109–127

Sehgal CB (1964) Artificial induction of polyembryony in *Foeniculum vulgare* Mill. Curr Sci 33: 373–374

Shah CS, Qadry JS, Chauhan MG (1969) Chemical races in *Foeniculum vulgare* Miller. Curr Sci 38: 365–366

Shah CS, Qadry JS, Chauhan MG (1970) Chemical races in fennel. Planta Med 18: 285–289

Shah RR, Dalal KC (1980) In vitro multiplication of *Glycyrrhiza*. Curr Sci 49: 69–71

Simard MH (1985) Méthodologies de sélection de plants de Fenouil amer (*Foeniculum vulgare*) tolérants au *Phomopsis foeniculi* et à ses principes toxiques. Rapport de stage, Ecole Nationale Supérieure d'Horticulture, Versailles, France

Toth L (1967) Untersuchungen über das ätherische Öl von *Foeniculum vulgare*. I. Die Zusammensetzung des Frucht- und Wurzelöls. Planta Med 15: 157–172

Trenkle K (1972) Neuere Untersuchungen an Fenchel (*Foeniculum vulgare* M.) 2 Mitteilung: das ätherische Öl von Frucht, Kraut und Wurzel fruktifizierender Pflanzen. Pharmazie 27: 319–324

II.3 Micropropagation of Celery (*Apium graveolens* var. *dulce*)

K. F. TOTH and M. L. LACY[1]

1 Introduction and Botany

Celery (*Apium graveolens* L. var *dulce* (Mill.) Pers. family Umbelliferae) probably originated in marshlands around the Mediterranean Sea (Ryder 1979). It was first grown for medicinal purposes, most likely because its bitter flavor and odor suggested curative properties (Ryder 1979) and because celery seed contains an opiate (Hart 1977). During the late 1500s to early 1600s, celery began to be cultivated as a food product to add flavor to cooked foods (Ryder 1979; Ware and McCollum 1975).

Celery grows by producing a close rosette of petioles on a compressed stem, forming a tight or loose head. Commercial celery cultivars are biennials, but will bolt and flower in the first year with a vernalization treatment (6–10 °C for 6 to 8 weeks). Bolting occurs when the main stem elongates to form a branched flower stalk containing numerous umbels of small white perfect flowers which have five petals and five stamens. Celery seeds are actually fruits made up of two compressed carpels which contain the true seed.

1.1 Distribution and Importance

As celery began to be cultivated as a fresh vegetable in the 1700s, its horticultural characteristics began to change (Ryder 1979). Wild celery was a leafy plant, probably resembling the leafy vegetable smallage (*A. graveolens* var. *secalinum*), which is still planted in gardens as a cooking herb. Selection and breeding lead to the development of the large, succulent, and highly edible petioles for which commercial celery is grown today.

Commercial celery cultivars fall into three categories: yellow types which contain some chlorophyll when blanched, easy-blanching (or self-blanching) types which are white when blanched, and green types which are not blanched (Hart 1977). The yellow and easy-blanching cultivars were grown in the United States until the mid 1940s. Growers turned to green cultivars during World War II because of the lack of field hands for blanching. After World War II, celery growers continued growing the green cultivars because consumers had accepted them and because the newer green cultivars were highly resistant to the fungal disease

[1] Department of Botany and Plant Pathology, Michigan State University, Plant Biology Building, East Lansing, MI 48824, USA

Biotechnology in Agriculture and Forestry, Vol. 19
High-Tech and Micropropagation III (ed. by Y.P.S. Bajaj)

Fusarium yellows and some diseases caused by mineral deficiencies (Hart 1977). Yellow celery cultivars are still most common in the European countries, of which the leading producers are Germany, France, Italy, Belgium, and England (Ryder 1979). No worldwide production statistics are available for celery.

Green celery is one of the salad and cooking vegetables most commonly used in the United States. California and Florida are the two largest celery-producing states, accounting for approximately 57% and 20% of total production, respectively (Ryder 1979). Michigan remains the third largest celery-producing state, producing 6.4% of the total United States production in 1987 (Shapley and Dudek 1989). Other celery-producing states include Texas, Ohio, Wisconsin, and New York.

1.2 Diseases

Numerous diseases and mineral deficiencies can affect celery production. The two most important foliar diseases [*Septoria* leaf blight (late blight) caused by *S. apiicola* Speg. and *Cercospora* leaf blight (early blight) incited by *C. apii* Fres.] occur worldwide and can cause serious losses if not properly controlled (Sherf and MacNab 1986). *Septoria* blight is characterized by yellow to brown, irregularly shaped lesions on leaves, petioles, and flower stalks of infected plants (Lacy and Grafius 1980). Mature lesions contain small dark pycnidia which, when wetted, extrude their conidia. The spores can then be disseminated by splashing rain, or by machinery and people moving through wet fields. *S. apiicola* can infect the seed of diseased plants and develop pycnidia on the carpel. Seed-borne infection can be controlled by using 3-year-old seed (the spores do not survive more than 2 years on seed) and by hot water-treated or thiram-treated seed (Sherf and MacNab 1986). Blight on plants is controlled with timely fungicidal sprays and by keeping machinery and people out of wet seedbeds and celery fields (Sherf and MacNab 1986).

Cercospora blight lesions are roughly circular and yellow to tan in color. They can enlarge to 1 cm or more in diameter on leaves and may become elongate on petioles (Lacy and Grafius 1980). The lesions appear gray when spores are produced. *C. apii* infection is favored by warm days followed by heavy night dews. *C. apii* is also controlled with clean seed, sanitation, and fungicidal sprays (Sherf and MacNab 1986).

Bacterial blights, caused by *Pseudomonas syringae* pv. *apii* (Jagger) Young or *P. cichorii* (Swingle) Stapp, are common, but rarely destructive problems (Lacy and Grafius 1980). *P. syringae* pv. *apii* causes red-colored, irregularly shaped lesions on celery leaves. Young lesions have a chlorotic halo, and older lesions may reach 5 mm in diameter and crack open (Sherf and MacNab 1986). *P. cichorii* produces larger tan, papery lesions on celery leaves (Sherf and MacNab 1986; Thayer and Wehlburg 1965). These pathogens rarely infect petioles, and both may be partially controlled by crop rotation, clean seed, sanitation and fixed copper sprays (Sherf and MacNab 1986).

Sclerotinia sclerotiorum and *Rhizoctonia solani* are two soil-borne pathogens which can cause damping-off of seedlings as well as disease in older plants (Lacy and Grafius 1980; Sherf and MacNab 1986). *Sclerotinia* pink rot is favored by cool, wet conditions, and can be controlled by fungicides when necessary. *Rhizoctonia* damping off and crater rot may be problems during warm, wet conditions. Sterilization of seedbed soil and fungicide drenches and sprays may control this pathogen.

Other diseases which can cause regional problems include cucumber mosaic and common celery mosaic viruses, and aster yellows. These pathogens are all controlled by eliminating weed hosts around celery fields, and by keeping their insect vectors under control (Lacy and Grafius 1980; Sherf and MacNab 1986). Black heart caused by calcium deficiency and cracked stem due to boron deficiency are two nutritional problems common in celery in midwestern and southern muck soils (Lacy and Grafius 1980). Both are controlled by alleviating the mineral deficiency with topical sprays of appropriate calcium and boron salts.

Fusarium yellows, caused by the soil-borne pathogen *Fusarium* oxysporum f. sp. *apii* (R. Nels. & Sherb.) Snyd. and Hans., was an important disease of yellow and self-blanching celery cultivars grown during the first half of this century in the United States (Ryker 1935; Nelson et al. 1937) Fusarium yellows vanished in the United States after the intròduction of highly resistant green celery cultivars in the 1940–1950s (Opgenorth and Endo 1985). In 1978, a new race of *Fusarium oxysporum* f. sp. *apii* (race 2) appeared in California, which attacked previously resistant green cultivars as well as yellow and self-blanching celery cultivars (Hart and Endo 1978; Schneider and Norelli 1981). *F. oxysporum* f. sp. *apii* race 2 was identified in Michigan in 1981 (Elmer and Lacy 1984) and had spread to all the celery growing regions in Michigan by 1987 (Toth and Lacy, unpubl.). Race 2 had also been identified on celery grown in New York (Awuah et al. 1986), Ohio (R. Rowe, pers. commun.), Wisconsin (Elmer and Lacy 1984), Texas (Martyn 1987), and Ontario, Canada (Cerkauskas and McDonald 1989).

The most common diagnostic symptom of *Fusarium* yellows is a red to brown discoloration in the vascular tissues of the roots and crowns of infected plants (Hart and Endo 1978; Lacy and Elmer 1985; Awuah et al. 1986). The discoloration may extend into the petioles in severely infected plants, and be accompanied by rotting of the central crown area. As the disease progresses, plants become stunted and chlorotic. Severely infected plants can wilt and die, some of them because of bacterial soft rot which develops in the diseased crowns. Even slightly infected plants are of low quality because they are often stunted and bitter-tasting (Awuah et al. 1986; Toth and Lacy, unpubl).

F. oxysporum f. sp. *apii* race 2 is difficult to control. Soil fumigation is expensive and has not yet proven to be an effective means of controlling the pathogen (Otto et al. 1976; Awuah et al. 1986; Greathead 1988). Likewise, no commercially available fungicides have controlled this pathogen (Awuah et al. 1986). Host resistance is the most practical method of controlling *F. oxysporum* f. sp. *apii* race 2 (Opgenorth and Endo 1985; Lorenz and Maynard 1988). Field trials in California (Hart and Endo 1978; Opgenorth and Endo 1979, 1985), Michigan (Elmer et al. 1986), and New York (Awuah et al. 1986) demonstrated that most yellow and green celery cultivars are susceptible to *F. oxysporum* f. sp. *apii* race 2. Celery cultivars moderately to highly resistant to race 2 are now being released, although most are horticulturally inferior to susceptible cultivars.

1.3 Traditional Breeding Techniques

Until recently, new celery cultivars were developed via mass selection and random mating. While these methods produced cultivars with improvements in such traits as reduced ribbiness, slow flowering, and some disease resistance, producing new cultivars in this way can take 10 or more years. In the United States, the need to more quickly develop celery cultivars highly resistant to *F. oxysporum* f. sp. *apii* race 2 has led celery breeders to begin employing single plant selections along with selfs, sib-crosses, and backcrosses in developing new lines. With advances in the horticultural characteristics of male-sterile celery lines and increases in the price of celery seed, F_1 hybrids are becoming more feasible for celery improvement.

When developing cultivars with increased disease resistance via conventional breeding, celery plants must be crossed with a related resistant species such as celeriac, or with distantly related plants such as parsley. Not only are such crosses difficult to make (hand pollination is challenging in celery because the flowers are small and do not open simultaneously in a given umbel), but many undesirable characteristics (for example thin petioles and an open growth habit) may be inherited along with resistance (Honma and Lacy 1980; Orton et al. 1984; Opgenorth and Endo 1985). Several rounds of backcrossing and selection are necessary to produce a highly resistant cultivar which is horticulturally acceptable to celery growers. Even with a vernalization treatment to force celery plants to flower in the 1st year, it takes 12 to 14 months to go from seed to seed (Honma 1959), and developing a new celery cultivar requires a number of years.

Like most other vegetable and fruit crops, maintaining acceptable horticultural traits is essential for breeding programs. New cultivars must be very similar to those familiar to consumers for them to be marketed successfully. In celery, these characteristics of greatest concern are mostly related to petiole morphology, such as height, number, ribbiness, color, and shape. This need for certain traits adds to the problems encountered and the amount of time required for producing new celery cultivars.

Some of the problems with conventional breeding programs can be exemplified by the celery breeding line UC-1. UC-1 was developed at the University of California, Davis, in 1984 (Orton et al. 1984). Derived from a celery X celeriac cross, UC-1 contains a high level of resistance to *F. oxysporum* f. sp. *apii* race 2, yet is not horticulturally desirable. It is a tall, vigorous line, but has thin petioles and an open growth habit reminiscent of its celeriac parentage (Toth 1989). Many celery breeding programs in the United States are utilizing UC-1 by selecting plants which most closely resemble commercial celery to be used as parents to develop lines which are highly resistant and commercially acceptable. Although this germ plasm has been available for a number of years, highly resistant cultivars derived from UC-1 are only now beginning to appear.

2 In Vitro Studies and Micropropagation

Much of the early tissue culture work with celery was reviewed in Volume 2 of the *Biotechnology in Agriculture and Forestry* series (Browers and Orton 1986).

Regeneration of celery plants from callus or cell suspension cultures was first initiated for vegetative propagation of celery for breeding purposes (Chen 1976; Williams and Collin 1976a, b). Somatic embryoids were observed in callus and cell suspension cultures from celery (Al-Abta and Collin 1978a, b, 1979; Chen 1976; Dunstan et al. 1982; Williams and Collin 1976a), leading researchers to propose using embryoids as planting stock (Orton 1985).

Although much variability in chromosome number and structure was observed in cultured celery cells (Browers and Orton 1982a, b; Murata and Orton 1983; Orton 1987), cells with gross chromosomal changes rarely regenerated into plants (Browers and Orton 1982a; Orton 1987). Still, several kinds of variability were observed in regenerated somaclones of celery (Orton 1983, 1985). These changes ranged from gross phenotypic variability to differences in electrophoretic banding patterns for some enzymes (Orton 1983) to karyological changes (Orton 1985).

We have also discovered a number of different phenotypic variants in somaclonal regenerants. Some of the somaclones which were highly resistant to *Fusarium* yellows resembled celeriac in gross morphology and growth habit (Toth 1989). These somaclones had short, thin petioles and a very open heart. Other somaclones exhibited unique leaf morphologies (Fig. 1) (Toth 1989).

Other changes were due to recessive mutations because the characteristics were not expressed in the original somaclones but in their progenies. In the R_1 progeny from one self-fertilized somaclone, seven out of 56 seedlings were albinos (Toth 1989). The albino seedlings were completely white and did not grow past the cotyledon stage in the greenhouse. Approximately 24% of the progeny from another somaclone were dwarf mutants (Toth 1989), exhibiting stunted petioles

Fig. 1. Variability in leaflet morphologies seen in somaclones regenerated from cells of celery cultivar Tall Utah 52–70 HK. The leaflet in the *upper left corner* is most similar in morphology to leaflets of Tall Utah 52–70 HK plants from commercial seed

and shortened but abnormally wide leaflets. Orton (1985) reported on the presence of a recessive mutation which caused an abnormal leaflet morphology and which was expressed in the selfed progeny of some celery somaclones with normal-looking leaves.

Greater genetic changes may be induced by in vitro culturing than is evident in regenerated plants. Williams and Collin (1976b) reported that out of 44 "normal-looking" regenerated celery plants, 32% were polyploid or aneuploid. We found a greater number of abnormal-looking cells in suspensions derived from R_2 somaclonal progeny than in those derived from commercial cultivars, even though the original somaclone and the resulting R_2 plants were very similar in gross morphology to the commercial cultivar (Toth and Lacy, unpubl.). A larger number of abnormal-looking plants were also regenerated from cell suspensions from somaclonal progeny than from cell suspensions from commercial cultivars (Toth and Lacy, unpubl.). This may indicate that more genetic changes were induced in the original somaclones by the culturing techniques than were evident by their phenotype.

2.1 Increased Disease Resistance in Celery from In Vitro Culturing

Wright and Lacy (1988) identified somaclones regenerated from the celery cultivars Florida 683 and Tall Utah 52–70 HK which were highly resistant to the following pathogens: *Fusarium oxysporum* f. sp. *apii* race 2, *Septoria apiicola, Cercospora apii,* or *Pseudomonas cichorii* (Table 1). Florida 683 was highly susceptible to all three pathogens, and Tall Utah 52–70 HK contained a moderate level of resistance only to *F. oxysporum* f. sp. *apii* race 2 (Ireland et al. 1987). Resistance to each pathogen appeared to arise independently in vitro because most somaclonal variants contained resistance to only one pathogen, and no variants were found which were highly resistant to all four pathogens. Unfortunately, since all the somaclones eventually died from the diseases to which they lacked resistance, the inheritances of the resistances derived via tissue culture were not studied.

In another study, somaclones derived from Tall Utah 52–70 HK, which were highly resistant to *F. oxysporum* f. sp. *apii* race 2, were found to pass on resistance to their progeny via self-fertilization (Toth 1989) (Table 1). The progeny of those somaclones chosen for seed production closely resembled Tall Utah 52–70 HK in gross morphology and growth habits. Future generations will have to be examined for variability in other characteristics, such as petiole width, height, or tendency to form side shoots, which would affect yield and quality.

Likewise, Heath-Pagliuso et al. (1988) identified somaclonal variants derived from susceptible celery cultivar Tall Utah 52–70 R which were highly resistant to *F. oxysporum* f. sp. *apii* race 2 (Table 1). The induced resistance was heritable and coded for by more than one gene. They have recently released a highly resistant celery line (UC-T3 somaclone) which was derived from this tissue culture study (Heath-Pagliuso et al. 1989).

Table 1. Summary of increased resistance in celery derived from somaclonal variation

Cultivar	Pathogens tested	Resistance inherited	Reference
Tall Utah 52–70 R	*Fusarium oxysporum* f. sp. *apii* race 2	Yes	Heath-Pagliuso et al. (1988)
Florida 683	*F. oxysporum* f. sp. *apii* race 2 *Septoria apiicola* *Cercospora apii* *Pseudomonas cichorii*	Not tested	Wright (1985) Wright and Lacy (1988)
Tall Utah 52–70 HK	*F. oxysporum* f. sp. *apii* race 2 *Septoria apiicola* *Cercospora apii* *Pseudomonas cichorii*	Not tested	Wright (1985) Wright and Lacy (1988)
Tall Utah 52–70 HK	*F. oxysporum* f. sp. *apii* race 2	Yes	Toth (1989)

3 Protocol

The tissue culture protocol followed by many researchers regenerating celery was reviewed by Browers and Orton (1986). Murashige and Skoog salts (MS) with either Murashige and Skoog or White's vitamins with various hormone concentrations are the best choices for culturing celery (Browers and Orton 1986). The composition of the media used in the authors' laboratory was based on MS, but modified for celery (Heath-Pagliuso et al. 1988; Wright 1985; Wright and Lacy 1988). Celery researchers also differ in their choice of tissue for explants (Browers and Orton 1986). We chose axillary buds from greenhouse-grown plants because they readily formed callus on the medium chosen (Wright 1985; Wright and Lacy 1988).

1. Media Composition

All media contained 1.65 g NH_4NO_3, 1.9 g KNO_3, 0.44 g CaCl, 0.37 g $MgSO_4 \cdot 7H_2O$, 0.17 g KH_2PO_4, 0.62 mg H_3BO_3, 1.69 mg $MnSO_4$, 1.06 mg $ZnSO_4$, 0.83 mg KI, 27.9 mg $(Fe_2SO_4)_3$, 37.3 mg Na_2EDTA, 2 mg glycine, 100 mg myo-inositol, 0.5 mg nicotinic acid, 0.5 mg pyridoxine HCl, and 0.1 mg thiamine HCl, in 1 l glass distilled water. Media for callus initiation, callus maintenance, and plant regeneration varied in levels of the synthetic auxin 2,4-dichlorophenoxyacetic acid (2,4-D) and either kinetin or benzyladenine (discussed in appropriate sections). All tissue culture media were adjusted to a pH of 5.8 with 1 N NaOH prior to autoclaving. Agar media contained 9.5 g Difco Bacto agar (Difco Laboratories, Detroit, Michigan) per liter of medium.

2. Establishment of Callus and Cell Suspensions

Axillary buds from 8–11-week-old plants were excised, surface sterilized in 0.53% sodium hypochlorite (a 1:9 v/v dilution of household bleach with sterile distilled water) for 10 min and then rinsed three times with sterile distilled water. Surface-sterilized buds were placed on MS agar containing 1 mg 2,4-D, 2 mg benzyladenine, and 30 g sucrose/l medium, and were incubated at 22 °C under a 16-h photoperiod. After 9 days, buds showing callus formation were transferred to MS agar which contained 0.5 mg 2,4-D, 0.1 mg kinetin, and 30 g sucrose/l medium. Six weeks later the callus was transferred to 250-ml flasks containing 50 ml MS broth with 0.5 mg 2,4-D, 0.1 mg kinetin, and 30 g sucrose/l medium, and these cultures were incubated on a rotary shaker at 100 rpm in the dark.

After 2 months, the cell suspension cultures were supplied monthly with fresh MS broth with the same hormone and sucrose concentrations. Cell suspensions were diluted 1:1 (v/v) with fresh broth, and one half (50 ml) of the diluted cell suspension was poured into a sterile 250-ml flask and placed on the shaker.

4 Regeneration and Transfer of Plants

Celery regenerates by producing somatic embryoids which develop from single cells and arise on clumps of cells on callus cultures or within cell suspensions (Williams and Collin 1976a; Dunstan et al. 1982). These embryoids appear to go through the same developmental stages as zygotic embryos (Dunstan et al. 1982). Embryoids develop from globular to heart-shaped stages while becoming polarized with cotyledon initials at one end and radicle at the other (Fig. 2). Once transferred to a solid medium devoid of hormones, the embryoids mature and develop into plantlets (Williams and Collin 1976a; Dunstan et al. 1982; Wright 1985; Wright and Lacy 1988; Toth 1989).

When cell clusters indicative of embryoid formation were observed in the cell suspensions (8–12 weeks after initiation of cell suspensions), 2 ml aliquots from cell suspensions were plated onto MS agar with no hormones, 20 g sucrose, and 0.5 g

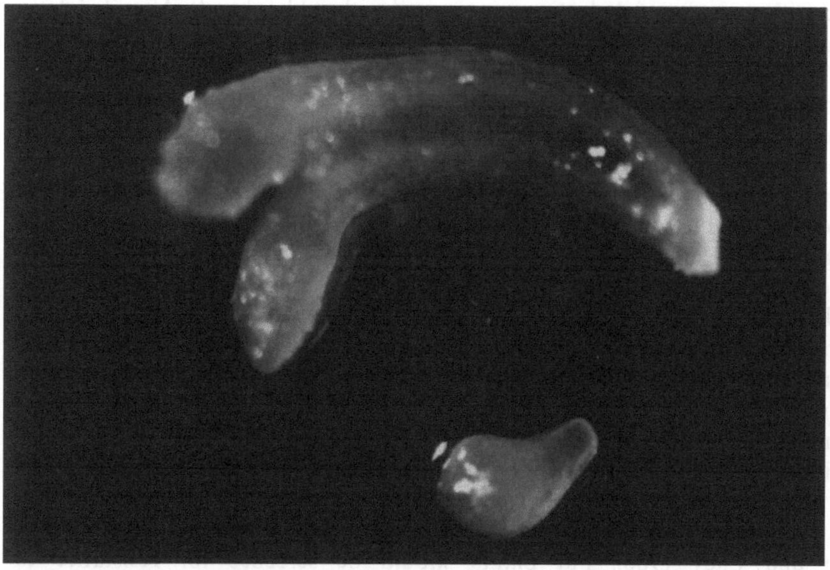

Fig. 2. Globular and heart-shaped embryoids (*from left to right*) from a suspension of celery cells developing on a medium lacking plant hormones

activated charcoal/l medium. The activated charcoal was necessary to absorb traces of hormones which could inhibit root development in celery somaclones. Two weeks later green plants were observed on the regeneration medium. The plants (somaclones) were transferred every 2–4 weeks to new MS agar with no hormones and 20 g sucrose in either Petri plates (10 × 15 cm) or culturing boxes (7.5 × 7.5 × 10 cm) (GA7 vessels, Magenta Corporation, Chicago, Illinois).

Vitrification, a condition thought to be caused by ethylene and denoted by broad, thick leaves which appear glassy and translucent and easily breakable (Daguin and Letouze 1986; Kevers et al. 1984), was a frequent problem in regenerated celery plants. Although the somaclones did tend to outgrow the symptoms, vitrification posed a problem when the crowns of somaclones became weakened and broke apart during transfer to fresh media, rendering the plants useless. In recent studies in our laboratory, vitrification was reduced by decreasing the NH_4NO_3 concentration to 0.825 g/l in the regeneration medium and by transferring the somaclones after 2 months of growth to culturing boxes (where air exchange is better) containing regeneration medium without charcoal (Toth and Lacy. unpubl.).

Hardening-off of somaclones of celery to enhance survival after they are transferred from agar media to soil can be a challenging procedure. Often as many as 50% or more do not survive the transfer from tissue culture medium to greenhouse conditions. For this reason, we tried different procedures for hardening-off of the somaclones. The procedure having the highest survival rates was based on a technique developed by Heath-Pagliuso et al. (1988) (Toth and Lacy, unpubl.). Medium-grade vermiculite was wetted with a Hoagland's solution and autoclaved for 45 mins. Disposable cups (8.5 cm high and 7 cm wide at top) were filled half way with the steamed vermiculite. The somaclones were transplanted into the cups, sprayed with a mist of sterile distilled water, and the cups were covered with a commercial plastic wrap secured with rubber bands. After 2–3 days, small holes were made in the plastic wrap. These holes were enlarged daily, and the plastic wrap removed in 1–2 weeks. Light intensity during this first week did not seem to play a role in survival, and the plants survived better if the vermiculite was kept well watered (Toth and Lacy, unpubl.). Somaclones which were 8 to 10 cm tall and had secondary leaflets (4 to 8 months after regeneration) survived at higher frequencies than smaller and younger plants (Toth 1989). After sufficient root and shoot development (Fig. 3), somaclones can be transplanted into a commercial potting mix in the greenhouse, or directly into the field, and screened for resistance to various pathogens (Wright 1985; Wright and Lacy 1988; Heath-Pagliuso et al. 1989; Toth 1989).

Most plant resistance mechanisms are tissue or whole-plant responses and these responses can only be adequately screened for, once plants have been regenerated from the in vitro cultures. When screening regenerants for disease resistance, care must be taken to provide an adequate inoculum and environmental conditions which are favorable for sufficient disease pressure. For example, in celery, plants can be satisfactorily screened for resistance to foliar diseases (*Septoria* blight, *Cercospora* blight, and bacterial blights) in the greenhouse (Wright 1985; Wright and Lacy 1988), but plants should be screened for resistance to *F. oxysporum* f. sp. *apii* race 2 in the field where the disease is more severe and escapes occur less frequently (Toth 1989).

Fig. 3. Five-month-old celery somaclones growing in the greenhouse. Plants are almost large enough to begin screening for resistance to foliar and soil-borne pathogens

5 Summary and Conclusions

With fewer conventional celery breeding programs available and because of the number of years needed to develop new cultivars via conventional breeding, other systems need to be employed for the development of improved celery cultivars. The genetics of many horticultural traits and of resistance to specific diseases in celery have not been well worked out. This, and the fact that very few genes in celery are mapped, would currently limit the application of some biotechnology techniques. In vitro culturing has proven to be a practical method for developing improved celery germplasm.

Tissue culture techniques have some disadvantages, however. Somaclonal variation is thought to result from several types of chromosomal changes (rearrangements, aneuploidy, deletions, and amplifications), from single gene mutations, and from epigenetic changes induced by the stresses of culturing conditions (Meins 1983; Orton 1984; Lee and Phillips 1988). A researcher has little or no control over the type and level of variability occurring in cultured cells. Tissue culture, especially without any selection pressure at the cellular level, is similar to other types of mutation breeding in that the desired genotypes may not be produced at all, or a large number of plants may have to be screened in order to find one of the desired type (Daub 1986).

The potential of tissue culturing for crop improvement far outweighs its problems. Tissue culture would not only provide a means of increasing disease resistance in crops, but it also has the potential for changing other traits, even quantitative characteristics for increasing yield. In vitro culturing is yet another tool for developing improved cultivars.

Acknowledgments. The authors express their appreciation to Raymond Hammerschmidt, John L. Lockwood, Lawrence Rappaport, and Peter P. Toth for their editorial contributions to this chapter.

References

Al-Abta S, Collin HA (1978a) Control of embryoid development in tissue cultures of celery. Ann Bot 42: 773–782

Al-Abta S, Collin HA (1978b) Cell differentiation in embryoids and plantlets of celery tissue cultures. New Phytol 80: 517–521

Al-Abta S, Collin HA (1979) Endogenous auxin and cytokinin changes during embryoid development in celery tissue cultures. New Phytol 82: 29–35

Awuah RT, Lorbeer JW, Ellerbrock LA (1986) Occurrence of Fusarium yellows of celery caused by *Fusarium oxysporum* f. sp. *apii* race 2 in New York and its control. Plant Dis 70: 1154–1158

Browers MA, Orton TJ (1982a) Transmission of gross chromosomal variability from suspension cultures into regenerated celery plants. J Hered 73: 159–162

Browers MA, Orton TJ (1982b) A factorial study of chromosomal variability in callus cultures of celery (*Apium graveolens*). Plant Sci Lett 26: 65–73

Browers MA, Orton TJ (1986) Celery (*Apium graveolens* L.). In: Bajaj YPS (ed) Biotechnology in agriculture and forestry, vol 2: Crops I. Springer, Berlin Heidelberg New York Tokyo, pp 405–420

Cerkauskas RF, McDonald MR (1989) Race 2 of *Fusarium oxysporum* f. sp. *apii* new to Ontario. Plant Dis 73: 859

Chen CH (1976) Vegetative propagation of the celery plant by tissue culture. Proc S D Acad Sci 55: 44–48

Daguin F, Letouze R (1986) Ammonium-induced vitrification in cultured tissues. Physiol Plant 66: 94–98

Daub ME (1986) Tissue culture and the selection of resistance to pathogens. Annu Rev Phytopathol 24: 159–186

Dunstan DI, Short KC, Merrick MA, Collin HA (1982) Origin and early growth of celery embryoids. New Phytol 91: 121–128

Elmer WH, Lacy ML (1984) Fusarium yellows (*Fusarium oxysporum* f. sp. *apii* race 2) of celery in Michigan. Plant Dis 68: 537

Elmer WH, Lacy ML, Honma S (1986) Evaluations of celery germ plasm for resistance to *Fusarium oxysporum* f. sp. *apii* race 2 in Michigan. Plant Dis. 70: 416–419

Greathead A (1988) Control of Fusarium yellows. California Celery. Calif Celery Res Adv Board, Dinuba, California 2: 3–5

Hart LP (1977) History of celery according to Gene Hill. (unpublished)

Hart LP, Endo RM (1978) The reappearance of *Fusarium* yellows of celery in California. Plant Dis Rep 62: 138–142

Heath-Pagliuso S, Pullman J, Rappaport L (1988) Somaclonal variation in celery: screening for resistance to *Fusarium oxysporum* f. sp. *apii*. Theor Appl Genet 75: 446–451

Heath-Pagliuso S, Pullman J, Rappaport L (1989) UC-T3 somaclone: celery somaclone resistant to *Fusarium oxysporum* f. sp. *apii* race 2. Hortic Sci 24: 711–712

Honma S (1959) A method for celery hybridization. Proc Am Soc Hortic Sci 73: 345–348

Honma S, Lacy ML (1980) Hybridization between pascal celery and parsley. Euphytica 29: 801–805

Ireland KF, Elmer WH, Lacy ML (1987) Field evaluation of celery germ plasm for resistance to *Fusarium oxysporum* f. sp. *apii* race 2. Phytopathology (Abstr) 77: 1712

Kevers C, Coumans M, Coumans-Gilles MF, Gaspar T (1984) Physiological and biochemical events leading to vitrification of plants cultured in vitro. Physiol Plant 61: 69–74

Lacy ML, Elmer WH (1985) Fusarium yellows of celery in Michigan. Michigan Extension Bull E-1823

Lacy ML, Grafius EJ (1980) Disease and insect pests of celery. Extension Bull E-1427. Cooperative Extension Service, Michigan State Univ

Lee M, Phillips RL (1988) The chromosomal basis of somaclonal variation. Annu Rev Physiol Plant Mol Biol 39: 413–437

Lorenz OA, Maynard DN (1988) Knott's handbook for vegetable growers, 3rd edn. Wiley, New York, 456 pp
Martyn RD (1987) *Fusarium* yellows of celery in Texas. Plant Dis 71: 651
Meins F (1983) Heritable variation in plant cell culture. Annu Rev Plant Physiol 34: 327–346
Murashige T, Skoog F (1962) A revised medium for rapid growth and bioassays with tobacco tissue cultures. Physiol Plant 15: 473–497
Murata M, Orton TJ (1983) Chromosome structural changes in cultured celery cells. In Vitro 19: 83–89
Nelson R, Coons GH, Cochran LC (1937) The *Fusarium* yellows disease of celery. Mich Agric Exp Sta Tech Bull 155, 74 pp
Opgenorth DC, Endo RM (1979) Sources of resistance to Fusarium yellows of celery in California. Plant Dis Rep 63: 165–169
Opgenorth DC, Endo RM (1985) Additional sources of resistance to race 2 of *Fusarium oxysporum* f. sp. *apii*. Plant Dis 69: 882–884
Orton TJ (1983) Spontaneous electrophoretic and chromosomal variability in callus cultures and regenerated plants of celery. Theor Appl Genet 67: 17–24
Orton TJ (1984) Genetic variation in somatic tissues: method or madness. pp 153–189 In: Ingram DS, Williams PH (eds.) Advances in plant pathology, vol 2. Academic Press, New York, 303 pp
Orton TJ (1985) Genetic instability during embryogenic cloning of celery. Plant Cell Tissue Organ Cult 4: 159–169
Orton TJ (1987) Genetic instability in celery tissue and cell cultures. Iowa State J Res 61: 481–489
Orton TJ, Hulbert SH, Durgan ME, Quiros CF (1984) UC1, *Fusarium* yellows-resistant celery breeding line. Hortic Sci 19: 594
Otto HW, Paulus AO, Snyder MJ, Endo RM, Hart LP, Nelson J (1976) A crown rot of celery. Calif Agric 30: 10–11
Ryder EJ (1979) Leafy salad vegetables. Avi, Westport, Connecticut, 266 pp
Ryker TC (1935) *Fusarium* yellows of celery. Phytopathology 25: 578–600
Schneider RW, Norelli JL (1981) A new race of *Fusarium oxysporum* f. sp. *apii* in California. Phytopathology (Abstr) 71: 108
Shapley AE, Dudek TA (1989) The cost of producing celery, West Central Michigan: 1990. Staff Paper, Dep Agricl Econ, Michigan State Univ, East Lansing, MI, 8 pp
Sherf AF, MacNab AA (1986) Vegetable diseases and their control. Wiley, New York, 728 pp
Thayer PL, Wehlburg C (1965) *Pseudomonas cichorii*; the cause of bacterial blight of celery in the Everglades. Phytopathology 55: 554–557
Toth KFI (1989) The Biology and Control of *Fusarium oxysporum* f. sp. *apii* race 2. Doctoral Dissertation. Michigan State Univ, 174 pp
Ware GW, McCollum JP (1975) Producing vegetable crops. Interstate, Danville, Illinois, 599 pp
Williams L, Collin HA (1976a) Embryogenesis and plantlet formation in tissue cultures of celery. Ann Bot 40: 325–332
Williams L, Collin HA (1976b) Growth and cytology of celery plants derived from tissue cultures. Ann Bot 40: 333–338
Wright JC (1985) Somaclonal variation occurring in disease response of regenerated celery plants from cell suspension and callus cultures. M. S. Thesis. Michigan State Univ, 90 pp
Wright JC, Lacy ML (1988) Increase of disease resistance in celery cultivars by regeneration of whole plants from cell suspension cultures. Plant Dis 72: 256–259

II.4 Micropropagation of Thyme (*Thymus vulgaris* L.)

M. Furmanowa and O. Olszowska[1]

1 Introduction

1.1 Botanical Aspects

1.1.1 Taxonomy and Distribution

Thymus vulgaris L. or common thyme belongs to the family Labiatae = Lamiaceae. Jalas (1972) in *Flora Europaea*, gives 66 species of *Thymus*, divided into two subgenera, *Coridothymus*, with only one species, and *Thymus* with 65 species. *Thymus vulgaris* belongs to the sixth section of *Thymus*. Classification of this species is difficult and therefore methods of comparative enzyme study may be useful for identification (Elena-Rosselló 1979). The variation in *T. vulgaris* is caused not only by genetic constitution, but also depends on ecological and environmental conditions. All of these factors influence the yield of volatile oil produced by thyme plants and the composition of the oil, which is very important.

Thymus vulgaris is an aromatic, perennial herb of the Mediterranean flora. Calyx 3–4 mm, with short hairs, the tube is campanulate, upper teeth as long as wide, not ciliate; corolla whitish to pale purple; 2n = 30. In many European countries this species is cultivated and propagated by seeds and stem cuttings. Pollination is mainly effected by bees (Brabant et al. 1980). Rumińska (1981) and Turowska et al. (1980) have described the methods of cultivation of *T. vulgaris* in the field.

1.1.2 Morphology and Biology

Apart from some morphological characters typical of the Labiatae family, *T. vulgaris* has another very important biological feature, gynodioecy. This means the occurrence of two kinds of individuals in natural populations: hermaphrodites or bisexuals also called, male fertiles (mF), which possess stamens and ovules, and females, called male steriles (mS), which do not bear functional stamens. The presence or absence of functional stamens leads to polymorphism. This situation presents a very interesting problem to growers (Rey 1990). The reproductive capacities of the sexual forms of *T. vulgaris* were studied by many authors

[1] Department of Biology and Pharmaceutical Botany, Medical Academy, Banacha 1, 02-097 Warsaw, Poland

Biotechnology in Agriculture and Forestry, Vol. 19
High-Tech and Micropropagation III (ed. by Y.P.S. Bajaj)

especially in France, where thyme is widely distributed in the south. Assouad et al. (1978) investigated the differences between male sterile and male fertile forms of *T. vulgaris*.

One of the most important conclusions was the observation that mS plants have almost twice as many seeds as mF plants. Size and weight of plants, germination, ease of pollination, and average number of pollen grains retained on the stigmas also belong to the advantageous characters in the mS form. The size of the flowers and the length of styles, however, appear to give mF plants an advantage. Assouad et al. (1978) gave the rate of germination for different types of pollinations: mF selfed, mF × mF, and mS × mF. The percent germination was highest in the crosses mS × mF (56.4%) and lowest in mF selfed (29.9%). Valdeyron et al. (1977) determined the rate of self-fertilization in male fertile plants of four thyme populations, homozygous for dominant chemical marker alleles. Dommée et al. (1978) observed that the percentage of mS varied from 5 to 95% in different natural populations, and depended on the environmental conditions.

1.2 Chemical Components

The thyme plant has been used in medicine since antiquity and has been the subject of numerous chemical and pharmacological studies in this century. Most characteristic for thyme is the production of volatile oil (0.7–6.5% dry wt. of leaves and flowering tops) (Rumińska 1981). Yamaura et al. (1989) suggested that peltate glandular trichomes (PGTs) are the primary site of monoterpene accumulation in thyme and the formation of PGTs is a prerequisite for the accumulation of essential oil in this plant.

Thyme contains also tannins, caffeic acid 1.8%, chlorogenic acid 1.1% (Barbawski 1965), oleanolic acid 0.6%, ursolic acid 1.7% (Rowe 1949), saponins (Garcia Marquina 1949), thymol and carvacrol glycosides (Skopp and Hörster 1976), luteolin (Awe et al. 1959), apigenin, apigenin-7-β-glucoside, luteolin-7-β-glucoside, luteolin-7-β-diglucoside, (Olechnowicz-Stępień and Lamer-Zarawska 1975), hydroxy-6-luteolin (Adzet and Martinez-Vergés 1980), and three methylated flavones isolated by van den Broucke et al. (1982): 5,4′-dihydroxy-6,7,8,3′-tetramethoxyflavone (8-methoxy-cirsilineol), 5,4′-dihydroxy-6,7,3′-trimethoxyflavone (cirsilineol), and 5,6,4′-trihydroxy-7,8,3′-trimethoxyflavone (thymonine) (Fig. 1).

Granger and Passet (1973) discovered six chemotypes in *T. vulgaris* growing in the south of France; their main oil components were geraniol, linalol, α-terpineol, thujanol-4 and terpineol-4, thymol, and carvacrol (Fig. 2). The seventh type, 1,8-cineol, is known to grow in Spain (Adzet et al. 1977). The above main components constitute 60–95% of the essential oil, depending on the type in question. Myrcene, limonene, α-pinene, β-pinene, borneol, and bornyl acetate are other components found in minor quantities. Granger and Passet (1973) found that natural populations are usually heterogeneous, composed of plants of different chemotypes. There is no correlation between morphological and anatomical features of plants, and the chemical composition of the oil. The chemotypes most frequently occurring in France are thymol, carvacrol, and linalol. Geraniol was found in only one locality in the Alps. Factors which could affect the distribution of chemotypes are external

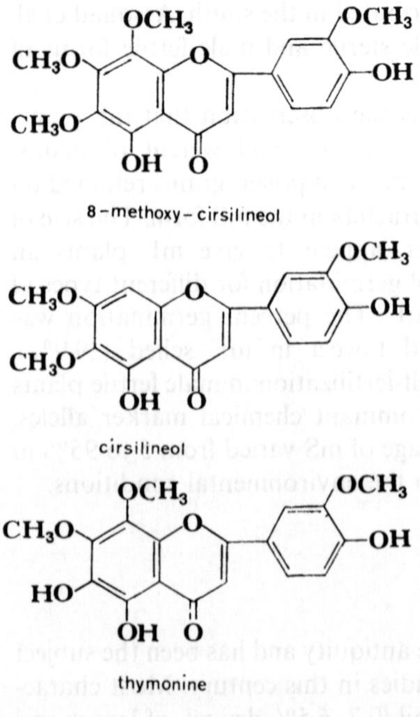

Fig. 1. Polymethoxyflavones isolated from leaves of *Thymus vulgaris* L. (van den Broucke et al. 1982)

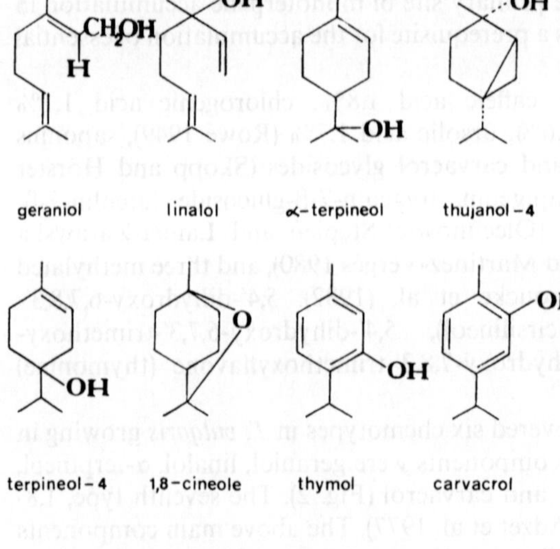

Fig. 2. Volatile oil components characteristic of chemotypes of *Thymus vulgaris* L.

(geographical, edaphic, climatic) and internal (sexual polymorphism and genetic factors) (Granger and Passet 1973).

More detailed investigations of chemotype distribution depending on ecological factors were carried out by Gouyon (1975), Vernet et al. (1977a, b), and Gouyon et al. (1986). It was found inter alia that phenolic populations (with carvacrol or thymol plants) then linalol, thujanol-4, and finally α-terpineol populations occurred in dry to relatively humid habitats. Gouyon et al. (1986) showed that correlations exist between chemotype polymorphism, sexual polymorphism, and the environment. Populations of thyme growing in stable and xeric environments are predominantly hermaphrodite (able to self-fertilize) and largely homogeneous in regard to chemotype (thymol or carvacrol). In more "difficult" environments, selection for genetic diversity may be greater. In these environments females can be selected for and may promote a high genetic diversity because of their obligate outcrossing. Nonphenolic chemotypes, which are favored in this environment, are capable of producing a wide range of chemotypes among their progeny.

The occurrence of geraniol (G), α-terpineol (A), thujanol-4 (U), linalol (L), carvacrol (C), and thymol (T) characteristic of the chemotypes is under genetic control (Valdeyron et al. 1977; Vernet 1977; Vernet et al. 1986). The nature of the monoterpenes is controlled by a series of loci. There are epistatic relationships between these loci and the general dominance relationship between different chemotypes is as follows:

$$G > A > U > L > C > T.$$

Thymol is recessive to all others. The genetic control of T and C; L; U and A chemotypes seems to be realized by one pair of alleles for each C/c, L/l, U/u, and A/a. The geraniol chemotype is controlled by two loci (G_1/g_1, G_2/g_2). When selfed, a T individual produces only T individuals in its progeny; a G individual heterozygous at all six loci will produce G, A, U, L, C, and T progenies when selfed. Unlike other types, the L-type individuals exhibit a qualitative variation of monoterpene content which is age-related. The first leaves produce phenols throughout their life, while leaves produced later will always produce linalol. A regulation system seems to be necessary to explain the behavior of the L type.

The existence of *T. vulgaris* chemotypes was tested in practice. Plants selected from natural populations were used to establish plantations supplying herbs for the raw materials containing volatile oil of fixed composition (Lamy 1983).

1.3 The Use of Active Principles

Thymus vulgaris is a medicinal plant, and it is also used in the food and cosmetic industries (Duke 1986). The antitussive, expectorant, and spasmolytic properties are commonly considered as the major action of thyme. This species is also used as an antiseptic, antifungal, antiphlogistic, rubefacient and anthelminthic (Duke 1986). Thyme preparations described in pharmacopoeias are: Herba Thymi, Oleum Thymi, Extractum Thymi Fluidum, Sirupus Thymi Compositus, Thymolum (Farmakopea Polska, *Pharmacopoea Polonica* Editio IV; *Arzneibuch* der Deutschen Demokratischen Republik 2. Ausgabe; *Deutsches Arzneibuch* 9. Ausgabe; *Pharmacopoea Bohemoslovaca* Editio quarta; *Pharmacopée Française* Xe Édition; Martindale *The Extra Pharmacopoeia* 29th Edition).

A minimum percentage of volatile oil (not less than 1% dry wt.) is usually required in Herba Thymi. Thyme oil and thyme extracts must also possess sufficient phenol (thymol and carvacrol).

The establishment of a fixed quantity of phenols is an attempt to standardize the crude drug, made necessary by the existence of thyme chemotypes containing very low concentrations of phenols. The phenols are generally regarded as the active agents of thyme. In some countries thyme liquid extracts are a fundamental component of many galenic preparations and drugs with antitussive activity (van den Broucke 1982). Van den Broucke and Lemli (1981) found that thymol and carvacrol cannot induce the spasmolytic activity of thyme liquid extracts because their concentration is too low ($\leqslant 0.01\%$). Some extracts showed high spasmolytic action, but no correlation between phenol content and activity was observed. Pharmacological studies (van den Broucke 1982, 1983) proved that the action of thyme extract may be almost completely explained by the flavone content. The concentration of polymethoxyflavones reported for the first time in *Thymus* species by van den Broucke et al. (1982) should be taken into account when evaluating thyme preparations. Van den Broucke (1982) and Brasseur (1983) gave detailed reviews of the literature on chemical compounds and medicinal properties of *Thymus vulgaris* and the use of this species in therapy in drugs used externally and internally. The antibacterial and antifungal properties were also discussed. Thyme is also used as a food preservative. Conner and Beuchat (1984) found that volatile thyme oil acts on several kinds of yeast that spoil food. Thymol inhibited the growth of toxigenic fungi and the production of mycotoxins of four *Aspergillus* species (Janssen et al. 1987).

1.4 Need for in Vitro Culture

The high morphological and chemical variability of *Thymus vulgaris* allows the selection of plants with features most valuable for growers. The development of tissue culture methods may contribute to the production of secondary metabolites in several ways. One of them is the rapid multiplication of selected plants by a procedure which allows plants to be obtained that are genetically identical to the donor plant. Another way could be the production of secondary metabolites by cell and tissue culture initiated from explants of selected plants. The volatile oil, however, is very difficult to obtain in plant cell and tissue cultures. It was found that the synthesis of the volatile oil occurred in morphologically differentiated cultures producing shoots and roots, or in suspension and callus cultures containing specialized cells (Collin and Watts 1983). Some volatiles were synthetized in vitro by callus and suspension cultures of the species of the Labiatae family: *Mentha* spp., *Melissa officinalis, Perilla frutescens, Ocimum basilicum, Pogostemon cablin, Thymus quinquecostatus*, and *Thymus vulgaris*. Bioconversion of monoterpenes by cultures of *Mentha* spp. was also studied (Koch-Heitzmann and Schultze 1988; Mulder-Krieger et al. 1988). Nerolidol, vanilin, ethyl benzoate, and isovaleramide were found in callus of *Thymus quinquecostatus* (Koch-Heitzmann and Schultze 1988). Production of volatiles in callus cultures derived from leaves of *Thymus vulgaris* was studied by Sugisawa et al. (1988). The effects of various constituents of culture media on callus induction and productivity of volatiles were studied using five kinds of basal media, Murashige and Skoog, B5, Nitsch and Nitsch, White's, and Linsmaier and Skoog. Callus induction was satisfactory on Murashige and Skoog medium with NAA (5 mg/l) and kinetin (10 mg/l). The yield of oil from callus was very low, corresponding to 1/500–1/1000 of that of the entire plant. Fifty-seven components including 21 aldehydes, 7 alcohols, 4 ketones, 8 terpenes, and others were identified. A remarkable feature of the essential oil obtained from callus was the presence of a variety of C15–C17 aldehydes and sesquiterpenes. Small amounts of thymol were also found. Geranylacetone, β-chamigrene, nerolidol, β-guaiene, cuparene, and γ-cuprenene, which were not detected in the intact plant, were found in callus. It was found that the rates of growth and essential oil production in callus reached a plateau after 30 days of subculture. The rate of oil production increased distinctly from the 25th to the 30th day. Growth regulators and precursors (mevalonic acid 1 g/10 ml) affected quantity and quality of the essential oil. The productivity of volatiles was increased by the addition of 2,4-D (10 mg/l) and kinetin (10 mg/l) to Murashige and Skoog medium; and was five times greater than that in the combination IAA (1 mg/l) and kinetin (1 mg/l). There were 109.3 and 554.3 μg of volatiles in 100 g fresh weight, respectively.

The production of essential oil in tissue culture is possible only in the range of laboratory studies and cannot be used for commercial purposes.

Regarding the great variability of thyme, micropropagation could contribute to the production of plants useful in breeding new varieties, thus supplying large quantities of valuable plant material.

2 Micropropagation

Various methods of micropropagation of medicinal and aromatic plants were presented by Bajaj et al. (1988). In this publication some medicinal plants from the Labiatae family are also cited. These plants were most frequently multiplied by axillary buds. Shoot tips or nodal segments were used to establish cultures. Plants were more rarely obtained as a result of formation of adventitious shoots directly from explants or from callus.

Thymus vulgaris plants were multiplied by axillary buds. Morphological structure of thyme makes it possible to obtain many explants from one plant. A preliminary report on the regeneration of *T. vulgaris* in vitro was presented by Furmanowa and Olszowska (1980). Detailed research on *T. vulgaris* micropropagation and phytochemical analysis were described by Olszowska (1982) and Olszowska and Furmanowa (1987).

2.1 Establishment of Tissue Cultures

Apical and axillary buds from plants grown in the field, and cotyledon segments with shoot tips of seedlings obtained under sterile conditions were used for the initiation of cultures. Shoots 5–7 cm long were rinsed in 70% alcohol and then three times in autoclaved distilled water. Apical and axillary buds (0.5–1 mm) containing apical meristem plus one or two pairs of leaf primordia were isolated under 34 × magnification and cultured in test tubes (16 × 1.5 cm) on filter paper bridges dipped in liquid medium. The seeds were surface sterilized with 5% calcium hypochlorite for 5 min, washed several times with sterile distilled water, and placed aseptically in Petri dishes on filter paper moistened with autoclaved tap water. Seeds germinated (90%) after 3 days at 25 °C in the light (12 h/day, 500 lx). Cotyledon segments with shoot tips were excised and placed on solid medium.

Plantlets developed from the explants were multiplied by taking nodal segments to solid medium.

2.2 Media

Media used for culture contained Nitsch and Nitsch (1969) (NN) nutrients, sucrose (2 or 4%), and 10 mg/l adenine sulfate (AS), and were supplemented with varying concentrations of auxins (NAA, IBA) and cytokinins [kinetin (K), BA, 2-iP]. The pH of the media was adjusted to 5.5. For solid media Difco-Bacto agar (0.6 or 0.8%) was added prior to autoclaving at 121 °C for 15 min. The cultures were kept under 2000 lx light for 12 h/day at 25 °C.

2.3 Plant Regeneration from Buds

Within 5 weeks plantlets about 2 cm long developed on NN medium with 2% sucrose, 10 mg/l AS and kinetin (0.05, 0.1, 0.25 mg/l) and NAA (0.05, 0.1, 0.25, 0.5 mg/l) in 12 combinations (Fig. 3). Similar results were obtained on media

Fig. 3. Plantlets developed from shoot buds on NN medium containing 0.1 mg/l K, 0.1 mg/l NAA, and 10 mg/l AS after 31 days, 62 days, and 77 days (from *left* to *right*)

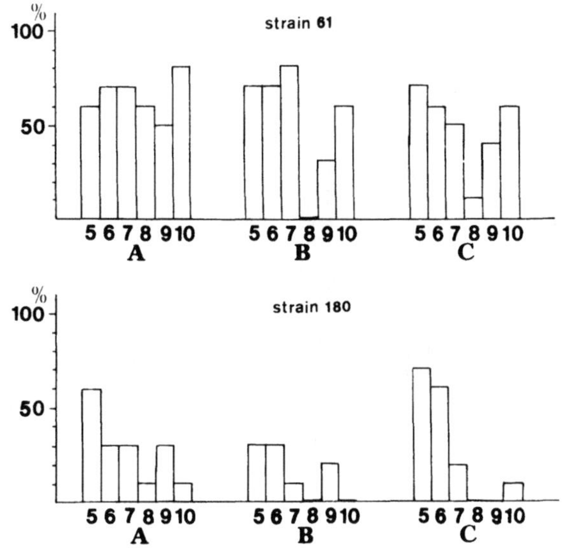

Fig. 4. Percentage of shoot buds developing into plantlets. The explants were isolated from May to October (*5–10*) from plants of two strains (*61, 180*) growing in the field, and transferred on NN medium with: *A* 10 mg/l AS, 0.1 mg/l K, 0.1 mg/l NAA; *B* 10 mg/l AS, 0.1 mg/l K, 0.3 mg/l IBA; *C* 10 mg/l AS, 0.5 mg/l IBA

containing: 0.3 mg/l IBA; 0.5 mg/l IBA; 0.1 mg/l K and 0.5 mg/l IBA. On media containing 0.5 mg/l K and NAA at concentrations of 1, 5, and 10 mg/l, buds produced shoots with callus at the base or only calli. On media with BA (0.05, 0.1, 0.2 mg/l) and 0.1 mg/l NAA, development of the main shoot was inhibited and proliferation of axillary shoots was observed. Part of the shoots was vitrified.

Media with 0.1 mg/l K and 0.1 mg/l NAA, 0.1 mg/l K and 0.3 mg/l IBA, and 0.5 mg/l IBA were chosen for subsequent experiments which showed that there was no difference in apical and axillary bud development. Development of plants from shoot tips depended on the intact plant which was the source of the explants and the time of year at which the buds were excised (Fig. 4). Shoot buds of two genotypes among 25 plants tested at random failed to grow in vitro. The best time to start in vitro culture from shoot buds was May to June; the percentage of buds producing plantlets was 60–70%.

2.4 Plant Development from Nodal Segments

Nodal segments of plants grown under sterile conditions were transferred on solid NN medium with 2% sucrose, 10 mg/l AS, 0.8% agar, supplemented with NAA (0.1, 0.3, 0.5 mg/l) or IBA (0.1, 0.3, 0.5 mg/l) and kinetin (0, 0.1 mg/l). Shoots developed well on all the media, but rooting depended on the kind and concentration of auxin, and was better on media containing IBA than NAA. The best rooting was observed on medium with 0.5 mg/l IBA without kinetin. Cytokinins (kinetin, BA, 2-iP) added at 0.05 mg/l to medium supplemented with 0.5 mg/l IBA had no effect or only little effect on plantlet development (Fig. 5). When increasing concentrations of the cytokinins were added (0.1, 0.5, 1, 5 mg/l), gradual inhibition of rooting and shoot growth was observed except for 1 mg/l 2-iP. Lower concentration of agar (0.6%) improved root development. Increased sucrose concentration (4%) improved both rooting and shoot development.

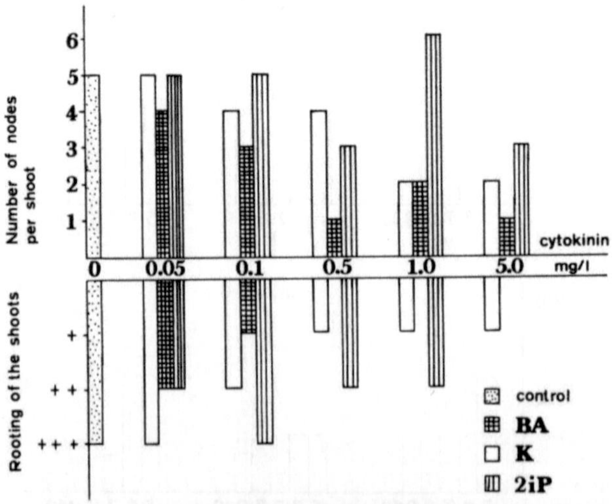

Fig. 5. Development of plants from nodal segments after 5 weeks of culture on solid NN medium (0.6% agar) with 10 mg/l AS, 0.5 mg/l IBA, 2% sucrose, and various concentrations of cytokinins: K, BA, and 2-iP. + = one or two roots about 1 cm long; + + = several roots about 2 cm long; + + + = several roots about 4 cm long

Fig. 6. Rooted plantlets regenerated from nodal segments on NN solid medium with 10 mg/l AS and 0.5 mg/l IBA, ready for transfer to pots

NN medium with 10 mg/l AS, 0.5 mg/l IBA, 4% sucrose, and 0.6% agar was chosen as standard medium for thyme multiplication by nodal segments and was also used for cotyledon segment culture. Within 4 weeks, an average of 90% of nodal segments developed into well-rooted plantlets (Fig. 6). The mean number of nodes per shoot was five. One shoot developed from the first node (apical segment). Both axillary buds of consecutive nodal segments developed into shoots. The position of the nodes (1–4) had no significant effect on plantlet development.

2.5 Transfer to Soil

Plantlets regenerated from buds after 6 weeks of culture and from nodal segments after 4 weeks of culture were transferred to pots filled with disinfected soil and maintained under beakers for the initial 7–10 days. One month after transplantation to pots, plants could be transferred to the field.

3 Summary and Conclusions

In the course of this research work 21 strains of *Thymus vulgaris* were obtained. They differ in morphological and physiological characters and essential oil content. Comparison of 11 strains is given in Table 1. Flowering shoots of three strains are shown in Fig. 7. Flowering plants of several strains are shown in Fig. 8. The composition of the volatile oil of all strains was similar, thymol being the main oil component (Fig. 9).

Thymus vulgaris can be propagated in vitro from shoot buds. The entire duration from initiation of culture to transfer to the field is about 67 days. Plants developed from buds excised in the late spring (May–June) can be transferred to the field in August. It is possible to obtain about 300 plants within 2.5 month taking buds from one 1–2-year-old thyme plant. This is more than from stem cuttings (about 50–60 plants) (Cybura 1956).

Table 1. Comparison of morphological features and volatile oil content in herb of 11 *Thymus vulgaris* L. strains obtained by micropropagation

Strain	Start of flowering	Sexual form[a]	Volatile oil content (% v/w)
2	20 V 83	mS	0.97
51	19 V 83	mS	1.52
61	17 V 83	mS	1.05
68	19 V 83	mS	1.57
180	19 V 83	mF	1.22
IP-10	23 V 83	mF	1.80
IP-12	19 V 83	mS	1.70
IP-23	19 V 83	mF	1.19
IP-24	17 V 83	mF	1.52
IP-26	23 V 83	mS	1.55
IP-28	26 V 83	mS	1.82

[a] mS = male sterile; mF = male fertile.

Fig. 7A–C. Flowering shoots of three strains of *Thymus vulgaris* L. **A–C** Strains 2, 68, and 180

Fig. 8. Flowering plants of several strains of *Thymus vulgaris* L.

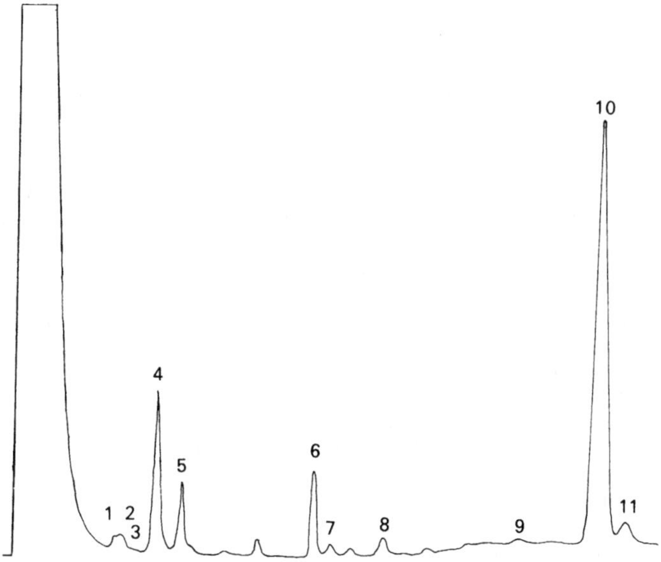

Fig. 9. Gas chromatogram of the volatile oil of *Thymus vulgaris* L. (strain 68) on a column (1.5 m × 3 mm) packed with 15% DEGS on 100–120 Gaschrom P. *1* Myrcene; *2* α-terpinene; *3* limonene; *4* γ-terpinene + 1,8-cineole; *5* p-cymene; *6* linalol; *7* bornyl acetate; *8* borneol; *9* geraniol; *10* thymol; *11* carvacrol

When plants derived in vitro are propagated by nodal segments, about 2200 plants can be obtained from one shoot bud after 5 months of culture.

The findings described here provide a simple procedure for rapid clonal multiplication of *T. vulgaris* which can be used for propagation of selected genotypes with biological and chemical characters suitable for breeding purposes.

4 Protocol

1. To start in vitro culture cut shoot buds 0.5–1 mm in length from a plant growing in the field, or take cotyledon segments with shoot tip of seedlings
2. For rapid multiplication divide plantlets growing in vitro into nodal segments
3 Place shoot tips on filter paper bridges dipped into liquid medium containing Nitsch and Nitsch (1969) (NN) nutrients, 10 mg/l AS, 2% sucrose and 0.5 mg/l IBA or 0.1 mg/l kinetin with 0.1 mg/l NAA
4. Place nodal segments and cotyledon segments with shoot tips on NN solid medium with 4% sucrose, 10 mg/l AS, 0.5 mg/l IBA, and 0.6% agar
5. Culture at 25 °C in 12-h photoperiod (2000 lx)
6. Plantlets developed from shoot buds after 6 weeks of culture, and plantlets regenerated from nodal segments after 4 weeks of culture are transferred to pots. Maintain the plants under glass beakers for the initial 7–10 days. One month after transplantation to pots, the plants can be transferred to the field
7. When nodal segments are used as explants, 2200 plants (in 5 months of culture) can be obtained from one plantlet, which was developed in vitro from one shoot bud

Acknowledgments. We wish to thank Prof. Dr. J. Kozłowski, Institute of Medicinal Plants, Poznań, for kindly supplying *Thymus vulgaris* seeds. We are also thankful to Mr. I. Rudnicki for the photographs.

References

Adzet T, Granger R, Passet J, San Martin R (1977) Le polymorphisme chimique dans le genre *Thymus*: sa signification taxonomique. Biochem Syst Ecol 5: 269–272

Adzet T, Martinez-Vergés F (1980) Luteolin and 6-hydroxyluteolin: taxonomically important flavones in the genus *Thymus*. Planta Med (Suppl) 52–55

Arzneibuch der Deutschen Demokratischen Republik (1983) 2. Ausg. Bd 3 Akademie Verlag, Berlin

Assouad MW, Dommée B, Lumaret R, Valdeyron G (1978) Reproductive capacities in the sexual forms of the gynodioecious species *Thymus vulgaris* L. Bot J Linn Soc 77: 29–39

Awe W, Schaler J, Kümmel H (1959) The flavones from *Thymus vulgaris*. Naturwissenschaften 46: 558

Bajaj YPS, Furmanowa M, Olszowska O (1988) Biotechnology of the micropropagation of medicinal and aromatic plants. In: Bajaj YPS (ed) Biotechnology in agriculture and forestry, vol 4. Medicinal and aromatic plants I. Springer, Berlin Heidelberg New York, pp 60–103

Barbawski B (1965) Colorimetric determination of caffeic and chlorogenic acids in plant material. Diss Pharm 17: 45–51

Brabant P, Gouyon PH, Lefort G, Valdeyron G, Vernet P (1980) Pollination studies in *Thymus vulgaris* L. (Labiatae). Acta Oecol Oecol Plant 1(15): 37–45

Brasseur T (1983) Études botaniques, phytochimiques et pharmacologiques consacrées au thym. J Pharm Belg 38: 261–272

Collin HA, Watts M (1983) Flavor production in culture. In: Evans DA, Sharp WR, Ammirato PV, Yamada Y (eds) Handbook of plant cell culture, vol 1. Techniques for propagation and breeding. Macmillan, New York, pp 729–747

Conner DE, Beuchat LR (1984) Effects of some essential oils from plants on growth of food spoilage yeasts. J Food Sci 49: 429–434

Cybura R (1956) Relation between morphological structure and yield of essential oil in the vegetative clones of *Thymus* species. Diss Pharm 7: 339–351

Deutsches Arzneibuch (1986) 9 Ausg. Dtsch Apotheker Verlag, Stuttgart, Govi-Verlag, Frankfurt

Dommée B, Assouad MW, Valdeyron G (1978) Natural selection and gynodioecy in *Thymus vulgaris* L. Bot J Linn Soc 77: 17–28

Duke JA (1986) Handbook of medicinal herbs. CRC, Boca Raton, Florida

Elena-Rosselló JA (1979) Identification of a species-specific enzyme marker in *Thymus* L. J Hered 70: 147–149

Farmakopea Polska IV wyd (1970) Pharmacopoea Polonica, 4th edn, vol 2. PZWL, Warszawa

Furmanowa M, Olszowska O (1980) *Thymus vulgaris* L. propagation through tissue culture. Acta Polon Pharm 37: 243–247

Garcia Marquina JM (1949) Saponins of *Thymus vulgaris*. Farmacognosia 9: 261–276

Gouyon PH (1975) Note sur la carte provisoire de la répartition des différentes formes chimiques de *Thymus vulgaris* L. dans le bassin de Saint-Martin-de-Londres (Hérault, France). Oecol Plant 10: 187–194

Gouyon PH, Vernet P, Guillerm JL, Valdeyron G (1986) Polymorphisms and environment: the adaptive value of the oil polymorphisms in *Thymus vulgaris* L. Heredity 57: 59–66

Granger R, Passet J (1973) *Thymus vulgaris* spontane de France: races chimiques et chemotaxonomie. Phytochemistry 12: 1683–1691

Jalas J (1972) *Thymus* L. In: Tutin TG, Heywood VH, Burges NA, Moore DM, Valentine DH, Walters SM, Webb DA (eds) Flora Europaea, vol 3. Univ Press, Cambridge, pp 172–182

Janssen AM, Scheffer JJC, Baerheim Svendsen A (1987) Antimicrobial activities of essential oils. A 1976–1986 literature review on possible applications. Pharm Weekblad Sci 9: 193–197

Koch-Heitzmann I, Schultze W (1988) Compilation of volatile compounds found in plant cell cultures. In: Schreier P (ed) Bioflavour '87. De Gruyter, Berlin, pp 365–395

Lamy J (1983) La culture du thym dans la Drome. Parf Cosm Aromes 51: 73–74

Martindale The extra pharmacopoeia (1989) 29th edn. Pharm Press, London

Mulder-Krieger T, Verpoorte R, Baerheim Svendsen A, Scheffer JJC (1988) Production of essential oils and flavours in plant cell and tissue cultures. A review. Plant Cell Tissue Organ Cult 13: 85–154

Nitsch JP, Nitsch C (1969) Haploid plants from pollen grains. Science 163: 85–87

Olechnowicz-Stępień W, Lamer-Zarawska E (1975) Investigation of flavonoid fraction of some crude drugs from the family Labiatae (*Herba Serpylli* L., *Herba Thymi* L., *Herba Majoranae* L. *Herba Origani* L.). Herba Pol 20: 347–356

Olszowska O (1982) Rozmnażanie klonalne *Thymus vulgaris* L. droga kultur tkankowych in vitro i ocena uzyskanych linii. (Vegetative propagation of *Thymus vulgaris* L. in tissue culture and estimation of selected strains). PhD Thesis, Med Acad, Warsaw

Olszowska O, Furmanowa M (1987) Micropropagation of thyme (*Thymus vulgaris* L.) from nodal segments. Herba Pol 33: 137–144

Pharmacopée Française (1986) 10th edn. L' ADRAPHARM, Paris

Pharmacopoea Bohemoslovaca (1987) 4th edn. Avicenum-Zdravotické Nakladatelstvi, NP, Praha

Rey C (1990) Selection of thyme (*Thymus vulgaris* L.) for extreme areas. Herba Hung 29: 30–33

Rowe E (1949) Isolation of oleanolic acid and ursolic acid of *Thymus vulgaris*. J Am Pharm Assoc 38: 12–14

Rumińska A (1981) Rośliny lecznicze (Medicinal plants). PWN, Warszawa

Skopp K, Hörster H (1976) An Zucker gebundene reguläre Monoterpene, Teil 1. Thymol- und Carvacrolglykoside in *Thymus vulgaris*. Planta Med 29: 208–215

Sugisawa H, Miwa K, Matsua T, Tamura H (1988) Volatile compounds produced from the cultured cells of thyme (*Thymus vulgaris* L.). In: Schreier P (ed) Bioflavour '87. De Gruyter, Berlin, pp 327–340

Turowska I, Kozłowski J, Golcz L (1980) Zarys zielarstwa (Medicinal plant cultivation in outline). PWN, Warszawa

Valdeyron G, Dommée B, Vernet P (1977) Self-fertilization in male-fertile plants of a gynodioecious species: *Thymus vulgaris* L. Heredity 39: 243–249

Van Den Broucke CO (1982) Thijm, voldoende gewaardeerd in de geneeskunde? Farm Tijdschr Belg 1: 49–60

Van Den Broucke CO (1983) The therapeutic value of *Thymus* species. Fitoterapia 4: 171–174

Van Den Broucke CO, Lemli JA (1981) Pharmacological and chemical investigation of thyme liquid extracts. Planta Med 41: 129–135

Van Den Broucke CO, Dommisse RA, Esmans EL, Lemli JA (1982) Three methylated flavones from *Thymus vulgaris*. Phytochemistry 21: 2581–2583

Vernet P (1977) Les variations de composition de l'essence de *Thymus vulgaris* L.: mode de transmission héréditaire de trois terpènes (le thymol, le carvacrol et le linalol). CR Acad Sci 284(D): 1289–1292

Vernet P, Guillerm JL, Gouyon PH (1977a) Le polymorphisme chimique de *Thymus vulgaris* L. (Labiée) I. Répartition des formes chimiques en relation avec certains facteurs écologiques. Oecol Plant 12: 159–179

Vernet P, Guillerm JL, Gouyon PH (1977b) Le polymorphisme chimique de *Thymus vulgaris* L. (Labiée) II. Carte à l'échelle 1/25000e des formes chimiques dans la région de Saint-Martin-de-Londres (Hérault-France). Oecol Plant 12: 181–194

Vernet P, Gouyon PH, Valdeyron G (1986) Genetic control of the oil content in *Thymus vulgaris* L.: a case of polymorphism in a biosynthetic chain. Genetica 69: 227–231

Yamaura T, Tanaka S, Tabata M (1989) Light-dependent formation of glandular trichomes and monoterpenes in thyme seedlings. Phytochemistry 28: 741–744

II.5 Micropropagation of Japanese Bunching Onion (*Allium fistulosum* L.) and Its Hybrid (*A. fistulosum* × *A. cepa*) Derivatives

E.B. PEFFLEY[1]

1 Introduction

Alliums belong to the family Liliaceae and are of considerable importance worldwide. Tremendous genetic diversity exists among the over 600 species (McCollum 1976), which are found mainly in northern temperate zones (Novak et al. 1986). *A. fistulosum* is widely grown in East and West Africa and Asia (Tindall 1983). *A. fistulosum* is most commonly known as the Japanese bunching onion, but is also called the Welsh onion (the word Welsh is derived from the Anglo-Saxon *welise* and the old German *welsche*, meaning foreign; it has no association with Wales) (Tindall 1983). Other less common names include the Chinese small onion (Herklots 1972), green bunching onion, spring onion, two-bladed onion, Ciboule (Fr.), Cebolla, Ceboletta (Sp.), Rohtenlauch (Ger.), Pijplook, Bieslook (Neth.), Taai Ts'ung (China), and Kikiyu onion (E. Afr.) (Tindall 1983).

The bunching onion is a diploid (x = 8) biennial herbaceous plant, flowering in the second season of growth. *A. fistulosum* has characteristics which distinguish it from other alliums (for review see Jones and Mann 1963). It has no well-developed bulb and a shortened stem-plate (basal plate) from which roots and shoots arise. The diameter increases with growth; when mature the basal plate appears dome-shaped. Lateral buds (in many cultivars) develop which give rise to multiple bulbs, tillers, or divisions. Leaves are produced from the apical meristem, pushing through the pseudostem formed by the sheath leaf bases of the older leaves (Yamaguchi 1983). Its leaves are foliar (nonstorage), hollow, erect, and fistulose (McCollum 1982). The scape is fistulose and uniformly straight; the spathe is short, without a long tip, and remains attached to the inflorescence base after flowering. Flowering is early (November through February in the greenhouse and February through May in the field) at Lubbock, Texas, USA. Flowering progresses uniformly from the apex of the umbel towards its base. The perianth is campanulate or lyriform, petals are translucent and pale greenish yellow (McCollum 1982), never open widely, and are without a pronounced mid-rib on the perianth segments. Florets are protandrous.

A. fistulosum can be separated by degree of tillering into three groups. The Hanegi (ha = leaf, negi = bunching) or Kujyo types are nondormant, continuously growing, relatively small, 2.5 cm in diameter and 30 to 38 cm in total length (Herklots 1972), and tend to tiller and form multiple stems; stems may be blanched

[1] Department of Agronomy, Horticulture and Entomology, Texas Tech University, Lubbock, TX 79409-2134, USA

Biotechnology in Agriculture and Forestry, Vol. 19
High-Tech and Micropropagation III (ed. by Y.P.S. Bajaj)

for 25 cm or more by placing earth up over the lower leaf bases of the growing plants; propagation is by tillers or seed. The Nebuka (ne = root, buka = deep) or Kaga types are larger plants, single-stemmed, go dormant in the winter, and are more cold-tolerant than the Hanegi types; propagation is by seed. The Senjyu types are intermediate (Haruhisa Inden, pers. commun.).

Commercial interest is greatest in Asia and North and Central America (Mexico). While there is not much commercial demand for this onion in Europe and the United States, onion breeders from these areas are interested in using *A. fistulosum* as a source of resistance genes for the improvement of the bulb onion, *A. cepa* (van der Meer and van Bennekom 1978). Many crosses between these species have been made (see Table 1) and introgression of *A. fistulosum* into *A. cepa* has recently been documented in backcross derivatives (Cryder et al. 1990; Peffley and Mangum 1990). Collections are found in the Federal Republic of Germany, Japan, The Netherlands, United Kingdom, and three locations in the United States.

A. fistulosum has been the main garden onion of China and Japan for centuries (McCollum 1982; Yamaguchi 1983) and is one of the most important vegetables in Japan (Tatebe 1961). Astley et al. (1982) cite China as the center of origin and Japan as the center of diversity, while Vavilov (1951) and most botanists attribute the origin of the onion to Persia and West Pakistan. The history of onions dates back to at least 3200–2800 B.C., from findings in Egyptian tombs (Yamaguchi 1983). Onions were found in India about 600 B.C. and the Greeks and Romans wrote about onions and garlic about 400–300 B.C. Onions had spread into northern Europe by the start of the Middle Ages. Alliums with onion-like flavor are found today in the wild in temperate regions of the Northern hemisphere. *A. fistulosum* is not known in the wild state.

A. fistulosum is grown primarily for the green leaves and white leaf bases. It is usually consumed fresh for the tops and white stems, is frequently sautéd, and is used in stir fry, soups, stews, and other dishes for enhanced flavor. The flavor of alliums is attributed to alliinase, an enzyme which acts on sulfur compounds when the tissues are disturbed. The volatile flavor compounds in onions are mainly propyl disulfide and methyl propyl disulfide. The tear- or lachrymator-causing

Table 1. Interspecific crosses involving *A. fistulosum*

Cross	Reference
A. cepa × *A. fistulosum*	Emsweller and Jones (1935)
	Levan (1936)
	Maeda (1937)
	Saini and Davis (1967)
	El-Gadi and Elkington (1975)
	Dolezel et al. (1980)
A. fistulosum × *A. ascalonicum*	Cochran (1950)
A. fistulosum × *A. cepa*	Dolezel et al. (1980)
	Corgan and Peffley (1986)
	Peters et al. (1984)
A. fistulosum × *A. galanthum*	El-Gadi and Elkington (1975)
A. fistulosum × *A. royeli*	McCollum (1982)

Table 2. World production of dry onion and garlic. (FAO 1979, as cited by Yamaguchi 1983)

	Onion bulb Area (10^3 ha)	Onion bulb Production (10^3 mt)		Garlic Area (10^3 ha)	Garlic Production (10^3 mt)
World	1500	19494		387	2446
Continent					
Africa	121	1241		8	174
North and Central America	65	1950		12	118
South America	127	1702		33	151
Asia	946	8632		237	1536
Europe	246	4293		87	446
Oceania	6	177			
USSR	170	1500		11	21
Leading countries					
1. China	208	2581	1. China	54	510
2. United States	50	1746	2. Korea, REP	50	350
3. India	215	1600	3. Thailand	42	300
4. USSR	170	1500	4. Spain	39	215
5. Japan	30	1120	5. Egypt	6	161
6. Turkey	70	1000	6. India	54	150
7. Spain	32	905	7. United States	6	84
8. Brazil	68	691	8. Argentina	17	81
9. Italy	22	540	9. Italy	6	60
10. Egypt	21	536	10. Turkey	12	60

Table 3. World production of green onions and garlic. (FAO 1975, as cited by Yamaguchi 1983)

	Area (10^3 ha)	Production (10^3 MT)	Yield (MT/ha)
World	110	1542	14.0
Continent			
Africa	12	153	12.6
North and Central America	26	296	11.4
South America	4	35	8.8
Asia	59	947	16.1
Europe	9	111	12.3
Leading countries			
1. Japan	25	610	24.4
2. Mexico	26	295	11.5
3. Turkey	14	85	6.1
4. Korea, REP	7	80	11.7
5. Nigeria	3	80	26.7
6. China	5	72	14.1
7. Iraq	5	70	13.6
8. Tunisia	2	44	22.0
9. Equador	4	35	8.8
10. Greece	3	31	12.1

compound is thiopropanyl sulfoxide, liberated by alliinase (Yamaguchi 1983). The strength of flavor increases with the age of the plant. Vitamin A content has been found to be 50 times higher in the bunching onion than the dry bulb onion (Nonnecke 1989).

Information on the economic importance of *A. fistulosum* is lacking. FAO world production figures for dry onions, garlic, green onions, and shallots have been cited by Yamaguchi (1983) and are shown in Tables 2 and 3. While the statistics for green onions and shallots, as given by Yamaguchi (1983), were not for *A. fistulosum* alone, many green onion cultivars are the species *A. fistulosum* and some shallots (Perkins et al. 1958) are interspecific *A. fistulosum* derivatives (Perkins et al. 1958); the figures in Tables 2 and 3 may provide comparative inference on the importance of this vegetable worldwide.

2 Need for Micropropagation

Utilization of male sterility in the breeding of crop plants has allowed for excellent improvement in the production of F_1 hybrid cultivars in *A. cepa*. Similar advancement in the development of *A. fistulosum* cultivars might be realized if a male sterile system could be established. The male sterility system in *A. cepa* has been traced back to a single male sterile (female) plant at the University of California at Davis in 1925 (Jones and Mann 1963) and since then genes for male sterility have been found in a number of crop species. Nishi and Hiraoka (1958, cited by Jones and Mann 1963) reported a monogenic, recessive type of sterility in *A. fistulosum* where the enlargement of the tapetal cells resulted in pollen abortion in the anthers. Male sterile mutants appearing spontaneously at high frequency in *A. fistulosum* were reported by Nishimura and Shino (1972 cited by Fujieda et al. 1977). Fujieda et al. (1977) wrote that the genetic mechanism controlling male sterility in *A. fistulosum* is more complicated than the system in *A. cepa* and that a technique for breeding male sterile strains had not been established. This remains so today. If sufficient numbers of male sterile seed parents could be generated, breeding of F_1 hybrid seed could be realized: vegetative reproduction avoids the maintenance of male sterile lines by a nonrestorer fertility line (Havel 1982). It is possible by tillering or divisions to propagate a sufficient number of male sterile plants for crossing, but multiplication is slow. Micropropagation offers a means to produce a large number of, theoretically, identical offspring from a limited amount of parent material, thus providing an efficient way to increase the number of female plants that are required for F_1 hybrid seed production.

Hybrids with *A. fistulosum* have been made in a number of species (Table 1). Crosses between *A. fistulosum* and *A. cepa* have been made with the goal of recovering a bulbing-type onion like *A. cepa* with the disease-resistant characteristics of *A. fistulosum* (for review see Jones and Mann 1963; van der Meer and van Bennekom 1978). Interspecific F_1 hybrids of *A. cepa* × *A. fistulosum* (and reciprocal) have been achieved with relative ease (Emsweller and Jones 1935; Levan 1936; Maeda 1937; Corgan and Peffley 1986), but because of high sterility, future generations are difficult to obtain. Tissue culture can be used for achieving these goals.

Tissue culture offers techniques that can be used in unconventional breeding methods. Firstly, high volumes of uniform individuals can be efficiently increased. Rapid clonal propagation provides a way to multiply unique and rare genotypes of interspecific derivation which possess elite traits, thereby generating sufficient numbers for use in breeding. Secondly, organ (i.e., embryo) culture offers a means to recover progeny from highly sterile interspecific F_1 hybrids and backcross progeny and has been employed in order to overcome incompatibility in interspecific (i.e., *A. cepa* × *A. fistulosum*) hybrids. In interspecific hybridization, pollination/fertilization occurs, but post-zygotic incompatibility may prevent development of the embryo: growth is prematurely arrested and abortion occurs. Before progeny can be recovered the barrier of cross incompatibility must be overcome. Thirdly, somatic embryogenesis offers a means whereby large numbers of individuals can be regenerated from a single progenitor; however, somaclonal variation negatively impacts the utility of this approach when clones are the desired product.

3 Review of Tissue Culture Literature

Much has been written on the tissue culture of selected species belonging to the genus *Allium* (i.e., *A. cepa* and *A. sativum*). In vitro culture techniques have been applied to various organs, tissues, and cells, giving rise to various kinds of morphogenic events (for review see Novak et al. 1986). This chapter concerns the in vitro culture of *Allium fistulosum* and its hybrid (*A. fistulosum* × *A. cepa* and reciprocal) derivatives; it emphasizes micropropagation and is limited to studies in which plant regeneration has been demonstrated. A survey of references concerning the in vitro culture of *A. fistulosum* and *A. cepa* interspecific derivatives is presented in Table 4.

Table 4. Summary of tissue culture studies on *Allium fistulosum*, *A. fistulosum* × *A. cepa*, and *A. cepa* × *A. fistulosum* interspecies crosses in which plants have been regenerated

Species	Type of tissue culture	Reference
A. fistulosum	Callus	Lin and Cui (1982); Phillips and Hubstenberger (1987); Shahin and Kaneko (1986); Tashiro et al. (1985); Yamane (1979)
	Micropropagation	Fujieda et al. (1977); Oosawa et al. (1981); Ohkoshi et al. (1987)
A. fistulosum × *A. cepa*	Callus	Lu et al. (1989); Phillips and Hubstenberger (1987)
A. cepa × *A. fistulosum*	Callus	Shahin and Kaneko (1986)
	Embryo	Dolezel et al. (1980); Gonzalez and Ford-Lloyd (1987)
	Micropropagation	Mar'yakhina et al. (1983)
	Ovary	Guan and Peffley (1989)

3.1 Micropropagation Techniques

Most micropropagation techniques utilize vegetative meristematic tissue. The use of vegetables regenerated from tissue culture for field production of the edible portion is usually not feasible or necessary, since most vegetables are seed-propagated and produce seed in relative abundance (Ng 1986). However, there are situations where micropropagation of vegetables is most useful in producing male sterile plants where an established cytoplasmic sterility system does not exist (Ng 1986). Increasing breeding lines of *A. fistulosum* by micropropagation was the objective of the research done by Fujieda et al. (1977). An elite male sterile female line was identified, but the male sterility could not be suitably exploited for lack of an efficient means of seed propagation. Shoot-tip explants of Welsh onion were cultured on MS medium (Murashige-Skoog 1962). Maximum proliferation of adventitious shoots were obtained when grown on MS medium supplemented with 2 mg/l 6-furfurylaminopurine (kinetin) and 0.5 mg/l 1-naphthylacetic acid (NAA), moderate temperature (20 °C), and long days. After 30 days on this regime plantlets were transferred onto MS medium containing no growth hormones. Roots formed after an additional 30 days. Primary cultures with many adventitious shoots were used as the source for more multiple shoots. Adventitious shooting was more active when the stock plants were more advanced in age. Fujieda et al. (1977) found that the presence of cytokinin stimulated the multiplication of adventitious shoots, but completely inhibited the formation of adventitious roots. If shoots were excised and placed back onto the same medium, roots formed. If the shoots were placed on the medium with the hormones omitted, root and leaf formation were stimulated so that a plantlet was formed. Others have reported the use of clonal multiplication through tissue culture (Mar'yakhina et al. 1983; Oosawa et al. 1981; Ohkoshi et al. 1987), but the Fujieda et al. (1977) report is the most thorough protocol published for the micropropagation of *A. fistulosum*. Plantlets were regenerated by Oosawa et al. (1981) that had been induced and cultured from callus derived from tissues immediately behind meristems and young leaf tips; regenerants from callus are usually not considered clonal propagules in the narrow sense. Recently, Ohkoshi et al. (1987) reported the cloning of meristem tips from the Welsh onion Bouzushirazu, but the protocol was not given. In vitro clonal propagation of *Allium* plants for generation of polyploid forms has been reported by Mar'yakhina et al. (1983). The explants were cultured on modified MS medium with and without added colchicine; optimum conditions were established for propagating the material studied but were not available in their report. The authors reported that polyploidization of the interspecific hybrids increased pollen viability.

3.2 Embryo Culture

Embryo culture of *A. fistulosum* genotypes is limited to the culture of interspecific material. Hybrids between *A. fistulosum* and *A. cepa* can be achieved by traditional methods, but backcross progeny are difficult to obtain since the F_1 hybrids are highly sterile and seed set is low. Dolezel et al. (1980) experienced poor seed set in *A. fistulosum* × *A. cepa* (and reciprocal) crosses. Imperfect development of endosperm was observed in most of the fertilized ovules. Only a small quantity of endosperm was present in the seeds. When embryos were cultured in vitro on B5 medium (Gamborg et al. 1968) without growth regulators, more hybrid plants were obtained than when seed was allowed to mature on the plant. Gonzales and Ford-Lloyd (1987) reported success in generating progeny from crosses between *A. cepa* and *A. fistulosum* when using an in ovulo embryo rescue technique. Embryos were inoculated onto B5 medium (after Dolezel et al. 1980). The greatest number of regenerants were achieved when embryos were rescued at between 15 and 20 days after pollination. Guan and Peffley (1989) cultured ovaries (after the protocol of Guha and Johri 1966) which had been excised from an interspecific triploid [*A. cepa* (2X = 16) × *A. fistulosum* (1X = 8)] fertilized with *A. cepa* (2X = 16) pollen. The greatest number of plants were recovered when ovaries were excised and cultured 5 days after pollination. Ten days after pollination the number of plants recovered was greatly reduced.

3.3 Callus

An interesting development in rapid propagation via tissue culture is that of somatic embryogenesis and the prospect of encapsulating embryos for use as synthetic seeds (Kitto and Janick 1985). Efficient use of this system will require high frequency and reliable initiation and recovery of somatic embryos. There are reports of plantlets regenerated from *A. fistulosum* callus (Lin and Cui 1982; Yamane 1979), but procedures and detailed results for these studies were not available. Somatic embryogenesis has been demonstrated in material from *A. fistulosum* and its interspecific hybrids (*A. fistulosum* × *A. cepa* and reciprocal). Phillips and Hubstenberger (1987) cultured explants taken from various parts of interspecific hybrid plants. Callus was initiated when inoculated onto BDS basal medium (Dunstan and Short 1977) supplemented with 0.75 mg/l 4-amino -3,5,6-trichloropicolinic acid (Picloram) and 2 mg/l N6-benzyladenine (BA). Plants were regenerated after transfer onto medium containing 0.03 mg/l Picloram and 0.35 mg/l BA (with and without 2.5 g/l proline). Regeneration of plants varied from very high to low among the different genotypes. Multiple shoots and somatic embryos were induced from callus of *A. fistulosum* cv. Japanese Bunching and Beltsville Bunching, an interspecific amphidiploid, in the work reported by Shahin and Kaneko (1986). Callus grew from radicles 2 weeks after seeds were germinated on media supplemented with various hormones. The presence of 2,4-dichlorophenoxyacetic acid (2,4-D) was sufficient for callus initiation and embryoid induction. After several transfers, callus became embryogenic. Kinetin was critical for shoot differentiation. Shoots were formed when transferred to medium with kinetin, and rooted on medium with NAA. From 3 to 35 shoots per callus were recovered. After drying the bulbs, plants were regrown in the greenhouse. Somatic embryos have also been recovered from callus of *A. fistulosum* × *A. cepa* interspecific diploid plants initiated from basal plates of nondormant, field-grown plants and from immature inflorescences (Lu et al. 1989). Explants were inoculated onto BDS-based medium supplemented with 0.75 mg/l Picloram and 2 mg/l BA (after Phillips and Luteyn 1983). Upon transfer to medium with reduced levels of growth hormones, N6-(2-isopentenyl) adenine (2-iP) substituted for BA, and supplemented with vitamins (Shahin 1984) and amino acids, large numbers of somatic embryos were initiated. Germinated embryos rooted upon transfer to hormone-free BDS-based medium and were transferred to the greenhouse. Several thousand plants have been established in the field. This work is reported in more detail in the following section.

4 Micropropagation

The objective of the research conducted in our laboratory was to establish a one-step process for the regeneration of multiple propagules.

4.1 Plant Material

Three genotypes of *A. fistulosum* were cultured: two plant introductions (PI), 433630 and 461401, obtained from the United States Department of Agriculture Regional Plant Introduction Station, Geneva, New York; var. Heshiko supplied by International Plant Breeders and four diploid populations of *A. fistulosum* × *A. cepa* F_1 hybrids (Corgan and Peffley 1986) were cultured. An *A. cepa* selection of New Mexico Yellow Grano line 80-20-1 supplied by Dr. J. N. Corgan of New Mexico State University, was the pollen parent in each of the hybrids. The *A. fistulosum* seed parents were Formosa Bunching, Ishikura Long White, Bunching No. 1, and Algara Bunching; all supplied by International Plant Breeders.

4.2 Media and Culture Conditions

Explants were inoculated onto either MS medium or onto B5 medium as modified by Dunstan and Short (BDS) (1977) with no supplemental plant growth hormones. Media were solidified with 10 g/l agar. Agar concentrations of 10 g/l were used in order to minimize vitreousness. Media were adjusted to pH 5.8 prior to autoclaving at 121 °C, 1.05 kg/cm^2 for 21 min and dispensed into Petri dishes. Cultures were grown under 18-h photoperiod at 20 °C. Subcultures onto fresh medium were done every 3–4 weeks.

4.3 Explants: Basal Plate

Greenhouse-grown *A. fistulosum* plants were harvested 2–3 days prior to culture. Plant material was taken to the laboratory and most of the root material was cut off, taking care not to injure the basal plate. Loose soil was removed with clean sudsy water under running tap water. Plants were drained upside down. Plant tops were removed close to the basal plate with at least two leaf bases remaining. Root stubs and outer portions of the basal plate were removed using a scalpel dipped in 70% ethanol. Scalpels were sterilized in alcohol and flamed after each cut. No material previously in contact with the soil remained. Multiple bulbs (those with several substantial growing points) were divided into two or more explants. Trimmed onion pieces were surface sterilized according to Lu et al. (1989). Onion pieces were stored overnight in sterile Petri dishes, and the following day the tissue surrounding the central portion of basal plate was removed. Explants were placed with the basal plate (root end) in contact with the medium. Petri dishes were sealed with Parafilm. No more than two explants were inoculated onto each plate. Explants which gave off a milky exudate were moved to clean medium as necessary.

4.4 Explants: Inflorescence

Unruptured, immature umbels of *A. fistulosum* genotypes were surface sterilized by passage through 70% ethanol for 5 min, 50% commercial sodium hypochlorite for 5 min, followed by three sterile distilled water rinses. The spathe was removed and florets were cut off, leaving a receptacle which was cut into two or four pieces (depending upon size of inflorescence) prior to inoculation onto the medium.

4.5 Regeneration: Basal Plate

Previous experiments in our laboratory showed that explants inoculated onto BDS medium without growth hormones produced shoots and roots within 1 month, whereas Fujieda et al. (1977) produced rooted plantlets only when explants were cultured on MS medium supplemented with kinetin and indolebutyric acid (IBA). A one-step propagation process is highly desirable and if it can be accomplished efficiently without supplements it will be more cost-efficient in terms of chemicals and labor. Therefore, a series of experiments were designed to test the effect of BDS and MS basal salts without growth regulators on the development of explants. Explants taken from Heshiko were inoculated onto medium containing either BDS or MS basal salts. Explants inoculated onto BDS medium were maintained on BDS medium throughout the duration of the experiment. Shoots which proliferated from explants inoculated onto MS medium elongated and turned green, but it was observed that if they were maintained on MS medium they became twisted and vitreous within 2 weeks. When explants were inoculated onto MS medium but transferred onto BDS medium 1 week after inoculation, the proliferating shoots were normal in appearance. This explantation and transfer system was followed for the remaining experiments and is called the MS regime (MSr). Explants inoculated onto BDS medium were transferred onto fresh BDS medium at the time explants were transferred from MS to BDS medium. Similar responses were observed in the number of multiple shoots (no roots) when explants were grown on either BDS or MS regimes, but only when basal plate explants of *A. fistulosum* plants were grown on BDS were plantlets (rooted shoots) produced (Table 5, Fig. 1A, B).

Plantlet regeneration from basal plates is dependent upon bulb physiology at time of excision: if plant material is developmentally reproductive (i.e., seed stalks initiated), vegetative, axillary shoot proliferation is inhibited. In our experiments basal plate explants that were reproductive produced only abnormal, dwarfed scapes which either ceased to develop or, in many cases, continued to mature and

Table 5. Regeneration from *Allium fistulosum* var. Heshiko basal plate explants after 30 days in culture

Response %	Basal salts	
	BDS	MSr[c]
Plantlets[a]	17	0
Shoots[b]	19.5	22

[a] 23 observations.
[b] 41 observations.
[c] Explantation onto MS for 1 week, then transfer to BDS.

formed sterile inflorescences in vitro (Fig. 1C). If the sterile inflorescences were severed from the basal plate and maintained in culture, vegetative shoots were generated from the receptacle tissue within 60 days. When the inflorescence remained on the scape, multiple shoots failed to form from the explant, but when scapes were removed from the basal plates, shoots developed from axillary buds (Fig. 1D). Multiple shoots were also produced from inflorescences that were initiated when reproductive basal plates were grown on MS medium for 1 month and then transferred to BDS medium (Fig. 1E).

4.6 Regeneration: Inflorescence

Experiments were conducted to test the effects of age, genotype, basal salts, and concentrations of the cytokinins 6-benzyl-aminopurine (BA) and kinetin on the development of shoots from inflorescence explants.

The first experiment tested the effect of cytokinin-supplemented media on shoot production. Unopened inflorescences of various ages were prepared from Heshiko and PI 418953, and explants were inoculated at random onto media which consisted of 0.0, 1 or 2 mg/l kinetin or BA and either MS or BDS basal salts. After 1 week, explants inoculated onto MS medium were transferred to BDS. All cultures were transferred to fresh BDS medium after 3 weeks. Numbers of shoots produced were recorded after 60 days in culture (Table 6). Genotypes responded differently. PI 418953 produced shoots only when cultured on MSr supplemented with 1 mg/l kinetin. Heshiko produced shoots on both the BDS and MS regimes and did not require exogenous cytokinins; in fact, no shoots were produced when cytokinins were supplied. Heshiko is a multiple-stemmed type of onion belonging to the Hanegi group (see Sect. 1) and exogenous cytokinins were not required in the culture media in order to form multiple shoots. PI 418953 is a single-stemmed, Nebuka type, and 1 mg/l kinetin was required in the culture medium to produce shoots. The genotypic differences may be explained in that multiple-stemmed types produce endogenous cytokinins (reflected in the tillering habit) and that exogenous applications of cytokinins are not necessary for shoot production in vitro. Whereas single-stemmed types may produce less cytokinins in vivo, (perhaps reflected in

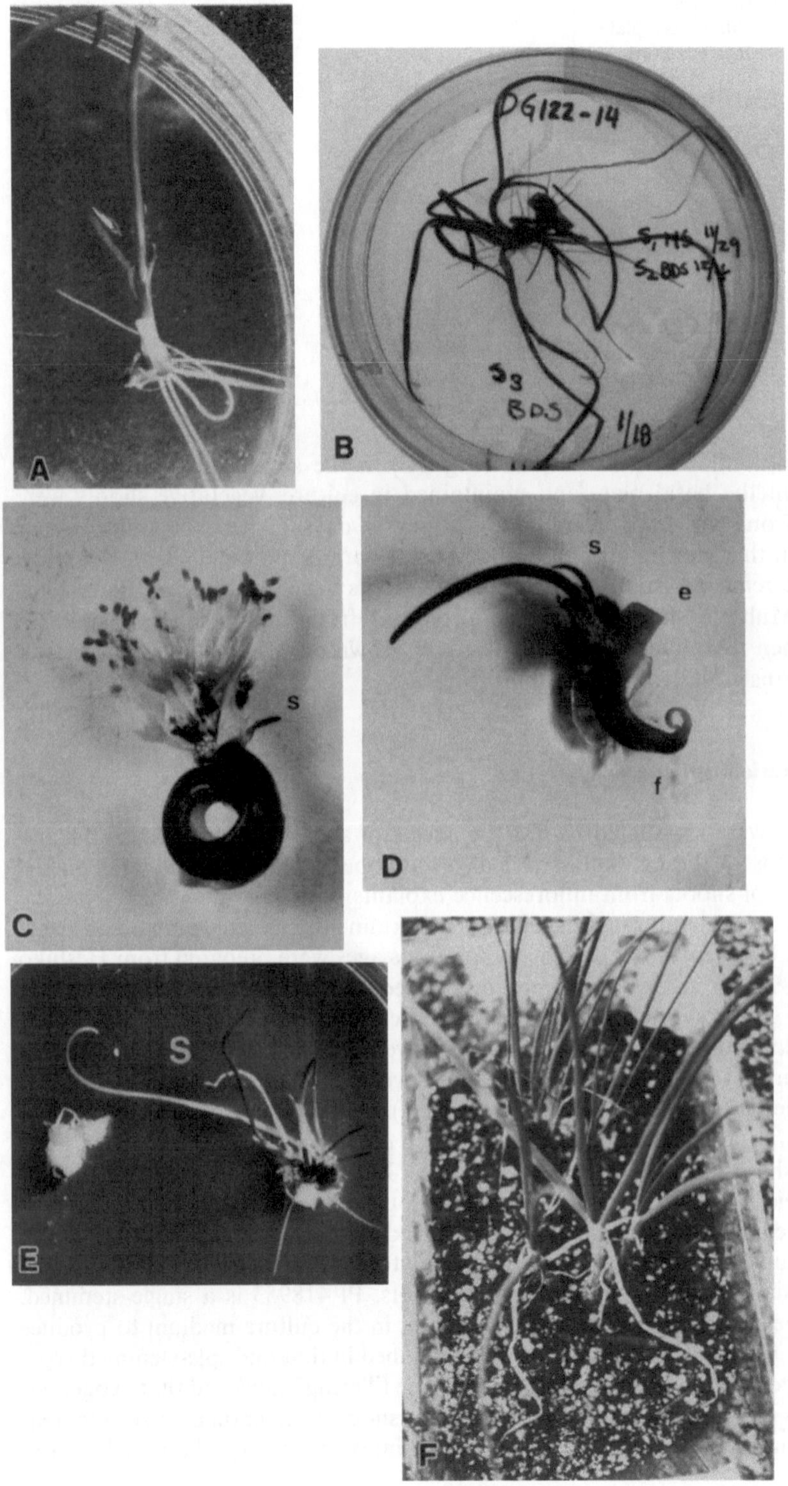
A
B
DG122-14
S3
BDS
1/18
C
s
D
s
e
f
E
S
F

Table 6. Effect of basal salts and cytokinin on shoot production from inflorescence explants of two *Allium fistulosum* genotypes after 60 days in culture

Genotype		PI 418953[a]		Heshiko[b]	
Cytokinin (mg/l)		Basal Salts			
		BDS	MSr	BDS	MSr[d]
Kinetin	1	–	+	–	–
	2	–	–	–	–
BA	1	–	–	–	–
	2	–	–	–	–
	0.0[c]	–	–	+	+

[a] 10 explants/treatment.
[b] 5 explants/treatment.
[c] 4 explants/treatment.
[d] Explantation onto MS for 1 week then transfer to BDS.

their lack of tillering, exhibiting strong apical dominance), and therefore exogenous cytokinins are required to allow multiple shooting.

A second experiment tested the age of inflorescences at time of explantation. Only Heshiko explants were used since similar shooting results were observed when cultured on either basal salt media with or without cytokinins. Every 2 days for 10 days, at least ten Heshiko umbels were tagged upon emergence from the leaf sheath. On the 11th day, each set of tagged (1, 4, 6, 8, 10 days after emergence) umbels was collected separately and ten new umbels were tagged for a total of six sets of umbels. Each set was prepared separately and inoculated at random onto BDS or MS media supplemented with 0, 1, or 2 mg/l kinetin or BA. The results, as shown in Table 7, corroborated earlier observations which showed that exogenous cytokinins did not enhance shooting in Heshiko. Shoot production did not vary with basal salts nor did the age or size of explant affect the production of shoots. The number of shoots produced from Heshiko explants was affected by basal salts and age of inflorescence but exogenous cytokinins were not necessary for shoot production. More shoots were produced from one- and same-day-old inflorescences on BDS media; on MS media more shoots were found on 8- and 10-day-old inflorescences.

Fig. 1A–F. Morphogenic events arising from cultures of *Allium fistulosum* var. Heshiko inoculated onto and maintained on BDS medium. **A, B** Plantlets regenerated from 7-week-old basal plate explants. **C** Flower stalk development arising from 4-week-old basal plate explant; shoot (*s*) arising from receptacle. **D** Shoots (*s*) proliferating on 4-week-old basal plate explant (*e*) from which the flower stalk (*f*) has been removed. **E** Multiple shoots (*s*) and roots arising from 4-week-old inflorescence. **F** 8-week-old plantlets derived from micropropagation and established in corsage box in the greenhouse

Table 7. Effect of basal salts and cytokinins on shoots produced from *Allium fistulosum* var. Heshiko inflorescence explants[a] taken 0, 1, 4, 6, 8, or 10 days after emergence from the leaf sheath

Cytokinin (mg/l)		Basal salts											
		BDS						MS					
		Inflorescence age (Days from emergence)											
		0	1	4	6	8	10	0	1	4	6	8	10
Kinetin	1	−	−	−	−	−	−	−	−	−	−	−	−
	2	−	−	−	−	−	−	−	−	−	−	−	−
BA	1	−	−	−	−	−	−	−	−	−	−	−	−
	2	−	−	−	−	−	−	−	−	−	−	−	−
	0	+	+	−	+	−	−	+	−	−	−	+	+
Number of shoots/treatment[b]		6	1	0	1	0	0	2	0	0	0	1	4

[a] 1 explant = 1/4 receptacle.
[b] 4 explants/treatment.

4.7 Other in Vitro Approaches: Somatic Embryogenesis

Regeneration of somatic embryos is now done routinely in our laboratory following the protocol established by Lu et al. (1989). Callus is established and maintained using both basal plate or inflorescence tissue on BDS medium supplemented with 0.75 mg/l Picloram, 2 mg/l BA, and 2.5 g/l proline. Transfers are done every 4–6 weeks. Generally after approximately 3 months in culture, callus becomes embryogenic, and following transfer to a lowered auxin concentration of 0.03 mg/l Picloram and changing the cytokinin to 0.32 N 6-(2-isopentenyl) adenine (2-iP) embryos germinate within 4 weeks (Fig. 2A, B). This system has produced as many as 100 plantlets per Petri dish.

High numbers of somatic embryos derived from callus cultures can be quickly multiplied in liquid suspension cultures and somatic embryos can be used for large-seed clonal propagation (Ammirato 1984). Somatic embryos are useful as clonal propagules only if genetic and chromosomal integrity has been maintained. Chromosomal alterations are accelerated with time in culture. Keeping culture time as brief as possible is necessary to maintain integrity.

No protocol has been established for a callus suspension culture in onion. We conducted experiments in order to develop a suspension culture system in onion which would allow for the quick recovery of somatic embryos from single or small cell aggregates. Onion callus was initiated and maintained on solid BDS medium with auxin, cytokinin, vitamin, and proline supplements (Lu et al. 1989), and then transferred to liquid BDS medium supplemented with 0.5, 1, or 2% Picloram, 2,4-D or NAA (Ou and Peffley 1989). Liquid cultures were agitated at 120 rpm at about 25 °C with 10/14 h light/dark photoperiod. The liquid medium was removed and replaced every 2 weeks.

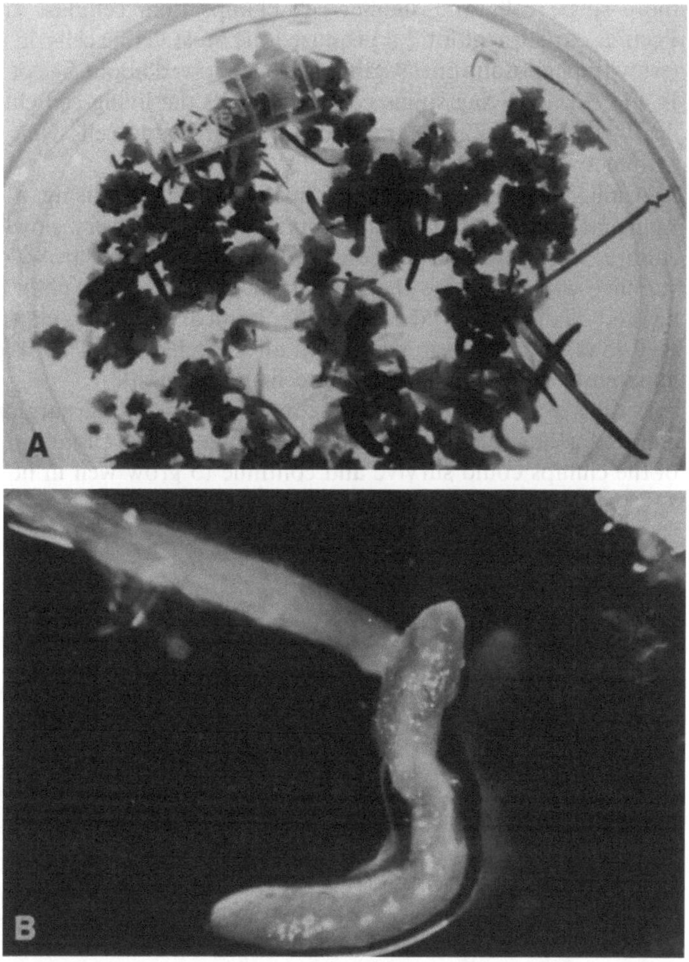

Fig. 2A, B. Somatic embryogenesis in diploid interspecific (*A. fistulosum* × *A. cepa*) F_1 hybrid onion. **A** Germinating embryos on solid BDS medium supplemented with 0.03 mg/l Picloram, 0.32 2-ip, vitamins (Shahin 1984), and 2.5 g/l proline. **B** Single germinating dipolar embryo

BDS liquid medium supplemented with Pectinase and Cellulase at 0.01, 0.05, 0.1, 0.15, and 0.2% was used to digest the cell aggregates into very small cell clumps with less than 30 cells in a clump. Enzyme treatments were conducted for 1, 2, 4, 6, 8, 16, and 24 h. Callus was either macerated into small clumps with a mortar and pestle, or cut into 1-2-mm^3 clumps using surgical blades. Clumps were cultured in liquid medium. Following transfer to liquid BDS medium supplemented with auxins at varying concentrations, callus enlarged and grew well. However, cell dispersion was not observed from suspensions of cultures after 6 months: cells had reaggregated into large callus clumps.

Pectinase and Cellulase digested cell aggregates into single cells and/or very small cell clumps with less than 30 cells in a clump. Pectinase treatments produced more single cells and smaller cell clumps than cellulase treatments. A 0.05% Pectinase treatment for 2 h produced the most viable cells, in which a cell viability test with phenosafranin (Widholm 1972) showed about 5% of the cells were viable. In Fig. 3A are living single cells, in Fig. 3B the living cell clumps, and in Fig. 3C dead cells after enzyme treatments. When viable cells were isolated, they grew extremely slowly in culture.

Callus was macerated into small cell clumps using a mortar and pestle. Aggregates larger than 2 mm^3 survived but grew very slowly in the subsequent culture. Macerated suspensions were sieved through an 860-μm screen, and cell viability was tested by the 2,3,5-triphenyltetrazolium chloride (TTC) reduction method (Towill and Mazur 1975). Immediately following maceration, 6.88% + 0.45 of the cells were viable, but all the cells were dead after 7 days in culture, apparently due to damage to the cells by the pestle.

Callus was also cut into 1–2-mm^3 clumps using surgical blades, and then cultured in liquid medium. Figure 3D shows the cell clumps in liquid culture. Most of the clumps could survive and continue to grow well in liquid culture.

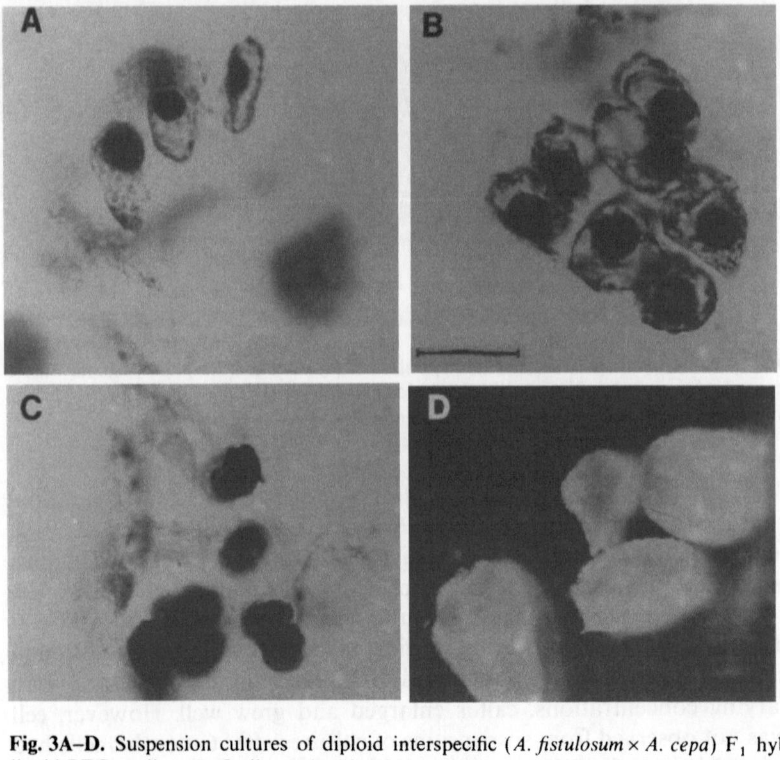

Fig. 3A–D. Suspension cultures of diploid interspecific (*A. fistulosum* × *A. cepa*) F_1 hybrid onion in liquid BDS medium **A**–**C**, digested with 0.05% pectinase. **A** Living single cells. **B** Living cell clumps. **C** Dead cells. **D** Living clumps after being cut by surgical blades

4.8 Plant Establishment

When two new leaves formed on plantlets the plants were transferred into sterilized potting media (peat moss: perlite) in commercial corsage boxes of clear plastic (Lu et al. 1989). Plantlets are placed in the dark for 2–3 days and then moved onto light shelves and given an 18-h photoperiod. When another new leaf has formed, the hardened plantlets are transferred to the greenhouse (Fig. 1f). Plants are transferred to the field after forming another new leaf.

5 Summary and Conclusions

Regeneration of plants from somatic embryogenesis has been reported earlier (Lu et al. 1989). Suspension culture systems are being investigated which will allow for rapid propagation of individual somatic embryos. In vitro clonal micropropagation of *Allium fistulosum* has been accomplished through multiple shoot regeneration from basal plate and inflorescence tissues. Rooted plantlets were recovered in a one-step process by growing on B5 medium as modified by Dunstan and Short (BDS) (1977) with no supplemental plant growth hormones: explants initiated and maintained on MS medium produced only vitreous twisted shoots. More rooted plants were obtained when explants were grown on BDS basal salts than when grown on the MSr system. Time of year is an important consideration when taking explants from mature greenhouse- or field-grown plants. Single-stemmed, nontillering Nebuka types required exogenous cytokinins for shoot production; Hanegi, tillering types, did not. Shoot production is not dependent upon age of inflorescences (up to 10 days of age). Plantlet regenerability from basal plate explants is dependent upon plant development at time of excision: if plants are developmentally reproductive (i.e., seed stalks initiated) axillary shoot proliferation is inhibited; when flower stalks are removed axillary bud development can proceed. Plants can be taken through each hardening-off stage when at least one new leaf has been formed.

6 Micropropagation Protocol

- Sterilization procedures for inflorescence and basal plate tissue were according to Lu et al. (1989)
- Explants were inoculated onto BDS-based medium with no supplemental hormones
- Rooted plantlets were ready for transfer out of in vitro culture after two true leaves had formed, usually in 4–6 weeks
- Plantlets were transferred to sterile potting media in covered corsage boxes and placed in the dark
- After 2–3 days plantlets were transferred to light shelves and provided with 18-h photoperiod
- When at least one new leaf had formed, plantlets were ready for transfer to the greenhouse

Acknowledgments. This work was supported in part by Vegetable Research, West Texas and by a grant through the Texas Advanced Technology Program. The technical assistance of Miss Judy Decker, Miss Chung-Cheng Lu, and Mr. Paul Mangum is appreciated.

References

Ammirato PV (1984) Induction, maintenance and manipulation of development in embryogenic cell suspensions. In: Vasil IK (ed) Cell culture and somatic cell genetics, vol 1. Laboratory procedures and their applications. Academic Press, Orlando, pp 139–151

Astley D, Innes NL, van der Meer QP (1982) Genetic resources of *Allium* species: a global report. Int Board Plant Genet Resour, IBPGR Secretariat, Rome, Italy

Cochran FD (1950) A study of the species hybrid, *Allium ascalonicum* × *Allium fistulosum* and its backcrossed progenies. J Am Soc Hortic Sci 55: 293–296

Corgan JN, Peffley EB (1986) Notice of release of *Allium* genetic materials (*Allium fistulosum* × *A. cepa*). New Mexico Agric Exp Stn Release

Cryder CM, Corgan JN, Urquhart NS, Clason D (1990) A study of progeny from the interspecific backcross (*A. fistulosum* × *A. cepa*) × *A. cepa*. In: Proc Natl Onion Res Conf, Dec 7–8, Boise, Idaho

Dolezel J, Novak FJ, Luzny J (1980) Embryo development and in vitro culture of *Allium cepa* and its interspecific hybrids. Z Pflanzenzücht 85: 177–184

Dunstan DI, Short KC (1977) Improved growth of tissue cultures of the onion, *Allium cepa* L. Physiol Plant 41: 70–72

El-Gadi A, Elkington TT (1975) Comparison of the Giemsa C-banding karyotypes and the relationships of *Allium cepa, Allium fistulosum* and *Allium galanthum*. Chromosoma 51: 19–23

Emsweller SL, Jones HA (1935) An interspecific hybrid in *Allium*. Hilgardia 9: 265–273

Fujieda K, Ando Y, Fujita Y (1977) Propagation of Welsh onion through shoot tip culture. J Fac Agric Kyushu Univ 22: 89–98

Gamborg OL, Miller RA, Ojima K (1968) Nutrient requirements of suspension cultures of soybean root cells. Exp Cell Res 50: 151–158

George EF, Sherrington FD (1984) Plant propagation by tissue culture. Handbook and Directory of Commercial Laboratories Exegetics Limited. Eastern P, England

Gonzalez LG, Ford-Lloyd BV (1987) Facilitation of wide-crossing through embryo rescue and pollen storage in interspecific hybridization of cultivated *Allium* species. Plant Breed 98: 318–322

Guan G, Peffley EB (1989) Megaspore culture of interspecific (*Allium cepa* × *A. fistulosum*) hyperploid onions. Hortic Sci, Proc 86th Annu Mtg Am Soc Hortic Sci

Guha S, Johri BM (1966) In vitro development of ovary and ovule of *Allium cepa* L. Phytomorphology 16: 353–364

Havel L (1982) Plant differentiation in tissue culture of some species of *Allium* genus. Thesis, Inst Exp Bot Czech Acad Sci, Praha

Herklots GAC (1972) Vegetables in southeast Asia. Hafner P, New York, pp 396–397

Jones HA, Mann LK (1963) Onions and their allies. Interscience, New York, 286 pp

Kitto SL, Janick J (1985) Production of synthetic seeds by encapsulating asexual embryos of carrot. J Am Soc Hortic 110(2): 277–282

Levan A (1936) Die Zytologie von *Allium cepa* × *A. fistulosum*. Hereditas 21: 185–214

Lin ZP, Cui QI (1982) Regeneration of plants from callus of *Allium fistulosum*. Acta Bot Sin 24(6): 586–587, Inst Bot, Beijing, China (in Chinese)

Lu C, Currah L, Peffley EB (1989) Somatic embryogenesis and plant regeneration in diploid *Allium fistulosum* × *A. cepa* F_1 hybrid onions. Plant Cell Rep 7: 696–700

Maeda T (1937) Chiasma studies in *Allium fistulosum, Allium cepa* and their F_1, F_2, and backcross hybrids. Jpn J Genet 13: 146–159

Mar'yakhina I Ya, Polumordvinova IV, Kozlova NM (1983) Clonal propagation of *Allium* plants and the development of polyploid forms in vitro. Skh Biol 6: 16–21, Vses Inst Prikladnoi Mol Biol Genet, Moscow, USSR (in Russian)

McCollum GD (1976) Onion and allies. In: Simmonds NW (ed) Evolution of crop plants. Longman, London, pp 186–190

McCollum GD (1982) Experimental hybrids between *Allium fistulosum* and *A, roylei*. Bot Gaz 143(2): 238–242

Murashige T, Skoog F (1962) A revised medium for rapid growth and bioassays with tobacco tissue cultures. Physiol Plant 15: 473–497

Ng TJ (1986) Use of tissue culture for micropropagation of vegetable crops. In: Zimmermann RH, Griesbach RJ, Hammerschlag FA, Lawson RH (eds) Tissue culture as a plant production system for horticultural crops. Nijhoff, Dordrecht, pp 259–270

Nishi S, Hiraoka T (1958) Histological studies on the degenerative process of male sterility in some vegetable crops. Bull Nat Inst Agric Sci, Jpn, Ser E, 6: 1–41
Nishimura J, Shino M (1972) Studies on the male sterility in Welsh onion. I. Cytological studies. Abstr, Spring Assembly, Jpn Soc Hortic Sci, pp 180–181 (in Japanese)
Nonnecke IL (1989) Vegetable production. Reinhold, New York, pp 294–319
Novak FJ, Havel L, Dolezel J (1986) Allium. In: Evans DA, Sharp WR, Ammirato PV (eds) Handbook of plant cell culture, vol. 4, Techniques and applications. Macmillan, New York, pp 419–456
Ohkoshi K, Ishii K, Jitsukawa S (1987) Vitrus elimination from the Welsh onion Bouzushirazu by means of the meristem tip culture method and its practical use, I: Examining the various conditions of the meristem tip culture and their variations in the regenerated plants. Bull Chiba-ken Foundation Seed and Stock Farm, Mar, 1987, 9: 1–10 (in Japanese)
Oosawa K, Kuriytama T, Sugahara Y (1981) Clonal multiplication of vegetatively propagated crops through tissue culture, I. Effective balance of auxin and cytokinin in the medium and suitable explant parts for mass production of plantlets in strawberry, garlic, scallion, Welsh onion, yan and taro. Bull Vegetable and Ornamental Crops Res Stn, 9: 1–46 Ano Mie, Jpn (in Japanese)
Ou G, Peffley EB (1989) Establishment of suspension culture in onions. HortScience. Proc 86th Annu Mtg Am Soc Hortic Sci
Peffley EB, Mangum PD (1990) Introgression of *Allium fistulosum* L. into *Allium cepa* L.: cytogenetic evidence. Theor Appl Genet 79: 113–118
Perkins DY, Kehr AE, Brown RT, Tims EC, Miller JC (1958) Delta Giant, a new long-season shallot. Circular 52. Louisiana State Univ Release
Peters RJ, Netzer D, Rabinowitch HD (1984) A progress report: pink root resistance in *Allium cepa* L. × *Allium fistulosum* L. hybrids and progeny. In: Proc 3rd Allium Eucarpia Symp, Wageningen, pp 70–73
Phillips GC, Hubstenberger JJF (1987) Plant regeneration in vitro of selected *Allium* species and interspecific hybrids. Hortic Sci 22(1): 124–125
Phillips GC, Luteyn KJ (1983) Effects of Picloram and other auxins on onion tissue culture. J Am Soc Hortic Sci 108: 948–953
Saini SS, Davis GN (1967) Compatibility in some *Allium* species. Proc Am Soc Hortic Sci 91: 401–409
Shahin EA (1984) Isolation, culture, and regeneration of potato leaf protoplasts from preconditioned in vitro. In: Vasil I (ed.) Cell culture and somatic cell genetics of plants, vol 1. Academic Press, New York, pp 381–390
Shahin EA Kaneko K (1986) Somatic embryogenesis and plant regeneration from callus cultures of nonbulbing onion. Hortic Sci 21(2): 294–295
Tashiro Y, Miyazaki S, Kanazawa K (1985) Effects of para-fluorophenylanine on the growth and chromosome number of *Allium* callus. Bull Fac Agric Saga Univ 59: 47–56 (in Japanese) *A. fistulosum* and *A. f.* × *A. ascalonicum* F_1 hybrids.
Tatebe T (1961) Genetic studies on the leaf variegation of *Allium fistulosum* Linnaeus. Jpn J Genet 36: 151–156
Tindall HD (1983) Vegetables of the tropics. Macmillan, London, pp 14–35
Towill LE, Mazur P (1975) Studies on the reduction of 2,3,5-triphenyltetrazolium chloride as a viability assay for plant tissue cultures. Can J Bot 53: 1097–1102
van der Meer QP, van Bennekom JL (1978) Improving the onion crop (*Allium cepa* L.) by transfer of characters from *Allium fistulosum* L. Biul Warzywniczy 22: 87–91
Vavilov NI (1951) The origin, immunity, and breeding of cultivated plants. Chron Bot 13: 1–6
Widholm JM (1972) The use of fluorescein diacetate and phenosafranine for determining the viability of cultured plant cells. Stain Technol 47: 189–194
Yamaguchi M (1983) World vegetables. AVI, Westport, Connecticut, pp 184–206
Yamane Y (1979) Regenerated plants derived from tissue culture of *Allium fistulosum*. Jpn J Genet (Abstr) 54(6): 475

II.6 Micropropagation of *Mentha* L.

E. ČELLÁROVÁ[1]

1 General Account

1.1 Botany/Morphology/Chromosome Number of *Mentha* L.

The genus *Mentha* L. (Lamiaceae) includes about 25 species that are mostly aromatic perennial (rarely annual) herbs with creeping rhizomes and scented foliage. Flowers are hermaphrodite or female, on the same or different plants, usually in dense, many-flowered verticillasters, sometimes forming a long-spiked inflorescence or a terminal head. The calyx is actinomorphic or weakly two-lipped, tubular or campanulate, 10- to 13-veined, with five (four) usually subequal lobes. Corolla weakly two-lipped with four subequal lobes, the upper lobe wider and usually emarginate, with the tube shorter than the calyx. Stamens about equal, divergent or ascending under the upper lip of corolla, exserted (except in *Mentha pulegium*, some hybrids and female flowers). Style branches are subequal. Nutlets are smooth, reticulate, or tuberculate.

Mentha L. is a taxonomically complex genus subdivided into four sections (Tutin et al. 1972): *Audibertia*, *Pulegium*, *Preslia*, and *Mentha*. In the section *Mentha* identification is frequently difficult, since in addition to much phenotypic plasticity and genetic variability, most of the species are capable of hybridization with each other. Hybrids are frequent in nature but can usually be recognized by their intermediate appearance and sterility, although fertile hybrid swarms occasionally occur.

The genus *Mentha*, that comprises two subgenera (*Menthastrum* and *Pulegium*), exhibits a variety of genetic forms with different sets of chromosomes. The basic chromosome number for the subgenus *Menthastrum* is $x = 6$ and for the subgenus *Pulegium* $x = 5$, as in the majority of the species belonging to the subgenus *Pulegium* the chromosome numbers are multiples of five (Löve and Löve 1974; Kundu and Sharma 1985).

The chromosome number in somatic cells of some species of the genus *Mentha* is shown in Table 1.

Earlier cytological investigations showed an important role of autopolyploidy and allopolyploidy in the evolution of the species within the genus *Mentha*. The polyploidy occurs at both intra- and interspecific levels. Polyploidy, aneuploidy, and aneusomaty associated with cryptic structural alterations of chromosomes

[1] Department of Experimental Botany and Genetics, Faculty of Science, P. J. Šafárik University Mánesova 23, 041 54 Košice, Czechoslovakia

Biotechnology in Agriculture and Forestry, Vol. 19
High-Tech and Micropropagation III (ed. by Y.P.S. Bajaj)

Table 1. Variation in chromosome number of the genus *Mentha* L.

Species	Tutin et al. (1972)	Löve and Löve (1974)	Kundu and Sharma (1985)
M. longifolia	24	24	—
M. microphylla	48	—	—
M. spicata (spearmint)	48	48	48
M. × *villosa*	36	—	—
M. × *rotundifolia*	24	24	—
M. suaveolens	24	—	—
M. × *pyramidalis*	72	—	—
M. × *maximilianea*	60 72–78 120	—	—
M. × *piperita* nm. *citrata*	84 120	—	—
M. × *piperita* (peppermint)	66 72	—	—
M. aquatica (water mint)	96	96	—
M. × *smithiana*	120	—	—
M. × *gentilis*	54 60 84 96 108 120	—	—
M. × *verticilata*	42 78 84 90 96 120 132	—	—
M. × *muellerana*	60	—	—
M. arvensis (wild mint)	24 72 90	72	96
M. arvensis, var. *javanica*	—	—	60
M. pulegium	20	20	—
M. requienii	18		

have been effective in the evolution of species as well as of intraspecific genotypes (Kundu and Sharma 1985). These authors also reported on the chromosome morphology of six populations and species of *Mentha* L. and concluded that chromosomes are characterized by having mostly median to nearly submedian constrictions, and the number of nucleolar chromosomes, which may give preliminary evidence about the ploidy level in the cell nucleus, varied from 4 to 12.

1.2 Distribution/Importance

Mentha is widely distributed, occurring on five continents. Several species are cultivated in different parts of the world for their medicinal and commercial importance. The representatives of the genus *Mentha* are aromatic plants with high essential oil content that is accumulated in the oil glands of mature leaves and flowers. Each of the *Mentha* species contains a specific composition of essential oil. The major components of *Mentha* oils are a principal ketone and the related alcohol and ester components that may be derived from it. These compounds account for 75 to 85% of the oil by weight, and give the oil and herbage their characteristic odor. Essential oil components from several *Mentha* species, especially from peppermint (*Mentha piperita*) are widely used in perfumery, in the food industry, and for pharmaceuticals. Among these components, one of the most important is 1-menthol, which is obtained by crystallization from the essential oil from the genus *Mentha* (Guenther 1966). The large-scale industrial manufacture of 1-menthol is based on the optical resolution of racemic d,1-menthol, which is achieved via esterification with benzoic acid derivatives and selective crystallization (Fleisher et al. 1971). Many microbial esterases preferentially hydrolyze 1-menthyl esters while leaving the d-menthyl ester untouched.

1.3 Conventional Propagation

The majority of the *Mentha* species are multiplied vegetatively through the vigorous rhizome system. Natural hybridization, which is rather common, is responsible for taxonomic complications of the species within the genus *Mentha*. Most of the hybrids are sterile or subfertile. Vegetative propagation enables a large number of such hybrids to establish themselves successfully along with their parents. Some of the plants behave as functionally female plants, due to abortion of the anthers, and thus cross-pollination is ensured (Kundu and Sharma 1985).

1.4 Need for in Vitro Culture of *Mentha* L.

Sexual propagation of the representatives of the genus *Mentha* is of no importance because the progeny is usually semi-fertile or sterile. In contrast to the conventional method of vegetative propagation through rhizomes that occur naturally, the use of in vitro propagation enables the production of a large number of uniform plants for planting in a small space and within a short time interval.

Another need for in vitro culture is the possible elimination of various pathogens such as viruses, mildew, etc., as chemical protection is not very effective.

Callus and cell cultures of *Mentha* are used as a model system for studying the biosynthesis and biotransformation of desirable secondary products.

2 Brief Review of Tissue Culture Work

Callus and cell cultures of several *Mentha* species were isolated under various conditions (Table 2), namely with respect to secondary products formation.

Earlier investigations with initiation of callus and cell suspension cultures of *Mentha piperita* showed that explants cultured on the medium with 2,4-D and coconut milk grew very slowly. For example, an addition of 15% coconut milk, 1 mg/l 2,4-D and 5 mg/l myo-inositol did not promote proliferation of calli (Lin and Staba 1961).

Table 2. Plant growth substances promoting callus formation in several *Mentha* species

Species	Plant growth substances	Reference
M. piperita *M. spicata*	6 mg/l 2,4-D	Lin and Staba (1961)
M. spicata	0.1 mg/l 2,4-D	Stohs and Staba (1965)
M. piperita *M. pulegium*	0.5 mg/l K 1 mg/l 2,4-D	Becker (1970)
M. pulegium	18% CM 1 mg/l 2,4-D	Becker (1970)
M. piperita	1 mg/l 2,4-D	Rodov and Reznikova (1982)
M. piperita	3 mg/l 2,4-D	Kireeva et al. (1978)
Six *Mentha* chemotypes	2 to 5 mg/l 2,4-D (callus induction) 0.2 mg/l 2,4-D 0.5 mg/l K (callus maintenance) 0.2 mg/l 2,4-D or 1 mg/l 2,4-D 0.2 mg/l K (cell suspension)	Aviv and Galun (1978)
M. arvensis, var. *piperascens*	0.02 mg/l NAA 0.1 mg/l K	Karasawa and Shimizu (1980)
M. aquatica *M. citrata* *M. longifolia* *M. pulegium* *M. spicata* *M. suaveolens* *M. troyleana* *M. spicata* × *suaveolens*	0.2 mg/l K 1 mg/l 2,4-D	Charlwood and Charlwood (1983)

Further investigations were focused on the formation of the components of essential oil in cultured plant cells. Stohs and Staba (1965) observed no accumulation of essential oil in *M. spicata* and *M. piperita* cell cultures either spontaneously, or after addition of precursors. They concluded that metabolic processes leading to the biosynthesis of essential oil depend on the presence of special morphological structures that are not found in a cell culture system. Similar data were published by Becker (1970).

Later results showed that callus cultures of an alloploid form of *M. piperita* ($4n = 144$) cultivated in the dark accumulated 1 to 1.8% v/w essential oil in dry matter and about the same amount has been found in the medium. Essential oil was accumulated in two types of parenchymatous cells, i.e., normal sized cells and giant cells. According to the qualitative and quantitative evaluation of the essential oil, it is evident that callus cultures accumulated pulegone and piperitone, but the further biosynthesis of menthone and menthol was not detectable (Kireeva et al. 1978). Callus and cell cultures of *M. piperita* synthesize early intermediates such as nerol, piperitone, and pulegone, but not the desirable final products (Bricout and Paupardin 1975; Rodov and Reznikova 1982). The biosynthesis of essential oil components synthesized by free suspension cultures is determined by genetic basis as well as by cell environment (Aviv and Galun 1978). Higher biotransformation effectiveness was achieved with *Mentha* cells immobilized in cross-linked polyacrylamide-hydrazide (Galun et al. 1983). These authors also found that *Mentha* cells after gamma-irradiation efficiently reduce (−)-menthone to (+)-neomenthol (Galun et al. 1985). Charlwood and Charlwood (1983), who isolated cell lines from several *Mentha* species, found a very variable qualitative and quantitative accumulation of monoterpenes both between taxa and different cell lines from the same species. Many of the calli accumulated only the early precursors of monoterpenes (namely geraniol, linalol, and lavandulol), but some species gave rise to cultures that accumulated monoterpenes characteristic of the intact plant (i.e., cell lines from *M. spicata* × *suaveolens* accumulate carvone). Maximum accumulation, which is appreciably higher in liquid suspension cultures than in the corresponding calli, occurred in late exponential phase cells.

3 Micropropagation

3.1 General

Within the genus *Mentha*, only limited information is available on micropropagation. The genetic variability and phenotypic plasticity within a given species may cause a specific morphogenetic response of individual genotypes to the exogenous promoting factors.

Among the *Mentha* species, interest has been focused especially on peppermint (*Mentha piperita* L.).

Up-to-date literature data show different morphogenetic patterns, depending on plant origin and culture conditions. In addition to the use of shoot apices for clonal propagation, also other parts may be successfully used following histological determination of meristem tissues in the plant organs, and can give rise to a great number of regenerants.

3.2 Explant

Several explants were successfully used for clonal propagation of *Mentha piperita* L.: terminal and axillary buds (Čellárová et al. 1984; Venkataraman and Ravishankar 1986), two to four youngest leaves and stem cuttings comprising the point of leaf insertion (Repčáková et al. 1986). Older leaves were not suitable for regeneration experiments due to their insufficient morphogenetic capacity.

The explants were chosen following the cytological characterization of young plants. Localization of meristem has been identified in squash preparations stained by the Giemsa method. The meristematic character of leaf mesophyll tissue was evident and characterized by the presence of cells in various stages of mitosis (Fig. 1). Leaf explants possessed epidermal secretory structures in various stages of development as secretory trichomes and glands, in which essential oil is accumulated. GLC analysis of isolated glandular hairs (three-celled) and glandular trichomes (ten-celled) showed the same qualitative composition of the essential oil, but the concentration of the individual compounds was different (Amelunxen et al. 1969).

According to morphological markers, simpler types of secretory structures are secretory trichomes consisting of three cells with one basal cell, one so-called stalk cell, and one larger secretory cell (Figs. 1a, 2).

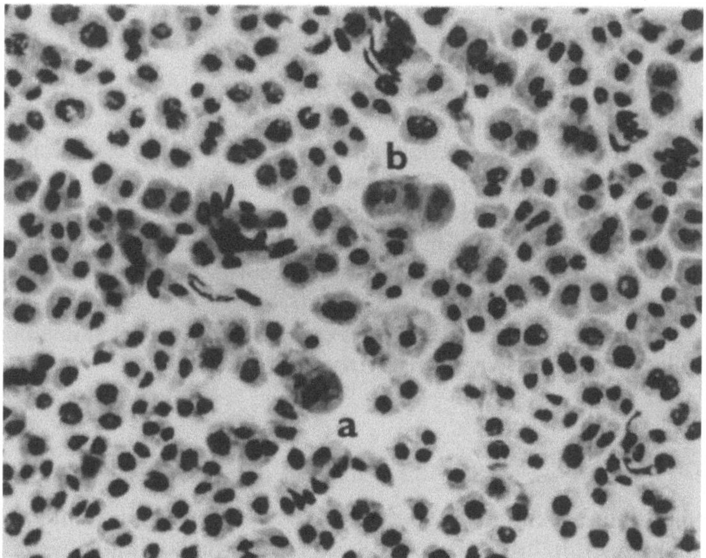

Fig. 1. Squash preparation of the meristem located in the first leaf pair. Development of secretory structures: *a* secretory trichome; *b* secretory gland. ($\times 310$). (Photo Dr. M. Rychlová)

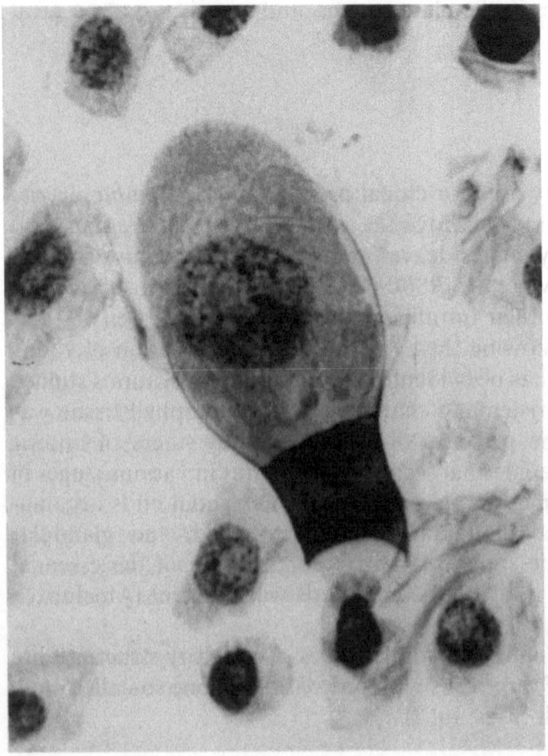

Fig. 2. Secretory trichome (three-celled) at the time of isolation of the explant. (× 760). (Photo Dr. M. Rychlová)

More complicated secretory structures are represented by a ten-celled gland which consists of one basal, one stalk cell, and eight secretory cells arranged in a circle (Figs. 1b, 3, 4). The surface of the secretory cells (trichomes and glands) is covered with the external cutinized layer of a cell wall. In the secretory cells, the connection between cutinized and cellulose layers of the cell wall is subsequently destroyed and results in subcuticular space formation, which is expanded and enlarged. Polyploidized nuclei of secretory cells may predict their high functional activity.

The third type of trichome is represented by a trichome consisting of several cells with protective and covering function.

3.3 Sterilization

Shoot apices containing two leaf pairs were rinsed in diluted detergent and then submerged in 70% ethanol for several seconds. Explants were then sterilized for 15 min with 5% aqueous solution of chloramine B containing several drops of Tween 80. The first 3 to 5 min of sterilization were performed under vacuum. Explants were then washed with sterile distilled water.

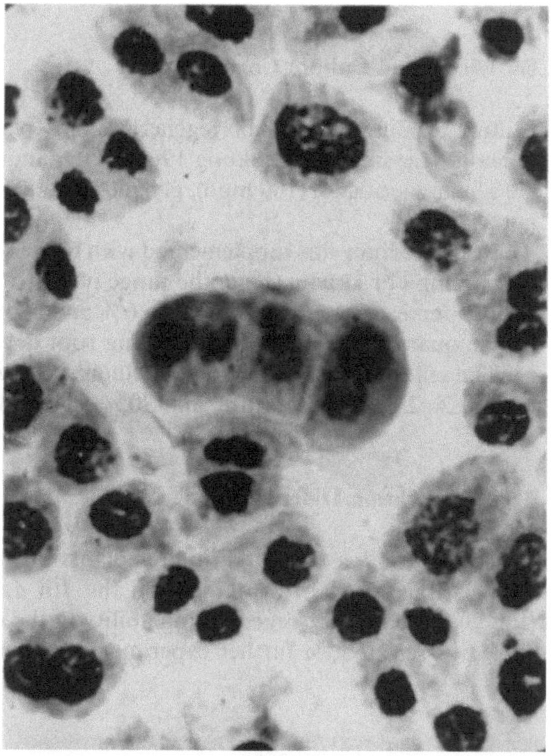

Fig. 3. Lateral view of the ten-celled secretory gland in the stage of differentiation. (× 760). (Photo Dr. M. Rychlová)

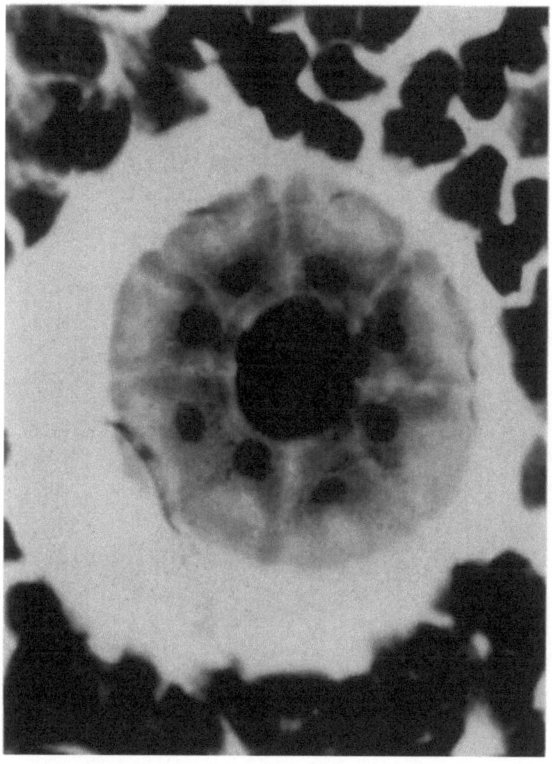

Fig. 4. Upper view of the differentiated secretory gland. (× 760). (Photo Dr. M. Rychlová)

3.4 Media and Culture Conditions

Culture medium for direct regeneration of plants from explants contains: LS minerals (Linsmaier and Skoog 1965), B5 vitamins (Gamborg et al. 1968), sucrose (30 g/l), myo-inositol (100 mg/l), glycine (2 mg/l), agar (0.4 or 0.8%), pH 5.6 before autoclaving.

Basal medium was supplemented with 6-benzylaminopurine (BAP) in the range 0.1 to 5 mg/l or kinetin (K) in the range of 0.1 to 1 mg/l. A modified basal medium with an increased sucrose content to 6% and a lowered inorganic nitrogen content to one quarter was used for promoting root development.

The cultures were grown under fluorescent lighting at 2500 lx, 16/8-h photoperiod, 26/22 °C temperature, and 70% relative humidity.

3.5 Regeneration/Differentiation

Differentiation was initiated in all explants on regeneration media supplemented with various levels of cytokinins on the 7th day after the isolation of explants (Fig. 5). Based on the regeneration ability of the explants, the medium with 2 mg/l BAP was selected for further experiments (Table 3).

Fig. 5. Early stage of direct regeneration of shoots from explants on Linsmaier-Skoog's modified culture medium. (Photo Mrs. P. Markušová)

Table 3. The influence of various levels of cytokinins on shoot differentiation in two *Mentha piperita* cultivars Bulgarian (B) and Krasnodar (K). (Repčáková unpubl.)

Plant growth regulator (mg/l)		No. of explants		No. of shoot-forming explants		No. of shoot-forming explants	
				14 d	10 d	30 d	30 d
		B	K	B	K	B	K
BAP	0.1	16	18	4	4	6 +	3 +
BAP	0.5	17	18	7	5	7 + +	7 + +
BAP	1.0	15	15	4	4	9 + +	4 +
BAP	1.5	18	18	4	4	9 + +	7 + + +
BAP	2.0	15	6	4	1	6 + + +	3 +
BAP	2.5	12	15	4	4	6 + + +	8 + +
BAP	3.0	18	15	4	5	6 + +	9 + +
BAP	4.0	18	12	2	2	5 + +	3 +
BAP	5.0	14	17	3	5	4 + +	8 +
K	0.1	14	12	3	3	1 +	5 +
K	0.5	18	14	2	3	3 +	2 +
K	1.0	17	12	7	4	10 +	2 +

+ , + + , + + + , intensity of regeneration.

The average number of regenerants increased when the agar content was lowered to 0.4%. After 2 months' cultivation the average number of shoots obtained on semi-solid medium was 24.9 ± 4.12, unlike that on the solid medium (0.8% agar), which was 7.52 ± 1.79 (Fig. 6). Some explants in the primary culture

Fig. 6. Regeneration of shoots on semi-solid culture medium. (Photo Dr. M. Repčák)

Table 4. The influence of modified culture medium on shoot regeneration. (Repčáková unpubl.)

RMB_2 medium with:	Average no. of regenerated shoots per explant
Sucrose lowered to 1%	21.5
Sucrose lowered to 2%	16
B5 vitamin complex omitted	14
Myo-inositol omitted	14
Glycine omitted	10

formed up to 50 shoots. Regeneration capacity was slightly decreased when certain components of the nutrient medium were omitted (Table 4). The number of shoots can be multiplied when regenerants are cut into smaller pieces (shoot apex with four leaves or stem cuttings with two leaves) as after rooting they serve as a source for repeating the process (Fig. 7). More intensive root development was achieved in stem cuttings.

Rhizogenesis was induced on the medium with decreased content of inorganic nitrogen to one quarter and sucrose increased to 6% with or without charcoal. The time needed for root system development was approximately 4 weeks. Sporadically, rhizogenesis was observed on the shoot regeneration medium before separating the regenerants from explants. In addition to the normal root system, adventive root formation was also observed. Despite this, typical rhizome structures were not found under in vitro conditions.

Venkataraman and Ravishankar (1986), who tested a range of growth substances for shoot differentiation, concluded that individual addition of IAA, 2,4-D, and NAA to MS medium induced plantlet formation in axillary and terminal buds. IBA induced multiple shoot formation that rooted on NAA medium. Efficient single-stage multiplication was obtained in 2,4-D and kinetin medium.

Using in vitro micropropagation, the number of regenerants is significantly higher than that obtained by conventional vegetative methods.

3.6 Transfer to the Greenhouse

Regenerated plants were transferred to nonsterile conditions (Fig. 8). A suitable substrate within the first days of adaptation is perlite. Plants were maintained under an environment of high relative humidity for several days. After 3 weeks the plants were transferred to the field (Figs. 9, 10). The regeneration cycle (explant – shoot differentiation – rooting – adaptation to field conditions) lasted 8 to 9 weeks.

3.7 Commercial Aspects of Micropropagation

The members of the genus *Mentha* are widely used in pharmacy, perfumery, and in the food industry. Conventional breeding methods are not suitable owing to the

Fig. 7. Rooting of the stem cuttings. (Photo Mrs. P. Markušová)

Fig. 8. Adaptation of *Mentha piperita* plants to nonsterile conditions. (Photo Mrs. P. Markušová)

great variability within the genus and the limited possibility of generative propagation. For these reasons, the biotechnological approach in selection and multiplication of plants with desirable properties is considered to be promising. It enables practically unlimited possibilities for cloning, shortening of the regeneration period, and potential elimination of various pathogens.

Fig. 9. Field cultivation of *Mentha piperita* regenerants. (Photo Dr. M. Repčák)

Fig. 10. Flowering plants of *Mentha piperita* regenerated in vitro. (Photo Dr. M. Repčák)

4 Summary and Conclusions

A great number of uniform and healthy plants of *Mentha piperita* L. can be obtained from terminal and axillary buds or leaf cuttings of the youngest leaf pairs by the following procedure. After surface sterilization, the explants are placed on the modified Linsmaier-Skoog medium with 2 mg/l BAP solidified with 0.4% agar. Cultures are maintained under 2500 lx illuminance for a 16-h day with an air temperature about 26 °C during the day and 22 °C at night. After 7 to 10 days the differentiation of shoots appears. When the shoots are about 5 cm tall, they can be cut into pieces comprising at least one leaf pair and transferred to the rooting medium with increased sugar and decreased inorganic nitrogen content. To multiply the number of regenerants, this process can be repeated indefinitely. The rooted plants may be moved into perlite for about 2 weeks, but during this period they must be protected against loss of moisture. Following adaptation, the regenerants can be planted in the field.

The above method can be used following the selection of individual *Mentha piperita* plants with optimal composition of secondary products to multiply desirable genotypes and to obtain uniform material for planting. With respect to the vegetative way of propagation through rhizomes, further multiplication is ensured along with the parent plants.

5 Protocol

1. *Best Explant*

Terminal and axillary buds, two to four youngest leaf cuttings.

2. *Best Medium*

For shoot differentiation: Linsmaier and Skoog minerals, sucrose 30 g/l, B5 vitamins, myo-inositol 100 mg/l, glycine 2 mg/l, BAP 2 mg/l, agar 0.4%, pH 5.6 before autoclaving.

For rooting: LS minerals with decreased inorganic nitrogen content to one quarter, sucrose 30 g/l, B5 vitamins, myo-inositol 100 mg/l, glycine 2 mg/l, agar 0.4%, pH 5.6 before autoclaving.

3. *Optimal Conditions*

In vitro: photoperiod 16/8 h, temperature 26/22 °C, relative humidity 70%, illuminance 2500 lx.

Before transferring to nonsterile conditions, an increased light intensity of about 10000 lx for 24 h is beneficial.

In vivo: transplantation to perlite, maintenance in high relative humidity environment, transfer to soil.

References

Amelunxen F, Wahlig T, Arbeiter H (1969) Über den Nachweis des ätherischen Öls in isolierten Drüsenhaaren und Drüsenschuppen von *Mentha piperita* L. Z Pflanzenphysiol 61: 68–72

Aviv D, Galun E (1978) Biotransformation of monoterpenes by *Mentha* cell lines: conversion of pulegone to isomenthone. Planta Med 33: 70–77

Becker H (1970) Untersuchungen zur Frage der Bildung flüchtiger Stoffwechselprodukte in Calluskulturen. Biochem Physiol Pflanz 161: 425–441

Bricout J, Paupardin C (1975) The composition of the essential oil of *Mentha piperita* L. cultured in vitro: influence of several factors on its synthesis. C R Acad Sci Paris D 281: 383–386

Charlwood BV, Charlwood KA (1983) The biosynthesis of mono- and sesquiterpenes in tissue culture. 63rd Meeting, Liverpool, Biochem Soc Trans 11: 592–593

Čellárová E, Repčáková K, Hončariv R (1984) Vegetative propagation of some medicinal plants through tissue cultures. Proc Int Symp Plant tissue and cell culture application to crop improvement. Olomouc, Czechoslovakia, pp 515–516

Fleischer J, Bauer K, Hopp R (1971) German patent 2109456

Galun E, Aviv D, Dantes A, Freeman A (1983) Biotransformation by plant cells immobilized in cross-linked polyacrylamidehydrazide. Planta Med 49: 9–13

Galun E, Aviv D, Dantes A, Freeman A (1985) Biotransformation by division-arrested and immobilized plant cells: Bioconversion of monoterpenes by gamma-irradiated suspended and entrapped cells of *Mentha* and *Nicotiana*. Planta Med 51: 511–514

Gamborg OL, Miller RA, Ojima K (1968) Nutrient requirements of suspension cultures of soybean root cells. Exp Cell Res 50: 151–158

Guenther E (1966) The essential oils, vol 2, 5th edn. Nostrand, Princeton, USA

Karasawa D, Shimizu S (1980) Triterpene acids in callus tissues from *Mentha arvensis* var. *piperascens*. Mal Agric Biol Chem 44: 1203–1205

Kireeva SA, Mélnikov VN, Reznikova SA, Meshchriakova NI (1978) Accumulation of volatile oil in the callus tissue of peppermint. Fiziol Rast (USSR) 25: 564–570

Kundu AK, Sharma AK (1985) Chromosome characteristics and DNA content in *Mentha* L. Nucleus 28: 89–96

Lin ML, Staba EJ (1961) Peppermint and spearmint tissue cultures. I. Callus formation and submerged culture. Lloydia 24: 139–145

Linsmaier EM, Skoog F (1965) Organic growth factor requirements of tobacco tissue cultures. Physiol Plant 18: 100–127

Löve Á, Löve D (1974) Cytotaxonomical atlas of the Slovenian flora. Cramer, Leutershausen

Repčáková K, Rychlová M, Čellárová E, Hončariv R (1986) Micropropagation of *Mentha piperita* L. through tissue cultures. Herba Hung 25: 77–88

Rodov VS, Reznikova SA (1982) Initiation and characterization of peppermint (*Mentha piperita* L.) cell suspension culture in relation to terpene biosynthesis. Fiziol Rast (USSR) 29: 644–648

Stohs SJ, Staba EJ (1965) Production of cardiac glycosides by plant tissue cultures. IV. Biotransformation of digitoxigenin and related substances. J Pharm Sci 54: 56–58

Venkataraman LV, Ravishankar GA (1986) Clonal propagation of elite plants of *Mentha piperita* by tissue culture. Proc 6th Int IAPTC Congr, Univ Minn, Mineapolis, p 65

Tutin TG, Heywood VH, Burges NA, Moore DM, Valentine DH, Walters SM, Webb DA (eds) (1972) Flora Europea, vol 3. Diapensiaceae to Myoporaceae. Cambridge Univ Press, pp 183–186

II.7 Micropropagation of Turmeric (*Curcuma domestica* Valet) and Other *Curcuma* Species

A. SUGAYA[1]

1 Introduction

1.1 Botany of *Curcuma* L.

Curcuma (family Zingiberaceae) is a large perennial herb found in the tropical and subtropical regions of India, southern Asia, and southern China. About 50 *Curcuma* species have been found in the world. Many of these, which bear ellipsoid tubers, are used as aromatic stomachics and cholagogues in traditional medicine, and also as spices or for yellow dye (Kirtikar and Basu 1975; Li et al. 1978; Perry 1980; Chuyakudaijiten 1985).

The leafy shoots about 1 m tall, oblong and often very large leaves with long leaf sheaths are surrounded by very short stems. The leaf sheath forms a pseudostem. Leaf blades can have hairs (for example, *C. aromatica*) or no hairs (for example, *C. domestica*), and *C. zodoaria* has a purple band along the midrib. Compound cylindrical spikes are formed from the central leafy stalk (Fig. 1 right, or separate shoot, Fig. 1 left). The thick, broad bracts are colored purple or pink at the uppermost tip of the spike in some species. Two to seven flowers open in each bract.

Flowers have thin bracteoles and short, tubular calices with an irregular toothed margin. The lobes of the funnel-shaped corolla are thin-walled, ovate, and oblong. Staminoids are elliptic-oblong and petaloid, a perfect stamen is short and contracts at the top. Pollen grains are nonaperturate single grains. The ovary is trilocular; the fruit a capsule, from which the inflorescence emerges. The chromosome numbers of *Curcuma* species are reported as follows: *C. amada* (mango ginger) is $n = 42$, *C. angustifolia* (narrow-leaved turmeric) ($n = 42$, $n = 64$), *C. aromatica* (wild turmeric) ($n = 42$, $n = 63$, $n = 86$), *C. domestica* (turmeric) ($n = 32$, $n = 62$, $n = 63$, $n = 64$), and *C. zedoaria* (zedoary) ($n = 63$, $n = 64$, $n = 64$) (Bolkhovskikh 1974).

The following items are keys for the identification of the species of *Curcuma*: spikes from the center of the stalks or formed separately from rhizomes, the folium size, color, and hairy or nonhairy leaves; the color and form of rhizomes.

[1] Department of Pharmacognosy, Faculty of Pharmaceutical Sciences, Josai University, Sakado-shi, Saitama-ken, 350-02, Japan

Biotechnology in Agriculture and Forestry, Vol. 19
High-Tech and Micropropagation III (ed. by Y.P.S. Bajaj)

Fig. 1. Flowering plants of *Curcuma zedoaria* (**A**) and *C. domestica* (**B**). The spike of *C. zedoaria* is formed separately from the rhizome and that of *C. domestica* from the center of the stalk

1.2 Distribution, Chemistry, and Value as a Product

1.2.1 Chemical Components of Curcuma Species

Plants of *Curcuma* contain essential oil and curcuminoids as yellow pigment. The essential oils as well as the pharmacology of the biologically active substances have been extensively studied (Fukushima et al. 1970; Hikino et al. 1975; Gupta and Banerjee 1976; Inayama et al. 1984 and other reports are in Table 1). The main chemical components of *C. domestica*, *C. aromatica*, and *C. zedoaria*, which are widely used in Japan, are as follows.

C. domestica Valet (= *C. longa* koenig)
Curcuminoid ca. 0.3%
curcumin, p-hydroxycinnamyl-feruloyl-methane,
p,p'-dihydroxycinnamyl-feruloyl-methane
Essential oil 1–5%
trumerone, dihydrotumerone 50%,
zingiberene 2%,
d-, *d*-phellandrone, cineol 1%

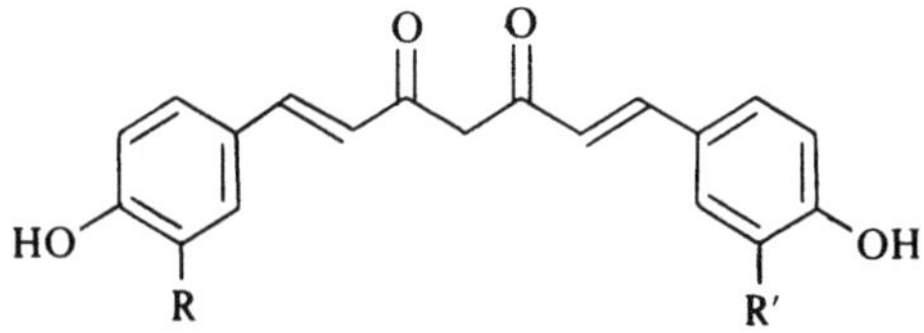

$R{=}OCH_3$, $R'{=}OCH_3$: curcumin
$R{=}OCH_3$, $R'{=}H$: *p*-hydroxycinnamoyl feruloyl methane
$R{=}H$, $R'{=}H$: *p,p'*-dihydroxy dicinnamoyl methane

C. aromatica Salisb.
Curcuminoid: small quantity
Essential oil: ca 6%
α-curcumene, β-curcumene: 65.5%
sesquiterpenalcohol: 22%
camphor: 2.5%
camphene: 0.8%

C. zedoaria Rosc.
Curcuminoid
curcumin
Essential oil: 1–1.5%
1.4.cineol, camphene, (+)borneol,
curzerenone, zederene, β-eleaene,
furanodenone, curdione,
curcumenol, curcumol, dihydrocurdione.

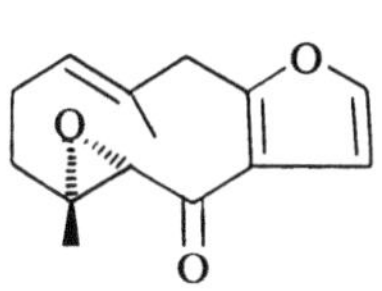

zederone

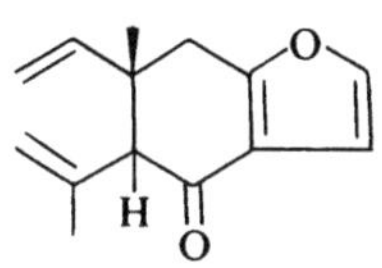

curzerenone

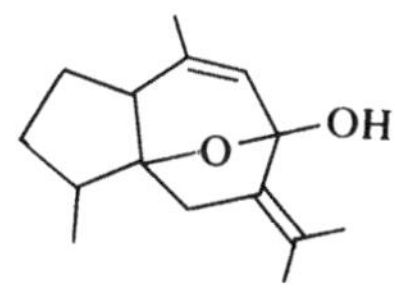

curcumenol

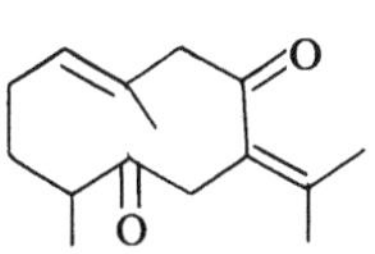

dihydrocurdione

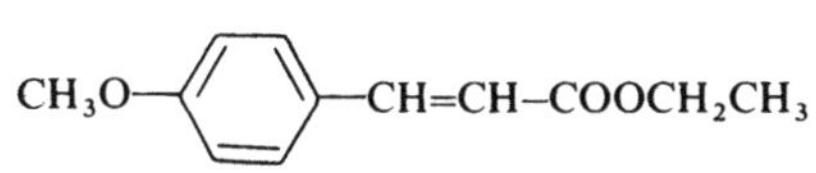

ethyl *p*-methoxy cinnamate

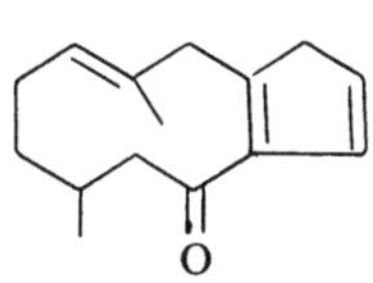

furanogermenone

For qualitative analysis in traditional medicine, the curcuminoid assay is used as a test for *Curcuma*, because *Curcuma* always contain curcuminoids to some extent (Asakawa et al. 1981). Among the essential oils, camphene is found in *C. zedoaria* and *C. aromatica* but never in *C. domestica*; camphene is used as a method for detection of admixtures of *C. domestica* (Sen et al. 1974; Shibuya et al. 1986).

1.2.2 Medicinal Use and Pharmacological and Biological Activities of Curcuma Species

In India and southeast Asia, tubers of *Curcuma*, mainly *C. domestica* (turmeric) are used as traditional medicine, spices, and dye. In Europe, *C. domestica* was used as a spice by the ancient Greeks. *C. aromatica* (wild turmeric) is used similarly to *C. domestica* and found or cultivated in India, China, and Japan. The tubers of *C. aromatica* pale yellow in color cannot be used as dyes, and are used only as medicine. In India, *C. angustifolia* Roxb. (narrow-leaved turmeric) and *C. lanceorrhiza* Roxb. are cultivated, and used as starch. In Indonesia, *C. xanthorrhiza* Roxb., with orange-yellow tubers, is called Temoe Lawak and used as medicine. All these *Curcuma* contain curcuminoid and are used as materials for curry and yellow dye for cotton, silk, and wool (Kirtikar and Basu 1975; Perry 1980). *C. zedoaria* (zedoary), with grayish brown-colored tubers, is another important *Curcuma* species. This plant was used as a spice, like *C. domestica*, in ancient times but is now seldom used in Europe. In China, southeast Asia, and Japan, this is used as an aromatic, analgetic, and Tsukeiyaku (drugs against Oketsu; Oketsu is a special concept in Japanese and Chinese traditional medicine which is probably due to microcirculatory disturbances such as stasis and thrombogenesis in arterioles) and registered in the Japanese Pharmacopeia. In China, *C. kwangsiensis* S.G. Lee et F. Liang is cultivated and sometimes imported to Japan under the name of zedoary.

The medical uses and pharmacological action of *Curcuma* species are summarized in Table 1.

1.3 Propagation of *Curcuma* Species

Conventional propagation of the *Curcuma* plants is by vegetative reproduction of the rhizomes. In the regions where *Curcuma* grows wild, artificial reproduction in cultures is easy. The seasonal changes of curcumin, essential oil, and starch were studied by Mathai (1979). The content of the yellow dye and essential oil decide the quality. Liu et al. (1974) studied the effects of nitrogen and phosphorus on the content of curcumin. Aoi et al. (1986) reported on the varietal differences in rhizome yield and curcuminoid content. In Japan, *Curcuma* is found in Okinawa and the southern region of Kyushu. In areas north of Okinawa, it is necessary to avoid frost.

For the production of *C. domestica*, easily drainable sandy soil with plenty of rain is necessary. Water is more important than nutrients (Miyazaki and Takagi 1980). According to Mathai, the best seasons are May and August. Like *C. domestica*, cold regions are not suitable for the production of *C. zedaria* and the nutrient soil is adequate but weak against dryness (Miyazaki and Takagi 1980). Rhizomes to be used for reproduction are frequently preserved until the following spring after being waxed overall to prevent drying.

1.4 Need for in Vitro Culture

The biological activity of the *Curcuma* species is gradually being clarified and the demand for the drug is increasing. The use of *C. domestica* as a spice is worldwide, and increased production of *C. domestica* is required. However, only tenfold reproduction from one rhizome is usual, since conditions for cultivation are difficult and the regenerative reproduction from the rhizomes is limited. In

Table 1. Distribution and uses of various species of *Curcuma*

Curcuma species	Region of cultivation and use in traditional medicine	Clinical treatment in ethnopharmacology (numbers correspond to the region in the left column)	Pharmacological and physiological research
C. domestica Val. = *C. longa*, *C. longa* var. *macrophylla* (turmeric)	India[1] Indonesia[2] Philippines[3] China Japan	Stomachic, stimulant, carminative, hematic or styptic in all kinds of hemorrhages certain types of jaundice and other liver disorders [1] Leucoderma, diseases of the blood, anthelmintic [2] External use: minor wounds and certain skin-eruptions burning sensation in the eyes gravel and to promote cleansing at parturition dispels itching and is gargled as mouthwash [3] Bronchial catarrh	*C. domestica* Anti-inflammatory (Arora et al. 1971; Yegnanarayan et al. 1976; Li 1985) Anti ulcer (Rafatullah et al. 1990) Anti-hepatotoxic (Kiso et al. 1983) Anti-microbial (Lutomski et al. 1974; Gupta and Banerjee 1976) Anti-oxidative (Toda et al. 1985; Tonnesen 1989) Experimental eye diseases (Mehra et al. 1984) Anti-neoplastic; anti-tumor; anti-cancer (Jiang et al. 1983; Kuttan et al. 1985, 1987; Nishino et al. 1987; Lee and Lin 1988) Allergic contact (Goh and Ng 1987) RES activity (Tomoda et al. 1990; Gonda et al. 1991) Review (Ammon and Wahl 1991)
C. aromatica Salisb. (wild turmeric)	India China Japan Indo-China[4]	Stomachic, carminative, emmenagogue abdominal pain, liver and stomach [4] Cut, crushed, and macerated rhizomes are placed in an embrocation to treat rheumatism	*C. aromatica* Anti-inflammatory (Li 1985) Anti-allergic (Dai et al. 1982) Anti-arrhythmia (Zeng et al. 1982) Effect on sleeping time (Yamazaki et al. 1988)
C. aeruginosa Rosb.	Burma (Malay Peninsula)[5] Java Indonesia[6]	For colic, [5] Asthma, cough, scurvy, mental derangement [6] Exanthem, poultice for skin irritation	

Table 1. (*Contd.*)

Curcuma species	Region of cultivation and use in traditional medicine	Clinical treatment in ethnopharmacology (numbers correspond to the region in the left column)	Pharmacological and physiological research
C. amada Roxb. (mango ginger)	India	Itching, skin diseases, asthma, inflammation	
C. angustifolia Roxb. (narrow-leaf turmeric)	India	Urinary discharge, ulcer, diseases of blood	
C. xanthorrhiza Roxb. (Temoe Lawak)	Indonesia Malay Peninsula	Abdominal diseases, liver disorders, liver, constipation, promotes lactation in neonates bloody diarrhea, dysentery, children's fevers, inflammation of the rectum, hemorrhoids, stomach disorders caused by cold, infected wounds, skin eruptions	*C. xanthorrhiza* Cholagogic action (Ozaki and Soedigdo 1988) Anti-tumor (Hirayama et al. 1984; Itokawa et al. 1985) Hypertermic effect (Yamazaki et al. 1988) Anti-inflammatory effect (Ozaki 1990)
C. zedoaria Rosc. = *C. pallida* (zedoary)	India[7] (Himalayas to Chittagong) Indonesia Indo-China[8] China Japan Taiwan[9] Malay Peninsula[10] Philippines[11]	Tonic nutrient, dissolves blood clots, promotes circulation abdominal pain during menstruation, women's illnesses. [7] Anthelmintic, antipyretic, leucoderma, asthma, tumors, spleen, epileptic seizure [8] Emmeagogue, condiment [9] Heart complaints, cholera, gonorrhea, irregular menstruation, snake bite [10] Tonic only external use [11] Wounds, ulcers	*C. zedoaria* Anti-hepatic (Yamahara et al. 1982) Gastrointestinal tract (cholagogic action) (Maeda et al. 1984) Anti-ulcer (Watanabe et al. 1986) Inhibition of contraction isolated ileum (Itokawa et al. 1983) Anti-tumor (Yokota 1986)
C. caesia Roxb. (black zedoary)	India[7]		
C. kwanglensis	China		

addition, the contents of the chemical component can vary with each variety in the same *Curcuma* species. For example, in *C. domestica*, the higher the content of curcumin the better the quality, as shown in two samples of the same Java turmeric, which contained 2.32 and 0.12% curcumine, respectively (Miyazaki and Takagi 1980). For this reason it is necessary to promote clone reproduction or production of secondary metabolites in the callus.

2 Tissue Culture

Tissue culture of *Curcuma* species was first performed by Nadgauda et al. in 1978 on stem tips of *C. domestica*. They reported that shoot formation was successful but callus formation could not be obtained using Murashige Skoog (MS) medium with benzylaminopurine (BA) or Smith medium with coconut milk, BA, and inositol. Mukhri and Yamaguchi (1986) reported successful shoot, root, and callus formation from *C. domestica* and *C. xanthorrhiza* using Ring and Nitsch medium (Ring and Nitsch 1968). Sato et al. (1987) measured curcuminoid in plants cultured in pots and in the field originating from shoots and plantlets cultured in MS medium. We cultured *C. domestica*, *C. aromatica*, and *C. zedoaria* from rhizomes using MS medium with various growth regulators, and were successful in obtaining callus; we also had success with plantlet and callus formation from stem tips of *C. zedoaria*. The plantlets were cultured in pots and in the field, and we identified the same essential oils as in the mother plants (Yasuda et al. 1988). Details of medium conditions are shown in Table 2.

3 Micropropagation

Research on tissue culture of *Curcuma* species is still in the initial stages: the culture conditions already reported are not always optimal and it is necessary to find a better basic medium and growth regulators. This section describes detailed procedures for in vitro tissue culture propagation of *C. domestica* and *C. zedoaria*, in which the essential oils or curcuminoid were identified as the active pharmacological components.

3.1 *C. domestica*

Nadgauda et al. (1978) obtained rooted plantlets cultured from the stem tip of rhizome buds using Smith or MS medium with kinetin (0.1 mg/l) and coconut milk (10%). They reported the suitable length of the explant to be 2–5 mm; at shorter length, they showed very poor growth or died. However, Sato et al. (1987) used a length of 1 mm. We also used 1 mm explants (Yasuda et al. 1988). Root formation was adequate on reculture with the same medium, but root formation was better when transplanted onto White's medium.

Table 2. Summary of in vitro culture studies on *Curcuma*

Commencement of cultivation						Continuation				
Curcuma spp	Explant source	Basal medium	Growth regulators (mg/l)	Light condition	Response	Basal medium	Growth regulator (mg/l)	Light condition	Response	Reference
C. domestica	Stem tip	MS	CM (10%) Kinetin (0.1) BA (0.2)	Day/night	Plantlet	MS	CM (10%) Kinetin (0.1) BA (0.2)	Day/night	Root	Nadgauda et al. (1978)
				Day/night	Plantlet	White	Na_2MoO_4 (0.25) $CuCl_2$ (0.25)	Day/night	Root	Nadgauda et al. (1978)
	Stem tip	Smith	CM (10%) Kinetin (0.1) BA (0.2)	Day/night	Plantlet	MS	CM (10%) Kinetin (0.1) BA (0.2)	Day/night	Root	Nadgauda et al. (1978)
				Day/night	Plantlet	White	Na_2MoO_4 (0.25) $CuCl_2$ (0.25)	Day/night	Root	Nadgauda et al. (1978)
	Stem tip (bud of rhizome)	Ringe and Nitsch	BA (1)	Dark	Shoot and root					Mukhri and Yamaguchi (1986)
	Stem tip (bud of rhizome)	Ringe and Nitsch	BA (10) 2,4-D (1)	Dark	Callus	Ringe and Nitsch	BA (1) NAA (1)	Day/night	Shoot	Mukhri and Yamaguchi (1986)
	Stem tip (bud of rhizome)	Ringe and Nitsch	BA (10) NAA (15)	Dark	Callus, embryoid, root	Ringe and Nitsch	BA (10)	Day/night	Shoot	Mukhri and Yamaguchi (1986)
	Stem tip	MS	Kinetin (1) Phloroglusinol (1 mM)	Day/night	Shoot, plantlet	MS	BA (10)	Day/night	Shoot (A)	Sato et al. (1987)
						MS	NAA (0.1) Kinetin (1)	Day/night	Shoot (A)	Sato et al. (1987)
					Shoot (A)	1/2 MS	NAA (0.1)	Day/night	Root	Sato et al. (1987)
	Slice of rhizome	MS	NAA (1)	Dark	Callus	MS	NAA (0.1)	Dark	Callus	Yasuda et al. (1988)
		MS	NAA (1.0) Kinetin (0.1)	Dark	Callus		NAA (1) Kinetin (0.1)	Dark	Callus	Yasuda et al. (1988)
	Stem tip (bud of rhizome)	MS	BAP (3.0)	Day/night	Plantlet	MS	BAP (3.0)	Day/night	Root	Balachandran et al. (1990)

C. zedoaria	Slice of rhizome	MS	2,4-D (1) Kinetin (0.1)	Dark	Callus	MS	2,4-D (1) Kinetin (0.1)	Dark	Callus	Yasuda et al. (1988)
	Slice of rhizome	MS	NAA (1) Kinetin (0.1)	Dark	Callus	MS	NAA (1) Kinetin (0.1)	Dark	Callus	Yasuda et al. (1988)
	Slice of rhizome	MS	NAA (10) Kinetin (0.1)	Dark	Gallus	MS	NAA (10) Kinetin (0.1)	Dark	Callus	Yasuda et al. (1988)
	Stem tip	MS	BA (3.0)	Dark	Callus Plantlet	MS	BA (3)	Dark Day/night	Callus Plantlet	Yasuda et al. (1988)
	Stem tip	MS	BA (10) 2,4-D (1)	Dark	Callus	MS	BA (10) 2,4-D (1)	Dark	Callus	Yasuda et al. (1988)
	Stem tip	MS	BA (10) NAA (15)	Dark	Callus	MS	BA (10) NAA (15)	Dark	Callus	Yasuda et al. (1988)
	Stem tip	MS	NAA (10)	Dark	Callus	MS	NAA (10)	Dark	Callus	Yasuda et al. (1988)
	Stem tip	MS	NAA (10) Kinetin (0.1)	Dark	Callus	MS	NAA (10) Kinetin (0.1)	Dark	Callus	Yasuda et al. (1988)
	Stem tip	MS	NAA (1.0)	Dark	Callus	MS	NAA (1)	Dark	Callus	Yasuda et al. (1988)
	Stem tip	MS	BA (0.05–3)	Day/night	Plantlet	MS	BA (0.05–3)	Day/night	Plantlet	Yasuda et al. (1988)
	Stem tip	MS	NAA (0.1)	Day/night	Plantlet	MS	NAA (0.1)	Day/night	Plantlet	Yasuda et al. (1988)
	Stem tip	MS	NAA (0.1 or 0.5)	Day/night	Shoot and plantlet	MS	NAA (0.1 or 0.5)	Day/night	Shoot and plantlet	Yasuda et al. (1988)
	Stem tip	MS	NAA (1.0) Kinetin (0.1)	Day/night	Plantlet	MS	NAA (1) Kinetin (0.1)	Day/night	Plantlet	Yasuda et al. (1988)
	Stem tip	MS	NAA (10) Kinetin (0.1)	Dark Day/night	Callus and plantlet	MS	NAA (10) Kinetin (0.1 ppm)	Dark Day/night	Callus and plantlet	Yasuda et al. (1988)
C. aromatica	Slice of rhizome	MS	IAA (10) Kinetin (0.1)	Dark	Callus	MS	IAA (10) Kinetin (0.1)	Dark	Callus	Yasuda et al. (1988)
	Slice of rhizome	MS	NAA (1)	Dark	Callus	MS	NAA (1.0)	Dark	Callus	Yasuda et al. (1988)
	Slice of rhizome	MS	NAA (1) Kinetin (0.1)	Dark	Callus	MS	NAA (1) Kinetin (0.1)	Dark	Callus	Yasuda et al. (1988)
	Slice of rhizome	MS	NAA (1.0) Kinetin (0.1)	Dark	Callus	MS	NAA (1.0) Kinetin (0.1)	Dark	Callus	Yasuda et al. (1988)
	Slice of rhizome	MS	NAA (1.0) Kinetin (0.1) Yeast extract (0.1%)	Dark	Callus	MS	NAA (1.0) Kinetin (0.1) Yeast extract (0.1%)	Dark	Callus	Yasuda et al. (1988)

Table 2. (*Contd.*)

Commencement of cultivation						Continuation				
Curcuma spp	Explant source	Basal medium	Growth regulators (mg/l)	Light condition	Response	Basal medium	Growth regulator (mg/l)	Light condition	Response	Reference
	Slice of rhizome	MS	NAA (1) Kinetin (0.1) Coconut milk (10%)	Dark	Callus	MS	NAA (1) Kinetin (0.1) Coconut milk (10%)	Dark	Callus	Yasuda et al. (1988)
	Slice of rhizome	MS	NAA (1) Kinetin (0.1) Casamino acids (0.1%)	Dark	Callus	MS	NAA (1) Kinetin (0.1) Casamino acids (0.1%)	Dark	Callus	Yasuda et al. (1988)
	Slice of rhizome	MS	NAA (1) Kinetin (0.2)	Dark	Callus	MS	NAA (1) Kinetin (0.2)	Dark	Callus	Yasuda et al. (1988)
	Slice of rhizome	MS	NAA (1) Kinetin (0.5)	Dark	Callus	MS	NAA (1.0) Kinetin (0.5)	Dark	Callus	Yasuda et al. (1988)
	Slice of rhizome	MS	NAA (10) Kinetin (0.1)	Dark	Callus	MS	NAA (10) Kinetin (0.1)	Dark	Callus	Yasuda et al. (1988)
	Slice of rhizome	MS	NAA (10) Kinetin (0.2)	Dark	Callus	MS	NAA (10) Kinetin (0.2)	Dark	Callus	Yasuda et al. (1988)
	Slice of rhizome	MS	NAA (10) Kinetin (0.5)	Dark	Callus	MS	NAA (10) Kinetin (0.5)	Dark	Callus	Yasuda et al. (1988)
C. xanthorrhiza	Stem tip (bud of rhizome)	Ringe and Nitsch	BA (1) NAA (1.0)	Dark	Root and shoot					Mukhri and Yamaguchi (1986)
	Stem tip (bud of rhizome)	Ringe and Nitsch	BA (10)	Dark	Root and bud	Ringe and Nitsch	BA (10) NAA (15)		Callus	Mukhri and Yamaguchi (1986)
					Callus	Ringe and Nitsch			Root	Mukhri and Yamaguchi (1986)

According to Sato, buds of 2–5 cm in length were taken from the rhizome and sterilized. The procedure of sterilization was as follows: buds were immersed in 70% ethanol for 30 s, rinsed with distilled water, immersed in 3% sodium hydrochloride solution (Tween 20 one drop/50 ml) for 10 min and then rinsed three times with distilled water. Stem tips 1 mm in length were cultivated as explants. MS medium with kinetin (1 mg/l) and phloroglusinol (1 mM) (medium A) or with NAA (0.1 mg/l) and kinetin (1 mg/l) (medium B) was used. Cultures were stored at 25 ± 2 °C and 12-h photoperiods. The formation of shoots from the subculture of plantlets was performed in MS medium with 10 mg/l each of media B and A. After shoot proliferation, culture was continued with NAA (0.1 mg/l) added to the MS medium and roots were produced. Plantlets produced by clone culture were transplanted into pots containing Akadama soil (light, uneven surfaced clod which has good airing and good water-preserving characteristics), mold, and sand (5:2:1), covered with polyvinyl sheets for 3 months, cultivated in the greenhouse and then transplanted to the field, where plantlet growth was 100%. Mukhri and Yamaguchi (1986) reported another method: plantlet formation and method of sterilization were the same as that of Sato and root formation used Ringe and Nitsch medium with 1 mg/l 6-benzyladenine (BA). In the same Ringe and Nitsch medium with 10 mg/l BA and 1 mg/l 2,4-D or the medium with 10 mg/BA and 15 mg/l NAA, callus is obtained. This callus can produce shoots and auxiliary shoots in the medium with 1 mg/l BA and 1 mg/l NAA. Mukhri and Yamaguchi's method is not less effective than Satoh's method but, regrettably, they did not carry out transplants to the field.

We could produce calli with 1–2-mm-thick rhizomes as an explant with MS medium with 0.1 mg/l NAA or with 1 mg/l NAA and 0.1 mg/l kinetin. These calli could produce pigment even in the basic medium with added nutrients.

Sato et al. (1987) noted that the ground parts of *C. domestica* were almost the same as in the mother plant in form and size, but there was no rhizome development and little enlargement of the lower part of the stem where yellow pigment was found. The proliferation of shoots was fivefold during the 6-week culture, which means far better efficiency than by ordinary nutritive proliferation. Although further studies are necessary for the development of tubers, application to commercial micropropagation of *C. domestica* is hoped for in the near future.

Recently, Balachandran et al. (1990) reported on the in vitro clonal multiplication of the *Curcuma* species. This report is important together with the above-mentioned method of Sato. Their method of multiplication is as follows.

After sterilization with a 0.1% mercuric chloride solution for 15 min, explants (ca. 1 cm each) with buds were cultured in the MS medium with BAP (0, 0.2, 1, 3, or 5 mg/l) or with a combination of BAP (1 mg/l) and kinetin (1 mg/l) at 27 °C using a 16-h photoperiod. After 4 weeks, multiple plantlets were formed in all treatments. The highest rate of multiplication was observed in MS medium with BAP (3 mg/l). Thereafter, regenerative plantlets were transplanted onto the same fresh medium. Before subculture, all the roots and leaves were removed. After immersion in 0.2% Bavestin (a fungicide) for 15–20 min, plantlets were transplanted onto a 1:1 mixture of sand and soil in a tray. The trays were covered with a polythene sheet and kept at 27 °C with a 16-h photoperiod. After 10 days, the polythene sheets were removed, and plants were acclimatized for 1 week. The plants were then transplanted into pots and kept in the nethouse.

Rhizome buds of *Curcuma* species (*C. domestica*, *C. aeruginosa*, *C. cassia*) formed complete plants after 4 weeks of culture in MS medium with BAP (3 mg/l). The multiplication rate was lower than that of the method of Nadganda et al. (1978). However, shoot multiplication occurred four times because of development of axillary buds. The regenerated plant and their parental stokes showed genetic purity according to an examination of isozyme patterns. The authors discussed that the subculture method was useful for germplasm conservation.

Callus formation of *C. domestica*, which has been considered difficult to obtain, was successfully carried out by the Mukhri group and our group, using different media. The Mukhri group used stem tips and our group used slices of rhizomes. Induction from rhizomes is probably better because it is simple. If it becomes possible to produce curcuminoid from the callus, it is a promising method for the production of curcumin.

3.2 *C. aromatica*

Conventional propagation of *C. aromatica* is comparatively easy; however, the development and proliferation of *C. aromatica* is as slow as those of other *Curcuma* species. Our success in callus formation from rhizomes is the first step towards

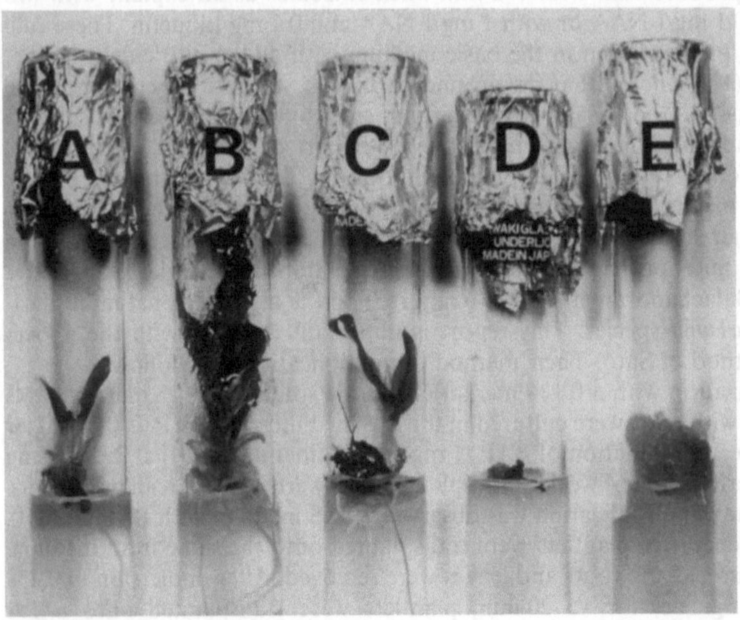

Fig. 2A–E. Callus or plantlet formation from stem tips of *C. zedoaria*, 4-week culture using MS medium containing BA and NAA. **A** Plantlet formation in MS medium containing 0.05 mg/l BA. **B** Plantlet and callus formation in MS medium containing 0.5 mg/l NAA. **C** Callus formation only in MS medium containing 10 mg/l NAA. (Yasuda et al. 1988)

mass production of *C. aromatica* (Yasuda et al. 1988). Callus formation of *C. aromatica* was observed in the MS medium with NAA (1 mg/l) and kinetin (0, 0.1, 0.2, and 0.5 mg/l). In MS medium with 1 mg/l NAA, 0.1 mg/l kinetin, or also the addition of 10% coconut milk, 0.1% casamino acid, or with 0.1% yeast added, callus production was observed; especially the addition of yeast resulted in 100% callus formation.

3.3 *C. zedoaria*

For tissue culture of *C. zedoaria*, our report is the only one (Yasuda et al. 1988). Callus formation was observed using MS medium with NAA (1 mg/l) and kinetin (0.1 mg/l) or NAA (10 mg/l) and kinetin (0, 0.1, 0.2 mg/l). With 2,4-D (1 mg/l) callus formation was observed only in combination with kinetin (0.1 mg/l). Coconut milk and casamino acid showed inhibitory effects on callus formation.

Formation of callus and plantlets from the stem tip of *C. zedoaria* is useful for the production of this plant, which is widely used as a medicine.

Fig. 3. Plantlet 9 months after transplantation to the field, following induction in MS medium containing 1 mg/l NAA and 0.1 mg/l kinetin. Note the adventitious rootlets, rhizome, and purple band along the midrib

For plantlets, rhizomes and buds were picked from April to June, immersed in 70% ethanol for 30 s, then in 10% chlorinated lime, and rinsed with distilled water twice. Terminal buds and lateral buds were excised and trimmed in a pyramid form 1 mm high and 1 mm wide at the base. Tissue culture was performed under conditions of darkness or 4000 lx under 12-h photoperiod and 26 ± 1 °C, and subcultured every 4 weeks. Primordia formation was observed in the MS medium with BA (0.05–3 mg/l) or MS medium with NAA (0.1 or 0.5 mg/l) (Fig. 2). Using MS medium with 1 mg/l NAA and 1 mg/l kinetin, the proliferative rate per year was calculated as 9.8×10^6.

The regenerated plantlets were successfully transplanted either to pots (vermiculite and Akadama soil, 1:1) or to the field, and showed 100% growth.

After transplantation of regenerated plantlets in the field, the midrib part became red in color, which is characteristic of the ground parts of *C. zedoaria*. After 9 months, a small rhizome with adventitious rootlets was observed in the underground parts (Fig. 3), and the sliced surfaces of the regenerated rhizomes were bluish green in color, resembling the mother plant. The chromosome number of *C. zedoaria* was reported as n = 63 or n = 64. Our tissue cultured plants showed no changes in either heteroploidy or polyploidy, so that this method of clone culture of *C. zedoaria* can be considered to be suitable for mass production.

In gas chromatographic analysis of the production of essential oil in *C. zedoaria*, the peak with retention time of cineol in the essential oil from *C. zedoaria* was observed both in tissue cultured plants and mother plants (comparison with the authentic substance), and other peaks had the same retention

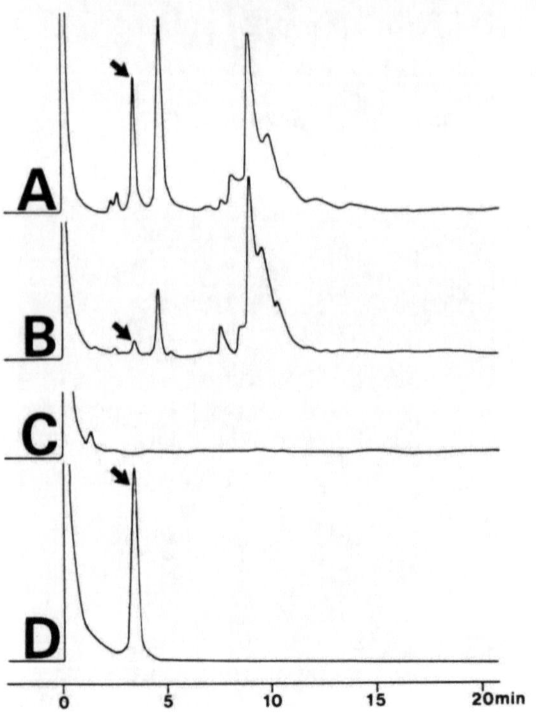

Fig. 4. Gas chromatograms of essential oils in *C. zedoaria*. *A* mother plants; *B* tissue cultured plants; *C* callus; *D* standard cineol. *Arrows* indicate position of cineol. (Yasuda et al. 1988)

times as those of the mother plants (Fig. 4). However, only a qualitative analysis could prove that the essential oil was the same as that of the mother plants.

These results suggest that multiplication by cloning is a useful method for the rapid propagation and production of the effective components of *C. zedoaria*.

Many mono- or sesquiterpenoids from the rhizomes of *C. zedoaria* were reported by several authors (Fukushima et al. 1970; Hikino et al. 1975; Gupta and Banerjee 1976). The monoterpenoids of *C. zedoaria*, cineol, camphene, and pinene, have an accelerating effect on bile secretion; and furanogermenone, one of the sesquiterpenoids of *C. zedoaria*, has been reported to have the same effect as ethyl *p*-methoxycinnamate (Yamahara et al. 1982). Considering that the essential oil produced by the cultured plants was probably the same as that of the mother plants, tissue culture of *Curcuma* species is highly successful and can be put to good use.

4 Summary and Conclusion

The basic conditions for in vitro propagation techniques were as follows.

1. For clone propagation of *Curcuma*, stem tips from rhizome buds are suitable for explants. MS medium or Ring and Nitsch medium are suitable as basic culture medium for shoot formation.
2. As growth regulators, BA and kinetin are suitable for *C. domestica.* A low concentration of BA (1 mg/l) is suitable for shoot formation and a higher one (10 mg/l) for callus formation. For the shoot formation of *C. zedoaria*, addition of a low concentration (3 mg/l and 0.5 mg/l respectively) of BA and NAA was effective, but a higher concentration accelerated callus formation.
3. Shoot formation by tissue culture is easy and the proliferation rate is high. From one rhizome 800 000 plantlets for *C. domestica* (Nadgauda et al. 1978), and 9 800 000 for *C. zedoaria* (Yasuda et al. 1988) were counted. The growth rate of these regenerative plants in the field is 100% (Sato et al. 1987) and 65% (Nadgauda et al. 1978) for *C. domestica*, and 100% (Yasuda et al. 1988) for *C. zedoaria.*
4. The chemical components of regenerative plants from stem tips are the same as those of the mother plants (Sato et al. 1987; Yasuda et al. 1988). These results demonstrate the usefulness of clone propagation from all *Curcuma* species.

Callus formation of *Curcuma* species was also successfully performed (Mukhri and Yamaguchi 1986; Yasuda et al. 1988), and this is the first step towards mass production of the effective components. The pigment contents of *Curcuma* are promising for dye production. Roughley and Whiting (1974) reported biosynthesis of curcumine in the plants. The addition of intermediate metabolites to the medium is probably effective for tissue culture.

5 Protocol

Tips of rhizomes were picked between April and July. Terminal buds were cultured in MS medium with BA (10 mg/l) and subcultured in half-strength MS medium with NAA (0.1 mg/l) (*C. domestica*) or cultured in MS medium with NAA (1 mg/l) and kinetin (0.1 mg/l) for 4 weeks, and subcultured for 4 weeks in the same medium (*C. zedoaria*) to obtain plantlets.

Acknowledgments. I wish to thank K. Yasuda, T. Tsuda, and H. Shimizu for their constant collaboration, and I am also grateful to Prof. T. Furuya and Associate Prof. T. Yoshikawa of Kitasato University for their helpful advice.

References

Ammon HPT, Wahl MA (1991) Pharmacology of *Curcuma longa*. Planta Med 57: 1–7

Aoi K, Kaburagi K, Seki T, Tabata T, Satake M, Kuroyanagi M (1986) Studies on the cultivation of turmeric (*Curcuma longa* L.) I. Varietal differences in rhizome yield and curcuminoid content. Eisei Shikenjo Hokoku 104: 124–128

Arora RB, Basu N, Kapoor V, Jain AP (1971) Anti-inflammatory studies on *Curcuma longa* (Turmeric). Indian J Med Res 59: 1289–1295

Asakawa N, Tsuno M, Hattori T, Ueyama M, Shinoda A, Miyake Y, Kagei K (1981) Determination of curcumin content of turmeric by high performance liquid chromatography. Yakugaku Zasshi 101: 374–377

Balachandran SM, Bhat SR, Chandel KPS (1990) In vitro clonal multiplication of turmeric (*Curcuma* spp.) and ginger (*Zingiber officinale* Rosc.). Plant Cell Rep 8: 521–524

Bolkhovskikh Z, Grif V, Matveieva T, Zakharyeva (1974) Chromosome numbers of flowering plants. Koelt, Koenigstein, 725 pp

Chuyakudaijiten (1985) Dictionary of Chinese medicinal plants. Shogakkan, Jpn, pp 57–60; 475–477; 2360–2362

Dai LM, Chen MZ, Xu SY (1982) The therapeutic effect of the chinese drug Yu-Jin (*Curcuma aromatica* Salisb.) on experimental allergic encephalomyelitis of the Guinea pig. Yao Hsueh Hsueh Pao 17: 692–695

Fukushima S, Kuroyanagi M, Ueno A, Akahori Y, Saiki Y (1970) Structure of curzerenone, a new sesquiterpene from *Curcuma zedoaria*. Yakugaku Zasshi 90: 863–869

Goh CL, NG SK (1987) Allergic contact dermatitis to *Curcuma longa* (Turmeric). In: Gupta B, Kulshrestha VK, Srivastava RK, Prasad DN (eds) Contact dermatitis 17: Indian J Med Res 71: 806–814

Gonda R, Tomoda M, Shimizu N, Kanari M (1991) Characterization of polysaccharides having activity on the reticuloendothelial system from the rhizome of *Curcuma longa*. Chem Pharm Bull 38: 482–486

Gupta SK, Banerjee AB (1976) Isolation of ethyl *p*-methoxycinnamate, the major antifungal principle of *Curcuma zedoaria*. Lloydia 39: 218–222

Hikino H, Konno C, Agatsuma K, Takemoto T (1975) Sesquiterpenoids. Part XLVII. Structure, configuration, conformation, and thermal rearrangement of furanodienone, isofuranodienone, curzerenone, epicurzerenone, and pyrocurzerenone, sesquiterpenoids of *Curcuma zedoaria*. J C S Perkin I: 1478–1484

Hirayama F (1984) Antitumor active substances from Indoneia Temu Lawak. Proc 104th Meeting of Jpn Pharm Soc, Sendai (Japan), p 134

Inayama S, Gao JF, Harimaya K, Kawamata T, Iitaka Y, Guo YT (1984) The absolute stereostructure of curcumol isolated from *Curcuma wenyujin*. Chem Pharm Bull 32: 3783–3786

Itokawa H, Mitsuhashi S, Watanabe K, Matumoto H, Hamanaka T (1983) Studies on the constituents of crude drugs having inhibitory activity against contraction of the ileum caused by histamine or barium chloride (I). Shoyakugaku Zasshi (1983) 37: 223–228

Itokawa H, Hirayama F, Funakoshi K, Takeya K (1985) Studies on the antitumor bisabolane sesquiterpenoids isolated from *Curcuma xanthorrhiza*. Chem Pharm Bull 33: 3488–3492

Jiang TL, Salmon SE, Liu RM (1983) Activity of camptothecin, harringtonin, cantharidin and curcumin in the human tumor stem cell assay. Eur J Cancer Clin Oncol 19: 263–270

Kirtikar KR, Basu BD (1975) Indian medicinal plants. M/S Periodical Experts, Delhi, 2417–2426

Kiso Y, Suzuki Y, Watanabe N, Oshima Y, Hikino H (1983) Antihepatotoxic principles of *Curcuma longa* rhizomes. Planta Med 49: 185–187

Kuttan R, Bhanumathy P, Nirmala K, George MC (1985) Potential anticancer activity of turmeric (*Curcuma longa*). Cancer Lett 29: 197–207

Kuttan R, Sudheeran PC, Joseph CD (1987) Turmeric and curcumin as topical agents in cancer therapy. Tumori 73: 29–31

Lee H, Lin JY (1988) Antimutagenic activity of extracts from anticancer drugs in Chinese medicine. Mutat Res 204: 229–234

Li CZ (1985) Anti-inflammatory effect of the volatile oil from *Curcuma aromatica*. Chung Yao Tung Pao 10: 38–40

Li HL, Hiu TS, Huang TC, Koyama T, De Vol CE (1978) Flora of Taiwan, Epoc, Taipei, pp 844–848

Liu LS, Yang YY, Chu YT (1974) The effect of N, P and K on yield and curcumin content of *Curcuma longa* Linn. Nung Yeh Yen Chiu 23: 284–291

Lutomski von J, Kedzia B, Debska W (1974) Wirkung des Athanolextraktes und aktiver Substanzen aus *Curcuma longa* auf Bakterien und Pilze. Planta Med 26: 8–18

Maeda H, Sunagane N, Kubota K (1984) Pharmacological effects of the powder from *Curcuma zedoaria* Roscoe on the gastrointestinal tract of experimental animals. Yakugaku Zasshi 104: 640–643

Mathai CK (1979) The pattern of rhizome yield and their accumulation of commercially important chemical constituents in turmeric (*Curcuma* species), during growth and development. Qual Plant Foods Hun Nutr 28: 219–226

Mehra KS, Gupta U, Gode KD (1984) *Curcuma longa* (Linn.) drops in corneal wound healing. Tokai J Exp Clin Med 9: 27–31

Miyazaki Y, Takagi M (1980) The new technique of medicinal plants cultivation. Hirokawa, Tokyo, pp 49–56

Mukhri Z, Yamaguchi H (1986) In vitro plant multiplication from rhizomes of turmeric (*Curcuma* domestica Val.) and Temoe Lawak (*C. xanthorrhiza* Roxb.). Plant Tissue Cult Lett 3: 28–30

Nadgauda R, Mascarenhas AF, Hendrs RR, Jagennathan V (1978) Rapid multiplication of turmeric (*Curcuma longa* L.) plants by tissue culture. Indian J Exp Biol 16: 120–122

Nishino H, Nishino A, Takayasu J (1987) Antitumor-promoting activity of curcumin, a major constituent of the food additive "turmeric yellow". Rapid communication. Kyoto Furitsuikadaigaku Zasshi 96: 725–728

Ozaki Y, Soedigdo S (1988) Cholagogic effect of *Zingiber* plants obtained from Indonesia. Shoyakugaku Zasshi 42: 333–336

Ozaki Y (1990) Antiinflammatory effect of *Curcuma xanthorrhiza* Roxb. and its active principles. Chem Pharm Bull 38: 1045–1048

Perry LM (1980) Medicinal plants of East and Southeast Asia. Mit Press, London, pp 439–440

Rafatullah S, Tariq M, Al-Yahya MA, Mossa JS, Ageel AM (1990) Evaluation of Turmeric (*Curcuma longa*) for gastric and duodenal antiulcer activity in rats. J Ethnopharmacol 29: 25–34

Ringe F, Nitsch JP (1968) Conditions leading to flower formation on excised *Begonia* fragment culture in vitro. Plant Cell Physiol 9: 639–652

Roughley PJ, Whiting DA (1974) Experiments in the biosynthesis of curcumin. J Chem Soc Perkin 20: 2379–2388

Sato M, Kuroyanagi M, Ueno A, Shimomura K, Satake M (1987) Plant tissue culture of *Zingiberaceae* (II) In vitro propagation of Turmeric (*Curcuma longa* Linn.). Plant Tissue Cult Lett 4: 86–88

Sen AR, Sen Gupta P, Dastidar G (1974) Detection of *Curcuma zedoaria* and *Curcuma aromatica* in *Curcuma longa* (turmeric) by thin-layer chromatography. Analyst 99: 153–155

Shibuya H, Yoshihara M, Kitano E, Nagasawa M, Kitagawa I (1986) Qualitative and quantitative analysis of essential oil constituents in various Zedoariae rhizome (Gajutsu) by means of gas liquid chromatography-mass spectrometry. Yakugaku Zasshi 106: 212–215

Tonnesen HH (1989) Studies on curcumin and curcuminoid 13. Catalytic effect of curcumin on the peroxidation of linoleic acid by 15-lipoxygenase. Int J Pharm 50: 67–70

Toda S, Miyase T, Arichi H, Tanizawa H (1985) Natural antioxidants III. Antioxidative components isolated from rhizome of *Curcuma longa* L. Chem Pharm Bull 33: 1725–1728

Tomoda M, Gonda R, Shimizu N, Kanari M, Kimura M (1990) A reticuloendothelial system activating glycan from the rhizomes of *Curcuma longa*. Phytochemistry 29: 1083–1086

Watanabe K, Shibata M, Yano S, Cai Y, Shibuya H, Kitagawa S (1986) Antiulcer activity of extracts and isolated compounds from Zedoary (Gajutsu) cultivated in Yakushima (Japan). Yakugaku Zasshi 106: 1137–1142

Yamahara J, Matsuda H, Sawada T, Sakurai H, Matsumura A, Toyama S, Suzuki I (1982) Effect of crude drugs on experimental liver damages. II. Effect of new sesquiterpenoid "furanogerenone". Yakugaku Zasshi 102: 272–277

Yamazaki M, Maebayashi Y, Iwse N, Kaneko T (1988) Studies on pharmacologically active principles from indonesian crude drugs I. Principle prolonging pentobarbital-induced sleeping time from *Curcuma xanthorrhiza* Roxb. Chem Pharm Bull 36: 2070–2074

Yasuda K, Tsuda T, Shimizu H, Sugaya A (1988) Multiplication of *Curucuma* species by tissue culture. Planta Med 1988: 75–79

Yegnanarayan R, Saraf AP, Balwani JH (1976) Comparison of anti-inflammatory activity of various extracts of *Curcuma longa* (Linn). Indian J Med Res 64: 601–608

Yokota M, Sugiyama K, Yamamoto F (1986) Studies on Chinese medicine used for cancer IV. Antitumor constituents in Rhizome Zedoariae, Penostracum Cicadae, Galla Wisteriae, and Radix Hostae. Yakugaku Zasshi 106: 425–426

Zeng LL, Wang GY, Wang DX, Chen CH, Xu SY (1982) Studies on the active ingredient of *Curcuma aromatica* Salisb. and its effects. Yao Hsueh Hsueh Pao 17: 946–950

II.8 Micropropagation of Lavandin (*Lavandula officinalis* Chaix × *Lavandula latifolia* Villars cv. Grosso)

M. Panizza and F. Tognoni[1]

1 Introduction

1.1 Characteristics and Importance of the Plant

Lavandin (Fig. 1) is a sterile hybrid between *Lavandula officinalis* Chaix (≡ *L. angustifolia* Mill. ≡ *L. vera* DC) and *Lavandula latifolia* Villars. *L. officinalis*

Fig. 1. Lavandin plants in bloom

[1] Dipartimento di Biologia delle Piante Agrarie, University of Pisa, Viale delle Piagge, 23, 56124 Pisa, Italy

Biotechnology in Agriculture and Forestry, Vol. 19
High-Tech and Micropropagation III (ed. by Y.P.S. Bajaj)

grows at higher altitudes than *L. latifolia* and blooms 10 to 20 days earlier but it frequently shows delayed blooming, hence allowing hybrid formation. Because of the incomplete homology of the chromosomes of the two species, problems arise during meiotic pairing in the hybrid, which is therefore sterile.

Lavandin is an essential oil-producing plant belonging to the family Labiatae. It is a perennial shrub with square young branches, opposite and entire leaves, greyish-green in color. The blue-violet flowers, crowded in cylindrical spikes, have a two-lipped corolla tubular at the base, a gamosepal calyx, and four stamens. Lavandin is characterized by hybrid vigor: the plant is higher and more robust than the parent plants; spike and flower numbers are more numerous than *L. officinalis* and chiefly than *L. latifolia*. The essential oil is obtained from flower spikes and more precisely it is produced in the flower calices rich in glandular hair. There are several cultivars of lavandin which vary in the composition of their essence (Abrialis, Super, Grosso). Lavandin cv. Grosso is a medium-sized plant characterized by straight inflorescences with quite important secondary spikes and a high oil yield (70–90 kg oil/ha). The essence (1.45% of drug fresh weight) is composed of monoterpenes with a high content of linalool (42%) and linalyl acetate (14.5%), and low content (7.5%) of camphor (Bonari et al. 1985). The essential oil is mainly used in industry for preparing perfumes, soaps, detergents, and insecticides. Further, it has medicinal properties as a sedative, diuretic, diaphoretic, antiseptic, healing, antirheumatic agent.

1.2 Need for Unconventional Practices for Propagation

Because of its sterility, lavandin is conventionally propagated by cuttings. The countries where it is mostly cultivated are France, Bulgaria, Hungary, and the USSR.

The major obstacle, which arises after repeated vegetative propagation, is clonal degeneration resulting in reduced life span, deterioration in vigor, low disease resistance, and decreased tolerance to environmental stress conditions. These problems have seriously affected the cv. Abrialis, intensely cultivated in France up to 1975, and it has progressively been replaced by the more resistant cultivar Grosso (Moutet 1980). Cv. Grosso, however, also suffers from deterioration, though to a lesser extent. Micropropagation through axillary budding may thus be proposed as a helpful technique to cope with this problem. Plant regeneration via callus can also be a source of genetic variability for selecting useful variants, and this is of importance in a sterile plant devoid of the genetic potentials of sexual propagation.

2 In Vitro Approaches

2.1 Review of Tissue Culture Studies on *Lavandula* spp.

To our knowledge, no study has previously been made on in vitro propagation of lavandin. The two parent plants, *L. officinalis* and *L. latifolia*, however, have been subjected to studies as regards micropropagation as well as secondary metabolite formation in cell cultures. In both species, the regeneration of somatic embryoids was obtained from leaf-bud calli (Quazi 1980) and complete plantlets were formed within 4 weeks of subculturing. More recently (Calvo and Segura 1988, 1989), in *L. latifolia*, adventitious bud formation was induced from seedling explants indirectly through callus.

Cell cultures of *L. angustifolia* were investigated for the ability to synthesize and accumulate lower terpenoids: no such compound could be detected in undifferentiated callus (Banthorpe et al. 1986) but callus-regenerated shoots accumulated monoterpenes (Webb et al. 1984). Moreover, cultured *L. officinalis* cells were able to produce secondary metabolites although different from those of the mother plant, such as biotin (Watanabe and Yamada 1982; Watanabe et al. 1982) and a blue pigment (Banthorpe et al. 1985; Watanabe et al. 1985; for more references see Segura and Calvo 1991).

2.2 Axillary Bud Proliferation

2.2.1 Growth Regulator Effect on Shoot Proliferation and Growth

Axillary bud proliferation of lavandin cv. Grosso (Panizza and Tognoni 1988) was performed on stem node explants. Shoots (8 cm in length) were excised from mature actively growing vegetative plants. Shoot proliferation was induced on Medium A (Table 1) supplemented with BAP and on Medium B (Table 1) provided with BAP and GA_3. The proliferation percentage was similar in all treatments (70% average). No difference could be found either in shoot number, as shown by the number of 2-cm-long shoots after 45 days in culture (Fig. 2A), or in shoot growth, as indicated by the number of 5-cm-long shoots after 75 days in culture (Fig. 2B).

In order to test whether a high cytokinin concentration might influence initial shoot proliferation, the explants were given 15 days in culture on Medium C (Table 1), containing BAP, followed by transfer to either Medium A (C/A) or

Table 1. Growth regulator composition of media used for in vitro culture of lavandin. (Panizza and Tognoni 1988)

Medium	Growth regulator (mg/l)		Response
A	BAP	(0.2)	Axillary budding
B	BAP GA_3	(0.2) + (0.5)	Axillary budding
C	BAP	(10.0)	Axillary budding and callus differentiation
D	BAP GA_3	(1.0) + (0.5)	Shoot elongation
E	2,4-D Kinetin	(1.0) + (0.5)	Callus formation and growth
F	NAA	(1.0)	Rooting
G	NAA	(0.5)	Rooting

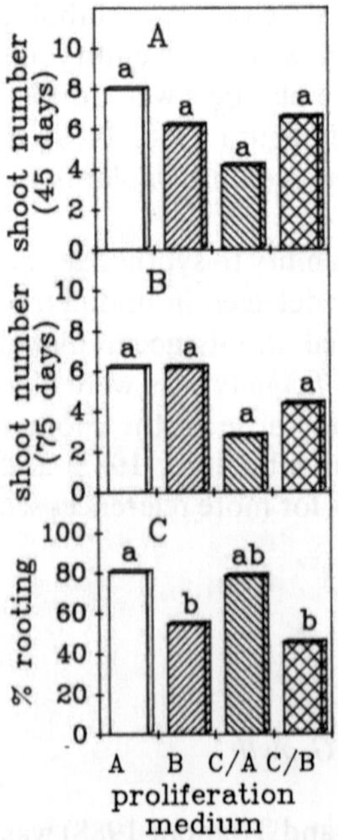

Fig. 2A–C. Axillary budding on media supplemented with various growth regulators (Table 1). *A* Medium A throughout, *B* Medium B throughout; *C/A* and *C/B* 15 days on Medium C followed by transfer to Medium A and B, respectively; 2- and 5-cm-long shoots were counted after 45 and 75 days in culture, respectively. Within each figure, values marked by *different small letters* differ significantly at 5% level. (Panizza and Tognoni 1988)

Medium B (C/B). These treatments, however, were ineffective in improving shoot number and growth (Fig. 2A, B).

2.2.2 Effect of Growth Regulators on Rooting

The plantlets regenerated on the above media (A, B, C/A, C/B) were transferred to Medium F (Table 1), supplemented with NAA, to promote rooting. As indicated in Fig. 2C, the hormonal treatments with GA_3 in the proliferation phase negatively influenced rooting, and the best response was found in the BAP treatment at low concentration (Medium A).

All plantlets, whatever their origin, acclimatized well first to the greenhouse and then to outdoor conditions. Establishment of micropropagated plants occurred at a high rate (90%) for rooted plants and at a low rate (20%) for unrooted ones.

2.2.3 Effect of Gelling Agents on Lavandin Micropropagation

The effect of the gelling agent on lavandin micropropagation was studied in relation to the kind of agent, Difco-bacto agar and Gelrite (Scott Lab. Inc.), and to its concentration (A7.5 = agar, 7.5 g/l; A5 = agar, 5 g/l; G2 = Gelrite, 2 g/l; G1 = Gelrite, 1 g/l) in Medium A (Table 1).

A reduced agar concentration as well as both Gelrite concentrations positively influenced shoot proliferation (Fig. 3A). As regards shoot length (Fig. 3B), it was improved only by Gelrite at the lower concentration. Nevertheless, these positive results obtained for shoot proliferation were reversed when we examined rooting on Medium F (Table 1) solidified with the above cited gelling agents (A7.5, A5, G2,

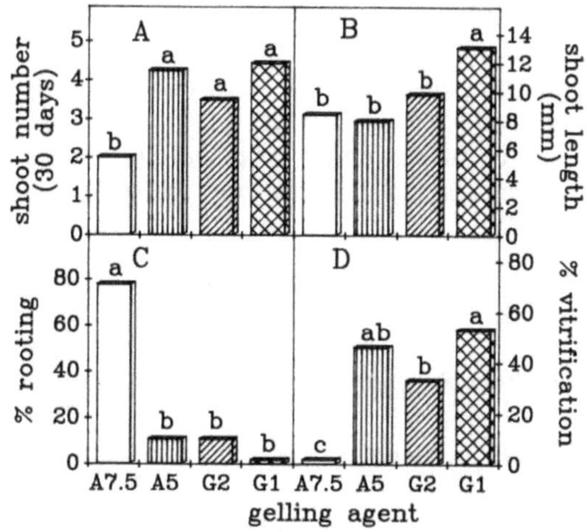

Fig. 3A–D. Axillary budding on Medium A (Table 1) supplemented with different amounts of agar (*A*) and Gelrite (*G*): *A7.5* (7.5 g/l); *A5* (5 g/l); *G2* (2 g/l); *G1* (1 g/l). See Fig. 2 for details

G1). In fact, all treatments but A7.5 strongly inhibited rooting (Fig. 3C). Moreover, the shoots regenerated on A5 and both Gelrite-supplemented media showed a high percentage of vitrification (Fig. 3D): they were bright green, with glassy consistency, large leaves, and highly hydrated tissues.

2.2.4 Morphological Changes in Micropropagated Plants

Micropropagated plants derived from Medium A and B (Table 1) were kept under observation in order to identify possible morphological changes induced by in vitro treatments (Panizza et al. 1989). Ten-cm-long cuttings excised from vegetative lateral shoots were rooted under mist and served as controls.

In vitro-propagated plants (both micropropagated and callus-regenerated ones) always bloomed later (16 months on average) than plants propagated by cuttings (8 months). At the first flowering, differences could be detected in the total spike number per plant (Fig. 4A), micropropagated plants showing in all cases

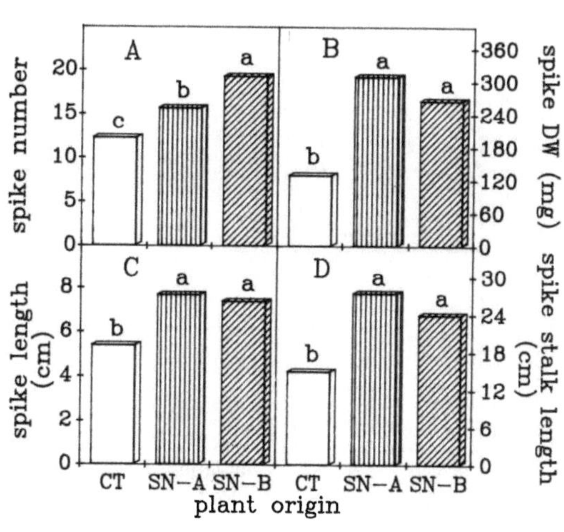

Fig. 4A–D. Morphological changes in micropropagated plants in comparison with the control (*CT* 10-cm-long cuttings). Average spike number is expressed on a per plant basis. Plants derived from node explants were cultured on Medium A, *SN-A*, and B, *SN-B* (Table 1). See Fig. 2 for details. (Panizza et al. 1989)

more spikes than the control. Such a variation was due to different numbers of secondary spikes. Both spike dry weight and length (Fig. 4B, C) were increased in micropropagated plants, but the number of flowers per spike was constant in all treatments (60 on average).

It is interesting to note that the length of the inflorescence stalk (distance between the last leaf pair and the spike) was greater in micropropagated plants (Fig. 4D).

2.3 Shoot Regeneration from Callus Cultures

2.3.1 Establishment of Callus Cultures

Callus formation from various explants was induced on Medium E supplemented with 2,4-D and kinetin (Table 1) (Panizza and Tognoni 1988). After 3 weeks in culture, fast-growing callus developed in explants from different sources, flower parts (bud and calyx), leaf, stem node, and shoot tip. The average percentage of callus formation was similar for all explants but stem node, which exhibited poor callus formation (Fig. 5A). Callus appeared along the cut edge of the explant and went on growing until the explant turned brown and died. When cultured on Medium E, calli from various origins were friable and pale yellow with a tendency to browning, which was greater the smaller the pieces of subcultured callus tissue. Lavandin callus was also characterized by the formation of a blue pigment that took place sporadically and independently of callus origin: neither callus growth nor the subsequent shoot regeneration were influenced by the appearance of the blue pigment.

Calli from different explants showed analogous growth patterns, reaching a stationary phase in 2 weeks, then they were subcultured.

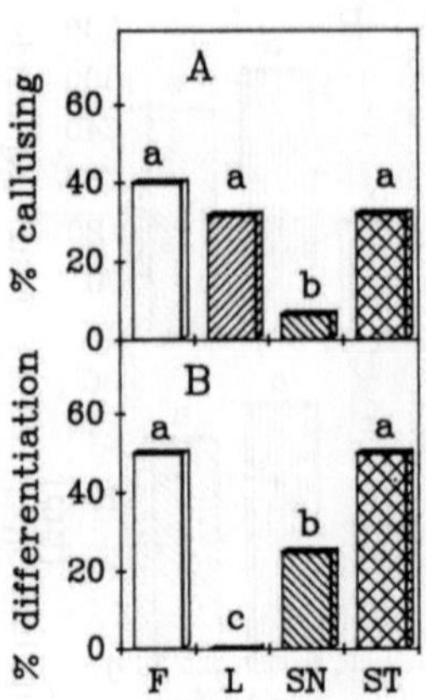

Fig. 5A, B. Influence of callus source on callus formation and differentiation. *F* flower parts (bud and calyx); *L* leaf; *SN* stem node; *ST* shoot tip. See Fig. 2 for details. (Panizza and Tognoni 1988)

2.3.2 Shoot Regeneration and Rooting

Calli from various origins were transferred to a medium with BAP (Medium C, Table 1) to promote shoot regeneration (Panizza and Tognoni 1988). Callus differentiation percentages were similar for flower parts (bud and calyx) and shoot tip and lower for stem node callus; leaf-derived callus never regenerated shoots under the described culture conditions (Fig. 5B). Time for regeneration varied for the different origins: 1 month for stem node callus, 2 months for calyx callus, 4 months for shoot tip callus, and about 1 year for flower bud callus. In this phase, too, the appearance of the blue pigment was noted; this occurred independently of callus origin and of its occurrence in the previous undifferentiated phase.

The first sign of shoot differentiation was the formation of green spots on the mass of the yellow callus; when the shoots became visible on the callus surface, shoot clusters were transferred to the shoot elongation Medium D (Table 1) supplemented with GA_3 and a reduced amount of BAP. After 45 days on Medium D, the maximum number of regenerated shoots was exhibited by stem node callus (Fig. 6A). Shoot growth was also maximum for stem-node-derived shoots, as shown by the number of 5-cm-long shoots counted after 75 days in culture on the elongation medium (Fig. 6B).

Rooting of the regenerated shoots was performed on Medium G (Table 1), supplemented with NAA. Stem node plants once again proved to give the best response as regards rooting (Fig. 6C).

The rate of acclimatization for callus-regenerated plants was similar to that for micropropagated ones.

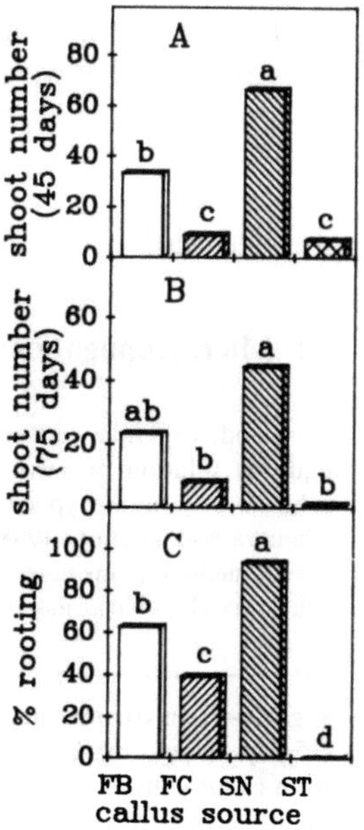

Fig. 6A–C. Influence of callus source on shoot regeneration (*FB* flower bud; *FC* flower calyx; *SN* stem node; *ST* shoot tip). 2- and 5-cm-long shoots were counted after 45 and 75 days on Medium D (Table 1), respectively. See Fig. 2 for details. (Panizza and Tognoni 1988)

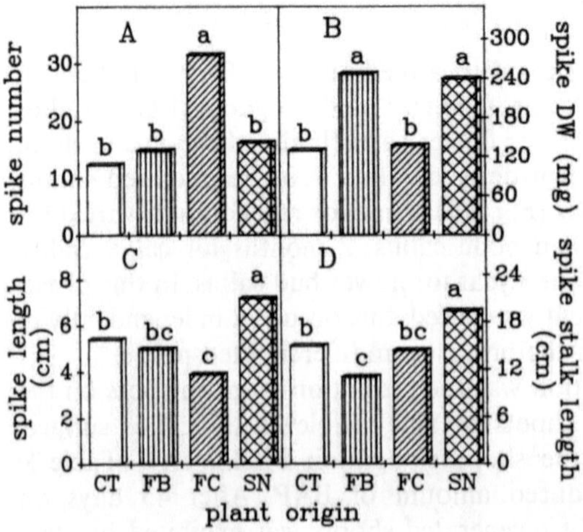

Fig. 7A–D. Morphological changes in callus-regenerated plants of various origins (*FB* flower bud; *FC* flower calyx; *SN* stem node) as compared with the control (*CT* 10-cm-long cuttings). Average spike number is expressed on a per plant basis. See Fig. 2 for details. (Panizza et al. 1989)

2.3.3 Morphological Changes in Callus-Regenerated Plants

Plants regenerated from calli of different origins (Panizza et al. 1989) exhibited morphological differences in comparison with the control (10-cm-long cuttings). At the first flowering, which, as described above, was delayed in time as compared with the control, spike number was enhanced in calyx-derived plants (Fig. 7A). This was due to the greater number of secondary spikes.

Spike dry weight (Fig. 7B) was maximum in bud and node-derived plants: in both cases the appearance of aberrant spikes larger in diameter than the control was noted. Spike and spike-stalk lengths were also increased in node-derived plants (Fig. 7C, D).

Furthermore, some plants regenerated from stem node callus showed increased leaf area and dry weight.

3 Protocol

3.1 Micropropagation Through Axillary Budding

Stem node explants were excised from mature vegetative plants. After an initial washing in 0.1% (v/v) aqueous solution of Tween 20 for 30 min, the explants were surface sterilized with 10% aqueous solution of sodium hypochlorite (7% chloro-active) followed by three rinses in sterile distilled water (Panizza and Tognoni 1988).

LS medium (Linsmaier and Skoog 1965), supplemented with 30 g/l sucrose and 7.5 g/l Difco-bacto agar, served as basal medium. The pH was adjusted to 5.7 and growth regulators were added before

Fig. 8A–F. In vitro propagation of lavandin. **A** Stem node explants. **B** Initial axillary bud proliferation (15 days in culture). **C** Advanced axillary bud proliferation (75 days in culture). **D** Shoot regeneration from callus cultures (*left*) and regenerated shoot elongation (*right*). **E** Rooting of in vitro propagated plants. **F** In vitro propagated plants 1 year after transfer outdoors

A
B
C
D
E
F

autoclaving the medium for 20 min at 120 °C. The cultures were placed in a growth chamber at 25 ± 1 °C and cool fluorescent light (10 W/m^2) was provided by Philips TL 40 W/33 RS lamps with a 16-h photoperiod.

Shoot proliferation was induced on Medium A (Table 1), with BAP. After 75 days, the shoots (5 cm in length) were rooted on Medium F (Table 1) supplemented with NAA. One month later, the rooted plantlets were transferred to potting soil in the greenhouse. Environmental high humidity was maintained by covering the pots with a plastic bag for 1 week. After 30 days of acclimatization the plantlets were transferred outdoors.

3.2 Shoot Regeneration via Callus

Callus formation was induced on stem node explants on Medium E (Table 1) supplemented with 2,4-D and kinetin. Explant sterilization and culture conditions were as described above. Fast-growing callus formed within 3 weeks and was subcultured every 2 weeks.

For shoot regeneration, lavandin callus was transferred to Medium C (Table 1) with BAP. One month later the regenerating callus was subcultured onto the shoot elongation Medium D (Table 1). After 75 days, 5-cm-long shoots were rooted on Medium G (Table 1), containing NAA.

Acclimatization procedures of callus regenerated plants were identical to those used for micropropagated plants.

4 Conclusions and Prospects

In vitro propagation of lavandin (Fig. 8) can be easily accomplished following these procedures. Axillary budding was feasible and showed a good proliferation rate. Rooting and acclimatization of micropropagated plants were also satisfactory. Moreover, micropropagation of lavandin by stem node explants enhanced the number of spikes per plant and spike stalk height, thus facilitating mechanical harvesting.

For all these reasons, we can propose lavandin micropropagation as a reliable method for rapid multiplication of elite clones as well as a potential useful tool for overcoming clonal degeneration problems. Shoot regeneration from callus, on the other hand, also proved to be feasible. Of the various explants tested, stem node was the best callus source as regards shoot formation. In fact, in spite of the low percentages of callusing and differentiation exhibited by stem node callus, shoot regeneration from node callus was much more abundant and rapid than the other sources and rooting was also adequate. Likewise, stem node callus-derived plants showed enhanced morphological features related to the spike (dry weight, length, stalk length).

We can thus conclude that the methodology of callus regeneration might become a useful means to create genetic variability in a plant devoid of sexual propagation.

References

Banthorpe DV, Bilyard HJ, Watson DG (1985) Pigment formation by callus of *Lavandula angustifolia*. Phytochemistry 24(11): 2677–2680

Banthorpe DV, Branch SA, Njar VCO, Osborne MG, Watson DG (1986) Ability of plant callus cultures to synthesize and accumulate lower terpenoids. Phytochemistry 25(3): 629–636

Bonari E, Cioni PL, Morelli I (1985) Effet de la densité de plantation sur une culture du lavandin (variété Grosso). Influence sur la qualité et la quantité d'essence. Plant Med Phytother 19(2): 98–108

Calvo MC, Segura J (1988) In vitro morphogenesis from explants of *Lavandula latifolia* and *Lavandula stoechas* seedlings. Sci Hortic 38: 131–137

Calvo MC, Segura J (1989) In vitro propagation of lavender. Hortic Sci 24(2): 375–376

Linsmaier EM, Skoog F (1965) Organic growth factor requirements of tobacco tissue cultures. Physiol Plant 18: 100–127

Moutet L (1980) Lavandin Abrialis, lavandin Grosso: what is their future? Perfumer and Flavorist 4(6): 27–29

Panizza M, Tognoni F (1988) Clonal propagation, callus formation and plant regeneration of lavandin. Sci Hortic 37: 157–163

Panizza M, Mensuali Sodi A, Tognoni F (1989) In vitro propagation of lavandin: morphological changes in regenerated plants. Proc Int Symp In vitro Culture and Horticultural Breeding, Cesena, Italy, May 30–June 3.

Quazi MH (1980) In vitro multiplication of *Lavandula* spp. Ann Bot 45(3): 361–362

Segura J, Calvo MC (1991) *Lavandula* spp. (lavender): in vitro culture, regeneration of plants, and the formation of essential oils and pigments. In: Bajaj YPS (ed) Biotechnology in agriculture and forestry, vol 15. Medicinal and aromatic plants III. Springer Berlin Heidelberg New York pp 283–310

Watanabe K, Yamada Y (1982) Selection of variants with high levels of biotin from cultured green *Lavandula vera* cells irradiated with gamma rays. Plant Cell Physiol 23(8): 1453–1456

Watanabe K, Yano SI, Yamada Y (1982) The selection of cultured plant cell lines producing high levels of biotin. Phytochemistry 21(3): 513–516

Watanabe K, Sato F, Furuta M, Yamada Y (1985) Induction of pigment by S-containing compounds in cultured *Lavandula vera* cells. Agric Biol Chem 49(2): 533–534

Webb JK, Banthorpe DV, Watson DG (1984) Monoterpene synthesis in shoots regenerated from callus cultures. Phytochemistry 23(4): 903–904

II.9 Micropropagation of *Agave* spp.

M.L. Robert, J.L. Herrera, J.L. Chan, and F. Contreras[1]

1 Introduction

Before the 1930s, plantation agaves were prominent crops in the economies of some Third World countries such as Mexico and Tanzania. The advent of synthetic fibers, however, severely reduced their economic value. Although their importance cannot be compared to that of basic food crops today, agaves are potentially useful plants for the future of semi-arid parts of the world because they are extremely hardy and well adapted to cope with water scarcity, as well as being the source of multiple products.

1.1 Botanical Classification and Description

The genus *Agave* L. is a member of the family Agavaceae, order Asparagales (Dahlgren et al. 1985). It is divided into two subgenera, *Littaea* and *Agave*, with 136 species that have been arranged into groups with close similarities. This system, however, is debatable and might not faithfully reflect natural relationships (Gentry 1982). The center of origin and diversity of the genus *Agave* is limited to Mexico. However, since the 17th century, agaves have been taken to practically all subtropical areas of the world, mainly for ornamental purposes.

With the exception of a few arborescent forms such as *Agave karwinski* Zucc., the stems of these plants are thick abbreviated shoots. The size and shape of the rosettes range from small, compact, and globose (*A. parryi* Engelm.) to the gigantic plants (*A. mapisaga* Tres.) that can reach a height of 2–2.5 m. The leaves have a spade shape with a sharp thorn at the tip, although some plants (*A. cantala* Roxb) have long, narrow and flexible leaves; they are generally thick and succulent with a spongy parenchyma for storing water and have a waxy cover that prevents water loss and gives them a glaucous look. The edges can be smooth, as in some forms of *A. cerulata* Trel., but are more frequently covered with lateral prickles or "teeth" of different shapes and numbers. The leaves arise from the meristematic region at the apex of the stem. Their formation in the bud takes several years and, when fully developed, they unfold outwards in a spiral arrangement forming a rosette.

[1] División de Biología Vegetal, Centro de Investigación Científica de Yucatán, A. C. Apdo. Postal 87 Cordemex, 97310 Mérida, Yucatán, México

Biotechnology in Agriculture and Forestry, Vol. 19
High-Tech and Micropropagation III (ed. by Y.P.S. Bajaj)

1.2 Life Cycle and Reproduction

Agaves live for periods of between 8 and 25 years according to the species. Until they reach sexual maturity they constantly produce new leaves while shedding the old ones.

Some agaves of the group Polycephalae (*A. celsii* Hook) are polycarpic perennials that flower repeatedly from the leaf axils excentric to the axis, while others, such as the agaves analyzed here, are monocarpic multiannuals that flower only once towards the end of their life. The flowers develop at the top of an inflorescence stalk that emerges from the apex of the stem. Many inflorescences are spectacular not only because of their size, which in *A. americana* L. can reach 9–10 m, but because of the speed at which they grow, causing the inflorescence to use up all the nutrients and water stored in the leaves and stem and finally resulting in the drying up and death of the plant.

Agaves reproduce vegetatively through the production of underground stems called rhizomes that form at the base of the rosette above the roots. The leaves of these shoots are modified, forming small bracteas that protect dormant axillary buds. The rhizomes elongate and eventually grow up to the surface, where the apex of the rhizome forms a new plantlet or sucker. The number of new individuals produced by this process is very variable; some agaves form new suckers only when the rosettes are young, others develop them throughout their life and still others only do so when they set flower. Another variety of vegetative propagation is the formation of bulbils, small buds that develop on the inflorescence after the fruits have fallen.

1.3 Adaptation to Arid Environments

Although the degree of adaptability varies considerably, most agaves are very well suited to arid environments (Burgess 1985). The form and distribution of the leaves help the plant capture the little rainfall available most efficiently (Gentry 1982). However, the main adaptive features to cope with dessication and high temperatures, widely studied by Nobel on *Agave deserti*, are the mechanisms that prevent the loss of water: the thick epicuticular layer of waxes and the crassulacean acid metabolism (CAM), a CO_2 fixation pathway in which stomatal opening and CO_2 uptake occur at night, thus preventing the loss of water through evapotranspiration that would normally occur if the stomata were open during day time (Nobel 1976; Nobel and Hartsoch 1979).

1.4 Uses and Economic Importance

Agaves have a wide range of uses, including landscaping and the prevention of soil erosion, but their main economic importance stems from being a source of raw material for hard fibers (sisal) and spirits (such as tequila). Because of their high sugar content, agave tissues have been chewed in Mesoamerica for hundreds of years. The soft, starchy white meristem at the top of the stem and the young inflorescence stalk are the most palatable. However, this use has been maintained as a local custom and has not been commercially exploited as in the case of the tender tips of many palms, which are considered a delicacy in many parts of the world. Animals eat the new leaves and suckers from agaves, and the fresh leaves of *A. salmiana* Otto ex Salm. have been used as cattle forage in some arid parts of Mexico. Nevertheless, these habits are probably more the result of necessity than an indication of the plant's potential as animal feed (Gentry 1982).

From an industrial point of view, the most important agaves are *A. fourcroydes* Lem (henequen) and its close relative *A. sisalana* Perrine (sisal), and to a lesser extent *A. lechuguilla* Torr. (Ixtle), whose leaves are harvested for hard fiber extraction, and *A. tequilana* Weber, which is used for the production of spirits.

At one stage, henequen cultivated in Mexico was the world's most important source of natural fiber. In 1921 it represented 88% of the world's hard-fiber production (Aznar Mendoza 1977) and was the basis of Yucatan's wealth. But from being one of the richest states in Mexico at the beginning of the century, Yucatan's fortunes have steadily declined along with its falling production of henequen. In 1988 sisal from Africa and South America represented 71% of the total world production (455 100 mt), while Mexico produced only 12.5% (FAO 1989). Although synthetic fibers have captured a large proportion of the world market, there is still a constant demand for hard, natural fibers for harvest twine, sacks, bags, carpets, and ropes (García de Fuentes and De Sicilia 1984). The short fiber is also used for padding materials (Baker 1985).

Another important commercial use is the production of alcoholic beverages such as tequila, mezcal, and pulque. Tequila is by far the most important and has become a well-known spirit all over the world mainly because of its use in the preparation of Margarita cocktails. However, when properly distilled and aged it is a very fine quality spirit with a distinctive flavor. Tequila is produced from *A. tequilana* grown around the town of Tequila in the State of Jalisco, Mexico. The mature "heads" (stems from which the leaves have been removed) are harvested and steamed to convert the stored starches into reducing sugars. After extraction, the juices are fermented, distilled in stainless steel vats, and bottled as tequila blanco or left to age in cedar vats. Different quality tequilas are then produced, depending on the length of time they are allowed to age, which ranges from 3 months (reposado), to 2 years (añejo). Tequila is a significant source of hard currency for Mexico, which exports around 30 million l of the spirit per year, mainly to the USA (95%), Germany, Japan, and Canada (Rico Tavera 1987).

Aguamiel, pulque, and mezcal are beverages that are consumed locally, made from *A. salmiana*, *A. atrovirens* Karw. ex Salm., and *A. potatorum*. Zucc. These agaves produce a syrup with a very high fructose content (70–90%) and could potentially be exploited in other forms by the food industry (Promotora del Maguey y del Nopal 1981).

Agave fibers have been used for the production of composite materials with increased strength and with heat-insulating properties that could replace asbestos as hot tops for steel ingot production (Cruz Ramos et al. 1985). Agave fibers have also been proposed as a source of cellulose for paper production. Unlike the woody species currently used, agaves would be far easier to propagate but, to be economically feasible, a paper production plant would require the planting of an additional 135 million agaves and would compete for raw material with the cordage industry (Villalvazo et al. 1985).

The juices generated as a waste product from the decortication of *A. fourcroydes* and *A. sisalana* are rich in sapogenins such as hecogenin (Blunden et al. 1980; Hertz 1985), which are precursors for the synthesis of corticoid hormones. The juices from sisal cultivated in Africa have been used by the British pharmaceutical company Glaxo as a raw material for the production of betametasone and dexametasone. The high concentration of tigogenin (which is difficult to separate from hecogenin) in henequen leaves has prevented its commercial exploitation.

Agaves are also used for ornamental purposes; some of the large ones, such as *A. angustifolia* Haw. (var. *marginata*) and *A. vilmoriniana* Berger, are very popular in gardens and parks, while others, such as *A. victoria* reginae T. Moore, are more adequate for small desert-like corners.

Large agaves are used to construct fences to keep cattle in or out of fields. Properly used, agaves could play an important role in the control of soil erosion, since they can be intercropped with many annual crops in traditional agricultural holdings.

2 Need for Micropropagation

Many cultivated agaves are sterile clones, sometimes polyploids that rarely produce viable seed. This is the case of the two species analyzed here. Furthermore, the little sexual reproduction that might occur in the fields is prevented by the producers themselves, who cut off the inflorescences as soon as they begin to develop to stop them using up the stored sugars and drying up the leaves. The genetic variability is therefore very limited; all the African and South American production of fiber comes from a few sisal clones collected from Mexico towards the end of the 19th century (Dewey 1973; Wienk 1976) and, although there are no records, it is quite likely that the present production of henequen and tequila is based on a few clones selected more than a century ago.

Probably due to the length of time it takes to cross-breed cultivated agaves, examples of programs to genetically improve them are rare, and even fewer have actually resulted in the successful production of improved varieties (Boulanger 1985). One such example is the celebrated diploid hybrid 11648, bred by Lock and collaborators in Tanzania from a cross between *A. angustifolia* × *A. amaniensis* Trel. & Nowell back-crossed with *A. amaniensis* (Lock 1962, 1985). The hybrid is a superb hard-fiber producer in terms of leaves, content, and quality of the fiber and has the added benefit of leaves with smooth edges; unfortunately, its extreme susceptibility to species of the fungus *Phytophthora* has prevented it from being successfully cultivated.

In spite of this, morphological variants with agronomically important characteristics such as larger heads or increased number of leaves are frequently observed in plantations and could be the basis of a genetic improvement program using tissue culture. Other characteristics such as resistance to disease, increased fiber content, or quality could also be selected for, cloned, and evaluated if a program with such aims were to be established (Boulanger 1985).

An additional problem is a shortage of plant material to maintain the production of both henequen and tequila at their present levels. In the case of tequila, the high international demand over the past two decades has led to overexploitation, that has resulted in a severe shortage of plants. This, in turn, has led to the indiscriminate planting of low-quality materials and, consequently, to a reduction in yields. Today, an average head weighs 25–30 kg, while it is reported that during the first half of the century the average weight was more than 70 kg (Tequila Cuervo, pers. commun.). The case of henequen is even more dramatic. Not only have productivity and absolute production figures greatly decreased, but in 1989 for the first time in its history Mexico had to import sisal fiber from Brazil to keep its cordage industry going.

Although agaves reproduce asexually, the number of new clonal individuals produced yearly is very limited and is of no use for improvement through selection. Micropropagation could play a very important role in re-establishing the productivity of the plantations, since the cloning of large numbers of elite individuals would not only increase the availability of "seed", but could also restore the high yields.

3 Review of Tissue Culture

There are a few reports on the micropropagation of *Agave* species from shoots regenerated from callus. Groenewald et al. (1977) were the first to regenerate plantlets of an unidentified species of *Agave*. Callus from seed embryos was induced and maintained in enriched Linmaier and Skoog medium supplemented with 1 mg/l 2,4-D and 5 mg/l kinetin. Organogenesis was obtained after 12 weeks and two transfers to medium with reduced concentrations of growth regulators (0.2 mg/l 2,4-D and 1 mg/l kinetin). Both shoots and roots formed in the same medium, but no indication is given of the number of regenerated plants. Hunault (1979) also reports the induction of callus from foliar discs of another unidentified agave, but no organogenesis was attained. Madrigal and collaborators (1981) regenerated plantlets of henequen (*A. fourcroydes*) from calli derived from stem tissues. Callus was induced in MS medium supplemented with unspecified concentrations of auxins (IAA, IBA, NAA, or 2,4-D), and adventitious shoots formed from this callus under the influence of high (unspecified) concentrations of cytokinin. Later, Robert et al. (1987) reported callus formation from rhizome internode tissues of *A. fourcroydes*. The formation of shoots from this callus was obtained in modified Hildebrandt and Shenck medium containing 0.25 mg/l 2,4-D and a high cytokinin concentration (10 mg/l BAP). Organogenesis is also influenced by the ratios of NO_3 and NH_4, and by 1 mM arginine, which shortened the period needed for adventitious shoots to appear by 2 weeks. Frydrich (1982) did not observe organogenesis in callus from either bulbil tissues or leaves of in vitro-cultivated plants of *A. sisalana*.

Recently, Powers and Backhaus (1989) regenerated plants of *A. arizonica* Gentry & Weber from callus derived from basal segments of bulbils. Callus induction, shoot proliferation, and rooting were carried out on Murashige and

Skoog basic salts supplemented with different concentrations of vitamins, growth regulators, NaH_2PO_4, and sucrose, optimal for each stage. The induction of callus occurs with 0.3 mg/l 2,4-D only, while organogenesis requires 10 mg/l (44.4 μM) BAP and NAA (1 mg/l). Callus of *A. wightii* Drum. & Prain has been cultivated in semi-solid and liquid medium by Sharma and Khanna (1979) to study the synthesis of steroids in undifferentiated tissues. Surprisingly, bulbil-like structures were formed in liquid medium supplemented with 2,4-D (2 mg/l).

Micropropagation utilizing axillary buds as explants was reported by Madrigal et al. (1981). The development of dormant buds from leaf axils cultivated in vitro was promoted by kinetin (1 mg/l) and IAA (0.3 mg/l). The shoots were rooted with IAA (2 mg/l) and transplanted to soil.

Propagation through the production of adventitious shoots from segments of vegetative tissue was first reported by Frydrich (1982), who used the base plate (stems) of bulbils of *A. sisalana*. The efficiency of this method is low because an average of only three shoots can be obtained from a single individual. The induction of adventitious shoots from stem segments of larger plants of *A. fourcroydes* (Robert et al. 1987) was only slightly more efficient. 130 shoots were obtained from explants from a single mother plant.

A far more efficient method is the induction by high cytokinin concentration (10 mg/l BAP) of adventitious shoots from in vitro-grown plants of *A. fourcroydes*, as reported by Robert et al. (1987). This procedure, that quadruples the number of plants every 4 weeks, can be repeated indefinitely, generating as many clonal individuals as wished. A similar procedure, producing five to ten new shoots/plant every 12 weeks, has been used with *A. arizonica* by Powers and Backhaus (1989).

In this Chapter we describe the work carried out at our laboratory to establish blueprints for efficient mass propagation of *A. fourcroydes* and *A. tequilana* (Robert et al., in prep.), which are the most widely cultivated agaves in Mexico due to the economic importance of the products extracted from them.

4 Micropropagation

4.1 Selection of Plant Material

One of the main difficulties in the selection of high-yielding individuals is the fact that their productivity cannot be assessed until they reach the productive stage (4–5 years for henequen and 8–9 years for tequila), and by then the tissues are not very suitable for micropropagation. In the case of agaves, it would appear that the greatest impact of tissue culture methods is achieved when they are incorporated into long-range integrated programs for genetic improvement. It is therefore of the utmost importance that a constant and methodical selection procedure is first carried out in the plantation. The rhizomes from promising individuals should be identified and grown to form a germplasm collection of potentially high-yielding individuals, while the parental yields are evaluated. Since the offshoots are the result of asexual reproduction, clonal lines of a few individuals (10–25) can be established that will serve as explant donors to start a mass propagation program.

4.2 Source of Explants

Several tissues, such as rhizome internodes, stem tissues, or axillary buds, can be used as a source of explants to initiate micropropagation. The choice will depend on the objectives and availability. Axillary buds from rhizomes offer the advantage that the mother plant does not have to be sacrificed as in the case of stem explants or axillary buds from leaves. However, rhizomes are not available all year round and sometimes they are difficult to disinfect.

Stem Segments. Plants 50–70 cm high (measured from the soil surface to the tip of the central leaves) are excised from the soil and all their leaves are removed, barring the central "head" or "piña". The heads are washed with plenty of running tap water before the surface is eliminated, cutting off all the fibrous parts and the base of leaves, leaving only a cube of white tissue (between 5 and 6 cm per side, depending on the size of the agave head). This tissue is surface sterilized by immersing it in a solution of 30% Cloralex commercial bleach (1.8% active chlorine) for 30 min before rinsing it four times with plenty of sterile distilled water. About 0.3 cm of the surfaces are cut and eliminated since this tissue has been severely damaged by the sterilizing treatment. The remaining tissue is cut into smaller cubes about 0.8 cm per side. Approximately 60 to 100 such explants can be obtained from each mother plant through this method. The small cubes are further sterilized by a much milder treatment in 10% Cloralex (0.6% active chloride) for 5 min, after which the cubes are rinsed thoroughly four times with sterile distilled water. It is not recommendable to leave the cubes in water too long, since the possibility exists of endogenous microorganisms from still-infected tissues diffusing to clean ones. The explants are first dried by placing them on top of sterile filter paper for a few seconds, and then planted (Fig. 1A). During the 1st week, these explants should be constantly checked for microbial contamination. If bacteria or fungi grow on any of the explants in a culture vessel, the rest should be transferred to another container.

Axillary Buds from Rhizome Scales. Rhizomes from 40–50 cm high plants (1-year-old tequila, 2-year-old henequen) are excavated and brought to the surface to green the etiolated tissues while remaining attached to the mother plant. Fifteen days later the rhizomes are cut into 15–20-cm-long segments and the cut surfaces are covered with Parafilm to protect them from contamination. The segments are immersed in distilled water and moved to the laboratory. The stem segments are washed with a solution of 2% v/v Extran before removing the bracteas with a scalpel. The axillary buds are then excised and washed with 2% v/v Extran for 10 min with continuous shaking. After rinsing thoroughly with sterile distilled water, the buds are washed again with a 2% v/v solution of Extran plus 15% (0.9% active chloride) Cloralex commercial bleach. The buds are rinsed at least three

Fig. 1A–F. Stages during the micropropagation of *A. tequilana*. **A** Stem explants just established in culture. **B** Adventitious shoots formed from the explants after 12 weeks. **C** Mature shoots in rooting medium ready to be transferred to polystyrene trays for weaning (**D**). **E** 18-month-old micropropagated plants in field trials, note the large rhizomes that develop at an early stage. **F** General view of the field trials at Tequila Herradura plantations in Amatitan, Jalisco. (Figure parts **D–F** see p. 314)

Fig. 1D–F

times in sterile, distilled water and dried before being planted in the culture vessels. The axillary buds from leaves are treated in exactly the same way.

4.3 Induction (Stage I)

To promote the development of axillary buds, the explants are put in baby-food jars containing 25 ml of induction medium (see Table 1) which contains a high cytokinin concentration (10 mg/l BAP). The jars are incubated for 2 weeks at 27 °C under continuous light. The explants from stems are also incubated in baby-food jars containing 25 ml of induction medium. The optimal physical and chemical parameters for adventitious shoot formation in this type of explants (Fig. 1B) have been extensively studied because of their plentiful supply all year round.

Nitrogen Balance. The total amount of nitrogen as well as the ratio between NO_3 and NH_4 are of primary importance for the induction and further development of adventitious shoots. Table 2 shows the number of shoots produced from stem explants under a variety of nitrogen regimes. The concentrations of 20 mM KNO_3 and 5 mM NH_4NO_3 were found to be optimal on the basis of the number of shoots produced, degree of vitrification, size, and general aspect of the shoots.

Temperature. The production of adventitious shoots is favored by high temperature (30 °C), so we recommend a minimum of 27 °C for good development (Fig. 2). Shoots that develop at higher temperature also have a less vitrified appearance (more opaque surface and sharper edges) than the ones grown at 25 °C.

Light. Adventitious shoots are produced under all lighting conditions. Continuous light (30–40 mol/m^2/s of photosynthetically active radiation) favors their initial formation from stem explants; however, further growth of the plantlets and the formation of adventitious shoots at their base is most efficient under a 16-h photoperiod. For maturation and rooting it is important to give plants kept in culture rooms the highest possible light intensity and to increase the content of red light with incandescent bulbs. The plantlets will develop faster and be better adapted for transplantation to soil under these conditions. However, ideally they should be indirectly exposed to natural sunlight.

Carbon Source. 20–30% sucrose produces optimal formation and growth of adventitious shoots, but higher concentrations (40%) completely inhibit the process (Fig. 3). In order to lower costs, we tested icing sugar as a carbon source with very satisfactory results. We found it to be considerably cheaper and, for reasons that we do not yet understand, it reduced the degree of vitrification in the newly formed adventitious shoots from stem explants. So far we have tested more than 50 batches of commercial icing sugar and have found no variation or negative effects at any stage of the in vitro process.

Variability of the Explant. In spite of the fact that they were derived from a single mother plant, these explants showed a great degree of variability, probably due to their differing locations in the large stem. The small cubes from the top (closer to

Table 1. Characteristics of the culture media and containers recommended for the different stages of agave micropropagation. In all cases the basic medium is MS with modified content of nitrogen salts: 20 mM KNO_3 and 5 mM NH_4NO_3

	Induction medium for stem explants	Induction medium for axillary buds	Growth medium	Maturation medium	Multiplication medium	Rooting medium
2,4-D mg/l	0.025	0.025	0.025	0.025	0.025	0.025
BAP mg/l	10	10	1	1	10	0
% Gelling agent	1 Agar	0.8 Agar	0.1 Agar 0.15 Gelrite	0.2 Agar 0.15 Gelrite	0.1 Agar 0.15 Gelrite	0.8 Agar
Containers	Baby-food jars	Baby-food jars	Magenta box	Magenta box	Magenta box	Magenta box
Approx. length of period in (weeks)	12	2	3	3	3	3

Table 2. Effect of the concentration and ratio of NO_3 and NH_4 ions on the induction of adventitious shoots from stem explants

KNO_3/NH_4NO_3 (mM)	Ratio NO_3/NH_4	Total nitrogen (mM)	Average Shoots/Explant
3.75/2.5	2.5	8.75	0.5
10.00/2.5	5.0	15.0	1.0
16.25/2.5	7.5	21.25	7.0
22.50/2.5	10.0	27.5	9.0
7.5/5	2.5	17.5	2.0
20.0/5	5.0	30.0	10.0
32.5/5	7.5	42.5	19.0
45.0/5	10.0	55.0	14.0
15.0/10	2.5	35.0	16.0
40.0/10	5.0	60.0	9.0
65.0/10	7.5	85.5	12.0
90.0/10	10.0	110.0	1.0
22.5/15	2.5	52.5	4.0
60.0/15	5.0	90.0	0
97.5/15	7.5	127.5	0
135.0/15	10.0	165.0	0

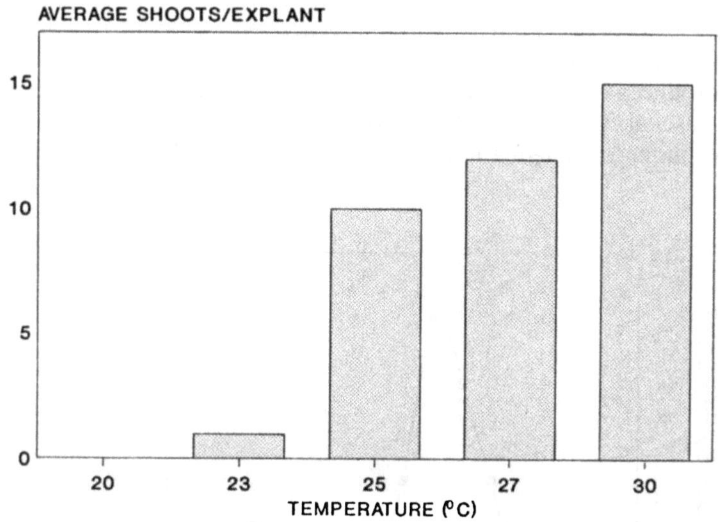

Fig. 2. Effect of temperature on the induction of adventitious shoots from stem explants of *A. tequilana*

the base of the new leaves) were always the first to produce shoots (6–8 weeks), while the ones from the bottom took 10–12 weeks to generate adventitious shoots though, in the end, they produced more than the former. The range of new shoot per explant varied between 0 to 30, but the averages were fairly constant. The production of adventitious shoots was analyzed in explants obtained from ten different mother plants cultivated together and kept under identical conditions.

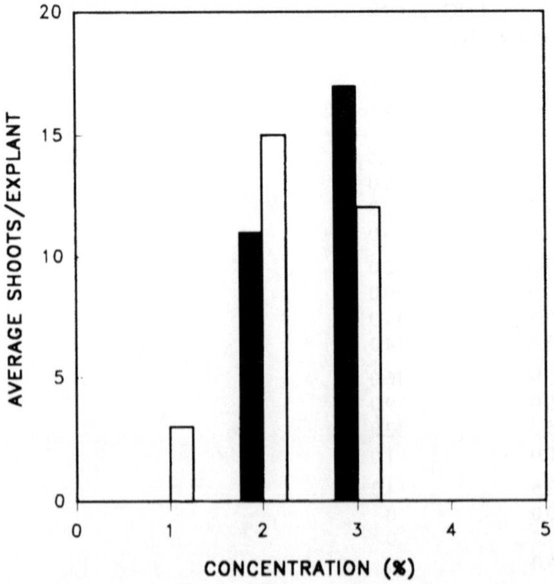

Fig. 3. Effect of different concentrations of sucrose (■) and icing sugar (□) on the induction of adventitious shoots from stem explants of *A. tequilana*

Eight out of ten plants produced an average of between 14 and 18 shoots per explant. Of the two remaining, one generated 23 shoots per explant while the other, which was clearly diseased (the explants were dark yellow instead of white), produced only four. Another source of variation in the response seems to be due to seasonal factors, since the mother plants were cultivated in the open and were exposed to changing natural conditions. As can be seen in Fig. 4, the number of

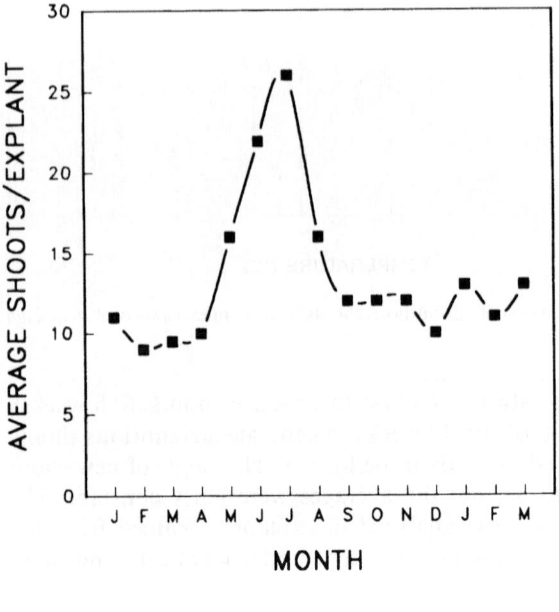

Fig. 4. Seasonal variation in the capacity of stem explants of *A. tequilana* to form adventitious shoots

adventitious shoots derived from stem explants from plants harvested during May and August is considerably higher than from the rest of the year. It is interesting to note that this corresponds to a parallel increase in the content of reducing sugars in the stems of plants growing in the fields.

4.4 Growth

Shoots developed from any type of explant are small and should be allowed to grow further before the multiplication stage is begun; firstly, because small plantlets are susceptible to vitrification and secondly, because the number of adventitious shoots they produce is very small. Both the containers (baby-food jars) and the induction medium are unsuitable for further growth and the small shoots should therefore be transferred to larger containers and a growth medium with reduced cytokinin (Table 1). The container type is very important: it should be high enough to allow the plantlets to grow without touching the top, since this stimulates ethylene production, which in turn causes the leaves to become elongated and their tips to dry out. We do not recommend the use of test tubes, where leaves grow abnormally long and no new leaves develop. In our experience, the Magenta box gives the best results for this purpose (Fig. 1C).

The characteristics of the gelling agent are also extremely important for good development (Table 3). In the large Magenta box with 0.8% agar, the small plants tend to lose water rapidly without being able to replenish it from the medium, and therefore die. Under various Gelrite concentrations the plantlets vitrify at great speed and cease growing. A mixture of both gelling agents, 0.1% agar plus 0.15% Gelrite, produced a satisfactory result which produced fast and healthy growth. The physiological effects of the differential water pressures produced by the various types of media are currently being studied in our laboratory (Castro Concha et al., 1990).

Table 3. Effect of the gelling concentration on the degree of vitrification and number of adventitious shoots formed in multiplication medium

Gelling agent		Continuous light		16-h Photoperiod		Status of mother plants	
Gelrite %	Agar %	Shoots/ plant	Vitrification	Shoots/ plant	Vitrification	Growth	Vitrification
0.5	0	0.22	Dry	0.44	–	–	Dry
0.25	0	0.28	–	0.63	+	+	+
0.15	0	0.39	+	2.39	+ + +	+ + +	+ + +
0.15	0.1	–		3.88	+	+ +	+
0.15	0.2	–		1.94	–	+	–
0.15	0.4	–		1.39	–	–	–
0	0.8	0	–	0.6	–	+	–
0	1.0	0	–	0.5	–	–	Dry

Table 4. Growth and capacity to produce adventitious shoots by shoot explants of different sizes in multiplication medium

Size of mother plant (cm)	No. of new adventitious shoots	Increase in height (cm)	Gain in new leaves
1–1.9	0.8	2.06	1.72
2–2.9	1.7	1.57	1.33
3–3.9	3.4	0.96	0.72

4.5 Multiplication (Stage II)

When the shoots (whether adventitious or from buds) reach a height of 4 cm, they are again transferred this time to a medium with increased cytokinin (10 mg/l BAP) to induce the formation of new adventitious shoots. The size of the plant is critical for multiplication: larger plants (3–4 cm high) produce more adventitious shoots than smaller ones (Table 4). Contrary to the conditions at the induction stage, a 16-h photoperiod also gives better results than continuous light.

The propagating capacity of the shoots grown in vitro can be maintained for long periods of time. We have kept lines of *A. fourcroydes* multiplying efficiently through the production of adventitious buds for more than 4 years, and the regenerated plants show no morphological changes and no signs of genetic instability at the chromosomal level.

4.6 Maturation

Plants originated in vitro under conditions of very high relative humidity and constant light (Fig. 5B) lack epicuticular waxes, present a reduced number of not fully developed stomata, and do not survive outside the culture dish. To preadapt them for transplantation to soil the shoots are placed in a maturation medium (Table 1) for 2 weeks before rooting. This medium has a reduced cytokinin concentration (1 mg/l BAP) and has double the amount of agar than the growth medium. Under these conditions the plants cease growing and turn dark green and more opaque due to an initial deposition of epicuticular waxes (Fig. 5C). Figure 5D shows the surface of a rooted plant already established in the nursery.

4.7 Rooting (Stage III)

In order to stimulate rooting, the micropropagated shoots are transferred to a rooting medium without cytokinin (Table 1). Shoots form better and larger roots when only agar 0.8% is used as a gelling agent, as it helps the plantlets to preadapt for the transfer to soil (Fig. 1C). As can be seen in Fig. 5B and C, there is a greater deposition of epicuticular waxes at this stage than in the leaves of shoots either at

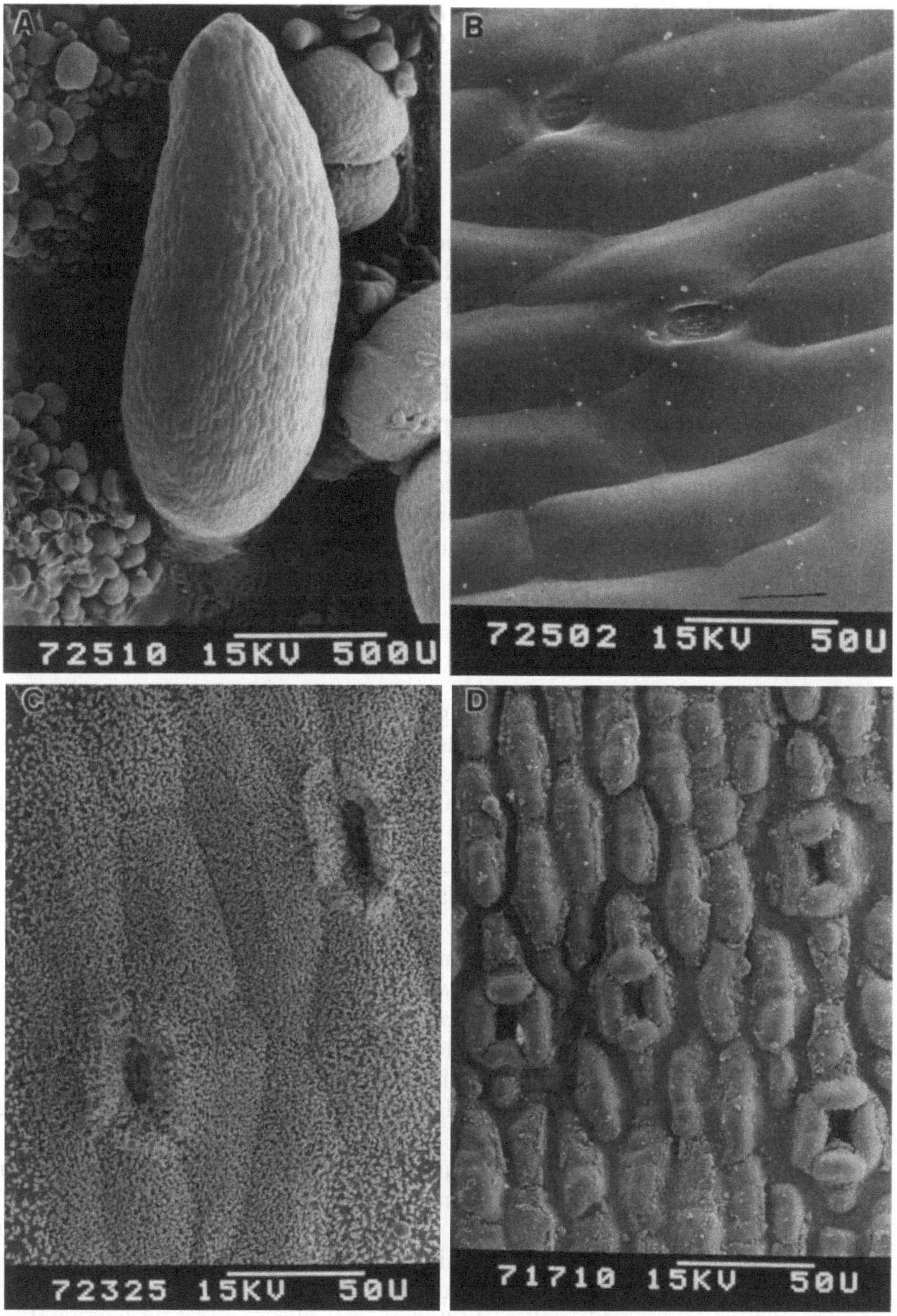

Fig. 5A–D. Scanning electron micrographs showing the surface of in vitro-cultured agave plants at different stages of development. **A** Adventitious shoot from a stem explant without any visible stomata. **B** Back surface of a leaf from an adventitious shoot during the multiplication stage showing a few not fully developed stomata. **C** Back surface of a leaf from a shoot during the maturation stage. **D** Back surface of a leaf from a rooted plant already established in soil the growth or multiplication stages. Plantlets rooted in vitro grow faster and present a much higher survival rate (85–90%) than in vivo-rooted ones (40–50%).

4.8 Weaning (Stage IV)

In vitro-rooted plants are planted in 200-well polystyrine trays (50 ml/well) (Fig. 1D) on a 2:1:1 mixture of compost, soil, and sand, and are kept under conditions of high humidity and high light intensity for 8 weeks. The trays are then moved to the nurseries but are still kept under 70% shading for 4 more weeks before planting them in soil at a distance of 15 cm between plants and between rows. By this time, the plantlets are 10 and 12 cm high and have an average of six to seven leaves. Although they are still smaller than the seedlings from rhizomes, they adapt rapidly to soil, and their development is equal to or better than that of naturally propagated plantlets. Under no circumstances should agave plants be placed in pots, since this prevents the development of the root system and produces severe abnormalities: the plants grow very slowly and produce a reduced number of very elongated leaves, while the formation of new rhizomes is totally inhibited.

4.9 Nursery and Field Performance

To be of real use in improvement programs, micropropagated agaves used for industrial purposes should produce yields as good or better than those of the plants from which they are derived. To evaluate this is, however, a lengthy task, since one has to wait 5 years before being able to assess how much fiber micropropagated henequen plants will produce, and 8–9 years before knowing the weight of the mezcal heads and their reducing sugar content for the production of tequila. It will not be until 1996 that the first tequila from micropropagated agaves will be produced. During the early stages, one can assess the performance of test tube plants only by comparing their growth and anatomical characteristics with those of traditionally produced plants. The growth of micropropagated henequen plants compared with that of rhizome-derived plantlets is illustrated in Fig. 6. It is particularly interesting that in one selected clonal line the number of new leaves is greater in micropropagated plants than in traditionally propagated ones (Fig. 6B), and that the other five clonal lines tested have performed as well or better than the control (Fig. 6D).

Field trials with *Agave tequilana* carried out by Tequila Herradura in their plantations in Amatitan, Jalisco, to compare agaves micropropagated at CICY with rhizome-derived plants, indicates that growth and rate of production of "hijuelos" are slightly higher (25–30%) in the former (Fig. 1E, F).

So far we have not observed any undesirable effects carried over from the in vitro culture stages, although early flowering is always a risk in perennial or multiannual crops. To prevent this, we recommend that micropropagation is used only for rapid clonal multiplication and that in vitro-produced agaves are employed as mother plants for the production of suckers that will be transferred to the plantations.

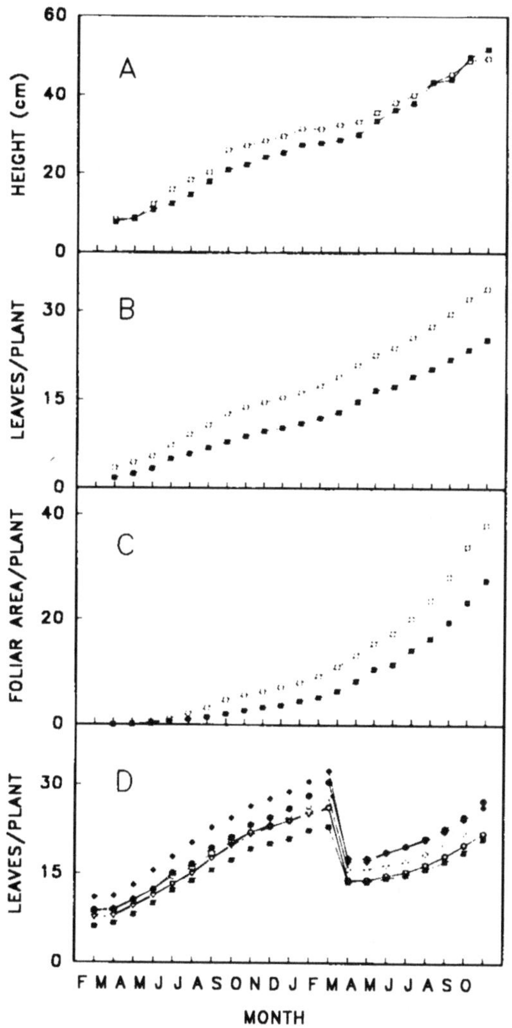

Fig. 6A–D. Comparative development of micropropagated (□) and naturally propagated (■) *A. fourcroydes*. **A, B,** and **C** show development in nurseries measured in terms of height, number of leaves, and foliar area. **D** The development of five clonal lines (1 = ○, 2 = ▽, 3 = ◇, 4 = ● and 5 = ◆) in the nursery and after transfer to field plantations

4.10 Genetic Stability

We have regularly screened the chromosome numbers of the micropropagated plants of both *A. fourcroydes* and *A. tequilana* and have found no variations whatsoever in plants multiplied through adventitious shoots. A very different picture is seen in plants derived from calli, where almost every single individual of *A. fourcroydes* that we have examined showed chromosomal losses (Castorena et al., unpubl.). This should not be a serious problem for henequen, since its pentaploid condition could neutralize the effects of chromosome losses. Although we have not observed any morphological alterations, these plants have not performed comparably to rhizome-propagated ones in the field, and a large proportion has been lost to fungal diseases.

4.11 Vitrification

Though not previously reported, vitrification seems to be a very common problem in in vitro-cultured agaves. We have observed that micropropagated shoots are very susceptible to changes in the balance of NO_3/NH_4 and particularly to the type and concentration of the gelling agent. In a recent study (Castro Concha et al., 1990), we discovered that the water potential of the medium (as determined by the concentration of the gelling agent) is the principal factor affecting both the number of shoots generated per explant and their degree of vitrification. The process of vitrification can be controlled by selecting an adequate gelling mixture (see Table 3).

5 Protocol

The methods described here were chosen not only for their multiplication efficiency but also for the high survival rates and the quality of the plants produced, as assessed by the rates of growth when transplanted to soil (Fig. 6). Although these methods are lengthy due to the biological characteristics of the species and the number of steps involved, they produce the best end results at a reasonable cost.

For *A. fourcroydes* we recommend the use of axillary buds as explants because of the small numbers of shoots that can be generated from stem explants. Previously (Robert et al. 1987), we suggested the induction of calli from these shoots followed by organogenesis as a suitable method for increasing the multiplication efficiency. We no longer recommend this procedure, as we have since learnt that callus-derived plants show a great incidence of chromosomal variation. Instead the procedure described below (Figs. 7 and 8), which is a very efficient method of micropropagating genetically stable plants is recommended.

1. Select mother plants.
2. Sterilize explants as previously described.
3. Induce the development of shoots either from axillary buds (from leaves or rhizomes) or stem segments planting them in baby-food jars with 25 ml of induction medium (Table 1). Incubate (2 weeks for axillary buds and 12 for stem segments) at 27–30 °C under continuous light.
4. Transfer the small shoots (S) to Magenta boxes containing 50 ml of growth medium (Table 1). Incubate for 3 weeks at 27–30 °C under a 16-h photoperiod until they are about 5 cm tall.
5. Transfer large shoots (L) to Magenta boxes containing 50 ml of multiplication medium (Table 1). Incubate at 27–30 °C under a 16-h photoperiod for 3 weeks.
6. Separate the newly formed small (S) adventitious shoots and transfer them to growth medium as in step 4. Transfer large shoots (L) to fresh multiplication medium as in step 5. Repeat until the desired number of plants is reached.
7. At the production stage select the largest and healthiest shoots and transfer a constant number per week to Magenta boxes containing 50 ml of maturation medium (Table 1). Incubate for 3 weeks under conditions of high light intensity, if possible natural sunlight. Continue multiplying the rest as indicated in steps 4 to 6.
8. Transfer large mature plants (LM) to Magenta boxes containing 50 ml of rooting medium (Table 1) for 3 weeks. Incubate as in step 7.
9. Transfer the rooted plants to polystyrene trays and keep them under fairly humid conditions and high light intensity (or indirect sunlight) for 8 weeks.
10. Transfer to the nurseries.

Figure 8 shows the mass production system that we use for the clonal propagation of elite agaves. Independently of the type of original explant, a multiplication stage is required to produce enough material for constant high production rates. During this stage, large shoots (L) derived from a single bud or from a number of buds from a single mother plant are kept in multiplication medium (×) where their

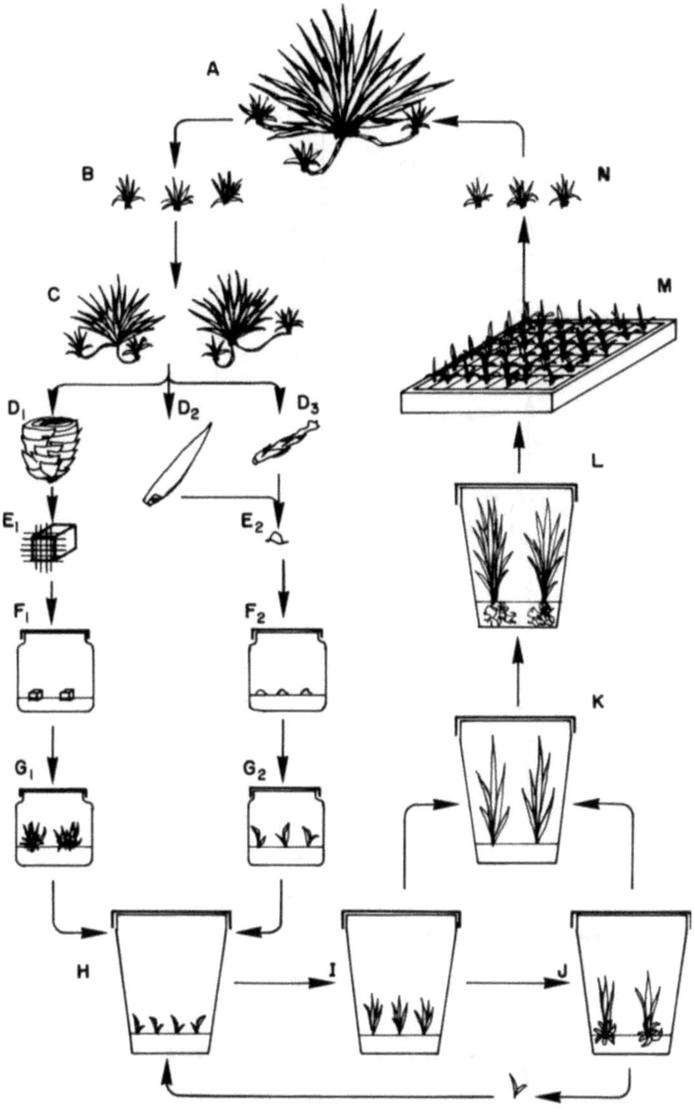

Fig. 7. Micropropagation of agaves: *A* Selection of elite plants; *B* establishment of a clonal line of rhizome derived plants to be used as a source of explants for micropropagation; *C* mother plants of the right age from which stem tissues (*D1*, *E1*), axillary buds from leaves (*D2*, *E2*), or axillary buds from rhizomes (*D3*, *E2*) can be taken as explants for in vitro culture. *F1*, *G1*, and *H* represent the induction of adventitious shoots from stem explants, and *F2*, *G2*, and *H* the development of axillary buds into shoots. The small shoots are grown (*I*) before they enter the multiplication stage (*J*). Small adventitious shoots are separated and grown (*H*, *I*) before repeating the cycle. Large shoots can be transferred to maturation (*K*) and to rooting medium (*L*) before weaning in polystyrene trays (*M*) and transferred to nurseries (*N*), where a new cycle of selection and cloning can be started or the production of suckers for plantations initiated

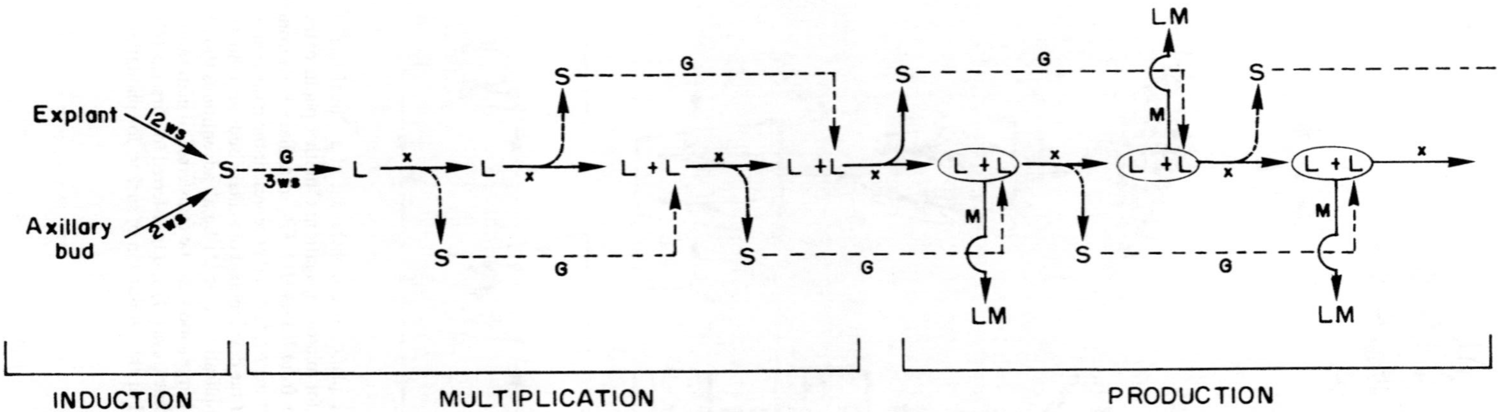

Fig. 8. Mass propagation system for elite agaves. Induction (stage I) is carried out with shoots derived from stem explants or axillary buds. The small shoots (*S*) grown for 3 weeks in growth medium before starting the multiplication stage (stage II). The small adventitious shoots (*S*) produced at the base of the large ones (*L*) grown in growth medium (*G*) before they return to the multiplication line. Once a critical mass has been reached the largest and healthiest shoots are transferred to maturation medium (*M*). The steps can be repeated as many times as desired to produce the required result

growth is minimal and the formation of adventitious shoots optimal. When a suitable number of individuals has been generated, the production can be stabilized, transferring a constant number of shoots to maturation medium (m) where they stop multiplying, resume growth, and start forming epicuticular waxes. These large, mature shoots (LM) are transferred to rooting medium and later to soil. It is important to note that the small adventitious shoots (S) have to reach a minimum size (4 cm) for the multiplication process to be efficient, which implies taking the small shoots out of the multiplication line in order to grow them (G) before putting them back into production. Hundreds of thousands of plants can be produced by this method from a single axillary bud in 1 year, as compared to the 25 to 35 suckers that a mother plant produces naturally over a period of 5 years.

6 Summary and Conclusion

The long life cycle of agaves has in the past been an obstacle to the improvement of these species. Now micropropagation techniques offer the possibility of reducing the time required to produce large numbers of clonal agaves from several years (through natural propagation) to a few months.

Plants can be established in vitro from various types of explants and then multiplied through repeated cycles of adventitious shoot production. This method not only provides large numbers of plants but ensures genetic stability for clonal propagation.

Survival rates through careful preadaptation are high, and field and nursery trials of micropropagated plants have shown them to be morphologically normal, with increased vigor manifested as rapid growth and earlier production of new rhizomes.

The techniques have been applied to several economically important agaves such as *A. fourcroydes*, *A. sisalana*, and *A. tequilana*, and could be used, with minor variations, for most other species. However, if large-scale propagation is attempted, careful adjustments should be made to ensure the highest possible efficiency.

There is no doubt that the productivity of plantation agaves can be improved through the utilization of micropropagation as part of field selection and multiplication programs. However, present production costs, even in large-scale schemes, are still too high to make the application of this technology economically attractive for the low-value, hard-fiber-producing agaves. Nevertheless, increases in multiplication efficiencies and higher survival rates promise to bring costs down to a level at which micropropagated plants could compete with traditionally propagated ones.

Biotechnological methods other than the clonal multiplication of promising variants could also help to improve agaves. For example, somaclonal and gametoclonal variants of the fertile diploid hybrid 11648 might yield individuals resistant to fungal infections and, in theory, could generate many other useful traits, such as thornless leaves, in other plantation agaves.

At present there is no obvious alternative to field selection of in vitro-generated variants (contrary to previous predictions which foresaw culture dish selection as a more efficient substitute), and therefore the lengthy periods needed for this process will continue to be a major obstacle. Additionally, in order to fully exploit this potential germplasm, collections of wild varieties and archetypal variants need to be established.

In spite of the limited use made of agaves today, they hold great potential for the future in the development of semi-arid zones where, under a more rational system of land use, they would undoubtedly play a far more prominent role.

Acknowledgments. The research work described here on *A. tequilana* was initially contracted by the company Tequila Cuervo. More recently, agaves have been mass propagated for Tequila Herradura, which is carrying out field trials in its plantations in Amatitan, Jalisco. We are grateful for both companies' financial support, their generous supply of plant materials, and the numerous discussions that proved so valuable for our work. Francisco Espadas, Agustín García, Andrés Quijano, and Carlos Talavera have also contributed to this work.

References

Aznar Mendoza E (1977) Historia de la industria henequenera desde 1919 hasta nuestros días. In: Encyclopedia Yucatanense, vol III, p 729

Baker DM (1985) Alternative uses for sisal fiber. In: Cruz C, Del Castillo L, Robert ML, Ondarza RN (eds) Biologia y aprovechamiento integral del henequén y otros agaves. Cent Invest Cient Yucatán, pp 177–186

Blunden G, Carabot A, Jewers K (1980) Steroidal sapogenins from leaves of some species of *Agave* and *Furcrea*. Phytochemistry 19: 2489

Boulanger J (1985) Mejoramiento de la variabilidad genética del Sisal (*A. sisalana* Perrine) y del henequen (*A. fourcroydes* Le maire). In: Cruz C, Del Castillo L, Robert ML, Ondarza RN (eds) Biología y aprovechamiento integral del henequén y otros agaves. Cent Invest Cient Yucatán, pp 75–81

Burgess TL (1985) Agave adaptation to aridity. In: Pinkava DJ, Gentry HS (eds) Desert Plants 7(2): 39–50

Castro Concha L, Loyola Vargas VM, Chan JL, Robert ML (1990) Glutamate dehydrogenase activity in normal and vitrified plants of *Agave tequilana* Weber propagated in vitro. Plant cell tissue organ cult 22: 147–151.

Cruz Ramos C, Orellana R, Robert ML (1985) Agave research progress in Yucatán. Desert Plants 7(2): 71–74, 89–93

Dahlgren RMT, Clifford HT, Yeo PF (1985) The families of the mono-cotyledons. Structure, evolution and taxonomy. Springer, Berlin Heidelberg New York Tokyo

Dewey LH (1973) Fiber production in the western hemisphere USA. Washington No 518: 1–95

FAO (1989) Estadísticas de sisal y henequén, 1984–1989. 23ª Reunión del Grupo Intergubernamental sobre Fibras Duras, Roma, 10–13 Oct

Frydrych D (1982) Induction in vitro de bourgeons adventifs à partir du sisal. Premiers résultats. Cot Fib Trop 27 Fasc 3: 295–304

García de Fuentes A, De Sicilia A (1984) El mercado mundial de las fibras duras. Cent Invest Cient Yucatán, AC

Gentry HS (1982) Agaves of continental North America. Tucson, Arizona, Univ Arizona Press, p 670

Groenewald EG, Wessels DCJ, Koeleman A (1977) Callus formation and subsequent plant regeneration from seed tissue of an *Agave* species (Agavaceae). Z Pflazenphysiol 21(4): 369–373

Hertz JE (1985) Modificación y recuperación de esteroides del henequén. In: Cruz C, Del Castillo L, Robert ML, Ondarza RN (eds) Bíologia y aprovechamiento integral del henequén y otros agaves. Cent Invest Cient Yucatán, pp 169–174

Hunault G (1979) Recherches sur le comportement des fragments d'organes et des tissus de Monocotyledones cultivés in vitro. II. Etude du cas de quelques Agavacées. Rev Cytol Biol Veget 2: 21–66

Lock GW (1962) Sisal. Twenty five years. In: Sisal research. Longman, London

Lock GW (1985) On the scientific and practical aspects of sisal (*Agave sisalana*) cultivation. In: Cruz C, Del Castillo L, Robert ML, Ondarza RN (eds) Biología y aprovechamiento integral del henequén y otros agaves. Centro de Investigacion Cientifica de Yucatán AC, pp 99–119

Madrigal LR, Dorantes GMR, Rodríguez de la O JL (1981) Propagación in vitro de henequén (*Agave fourcroydes* Lem). Mem Primer Simposio del Agave, Cordemex, Mérida

Nobel PS (1976) Water relations and photosynthesis of a desert CAM plant, *Agave deserti.* Plant Physiol 58: 576–582

Nobel PS, Hartsock TL (1979) Environmental influences on open stomates of a Crassulacean acid metabolism plant, *Agave deserti.* Plant Physiol 63: 63–66

Piña F, Herrera JL, Robert ML (in preparation) A computer programme for the planning and control of large-scale micropropagation

Powers DE, Backhaus (1989) In vitro propagation of *Agave arizonica* Gentry and Weber. Plant Cell Tissue Organ Cult 16: 57–60

Promotora del Maguey y del Nopal (1981) Anteproyecto de factibilidad técnica para la producción de jarabes de alto contenido de fructuosa a partir de diferentes variedades de Agave. México DF, p 46

Rico Tavera G (1987) De tequila es el tequila (informe especial). Expansión 19(474): 40–50

Robert ML, Herrera JL, Contreras F, Scorer KN (1987) In vitro propagation of *Agave fourcroydes* Lem. (Henequen). Plant Cell Tissue Organ Cult 8: 37–48

Sharma OP, Khanna P (1979) Studies on steroidal sapogenins from tissue cultures of *Agave wightii.* J Nat Prod 43(4): 459–462

Villalvazo J, Ramírez J, Frellmann KA (1985) Estudios técnicos y económicos para la instalación de una planta de celulosa y papel a partir del henequén. In: Cruz C, Del Castillo L, Robert ML, Ondarza RN (eds) Biología y aprovechamiento integral del henequén y otros agaves. Cent Invest Cient Yucatán, pp 213–222

Wienk JF (1976) Sisal and relatives. *Agave* (Agavaceae-Agaveae). In: Simmonds NW (ed) Evolution of crop plants. Logman, New York, pp 1–4

II.10 Micropropagation of Yucca (*Yucca glauca* Nutt.)

S.E. Bentz[1] and H.-J. Talbott[2]

1 Introduction

1.1 Botanical Description

Approximately 30 species comprise the genus *Yucca*, family Agavaceae (Alexander and Pond 1974). Vegetative characteristics include sword-shaped, evergreen leaves that are borne rosette-like on acaulescent to caulescent stems to 30 feet or more in height. Generally slow-growing, yuccas may require 3 to 7 years to flower from seed (Webber 1953). The clump-forming or rhizomatous habit of many yuccas is also an important means of reproduction because lack of rainfall may limit successful natural reproduction by seed (Webber 1953; Alexander and Pond 1974). Flowers are greenish to creamy white, sometimes tinged pink or purple, and borne erect in racemes or panicles. Most *Yucca* sp. are dependent upon moths for pollination and seed set. Fruits are dehiscent or indehiscent and contain hard black seeds. The basic chromosome number is reported to be $x = 30$ ($2n = 60$) (Fedorov 1974).

1.2 Distribution

The range of the genus *Yucca* extends east to west across the North American continent and north to south from North Dakota (49° north latitude) to Central America and the West Indies (22° north latitude) (Pina-Lujan 1980). Within this range yuccas are found in diverse habitats including plains, valleys, high mountain elevations, arid desert, or semi-desert, and regions receiving 20 or more inches of annual precipitation (Webber 1953; Alexander and Pond 1974). Soils typically are sandy or gravelly and well-drained (Webber 1953).

1.3 Importance and Role of In vitro Culture

Although traditionally utilized for fiber and animal feed, the use of *Yucca* species as ornamental plants and as sources of valuable biochemicals has received increased

[1] USDA-ARS, US National Arboretum, 3501 New York Avenue, N.E., Washington, DC 20002, USA
[2] USDA-ARS, National Plant Germplasm Quarantine Laboratory, 11601 Old Pond Drive, Glenn Dale, MD 20769, USA

Biotechnology in Agriculture and Forestry, Vol. 19
High-Tech and Micropropagation III (ed. by Y.P.S. Bajaj)

attention recently. Slow initial growth and minimal production by traditional propagation methods of offsets, rhizomes, and cuttings has led to the investigation of tissue culture techniques as a means of mass production. In vitro studies have been conducted on a chimeral *Y. elephantipes* (Pierik and Steegmans 1983) and an unidentified *Yucca* species (Litz and Conover 1977) grown as interior foliage plants. Our work (Bentz et al. 1988) examined micropropagation of selected pink- or white-flowered and potentially cold-hardy *Y. glauca* to improve distribution and evaluation of new introductions through US National Arboretum programs.

Many plant parts (leaves, seeds, and rhizomes) of yuccas are sources of steroids and sapogenins, whose uses include combating arthritis, relieving stress, and augmenting cattle feed (Kaneda et al. 1987; Webber 1953). Interest in the study and in vitro production of these chemicals resulted in development of callus/suspension culture techniques for *Y. aloefolia, Y. filifera* (Quintero et al. 1982), *Y. glauca* (Stohs et al. 1974), *Y. gloriosa* (Meskhi et al. 1978), and *Y. schidigera* (Kaneda et al. 1987). The effect of light on morphogenesis of *Y. schidigera* root organ cultures (MacCarthy and Staba 1985) and the regeneration of plants from callus in *Y. gloriosa* (Durmashidze et al. 1983; Stohs et al. 1974) were reported in conjunction with these studies.

2 Micropropagation

2.1 General Procedures

Micropropagation techniques have been developed as a result of investigations into the effects of wounding, genotype, daylength, temperature, time in culture, and growth regulators on in vitro shoot proliferation and rooting (Table 1). For

Table 1. Review of micropropagation studies in *Yucca*

Species	Explant	Result
Y. elephantipes	Shoot tip	Shoot proliferation, rooting (Pierik and Steegmans 1983)
Y. filifera	Seedling tissue, coleoptile	Callus, suspension cultures cultures (Quintero 1989; Quintero et al. 1982)
Y. glauca	Shoot tip, flower bud, rhizome	Shoot proliferation, rooting (Bentz and Parliman 1985; Bentz et al. 1988)
Y. glauca	Seedling tissue	Callus, suspension culture (Stohs et al. 1974)
Y. gloriosa	Flower bud	Shoot and root morphogenesis from callus (Durmishidze et al. 1983)
Y. gloriosa	Flower bud	Callus, rhizogenesis, organogenesis (Meshki et al. 1978)
Y. schidigera	Seedling tissue	Callus, root, and shoot cultures (Kaneda et al. 1987)
Y. schidigera	Hypocotyl	Callus, suspension cultures; root morphogenesis (MacCarthy and Staba 1985)
Y. species	Lateral stem bud	Shoot proliferation, rooting (Litz and Conover 1977)
Y. species	Shoot tip	Adventitious shoots, asexual embryos (Murashige 1974)

additional approaches, readers are referred to work on *Yucca* relatives, particularly in the genus *Agave* (Robert et al. 1987).

In vitro cultures of *Yucca* spp. have been successfully established using shoot tips, rhizomes, or flower tissue (Bentz and Parliman 1985; Bentz et al. 1988; Litz and Conover 1977; Pierik and Steegmans 1983). Shoot tips are readily available and easy to disinfest. Following disinfestation, explants have been cultured most successfully on MS (Murashige and Skoog 1962) medium containing 3% sucrose, 0.6% to 0.8% Difco-bacto agar and supplemented with various combinations of 6-benzyladenine (BA), and α-naphthaleneacetic acid (NAA). Media pH range from 5.7–6. Cultures generally have been maintained on a 16 h light/8 h dark schedule under cool-white fluorescent lights at 25–28 °C. Specific procedures outlined below are results of our work with *Y. glauca*.

2.2 Establishment of Cultures

Vegetative shoots of mature *Y. glauca* were excised and outer leaves removed until a leafy shoot tip explant remained, approximately 1 cm long and 0.5 cm wide. Explants then were disinfested in a 1.05% sodium hypochlorite solution for 15 to 20 min and rinsed in sterile distilled water. Culture medium used consisted of MS salts and vitamins (Murashige and Skoog 1962) solidified with 0.6% Difco bacto agar and supplemented with growth regulators. Combinations of NAA (0.0–3.2 μM) and BA (0.0–45 μM) or N^6-[2-isopentenyl]adenine (2-iP) (10–32 μM) were tested. Components were added and the medium pH adjusted to 5.7 with 1 N KOH or 1 N HCl prior to autoclaving at 120 °C at 1.05 kg/cm^2 for 15 min. Cultures were maintained at 25 ± 2 °C under cool-white fluorescent lights (30–55 μmol s/m^2) on a 16-h light/8-h dark cycle, and subcultured every 4 weeks. Once established, cultures could be stored in the dark for up to 8 weeks at 4 °C.

2.3 Regeneration and Differentiation

Cytokinins were necessary for shoot proliferation in the *Yucca* spp. tested (Bentz et al. 1988; Litz and Conover 1977; Pierik and Steegmans 1983). In *Y. glauca*, shoots were found to proliferate with increasing concentration of BA and to produce callus and poorer quality shoots in the presence of NAA and the absence of BA (Fig. 1). Shoot numbers increased and shoot length decreased with increasing concentrations of BA tested (Fig. 2). Shoot quality and callus production in *Y. glauca* were dependent on both NAA and BA, with the highest quality shoots found over a range of concentrations (0.32 or 1 μM NAA with 3.2 or 10 μM BA). In a chimeral *Y. elephantipes* (Pierik and Steegmans 1983) and another *Yucca* species (Litz and Conover 1977), BA, in the presence of NAA, also was effective in inducing shoot proliferation. Optimal rates reported were 4.4 μM BA with 0.5 μM NAA and 3 μM BA with 1 μM NAA, respectively. In the chimeral *Y. elephantipes* (Pierik and Steegmans 1983), exposure to high levels of BA (9–18 μM) resulted in stunting, callus formation, and loss of chimeral character. Experiments comparing 2-iP with BA showed that 2-iP was not effective in promoting shoot proliferation in *Y. glauca* (Table 2) (Bentz et al. 1988) or *Y. elephantipes* (Pierik and Steegmans

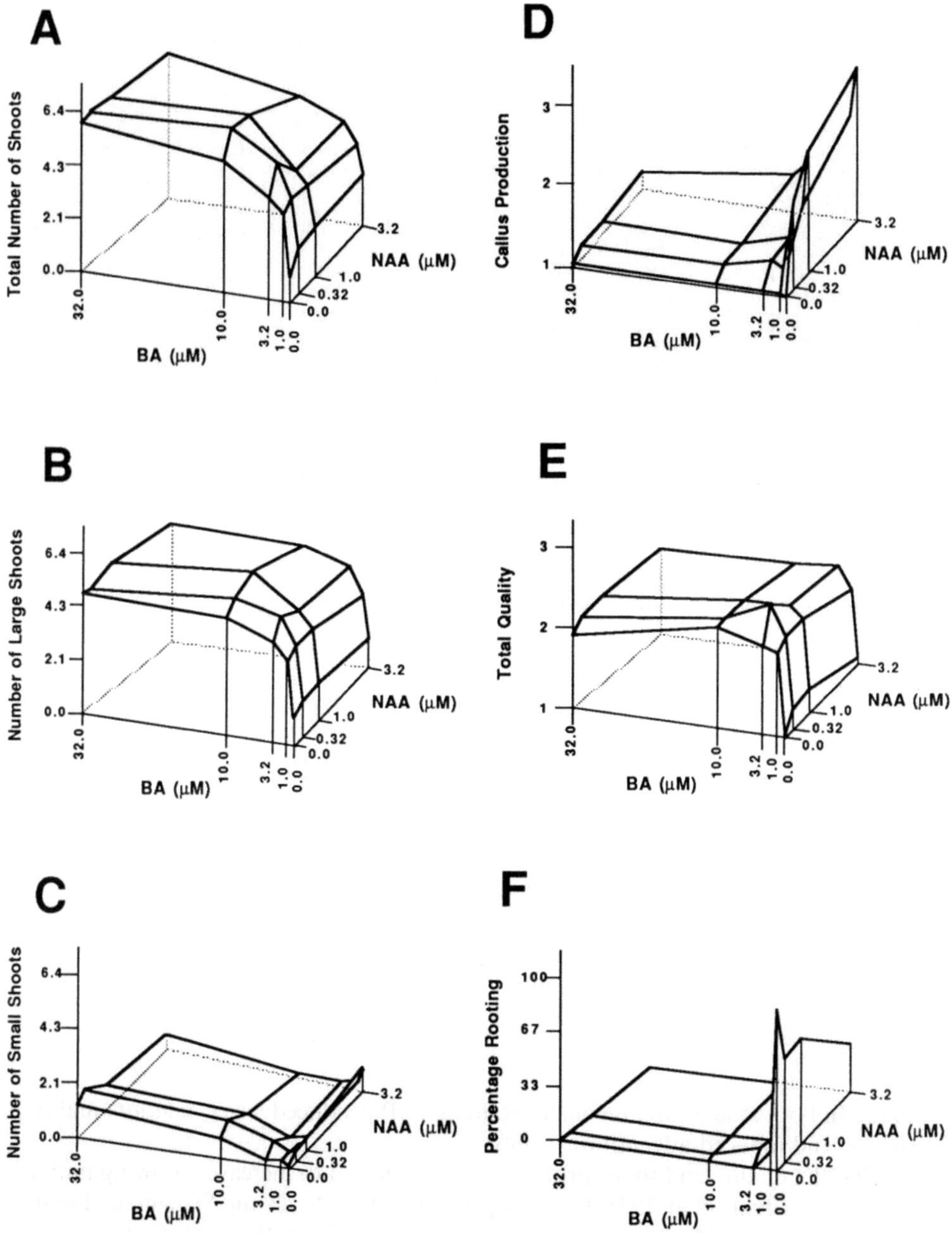

Fig. 1A–F. Influence of BA and NAA on explants of *Yucca glauca*. Significant effects ($P < 0.05$, statistical analysis by log-linear models) were: **A** BA, **B** BA, wounding, **C, D** BA, NAA, NAA × BA, **E** BA, NAA × BA. Data presented are means of nonwounded treatments. (Bentz et al. 1988)

Fig. 2. Response of *Yucca glauca* to increasing levels of cytokinin on MS medium with 0.32 μM NAA. *From left to right* BA concentrations: 0.0, 1, 3.2, 10, 32 μM

Table 2. Effects of BA and 2-iP concentration on shoot production and rooting in *Yucca glauca*.[a] (Bentz et al. 1988)

Cytokinin (μM)		NAA (μM)	Total no. of shoots	Cultures rooting (%)
BA	10	0.32	5.0a	0b
	32	0.32	5.4a	0b
2-iP	10	0.32	2.1b	50a
	32	0.32	2.6b	53a
	10	1.00	1.8b	68a
	32	1.00	2.3b	45a

[a] Treatment means within columns followed by the same letter are not significantly different ($P < 0.05$) by pairwise chi-square tests with Type I error adjusted to match that for Duncan's multiple range test.

1983). In both studies, treatments containing 2-iP produced fewer, inferior-quality shoots, and induced adventitious roots.

Plantlet quality and shoot proliferation were found to increase following initial subcultures (Bentz et al. 1988; Litz and Conover 1977; Pierik and Steegmans 1983). In *Y. glauca*, proliferation and quality increased from the first to the third subculture. Wounding, however, was ineffective in promoting shoot proliferation. In contrast, in *Y. elephantipes*, Pierik and Steegmans (1983) found that wounding, by longitudinal splitting in half, doubled shoot production.

Genotype was shown to be a source of variation in tissue cultured *Y. glauca*. Plants of three genotypes responded similarly to BA and NAA, but differed significantly in shoot number, length, quality, and in adventitious shoot production on attached leaves (Bentz et al. 1988).

Durmishidze et al. (1983) reported plantlet regeneration from callus derived from flower buds of *Y. gloriosa* cultured on modified MS media containing Staba vitamins (Kaul and Staba 1967) and 5.4 μM 2,4-dichlorophenoxyacetic acid (2,4-D) at 25 °C in the dark. Culture of callus at 20 °C enhanced the regenerative capacity of the callus tissue. Following exposure to light, plantlets became green and were transferred successfully to soil.

2.4 Rooting

Ninety percent rooting occurred on growth regulator-free media within 4 weeks of transfer in the *Yucca* spp. discussed above. In *Y. glauca*, BA was clearly inhibitory to rooting (Fig. 2) and the use of the auxins NAA or indole-3-butyric acid (IBA) either as medium components or as plantlet dips failed to improve rooting over growth regulator-free controls (Bentz et al. 1988). Pierik and Steegmans (1983) reported optimal root primordia number and shoot growth of *Y. elephantipes* using MS medium containing 4.9 μM IBA. Litz and Conover (1977) reported 0–0.5 μM NAA as satisfactory for root induction, depending upon genotype.

Rooted plantlets of all species can be transferred and acclimated to greenhouse conditions with approximately 90% survival; however, overwatering must be

Fig. 3. Micropropagated *Yucca glauca*, field-grown for 5 years, height 60 cm, width 90 cm

avoided. Use of plastic tenting and a well-drained soilless growing medium under greenhouse conditions has been very successful (Bentz et al. 1988). Shoots also may be rooted in high humidity conditions ex vitro (Pierik and Steegmans 1983; Bentz et al. 1988).

3 Conclusions

Micropropagation is a suitable propagation method for *Y. glauca* and other *Yucca* species. Following an initial culture period of 5–12 weeks, as many as six to eight shoots per subculture can be produced and subsequently rooted and acclimated to greenhouse or field conditions. Methodology and media are documented and routine. *Y. glauca*'s amenability to cold/dark storage (see above) suggests flexibility in handling for research or commercial purposes. Herbaceous growth of in vitro propagated *Yucca* spp. from several months to several years has been reported to be uniform and normal (Bentz et al. 1988, Pierik and Steegmans 1983). Sporadic flowering of in vitro-propagated plants has been observed at 5 and 6 years from field planting (Fig. 3).

4 Protocol

Excise and disinfest shoot tips following the procedures outlined in Section 2.2. Place explant on MS medium containing 0.32 to 1 μM NAA and 3.2 to 1 μM BA and 0.6% Difco-bacto agar. Maintain cultures at 25 °C under cool-white fluorescent illumination on a 16-h light/8-h dark cycle. Subculture every 4 weeks. Root propagules on growth regulator-free MS medium in vitro or under high humidity ex vitro.

References

Alexander RR, Pond FWA (1974) *Yucca* L. In: Seeds of woody plants in the United States. USDA Washington, DC, Agric Handb 450: 857–858

Bentz SE, Parliman BJ (1985) Sources of explants for in vitro propagation of *Yucca glauca* Nutt. Hortic Sci 20: 540

Bentz SE, Parliman BJ, Talbott H-J, Ackerman WL (1988) Factors affecting in vitro propagation of *Yucca glauca*. Plant Cell Tiss Organ Cult 14: 111–120

Durmishidze SV, Gogoberidze MK, Mamaladze MN (1983) Regeneration of plants from callus tissue of *Yucca gloriosa* buds. Z Pflanzenphysiol 11: 179–182

Fedorov A (ed) (1974) Chromosome numbers of flowering plants. Koeltz, Koenigstein, West Germany

Kaneda N, Nakanishi H, Staba EJ (1987) Steroidal constituents of *Yucca schidigera* plants and tissue cultures. Phytochemistry 26: 1425–1429

Kaul B, Staba EI (1967) *Dioscorea deltoidea* tissue cultures. I. Production and isolation of diostenin. Lloydia 30: 285

Litz RE, Conover RA (1977) Tissue culture propagation of some foliage plants. Proc Fla State Hortic Soc 90: 301–303

MacCarthy JJ, Staba EJ (1985) Morphogenesis in *Yucca schidigera* Roezl. root organ cultures. Ann Bot 56: 205–210

Meskhi AB, Gogoberidze MK, Katsitadze KP (1978) Tissue culture of *Yucca gloriosa*. Izv Akad Nauk Gruz SSR Ser Biol 4: 79–82

Murashige T (1974) Plant propagation through tissue cultures. Annu Rev Plant Physiol 25: 135–166

Murashige T, Skoog F (1962) A revised medium for rapid growth and bioassays with tobacco tissue cultures. Physiol Plant 15: 473–497

Pierik RLM, Steegmans HHM (1983) Vegetative propagation of a chimerical *Yucca elephantipes* Regel in vitro. Sci Hortic 21: 261–272

Pina-Lujan I (1980) Geographic distribution of the genus *Yucca*. Cactus Succulent J 52: 277–281

Quintero A (1989) Yucca. In: Ammirato PV, Evans DA, Sharp WR, Bajaj YPS (eds) Handbook of plant cell culture, vol 5. McGraw Hill, New York, pp 783–799

Quintero A, Rosas V, Zamudio F, Capella S, Romo A (1982) Tissue culture of *Yucca filifera* cells. Identification of steroidal precursors. In: Fujiwara A (ed) Plant tissue culture 1982. Maruzen, Tokyo, pp 295–296

Robert ML, Herrera JL, Contreras F, Scorer KN (1987) In vitro propagation of *Agave fourcroydes* (Henequen). Plant Cell Tissue Organ Cult 8: 37–48

Stohs SJ, Sabatka JJ, Obrist JJ, Rosenberg H (1974) Sapogenins of *Yucca glauca* tissue cultures. Lloydia 37: 504–505

Webber JM (1953) Yuccas of the southwest. US Dept Agr Monog 17: 97

II.11 Micropropagation of Monkshood (*Aconitum* spp.)

R.L. CERVELLI[1]

1 General Account

The genus *Aconitum*, commonly known as the monkshood, aconite, or wolfsbane, is composed of herbaceous perennials generally recognized as the most advanced group within the Ranunculaceae (Leppik 1964). Growing 60 cm to 1.5 m tall from short tuberous roots, the plants display deep palmately dissected basal or cauline leaves. The blue to white to yellow flowers, borne on a terminal raceme, are distinguished by the large arched or hood-shaped upper sepal, hence the name monkshood (Fig. 1). The two true petals are developed into nectaries contained

Fig. 1. The threatened species, Northern monkshood, *Aconitum noveboracense*, in the Kickapoo River Valley, Vernon County, Wisconsin, USA, August 2, 1984 (Cervelli 1985)

[1] Nova Scotia Research Foundation Corporation, P.O. Box 790, Dartmouth, NS, Canada B2 327

Biotechnology in Agriculture and Forestry, Vol. 19
High-Tech and Micropropagation III (ed. by Y.P.S. Bajaj)

within the hood. Seeds are numerous, three-cornered, and dispersed from follicles. The chromosomes are large and karyotypic analysis can be made without difficulty. The chromosome number of the genus is n = 8, somatic number (2n) = 16 (Jensen 1950). Vesicular-arbuscular mycorrhizae have been documented in the roots of at least two species, *A. napellus* and *A. noveboracense* (Cervelli, unpubl.).

Considerable phenotypic plasticity exists among the approximately 500 species distributed worldwide (Munz 1945). Greatest diversity is in the mountains of eastern Asia, the next in those of central Europe, with a smaller group in western N. America. Habitat is primarily at higher elevations, in shaded woods or subalpine meadows, although some species can be found on rich bottomland or sheltered topography in the central United States.

The importance of aconite extends from its use in medicine to ornamentals. The potent diterpene alkaloids have given the aconite a reputation since antiquity. There is early evidence (Cook 1784) that aconite common to the Kamchatka Peninsula and Aleutian Islands was used as whale poison. Species in Europe came to be known as wolfsbane, presumably for its use as a wolf poison (Pelletier 1983). *Aconitum lycoctonum* L. derived its name from the Greek word for 'wolf-slayer.'

The aconite alkaloids are neurotoxins. The molecule binds to the nerve-cell membrane and prevents the normal closing of sodium channels (Pelletier 1983). Pharmacologically, there is a narrow range between a therapeutic dose and a toxic dose of the crude drug preparations. Aconite preparations have been used as sedatives (Banerji and Mukherjee 1984), analgesics (Murayama et al. 1984), anti-inflammatory agents (Tang et al. 1984), anti-arrhythmic agents (Chen et al. 1983), and cardiotonics (Pelletier 1983). Anticancer potential has been reported, but is as yet unproved (Pelletier 1983).

Ornamental usage of the genus, cultivated in Europe since the 16th century, is recently gaining popularity in North America because of its showy flower stem, which blooms mid-season. Cut flower use is increasing. Warnings about mistaken consumption of the tubers by children or pets are frequently provided by commercial growers. The species of greatest ornamental use are *A. napellus*, *A. carmichaeli*, and *A. fischeri*.

Conventional propagation is by seed, which are produced prolifically. Germination rates can be high if care is taken to overwinter the seed in the imbibed state (Cervelli et al. 1986). Imbibition may be necessary to provide development of a rudimentary embryo prior to germination (Cervelli et al. unpubl.). For this reason, the seed may not store long in a dry condition. Mature plants can be propagated by root divisions, but plants tend to resist disturbance. Little breeding work has been done within the genus and there are few hybrids.

Tissue culture may be beneficial to facilitate the breeding and multiplication of particular species. In vitro study and production of diterpene alkaloids may provide additional knowledge and control for pharmaceutical use. Micropropagation can be beneficial for the commercial production of ornamental species, or for the preservation of rare or threatened species.

Table 1. Summary of the in vitro work done on *Aconitum* spp. (see also Shoyama et al. 1991)

Species	Culture and response	Medium	Reference
A. napellus	Somatic embryogenesis from leaf-derived calli	MS + 2,4-D	Cervelli (1986)
A. noveboracense	Somatic embryogenesis from leaf-derived calli	MS + 2,4-D	Cervelli (1986)
A. carmichaeli	Somatic embryogenesis from anther-derived calli	MS + 2,4-D + K	Hatano et al. (1987)
A. napellus	Shoot multiplication from axillary buds	MS + BA + PVPP	Cervelli (1987)
A. noveboracense	Shoot multiplication from axillary buds	MS + BA + PVPP	Cervelli (1987)
A. carmichaeli	Shoot multiplication from stem tips and axillary buds	MS + BA	Hatano et al. (1988)

2 Review of Tissue Culture Work

The in vitro work to date has been summarized in Table 1. Shoot multiplication and leaf callus-derived somatic embryogenesis protocol has been established for common European monkshood, *A. napellus*, and for a threatened species, Northern monkshood, *A. noveboracense* (Cervelli 1986, 1987). Hatano et al. (1987, 1988) successfully produced plants of *A. carmichaeli* from anther-derived somatic embryos and from multiplication of shoot tips and axillary buds. Little other published in vitro work appears to have been done at present (see Shoyama et al. 1991), although the genus is receiving considerable attention from ornamental and pharmaceutical researchers.

3 Micropropagation

Explants can be taken from several tissue sources, including in vitro-germinated seeds (Cervelli, unpubl.), axillary buds and shoot tips (Cervelli 1986, 1987, Hatano et al. 1988), and immature anthers (Hatano et al. 1987). Use of tuberous roots as an explant source results in high contamination rates (Cervelli 1985). It appears that somatic embryos can be obtained from either leaf (Cervelli 1986) or anther-derived calli (Hatano et al. 1987). Although successful rooting, acclimation, and transfer to the greenhouse has been reported, considerable work remains to be done on commercial-scale micropropagation.

3.1 Explant Sterilization and Culture Establishment

Axillary bud nodal segments (Fig. 2) isolated from stems produced from chamber-grown first year tubers were sterilized with successive immersion in 70% ethanol for 1 min, 0.5% sodium hypochlorite plus two drops Tween 20 per 100 ml for 20 min, followed by two rinses of sterile distilled water. Alternatively, sterilization occurred with a 2% solution of antiformin for 10 min and 70% alcohol for 1 min. Young flower buds, 7–12 mm in diameter, were taken from field-grown plants and sterilized with 2% sodium hypochlorite plus 0.1% Tween 80 for 10 min, then with 70% alcohol for 30 s and double-rinsed with sterile water.

Unimbibed seeds were also used as an explant source but required standard sterilization techniques as outlined above, followed by 9 weeks of warm (17 °C) stratification and 6 weeks of cold (– 5 °C) stratification (Cervelli, unpubl.). Seeds were placed individually in sterile 5-ml vials containing one-eighth strength MS salts and 10 g/l agar. Following axenic germination (Fig. 3), with rates as high as 95%, epicotyls were transferred to shoot multiplication medium.

3.2 Media and Culture Conditions

Murashige and Skoog's basal medium (1962) was supplemented with 3% sucrose and different hormones depending on tissue and intended growth response, as outlined in Section 3.5. Phenolic oxidation of shoot cultures of *A. napellus* and *A. noveboracense*, and diffusion into the medium were controlled with media supplements of 25 mg/l citric acid (CA) and 500 mg/l polyvinylpolypyrolidone (PVPP). The occurrence of nonchimeric albino foliage and reduced growth rates for these two species was alleviated by switching from 8 g/l agar to 2 g/l gellan gum in the medium. Rifampicin was used successfully at 50 mg/l to control contamination in some stem node explants (Cervelli 1987).

Culture growth appeared best at 25 °C, with 40 μmol/s/m^2 cool white fluorescent light at either 16 or 24 h per day. Subculturing varied from 4 to 8 weeks.

3.3 Regeneration and Differentiation

Optimum shoot multiplication (Fig. 4a), which occurred in the form of basal rosettes (Fig. 4b), from stem tips or axillary buds was achieved on medium supplemented with 1 mg/l (for *A. napellus* and *A. noveboracense*) to 5 mg/l (for *A. carmichaeli*) benzyladenine (BA). There did not appear to be significant differences in rosette production between genotypes and species. No significant decline in rosette production occurred through ten subcultures of 4 weeks each for *A. napellus* and *A. noveboracense* (Table 2).

Anther-derived calli of *A. carmichaeli* were produced in the dark on medium containing from 0.1 to 10 mg/l 2,4-D and with or without 1 mg/l kinetin. Optimum callus growth occurred on medium containing 5 mg/l 2,4-D and 1 mg/l kinetin (Hatano et al. 1987). Optimum leaf-derived callus growth for *A. napellus* and *A. noveboracense* occurred in the light on 1 mg/l 2,4-D (Cervelli 1986). Calli from all systems could be induced to form somatic embryos (Fig. 5) by repeated subculture for at least 10 weeks on 1 mg/l 2,4-D, although the extent of embryogeny varied considerably with genotype. Transfer of embryos to medium without hormones resulted in the establishment of independent plantlets (Cervelli 1987).

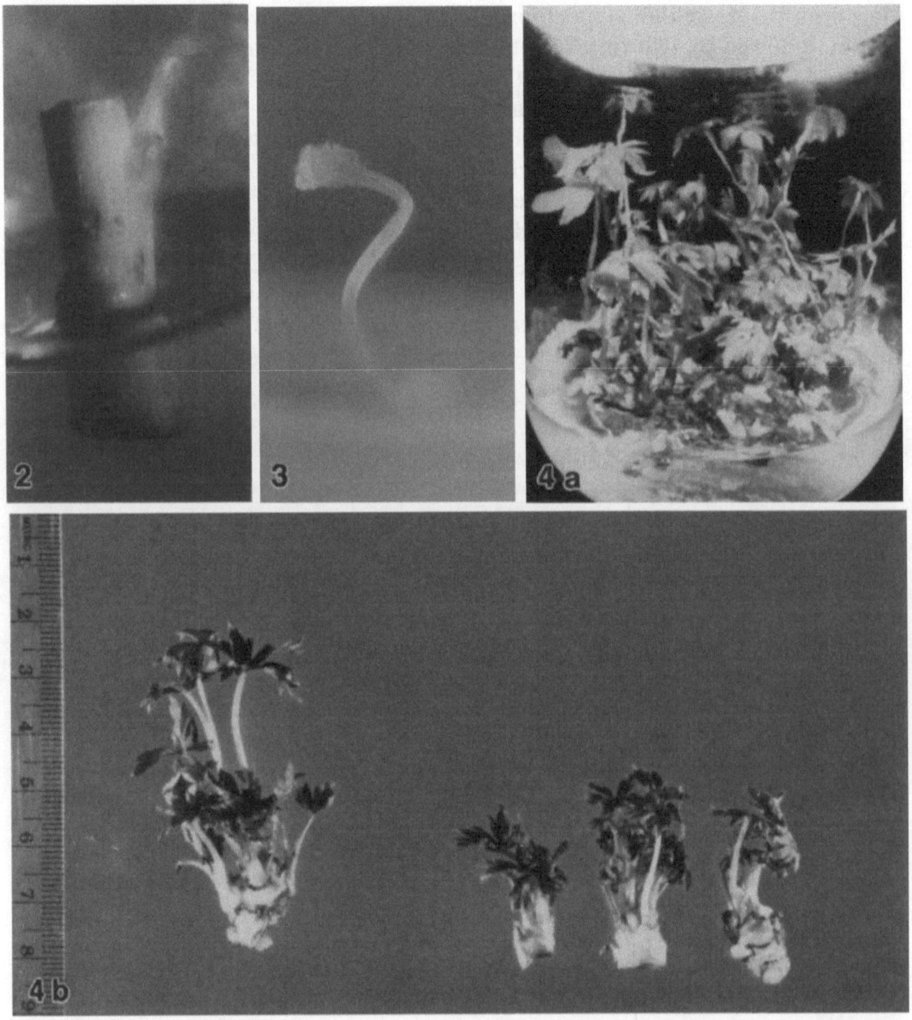

Figs. 2–4. Shoot initiation and multiplication of *A. napellus* and *A. noveboracense*. **2** Nodal explant with axillary bud breaking dormancy after 2 weeks on MS medium with 1 mg/l BA. **3** Axenic seed germination after 15 weeks stratification on one-eighth strength MS medium without hormones. **4a** Four explants per container after 4 weeks as propagated in vitro on MS medium with 1 mg/l BA. **4b** Rosette cluster and excised individual rosettes taken from micropropagation container. (After Cervelli 1987)

Embryos of *A. carmichaeli* developed normal shoots after transfer to medium containing 1 mg/l gibberellic acid (GA) and 5 mg/l BA (Hatano et al. 1987).

The in vitro development of incipient axillary bulbils from stem node explants taken from field-grown stems of *A. noveboracense* was noted in the work of Cervelli (1985) (Fig. 6). Bulbil development was expressed as simultaneous growth of root and shoot within 2 weeks of explant inoculation. There was a greater tendency towards in vitro bulbil expression in nodal explants taken from the basipetal

Table 2. Rosette production of *Aconitum* as a function of subculture.[a] (Cervelli 1987)

Genotype	Mean no. of rosettes[b]							
	Propagative cycle							
	2	3	4	5	6	7	8	9
A. noveboracense								
WCM[c]	2.5 ± 0.3	2.4 ± 0.4	3.2 ± 0.2	2.2 ± 0.2	2.1 ± 0.2	2.3 ± 0.2	2.0 ± 0.2	NA
HV	2.5	3.0 ± 0.3	3.2 ± 0.4	2.5 ± 0.3	2.6 ± 0.2	2.3 ± 0.3	1.8 ± 0.2	NA
WC	2.5	1.5 ± 0.3	3.6 ± 0.4	1.7 ± 0.3	2.9 ± 0.3	3.0 ± 0	2.7 ± 0.2	4.0 ± 0.6
A. napellus	NA	NA	NA	3.4 ± 0.2	2.2 ± 0.2	2.8 ± 0.3	2.8 ± 0.2	3.6 ± 0.4

[a] Stage culture, 4.5 μM BA, 4-week cycles.
[b] ± SE; NA = data not available.
[c] Genotypes from Vernon County, Wis.: WCM = Wildcat Mountain, HV = Hay Valley, and WC = Weister Creek.

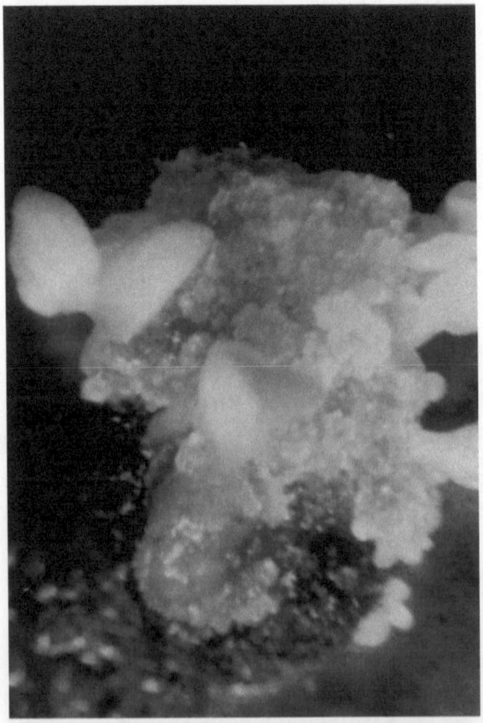

Fig. 5. Embryos proliferating from leaf-derived callus of *A. noveboracense* after 10 weeks on MS medium with 1 mg/l 2,4-D (Cervelli 1986)

Fig. 6. Development of quiescent axillary bulbil of *A. noveboracense* after 3 weeks on MS medium with 1 mg/l BA. (Cervelli 1985)

portion of the stem, which correlated with field observations of in situ bulbil development.

Cold (4 °C) storage of excised rosettes of *A. napellus* and *A. noveboracense* was attempted for 4 months on shoot multiplication medium containing 1 mg/l BA (Cervelli, unpubl.). Although the original tissue tended to senesce, the development of viable bulbils was observed in all *A. noveboracense* genotypes of cold-stored rosettes after 4 months, possibly mimicking an in situ vegetative propagation response to seasonal changes (Cervelli 1985).

3.4 Transfer to Greenhouse and Field

Rooting of rosettes and shoots occurred in vitro after 6 weeks on media either lacking hormones or supplemented with 0.5 mg/l indole-3-acetic acid (IAA). Rosettes could also be rooted by transfer directly to peat plugs. Survival in direct rooting depended on genotype and plantlet size. For the most successful genotype, 74% of the rosettes rooted in a high humidity chamber with 16 h daylength after 10 weeks. *Aconitum* produces thick, tuberous roots which appear to be functional whether plantlets are rooted in vitro or directly in peat plugs, but the rooting response varies with genotype.

Rooted plantlets of *A. napellus* and *A. noveboracense* tended to go dormant after 2 to 3 months' growth (Fig. 7), and flowering was stimulated by a 10-week

Fig. 7. *A. noveboracense* plantlet obtained from shoot multiplication culture. (Cervelli 1985)

dormancy treatment (3 °C, dark). Subsequent growth morphology appeared normal (Cervelli 1987). Dormancy problems do not appear to have been encountered in micropropagated plantlets of *A. carmichaeli*, which were rooted in vitro on 0.5 to 1 mg/l IAA and grown in 100% vermiculite at 20 °C for 4 weeks (Hatano et al. 1988). Flowering occurred within 5 months. Further work is required for *Aconitum* species in optimizing commercial-scale techniques for rooting and cultivation to plant maturity.

4 Summary and Conclusions

The genus *Aconitum* is responsive to shoot multiplication and somatic embryogenesis techniques. Responses between species appear to be consistent for shoot multiplication protocols, although large inter-specific and genotypic differences were observed with somatic embryogenesis protocol. Rosette production offers a usable plantlet for large-scale handling which acclimates well. Large tuberous roots appear functional whether grown in vitro or ex vitro. Dormancy problems exist in young rooted plantlets which can be overcome by cold treatment. Foliar sprays of GA_{4+7} have been reported to break dormancy in some woody micropropagation systems (Pruski et al. 1985) but were not tried with these species. Both shoot multiplication and somatic embryogenesis systems of *Aconitum* require refinement before commercial-scale micropropagation is viable.

5 Protocol

1. Best Explant
   Axillary buds or shoot tips, given standard sterilization treatments in 70% ethanol and 0.5% sodium hypochlorite
2. Best Multiplication Medium
   MS basal medium + 1 to 5 mg/l BA, 25 mg/l CA and 500 mg/l PVPP solidified with 2 mg/l gellan gum
3. Best Rooting Medium
   MS basal medium with 0.5 to 1 mg/l IAA
4. Best Acclimation
   4 weeks at high humidity at 20 °C in 100% peat moss or vermiculite

References

Banerji P, Mukherjee S (1984) The effect of *Aconitum napellus*, *Arnica montana* and *Rhus toxicodendron* extracts on hypertension. Indian J Pharm Sci 46(1)

Cervelli RL (1985) Vegetative propagation of the threatened perennial, northern monkshood, *Aconitum noveboracense*: natural and in vitro techniques. Masters Thesis, Univ Wisconsin, Milwaukee

Cervelli RL (1986) Somatic embryogenesis and shoot multiplication of a threatened perennial: Northern monkshood, *Aconitum noveboracense*. In: Abstr 6th Int IAPTC Congr, Univ Minn, Minneapolis, p 189

Cervelli RL (1987) In vitro propagation of *Aconitum napellus* and *Aconitum noveboracense*. HortScience 22(2): 304–5

Cervelli RL, Kuchenreuther M, Kessenich T, Stearns F (1986) Germination protocol for *Aconitum noveboracense*, a threatened species. Restoration Manag Notes UW-Arboretum 4(1): 36

Chen WZ, Dong YL, Zhang YF, Ding GS (1983) Anti-arrhythmic effects of guan-fu base A. Acta Pharm Sin 4: 247–250

Cook J (1784) A voyage to the Pacific Ocean undertaken by command of his majesty for making discoveries in the northern hemisphere, vol 4. Stockdale, Scratcherd, Whitaker, Fielding and Hardy, London

Hatano K, Shoyama Y, Nishioka I (1987) Somatic embryogenesis and plant regeneration from the anther of *Aconitum carmichaeli* Debx. Plant Cell Rep 6: 446–448

Hatano K, Kamura K, Shoyama Y, Nishioka I (1988) Clonal multiplication of *Aconitum carmichaeli* by tip tissue culture and alkaloid contents of clonally propagated plant. Planta Med 2: 152–155

Jensen H (1950) Meiosis in an unusual form of *Aconitum uncinatum* L. Am Nat 84: 17–22

Leppik EE (1964) Floral evolution in the Ranunculaceae. Iowa State J Sci 39: 1–101

Munz P (1945) The cultivated aconites. Gentes Herb 6

Murashige T, Skoog F (1962) A revised medium for rapid growth and bioassays with tobacco tissue cultures. Physiol Plant 15: 473–497

Murayama MT, Ito T, Konno C, Hikino H (1984) Mechanism of analgesic action of mesoaconitine 1. Relationship between analgesic effect and central monoamines or opiate receptors. Eur J Pharm 101: 29–36

Pelletier SW (1983) Alkaloids: chemical and biological perspectives. Wiley, New York

Pruski K, Grainger G and Nowak J (1985) Micropropagation of Saskatoons. In: Zimmerman RH et al. (eds) Current plant science and biotechnology in agriculture: tissue culture as a plant production system for horticultural crops. Conference, Beltsville, Maryland, USA, Oct 20–23. Nijhoff, Dordrecht, Abstr 26

Shoyama T, Nishioka I, Hatano K (1991) *Aconitum* spp. (Monkshood): Somatic embryogenesis, plant regeneration, and the production of aconitine and other alkaloids. In: Bajaj YPS (ed) Biotechnology in agriculture and forestry, vol 15: Medicinal and Aromatic Plants III, pp 58–72

Tang XC, Lin ZG, Cai W, Chen N, Shen L (1984) Anti-inflammatory effect of 3 acetyl aconitine. Acta Pharm Sin 5: 85–89

II.12 Micropropagation of French Tarragon (*Artemisia dracunculus* L. var. *sativa*)

S.L. Kitto[1] and W.A. Mackay[2]

1 Introduction

Artemisia dracunculus L. var. *sativa*, French tarragon, is a sterile perennial herb that reaches 2 feet in height and has erect branching stems having glossy, dark green, lanceolate to linear leaves 1 to 4 inches long (Foster 1974). French tarragon should not be confused with Russian tarragon, the fertile, seed-bearing variety, *A. dracunculus* L., which has erect, branching stems that reach 4 to 5 feet in height and have wider, light-green leaves. Most importantly, the leaves lack the pronounced flavor characteristic of French tarragon (Foster 1974). French tarragon

Fig. 1. Greenhouse-grown plant of French tarragon in flower

[1] Delaware Agricultural Experiment Station, Department of Plant Science, College of Agricultural Sciences, University of Delaware, Newark, DE 19717-1303, USA
[2] Former Research Assistant. Currently Research Assistant, Department of Horticulture, University of Maryland, College Park, MD 20742, USA

Biotechnology in Agriculture and Forestry, Vol. 19
High-Tech and Micropropagation III (ed. by Y.P.S. Bajaj)

leaves contain 0.25 to 1% aromatic oils, of which 75% is methyl chavicol, the origin of the licorice flavor. Other aromatic oils which give French tarragon its characteristic flavor are ocimenes and capillenes that provide the floral tarragon flavor (Zarghami and Russel 1973). The essential oils are used by the fragrance and food industries as a fragrance component in soaps, detergents, creams, lotions, and perfumes, and as a flavor component in beverages, candies, processed meats, fats and oils, and gravies. French tarragon leaves are used both fresh and dried as a seasoning.

It is commonly reported that French tarragon does not flower; however, our greenhouse stock plants flowered for several months each year starting in early spring. The whitish-green flowers are borne in a spreading panicle (Fig. 1), and are of two forms, pistillate and bisexual, the latter being sterile. Rousi (1968) has attributed one cause of infertility to disturbances in meiosis of the pollen mother cells.

2 Conventional Propagation

Propagation of French tarragon is achieved either by crown divisions, approximately once every 3 years, or by stem or root cuttings. However, there are problems with these propagation methods. Problems associated with propagation by division include the limited number of plants that can be obtained from one mother plant in a year, and the spread of northern root-knot nematode, *Meloidogyne hapla*, by introducing the pest into uninfested fields. Root cuttings also may spread northern root-knot nematode. Problems reported to be associated with propagation by stem cuttings include either a failure to root or a seasonal responsiveness to rooting. However, we had no problem in rooting cuttings year-round from greenhouse-grown stock plants. Quinn (1987) has found that if cuttings originate from healthy main shoots and are 2 inches in length they will root within 14–19 days. For best results, the cuttings were washed, treated with a rooting hormone, stuck in Techniculture plugs and placed under mist. Other potential problems with propagation by division or cuttings are the spread of unidentified viruses and pathogens that may reduce yields.

Greenhouse production of fresh tarragon during winter months is limited by the cessation of growth. It has been reported that French tarragon growth decreases during the winter as the natural light levels decrease. Providing a long day length with supplemental light has been reported to eliminate this "dormancy" period allowing year-round production of fresh tarragon in the greenhouse (Bassuk 1986).

3 Need for Micropropagation

Demand for dried French tarragon in the United States has increased approximately 2.5 times in volume and 5.4 times in value during the last decade (Anon

1973–83). There are no available nationwide statistics for the volume of fresh French tarragon utilized in the United States. The primary suppliers of French tarragon have been European growers with labor and production costs comparable to the United States, suggesting that there may be a market that could be filled by domestic producers. Production costs in the United States could be lowered with the introduction of improved cultivars of tarragon that are resistant to *Fusarium* spp., are higher yielding, or are capable of sexual reproduction. Tarragon plants are sterile, and elite cultivars are currently obtained via sports. However, it would be desirable to produce variants using tissue culture, i.e., somaclonal variants. One method of producing somaclonal variants is by initiating callus, maintaining the callus over a period of time, and in some cases screening the callus with the appropriate selection pressure, then regenerating shoots which are rooted to form new plants. The regenerated plants are then screened for desirable characteristics. This method takes advantage of variation that is either already present in the plant or induced during tissue culture. Once a desirable variant is identified, then it is necessary to have a system that allows for rapid micropropagation.

4 Micropropagation

French tarragon has been previously micropropagated using leaves as the explant source from 3-year-old plants grown outdoors in a herb garden (Garland and Stoltz 1980). However, shoot tips are the most common explant used in commercial micropropagation. A shoot tip has several advantages: it contains numerous axillary buds, starts growth rapidly, has higher survivability during transfer to in vitro conditions than a smaller explant (George and Sherrington 1984), and also ensures clonal uniformity.

Our studies have shown that French tarragon can be successfully micropropagated in vitro using shoot tips as the explant source (Fig. 2) on medium containing Murashige and Skoog's (1962) salts and vitamins (Mackay 1987, Mackay and Kitto 1987, 1988). Sucrose was found to be essential for proliferation with 30 g/l optimal. Shoot-tip position on the greenhouse-grown stock plant, axillary or apical in origin, did not affect proliferation or growth. However, shoot tip explants placed horizontally on the medium had greater proliferation than those placed vertically on the medium (Fig. 3). Horizontal placement of unpinched shoots produced the most axillary shoots, particularly those > 5 mm (Fig. 3, Table 1). Pinching the shoot tip apex significantly reduced the number of axillary shoots, but average shoot length was increased significantly (Table 1).

Two stock plant pretreatments, potting media and lighting, were investigated as a means of maximizing proliferation in culture. Stock plants were produced by potting rooted cuttings in one of four media: sand, vermiculite, peatlite, or peatmoss. Those maintained in peatmoss died within 2 weeks. Explants were obtained from the remaining media stock plants and cultured. Analysis of the data indicated that there was no difference for the main shoot total length, the number of axillary shoots, or the average length of the axillary shoots for the remaining stock plant media (Mackay 1987).

Table 1. Effect of shoot-tip treatment on shoot proliferation of French tarragon. (Mackay and Kitto 1988)

Shoot tip treatment	Main shoot length (mm)	No. of axillary shoots	Av. length of axillary shoots (mm)
Vertical unpinched	24.4a[a]	9.4a	3.7b
Horizontal unpinched	25.0a	10.6a	3.9b
Vertical pinched	20.7b	6.8b	4.8a

[a] Mean separation in columns by Duncan's multiple range test, 5% level.

Fig. 2. Greenhouse shoots 2 cm in length were collected, leaves removed and surface disinfected for 10 min in a 0.5% sodium hypochlorite solution containing 0.1% Tween 20 (v/v). After rinsing three times in sterile water, shoots were trimmed to 1.5 cm in length and cultured. Proliferated cultures were terminated after 4 weeks

An equal number of stock plants were maintained under either natural daylength (10 h, 18 min to 12 h, 10 min) or 24 h supplemental light provided by cool-white fluorescent lamps (65 μmol/s/m^2). Main shoot length and number of axillary shoots were unaffected by light treatment (Table 2). Natural daylength produced higher average shoot lengths in vitro.

The best combined response between axillary shoot number (15.1) and average shoot length (5.1 mm) was achieved on medium containing a combination of 0.6 mg/l BA and 0.05 mg/l NAA (Mackay and Kitto 1988).

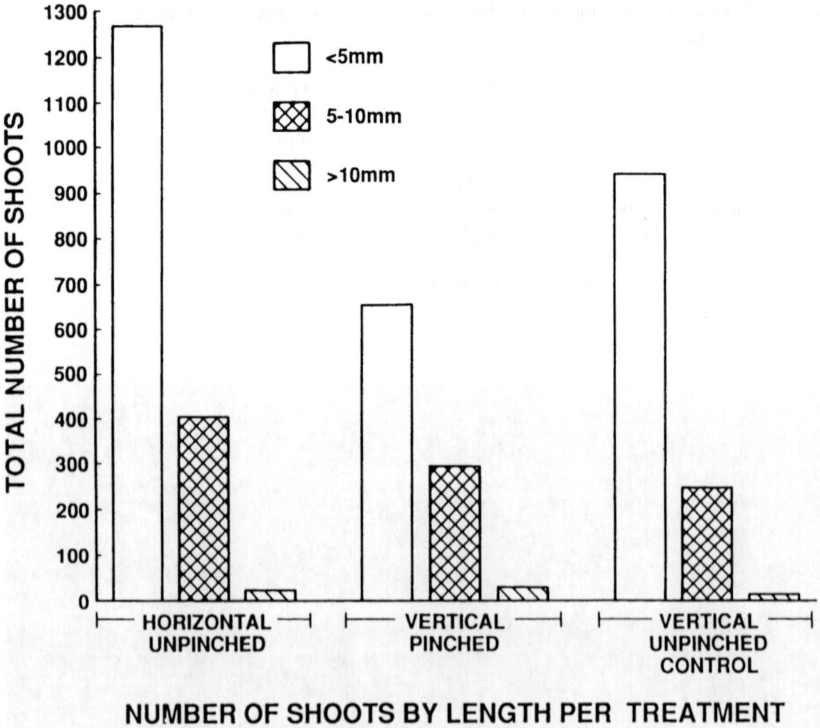

Fig. 3. Axillary shoot number and length in French tarragon as influenced by explant orientation and pinching of the apex. (Mackay and Kitto 1988)

Table 2. Effect of light regime pretreatment of stock plants on shoot proliferation of French tarragon. (Mackay and Kitto 1988)

Light regime	Av. main shoot length (mm)	Av. no. of axillary shoots	Av. length of axillary shoots (mm)
Supplemental[a]	30.3*[b]	12.8	3.5*
Natural daylength	30.9	14.3	3.8

[a] Twenty-four-hour cool-white fluorescent lamps (65 μmol/s/m^2).
[b] Natural daylength ranged from 10 h, 18 min to 12 h, 10 min.
* Significant at the 5% level.

Preliminary experiments suggested that subculturing the proliferated explants onto medium without growth regulators would promote axillary shoot extension. Therefore, the presence or absence of the growth regulators BA (0.6 mg/l) and NAA (0.05 mg/l) on proliferation was investigated using four treatments: 4 weeks with growth regulators (the control); 2 weeks with growth regulators and 2 weeks without growth regulators; 8 weeks with growth regulators (subcultured after 4 weeks); and 4 weeks with growth regulators and 4 weeks without growth regulators. A significantly greater number of axillary shoots was produced with the

8-week treatment; however, there was no difference for main shoot length or average length of the axillary shoots (Table 3).

Gelling agent material and concentration affected proliferation. As the gelling agent concentration increased (washed Difco Bacto agar; 2, 4, 6, 8, 10, 12 g/l or Gelrite; 0.5, 1, 1.5, 2, 2.5, 3 g/l) there was a decrease in the number and length of the axillary shoots produced (Fig. 4, Table 4). Shoot proliferation was greater on

Table 3. Effect of the presence or absence of BA and NAA on shoot proliferation of French tarragon. (Mackay and Kitto 1988)

Growth regulators[a] (weeks) +	–	Main shoot length (mm)	No. of axillary shoots	Av. length of axillary shoots (mm)
4	–	23.2a[b]	7.1b	4.0a
2	2	23.3a	6.4b	4.2a
8	0	23.3a	14.9a	4.1a
4	4	23.7a	7.8b	4.2a

[a] Growth regulators: BA (0.6 mg/l) and NAA (0.05 mg/l).
[b] Mean separation in columns by Duncan's multiple range test, 5% level.

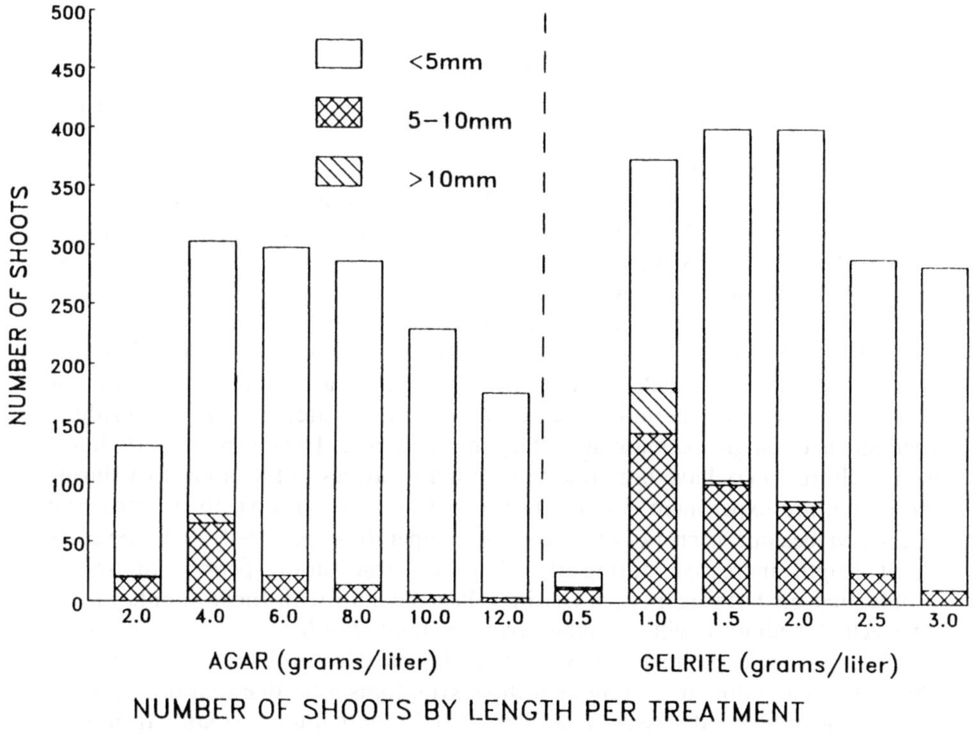

Fig. 4. Axillary shoot number and length in French tarragon as influenced by the medium gelling agents, agar and Gelrite. (Mackay and Kitto 1987)

Table 4. Effect of type and concentration of gelling agent on shoot proliferation of French tarragon. (Mackay and Kitto 1988)

Gelling agent (g/l)	Gel strength (MPa × 10^6)	Av. main shoot length (mm)	Av. no. of axillary shoots	Av. length of axillary shoots (mm)
Agar[a]				
2.0	10	25.1 ± 1.1[b]	3.5 ± 0.5	3.9 ± 0.1
4.0	210	27.7 ± 0.8	8.3 ± 1.0	3.9 ± 0.2
6.0	791	25.0 ± 0.5	7.8 ± 0.5	3.1 ± 0.1
8.0	2059	23.5 ± 0.4	7.6 ± 0.3	2.6 ± 0.1
10.0	4393	23.3 ± 0.4	6.2 ± 0.3	2.5 ± 0.1
12.0	5990	21.0 ± 0.8	4.6 ± 0.3	2.6 ± 0.1
Gelrite				
0.5	50	29.0 ± 0.9	0.7 ± 0.6	5.4 ± 0.5
1.0	790	32.6 ± 1.3	12.0 ± 1.6	5.5 ± 0.2
1.5	1722	28.5 ± 0.9	10.7 ± 0.8	4.0 ± 0.2
2.0	1847	27.9 ± 0.8	11.1 ± 0.7	3.7 ± 0.2
2.5	2745	26.2 ± 0.7	7.7 ± 0.5	3.0 ± 0.1
3.0	3582	23.9 ± 1.0	7.5 ± 0.4	2.7 ± 0.1
Agar		24.3*	6.3*	3.0*
Gelrite		28.0	8.3	3.7

[a] Washed Difco Bacto agar.
[b] Mean ± SE.
* Significant at the 5% level.

medium gelled with Gelrite than washed Difco Bacto agar if the shoots were not subcultured. At the lowest concentration of both gelling agents, vitrification was increased.

With all other concentrations, average axillary shoot number increased as gelling agent concentration decreased and was greatest in medium containing 4 g/l agar or 1 g/l Gelrite (Table 4). However, 1.5 g/l Gelrite was considered the best treatment if the shoots were not subcultured, based on number and length of axillary shoots and the absence of vitrification.

However, if the shoots were subcultured, greater proliferation was achieved on medium gelled with the washed Difco Bacto agar. The average number of axillary shoots (Fig. 5) proliferated on agar-gelled medium generally increased over the 7-month observation period, whereas shoots proliferated on Gelrite medium generally had a slight decrease in axillary shoot number. The low number of shoots at subculture 3 may have been due to high temperatures in the laboratory during the subculture experiment. French tarragon was very sensitive to in vitro heat stress and became necrotic whenever the temperature exceeded 30 °C. Average shoot length increased during the 7-month observation period for shoots proliferated on both agar- and Gelrite-gelled medium with those shoots on the agar-gelled medium having a higher average shoot length.

During early experiments it was noted that in vitro-proliferated shoots were chlorotic. In an effort to determine if these symptoms were due to a deficiency in essential plant nutrients, tissue was collected at 3-month intervals (Jan., April, July, and Oct.) from greenhouse stock plants and from shoots proliferated in vitro.

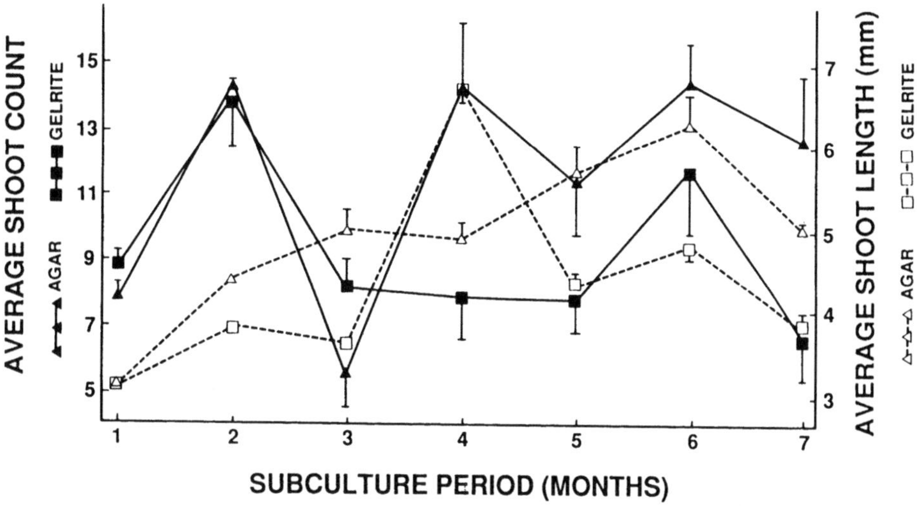

Fig. 5. Axillary shoot number and length over a 7-month subculture period as influenced by gelling agent (0.4% agar vs. 0.15% Gelrite). (Mackay and Kitto 1988)

Analysis of the plant tissue indicated differences in the mineral nutrient status between the stock plants and the in vitro-proliferated shoots. In vitro-proliferated shoots had lower levels of Ca, Mg, Cu, Fe, P, and S but had higher levels of K, Mn, Na, and Zn than the greenhouse stock plants (Table 5).

It was noted that control treatment proliferation varied between experiments when using containers other than 25 × 150-mm culture tubes. Container size was investigated using three treatments: (1) 50 × 70-mm jars containing three explants; (2) 90 × 95-mm jars containing five explants; (3) 25 × 150-mm test tubes containing one explant. Treatments did not affect main shoot length (Table 6). Treatment 1 had significantly more axillary shoots than treatment 3, but was not significantly different from treatment 2. Treatment 1 had significantly greater average axillary shoot length than either treatment 2 or 3. Several factors could be responsible for this effect including shoot density in the container, the type of container closure, or the atmospheric volume.

Table 5. Mineral nutrient status of in vitro and greenhouse-grown stock plants. (Mackay and Kitto 1987)

Source	Elements (mg/kg dry weight of tissue)									
	Ca	Cu	Fe	K	Mn	Mg	Na	P	S	Z
Greenhouse	10462[a]	12	94	33 192	62	2336	130	6327	3556	69
In vitro	5142	—	40	57 192	141	1803	1382	6010	3164	91

[a] Mean of four samples.

Table 6. Effect of shoot density in different sized containers on shoot proliferation of French tarragon. (Mackay and Kitto 1988)

Container size (mm)	No. of shoots	Volume			Ratios			Main shoot length (mm)	No. of axillary shoots	Av. length of axillary shoots (mm)
		Medium	Air	Total	Shoot: medium	Shoot: air	Shoot: total			
55 × 70	3	25	121	146	1:8	1:40	1:49	30.1a[a]	11.7a	3.9a
90 × 95	5	50	440	490	1:10	1:88	1:98	30.2a	10.9ab	3.5b
25 × 150	1	20	27	57	1:20	1:27	1:57	31.6a	8.7b	3.4b

[a] Mean separation in columns by Duncan's multiple range test, 5% level.

5 Induction of Rooting

Rooting was accomplished both in vitro and in vivo. Rooting media included perlite, Techniculture plugs, or a proprietary peat-lite (Redi-Earth) (Fig. 6). Containers used during the rooting studies in addition to the Techniculture plug trays included seedling plug trays and community packs. Rooting success was greatest in community packs containing Redi-Earth (Mackay and Kitto, unpubl.). Microcuttings rooted in the laboratory had significantly more roots, significantly longer main shoots and greater rooting and survival percentages than the greenhouse-rooted cuttings (Table 7). There was no difference in the number of roots produced in the greenhouse or the laboratory. Rooting success was maximal when microcuttings > 10 mm in length derived from shoots proliferated on agar-gelled medium. Rooting percentage and root numbers increased when the basal portion of the microcuttings were dipped in auxin (NAA or IBA, dissolved in 50% ethanol) for 5 s (data not shown). Once microcuttings were rooted they had high survival rates.

As auxin concentration increased from 0.1 to 1 g/l, the number of roots produced increased; however, the average length of the roots and main shoot length decreased (MacKay and Kitto 1988). The presence of auxin increased the percentage rooting but did not affect survival percentage.

Proliferation medium gelling agent (agar or Gelrite) had no effect on the number and length of roots produced, but the microcuttings from explants proliferated on Gelrite-gelled medium had significantly more shoots of greater length than microcuttings from explants proliferated on agar-gelled medium. Although microcuttings from explants proliferated on Gelrite-gelled medium had

Table 7. Effect of cutting size, rooting environment, and proliferation media gelling agent on rooting of French tarragon. (Mackay and Kitto 1988)

Treatment	Rooting (%)	Survival (%)	Av. no. of roots/ explant	Av.[a] root length (mm)	Av. no. of shoots/ explant	Av. main shoot length (mm)
		Cutting size				
5 to 10 mm	77	96	1.5*	2.4	1.1	1.2*
> 10 mm	78	97	1.9	2.4	1.1	1.9
		Rooting environment				
Laboratory	81	99	1.8	2.2*	1.1	1.7*
Greenhouse	73	93	1.5	2.7	1.1	1.2
		Gelling agent				
Agar[b]	80	95	1.7	2.3	1.1*	1.3*
Gelrite	73	97	1.6	2.5	1.2	1.6

[a] Primary roots only.
[b] Washed Difco Bacto agar.
* Significant at the 5% level.

Fig. 6. French tarragon plants 4 weeks after sticking shoots in Redi-Earth

a slightly higher survival percentage, rooting percentage was much less than on microcuttings from explants proliferated on agar-gelled medium (Table 7).

French tarragon wilts rapidly when removed from the high humidity environment employed during the proliferation stage. If the in vitro-proliferated microcuttings are rooted in a high humidity environment, an additional period of acclimation is important for the survival of the in vitro-produced plant. The total time required for proliferation and rooting is 8 weeks, with an additional week of acclimation required when rooting took place in a 100% relative humidity environment (Fig. 7). During a 7-month subculture experiment, an average of 11.5 axillary shoots per explant were produced in each 4-week subculture period. If all the axillary shoots had been subcultured, approximately 24 million plants would have been produced. Tissue culture of French tarragon is an economical method of production of large numbers of desirable cultivars and can eliminate several of the problems associated with conventional propagation methods.

Fig. 7. Axillary shoot rooted in Techniculture plug. Plant shown is 9 weeks old (4 weeks for in vitro proliferation and 5 weeks for in vivo rooting). (Mackay and Kitto 1987b)

6 Summary and Conclusions

The objective of this research was to develop a micropropagation system for French tarragon, *Artemisia dracunculus* L. var. *sativa*. Optimum proliferation of shoot tips was obtained on medium composed of MS salts and vitamins supplemented with 0.6 mg/l BA, 0.05 mg/l NAA, and 3% sucrose. Proliferation was maximal after 4 weeks of culture. Stock plant environment (natural daylength or 24-h supplemental light) did not influence proliferation in vitro. Horizontally placed shoot-tip explants produced more axillary shoots than vertically placed explants. Significantly greater proliferation occurred on medium gelled with Gelrite than with washed Difco Bacto agar. Increasing concentrations of both gelling agents decreased proliferation; however, long-term subculture (6 months) on agar-gelled medium (4 g/l) resulted in greater proliferation than those cultured on Gelrite-gelled medium (1.5 g/l). Container size had an effect on proliferation, with shoots in 55 × 70 mm jars having significantly greater axillary shoot length than those in 25 × 150 mm culture tubes or 90 × 95 mm jars and a significantly greater number of axillary shoots than those in 25 × 150 mm culture tubes. Rooting was maximized by placing microcuttings > 10 mm in length, derived from shoots proliferated on agar-gelled medium, in clear plastic containers in the laboratory. A 5-s dip of the basal portion of the microcuttings in either NAA (90, 450, or 900 mg/l) or IBA (90, 450, or 900 mg/l) increased rooting percentage and root numbers.

7 Protocol

1. *Proliferation*
Shoots, 2 cm in length, should be collected from actively growing stock plants. After removing the leaves, surface disinfect shoots in a 0.5% sodium hypochlorite + 0.1% surfactant solution. Prior to culturing, rinse the shoots three times with sterile water and trim to 1 cm. Culture explants horizontally onto proliferation medium composed of Murashige-Skoog salts and vitamins, 3% sucrose, 4 g Difco Bacto agar/l, 0.6 mg/l BA and 0.05 mg/l NAA. Proliferate in 50 × 70 mm jars (baby-food jars) containing 25 ml of medium and subculture stocks every 4 weeks.

2. *Rooting*
Dip microcuttings > 5 mm in length in an auxin solution (NAA 90, 450, or 900 or IBA 90, 450, or 900 mg/l; 50% ethanol) prior to sticking in a 1 peat-lite : 1 perlite (v : v) soilless mix. Microcuttings root in 4 weeks when placed under intermittent mist in a greenhouse or into clear plastic trays under lights (50 $\mu M/s/m^2$) in the laboratory.

References

Anon USDA (1973–1983) Economic Research Service, U.S. Foreign Agricultural Trade Statistical Report. USDA Econ Res Serv

Bassuk N (1986) Year-round production of greenhouse-grown French tarragon. HortSci 21: 258–259

Foster GB (1974) Tarragon twins. Herb Grower Mag 27: 2–4

Garland P, Stoltz LP (1980) In vitro propagation of tarragon. HortSci 15: 739

George EF, Sherrington PD (1984) Plant propagation by tissue culture. Handbook and directory of commercial laboratories. Exegetics Limited, Reading, Berks, Great Britain

Mackay WA (1987) Factors affecting the micropropagation of *Artemisia dracunculus* L. var. *sativa*, French tarragon, using shoot tips as the explant source. MS Thesis, Univ Delaware

Mackay WA, Kitto SL (1987) Rapid propagation of French tarragon using in vitro techniques. Acta Hortic 208: 251–257

Mackay WA, Kitto SL (1988) Factors affecting in vitro shoot proliferation of French tarragon. J Am Soc Hortic Sci 113(2): 282–287

Murashige T, Skoog F (1962) A revised medium for rapid growth and bio assays with tobacco cultures. Physiol Plant 15: 473–497

Quinn J (1987) Propagation of French tarragon. In: Simon JE, Grant L (eds) Proc 2nd Nat Herb Growing and Marketing Conf. Purdue Res Found, West Lafayette, Indiana, pp 88–90

Rousi A (1968) Cytogenetic comparison between two kinds of cultivated tarragon (*Artemisia dracunculus*). Hereitas 10: 194–213

Zarghami NS, Russell GF (1973) The volatile constituents of tarragon (*Artemisia dracunculus*). Chem Mikrobiol Technol Lebensm 2: 184–187

II.13 Micropropagation of *Camellia* spp.

A.M. VIEITEZ[1], M.L. VIEITEZ[2], A. BALLESTER[1], and E. VIEITEZ[2]

1 General Account

1.1 Origin and Distribution

In 1689 the Dutch physician and botanist André Cley published a Latin work mentioning a plant called the "tzumacky", a transcription of the Japanese name "tsubaki" (Odriozola 1986). The same name was used in association with the descriptions and drawings of James Petiver in 1702 and Engelbert Kaempfer in 1712. In Linnaeus' *Sistema Natura* of 1735, the plant figures as *Camellia tsubaki*, the genus being dedicated to the Moravian Jesuit Kamel, and it is not until the appearance of his *Species Plantarum* (which also features the tea-plant *Thea sinensis*, now *C. sinensis*) that it acquires its present systematic name, *Camellia japonica*.

Today the most widespread species of the genus, *C. japonica* (Fig. 1A, B) did indeed originate in Japan, as did three other species. Most species of *Camellia*, some 312, are nevertheless distributed wholly within China (in the provinces of Guagxi, Yunnan, Guangdon, Chekiang, Kweichow, Sichuan, Guizhou, Hunan, Fujian, Jiangxii, Anhui, and Hainan), and 54 others in Japan, Vietnam, Burma, Laos, India, Nepal, Thailand, Indonesia, the Phillipines, Borneo, and Taiwan.

In spite of its long history in Asia, camellia was introduced into Europe only some 250 years ago, when in 1739 James Lord Petre established a *C. japonica* plant at Thordon Hall in Essex (England); and it was not until 1792 that Captain Connor of the East India Company's merchantman *Carnatic* brought home the most successful varieties *C. japonica* cvs. Alba Plena and Variegata, although by 1812 there were 11 varieties recorded in Kew Gardens. The rapidly increasing popularity of tea among the English had led to *C. sinensis* being brought in 1740 by the East India Company's Captain Goff, but these plants died (Sealy 1958), and the first successful introduction was achieved by the merchant and naturalist John Ellis in 1768 (Aiton 1789; Booth 1830). *C. oleifera* Abel was brought to Kew from Canton in 1803 by the collector William Kerr (Bretschneider 1898). *C. reticulata* Lindley cv. Captain Rawes, one of the most beautiful of shrubs, was first brought to England for Palmer by the merchant Captain Rawes, after whom it is named, and shortly afterwards appeared in other European countries, but the other variety of *C. reticulata*, Rober Fortune, seems to have died out in England soon after its

[1] Plant Physiology, CSIC, Apartado 122, 15080 Santiago de Compostela, Spain
[2] Department of Biology, Faculty of Biology, University Santiago de Compostela, 15080 Santiago de Compostela, Spain

Biotechnology in Agriculture and Forestry, Vol. 19
High-Tech and Micropropagation III (ed. by Y.P.S. Bajaj)

Fig. 1A, B. *Camellia japonica* trees showing the typical habit growth. **C** Flower of *C. japonica* cv. Alba Plena

introduction, and to have met with the same fate throughout the rest of Europe except Portugal; *C. reticulata* was recorded in China as early as the first T'ang emperor in the 9th century B.C. The most important yellow species is *C. chrysantha*.

1.2 Taxonomy and Chromosome Number of the Genus *Camellia*

The family Theaceae D. Don comprises 16–25 genera with over 200 species. All are tropical or subtropical trees or shrubs. Many species are of economic importance for tea or edible oil, or as ornamental plants. Chang and Bartolomew (1984) recognized more than 200 species in the genus *Camellia* alone, which they divided into four subgenera and 20 sections. The subgenus Camellia comprises seven sections containing 73 species: the most widely cultivated species of the section Oleifera are *C. oleifera* and *C. sasanqua*, while the most notable of the 33 species of the section Camellia are *C. reticulata*, *C. pitardii*, *C. hongkongensis*, *C. saluenensis*, and *C. japonica*, which together have some 500 varieties. Of the eight sections of the subgenus *Thea*, the most noteworthy is *Thea*, the most widely cultivated of whose 18 species are *C. sinensis* and *C. irrawadiensis*.

1.2.1 Hybrids and New Varieties

In 1930, Williams crossed *C. saluenensis* with *C. japonica* cultivars to produce the hybrids named *C. williamsii* (Gimson 1986). Of the many hybrids since obtained, the most noteworthy include *C. japonica* × *C. saluenensis*, *C. cuspidata* × *C. saluenensis* (type Cornish snow), *C. saluenensis* × *C. reticulata* (type Lamertsii), *C. japonica* × *C. reticulata* Crimson Robe (type Crimson Robe), *C. williamsii* × *C. reticulata* (type Crimson Robe), *C. williamsii* × *C. reticulata* (type Black Lace), *C. pitardii* × *C. reticulata* and *C. japonica* × *C. fraterna* (type Swadii). In the USA, the hybrid known as Brigadoon is particularly well known, as are G.E.G. Waterhouse in Australia, Citation, Donation, and C.S. Williams in Europe and Anticipation, Daintiness, and Grand Jury in New Zealand.

The dream of all camellia growers is to achieve a scented yellow double camellia. The problem is to introduce the yellow of *C. chrysantha* into the reservoirs of the cultivars of *C. japonica.* The Botanic Gardens of Kunmig have crossed *C. chrysantha* with *C. pitardii* cv. Yunnanica, but the hybrids are sterile. *C. chrysantha* is diploid, whereas *C. reticulata* and *C. pitardii* cv. Yunnanica are hexaploid, and it might be more feasible to cross *C. chrysantha* with other diploid camellias such as some cultivars of *C. japonica. C. semiserrata, C. polydont, C. gigantocarpa,* and *C. octopetala* are also all diploid, and like *C. chrysantha* have a total of 30 chromosomes, the predominant number in the genus *Camellia.*

1.3 Economic Importance

Although very few *Camellia* species are marketed for horticulture, of those that are of great importance, the foremost is, of course, *C. japonica,* followed by *C. reticulata* and *C. sasanqua.* By far the majority of existing ornamental camellia cultivars belong to these three species and their hybrids. The economic importance of the genus is nevertheless due largely to *C. sinensis* cvs. Sinensis and Assamica, from whose young leaves tea is made. Another important product is camellia seed oil, which is greatly used in Chinese cookery, especially in the southern provinces. This is obtained mainly from three species of the section Oleifera, *C. oleifera, C. semiserrata,* and *C. chekiangolomy.* Finally, camellia leaves contain a number of substances used in the pharmaceutical industry, including xanthine, teophiline, teabromine, adenine, tearine, and oleic acid.

1.4 Conventional Propagation

For several decades, the usual method of propagating camellias was to use cuttings, since the advent of auxins and misting has enabled conditions suitable for rooting to be maintained for long periods. The most widely used auxins are indolebutyric acid (IBA), naphthaleneacetic acid (NAA), and combinations of the two, in concentrations ranging from 5 to 12 g/l. Rooting response nevertheless varies greatly from one species or variety to another. With *C. japonica* cv. Bella Romana, 100% rooting rates are achieved, and rates of 85–100% with *C. sasanqua* (Thiman and Behnke-Rogers 1950), but with *C. japonica* cv. Alba Plena the rates are only

20–65% and only 30% with *C. japonica* cv. Camursa (Salinero and Silva-Pando 1986), while the beautiful *C. reticulata* fails to root at all (Laborey 1986) and must be propagated vegetatively by other methods.

Stooling and air-layering are also both used for camellia propagation. In stooling, the branch is bent down to the ground and buried after the bark has been stripped from 3–4 cm (auxin may or may not be applied). In air-layering, auxins are applied to the stripped branch, which is then covered in sphagnum and aluminum foil.

In the 19th century, when modern methods of rooting cuttings had not yet been developed, it was common to propagate camellias by grafting. Normally, crown grafts were made at the base of young *C. japonica* stocks, though different kinds of side cleft graft and approach graft can also be used (Laborey 1986). Juvenile grafting onto the hypocotyls of seedlings has proved to be both very effective and simple to practise (Vieitez and Vieitez 1983). In Japan, particularly in the camellia center of Kyuohu Island, it is common for highly prized bonsai camellias to be created by grafting onto segments of root 50–60 cm long and 2–3 cm in diameter taken from wild *C. japonica*; flowering begins 3–4 years later.

1.5 Need for In Vitro Culture

Since the propagation of some species of *Camellia* by conventional methods is inefficient, other means have been sought. Much work has been done in recent years on the extension of in vitro techniques to woody species (Bajaj 1986; Bonga and Durzan 1987). It has been suggested that it may be possible and desirable to apply these techniques to rapid clonal propagation of *Camellia* (Bennett 1977, 1978). This would have several advantages over conventional methods (Mantell 1982): clonal propagation would allow rapid multiplication of selected mother plants and is independent of season; in vitro culture of very small meristems or shoot tips would hopefully rid several highly prized existing clones of the viruses with which they are now naturally infected; and in the not-too-distant future it would be possible to approach the genetic improvement of existing clones to produce new types of camellia. The improvement of plant quality would make the camellia more generally popular, and the possibility of all-year-round production would make it easier for growers to respond to increased demand.

2 Brief Review of Work on Tissue Culture

The species whose in vitro propagation has been most intensively investigated are naturally those of greatest economic importance: *C. sinensis* (L.) O. Kuntze *C. oleifera* Abel., *C. japonica* L., *C. reticulata* Lindlay, *C. chrysantha* (Hu) Tuyama, and some of their hybrids. One of the earliest studies was Lammerts' (1958) work on embryo culture, a technique that is very useful for propagating hybrids in which the embryo may be incompletely developed within the seed (Ackerman 1978). Lammerts removed the seed coat and cultured the embryos in the nutrient

Table 1. Summary of the results obtained with in vitro culture of *Camellia*

Species	Explant source	Medium[a]	Growth response	Reference
C. chrysantha	Cotyledons	MS + BA + NAA	Somatic embryogenesis and plantlet regeneration	Zhuang and Liang (1985a)
C. japonica	Embryos	T	Seedling growth	Ackerman (1978)
	Shoot tips and nodes of 3–4-year-old seedlings	MS + K(1) + 2-iP(1) + GA_3(1) + BA(1) + IAA(1) + Ad(20) + PVP(10 g/l)	Axillary shoot development	Crézé and Beauchesne (1980)
	Cotyledons	Knop + BA(0.5–5.0)	Somatic embryogenesis and plantlet regeneration	Bennett and Scheibert (1982)
	Seeds and shoot tips of adult trees	T	Micrografted plant	Crézé (1983, 1985)
	Shoot tips from 4-month-old seedlings	MS + BA(1) MS + dipping IBA(1 g/l) 20–30 min	Axillary shoot development Root induction and plantlet regeneration	Samartín et al. (1984, 1986)
	Shoot tips and nodes of 10-year-old trees	MS1/2 + BA(2.2–4.4 μM) + GA_3(0.3–14.4 μM)	Axillary shoot development	Carlisi and Torres (1986)
	Cotyledon slices	MS + BA(0–10) + IBA (0–2) MS + GA_3(1)	Somatic embryogenesis Somatic embryos germination	Kato (1986)
	Shoot tips and nodes of adult trees	WPM + BA(2) + Z(2) + 2-iP(2) + IBA(0.01) WPM1/2 + Dipping IBA(1 g/l) 15 min	Axillary shoot development Root induction and plant development	Vieitez et al. (1989a, b)
	Immature and mature zygotic embryos	MS + BA(0–2) + IBA(0–2)	Somatic embryogenesis and plantlet regeneration	Vieitez and Barciela (1990)
C. oleifera	Immature cotyledons and embryos	MS + BA(4) + NAA(2)	Adventitious buds and plantlet regeneration	Tian-Ling (1982)
	Lateral buds of adult trees	—	Axillary shoot development	Yan et al. (1984)
C. reticulata	Cotyledons	MS + BA(1) + NAA(0.2)	Somatic embryogenesis	Zuang and Liang (1985b)
	Shoot tips and nodes of adult trees	WPM + BA(2) + Z(2) + 2-iP(2) + IBA(0.01)	Axillary shoot development and plantlet regeneration	San-José and Vieitez (1990)

Table 1. (*Contd.*)

Species	Explant source	Medium[a]	Growth response	Reference
C. sasanqua	Shoot tips	MS1/2 + BA (2.2–4.4 μM)	Axillary shoot and plantlet regeneration	Torres and Carlisi (1986)
	Cotyledons	MS + BA (1–2) + NAA (0.2–0.5)	Somatic embryogenesis and plantlet regeneration	Zhuang et al. (1988)
C. sinensis	Stem segments, apical buds	—	Callus	Forrest (1969)
	Stem segments	H + 2,4-D (6) + CM (20%)	Callus	Ogutuga and Northcote (1970a,b)
	Leaves, stems	K (5 μM) + 2,4-D (20 μM)	Callus	Korestkaya and Zaprometov (1975) Zaprometov et al. (1976, 1979, 1986) Zagoskina and Zaprometov (1979, 1983, 1986)
	Cotyledons, anthers, shoot tips	MS + 2,4-D (2–8) + IAA (2–4) + K (2)	Callus	Wu (1976)
	Cotyledons	MS + IAA (1) + K (10)	Plantlet regeneration	Wu et al. (1981)
	Stem segments, anthers	MS + 2,4-D (1 μM) + K (10 μM)	Callus	Doy (1980, 1981)
	Anthers	MS + NAA (10 μM) + K (10 μM)	Roots	
	Cotyledons		Embryoids	Yan and Ping (1983)
	Nodal segments, shoot tips of seedlings and mature trees	MS + NAA (1) + BA (3) MS + IAA (2) + K (8)	Callus, adventitious buds Axillary shoot development	Phukan and Mitra (1984)
	Stem pieces, leaf petioles	MS + 2,4-D (45 μM) + YE (0.2%) MS + BA (10 μM)	Callus Axillary shoot development	Sarwar (1985)
	Seedling stem segments	MS + IBA (2–4) + BA (4–2) MS + IBA (0.5) + BA (10) MS1/2 + IBA (0.5–1)	Callus Adventitious buds Root induction and plantlet regeneration	Kato (1985)
	Cotyledon slices	MS + BA (0–10) + IBA (0–2) MS + IBA (0.5) + BA (10)	Somatic embryogenesis Somatic embryos germination	Kato (1986)

	Anthers, leaves, roots and stem segments	MS + K(5) + 2,4-D(3)	Callus and roots	Kim (1986)
	Anthers	N6 + 2,4-D(0.05) + K(0.2)	Adventitious embryos	Shimokado et al. (1986)
	Shoot tips, nodal segments	MS + BA(2)	Axillary shoot development	Arulpragasam and Lattif (1986)
	Green wood stem segments	MS + 2,4,5–T(0.1 μM) + K(10 μM)	Callus	Frisch and Camper (1987)
C. japonica × *C. cuspidata*	Embryos	T	Seedling growth	Lammerts (1958)
C. japonica × *C. reticulata*	Embryos	T	Seedling growth	
C. japonica × *C. saluenensis*	Embryos	T	Seedling growth	
C. japonica × *C. chrysantha*	Immature embryos	MS or A + NAA(0.5–1) + K(0.1–0.5) + YE(1 g/l)	Somatic embryogenesis	Yamaguchi et al. (1987)
C. saluenensis × *C. chrysantha*	Shoot tips of seedlings	MS + K(1) + BA(1) + 2-iP(1) + GA_3(1) + IAA(1) + Ad(20) + PVP(10 g/l)	Axillary shoot development	Crézé and Beauchesne (1980)
C. saluenensis × *C. japonica*	Shoot cultures	MS1/2 + dipping IBA(1–2 g/l) 15 min	Root induction and plantlet regeneration	Beretta et al. (1987)
C. vietnamensis × *C. chrysantha*	Cotyledons	MS + BA(3) + NAA(1)	Somatic embryogenesis and adventitious buds	Nadamitsu et al. (1986)
		MS + GA_3(1) + CM(10%)	Somatic embryos germination	

Abbreviations

Mineral media: A: Anderson (1984); H: Heller (1953); MS: Murashige and Skoog (1962); N6: Chu (1976); T: Tukey (1934); WPM: Woody Plant Medium (McCown and Lloyd 1981)

Supplements: IAA: indole-3-acetic acid; IBA: 3-indolebutyric acid; NAA: naphthaleneacetic acid; 2,4-D: 2,4-Dichlorophenoxyacetic acid; 2,4,5-T: 2,4,5-Trichlorophenoxyacetic acid; Ad: Adenine; BA: Benzylaminopurine; 2-iP: iso-pentenylaminopurine; K: Kinetin; Z: Zeatin; GA_3: Gibberellin; CM: Coconut milk; PVP: Polyvinylpyrrolidone; YE: Yeast extract.

[a] Quantities in brackets are expressed in mg/l unless otherwise stated.

solution of Tukey (1934). Good results were obtained with interspecific hybrids such as *C. japonica* × *C. reticulata*, *C. japonica* × *C. cuspidata*, and *C. japonica* × *C. saluenensis*, fully developed plants being obtained within 4 weeks.

Little more research on in vitro culture of *Camellia* was carried out before 1980, most of what was done being concerned with the establishment of callus cultures or cell suspensions of tea-producing varieties as a very convenient model system for studies of the biosynthesis of the phenols that give tea a dark brown color (Forrest 1969; Ogutuga and Northcote 1970a, b; Koretskaya and Zaprometov 1975; Zaprometov et al. 1979; Zagoskina and Zaprometov 1979, 1983). Not until the 1980s was the regeneration and micropropagation of plantlets obtained, generally by somatic embryogenesis or shoot tip culture. These studies are summarized in Table 1 (see also Kato 1989).

3 Micropropagation

According to the literature, *Camellia* has been regenerated in vitro by three methods: induction of adventitious buds; somatic embryogenesis; and development of axillary buds from shoot tips and nodal explants. Plantlet regeneration obtained through the three methods mentioned will be described.

3.1 Explants and Sterilization

The sterilizing agents most commonly used for *Camellia* explants are sodium- or calcium hypochlorite and ethanol. When Kato (1986) used cotyledons of *C. sinensis* and *C. japonica* as explants, she first sterilized the decorticated seeds with 70% alcohol and then soaked them in 7% w/v calcium hypochlorite for 1 h, whereas Crézé (1980) sterilized seeds in 10% sodium hypochlorite for 10 min. In our laboratory we have found that *C. japonica* seeds treated with 70% alcohol for 90 s followed by commercial bleach (having 4% of chlorine) 20% for 20 min had a 12% residual contamination rate.

For sterilizing the stems of glasshouse-grown seedlings, Kato (1985) used immersion in 7% calcium hypochlorite for 20 min. In our laboratory, shoots of glasshouse-grown plants from which shoot tips and nodal segments were to be obtained as explants have been sterilized by treating them with 70% alcohol for 2 min followed by 5% calcium hypochlorite for 6 min and three rinses in sterile water (Samartín et al. 1984). Carlisi and Torres (1986) sterilized herbaceous stem segments excised from the current season's growth of 10-year-old stock plants with a solution containing 1.05% of sodium hypochlorite and 0.1% of Tween 20. According to Crézé and Beauchesne (1980), it is not necessary to disinfect buds prior to dissection when very small (0.2–0.5 mm) shoot tips are to be used as explants, since these portions of the plant are protected by their scales and final infection rate is less than 10%.

Sterilizing material from field-grown plants can be somewhat more complicated, since joint bacterial and fungal contamination rates of at least 90% are typical in field-grown shoots of *C. sinensis* and *C. japonica* (Haldeman et al. 1987). Haldeman proposes, following traditional disinfestation, a 24-h culture in liquid medium containing the fungicide benomyl and the antibiotic rifampicin: with 1 g/l of benomyl and 10 mg/l of rifampicin, 95% of *C. sinensis* shoots were contamination-free, while 4 g/l of benomyl and 10 mg/l of rifampicin gave an average 85.5% of uncontaminated *C. japonica* shoots with no phytotoxic effects. Frisch and Camper (1987) treated greenwood stems in a series of stirred baths in a procedure comprising an initial water rinse, 1.5–2 min in 0.1 HCl, 5 min in autoclaved distilled water, 10 min in 3.75% sodium hypochlorite (7.5% Clorox) with three drops of Tween 20, 5 min in autoclaved distilled water, 10 min in 7.5% $CaCl_2$ and three final 5-min rinses in autoclaved distilled water.

We have sterilized shoots forced to develop from the buds of a 50-year-old tree by placing the base of collected branchlets in water in a growth chamber with controlled temperature and humidity. The new shoots were collected when 2–4 cm long and sterilized by successive immersion in 70% alcohol for 30 s followed by either 15 min in 15% commercial bleach (having 4% chlorine) or 5 min in 5% calcium hypochlorite. The contamination rate among the shoot tips and nodes subsequently excised was 5% (Vieitez et al. 1989a). In general, it is clear that, whenever possible, mother plants should be maintained in a clean atmosphere in a glasshouse or phytotron.

3.2 Culture Media

Table 1 shows that the mineral medium most widely used for *Camellia* cultures, whether for callus culture for secondary product biosynthesis research (Frisch and Camper 1987) or for plant regeneration by embryogenesis, induction of adventitious buds, or shoot tip and bud culture (Crézé and Beauchesne 1980; Phukan and Mitra 1984; Arulpragasam and Lattif 1980) has been Murashige and Skoog's medium (1962) (MS) or some modifications of it. For the shoot multiplication stage of *C. japonica* regeneration from buds of juvenile origin in our laboratory (Samartín 1984), MS consistently gave the best results in terms of both growth and vigor when compared with the macronutrient formula of Lepoivre (Quoirin and Lepoivre 1977), Knop (Tabachnik and Kester 1977), Schenk and Hildebrandt (1972), and Heller (1953) modified this with the concentration of all macronutrients increased by a factor of 1.25 and the inclusion of 1 mM $(NH_4)_2SO_4$(H + SO_4). MS nevertheless failed to give satisfactory results for regeneration from adult *C. japonica* cv. Alba Plena material (Vieitez et al. 1989a); in a series of shoot multiplication experiments, the effect of six macronutrient formulas were investigated (Table 2): H + SO_4, MS, half-strength MS (MS 1/2), Woody Plant Medium (McCown and Lloyd 1981) (WPM), Gresshoff and Doy (1972) (GD), and Anderson (Anderson 1984). The best multiplication rates and elongation were achieved with WPM and H + SO_4 and the worst with MS and MS 1/2. The optimal medium in these experiments was WPM, which is also very efficient for culturing other woody dicotyledoneous species (Read et al. 1982; McCown and Sellmer 1987), since although the differences between WPM and H + SO_4 as regards multiplication rate and elongation were not significant, the H + SO_4 cultures were more chlorotic and less vigorous. The poor performance of MS and MS 1/2 contrasts with our own results with material of juvenile origin and with those obtained by Carlisi and Torres (1986) when they successfully used MS 1/2 to culture *C. japonica* cv. Purple Dawn of adult origin. These apparent contradictions may have an explanation in terms of genotype differences.

In the rooting stage, macronutrient formulas should be used half-strength, as has been normal practice with both *C. sinensis* (Kato 1985) and *C. japonica* (Samartín et al. 1986). In the latter study, MS 1/2 was found to produce significantly more roots per rooted shoot than undiluted MS medium. The concentration of sucrose was also found to affect rooting, the optimal level being 30–50 g/l (depending on clone), while with levels of 5–10 g/l the rooting rate was considerably smaller. The important role of glucose was likewise stressed by Beretta et al. (1987), who found that 30 g/l increased both the number and length of roots.

As Table 1 shows, growth regulators are usually used at high concentrations in *Camellia* cultures. Auxin doses such as 2–4 mg/l of IBA were used by Kato (1985) with 2–4 mg/l of 6-benzyladenine (BA) to initiate callus from which buds of *C. sinensis* subsequently differentiated. NAA has often been used to induce somatic embryos from cotyledon explants, while 2,4-dichlorophenoxyacetic acid (2,4-D) appears to be essential for callus induction in tea cotyledons, but was ineffective for the regeneration of plantlets (Wu et al. 1981) and detrimental for embryogenesis from *C. reticulata* cotyledons (Zhuang and Liang 1985b). The cytokinin most widely used for somatic embryogenesis is BA, but kinetin has also been found to induce the differentiation of plantlets from cotyledon tissues when used at concentrations of 10 mg/l for *C. sinensis* (Wu 1976; Wu et al. 1981) and 0.1–0.5 mg/l for the hybrid *C. japonica* × *C. chrysantha* (Yamaguchi et al. 1987). One or two mg/l of BA have generally been effective for induction of axillary bud growth in both initial explant cultures and subsequent shoot multiplication stages (Crézé and Beauchesne 1980; Samartín et al. 1984; Carlisi and Torres 1986; Arulpragasam and Lattif 1986). Zeatin (1–2 mg/l), together with N^6-Δ^2-isopentenyl adenine (2-iP) is also beneficial in the shoot multiplication stage for *C. japonica* cv. Alba Plena (Vieitez et al. 1989a), *C. japonica* cv. Fimbriata Alba, and *C. reticulata* (San-José and Vieitez 1990). In the rooting stage, the auxin yielding the best results was IBA.

There have also been a few reports of the use of growth substances of undefined chemical composition. Addition of 10% v/v coconut milk has been reported to favor both the growth of callus in *C. sinensis* (Wu et al. 1981) and the induction of adventitious buds and roots on cotyledons of the hybrid *C. vietnamensis* × *C. chrysantha* (Nadamitsu et al. 1986), while 1 g/l of yeast extract promoted embryogenesis during culture of immature *C. japonica* × *C. chrysantha* embryos (Yamaguchi et al. 1987). Further details on the growth regulators used to induce morphogenesis in callus, embryos, buds, or shoot tip cultures are given in the corresponding sections below.

3.3 Micropropagation by Neoformation of Adventitious Buds

Most efforts of this kind have been made using *C. sinensis*. Wu (1976) and Wu et al. (1981) reported successful regeneration from callus obtained from *C. sinensis* cotyledons using medium containing 1 mg/l of indole-3-acetic acid (IAA) and 10 mg/l of kinetin, though for some varieties it was necessary to raise these dosages to 4 and 20 mg/l respectively, and it is not made clear whether what occurred in the cotyledon callus was somatic embryogenesis or the differentiation of adventitious buds that subsequently root. Tian-Ling (1982) does specify that in his work it was a large number of buds that were obtained from callus differentiated from immature cotyledons and embryos of *C. oleifera*, and that subsequently became plantlets. Indirect regeneration of shoot buds from callus initiated on twig explants was achieved using 1 mg/l of NAA and 3 mg/l of BA by Phukan and Mitra (1984). Kato (1985) reported induction of callus tissue from intact stem segments of tea seedlings, from segments stripped of their epidermal layers, and from the epidermal layers alone (the epidermis and cortex). Of the callus produced from epidermal layers alone, 20% differentiated buds when transferred to a medium containing 0.5 mg/l of IBA and 10 mg/l of BA; the shoot buds subsequently developed roots on a filter paper bridge on a liquid medium supplemented with 0.5–1 mg/l of IBA, and the plantlets continued to grow successfully in pots.

3.4 Micropropagation by Somatic Embryogenesis

The *Camellia* seed's two large cotyledons appear to have great capacity for morphogenesis of both adventitious buds and somatic embryos, either indirectly via callus, as in Bennett and Scheibert's (1982) work on *C. japonica*, or directly on the cotyledon surface. The direct technique has been employed with *C. sinensis* and *C. japonica* (Kato 1982, 1986; Yan and Ping 1983), *C. chrysantha* (Zhuang and Liang 1985a), *C. reticulata* (Zhuang and Liang 1985b), and *C. sasanqua* (Zhuang et al. 1988). For *C. reticulata* the average embryoid induction rate was about 43%, and normal plantlets were obtained when a modified MS medium containing 1 mg/l of BA and 0.2 mg/l of NAA was used. For *C. sasanqua* the somatic embryoids were also initiated on the MS medium supplemented with BA and NAA. The growth regulator test demonstrated that gibberellic acid (GA_3) has an inhibitory effect on formation of embryoids, but GA_3 could stimulate shoot growth and root development of the mature embryoids (Zhuang et al. 1988).

Kato (1986) described a useful method of micropropagation by mature cotyledon slice culture: somatic embryos of *C. japonica* and, less abundantly, *C. sinensis* formed directly on swollen parts of the cotyledon surface without an intermediate callus, regardless of the presence or concentrations of the growth regulators used (BA, IBA, and NAA). Somatic embryos of *C. japonica* germinated on transfer to a medium supplemented with 1 mg/l of GA_3, where additional embryos arose from the hypocotyl during successive subcultures, whereas *C. sinensis* embryos germinated in the presence of 10 mg/l of BA and 0.5 mg/l of IBA but not on GA_3 medium. The same Kato's method has been used to obtain plantlets of *C. vietnamensis* × *C. chrysantha* on MS medium supplemented with 3 mg/l of BA and 1 mg/l of NAA, the germination of embryos being achieved on a GA_3 medium with 10% of coconut milk (Nadamitsu et al. 1986). Similarly, Yamaguchi et al. (1987) obtained somatic embryogenesis working with immature embryos of *C. japonica* × *C. chrysantha* just prior to their degeneration. In this case, germination was promoted by 5 g/l of polyvinylpyrrolidone (K-90), 5 mg/l of GA_3 and 0.2–0.5 mg/l of 2-iP.

Somatic embryogenesis and plantlet regeneration were achieved from immature and mature zygotic *C. japonica* embryos cultured on MS mineral medium without growth regulators or with various combinations of IBA (0–2 mg/l) and BA (0–4 mg/l), though high concentrations such as 4 mg/l BA plus 2 mg/l IBA were inhibitory (Vieitez and Barciela 1990). This study showed that somatic embryogenesis readily occurs in vitro on the cotyledons, cotyledonary petioles, and hypocotyl of the zygotic embryos. Success depends on both the physiological maturity of the zygotic embryo and the kind of initial explant used (embryonic axis or cotyledon segments). It is important to emphasize the positive response of the embryonic axes, especially with the immature September embryos, 94% of which were embryogenic-competent. This result improves the efficiency of the embryogenic system described by Kato (1986) in which only cotyledons explants were used. The somatic embryos produced were generally one of the two clearly distinct types: (1) "seed-like" embryos, which were yellowish white, with large cotyledons like those of zygotic *Camellia* embryos, were by far the most numerous; (2) "bud-like" embryos which were green, with cotyledons like true leaves. They were genuinely bipolar, developing whole plantlets with shoot and root poles, and they generally occurred in media with a relatively high concentration of BA. Embryogenic potential was maintained by secondary embryogenesis through the successive generations of embryos. Indirect somatic embryogenesis via callus was also evident.

In *C. reticulata* cv. Mouchang somatic embryo formation was also performed through cotyledon sections and embryonic axes culture (Plata and Vieitez 1990) (Fig. 2). Embryogenic competence was increased by 0.5–1 mg/l IBA and decreased by 0.2–1 mg/l NAA, though in the cv. Early Crimson Zhuang and Liang (1985b) induced embryogenesis with 0.2 mg/l NAA plus 1 mg/l BA, but did not consider the effect of IBA. We have obtained up to 40% conversion to plantlets when mature somatic embryos were isolated and incubated in medium supplemented with 5 mg/l of GA_3 and 2 mg/l of IAA.

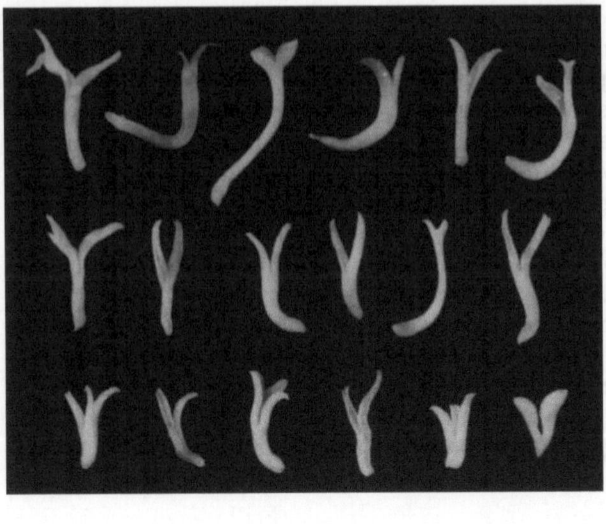

Fig. 2. Somatic embryos of *C. reticulata* cv. Mouchang, some of them showing expanded cotyledons and elongated radicle. (Courtesy E. Plata)

Indirect somatic embryogenesis was induced in callus of *C. reticulata* differentiated on cotyledon explants after 3 months' culture in media containing IBA and/or BA, embryogenic capacity being retained by callus subcultured on 0.5 mg/l IBA and 1 mg/l BA.

3.5 Micropropagation from Shoot Tips and Nodal Explants

Micropropagation of *Camellia* by somatic embryogenesis from cotyledons or neoformation of buds in callus culture has only been achieved with material of juvenile origin. For in vitro propagation of adult material, other techniques must be sought. Since micropropagation by induction of axillary shoots and subsequent transfer to a rooting medium is currently recognized as a valuable method for many kinds of woody species, the potential of bud culture has been explored for *Camellia*, too.

3.5.1 Material from Juvenile Origin

The first reported attempts to regenerate plants from shoot tips and axillary buds were made by Crézé and Beauchesne (1980), who took meristems with one or two leaf primordia, 0.5 mm long, from 1-year-old rooted cuttings or (more successfully) from 3–4 year-old seedlings that had not yet flowered, and cultured them on a modified MS medium supplemented with 20 mg/l of adenine, 0.1 mg/l of IAA, 1 mg/l each of kinetin, 2-iP, BA, and GA_3, and 10 g/l of polyvinylpyrrolidone. Although the cultures were established and elongated, and node cultures were observed to produce shoots more rapidly than shoot tips, no multiplication rates were reported, and rooting and transfer to soil were not described.

In our work on in vitro propagation of *C. japonica* seedlings (Samartín et al. 1984), camellia seeds were germinated in pots in the glasshouse and shoots from the plants thus produced were used as the starting material for in vitro cultures when the plants were 3 or 4 months old. Apex and nodal explants 1 cm long were aseptically placed on the basal medium, a modified form of that used by Crézé and Beauchesne but without polyvinylpyrrolidone, 2-iP, and GA_3, and were incubated under 30 $\mu E/m^2/s$ delivered during a 16-h photoperiod by cool white fluorescent lamps with day and night temperatures of 25 and 18 °C, respectively. Shoot apex explants began to develop after 4 weeks, and on the transfer to fresh medium produced single shoots whose axillary buds began to grow about 6 weeks after the transfer, making a total of 10 weeks from the cutting of explants to the establishment of shoot multiplication cultures. Nodal explants formed callus but, unlike Crézé and Beauchesne's nodal explants, developed no buds, a discrepancy that may probably be attributed to the age difference between Crézé and Beauchesne's starting material and our own, and the consequent difference in the physiological state of our lateral buds.

Our successful apex-derived cultures were successively subcultured every 8 weeks to maintain the clones (Fig. 3). As mentioned in Section 3.2, the most suitable mineral medium for this was MS. In order to simplify the growth regulation formula and determine the optimal level of cytokinins in the shoot

Fig. 3. Shoot multiplication culture of a juvenile *C. japonica* clone after 8 weeks on shoot multiplication medium

multiplication stage, the effects of BA and kinetin (0–2 mg/l) were studied independently and in combination. The cytokinin experiments showed that BA is necessary for axillary bud development. The highest number of shoots was obtained with 2 mg/l of BA, but their length dwindled with increasing BA concentration; in earlier experiments, shoots produced using 5 mg/l of BA had grown to no more than 1 cm in length, and with 2 mg/l most shoots were less than 2 cm long and became succulent, with thick, brittle leaves. We therefore prefer to use a dosage of 1 mg/l of BA, which affords a lower multiplication rate but ensures the production of shoots with a length of 2–4 cm and a vigorous constitution more suitable for rooting. The overall multiplication rate can be raised by obtaining further axillary shoots from subcultures of nodal segments of the shoots produced in vitro. In the absence of cytokinins, or with kinetin alone, cultured shoots elongated to 4–5 cm but produced no axillary shoots. Since the joint use of kinetin and BA gave results no better than those obtained with 1 mg/l of BA alone, simplifying the initial medium by eliminating kinetin from the shoot multiplication stage seems reasonable. Subsequent experiments showed that the IAA and adenine supplements were also superfluous for culture development and condition. The only growth regulator now used in the shoot multiplication medium is therefore BA.

3.5.2 Root Induction

The production of complete cultured plantlets requires that roots be induced to form on shoots isolated from the multiplication cultures. In *Camellia*, roots are generally induced by the application of auxin, either in low concentrations supplied in the medium, or by dipping the basal ends of the shoots in concentrated auxin before transferring them to an auxin-free medium for the roots to develop. Untreated shoots never root at all.

In our first rooting experiment (Samartín et al. 1986), cultures were placed for 11 days in half-strength MS medium to which 1–10 mg/l of IBA had been added, after which they were transferred to an IBA-free root development medium. Since rooting rates of only 8–16% were achieved (depending on the clone), in subsequent experiments we employed the immersion technique, in which the basal 1 cm of the shoot is dipped for a short period in concentrated auxin solution. Trials involving dipping for 10, 20, or 30 min in 1 g/l IBA or NAA showed that the two auxins do not differ significantly as regards rooting rates or the number of roots per rooted shoot, but do differ as regards the speed of root induction and the length of the roots produced, the latter especially. NAA-induced roots emerged 2–4 days earlier than roots induced by IBA, but were consistently short and thick, growing no further than 3–4 mm attained when measured after 4 weeks' culture, whereas IBA-induced roots were longer (about 10 mm), fibrous, and continued growing. NAA also produced more callus on the base of shoots than IBA, which is undesirable for transfer to soil. IBA was accordingly used to induce rooting in all subsequent work.

Dipping in 2 or 5 g/l IBA was found invariably to produce necrosis, even with dipping times as short as 15 s. The results obtained using 0.5 and 1 g/l IBA were compared for dipping times ranging from 15 s to 1 h (Fig. 4). The 1 g/l concentration generally gave better results as regards both rooting rates and the number of roots per rooted shoot, regardless of dipping time or clone. Very short dipping times afforded low rooting rates, but times of 40 min or more caused large swellings to appear at the base of the shoots, accompanied by apical necrosis. However, it must be emphasized that the genotype also strongly influences rooting ability in vitro, with figures ranging from 96% after 20 min dipping for clone 20 to 43% after 30 min for clone 50. This variability in the different clones' sensitivity to exogenous auxin may reflect genetically determined differences in endogenous auxin content, rooting cofactors, or auxin metabolism (Samartín et al. 1986).

In comparison with other species, the camellia requires higher IBA concentrations and longer immersion times if acceptable rooting rates are to be achieved (see Sriskandarajah et al. 1982; San-José et al. 1988). Such vigorous rooting treatment is not as deleterious to the camellia as it is to woody species such as *Prunus avium* (Riffaud and Cornu 1981).

It is well known that complete darkness at the beginning of the rooting stage may be beneficial (Druart et al. 1982). Increased rooting rates when an initial dark period is used have been reported for apple (James 1983), raspberry (Welander 1985), and plum (Druart and Gruselle 1986). For *Camellia*, we studied the influence on rooting of various periods of post-auxin-dipping darkness in two clones with low rooting rates under normal photoperiod. Both the rooting rate and the number of roots per rooted shoot increased after 20 min treatment with 1 g/l IBA followed by 9, 18, or 27 days' darkness (Fig. 5), an effect that was statistically significant after

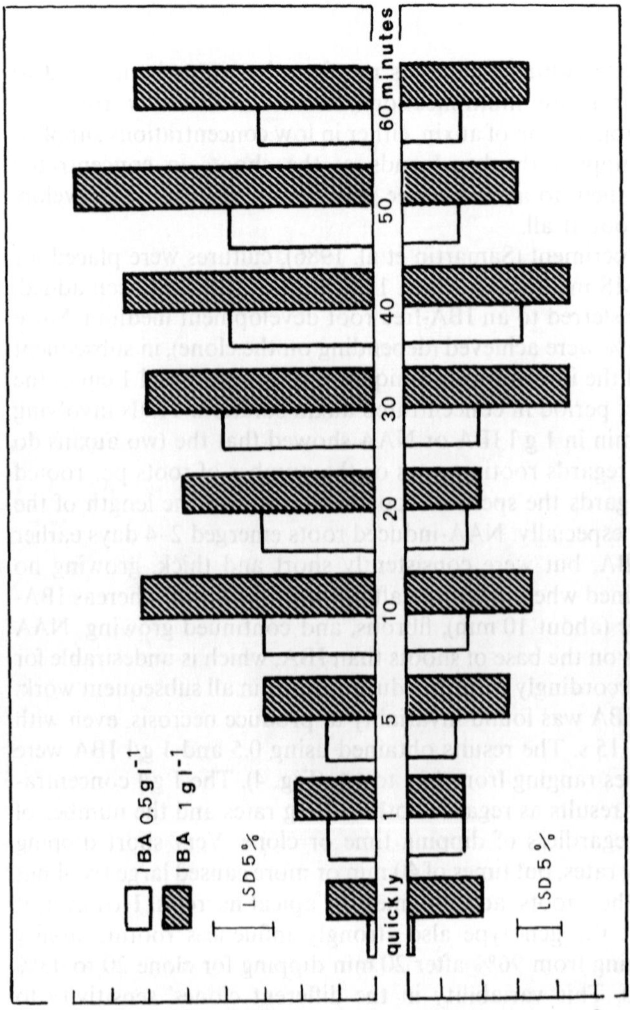

Fig. 4. Effect of basal treatment with IBA for periods ranging from 15 s to 60 min on rooting percentage and roots per rooted culture. (Samartin et al. 1986)

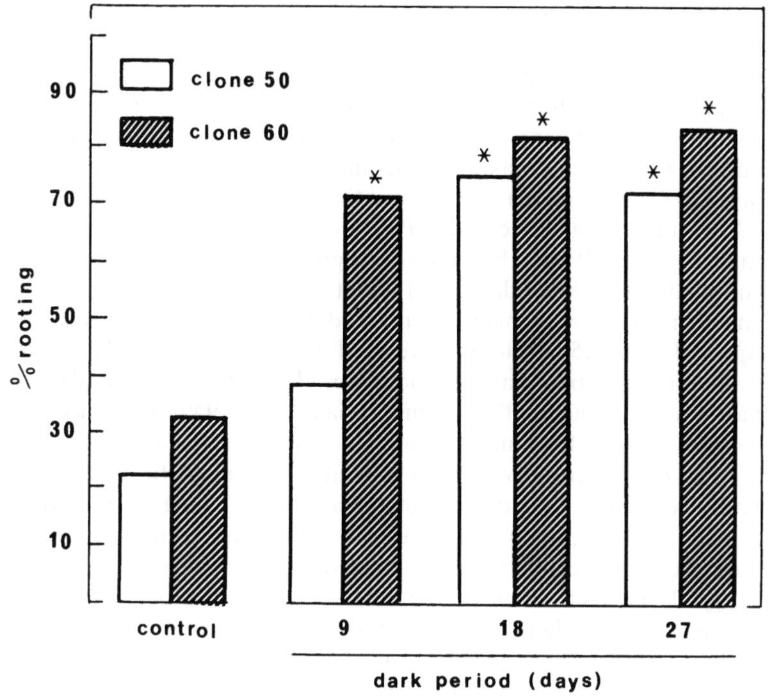

Fig. 5. Rooting of two clones of *C. japonica* under varying lengths of dark treatment. Control = 5 weeks under 16-h photoperiod. *Dark period* days in complete darkness followed by incubation under 16-h photoperiod. *Significantly higher (P = 0.05) than corresponding control using the chi-square test. (Samartin et al. 1986)

9 days for clone 60 and after 18 days for clone 50. Dark treatment also accelerated root emergence by 2–4 days, depending on the clone. The root-promoting effect of darkness may be due to the plantlet's acquiring higher total auxin levels than in shoots grown under light, which decomposes endogenous IAA. Alternatively, the light may bring about photoinactivation of root-promoting cofactors (Mosella-Chancel et al. 1980).

In an anatomical study of in vitro root formation in a clone with high rooting capacity (93%), it was found that small meristematic areas appeared in the phloem about 7–8 days after auxin treatment (Samartín et al. 1986). Mitotic activity continued during the following days, and about day 15 there appeared what Thorpe (1978) called "meristemoids", from which the root primordia originate. By day 18 the root primordia had a differentiated vascular system and emerged. They were occasionally observed to have differentiated in the pith, without further growth.

3.5.3 *Material of Adult Origin*

In vitro culture of shoot tips or buds of tea has scarcely been studied. However, Phukan and Mitra (1984) used a medium with 2 mg/l of IAA and 8 mg/l of kinetin to produce shoots from nodal explants obtained from both seedlings and mature twigs of *C. sinensis*, and Arulpragasam and Lattif (1986) have described the proliferation of shoots of *C. sinensis* clones of adult origin using MS medium supplemented with 2 mg/l of BA, though they mention no attempt to induce rooting. Our own work on adult *C. japonica* cvs. Alba Plena and Fimbriata Alba and on *C. reticulata* has followed on the success of our preliminary work with juvenile *C. japonica* material (Samartín et al. 1984, 1986).

Mature *C. japonica* cv. Alba Plena material was obtained from a tree aged more than 50 years. When the new year's growth began in February, branches were collected whose buds developed shoots after being placed in water in a growth chamber. The new shoots were collected when 2–4 cm long and the terminal shoot tips and nodes (5 mm long) were used as explants; the axillary buds were found to be more reactive than the shoot tips. Three monthly transfers to fresh WPM medium with Jacquiot's vitamins (Gautheret 1959) were necessary to obtain cultures suitable for multiplication (Vieitez et al. 1989a).

The combination of growth regulators found to be most suitable for shoot multiplication by development of axillary buds is 2 mg/l BA, 2 mg/l zeatin, 2 mg/l 2-iP, and 0.01 mg/l IBA. After 4 weeks, cultures are transferred to fresh medium of the same composition except for the cytokinins, which are all reduced in strength to 1 mg/l. Four weeks later, shoots are harvested and explants obtained for the next multiplication cycle (Fig. 6A, B). A similar shoot multiplication system has also been established for adult *C. japonica* cv. Fimbriata Alba (Fig. 6C) (unpubl.) and *C. reticulata* (Fig. 6D) (San-José and Vieitez 1990), with the difference that for these species the shoots established initially were obtained by hypocotyl grafting (Vieitez and Vieitez 1983) instead of by inducing the flushing of bud branches in the phytotron. The fact that *Camellia* cultures generally need high concentrations of cytokinins may be due to the penetration of growth substances into the heavily cutinized shoots being very slow otherwise.

In the shoot multiplication stage, three kinds of explant with very different behaviors were distinguished (Fig. 7) (Vieitez et al. 1989a): ST1, the shoot tips of harvested shoots longer than 13–14 mm; n, the nodal segments of constituting the lower parts of the shoots from which the ST1 were taken; and ST2, whole harvested shoots 5–10 mm long. Table 2 lists the results achieved with each of the three kinds of explant in each of the six macronutrient media tested (see above, Section 3.2). The ST1 explants hardly produced axillary shoots in any of the media and elongated poorly. ST2 and n explants both had better multiplication rates, and attained lengths allowing them to be used directly for rooting. The production of fewer axillary shoots by shoot tips than by nodal explants has also been reported in the case of other woody species (Hutchinson 1984; Mascarenhas et al. 1987; Amin and Jaiswal 1987; San-José et al. 1988), and has been attributed to the effects of apical dominance in the shoot tips and to the presence of axillary buds at a more advanced stage of development. Why the inhibitory effect of the apical meristem is different in ST1 and ST2 explants is not known; it may be due to the apical meristems of dominant (long) shoots exerting a greater inhibitory effect than those of shorter shoots, perhaps because of greater physiological aging. Similar results have been reported for rose by Mederos and Rodríguez-Enríquez (1985).

To increase the efficiency of the shoot multiplication procedure in *C. reticulata* cv. Captain Rawes, decapitated shoots 20–25 mm in length were cultured in

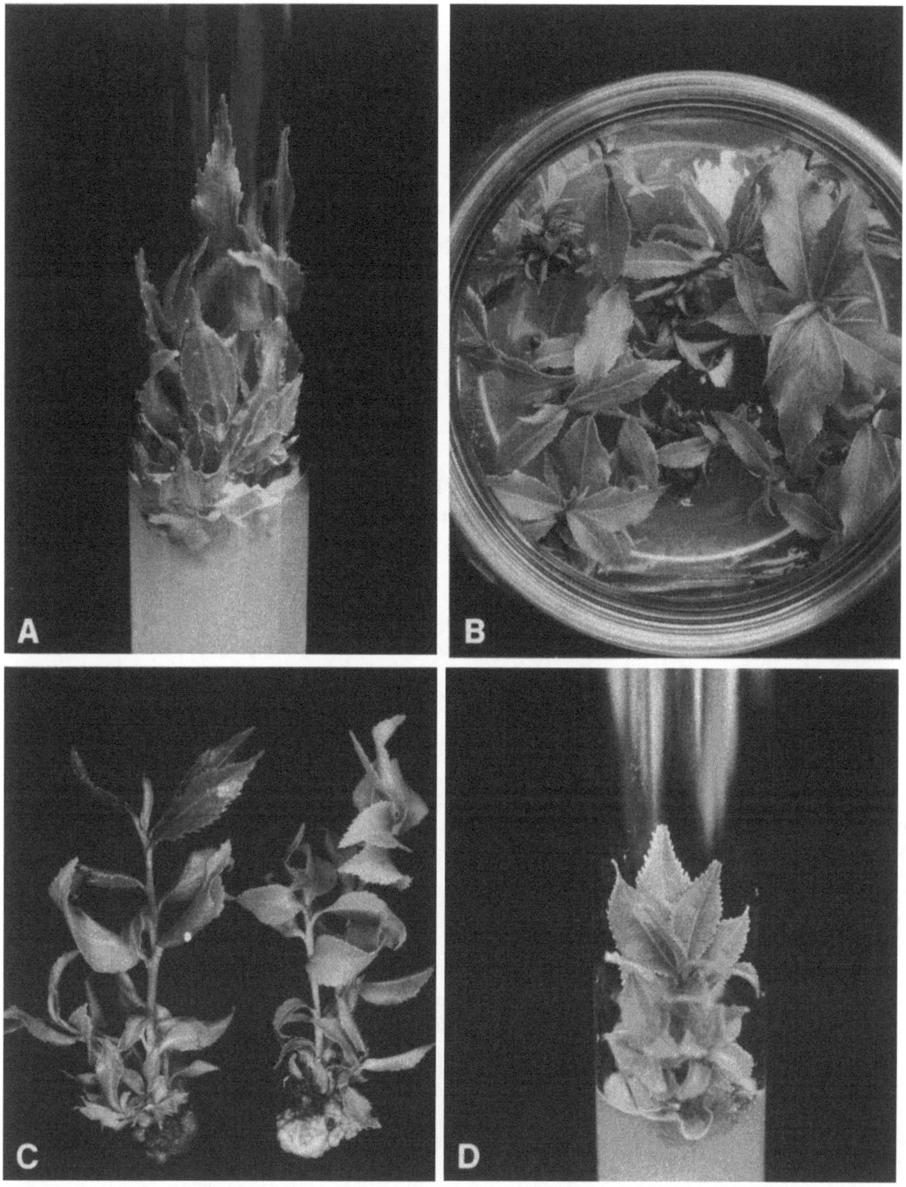

Fig. 6A–D. Shoot cultures 8-week-old of: *C. japonica* cv. Alba Plena (**A, B**), *C. japonica* cv. Fimbriata Alba (**C**) and *C. reticulata* cv. Captain Rawes (**D**)

a horizontal position (San-José and Vieitez 1990). It was possible for donor shoots so cultured to be recultured on fresh medium several times and the newly formed shoots harvested before each transfer.

Interesting findings were obtained in experiments to compare the results of transferring 4-week-old multiplication cultures to fresh WPM agar medium with

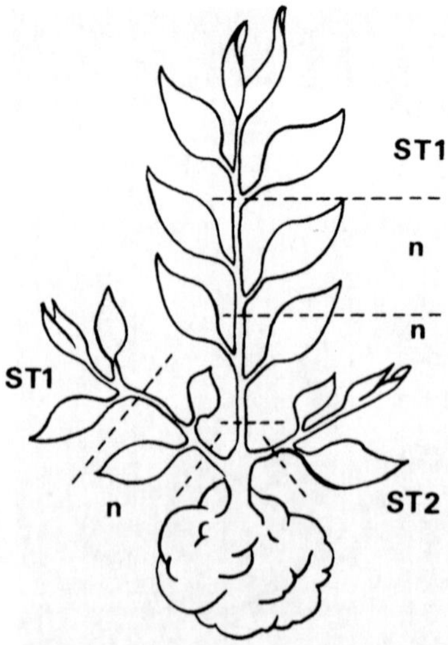

Fig. 7. Explant type in shoot multiplication culture of *C. japonica* cv. Alba Plena. (Vieitez et al. 1989a)

Table 2a. Survival and shoot production performance by three kinds of explant (ST1, ST2 and n) after 8 weeks' culture in one of six mineral media. (Vieitez et al. 1989a).

	ST1				ST2				n			
	SH	SE	L	%	SH	SE	L	%	SH	SE	L	%
$H + SO_4$	1.5	1.8	12.1	100.0	3.1	5.2	24.3	100.0	3.9	5.0	17.5	100.0
GD	1.4	1.9	10.6	100.0	3.5	4.4	14.6	93.3	3.8	4.9	14.8	100.0
MS	0.9	1.0	7.5	73.7	1.5	1.7	7.7	73.3	0.8	0.8	3.2	42.9
MS1/2	0.9	0.9	4.9	53.3	1.4	1.6	9.5	84.2	1.9	2.4	9.1	78.6
WPM	1.1	1.4	11.5	100.0	3.9	6.2	25.3	100.0	4.7	6.3	19.0	100.0
Anderson	1.1	1.2	7.4	87.5	2.7	3.4	13.6	93.8	3.6	4.7	14.6	100.0

SH, number of shoots per explant; SE, number of segments (including shoot tips) per explant; L, length (mm) of the tallest shoot; %, percentage of cultures surviving.

Table 2b. Table of variance of the two factors studied (two-way ANOVA)

	LSD 1%	F-test Factor A	F-test Factor B
SH	1.7	17.98**	49.46**
SE	2.6	26.50**	59.49**
L	8.0	30.18**	27.45**

** Significant at $p = 0.01$.
LSD: least significant difference.
Factor A: mineral medium.
Factor B: kind of explant.

those of replenishing the original medium with a liquid solution of the same composition as the second medium (Vieitez et al. 1989a). The number and length of shoots were generally increased by replenishing with liquid medium, especially in the case of ST1 explants. Since this "double phase medium" technique (Maene and Debergh 1985) is at least as effective as transfer to fresh agar medium, it has been adopted to save labor. Carlisi and Torres (1986) have likewise reported increased shoot length and number when liquid medium with GA_3 was used for shoot cultures established from 10-year-old *C. japonica* cv. Purple Dawn, though their shoots displayed some symptoms of vitrification, a phenomenon that was not observed in Alba Plena cultures when liquid medium is used only during the second month of each cycle.

As can be seen, the culture medium optimized for shoot multiplication of cultures established from juvenile material (Samartín et al. 1984) proved to be rather unsuitable for adult material, since several changes had to be made.

3.5.4 Root Induction

Basal dipping in 1 g/l IBA for 15 min, which had given good results in the rooting stage of juvenile cultures, also proved to be satisfactory for the adult Alba Plena cultivar. As for some recalcitrant juvenile clones, however, a dark period was found to be essential for root induction in the adult material: the 10% rooting rate achieved under the standard 16-h photoperiod increased to 70–90% when an initial period of 12 days' darkness was imposed (Vieitez et al. 1989b) (Fig. 8A). The partial shading achieved by wrapping the culture tubes in aluminum foil up to the height of the agar failed to improve on the results obtained without any dark period; 9 days' darkness afforded lower rooting rates than with 12 days; and 15 days' darkness afforded no better rooting rates than with 12 days. The apparent discrepancy between these findings and those of Beretta et al. (1987), who found darkness to be of no advantage for rooting *C. saluenensis* × *C. japonica* cultures, is presumably attributable to genetic differences; it is possible that clones rooting poorly under light may have lower levels of endogenous IAA or auxin protectors.

In experiments to optimize IBA concentration, dipping time, and medium (Vieitez et al. 1989b) it was found that the 87% rooting rate achieved by dipping in 1 g/l IBA for 15 min fell to 52% when the IBA concentration was reduced to 0.5 g/l, and to 72% when the dipping (in 1 g/l IBA) time was reduced to 7 min (Table 3). WPM and $H + SO_4$ media afforded similar rooting rates, but the number of roots per rooted shoot was 9.3 for WPM against only 5.7 for $H + SO_4$. The $H + SO_4$ cultures also produced less vigorous plantlets that would presumably be more difficult to transfer to soil, and some exhibited apical necrosis.

Tian-Ling (1982) and Kato (1986) induced root formation on a filter paper bridge over liquid medium, whereas Beretta et al. (1987) claim that agar medium is definitely superior to liquid medium.

Our result with the cv. Alba Plena (Vieitez et al. 1989b) showed that the type of support (agar or paper bridges) did not significantly affect the rooting percentage or the number of roots per rooted shoot. However, paper supports on liquid media induced greater root elongation.

Fig. 8. In vitro rooting (**A**) and potted plants (**B**) of *C. japonica* cv. Alba Plena (Vieitez et al. 1989b). **C** Micropropagated camellias from juvenile origin 4 years after field planting

Table 3. Effect of IBA concentration and dipping time on rooting percentage (%) and number of roots per rooted shoot (N ± S.D.). (Vieitez et al. 1989b)

Treatment (g/l)	%	N
IBA 0.5–7 min	44.4a[a]	4.3 ± 3.5a
IBA 0.5–15 min	51.9a	5.7 ± 3.8ab
IBA 1–7 min	72.2b	7.4 ± 4.6b
IBA 1–15 min	87.0c	7.4 ± 5.6b

[a] Figures followed by the same letter are not significantly different ($p < 0.05$) using the chi-square test.

In *C. reticulata* cv. Captain Rawes, which is notoriously difficult to propagate by cuttings, rooting rates of 60% were achieved among shoots excised from the second crop of horizontally cultured donor shoots, by dipping their basal ends in 1 g/l IBA solution for 30 min and keeping them in the dark for a further 18 days in WPM with half-strength macronutrients and no growth regulators.

3.6 Transfer to Soil

Few reports on the regeneration of *Camellia* plantlets mention conditions and preparation for transfer to soil. Among those who have obtained plantlets by somatic embryogenesis or induction of adventitious buds, Tian-Ling (1982) and Kato (1985) used a mixture of soil and vermiculite for plantlet acclimatization. Wu et al. (1981) reported that tea plantlets derived from cotyledon callus grew vigorously in soil, but were morphologically considerably different from naturally grown plants; plantlets regenerated in vitro flowered in the year following their transfer to soil, and new F_1 hybrids were obtained from them. Phukan and Mitra (1984) observed genetic variability among plants produced by indirect regeneration via callus tissue, whereas plantlets obtained by axillary bud culture were true to type.

We have achieved a survival rate of 70–90% after transfer of 4-week-old rooted juvenile or adult plantlets to 1:1 mixture of soil and quarz sand in pots placed under a plastic tunnel cover with mist and fog systems in a greenhouse (Samartín et al. 1986; Vieitez et al. 1989b). We have found that acclimatization can also be performed by transferring the plantlets to the tunnel just before roots appear (12–14 days after auxin treatment); roots thus prevented from developing in agar look and work better. Acclimatization also proves to be favored by the use of a filter paper bridge system during the rooting stage, possibly because of an improvement of root quality.

After 6–8 weeks, when new growth has begun, the plantlets can be removed from the tunnel and placed in a shade house for further acclimatization before planting out (Fig. 8B). Regenerated camellia plantlets, which are heavily cutinized, are not very prone to desiccation; the longish time spent in the acclimatization tunnel is due more to camellia's slow growth than to plantlets' sensitivity. Plants regenerated from cultures established from adult Alba Plena material are still in their 2nd year of outdoor growth, and have not yet flowered; those obtained from cultures established from seedling material flowered during their 4th year outdoors (Fig. 8C).

4 Summary and Conclusions

The capacity of regeneration of embryos, cotyledons, and other juvenile tissues of *Camellia* species is of considerable interest, as since the embryo forms after completion of fertilization, it is heterogenic in hybrid seed, and somatic embryogenesis by cotyledon and embryo culture may constitute a viable means of rapid clonal propagation of crossing materials or selected *Camellia* clones (Kato 1989).

This is particularly desirable for the propagation of hybrids whose seeds, as in many cases of interspecific cross-pollination, have only partially developed embryos or low viability rates. Examples of this application are provided by the success of embryo culture of various hybrids of *C. chrysantha* (Nadamitsu et al. 1986; Yamaguchi et al. 1987). Further studies are nevertheless required to optimize culture conditions for the development of somatic embryos.

One of the most obvious applications of in vitro techniques is the propagation of selected trees with desired characteristics. It is still difficult to induce morphogenesis in adult material, but good results have been obtained for some varieties by stimulation of axillary buds. Rooting shoots obtained by this method also seems to be possible, though it is necessary to adapt the technique employed to the individual cultivar being propagated. For advances on this front, it is therefore necessary for more time and research staff to be devoted to the task. A crucial question that remains to be answered is whether plants produced in vitro are true to type, i.e., whether the genotypes are really cloned. In vitro genetic stability varies from one micropropagation method to another, and must therefore be investigated separately for each.

Another objective is to obtain disease-free old cultivars by grafting their shoot tips onto root-stocks obtained from seed. Crézé (1983, 1985) reported that 79% of plants so grafted showed no signs of virosis after more than 2 years of culture, though no serological methods have been developed to check this.

5 Protocol of Micropropagation by Development of Axillary Shoots

1. Best Explant
   Shoot tips and nodal segments of the current season's growth.

2. Best Medium
   Initial and multiplication stages: the Woody Plant Medium mineral formula with Jacquiot's vitamins, 100 mg/l of m-inositol, 2 mg/l of BA, 2 mg/l of 2-iP, 2 mg/l of zeatin, 0.01 mg/l of IBA, 30 g/l of sucrose, and 6 g/l of agar, at pH 5.5.

   Rooting stage: Root induction by dipping the basal ends of shoots in 1 g/l of IBA for 15 min, followed by root development in the same medium as for the multiplication stage but with macronutrients reduced to half strength and with no growth regulators.

3. Optimal Conditions
   16-h photoperiod with day and night temperatures of 25 and 18 °C respectively. 8-week subculture period in the multiplication stage, with liquid medium of the same composition (but with cytokinins reduced to 1 mg/l) being added after 4 weeks. 12 days darkness at the beginning of rooting stage, followed by development under the standard photoperiod regime. Acclimatization by transfer to a 1:1 mixture of peat and sand in a tunnel with programmable mist and fog system.

References

Ackerman WL (1978) Handling *Camellia* seed by aseptic culture. Am Camellia Yearb 1978: 70–74

Aiton W (1789) Hortus Kewensis, or a catalogue of the plants at the Royal Botanical Garden at Kew. London

Amin MN, Jaiswal VS (1987) Rapid clonal propagation of guava through in vitro shoot proliferation on nodal explants of mature trees. Plant Cell Tissue Organ Cult 9: 235–243

Anderson WC (1984) A revised tissue culture medium for shoot multiplication of *Rhododendron*. J Am Soc Hortic Sci 109: 343–347

Arulpragasam PV, Latiff R (1986) Studies on the tissue culture of tea [*Camellia sinensis* (L.) O. Kuntze]. 1. Development of a culture method for the multiplication of shoots. Sri Lanka J Tea Sci 55: 44–47

Bajaj YPS (1986) Biotechnology in Agriculture and Forestry, vol 1. Trees I. Springer, Berlin Heidelberg New York Tokyo

Bennett WY (1977) Tissue culture for camellias? Am Camellia Yearb, 1977: 188–190

Bennett WY (1978) Tissue culture for Camellias II. Am Camellia Yearb 1978: 102–104

Bennett WY, Scheibert P (1982) In vitro generation of callus and plantlets from cotyledons of *C. japonica*. Camellia J 37: 12–15

Beretta D, Vanoli M, Eccher T (1987) The influence of glucose, vitamins and IBA on rooting of camellia shoots in vitro. Abstr Symp on Vegetative Propagation of Woody Species, Pisa (Italy), p 105

Bonga JM, Durzan DJ (ed) (1987) Cell and tissue culture in forestry: Vols 1, 2, and 3. Martinus Nijhoff, Dordrecht

Booth WB (1830) History and description of the species of *Camellia* and *Thea*. Trans Hort Soc Lond 7: 519–562

Bretschneider E (1898) Early european researches into the flora of China. J N China Branch R Asiat Soc 15: 1–192

Carlisi JA, Torres KC (1986) In vitro shoot proliferation of camellia Purple Down. HortScience 21: 314

Chang HT, Bartolomew B (1984) Camelliar. Batsford, London

Chu CC (1976) The N_6 medium and its applications to anther culture of cereal crops. In: Proc Symp on Plant tissue culture. Science Press, Peking, pp 45–54

Crézé J (1980) The graft of *Camellia* meristem in vitro. Camellia Rev 42: 11–12

Crézé J (1983) Where do we stand in regard to the grafting of apex in camellia? Int Camellia J 15: 56–60

Crézé J (1985) Elimination des virus de camellia par greffe d'apex in vitro. C R Seances Acad Agric Fr 71: 572–576

Crézé J, Beauchesne MG (1980) Camellia cultivation in vitro. Int Camellia J 12: 31–34

Doi Y (1980) Suitable culture conditions for callus induction from stem segments and their application to anther culture in tea. Stud Tea 58: 1–9

Doy Y (1981) Frequency of root differentiation in anther culture of tea. Stud Tea 60: 1–3

Druart P, Gruselle R (1986) Plum (*Prunus domestica*). In: Bajaj YPS (ed) Biotechnology in agriculture and forestry, vol 1: Trees I. Springer, Berlin Heidelberg New York Tokyo, pp 130–154

Druart P, Kevers C, Boxus P, Gaspar T (1982) In vitro promotion of root formation by apple shoots through darkness effect on endogenous phenols and peroxidases. Z Pflanzenphysiol 108: 429–436

Forrest GI (1969) Studies on the polyphenol metabolism of tissues cultures derived from the tea plant (*Camellia sinensis* L.). Biochem J 113: 765–772

Frisch CH, Camper ND (1987) Effect of synthetic auxins on callus induction from tea stem tissue. Plant Cell Tissue Organ Cult 8: 207–213

Gautheret RJ (1959) La culture des tissus végétaux: techniques et réalisations. Masson, Paris

Gimson R (1986) Nuevos híbridos y nuevas especies. In: Diputación Pontevedra (ed) La Camelia. Peón, Pontevedra, Spain, pp 91–97

Gresshoff PM, Doy CH (1972) Development and differentiation of haploid *Lycopersicon esculentum*. Planta 107: 161–170

Haldeman JH, Thomas RL, McKamy DL (1987) Use of benomyl and rifampicin for in vitro shoot tip culture of *Camellia sinensis* and *C. japonica*. HortScience 22: 306–307

Heller R (1953) Recherches sur la nutrition minérale des tissus végétaux cultivés in vitro. Ann Sci Nat Bot Biol Veg 14: 1–223

Hutchinson JF (1984) Factors affecting shoot proliferation and root initiation in organ cultures of the apple Northern Spy. Sci Hortic 22: 347–358

James DJ (1983) Adventitious root formation in vitro in apple rootstocks (*Malus pumila*) I. Factors affecting the length of auxin-sensitive phase in M.9. Physiol Plant 57: 149–153

Kaempfer E (1712) Amoenitatum Exoticorum politico-physico-medicarum. Lemgo

Kato M (1982) Results of organ culture on *Camellia japonica* and *C. sinensis*. Jpn J Breed 32: 276–277

Kato M (1985) Regeneration of plantlets from tea stem callus. Jpn J Breed 35: 317–322

Kato M (1986) Micropropagation through cotyledon culture in *Camellia japonica* L. and *C. sinensis* L. Jpn J Breed 36: 31–38

Kato M (1989) *Camellia sinensis* L. (Tea): In vitro regeneration. In: Bajaj YPS (ed) Biotechnology in agriculture and forestry, vol 7: Medicinal and aromatic plants II. Springer, Berlin Heidelberg New York Tokyo, pp 82–98

Kim JS (1986) Studies on the propagation of Korean tea plants by tissue culture. J Korean For Soc 75: 25–31

Korestkaya TF, Zaprometov MN (1975) Cultivation of tissue of the tea plant (*Camellia sinensis*) as a model for studying conditions of phenolic compounds synthesis. Fiziol Rast 22: 282–288

Laborey J (1986) Les camelias. La Maison Rustique, Flamarion

Lammerts WE (1958) Embryo culture in camellia seed germination. In: Tourje EC (ed) Camellia culture. Southern California Camellia Society, Pasadena, California, pp 171–174

Linneaus C (1735) Sistema Naturae. Leiden

Maene L, Debergh P (1985) Liquid medium additions to established tissue cultures to improve elongation and rooting in vitro. Plant Cell Tissue Organ Cult 5: 23–33

Mantell SH (1982) Tissue culture techniques. Camellia Int Soc 1982: 61–64

Mascarenhas AF, Kendurkar SV, Gupta PR, Khuspe SS, Agrawa DC (1987) Teak. In: Bonga JM, Durzan DJ (eds) Cell and tissue culture in forestry, vol 3. Martinus Nijhoff, Dordrecht, pp 300–315

McCown BH, Lloyd G (1981) Woody plant medium (WPM). A mineral nutrient formulation for microculture of woody plant species. HortScience 16: 453

McCown BH, Sellmer JC (1987) General media and vessels suitable for woody plant culture. In: Bonga JM, Durzan DJ (eds) Cell and tissue culture in forestry, vol 1. Martinus Nijhoff, Dordrecht, pp 4–16

Mederos S, Rodriguez-Enriquez MJ (1985) In vitro propagation of Golden Times rose. Factors affecting shoot tips and axillary buds growth and morphogenesis. Abstr Symp on In vitro problems related to mass propagation of horticultural plants. Gembloux, Belgium, p 27

Mosella-Chancel L, Macheix JJ, Jonard R (1980) Les conditions du microbouturage in vitro du pêcher (*Prunus persica* Batsch): influences combinées des substances de croissance et de divers composes phénoliques. Physiol Vég 18: 597–608

Murashige T, Skoog F (1962) A revised medium for rapid growth and bioassays with tobacco tissue cultures. Physiol Plant 15: 473–497

Nadamitsu S, Andoh Y, Kondo K, Segawa M (1986) Interspecific hybrids between *Camellia vietnamensis* and *C. chrysantha* by cotyledon culture. Jpn J Breed 36: 309–313

Odriozola A (1986) Lo que conviene saber sobre la camelia. In: Diputación Pontevedra (ed) La camelia. Péon, Pontevedra, Spain, pp 23–30

Ogutuga DB, Northcote DH (1970a) Caffeine formation in tea callus tissue. J Exp Bot 21: 258–273

Ogutuga DB, Northcote DH (1970b) Biosynthesis of caffeine in tea callus tissue. Biochem J 117: 715–720

Petiver J (1702) Gazophylacill Naturae et artis Decades X. London

Phukan MK, Mitra GC (1984) Regeneration of tea shoots from nodal explants in tissue culture. Curr Sci (India) 53: 874–876

Plata E, Vieitez AM (1990) In vitro regeneration of *Camellia reticulata* by somatic embryogenesis. J Hortic Sci 65: 707–714

Quoirin M, Lepoivre P (1977) Improved medium for in vitro culture of *Prunus* sp. Acta Hortic 78: 437–442

Read PE, Garton SM, Louis KA, Zimmermann ES, Farnham RS (1982) In vitro propagation of species for bioenergy plantations. In: Fujiwara A (ed) Plant tissue culture 1982. Maruzen, Tokyo, pp 757–758

Riffaud JL, Cornu D (1981) Utilisation de la culture in vitro pour la multiplication de merisiérs adultes (*Prunus avium* L.) sélectionnés en fôret. Agronomie 1: 633–640

Salinero MC, Silva-Pando F (1986) La multiplicación de las camelias. In: Diputación Pontevedra (ed) La camelia. Peón, Pontevedra, Spain, pp 175–184

Samartín A (1984) Regeneración de plantas de *Camellia japonica* L. mediante cultivo in vitro. Doctoral Thesis, Fac. Biol, Univ Santiago de Compostela, Spain

Samartín A, Vieitez AM, Vieitez E (1984) In vitro propagation of *Camellia japonica* seedlings. HortScience 19: 225–226

Samartín A, Vieitez AM, Vieitez E (1986) Rooting of tissue cultured camellias. J Hortic Sci 61: 113–120

San-José MC, Vieitez AM (1990) In vitro regeneration of *Camellia reticulata* cv. Captain Rawes from adult material. Sci Hortic 43: 155–162

San-José MC, Ballester A, Vieitez AM (1988) Factors affecting in vitro propagation of *Quercus robur* L. Tree Physiol 4: 281–290

Sarwar M (1985) Callus formation from explanted organs of tea (*Camellia sinensis* L.) Sri Lanka J Tea Sci 54: 18–22
Schenk RU, Hildebrandt AC (1972) Medium and techniques for induction and growth of monocotyledonous and dicotyledonous plant cell cultures. Can J Bot 50: 199–204
Sealy JR (1958) A revision of the genus *Camellia*. R Hortic Soc London
Shimokado T, Murata T, Miyaji Y (1986) Formation of embryoid by anther culture of tea. Jpn J Breed (Suppl 2) 36: 282–283
Sriskandarajah S, Mullins MG, Nair Y (1982) Induction of adventitious rooting in vitro in difficult to propagate cultivars of apple. Plant Sci Lett: 24: 1–9
Tabachnik L, Kester DE (1977) Shoot cultures for almond and almond-peach hybrid clones in vitro. HortScience 12: 545–547
Thimann KV, Behnke-Rogers J (1950) The use of auxins in the rooting of woody cuttings. Cabot Found, Massachussetts
Thorpe TA (1978) Regulation of organogenesis in vitro. In: Proc Propagation of higher plants through tissue culture. A bridge between research and applications. Univ Tennessee, Knoxville, pp 87–101
Tian-Ling L (1982) Regeneration of plantlets in cultures of immature cotyledons and young embryos of *Camellia oleafera* Abel. Acta Biol Exp Sin 15: 393–403
Torres KC, Carlisi JA (1986) Shoot and root organogenesis of *Camellia sasanqua*. Plant Cell Rep 5: 381–384
Tukey HB (1934) Artificial culture methods for isolated embryos of deciduous fruits. Am Soc Hortic Sci 32: 303–322
Vieitez AM, Barciela J (1990) Somatic embryogenesis and plant regeneration from embryonic tissues of *Camellia japonica* L. Plant Cell Tissue Organ Cult 21: 267–274
Vieitez AM, Barciela J, Ballester A (1989a) Tissue culture propagation of *Camellia japonica* cv. Alba Plena. J Hortic Sci 64: 177–182
Vieitez AM, San-José MC, Ballester A (1989b) Progress towards clonal propagation of *Camellia japonica* cv. Alba Plena by tissue culture techniques. J Hortic Sci 64: 605–610
Vieitez ML, Vieitez AM (1983) Propagation of camellia by the hypocotyl gafting. Am Camellia Yearb 1983: 1–4
Welander M (1985) In vitro culture of raspberry (*Rubus ideaus*) for mass propagation. J Hortic Sci 60: 493–499
Wu CT (1976) Studies on the tissue culture of tea plant. J Agric Assoc China 93: 30–42
Wu CT, Huang TK, Chen GR, Chen SY (1981) A review on the tissue culture of tea plants and on the utilization of callus derived plantlets. In: Rao AN (ed) Tissue cultured of economically important plants. COSTED, Singapore, pp 104–106
Yamaguchi S, Kunitake T, Hisatomi S (1987) Interspecific hybrid between *Camellia japonica* cv. Choclidori and *C. chrysantha* produced by embryo culture. Jpn J Breed 37: 203–206
Yan MQ, Ping C (1983) Studies on development of embryoid from the culture cotyledons of *Thea sinensis* L. Sci Silvae Sin 19: 25–29
Yan MQ, Ping C, Wei M, Wang YH (1984) Tissue culture and transplanting of *Camellia oleifera*. Sci Silvae Sin 20: 341–350
Zagoskina NV, Zaprometov MN (1979) Effect of 1-naphthylacetic acid on growth and phenolic compound formation in tea tissue culture. Fiziol Rast 26: 681–687
Zagoskina NV, Zaprometov MN (1983) Effect of kinetin on formation of phenolic compounds in the long passaged tea plant tissue culture. Fiziol Biokhim Kul't Ras 15: 250–253
Zagoskina NV, Zaprometov MN (1986) Phenolic compounds callus cultures of the tea plant and the possibility of regulating their formation. Kul't Kletok Rast Biotek 49–52
Zaprometov MN, Zagoskina NV, Koretskaya TF (1976) Effects of certain precursors on synthesis of phenolic compounds in tea plant tissue cultures. Fiziol Rast 23: 1274–1279
Zaprometov MN, Zagoskina NV, Strekova UY, Morozova GA (1979) Formation of phenolic compounds and the process of differentiation in tea callus culture. Fiziol Rast 26: 485–491
Zaprometov MN, Strekova UY, Subbotina GA, Zagoskina NV (1986) Effect of kinetin on differentiation and on phenolic compound formation in tea tissue culture. Fiziol Rast 33: 356–364
Zhuang C, Liang H (1985a) Somatic embryogenesis and plantlets formation in cotyledon culture of *Camellia chrysantha*. Act Bot Yunn 7: 446–450
Zhuang C, Liang H (1985b) In vitro embryoid formation of *Camellia reticulata* L. Acta Biol Exp Sin 18: 275–281
Zhuang C, Duan J, Zhou J (1988) Somatic embryogenesis and plantlets regeneration of *Camellia sasanqua*. Acta Bot Yunn 10: 241–244

II.14 Micropropagation of Common Centaury (*Centaurium erythraea* Rafn.)

H. VÁGNEROVÁ[1]

1 General Account

1.1 Botany and Morphology

Centaurium erythraea Rafn (Synonym: *C. umbellatum* Gilib., *C. minus* Moench, *Erythraea centaurium* Pers., common centaury), is a medicinal plant known for centuries; it was described by Dioscorides as early as in the 1st century A.D. It is a member of the family Gentianaceae (Fig. 1). It is a herbaceous winter-annual plant, 2–50 cm high, with a basal rosette of leaves and usually solitary, but sometimes several erect stems branched at the top. Basal leaves are obovate or elliptic (10–50 × 8–20 mm), prominently three- to seven-veined. Cauline leaves are shorter, narrow, and acute, but never parallel-sided. Flowers are sessile or subsessile, pink, often clustered, forming a dense corymb-like cyme.

Chromosome number is n = 40 (Oberdorfer 1970) or n = 42 (Zeltner 1970). Diploids and tetraploids exist in wild populations.

1.2 Distribution and Importance

Centaurium erythraea is distributed in oceanic Europe and the Mediterranean (Meusel et al. 1978). Diploids occur in the Eumediterranean and Atlantic areas to the East, tetraploids spread from the Mediterranean to the northern border of *Centaurium erythraea* distribution.

For medicinal purposes its aerial flowering part, mostly called *Centaurii herba* or *Herba centaurii minoris*, is used. According to some pharmacopoeias, this drug may include also *C. pulchellum* (Sw.) Druce, *C. uliginosum* (W. et K.) Beck [synonym: *C. littorale* ssp. *uliginosum* (W. et K.) Melderis] and *C. majus* (Hoffm. et Link) Zeltner [synonym: *Erythraea major* Hoffm. et Link]. Comprehensive reports on the taxonomy of the genus *Centaurium* were published inter alia by Zeltner (1970) and Melderis (1972).

All species of the genus *Centaurium* synthesize a similar spectrum of secondary products studied in detail, together with their chemotaxonomic analysis, by Van der Sluis (1985). The main group of secondary products of *C. erythraea* are bitter secoiridoid glucosides, which are very important medicinally, mainly in the

[1] Institute of Experimental Botany, Czechoslovak Academy of Science, Ke dvoru 15, 166 30 Praha 6, Czechoslovakia

Biotechnology in Agriculture and Forestry, Vol. 19
High-Tech and Micropropagation III (ed. by Y.P.S. Bajaj)

Fig. 1. Flowering *Centaurium erythraea* plant from the field. (Barešová 1983)

treatment of gastric complaints, hypertension, etc. *C. erythraea* has been widely used in folk herbal medicine for treating high temperature, anemia, scrofula, gout, etc. and is also used in homeopathy. The crude drug is utilized for preparing tinctures, liquid extracts, syrups, wines, and powders. Since its medicinal application mostly makes use of its high content of bitter substances, it is part of many well-known preparations, such as *Tinctura amara*, Species *amarae*, and *Vinum amarum*, to name a few. The pure extract of the drug is known under the name *Extractum centaurii minoris* and it is applied in the treatment of diseases of the gastrointestinal tract, eczema, wounds, as well being used as a sedative, antihelminthic, etc. A comprehensive account of its pharmacology was published by Van der Sluis (1985). The drug is further used for preparing commercial beverages by some companies (Jan Becher, Cinzano, Martini Rossi).

1.3 Conventional Propagation

Centaurium erythraea is one of medicinal plants which appears to be suitable for cultivation at least on small-plot areas, instead of being collected from local wild populations throughout Europe, today mostly from those in the Mediterranean region, mainly in Yugoslavia, Algeria, and Morocco. The populations of this winter-annual plant are greatly endangered by collecting unless at least one-quarter of their population is kept intact to ensure reproduction. *Centaurium erythraea* appears on many red lists of endangered plants (e.g., Holub et al. 1979). Some attempts at field cultivation have already been made (Macková 1988), but the germination of the seeds and development of the seedlings in field conditions is unpredictable, and strongly depends on mycorrhiza (McGee 1985). Another difficulties of field cultivation are its inability to grow in dense stands and to compete with common weeds. Macková (1988) tried to overcome these problems by mixing the seeds with those of *Chamomilla recutita* in the autumn. Camille plants shadowed the young rosettes, thereby helping them to overwinter.

2 Micropropagation

Micropropagation of *C. erythraea* can be accomplished by three methods: (1) induction and rooting of shoots from primary explants, (2) induction of callus and subsequent organogenesis of shoots which are then rooted, (3) induction of somatic embryogenesis in suspension culture derived from callus. Data without reference are my own results (Barešová 1988).

2.1 Explant and Establishment of Cultures

2.1.1 Micropropagation from Leaf Segments

The easiest method of micropropagation of *Centaurium erythraea* was to induce organogenesis in leaf segments, which under appropriate conditions formed abundant easily rooting shoots directly. Leaf segments used for shoot induction were of various sizes and origin. Čellárová et al. (1984) used small segments (2 mm^2) from the leaves of seedlings (3–8 weeks old) or from those of wild plants (3 months old). Laurová et al. (1986a) used segments (3 × 3 mm) from the base of leaves of plants cultivated in vitro. Barešová et al. (1985) worked with transversally cut rosette leaves (basal, central, apical segment; the central being approx. 7 × 7 mm) from the seventh to tenth pairs of leaves from rosettes grown in vitro.

2.1.2 Micropropagation from Callus

Callus cultures were established from all tested organs of mature plants collected in the field (leaves, roots, stems, axillary buds, flower stalks, calyces, corollae, and anthers) and roots and shoots of seedlings grown from surface sterilized seeds. The calli formed on primary explants were carefully removed and subcultured on fresh medium. With the exception of calli derived from corollae and anthers, all callus strains formed adventive organs in the second subculture on a standard medium with IAA and kinetin, and could be used for plant regeneration.

In long-term cultures (after 1 year of cultivation), calli still required PGR for their growth. Nevertheless, after $1\frac{1}{2}$ years, several strains (6 out of 43) grew without kinetin; three of them did not require auxin (IAA) either. Simultaneously, these three had lost their ability to form adventive organs, but they did grow faster than the organogenic strains and formed more friable calli.

The PGR requirement was also eliminated by cultivation of crown-gall tumors of *C. erythraea*. These tumors were induced by strain C58M of *Agrobacterium tumefaciens* (Dusbábková et al. 1985), but they did not form teratomas and regenerants.

2.1.3 Embryogenesis

Calli were subcultured at least seven times until suspension cultures were established. Irrespective of the initial organ source, the cells of the calli suspended in a liquid medium formed abundant embryos.

Callus or suspension strains differed in their embryogenic capacity. These differences did not seem to be related to the strain origin (root or shoot), but correlated with the medium used for callus cultivation. Some strains consistently lacked any embryogenic capacity (3 out of 15), and responded neither to auxin level nor to illumination. Other strains, although capable of embryogenesis, differed in their response to embryogenic factors (Barešová and Kamínek 1984).

2.2 Sterilization

Seeds of wild plants were thoroughly washed with a detergent and water, and subsequently immersed in 70% (v/v) ethanol for 2 min and in 5% (w/v) sodium hypochlorite for 7 min. The sterilized seeds were washed with sterile distilled water and germinated on filter paper in Knop solution in light. Wild plants were partly cut into organs and sterilized in the same way, but they were immersed in ethanol for 1 min and in 2% sodium hypochlorite for 5 min.

2.3 Media

Tissue cultures, i.e., primary explants, calli and suspensions, and plants, were cultivated on a basal LS medium (Linsmaier and Skoog 1965) supplemented with plant growth regulators (PGR).

2.3.1 Media for Micropropagation from Leaf Segments

Leaf segments regenerated easily, forming shoots, roots, and calli separately or together according to PGR, sucrose concentration, and salts in their medium. Shoot formation was induced by a low auxin/cytokinin ratio. The highest number of shoots formed on the leaf segments cultivated on the medium containing 0.22–0.66 mg/l IAA and 0.2 mg/l kinetin (Barešová et al. 1985) or 0.5–3 mg/l benzyladenine only (Čellárová et al. 1984). The absence of kinetin or the substitution of NAA or 2,4-D for IAA, did not lead to shoot formation at any of the concentrations employed, and only callus and roots were formed in these cases.

Sucrose content of the medium deeply influenced the type of morphogenesis. Lowering the sucrose content (up to 7.5 g/l, with the optimum at 15 g/l) strongly stimulated shoot formation; conversely, increasing the sucrose concentration (up to 70 g/l, with the optimum at 50–60 g/l) stimulated root formation. This effect is especially strong at intermediate IAA concentrations (ca. 2 mg/l). At other IAA concentrations, sucrose content influenced mainly the number of both organ types. The lowest (5 g/l) and highest (90–120 g/l) concentrations tested reduced organogenesis.

Callus formation on leaf segments depended on the nutritional effect of sucrose: an increase of sucrose content synergistically stimulated callus growth with IAA content.

Reduction of inorganic salts in an LS medium (to one-quarter or one-eighth) lowered the number of regenerated shoots (Čellárová et al. 1984). Laurová et al. (1986a) observed that NaCl 100 mg/l stimulated plant regeneration on leaf segments. Other salts (KNO_3 250 mg/l, NaCl 2250 mg/l, $CaCl_2 \cdot 2H_2O$ 1400 mg/l, $MgSO_4 \cdot 7H_2O$ 550 mg/l, supplemented in LS medium), chosen according to average anion and cation content in the East-Slovakia Lowland, had no negative effect on leaf segment regeneration and development of young shoots.

2.3.2 Micropropagation from Callus

Media containing only kinetin failed to induce callogenesis. Media containing only IAA induce callogenesis; optimal concentration range was 2–6 mg/l. Addition of kinetin to IAA-containing medium had a pronounced stimulative effect on callogenesis. Other auxins (NAA, 2,4-D) were capable of inducing higher callogenesis than IAA, and their optimum concentration for callogenesis was lower than that of IAA (NAA 0.19–1.67 mg/l, 2,4-D 0.07–0.66 mg/l). Čellárová et al. (1983) combined 2,4-D (1 mg/l) + BA (0.5 mg/l) + NAA (0.5 mg/l). Laurová et al. (1986b) found that media supplemented with 2,4-D (3 mg/l) best enhanced the growth of calli. Further, the effect of amino acids on callus growth and regeneration was studied (Laurová et al. 1986b). The growth of calli cultivated on LS medium with 2,4-D (3 mg/l) and GLY (1 mg/l) was strongly stimulated by LYS (1 mg/l), whereas other amino acids tested (ASP, GLU, MET, SER, β-ALA, VAL) changed the growth curve only slightly and did not change (or even slightly reduced) the resulting fresh weight. None of them induced organogenesis or changed the callus consistency.

2.3.3 Media for Embryogenic Suspension

For suspension initiation the callus cells cultivated on a solid medium were transferred to the same liquid medium with IAA (2 mg/l) and kinetin (0.21 mg/l) and became embryogenic when cultivated under illumination. Application of 2,4-D

(0.22 mg/l) in both solid and liquid medium suppressed embryogenesis, partly or completely, depending on the cell strain (Barešová and Kamínek 1984).

2.4 Optimal Conditions

2.4.1 Effect of Light

Centaurium erythraea seeds already require light for germination. Derived tissue cultures were cultivated (in both research groups) either in dark or in light (2500 lx, 16 h/d).

Effect of Light on Leaf Segment Regeneration. Regeneration of leaf segments was influenced more by PGR than by illumination. Under illumination, all the morphogenetic responses shifted on the IAA concentration scale toward higher levels (Barešová et al. 1985) and the number of organs formed was significantly lower. Shoots were etiolated and green in darkness and light, respectively (Fig. 2).

Effect of Light on Callus Regeneration. The type of organogenesis and the rate of callus growth were not influenced by illumination. The only difference was the formation of etiolated sprouts in darkness and small green rosettes in about half the amount of that formed in darkness. Further, light induced chlorophyll formation in the callus itself, more on media with IAA than on those with 2,4-D.

Effect of Light on Embryogenesis. In most strains light appeared to be the main embryogenic factor. Since IAA is photolabile, the effect of light may be partly

Fig. 2. Regenerating leaf segments, after 4 weeks, in light (*left*) and in darkness (*right*), LS medium + IAA (0.6 mg/l) + kinetin (0.2 mg/l). (Barešová 1983)

explained in terms of IAA photodestruction, thus making a change of concentration promoting the embryogenesis. Nevertheless, other factors may be involved, because light-induced embryogenesis also occurred (though more rarely) in suspensions cultivated on light-stable 2,4-D as auxin (Barešová and Kamínek 1984).

2.4.2 Effect of Temperature and Other Factors

Morphogenetic response to temperature was not investigated. Čellárová maintained the day/night temperature in the cultivation room at 26/22 °C; we kept the temperature constant at 25 °C. Regarding the other factors the number of shoots regenerated on leaf segments increased by lowering the volume of medium per segment in one dish. In the standard experiment we cultivated three leaves (i.e., nine segments) in a deep Petri dish (3 cm high, 9 cm in diameter) with 30 ml of the medium. Shoot formation was enhanced by lowering the medium volume to 15 ml or by increasing the number of leaves to five. Rhizogenesis and callogenesis were stimulated by increasing the medium volume per leaf. To achieve successful regeneration, the leaf segments must lie on their abaxial side. Lying on the adaxial side, the leaf segments swelled and formed calli over the whole area, while organogenesis was suppressed.

2.5 Regeneration

2.5.1 Regeneration of Plants from Leaf Segments

In general, the formation of either shoots or calli with roots seems to be the two different patterns of organogenesis. Organ types differed in their mode of origin: shoots initiated in the primary explant, whereas the roots formed within the callus tissue and were contingent on its growth.

Further, regeneration of leaf segments had a different time course with respect to the PGR content of the medium. Although both organ types formed at a rather wide range of IAA concentrations, they appeared earlier in the conditions optimal for their formation. Therefore, at the end of the third week of culture, the organogenesis was in agreement with the Skoog and Miller (1957) model: shoots formed on leaves exposed to a medium with low auxin/cytokinin ratio (IAA 0.07–0.66 mg/l, kinetin was kept constant at 0.21 mg/l), roots in the medium with high auxin/cytokinin ratio (IAA range 2–18 mg/l). However, in the 4th week, both organ types began to form in a broader range of IAA concentrations (shoots 0.07–6 mg/l, roots 0.66–18 mg/l), i.e., their concentration ranges began to overlap. The roots and shoots formed on one leaf cut were not anatomically connected.

The most efficient PGR supplement (IAA 0.2–0.6 mg/l + kinetin 0.2 mg/l or BA 1 mg/l) caused about 90–95% of the leaf segments to produce normal shoots. In this case, the average number of strong, well-developed shoots produced, for example per central leaf segment, was 15–20.

The rooting of shoots (Fig. 3) derived from leaf segments is very fast on LS medium without PGR. Higher sucrose concentration stimulated rooting (50–60 g/l). Čellárová et al. (1984) reported stimulation of rhizogenesis by lowering

Fig. 3. Rooting of shoots regenerated on leaf segments, 6 weeks from the start of culture, LS medium + sucrose (50 g/l). (Čellárová 1985)

the nitrogen content of the medium (by a factor of 4). A high light intensity of at least 2500 lx was necessary for good growth of rosettes.

2.5.2 Regeneration of Plants from Callus

The best media for callus growth contained NAA + BA + 2,4-D (Čellárová et al. 1983) or 2,4-D (Laurová et al. 1986b), but only those with IAA + kinetin stimulated the formation of adventive organs. Under optimal conditions (IAA 0.6–2 mg/l + kinetin 0.2 mg/l, in darkness), about 30 shoots regenerated in some callus strains in one 100-ml Erlenmayer flask after 3 weeks. Callus strains grow fast, and with a few exceptions they never cease to form adventive organs. The shoots were etiolated and fragile and their subculture on fresh media to induce rhizogenesis was more difficult than in shoots from leaf explants.

2.5.3 Plants from Embryogenic Suspensions

At the beginning, after 2–3 weeks, a large number of small embryoids (1 mm in diameter) were formed in the suspensions, but only a few of them completed typical embryonic development. Therefore only a small number of large embryos (5 mm) were found in the later stages. Sometimes small calli began to form on the embryo and released free cells, which eventually formed further embryos. Although some embryos were deformed (shoot part prevailed over the root part), normal plants could be obtained after their transfer to LS medium without PGR when cultivated under light intensity of 2500 lx. Under these conditions, well-developed rosettes with an extensive root system formed. Light intensity reduced to 500 lx resulted in growth of weak sprouts, which rooted only to a very limited extent and in addition, vitrification and callus formation were often observed, especially in the presence of IAA. In general, auxins could not be used to enhance the root formation of the shoots or embryos, since they mainly induce callogenesis.

2.6 Transfer to Field

The plants obtained from leaf segments, calli, or embryos were cultivated on LS medium without PGR in light and were used as explant source or transferred to the field (Fig. 4). About 3 days before transfer the flasks were opened and covered with a wet cotton sheet and slowly dried. Then the roots were thoroughly but carefully rinsed in water and the plants transferred to perlite substrate in the greenhouse and eventually to soil (Fig. 5). In the 1st week, high air humidity of at least 70% was necessary to prevent dessication of the plants (Čellárová et al. 1984).

Fig. 4. In vitro-regenerated *Centaurium erythraea* rosette plant, 9 weeks after culture initiation. (Barešová 1983)

Fig. 5. *Centaurium erythraea* plants regenerated in vitro cultured in pots in the greenhouse, 3 months old. (Čellárová 1985)

2.7 Commercial Aspects of Micropropagation

So far, micropropagation of *Centaurium erythraea* has not been exploited commercially. Its prospects in the future are vague; however, the demand for *Herba centaurii* exceeds the supply, and *C. erythraea* as a salt-tolerant plant could be grown under poor soil conditions. Economically, it is preferable to regenerate shoots directly from leaf segments than via embryogenesis because it does not require the technical equipment (roller or horizontal shaker) necessary for the cultivation of suspensions. The plants regenerated from leaf segments are ready for potting earlier than those grown from embryos. In addition, after the first subculture, selection of the best-developed shoots on the leaf cuts is clearer than choosing between well and poorly developed embryos, often in wet clusters, which make the handling difficult.

3 Summary and Conclusions

Centaurium erythraea is suitable for relatively easy tissue culture establishment. Both methods, multiplication of plants in vitro and growing of callus biomass, have been described and tested sufficiently. This would permit their commercial exploitation without much delay, should demand for this plant increase.

Fig. 6. Flowering *Centaurium erythraea* in the field. (Čellárová 1985)

Regeneration of this plant can be manipulated easily by changing the hormonal composition of the medium. Plants under repeated in vitro culture rarely vitrify and can be transferred back into nonsterile conditions (even to saline soil, Laurová et al. 1986a). This method does not eliminate the necessity of field cultivation, but may greatly increase its reliability by eliminating plant propagation as a limiting factor. Moreover, these plants can reach flowering in the first year of cultivation (Fig. 6).

4 Protocol

1. *Best Explant*
   Segments of rosette leaves from plants grown in vitro under light, on LS.

2. *Best Medium for Shoot Induction*
   LS + IAA (0.2–0.6 mg/l) + kinetin (0.2 mg/l) or
   LS + BA (1 mg/l); low sucrose content (15–20 g/l)

3. *Best Medium for Rooting of Shoots*
   LS with a high sucrose concentration (50 g/l)

4. *Optimal Conditions for Shoot Induction*
In dishes with a low medium volume per segment (15 segments placed on 30 ml of medium on abaxial side), 25 °C, 4 weeks in dark, 5th and 6th weeks in light (2500 lx, 16 h)

References

Barešová H (1988) *Centaurium erythraea* Rafn: micropropagation and the production of secoiridoid glucosides. In: Bajaj YPS (ed) Biotechnology in agriculture and forestry, vol 4: Medicinal and aromatic plants I. Springer, Berlin Heidelberg New York Tokyo, pp 350–366

Barešová H, Kamínek M (1984) Light-induced embryogenesis in suspension cultures of *Centaurium erythraea*. In: Novák FJ, Havel L, Doležel J (eds) Plant tissue and cell culture application to crop improvement. Czech Acad Sci, Prague, pp 163–164

Barešová H, Herben T, Kamínek M, Krekule J (1985) Hormonal control of morphogenesis in leaf segments of *Centaurium erythraea*. Biol Plant 27: 286–291

Čellárová E, Repčáková K, Hončariv R (1983) Some physiological aspects of callus induction and growth in *Matricaria chamomilla* L. and *Centaurium erythraea* Rafn. In: Herben T, Šrámek F (eds) Dny rostlinné fyziologie III. Czech Acad Sci, Prague, pp 177–178

Čellárová E, Repčáková K, Hončariv R (1984) Vegetative propagation of some medicinal plants through tissue cultures. In: Novák FJ, Havel L, Doležel J (eds) Plant tissue and cell culture application to crop improvement. Czech Acad Sci, Prague, pp 515–516

Dusbábková J, Nečásek J, Pešina K (1985) Crown gall tumors in *Centaurium*. Biol Plant 27: 463–467

Holub J, Procházka F, Čeřovský J (1979) List of vanished, endemic and endangered higher plant taxa of the flora of the ČSSR. Preslia 51: 213–237

Laurová D, Čellárová E, Hončariv R (1986a) Tolerance of plant tissue of *Centaurium erythraea* Rafn to increased concentration of ions present in soils of the East-Slovakia Lowland. In:Ješko T, Repčák M (eds) Dni rastlinnej fyziológie IV. Slovenská botanická spoločnosť pri SAV, Košice, pp 221–222

Laurová D, Čellárová E, Hončariv R (1986b) Effect of some aminoacids on growth of callus culture of *Centaurium erythraea* Rafn. In:Ješko T, Repčák M (eds) Dni rastlinnej fyziológie IV. Slovenská botanická spoločnosť pri SAV, Košice, pp 223–224

Linsmaier EM, Skoog F (1965) Organic growth requirements of tobacco tissue culture. Physiol Plant 18: 100–127

Macková A (1988) Small-plot cultivation of common centaury (*Centaurium erythraea* Rafn, Gentianaceae) in East-Slovakia. Herba Hung 27: 67–73

McGee PA (1985) Lack of spread of endomycorrhizas of *Centaurium* (Gentianaceae). New Phytol 101: 451–458

Melderis A (1972) *Centaurium* Hill. In: Tutin TG, Heywood VH, Burges NA, Moore DM, Valentine DH, Walters SM, Webb DA (eds) Flora europaea, vol 3. Univ Press, Cambridge, pp 56–59

Meusel H, Jäger E, Weinert E (1978) Vergleichende Chorologie der zentraleuropäischen Flora, vol 2. Fischer, Jena

Oberdorfer E (1970) Pflanzensoziologische Exkursionsflora für Süddeutschland und die angrenzenden Gebiete. Stuttgart

Skoog F, Miller CO (1957) Chemical regulation of growth and organ formation in plant tissues cultured in vitro. Symp Soc Exp Biol 11: 118–131

Van der Sluis WG (1985) Secoiridoids and xanthones in the genus *Centaurium* Hill (Gentianaceae) – a pharmacognostical study. Ph.D. Thesis, Rijksuniv Utrecht

Zeltner L (1970) Recherches de biosystématique sur les genres *Blackstonia* Huds. et *Centaurium* Hill. Bull Soc Neuchâtel Sci Nat 93: 1–164

II.15 Micropropagation of Woolly Foxglove (*Digitalis lanata*) by Shoot Tip Culture

M. Luckner and B. Diettrich[1,2]

1 Introduction

1.1 *Digitalis lanata*, a Biannual Herb from the Mediterranean Countries

D. lanata (family Scrophulariaceae) grows on poor soil in southeast Europe, with its dry and warm climate. It may, however, be cultivated in other parts of the world,

Fig. 1. Flowering *D. lanata* clonal plants

[1] Institute of Pharmaceutical Biology, Martin-Luther-University Halle-Wittenberg, Weinbergweg 15, 0-4050 Halle, FRG
[2] Dedicated to Prof. Dr. E. Reinhard, Tübingen, on the occasion of his 65th birthday.

Biotechnology in Agriculture and Forestry, Vol. 19
High-Tech and Micropropagation III (ed. by Y.P.S. Bajaj)

e.g., in Central and Western Europe (Heeger 1956; Mastenbroek 1980, 1985), and in North and South America (Claus et al. 1970).

D. lanata forms a rosette in the first year of vegetation, starts flowering in May–June of the second year, and dies after setting seeds in the autumn of that year. Before flowering, the plant forms an elongated stem which may reach a height of 120 cm (Fig. 1). The flowers are light brown to violet. The fruit is a capsule which contains a large number of seeds.

1.2 *Digitalis lanata* as a Source of Cardiac Glycosides

Most parts of *D. lanata* contain cardiac glycosides (cardenolides) (Baumgarten 1963; Vogel and Luckner 1981). These compounds are used in medicine as a remedy for heart diseases. The formulas of the most important cardiac glycosides occurring in *D. lanata*, the lanatosides A and C, are shown in Fig. 2.

The main source of the cardenolides are the leaves, which are harvested at the end of the first vegetation period. The leaves are either dried immediately or are fermented before drying to increase the content of digoxin derivatives (Bärisch et al. 1972a, b; Luckner and Diettrich 1979). They are processed for the pharmaceutical industry.

1.3 Conventional Breeding

Important for the economy of the production of *D. lanata* glycosides is the cardenolide content of the leaves. Starting from populations of wild plants, genotypes with a high cardenolide content and an interesting pattern of cardenolides were selected (cf. Weiler and Westekemper 1979). Cultivars with high cardenolide content were obtained by inbreeding of suitable genotypes and crossing of the inbred lines (Ligeti 1959; Tetenyi 1959; Schwerdtfeger 1961; Balint et al. 1980). These cultivars, however, lack stability. Hence, for maintenance and improvement of the characteristic properties, expensive breeding programs have to be repeated each year. The most important steps in these are the selection of the best plants with respect to quality and quantity of cardenolide content, and the crossing by hand-pollination of the elite plants selected (Heinrich et al. 1986). Spatial isolation of the elite plants from less valuable plants and pollination are laborious and cause a high physical strain on the workers involved.

1.4 Impact of Tissue Culture on Breeding

D. lanata cell and tissue cultures have been extensively investigated in the last few years (for reviews, see Luckner and Diettrich 1988, 1989; Rücker 1988). Several papers describe methods for the micropropagation of elite genotypes by shoot tip culture (Erdei et al. 1981, 1982; Dobos et al. 1982; Schöner and Reinhard 1982, 1986; Breuel et al. 1984; Luckner et al. 1984; Springer et al. 1986; Diettrich et al. 1987b, 1990; Kubalakova et al. 1987). The basic features (Fig. 3) will be discussed in

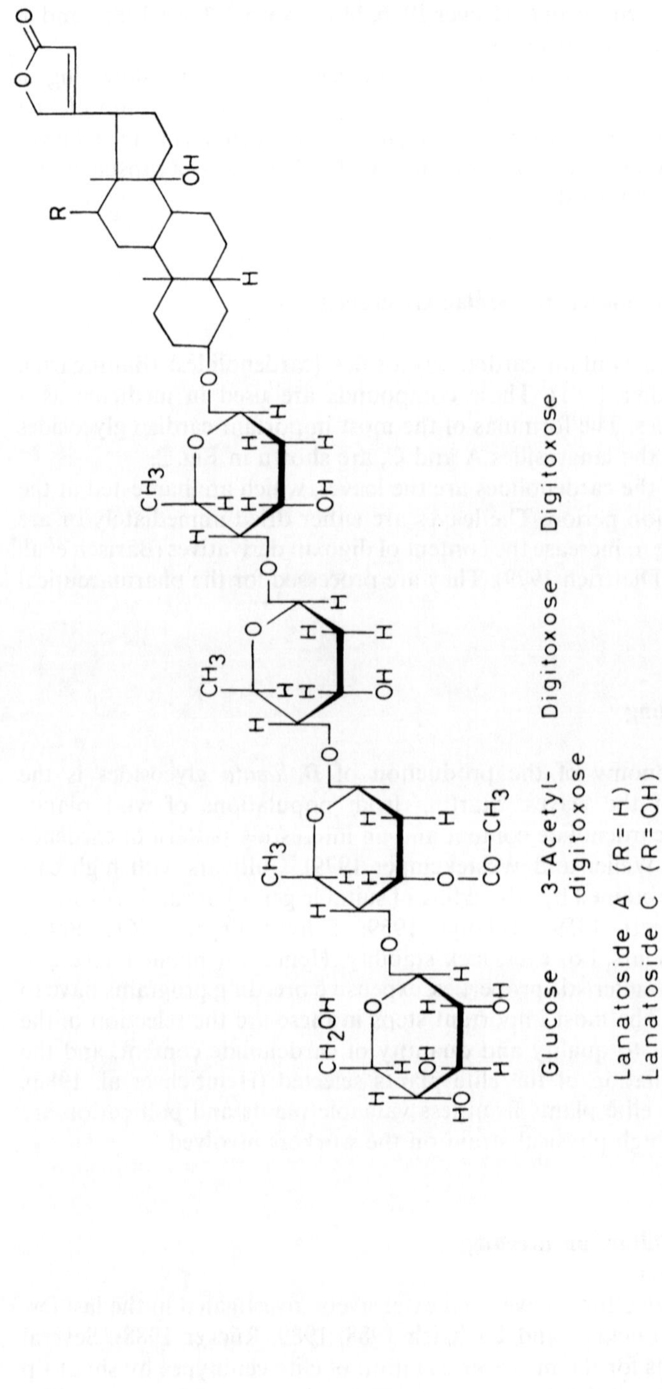

Fig. 2. Structural formula of the main cardenolides of *D. lanata*

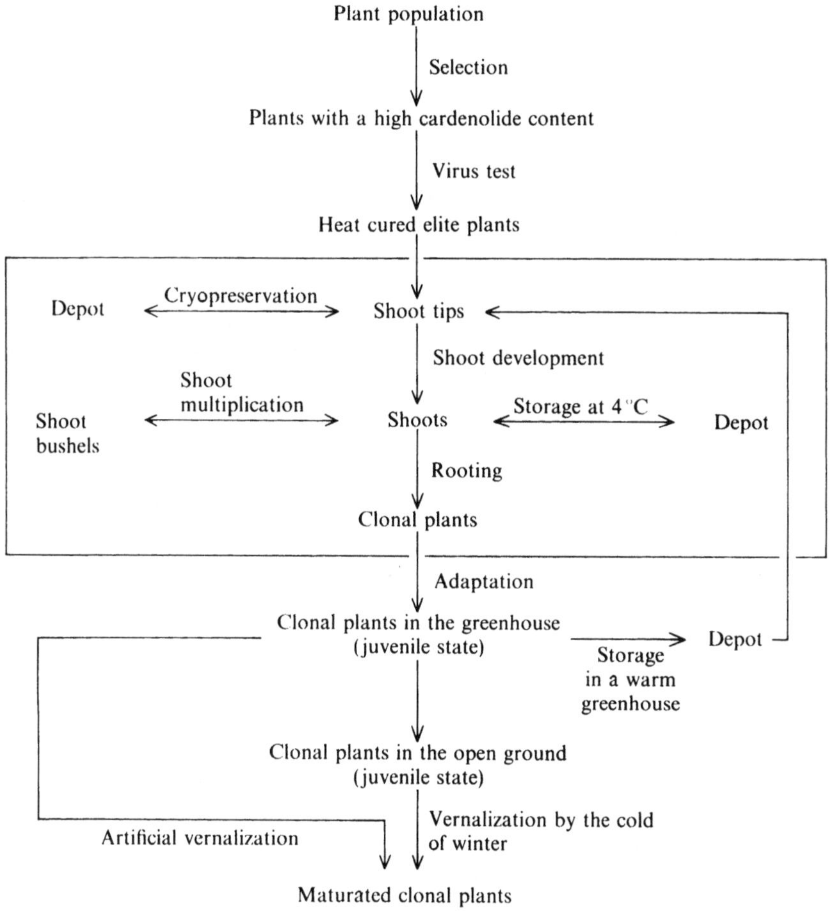

Fig. 3. Main steps in the formation of *D. lanata* plant clones by shoot tip culture. The part in *box* has to be carried out under axenic conditions

the following sections, based on experiments described by Diettrich et al. (1985, 1987b, c, 1990, 1991) and Mertinat et al. (1989).

2 Micropropagation of Valuable Genotypes by Shoot Tip Culture

2.1 The Growth of Juvenile Shoots from Shoot Tips

Healthy, virus-free shoot tips, which may be obtained from heat-cured plants (Oertel and Breuel 1987), are suitable explants for the establishment of shoot cultures. Heat curing is recommended because several authors have described virus infections of *D. lanata* (Schumann 1963a, b; Silva and Pop 1965). Recently, the broad bean wilt virus has been identified (Oertel and Breuel 1987).

Terminal or lateral buds may be excised from the rosette as well as from plants in the second year of vegetation with already elongated stem before or after the beginning of flowering. In routine experiments, the buds were excised from plants in the second year of vegetation before setting flowers. At that time, the chemical evaluation of the plants (based on the determination of the cardenolide content of the leaves, which had been harvested in the autumn of the first year of vegetation) was already finished, and the most valuable mother plants had been selected.

Shoot tips of immature plants from the first year of vegetation, as well as shoot tips of mature plants in the second year of vegetation, formed shoots with a short stem (rosettes) which were in the juvenile state, i.e., were unable to start flowering without vernalization (cf. Sect. 2.5). This phenomenon, first described by Luckner et al. (1984), is the basis of the long-term cultivation of *D. lanata* shoots in vitro. Rejuvenilization is probably brought about by altering the hormone status of the shoot tips by excision from the plant body and by the BA present in the nutrient medium used for shoot development (Zimmerman et al. 1985).

Buds from the middle region of the elongated stem of mature plants were more suitable as explants than buds from the top or ground regions. Buds near the ground were frequently severely infected by microorganisms, and buds from the terminal part produced shoots with an elongated stem, which formed rosettes only after the first or second subculture. Shoot tips of flower buds developed flowers and thus were not suitable as explants. Whole flower buds or flowers formed callus and adventitious shoots at the base where they were cut off. The adventitious shoots developed normally. However, since they were probably derived from the callus, they might have been genetically heterogeneous and were not used for further cultivation. From pieces of the stem or from root tips no shoots were obtained.

The rate and speed of shoot formation depended on the size of the shoot tips used as explants (Table 1). Meristems without leaf primordia (length of the shoot tips ca. 0.3 mm) did not develop into shoots. Instead they formed a green, nodular, durable tissue (see below). Large explants (length $\geq$ 1 mm), which were also used by Erdei et al. (1981) and Kubalakova et al. (1987), developed rapidly into shoots. These, however, were frequently infected with systemic bacteria probably living in the xylem vessels of the mother plants. These infections did not occur if smaller explants were used. Most suitable were shoot tips with a length of about 0.5 mm containing the meristem, the first leaf primordia, and a small piece of subapical tissue (Fig. 4). The use of shoot tips with a length of 0.5–1 mm was also recommended by Schöner and Reinhard (1982, 1986).

The presence of a cytokinin (e.g., BA), but not of an auxin, was necessary for the development of the shoot tips. Without cytokinins the shoot tips formed only small chlorotic shoots which did not develop further. With low cytokinin concentrations green healthy shoots were formed, whereas higher concentrations suppressed the formation of shoots and caused the formation of an increasing amount of a durable, green, nodular tissue. This tissue consisted of meristematic cells as well as parenchyma cells and xylem vessels. It was probably formed by the non-coordinated formation of daughter meristems. The ratio of the weight of shoots to that of the green nodular tissue was optimum with 2.5×10^{-3} mmol BA/l nutrient medium (Nm 1).

Table 1. Influence of shoot tip size on shoot formation. (Diettrich et al. 1990)

Explant (grown on solidified Nm 1)	Axenic cultures (%)	Cultures forming normal shoots (% of the axenic cultures)	Period necessary for shoot formation (weeks)
Length of shoot tips of apical buds (mm)			
0.5	96	50	5-8
1.0	88	88	4-6
3.0[a]	82	93	3-4
5.0[a]	46	91	2-3
Length of shoot tips of lateral buds (mm)			
0.3[b]	99	33	5-8
0.5	95	52	5-8
1.0	85	87	4-6
3.0[a]	63	91	3-4

[a] Frequently containing systemic bacterial infections.
[b] From most explants only a green nodular tissue was formed.

Fig. 4. *D. lanata* shoot tip (meristem with two leaf primordia) immediately after excision

2.2 Propagation of Shoots

The dominance of the apical shoot meristem may be overcome by a small increase in the BA concentration of the nutrient medium (Nm 2). In Nm 2, daughter shoots developed from lateral buds. The number of shoots formed depended on the BA concentration and the intensity of illumination (Tables 2, 3). Most favorable illumination was with light possessing high intensities in the blue and red regions of the spectrum; optimal illumination was with a rather high light intensity and

Table 2. Influence of BA concentration in nutrient medium 2 on daughter shoot formation. (Diettrich et al. 1990)

BA (mmol × 10^{-3}/l)	No. of shoots formed within 4 weeks	Appearance of shoots
0	0	–
5	4.3 ± 2.2	Normal, green
10	9.0 ± 2.6	Normal, green
15	10.8 ± 3.5	Glassy

Table 3. Dependence of daughter shoot formation on intensity of illumination on nutrient medium 2. (Diettrich et al. 1990)

Illumination (W/m^2)[a]	No. of shoots formed within 4 weeks	Appearance of shoots
1	4.7 ± 2.4	Small, yellow
3	5.3 ± 2.2	Normal, yellow
4	5.9 ± 2.6	Normal, green
8	8.5 ± 2.5	Normal, green
10	9.4 ± 3.1	Normal, green

[a] Measured at bench level. Within the vessels illumination was only about 20% of the given values.

a cytokinin concentration of 10×10^{-3} mmol BA/l in the nutrient medium. For the development of shoots from shoot tips (see above), and in contrast to the results of other groups (Erdei et al. 1981; Dobos et al. 1982; Schöner and Reinhard 1982, 1986; Kubalakova 1987), no external auxin was necessary for shoot multiplication.

The rate of shoot multiplication was slightly different with respect to different *D. lanata* genotypes, and deviated from the mean at repeated subcultures even with the same genotype. With suitable genotypes, within 4 weeks, from one shoot a shoot bushel was obtained (Fig. 5) from which about 8–10 daughter shoots could be cut (Fig. 6). Similar propagation rates (formation of 15–30 daughter shoots within 6 weeks) were also reported by Erdei et al. (1981).

Nm 2 was specifically designed for the propagation of *D. lanata* shoots. It is based on the nutrient medium of Murashige and Skoog (1962) and had a simpler composition than the nutrient media used for *D. lanata* shoot cultures by Erdei et al. (1981), Dobos et al. (1982), Schöner and Reinhard (1982, 1986), and Kubalakova (1987), but contains all nutrients necessary for the development of the *D. lanata* shoots in optimum concentrations.

A special problem in the in vitro clonation of *D. lanata* was the formation of vitrified shoots. The leaves of these shoots were much thicker than normal leaves and had a much smaller blade (Fig. 7). They were hard, translucent, and contained a higher percentage of water than normal leaves. The stomata of vitrified shoots did not respond to stimuli influencing the aperture of nonvitrified in vitro leaves (cf. Sect. 2.4). Vitrified shoots were unsuitable for rooting and adaptation to the greenhouse. Vitrification was relatively low if shoots were cultivated on nutrient

Fig. 5. Shoot bushel after a 4-week cultivation period on Nm 2

media solidified with agar. Increased agar concentrations decreased the extent of vitrification. In routine experiments, therefore, solidified media were applied (cf. also Erdei et al. 1981; Dobos et al. 1982). The smooth surface of the agar medium also had the advantage that infections of the shoots by systemic bacteria could be seen easily. These bacteria formed a white or yellowish layer on the surface of the medium near the point where the shoot was cut off and planted into the agar medium.

2.3 Rooting of Shoots

Auxins such as NAA caused the formation of roots at the base of the in vitro-grown shoots. Induction of rooting was brought about either by submerging shoots for 1 h in a concentrated solution of NAA containing a small amount of BA (root induction bath; optimal concentrations 125×10^{-3} mmol NAA/l and 10×10^{-3} mmol BA/l, Table 4), or was originated by cultivation of the shoots for

Fig. 6. Daughter shoot cut from a shoot bushel shown in Fig. 5

2 weeks on solidified Nm 3 supplemented with 2.5×10^{-3} mmol NAA/l and 1×10^{-3} mmol BA/l (cf. also Erdei et al. 1981; Dobos et al. 1982; Schöner and Reinhard 1982, 1986; Kubalakova et al. 1987). The root induction bath, as well as cultivation on the growth factor-supplemented Nm 3, caused a rooting efficiency of 100% (Table 5; cf. also Erdei et al. 1981; Schöner and Reinhard 1986; Kubalakova et al. 1987). Without the addition of auxins, only a small percentage of the shoots formed roots (see also Erdei et al. 1981). This fraction was only slightly increased if auxins (0.5×10^{-3} mmol NAA/l) were added to nutrient medium 2 during shoot propagation. In no case was the growth of callus observed.

The full development of the root system was hindered by auxins, as is known also with other plant species (cf. Maene and Debergh 1983). After treatment with the root induction bath or growth factor-supplemented Nm 3, the *D. lanata* shoots were grown on Nm 3, which was auxin-free. Using the root induction bath, plants with a fully developed root system (cf. Fig. 8; dry weight about 50–100 mg per plant) were obtained already after 4 weeks. The overall time necessary for rooting was thus shortened by at least 2 weeks when compared with the use of the growth factor containing Nm 3. Nevertheless, in routine experiments, the latter variant may be advantageous for rooting since infections are easily spread by the root induction bath.

Root development was optimal in perlite hydroculture. Within 4 weeks, a yellowish hairy, dense root system developed, whereas the roots of plants formed in medium solidified with agar (as recommended, for instance, by Schöner and Reinhard 1982, 1986) were brown and showed less branching. Root development

Fig. 7. Vitrified *D. lanata* shoot formed during cultivation on Nm 2

Table 4. Dependence of root development on hormone content in the root induction bath. (Diettrich et al. 1990)

Hormone content in bath (mmol × 10^{-3}/l) NAA	BA	Roots formed (mg dry weight per plant)[a]	Hormone content in bath (mmol × 10^{-3}/l) NAA	BA	Roots formed (mg dry weight per plant)[a]
0	0	4 ± 1	125	5	48 ± 10
50	0	18 ± 5	125	10	60 ± 10
100	0	27 ± 8	125	15	43 ± 12
125	0	41 ± 8	125	20	39 ± 8
150	0	35 ± 10	125	25	38 ± 8
200	0	36 ± 10	125	40	36 ± 8
250	0	20 ± 6	125	50	34 ± 10

[a] 100% of the shoots formed roots at all experimental variants. The dry weight of roots was determined after a 4-week cultivation period on Nm 3 solidified with perlite.

Table 5. Rooting on medium solidified with agar and on perlite hydroculture. (Diettrich et al. 1990)

Root induction	Root growth	Rooted shoots (%)	Dry weight of roots per plant (mg)	(%)
Root induction bath[a]	Nm 3; perlite[b]	100	58 ± 8	100
Nm 3 with 1×10^{-3} mmol kinetin and 2.5×10^{-3} mmol NAA/l; agar[c]	Nm 3; perlite[c]	100	23 ± 4	40
Nm 3 with 1×10^{-3} mmol kinetin and 2.5×10^{-3} mmol NAA/l; agar[c]	Nm 3; agar[c]	100	13 ± 2	22
Nm 3; agar[c]	Nm 3; agar[c]	20	0.25 ± 0.07	0.4

[a] Treatment: 1 h.
[b] Cultivation period: 4 weeks.
[c] Cultivation period: 2 weeks.

was optimum at 25 ± 2 °C. A reduction of the temperature as described by Schöner and Reinhard (1982, 1986) reduced the growth rate of the roots. At 20 ± 2 °C the root weight was only about 80% of that of the control experiments carried out at 25 ± 2 °C.

Nm 3 was optimized for root formation; its composition was simpler than that used for shoot multiplication (thiamine, inositol, and kinetin were omitted) and it contained less ingredients than the nutrient media described for rooting of *D. lanata* shoots by other authors (Erdei et al. 1981; Dobos et al. 1982; Schöner and Reinhard 1982, 1986).

A decrease in the sucrose content in Nm 3 or the complete omission of sucrose reduced the growth of roots to an extent which negatively influenced the adaptation of the regenerated plants to the conditions of the greenhouse (see below). Rooting of *D. lanata* shoots under nonaxenic conditions in media devoid of a carbon source (as recommended, e.g., by Maene and Debergh 1983 for ornamental plant species), was therefore not favorable. An increased nitrogen content in nutrient medium 3 (60 mmol/l as present in nutrient medium 1 and 2 instead of 30 mmol/l) decreased the amount of roots built. The positive effect of a low nitrogen content on root formation in the medium was also described by Erdei et al. (1981).

2.4 Adaptation of the Rooted Plants to the Greenhouse and Open Ground

A critical step was the adaptation of the in vitro-grown plants to the conditions of the greenhouse and open ground. During this process, the plants cultivated in vitro in an atmosphere nearly saturated with water were confronted with a much lower

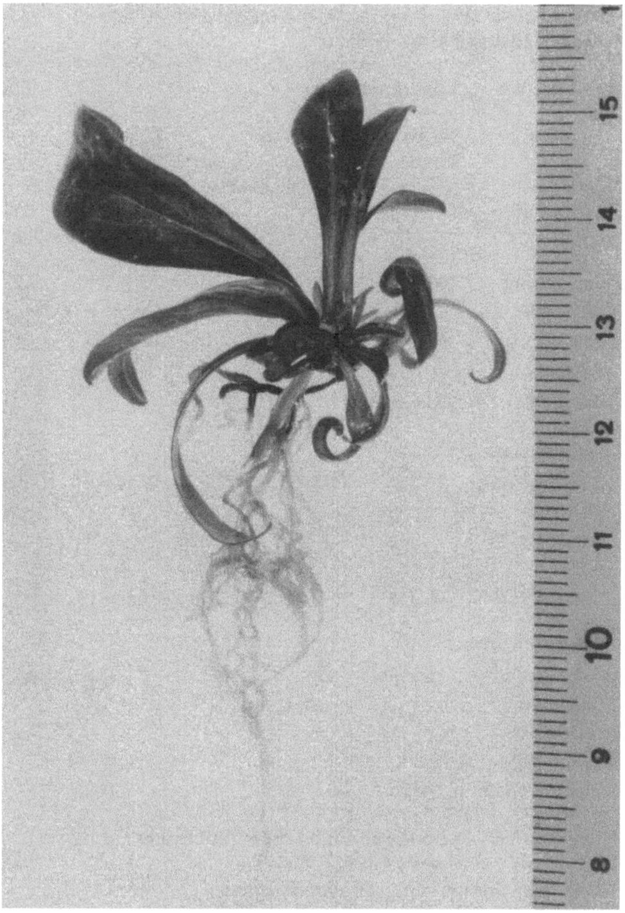

Fig. 8. *D. lanata* clonal plant at the end of the rooting period

relative water content in the gas phase. Of prime importance in the process of adaptation to the lower moisture content was the reduction of transpiration by formation of a water-reluctant cuticle covered by a layer of lipids (cf. Schönherr 1982) and by reduction of the aperture of the stomata (cf. Raschke 1979; Zeiger 1983).

In vitro-grown leaves of *D. lanata* showed extremely high rates of transpiration if kept in a stream of air with low water content (Table 6). The loss of water was about 30 times higher than that of leaves from plants cultivated in the open ground. During rooting and adaptation to greenhouse conditions, the loss of water gradually became less. Plants adapted to the greenhouse with newly formed leaves met the values obtained with plants originating from seed grown in the open ground.

Leaves formed in the open ground or in the greenhouse were covered by a thick cuticle and a layer of epicuticular lipids (Fig. 9), which may be dissolved in $CHCl_3$

Table 6. Relationship between the foliar transpiration rate and the epidermal wax layer of *D. lanata*. (Diettrich et al. 1991)

Leaves derived from	Loss of water[a] by			Epicuticular lipids dissolved in $CHCl_3$[b] (μg/cm² leaf area) of shoots and plants	
	Native leaves	Leaves without the epicuticular lipids[b]	Leaves treated with paraffins[c]	Treated normally	Water stressed
Shoots during proliferation	100 ± 9	- -	59 ± 7	2 ± 1	-
Shoots during rooting	66 ± 8	82 ± 10	44 ± 6	5 ± 1	-
Plants after adaptation to the greenhouse					
– Leaves formed in vitro (= persistent leaves)	37 ± 8	60 ± 9	25 ± 4	8 ± 1	- -
– Leaves formed ex vitro (= new leaves)	9 ± 6	30 ± 6	10 ± 2	11 ± 1	14 ± 2
Seed-derived plants grown in the greenhouse	3 ± 1	23 ± 3	2 ± 1	17 ± 2	20 ± 2
Seedlings grown in the open ground	3 ± 1	25 ± 5	2 ± 1	22 ± 2	26 ± 3

[a] Relative values; native leaves during shoot proliferation = 100%.
[b] Leaves were submerged in $CHCl_3$ for 60 s.
[c] Leaves were dipped into a paraffin emulsion.

(Table 6). In contrast, during the stage of shoot propagation in vitro, the cuticle of the leaves was thin and nearly no wax was obtained. During rooting, the thickness of the cuticle and the amount of wax increased. The cuticle and wax layer were, however, still smaller than those of the leaves of greenhouse plants or plants grown in the open ground.

There are several reports in the literature describing the deposition of increased amounts of surface lipids if plants undergo water stress (cf. Wardle et al. 1983). The *D. lanata* plants grown in vitro, however, showed only a slightly higher lipid amount if scarcely watered for several weeks (Table 6). Transpiration could be reduced by covering the leaves with a layer of paraffins, e.g., by dipping the leaves in a wax emulsion as is used in the preparation of fruits for long-term storage (cf. Sutter and Hutzell 1984, Table 6). However, the transpiration of the in vitro-grown leaves after treatment with paraffins was still much higher than transpiration of the untreated leaves of plants in the greenhouse.

The stomata of leaves from shoots grown in vitro were physiologically abnormal. They were completely open in the light as well as in the dark (Fig. 10A). In contrast, the stomata of new leaves formed in the greenhouse showed normal

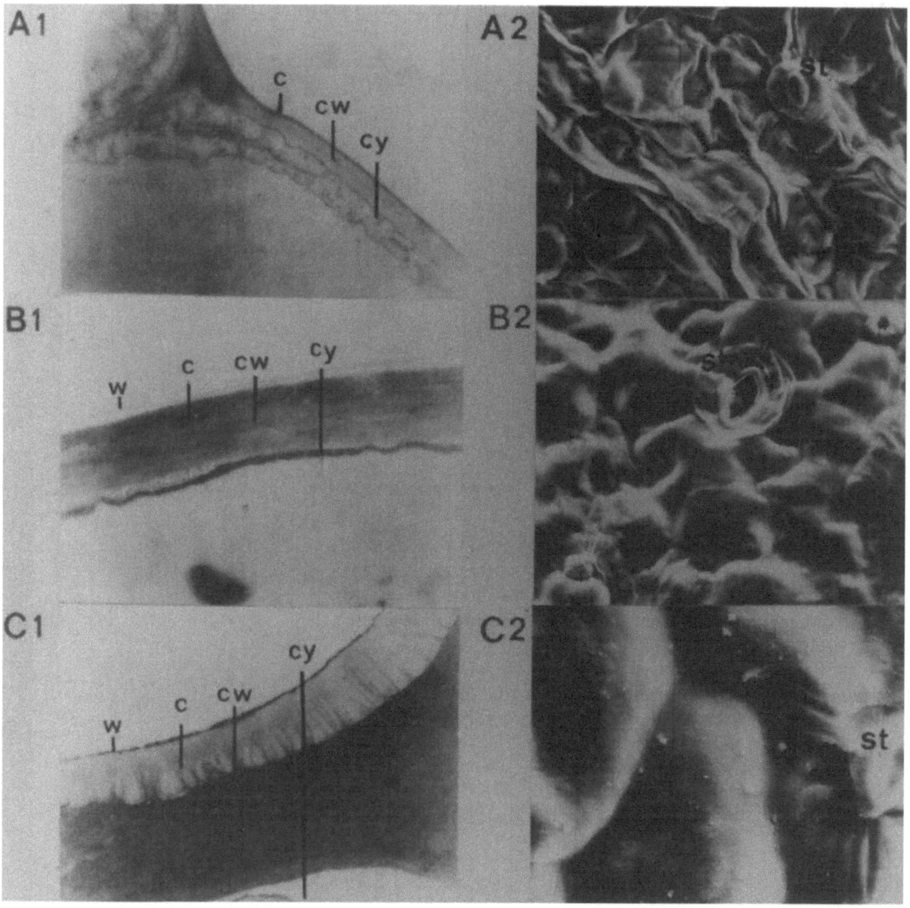

Fig. 9. Cuticle and epicuticular wax layer covering the upper surface of *D. lanata* leaves. A_1, B_1, C_1. Cross-sections through the upper epidermis of leaves; A_2, B_2, C_2 scanning electron micrographs of the upper surface of leaves. **A** Leaf from the stage of shoot propagation; **B** leaf from a plant at the end of the rooting stage; **C** leaf from a plant cultivated in the open ground; **w** wax layer; **c** cuticle; **cw** cell wall; **cy** cytoplasm

physiological reactions. They were partially open if the leaves were illuminated and were closed in the dark (Fig. 10B).

It is of interest that the aperture of the stomata of shoots at the stage of daughter shoot formation was larger than in shoots at the rooting stage, though both processes were carried out in an atmosphere nearly saturated with water (Table 7). This difference was shown to be caused by the high concentrations of BA and K^+ ions present in Nm 2 used for shoot multiplication which both enlarged the aperture of the stomata (Table 8, cf. Göring and Mardanov 1976; Zeiger 1983). Increased levels of CO_2 and abscisic acid caused closure of the physiologically active stomata of greenhouse plants even if the leaves were kept in the light, but originated only a partial decrease in the aperture of stomata of in vitro-grown leaves (Tables 7, 8).

Table 7. Aperture of the stomata of greenhouse and in vitro-grown *D. lanata* plants. (Diettrich et al. 1991)

Origin of the leaves	Aperture (%)[a] Without administration of CO_2[b]		After administration of CO_2[b]	
	In the light	In the dark	In the light	In the dark
Shoots during proliferation	100 ± 9	100 ± 10	100 ± 9	100 ± 10
Shoots during rooting stage	63 ± 6	59 ± 6	–	–
Plants grown in vitro	50 ± 5	40 ± 5	40 ± 4	30 ± 5
Seed-derived plants grown in the greenhouse	40 ± 5	6 ± 3	0 ± 2	4 ± 2

[a] The aperture of fully open stomata (Fig. 10A) was assigned 100%, that of completely closed stomata (Fig. 10B) 0%.
[b] The gas phase was either air or air enriched with 0.5% CO_2.

Table 8. Aperture of the stomata of *D. lanata* shoots grown in vitro on different nutrient media after 12 h in the dark. (Diettrich et al. 1991)

Compounds added	Concentration (mmol/l)	Stomatal aperture (%); period of incubation 2 h	1 week
Control[a]		70 ± 5	75 ± 6
K^+	0	50 ± 4	55 ± 4
	6	50 ± 3	55 ± 4
	12	70 ± 4	70 ± 5
	24	70 ± 5	75 ± 6
	48	85 ± 8	80 ± 5
Al^{3+}	0.2	62 ± 6	55 ± 4
	1	50 ± 6	45 ± 5
	5	45 ± 8	40 ± 4
Benzyladenine	2.5×10^{-3}	80 ± 7	85 ± 8
	5×10^{-3}	85 ± 8	95 ± 6
	25×10^{-3}	85 ± 5	100 ± 7
Kinetin	2.5×10^{-3}	90 ± 9	90 ± 9
	5×10^{-3}	85 ± 6	85 ± 11
	25×10^{-3}	95 ± 8	100 ± 7
ABA	2×10^{-3}	70 ± 5	35 ± 6
	4×10^{-3}	55 ± 9	30 ± 2
	10×10^{-3}	25 ± 3	20 ± 4

[a] Nutrient medium 2 without BA.

Microscope examination showed that the guard cells of *D. lanata* stomata were rich in starch granules. It may thus be expected that malate, which is formed as a product of starch degradation, acts as counter anion for the K^+ ions in *D. lanata* guard cells (Schnabl et al. 1982; Schnabl 1983). Al^{3+} ions, which inhibit the degradation of starch to malate (Schnabl 1976) and are used commercially in the preservation of cut flowers, decreased the aperture of the stomata of the in vitro-grown *D. lanata* shoots (Table 8) as well as of field plants. The auxins indolylacetic acid and NAA, which may be used in rooting of *D. lanata* shoots (cf. Sect. 2.3), showed no influence on the stomata aperture of in vitro-grown leaves.

In summary, these results demonstrate that the stomata of the in vitro-grown shoots of *D. lanata* are able to respond to several of the stimuli which influence the aperture of the physiologically active stomata of *D. lanata* plants. However, under all conditions, the stomata were only partially closed, the reason for this difference probably being an abnormally rigid cell wall (Ziv et al. 1987; Koshuchowa 1988) formed under the artificial conditions of in vitro cultivation. This cell wall prevents proper guard cell movements and closure of the stomata even if the turgor is lost completely, as was shown by plasmolysis of the guard cells (Fig. 10C).

The abnormality of the wall of the guard cells was not reversible during further cultivation. Hence, the leaves formed in vitro were unable to develop completely normal physiological functions. This is in agreement with the finding that they died about 2 months after the beginning of cultivation in the greenhouse and were replaced by leaves developing during the process of adaptation, i.e., at already reduced water content in the atmosphere, or in the greenhouse itself.

In contrast to the adaptation to reduced moisture, the clonal plants easily became used to nonaxenic conditions. During the period of adaptation to the conditions of the greenhouse, the plants were therefore not protected against bacteria and fungi. For adaptation, groups of plants were covered with translucent

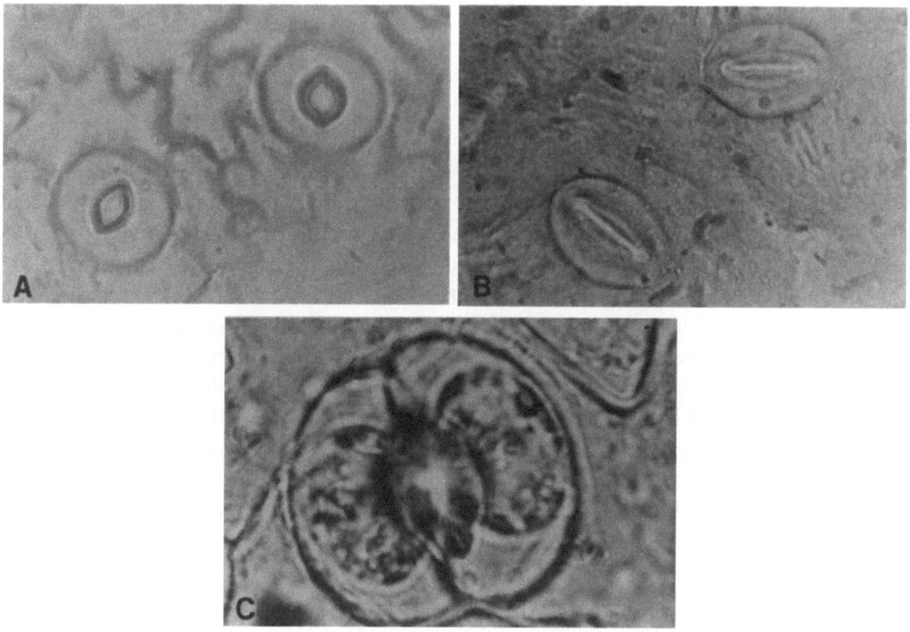

Fig. 10. A, B Photographs of replicas of the lower surface of leaves of an in vitro-grown shoot with fully open stomata and of plants cultivated in the open ground with completely closed stomata, respectively; **C** photograph of a stoma of an in vitro-grown leaf after plasmolysis

Fig. 11. *D. lanata* clonal plants in the greenhouse

plastic foil and kept for 1 week at about 20 °C and permanent illumination (intensity 10 W/m^2) in a climatized chamber. Then the foil was partially removed for increasing periods. After 3 weeks the foil was withdrawn completely and the plants were transferred to the greenhouse (Fig. 11) and later to the open ground.

The success of adaptation of the plants to greenhouse conditions depended on the ratio of roots to leaves. Important was also the physiological state of the plants. Plants with dark green leaves, length < 4 cm, and with a high dry mass of roots were most suitable. Higher temperatures with largely deviating extremes (25 ± 5 °C, as occurring in the greenhouse) as well as low illumination reduced the rate of adapted plants. If root development was carried out using perlite hydroculture, the rate of survival during adaptation was 90–100%. Similar rates were also described by Erdei et al. (1981) and Schöner and Reinhard (1986).

2.5 Vernalization of Clonal Plants

Flowering could be induced in *D. lanata* clonal plants by cultivation at reduced temperature. In this respect they resembled *D. purpurea* plants (Lang 1965; Halevy 1985). The minimum age for successful vernalization was 3 months after transfer of the in vitro-grown plants to the greenhouse (Table 9). In agreement with results obtained on other plant species (Hänsel 1953; Lang 1965; Pierik 1967), younger individuals did not become vernalized under all experimental conditions tested.

Cultivation of the plants at 1–2 or 3–5 °C resulted in a similar degree of vernalization. The time period necessary in both cases was relatively long (at least 3 months with younger plants and 2 months with older plants). Light intensity (between 4.5 and 19.0 W/m^2) or photoperiod (8 h illumination per day or continuous illumination) had no influence on vernalization. Lower temperatures for shorter periods (5 days at −4 °C after 1 month at 1 °C) did not cause flowering.

The artificially vernalized plants formed one or two inflorescences with 30–40 flowers each. The flowers showed normal morphology and contained viable pollen and ovules. They developed seeds that germinated at high rates (> 90%). Pollen and ovules were used successfully in breeding experiments. There was no difference between the plants vernalized in the laboratory and plants cold-treated "naturally" in the open ground during the winter.

Treatment with GA_3 (application of up to 100 μg per plant by spraying the surface of the leaves with an aqueous solution of GA_3) caused elongation of leaves and stems, but no induction of flowering. This is in contrast to the results of

Table 9. Vernalization of *D. lanata* clonal plants. (Mertinat et al. 1989)

Age (months)[a]	Leaves (no.)	Size of the leaves		Period of cold treatment (months)[c]	Flowering plants (%)[d]
		Length[b] (cm)	Width[b] (cm)		
1.5	4–9	3.5–7	0.5–1.2	1	0
				2	0
				3	0
2.5	7–10	6–11	0.9–1.7	1	0
				2	0
				3	0
				4	0
3	14–23	10–20	1.4–2.4	1	0[e]
				2	0[e]
				3	60
				4	80
				8	100
4	15–35	20–26	1.8–2.8	1	0[e]
				2	10
				3	80
				4	90
				8	100
7	20–40	15–20	2.1–3.4	2	100
				3	100

[a] After transfer of the in vitro-grown clonal plants to the greenhouse.
[b] Mean of the three largest leaves.
[c] Temperature 3–5 C, illumination 10 W/m^2 8 h per day.
[d] n = 10.
[e] Plants produced elongated leaves, but showed no formation of elongated axes or flowers.

Wittwer and Bukovac (1957) with *D. purpurea*. GA_3 also did not reduce the period of cold treatment necessary for flower induction. During further cultivation without GA_3 in the greenhouse, the gibberellin-treated plants formed a new rosette at the top of the elongated stem.

Flowering could not be induced in vitro (in axenic shoots or plants) under various conditions, e.g., increased sugar content of the medium, different concentrations of auxins and cytokinins, administration of GA_3, benzoic acid, orotic acid, and adenine. Even a cultivation period of up to 2 years in vitro at 4 °C (see Sect. 2.6) did not result in flowering.

2.6 Establishment of a Genotype Bank

2.6.1 Storage of Mother Plants in the Greenhouse

Important for the use of cloning in *D. lanata* breeding was labor-saving maintenance of valuable genotypes for long periods. For medium-term storage (several years) the maintenance of shoots at reduced metabolism or the cultivation of regenerated (juvenile) plants under conditions which prevented maturation was most suitable.

Continuous growth in a warm greenhouse kept *D. lanata* plants in the vegetative state for indefinite periods. After 5 years, about 40% of the plants were still alive, most of them possessing several side axes. Every year in May–June a few plants formed flowering shoots, the individual clones responding differently, showing 0–20% flowering individuals. Unexpectedly, the flowering plants did not die after setting fruits, as is obligatory under normal physiological conditions. They continued to grow further at the nonflowering shoots.

Although the stem of the juvenile *D. lanata* plants kept in the greenhouse was short, there was a slow increase in length during the formation of new leaves. Because old dead leaves were cut off to reduce infections, after some time the plants assumed a "palmlike" shape. The buds present at the points of insertion of the dead leaves were suitable for explants. From these, about 30% axenic shoots were obtained. In addition, the buds of side axes still carrying leaves could be used as explants without destroying the mother plant. In all cases the shoots obtained in vitro showed no abnormalities.

2.6.2 Storage of Axenic Shoots at Reduced Metabolism

As an alternative, valuable genotypes were stored as axenic shoots. Most important for prolongation of the storage periods was the retardation of metabolism by increasing the osmolality of the nutrient medium and the decrease of the storage temperature (Table 10). Under optimal conditions (200 mmol sucrose/l, dim illumination, and 4 °C) shoot cultures could be kept alive for 2 years. Shoots stored in the dark turned brown and died much earlier than the dimly illuminated shoots. An increase in illumination intensity also shortened the period of survival.

Shoots were best stored in test tubes because of their small volume and opening. In these vessels the rate of infection during the 2-year storage period was less than

Table 10. Storage of axenic shoots at reduced metabolism. (Diettrich et al. 1990)

Storage regime	Half-life of the stored shoots (weeks)[a]
Nm 1 with 10×10^{-3} mmol BA/l (Nm 2), 24 °C	10
Nm 1 – BA, 24 °C	20
Nm 1 – BA, half-concentrated, with 100 mmol sucrose, 24 °C	24
Nm 1 – BA, half-concentrated, with 200 mmol sucrose, 24 °C[b]	30
Nm 1 – BA, half-concentrated, with 200 mmol sucrose, 4 °C	100

[a] Period after which 50% of the shoots died.
[b] An increase in the sucrose concentration to 300 and 400 mmol/l caused the formation of abnormal shoots from which the regeneration of normal shoots was difficult.

5%. The surviving shoots reached the standard rate of propagation after one or two subcultures. There were no significant differences between the shoot clones tested relative to survival during storage or subsequent recultivation. In no case were deviations from normal shoot and plant development observed. In contrast to the storage of clonal plants in the greenhouse, the storage of axenic shoots had the advantage that the shoots could be rapidly reintegrated into the process of shoot propagation.

2.6.3 Cryopreservation of Shoot Tips

For the long-term storage of large numbers of genotypes, however, the methods given above proved to be too laborious. The establishment of a gene bank in which shoot tips of the desired *D. lanata* genotypes were stored in the deeply frozen state, i.e., in liquid nitrogen at – 196 °C, was carried out instead. In the deeply frozen (cryopreserved) state any metabolic activity was inhibited and the genetic material was nearly completely stable.

Tips of cold-hardened *D. lanata* shoots (8 weeks at 4 °C) were the most suitable explants for cryopreservation. These shoot tips were treated with a suitable cryoprotectant solution. The best results were obtained with DMSO (2 mol/l, incubation period 2 h; or 1 mol/l, incubation period 18 h), as well as glycerol-sucrose (1 mol/l, incubation period 18 h) (cf. Diettrich et al. 1982, 1985, 1986, 1987a).

For cooling, the shoot tips were mounted on filter paper bridges and then placed together with a small amount of cryoprotectant into plastic ampules, which were sealed to prevent contamination. The shoot tips showed optimal survival and shoot regeneration if they were cooled slowly (0.25–0.5 K/min) to – 40 °C. Artificial crystallization by dipping the ampules at – 5 °C for about 1 s in liquid nitrogen increased survival rates and to some extent also shoot regeneration. With this method > 80% of the shoot tips survived cryopreservation and > 25% regenerated shoots.

The ampules stored at − 196 °C were rapidly rewarmed in a water bath at 40 °C before use. The rewarmed shoot tips were cultivated on solidified Nm 1, the concentration of the cryoprotectant decreasing slowly by diffusion into the medium. During the first 14 days of culture, most of the surviving shoot tips formed a mass of durable, green nodular tissue. From a certain fraction of this material bushels of shoots developed. In contrast, from unfrozen shoot tips (and in rare cases also from frozen shoot tips) one unbranched shoot was formed. The reason for this difference is probably the destruction of the apical dome during cooling and/or rewarming and its step-by-step regeneration in the surviving shoot-forming tips (Withers 1983).

There was no decrease in survival rates or shoot regeneration even if the shoot tips were kept at − 196 °C for several years.

2.7 Summary and Conclusions

The results mentioned above show that the propagation of *D. lanata* by shoot tip culture made possible (1) the rapid multiplication of elite plants with the formation of plant clones, and (2) long-term cultivation of plants which under normal growth conditions would have died at the end of the second vegetation period.

Optimal conditions were established for the regeneration of shoots from shoot tips, for daughter shoot formation, and rooting, as well as for the adaptation of the regenerated plants to the open ground. Gene banks of valuable clones were built by keeping shoots at 4 °C on media with high sucrose concentration (maximum period of storage 2 years), by growing clonal plants in the greenhouse at temperatures preventing the induction of flowering, or by cryopreservation of shoot tips (storage at − 196 °C). The clonal plants were in the juvenile state even if they were derived from flowering mother plants. Flowering could be induced by vernalization or by keeping the clonal plants in the open ground during the winter.

In field tests the clonal plants showed normal development, set flowers in the 2nd year of vegetation, and formed fertile seeds. The variability of the morphological and chemical traits within the clones was reduced if compared with a population of plants originating from seeds (Luckner et al. 1984; cf. also Schöner and Reinhard 1986; Kubalakova et al. 1987). There was no decrease in the homogeneity of clonal plants produced during a period of 7 years, the investigated shoot clones being kept in vitro (see also Schöner and Reinhard 1986; Kubalakova et al. 1987). In several cases, however, the mean cardenolide content of the clonal plants was somewhat lower than in the mother plants (Breuel, pers. commun.). This indicates that the high cardenolide content of the mother plants was only in part genetically determined and reflected also a positive epigenetic deviation.

Daughter plants obtained by crossing of different clonal lines contained at least the same cardenolide contents as the plants of a standard clone (Breuel et al. 1988). This result showed that free pollination between members of suitable clones allowed reproduction of the high cardenolide content characteristic for the cultivars under investigation, and demonstrated that clonation of elite mother plants by shoot tip culture is indeed a valuable step in the process of *D. lanata* breeding.

3 Protocol

1. Nutrient Media (concentrations in mmol/l)

   Nm 1: sucrose 100, KNO_3 20, NH_4NO_3 20, KH_2PO_4 5, $CaCl_2$ 3, $MgSO_4$ 1.5, $FeSO_4$ 0.2, H_3BO_3 0.1, $MnSO_4$ 0.1, Na_2 EDTA 0.1, $ZnSO_4$ 0.03, KI 5×10^{-3}, $NaMoO_4$ 1×10^{-3}, myo-inositol 0.5, thiamine 1.5×10^{-3}, BA 2.5×10^{-3}, agar 12 g.

   Nm 2: Nm 1 with BA 10×10^{-3} and agar 8 g.

   Nm 3: Nm 1 without myo-inositol, thiamine, BA and agar; KNO_3 10, NH_4NO_3 10.

   Nm 4: Nm 1 diluted with water 1:1, sucrose 200, agar 8 g.

2. Plant Material. Plants with elongated stem in the second year of vegetation or rosettes in the first year of vegetation were used in the establishment of shoot tips cultures.

3. Preparation of Explants. Buds were sterilized with ethanol (70%), tosylchloramide sodium/HCl, and again with ethanol (70%). Then the shoot tips (length 0.5 mm), including the meristem and the first leaf primordia, were excised.

4. Cultivation of Explants. Four shoot tips each were cultivated on 15 ml solidified Nm 1 in a 50-ml beaker for 4 weeks.

5. Propagation of Shoots. Shoots were cultivated individually in 100-ml beakers containing 25 ml solidified Nm 2 for 4 weeks.

6. Root Induction Bath. Shoots were submerged for 1 h in a solution containing salts, sucrose, 150 $\times 10^{-3}$ mol NAA/l, and 10×10^{-3} mol BA/l.

   Root Induction by Auxin-Containing Nutrient Medium. Shoots were cultivated for 2 weeks in 100-ml beakers on 15 ml Nm 3 solidified with agar (8 g/l) containing 2.5×10^{-3} mmol NAA/l and 1×10^{-3} mol kinetin/l.

7. Root Development. After root induction, the shoots were cultivated for 4 weeks individually in 100-ml beakers on 15 ml Nm 3 without hormones, solidified with 4 g perlite.

8. Adaptation to the Greenhouse. In vitro-grown plants were covered with plastic foil and kept under nonaxenic conditions. Adaptation to reduced moisture was brought about by partial removal of the foil for increasing periods of time.

9. Storage of Shoots at 4 °C. Shoots were transferred into test tubes containing 3 ml solidified Nm 4 and kept at 4 °C under illumination with dim light.

10. Cryopreservation of Shoot Tips. Cryopreservation included the following steps: (a) hardening of shoots by cultivation at 4 °C for 8 weeks, (b) treatment of the explanted shoot tips with cryoprotectants, e.g., 2 mol DMSO/l for 2 h, (c) slow cooling (ca. 0.5 K/min) of the shoot tips to − 40 °C using a programable freezer, (d) storage of the shoot tips at − 196 °C in liquid nitrogen, (e) rapid rewarming of the ampules containing the shoot tips with water at 40 °C, and (f) recultivation of the shoot tips on the surface of solidified Nm 1.

11. Vernalization of Clonal Plants. Plants (age 3 months) were kept for 3 months at 3–5 °C (illumination 5–9 W/m^2). Then the cold-treated plants were cultivated for development of the inflorescences in the greenhouse at photoperiods $>$ 16 h/day.

References

Bärisch H, Fischer F, Pötter H, Schmidt HJ (1972a) Zur Umwandlung von Digitoxigeninglykosiden in Digoxin. Pharmazie 27: 415–416

Bärisch H, Fischer F, Schmidt HJ, Pötter H (1972b) Über den Einfluß von Luft auf fermentative Vorgänge in *Digitalis lanata* Erh. Pharmazie 25: 415

Balint A, Nyaradi-Szabady J, Tetenyi P (1980) Untersuchung über die Heterosis an *Digitalis lanata* Ehrh. Acta Hortic 96: 175–184

Baumgarten G (1963) Die herzwirksamen Glykoside. Thieme, Leipzig

Breuel K, Poetter H, Luckner M, Diettrich B, Springer M, Oertel C (1984) Verfahren zur virusfreien vegetativen Vermehrung und Erhaltung von Digitalis-Hochleistungspflanzen. DDR-Patent DD 207 731

Breuel K, Hess A, Feist M (1988) Zur Leistung generativer Nachkommen von *Digitalis-lanata*-Klonen. In: Potsdamer Forschungen, Reihe B, Heft 57: Wissenschaftliche Vortragstagung. Verbesserung von Kulturpflanzen durch biotechnologische Verfahren. Tagungsberichte. Pädagogische Hochschule Potsdam, Potsdam, pp 187–189

Claus EP, Tyler VE, Brady LR (1970) Pharmacognosy. Lea and Febiger, Philadelphia

Diettrich B, Popov AS, Pfeiffer B, Neumann D, Butenko RG, Luckner M (1982) Cryopreservation of *Digitalis lanata* cell cultures. Planta Med 46: 82–87

Diettrich B, Haack U, Popov AS, Butenko RG, Luckner M (1985) Long-term storage in liquid nitrogen of an embryogenic cell strain of *Digitalis lanata*. Biochem Physiol Pflanz 180: 33–43

Diettrich B, Haack U, Luckner M (1986) Cryopreservation of *Digitalis lanata* cells grown in vitro. Precultivation and Recultivation. J Plant Physiol 126: 63–73

Diettrich B, Haack U, Thom F, Matthes G, Luckner M (1987a) Influence of intracellular ice formation on the survival of *Digitalis lanata* cells grown in vitro. Cryo-Lett 8: 98–107

Diettrich B, Mertinat H, Luckner M, Breuel K, Dauth C (1987b) Die Gewinnung von Klonlinien aus *Digitalis-lanata*-Hochleistungspflanzen durch Sproßspitzenkultur. Wiss Z Univ Halle 36 M, 5: 90–102

Diettrich B, Wolf T, Bormann A, Popov AS, Butenko RG, Luckner M (1987c) Cryopreservation of *Digitalis lanata* shoot tips. Planta Med 53: 359–363

Diettrich B, Mertinat H, Luckner M (1990) Formation of *Digitalis lanata* clone lines by shoot tip culture. Planta Med 56: 53–58

Diettrich B, Mertinat H, Luckner M (1992) Reduction of water loss during ex vitro acclimatization of micropropagated *Digitalis lanata* clone plants. Biochem Physiol Pflanz 188: 23–31

Dobos E, Tetenyi P, Molnar G, Bernath J (1982) A method developed for the vegetative propagation of *Digitalis lanata* Ehrh. by tissue culture. Herba Hung 21: 49–57

Erdei I, Kiss Z, Maliga P (1981) Rapid clonal multiplication of *Digitalis lanata* in tissue culture. Plant Cell Rep 1: 34–35

Erdei I, Dobos E, Bernath J, Alörincz C, Molnar G, Nyaradi J, Reichert P, Tetenyi P, Maliga P (1982) Verfahren zur Gewinnung von Vermehrungsmaterial (Setzlingen) des Fingerhuts (*Digitalis lanata* Ehrh.) durch vegetative Mikrovermehrung in Gewebekulturen und Verwendung des nach diesem erhaltenen an die Freilandbedingungen angepaßten Vermehrungsmaterials (Setzlinge). BRD-Patent DE 3217857 A1

Göring H, Mardanov AA (1976) Beziehungen zwischen dem K^+/Ca^{++}-Verhältnis im Gewebe und der Wirkung von Zytokininen in höheren Pflanzen. Biol Rundsch 14: 177–189

Hänsel H (1953) Vernalization (Jarovisation, Kältestimmung). Z Pflanzenzücht 32: 233–274

Halevy HA (ed) (1985) CRC Handbook of Flowering, vol 2. CRC, Boca Raton

Heeger EF (1956) Handbuch des Arznei- und Gewürzpflanzenbaues. Dtsch Bauernverlag, Berlin

Heinrich C, Breuel K, Hahn H-J (1986) Die Aufwand-Nutzen-Analyse – eine Methode zur Bewertung der Effektivität von Vorhaben im Bereich Forschung und Entwicklung. Pharmazie 41: 409–414

Koshuchowa S (1988) Cytologische und histologische Veränderungen bei der Vitrifikation von Pflanzen unter Bedingungen der In-vitro-Kultur. In: Potsdamer Forschungen, Reihe B, Heft 57: Wissenschaftliche Vortragstagung. Verbesserung von Kulturpflanzen durch biotechnologische Verfahren. Tagungsberichte, Pädagogische Hochschule Potsdam, Potsdam, pp 195–196

Kubalakova M, Spitzova I, Novak FJ (1987) Stability of lanatoside C content in the in vitro propagated *Digitalis lanata* clones. Biol Plant (Praha) 29: 7–9

Lang A (1965) Physiology of flower initiation. In: Ruhland W (ed) Handbuch der Pflanzenphysiologie, vol. XV/1, Springer, Berlin Heidelberg New York, pp 1380–1536

Ligeti G (1959) Die Vererbung des Glykosidgehalts in den Blättern bei *Digitalis lanata* Ehrh. Pharmazie 14: 162–166

Luckner M, Diettrich B (1979) Herzglykoside – Struktur, Gewinnung, Verwendung in der Therapie. Pharmazie 34: 477–481

Luckner M, Diettrich B (1988) Cardenolides. In: Constabel F, Vasil IK (eds) Cell culture and somatic cell genetics of plants, vol 5. Phytochemicals in plant cell cultures. Academic Press, San Diego, pp 193–212

Luckner M, Diettrich B (1989) The integration of cardenolide biosynthesis in the differentiation program "somatic embryogenesis" of *Digitalis lanata*. In: Kurz WGW (ed) Primary and secondary metabolism of plant cell cultures II. Springer, Berlin Heidelberg New York, pp 117–124

Luckner M, Diettrich B, Springer M, Breuel K, Oertel C (1984) Verklonung von *Digitalis lanata* Hochleistungspflanzen durch Sproßspitzenkultur. In: Methoden und Verfahren der Züchtung, des Anbaus, der Sammlung und der industriellen Verarbeitung von Arznei- und Gewürzpflanzen. Artern 18–22 June 1984. Vortragstexte, Teil 1, VEB Pharmazeutisches Werk Halle, Halle, pp 113–127

Lui JHC, Staba EJ (1979) Effects of precursors on serially propagated *Digitalis lanata* leaf and root cultures. Phytochemistry 18: 1913–1916

Maene LM, Debergh PC (1983) Rooting of tissue cultured plants under in vivo conditions. Acta Hortic 131: 201–208

Mastenbroek C (1980) Some experiences in breeding *Digitalis lanata*. Acta Hortic 96: 167–173

Mastenbroek C (1985) Cultivation and breeding of *Digitalis lanata* in the Netherlands. Br Heart J 54: 262–268

Mertinat H, Diettrich B, Luckner M (1989) Vernalization of in vitro-grown clone plants of *Digitalis lanata*. J Plant Physiol 135: 378–379

Murashige T, Skoog F (1962) A revised medium for rapid growth and bioassays with tobacco tissue cultures. Physiol Plant 15: 473–497

Oertel C, Breuel K (1987) Virusfreie Klonpflanzen für die Digitaliszüchtung. Pharmazie 42: 217

Pierik RLM (1967) Effect of light and temperature on flowering in *Cardamine pratensis* L. Z Pflanzenphysiol 56: 141–152

Raschke K (1979) Movements of stomata. In: Haupt W, Feinleib ME (eds) Encyclopedia of plant physiology. New Series, vol 7: Physiology of movements. Springer, Berlin Heidelberg New York, pp 383–441

Rücker W (1988) *Digitalis* spp.: in vitro culture, regeneration, and the production of cardenolides and other secondary products. In: Bajaj YPS (ed) Biotechnology in agriculture and forestry, vol 4. Medicinal and aromatic plants I. Springer, Berlin Heidelberg New York, pp 388–418

Schnabl H (1976) Der Einfluß von Aluminiumionen auf den Stärkemetabolismus von *Vicia faba*-Epidermen. Z Pflanzenphysiol 77: 167–173

Schnabl H (1983) The key role of phosphoenolpyruvate carboxylase during the volume changes of guard cell protoplasts. Physiol Vég 21: 955–962

Schnabl H, Elbert C, Krämer G (1982) The regulation of the starch-malate balances during volume changes of guard cell protoplasts. J Exp Bot 33: 996–1003

Schöner S, Reinhard E (1982) Clonal multiplication of *Digitalis lanata* by meristem culture. Planta Med 45: 155

Schöner S, Reinhard E (1986) Long-term cultivation of *Digitalis lanata* clones propagated in vitro: cardenolide content of the regenerated plants. Planta Med 478–481

Schönherr J (1982) Resistance of plant surfaces to water loss: transport properties of cutin, suberin and associated lipids. In: Lange OL, Nobel PS, Osmond CB, Ziegler H (eds) Encyclopedia of plant physiology. New Series, vol 12 B: Physiological plant ecology II. Springer, Berlin Heidelberg New York, pp 153–179

Schumann K (1963a) Das 'Digitalis-Mosaik' – eine Viruskrankheit an *Digitalis lanata* Ehrh. Pharmazie 18: 497–501

Schumann K (1963b) Untersuchungen zur Charakterisierung und Identifizierung der Erreger des 'Digitalis Mosaik'. II. Das Ringmosaik-Virus der Kapuzinerkresse. Phytopathol Z 48: 135–148

Schwerdtfeger G (1961) Ein Beitrag zur Züchtung von *Digitalis lanata* Ehrh. Züchter 31: 202–211

Silva F, Pop I (1965) Mosaikvirus an *Digitalis lanata* Ehrh. und sein Einfluß auf den Gehalt an Wirkstoffen. Pharmazie 20: 110–112

Springer M, Mertinat H, Diettrich B, Luckner M, Hess A, Breuel K (1986) Verfahren zur vegetativen Vermehrung von Digitalis-Hochleistungspflanzen. DDR-Patent DD 236 550

Stary F (1971) Unsere Erkenntnisse bei der Züchtung von *Digitalis lanata* subsp. *lanata*. Planta Med (Suppl) 4: 166–168

Sutter EG, Hutzell M (1984) Use of humidity tents and antitranspirants in the acclimatization of tissue-cultured plants to the greenhouse. Sci Hortic 23: 303–312

Tetenyi P (1959) Zur Frage der chemischen Taxa der *Digitalis lanata* Ehrh. Pharmazie 14: 690–691

Vogel E, Luckner M (1981) Distribution of cardenolides in *Digitalis lanata*. Planta Med 41: 161–165

Wardle K, Dobbs EB, Short KC (1983) In vitro acclimatization of aseptically cultured plantlets to humidity. J Am Soc Hortic Sci 108: 386–389
Weiler EW, Westekemper P (1979) Rapid selection of strains of *Digitalis lanata* Ehrh. with high digoxin content. Planta Med 35: 316–322
Withers LA (1983) Germplasm storage in plant biotechnology. In: Mantell SH, Smith J (eds) Plant Biotechnology. Cambridge Univ Press, Cambridge, pp 187–218
Wittwer SH, Bukovac MJ (1957) Gibberellin effects on temperature and photoperiodic requirements for flowering of some plants. Science 126: 30–31
Zeiger E (1983) The biology of stomatal guard cells. Annu Rev Plant Physiol 34: 441–475
Zimmerman RH, Hackett WP, Pharis RP (1985) Hormonal aspects of phase change and precocious flowering. In: Pharis RP, Reid DM (eds) Encyclopedia of plant physiology. New Series, vol 11: Hormonal regulation of development, III. Role of environmental factors. Springer, Berlin Heidelberg New York, pp 79–115
Ziv M, Schwartz A, Fleminger D (1987) Malfunctioning stomata in vitreous leaves of carnation (*Dianthus caryophyllus*) plants propagated in vitro; implications for hardening. Plant Sci 52: 127–134

II.16 Micropropagation of Medicinal *Dioscorea* Species

J. Van Staden and D.L. Fowlds[1]

1 Introduction

The genus *Dioscorea* L. belongs to the family Dioscoreaceae in the order Liliales of the Class Monocotyledoneae. It contains six genera with 650 spp., of which *Dioscorea* is the largest with ca. 600 spp, which occur mainly in the tropics and subtropics (Purseglove 1972; Takeda 1972).

1.1 Morphology and Karyology

Most *Dioscorea* species are herbaceous climbers with rhizomes or tubers. In most species the rhizome which annually produces shoots is enlarged to a storage organ (Purseglove 1972). The shoots twine to the right or the left and the leaves are alternate or sometimes opposite, petiolated, entire or digitately compound. The inflorescence is a spike, raceme, or panicle, dioecious, or rarely monoecious. The flowers are small and dull (white, green, or red) and pollination is entomophilous. The fruit is a trilocular capsule. The seeds are often winged (Purseglove 1972; Ross 1972; O'Hair 1984). In many species bulbils or aerial tubers form in the leaf axils (Rao 1969).

Dioscorea has either a basic set of nine (*D. composita* Hemsl., *D. floribunda* Mart. and Gal., *D. mexicana* Guillem) or ten chromosomes (*D. deltoidea* Wall., *D. prazeri* Prain and Burk.) (Pal and Sharma 1977). Some authors indicate $x = 10$ with various degrees of polyploidy between and within species, e.g., *D. alata* L. 2n = 30, 40, 50, 60, 70, 80, *D. bulbifera* L. 2n = 40, 60, 80, 100; *D. cayenensis* Lam., and *D. rotundata* Poir. 2n = 36, 54, 140 (Purseglove 1972; O'Hair 1984).

1.2 Distribution and Importance

Dioscorea species cannot tolerate frost, although *D. opposita* Thunb. does survive at colder temperatures. Ideal conditions are a temperature of 30 °C, a dry season of not more than 2 to 4 months, 1500-mm rain distributed evenly throughout the remaining months and loose, deep, free-draining, fertile soils (Purseglove 1972).

[1] UN/FRD Research Unit for Plant Growth and Development, Department of Botany, University of Natal, P.O. Box 375, Pietermaritzburg 3200, Republic of South Africa

Biotechnology in Agriculture and Forestry, Vol. 19
High-Tech and Micropropagation III (ed. by Y.P.S. Bajaj)

The most important cultivated species are *D. alata*, *D. cayenensis*, *D. rotundata*, and *D. trifida* L. These are grown widely throughout tropical Africa (O'Hair 1984). Tubers of wild yams are used in times of famine, but they usually contain toxic alkaloids which have to be removed by boiling or soaking in water.

The tubers of *Dioscorea* contain steroidal sapogenins, the most important being diosgenin, which is present in concentrations averaging between 4 to 5% dry weight (range 0 to 14%) (Staba 1977). Tubers of wild yams are amongst the principle sources of diosgenin (Table 1).

Diosgenin is industrially important because it is either microbially or chemically converted to medicinally useful steroids. Products derived from these plant steroids include corticosteroids, fertility control compounds, sex hormones, anabolic agents, and cardiatonic glycosides. With the increasing need, especially of oral contraceptives, it is predictable that the world's raw material requirement, of which 85% comes from plants in the wild, needs to be supplemented by cultivation.

Table 1. List of important medicinal *Dioscorea* species

Species name	Common name	Origin	Diosgenin levels
D. bulbifera	Air potato (Fo) Potato or aerial yam (P)	Tropical Asia and Africa (F; P)	0–4.5% (F)
[a] *D. Composita*	—	Mexico (F; P)	4–6% (M) 0–13% (F) 5% (dry wt.) (H; P)
D. deltoidea	Himalayan yam (D)	Himalayas; India (F; M; P)	up to 10% dry wt. (P) 0–6% (F) 5% dry wt. (H; P)
D. elephantipes	Elephant's foot (Fo; P)	South Africa (P)	-
[a] *D. floribunda*	Medicinal yam	Mexico; Central America. (F; M; P)	6–8% (M) 0.2–4% (F) 6% dry wt. (H; P)
D. mexicana		Mexico; Central America (F; M; P)	0.3–0.8% (F) 6% dry wt. (H; P)
D. opposita (syn. *D. batatas*)	Chinese yam (P) Cinnamon yam (P)	China; Korea; Japan; Taiwan (P)	—
D. prazeri	—	Tropical Africa (F) India (F; M)	2.1% (F) 4% dry wt. (H; P)
D. sylvatica	Elephant's foot (Fo; P)	South Africa (F; Fo; M; P)	0.2–3.4% (F) 5% dry wt. (H; P)

[a] Especially suited to cultivation.
References: D = Drapeau et al. (1986b); F = Furmanowa and Guzewska (1989); Fo = Forsyth (1982); H = Hardman (1969); M = Martin and Gaskins (1968); P = Purseglove (1972)

1.3 Conventional Propagation and Breeding

Edible *Dioscorea* species are treated as annual crops. On the other hand, wild yams are perennial. Experiments on the feasibility of using wild *Dioscorea* as cultivated crops have been carried out in Puerto Rico and Kenya (Purseglove 1972). The most promising species is *D. floribunda*, which yielded 10% diosgenin on a dry weight basis. Also tested were *D. deltoidea*, *D. sylvatica*, and *D. balcanica* Kosanin. For the commerical production of diosgenin, *Dioscorea* plants must be grown for a minimum of 3 years before harvest. Four years is perferable, as in this last year the sapogenin level almost doubles. The tubers are best harvested in the dry winter months when the diosgenin levels are stable. In spring, when the plants start to regrow, the diosgenin levels decrease. These levels again increase during the growing season (Martin and Gaskins 1968).

A major problem in the development of high-yielding *Dioscorea* species as crops is a lack of rapid propagation methods.

1.3.1 Propagation by Seed

The dioecious character, inconsistent flowering and seed production, and the ploidy levels make breeding of *Dioscorea* difficult, if not impossible. Some cultivars, usually the cultivated edible yams, flower only rarely and even more rarely set fertile seed. On the other hand, many wild yams reproduce freely in nature by means of winged seeds. (Hardman 1969; Purseglove 1972)

Seed germination has been studied in a few species. Freshly collected *D. composita* seeds are positively photoblastic. This sensitivity is lost with time. Maximum germination was 70% at 25 °C. The viability of the seed decreased rapidly and was low after 5 years in storage (Viana and Felippe 1986, 1988). *D. bulbifera* seeds germinate on moist filter paper at 30 °C within 2 to 3 weeks after harvest (Purseglove 1972). *D. tokoro* Makino and *D. tenuipes* Franch. and Sav. seed germinated well at 20 °C both in the light and dark after a short period of chilling at 0 °C (Okagami and Kawai 1982). However, because of the variation in diosgenin levels between clones, sexual reproduction is not the best option, and vegetative reproduction is preferable.

1.3.2 Propagation by Leaf and Stem Cuttings

Propagation by leaf and stem cuttings has been attempted, but the success rate was low (Lakshmi Sita et al. 1976). Application of Rootone F to stem cuttings of *D. composita* resulted in 43% rooting (Martin and Delpin 1969). Two-year-old stem cuttings of *D. alata* containing a 2-cm length of internode on either side of the node plus the leaves were treated with 0.1% 3-indolebutyric acid (IBA) powder. After 5 weeks, between 16 and 88% of the cuttings rooted and developed shoots. Plantlets were successfully planted in the field (Van der Zaag and Fox 1981). Stem cuttings of *D. composita* produce callus and roots if treated with IBA. However, if the IBA treatment was followed by 6-benzylaminopurine (BA) applications at specific intervals after the IBA treatment, a high rate of rooting was achieved. In general, it took about 50 days to produce *D. composita* plants, compared to 15 days for *D. spiculiflora* Hemsl. (Viana and Felippe 1988). Rooting was poor during winter. The best leaf cuttings are those taken from the median part of leaves on vegetative stems (Viana and Felippe 1987).

1.3.3 Propagation by Tubers

The usual method of yam propagation is by using small tubers, tuber pieces or setts, or aerial tubers (bulbils). When the tuber is used, 10 to 20% of the previous crop is required for this purpose (Van der Zaag and Fox 1981). The usual method is by using tuber pieces ranging from 0.1 to 2 Kg. Setts taken from the proximal (top) portion of the tuber (or head) usually sprout more quickly and give higher yields (Purseglove 1972). As *Dioscorea* produce few new tubers per growing season (Okoli 1978), this system is expensive, results in low multiplication ratios, and involves high losses of produce meant for food. Such a low multiplication ratio is a handicap in breeding programs and there is a need for improvement so as to increase the amount of "seed" yam and to shorten the time for developing lines in breeding programs. In a recent study, it has been shown that setts weighing as little as 25 to 30 g can be used to produce seed- and ware-tubers respectively in *D. alata* and *D. rotundata.* In *D. cayenensis,* these size setts are only acceptable if taken from the head region of the tuber. This technique has in some cases resulted in a ten fold increase in the multiplication rate compared to the traditional methods used (Kalu 1989).

The multiplication rate can be increased further and ratios of 1:58 and 1:62 have been achieved by "segmenting" the tubers. This method induces "eyes" on the tuber to sprout and the sprout-bearing segments are then removed from the tuber and used as propagules for new plants. This method holds promise for both breeding and the commercial production of yams (Okoli 1987).

1.4 The Need for In Vitro Culture

The annual world demand for steroid-based hormones and oral contraceptives has risen dramatically. Diosgenin, the most common precursor, is not derived from cultivated *Dioscorea* plants, even though *D. alata* and *D. bulbifera* are grown as vegetables. Thus, to satisfy the pharmaceutical requirements, wild plants have been collected, leading to their depletion. In spite of recent advances, propagation from seed or conventional vegetative methods is too slow to provide sufficient plants. An efficient tissue culture method would be the answer.

2 In Vitro Production of Diosgenin

Most in vitro investigations have been directed towards rapid clonal multiplication of the commercial diosgenin-yielding species either by the proliferation of axillary buds, as with *D. bulbifera* (Uduebo 1971) and *D. floribunda* (Lakshmi Sita et al. 1976), or the regeneration of plantlets from callus cultures of *D. deltoidea* (Grewal and Atal 1976). In vitro tuber formation has also been studied (Rao 1969; Sengupta et al. 1984; Forsyth and Van Staden 1984; Ng 1988). Increases in diosgenin production in callus and suspension cultures have been achieved and are potentially another method for the production of sufficient quantities of diosgenin (Rokem et al. 1985; Drapeau et al. 1986b).

2.1 Diosgenin Levels In Vivo and In Vitro

In vivo the diosgenin content in different populations of the same species, and between species, is highly variable. This difference could originate from the polyploid nature of *Dioscorea* (Viana and Felippe 1988). Diosgenin levels in leaves are much lower than in tubers. Leaves from male plants have a higher level (Datta and Datta 1984; Viana et al. 1987). The levels of diosgenin produced in vitro can approach or exceed those in the plant under natural conditions (Staba 1977).

Table 2. Diosgenin content in callus cultures of *Dioscorea* spp.

Species	Explant	Hormones in medium for callus production	Diosgenin content	Reference
D. composita	Stem nodes	MS + 0.5 mg/l BA	About 2033 μg/g dry wt.	Heble and Staba (1980b)
	Internodal stem segments	MS + 0.5 mg/l BA	Traces	Heble and Staba (1980b)
D. floribunda	Shoot tip	Revised tobacco medium + 1–2 mg/l 2,4-D	0.295%	Aminuddin and Chowdhury (1983)
	Tuber	Revised tobacco medium + 1–2 mg/l 2,4-D	1.33%	
D. zingiberensis	Hypocotyl	Basal medium + 4–6 mg/l 2,4-D	0.42% dry wt.	Research group –Kiangsu Bot. Inst. (1981)
D. deltoidea			0.3–2.5%	Chaturvedi and Srivastava (1976)
D. bulbifera	Tuber	Miller's + BA + NAA (optimum callus growth with 5 mg/l BA + 2 mg/l NAA)		
	Leaf stem petiole		0.1%/g dry wt.	Forsyth (1982)
D. composita	Stem nodes	30-day-old callus 90-day-old callus	1.61% diosgenin dry wt; emergent shoot 2.5%; regenerated roots 0.09%	Datta and Datta (1984)

2.2 Undifferentiated Cultures

Callus or cell cultures of *Dioscorea* species produce diosgenin (Table 2). Diosgenin production is increased by using 2,4-dichlorophenoxyacetic acid (2,4-D) (Kaul et al. 1969; Heble and Staba 1980a; Forsyth 1982); by continuous cool-white fluorescent light (Seibert and Kadkade 1980); and by optimizing the ratio of carbon to nitrogen (Forsyth 1982; Tal et al. 1982). The effects of temperature, pH, and aeration have not been thoroughly investigated (Forsyth 1982).

It is apparently not essential that the tissue be organized. Studies with *D. deltoidea* showed diosgenin biosynthesis to be greater in unorganized tissue than in organized root cultures, and that the presence of BA ultimately inhibits growth and diosgenin production (Heble and Staba 1980b). Continuous cultures of excised roots of *D. floribunda* showed no production of diosgenin (Sengupta et al. 1985).

D. deltoidea cells will incorporate (14-^{14}C)- and (26-^{14}C)-cholesterol and sitosterol into diosgenin, and will biotransform progesterone (Staba 1980). *D. deltoidea* and *D. composita* cultures utilize cholesterol, at levels of at least 0.005%, to produce increased amounts of diosgenin. In the presence of cholesterol and 2,4-D the diosgenin yield is 1 to 2.5% dry weight, compared to less than 1% dry weight in the controls (Kaul et al. 1969; Staba and Kaul 1971; Staba 1977, 1980).

2.3 Suspension Cultures

Diosgenin is a nongrowth-associated product that is produced by nondividing cells of *Dioscorea* suspension cultures via an intermediate metabolite precursor (Furostanol I). This intermediate accumulates during the logarithmic growth phase and is biotransformed into diosgenin during the stationary phase. Diosgenin, which is exclusively intracellular, accumulates rapidly in the days following sugar depletion. Diosgenin biosynthesis is influenced by the initial sucrose level, hormone levels, phosphates, nitrogen sources, and some fungal elicitors. Initially 0.9 to 1% diosgenin (dry weight) or 98–114 mg/l was obtained after 28 days in culture. This was increased to 7.2% dry weight (230 mg/l) within 12 days of culture by using microbial and biochemical techniques (Heble and Staba 1980a; Rokem et al. 1984, 1985; Tal et al. 1984a, b; Drapeau et al. 1986a). Thus diosgenin need no longer be considered as a nongrowth-associated product, as diosgenin can be obtained directly from cultures in an exponential growth phase, and it will no longer be necessary to develop nongrowth (stationary) phases by incubating exponential cultures in sugar-free meidum to obtain optimum results (Drapeau et al. 1986b).

3 Micropropagation

There are several methods to micropropagate *Dioscorea* in vitro, including undifferentiated callus cultures, somatic embryogenesis, in vitro tuberization, embryo culture, and axillary bud or nodal culture.

The explants most frequently used are nodal stem segments. In a few cases, tuber or bulbil material has been utilized (Rao 1969). In all the recorded sterilization methods for *Dioscorea*, a presterilization routine is included. This usually involves a quick immersion in 70 to 80% ethanol (Asahira and Nitsch 1969; Chaturvedi 1975; Mantell et al. 1978; Forsyth 1982; Forsyth and Van Staden 1982, 1984); a wash in detergent, e.g., 5% Teepol (Chaturvedi 1975; Lakshmi Sita et al. 1976; Sengupta et al. 1984); or treatment with a fungicide, e.g., 0.1 to 1% benzimidazole (Lakshmi Sita et al. 1976; Forsyth 1982; Forsyth and Van Staden 1982). The majority of authors utilized active chlorine as the sterilant although a few used $HgCl_2$ (Chaturvedi 1975; Sengupta et al. 1984).

The most common medium used is modified MS medium (Murashige and Skoog 1962). In order to establish plantlets from callus, Rao (1969) utilized White's (1954) medium. The carbohydrate requirement is generally met with 1 to 3% sucrose. *Dioscorea* may be grown in liquid culture (Heble and Staba 1980b) but most media are solidified with 0.7 to 1% (w/v) agar. The pH of the media used was between pH 5.5 and 6.

3.1 Plant Regeneration from Callus Cultures

Hypocotyl material from aseptically germinated seeds of *D. deltoidea* has been used to initiate callus from which whole plants could be generated. The callus was subcultured for 12 months, after which it still retained its generation capacity (Grewal and Atal 1976; Osifo 1988). Callus from *D. deltoidea* tubers was used to study the effect of nitrogen sources and cytokinins on shoot bud initiation (Singh 1978).

Plants have been established from callus derived from young developing bulbils of *D. sansibarensis* Pax. In this case, rooting was common but only 2% of the cultures developed bud initials (Rao 1969). Callus established from node and internode segments of *D. floribunda* developed plants through a series of subcultures onto media containing various supplements. The initial subculture resulted in shoot development. Several "crops" of shoots were obtained from a single differentiating callus culture by excising the shoots and subculturing the residual callus onto the same medium. The excised shoots were rooted on a third medium. Seventy percent of the plantlets survived rooting and subsequent transfer to the soil (Table 3) (Sengupta et al. 1984).

Yellow friable callus was obtained from nodes and in vitro hypocotyls of *D. deltoidea* on RT medium containing 2,4-D. Shoot induction occurred if 2,4-D was replaced with BA and roots were established by changing the medium. Plantlets obtained in this way grew vigorously and after 4 weeks were planted in pots (Table 3) (Furmanowa and Guzewska 1989).

3.2 Somatic Embryogenesis and Embryo Culture

Callus tissue derived from dormant zygotic embryos of *D. rotundata* formed somatic embryos when cultured on MS medium supplemented with NAA and casein hydrolysate, in the light. When transferred to a medium with lower concentrations of NAA and casein hydrolysate, complete plantlets developed that

Table 3. Plantlet formation from *Dioscorea* nodal or axillary buds in vitro

Species	Source of explants	Medium	Hormones	Morphogenesis/ regeneration	Plantlets established	Reference
D. batatas and *D. japonica*	Stem segments	—	—	—	—	Asahira and Nitsch (1968, 1969)
D. floribunda	Single-node stem segments (in vivo)	Modified MS	15 mg/l adenine sulfate + 0.1 mg/l NAA	Plantlets with one or two shoots	No	Chaturvedi (1975)
	Single-node stem segments; shoot apices (in vitro)	Modified MS	2 mg/l BA followed by subculture onto 0.5 mg/l NAA	Five or more shoots developed initially. After subculturing prolific rooting	Yes	Chaturvedi (1975)
	Single-node stem segments	MS	2 mg/l kinetin with or without 0.2 mg/l NAA	Axillary bud produced shoots; roots arose from nodal region, i.e., plantlets obtained	Yes	Lakshmi Sita et al. (1976)
	Single-node stem segments	MS	As above, 0.5 mg/l NAA with or without kinetin	Callus	—	Lakshmi Sita et al. (1976)
	Internodal segments; shoot apex; callus	MS		No plantlets induced	—	Lakshmi Sita et al (1976)
D. floribunda	Node and internode	a) MS or modified Whites	2,4-D or NAA BA or Kinetin	Callus and shoot primordia	Yes	Sengupta et al. (1984)
		b) Half-strength MS	None	Increased shoot development		
		c) Half-strength MS	0.5 mg/l NAA or 0.25 mg/l IBA	Root development		
		d) As above	As above	In vitro tubers	Yes	
D. alata and *D. rotundata*	Nodal segments	MS grown under 16-h photoperiod	None	Callus; roots and shoots (plantlets). Plantlets in culture multibranched bearing numerous nodes which could be subcultured	Yes	Mantell et al. (1978)
			Different levels of NAA and kinetin	Callus	—	Mantell et al. (1978)

D. bulbifera	Nodal segments	MS	5-10 mg/l IAA + 0.05; 0.5 mg/l kinetin	Bulbils, shoots, roots	No	Uduebo (1971)
			5; 10 mg/l IAA + 2.5; 5 mg/l kinetin	Callus, unexpanded shoots	—	Uduebo (1971)
			15 mg/l IAA + kinetin	Callus	—	Uduebo (1971)
	Nodal segments	Revised MS	a) 5 mg/l kinetin or 1 mg/l BA	a) Elongated shoots initiated on nodes	—	Forsyth (1982)
			b) 5 mg/l IBA	b) Roots initiated		
			c) No hormones	c) Roots elongated	Yes	Forsyth and Van Staden (1982)
D. composita	Young stems and nodal segments	Revised MS	0.1 to 1 mg/l BA	Shoot and bud initiation (highest) with 0.5 and 0.75 mg/l BA	No	Heble and Staba (1980b)
D. sylvatica	Nodal segments	Revised MS	1 mg/l BA	Five shoots initiated per node. No rooting	No	Forsyth (1982); Forsyth and Van Staden (1982)
D. rotundata	In vitro single node cuttings	Revised MS	0.5 mg/l kinetin (+ 20 mg/l cysteine)	Tuberization mainly at base of stem/node. Frequency influenced by sucrose and day-length (optimum 12–16 h light)	About 80% of the microtubers sprouted after 2–3 months in aseptic conditions. Transferred in seedbed. After 7 months, 50% tuber weighed + 100 g	Ng (1988)
D. deltoidea	Nodes (also in vitro)	a) RT	1 mg/l 2,4-D	Friable yellow callus in 6 weeks Shoot induction Adventitious shoots rooted	Yes	Furmanowa and Guzewska (1989)
		b) RT c) i) SH (modified) ii) CS (Gibco Bio-Cult)	1 mg/l BA 0.5 mg/l 2,4-D 2 mg/l IAA 1 mg/l IAA supplemented with 170 mg/l Na_2HPO_4			

were established in the soil without difficulty. In the dark, only roots developed. This method is useful in that it circumvents the seed dormancy problem associated with conventional breeding of yams (Osifo 1988).

Callus of *D. floribunda* and *D. bulbifera* initiated on MS containing 2,4-D and/or BA produced somatic embryos if the cultures were aged or if the callus was transferred to media lacking 2,4-D but containing glutamine and zeatin or abscisic acid (ABA). These embryos germinated without growth hormones if at least five were subcultured together. The presence of zeatin and glutamine was required if subcultured individually. The embryos had to be transferred from a liquid to a semi-solid medium to obtain plantlet development (Ammirato 1982; George and Sherrington 1984). Germination of detached zygotic embryos from dormant *D. tokoro* seeds was studied under various nutritional and temperature regimes in vitro. The intact dormant seeds germinate slowly, if at all, between 11 to 23 °C. For rapid germination, prior chilling treatments were required. Detached embryos were grown in vitro on MS medium with the essential ingredients for germination being sucrose, NO_3^-, and $SO_4^=$. This requirement decreased with seed age. After 1 year's storage, 65% of the seed germinated in the absence of NO_3^- and $SO_4^=$. Detached embryos incubated with sucrose and KNO_3 germinated more rapidly with increasing temperature up to 35 °C, whereas the intact dormant seeds could not be induced to germinate above 26 °C. Germination was found to be more rapid, but overall germination percentage increased only slightly, if the intact seed was cut near the radicle. This suggests that the endosperm exerts an inhibitory effect on germination by means of a mechanical barrier or its impermeability to water or gases (Terui and Okagami 1989). To be able to germinate embryos from otherwise dormant seed will aid in the propagation of *Dioscorea*.

3.3 In Vitro Tuberization

Another method of propagation of *Dioscorea* is by in vitro tuber formation. Ammirato (1982) obtained in vitro aerial tubers or bulbils on plantlets of *D. alata* and *D. bulbifera* growing in vitro on MS medium. Single-node segments formed tubers directly in continuous light with NAA, ABA, or zeatin supplements. The in vitro tubers could be harvested or stored on the plants for up to a year. Tubers weighing more than 90 mg were planted in the soil and produced normal plants (George and Sherrington 1984).

Plantlets have been established from callus of *D. floribunda* node and internode segments (Table 3). The shoots of these were rooted on MS supplemented with 0.5 mg/l NAA or 0.25 mg/l IBA. If the plantlets were maintained on this medium, 70% of them produced one to five aerial tubers at the nodes in the axils of the leaves. These tubers, when separated from the mother plant, developed roots and shoots and could be transferred to soil with 100% survival. Aerial tubers develop more readily in liquid than on solid media and the tubers ranged in length from 2 to 15 mm (Sengupta et al. 1984).

D. bulbifera produces aerial tubers or bulbils, in response to short-day conditions in vivo. Nodal segments collected in spring, i.e., noninducing long-day conditions, were cultured on MS containing 2% sucrose and 5 mg/l kinetin. Under a long-day regime (16 h light; 25 °C constant) no tuberization occurred. Under a short-day regime (8 h light, 25 °C; 16 h dark, 18 °C), 80% of the cultures tuberized within 30 days. Thus in vitro tuberization in *D. bulbifera* is promoted by short days together with low night temperatures (Forsyth 1982; Forsyth and Van Staden 1984).

Cytokinins and carbohydrates are known to affect tuberization. The effect of sucrose and kinetin was studied on *D. bulbifera*. In the absence of kinetin tuberization occurred at all levels of sucrose concentrations tested, i.e., 2 to 8%. At 2% sucrose the tubers developed within 4 weeks. Using 5 mg/l kinetin, all plants tuberized within 6 weeks if also grown on 8% sucrose. If 5% sucrose was used, 65% of the plantlets tuberized, the remainder formed shoots. With 2% sucrose only shoots were produced. With higher kinetin levels (10 mg/l) all plantlets formed shoots on 2 and 5% sucrose. Using 8% sucrose resulted in tuber formation. Thus tuberization is induced by high sucrose levels (all plantlets on 8% tuberized irrespective of the kinetin concentration) and/or low kinetin concentrations. However, the effect of sucrose on tuberization was observed to be overcome by increasing the kinetin concentration in the medium (Forsyth 1982; Forsyth and Van Staden 1984).

Nodal cuttings of *D. rotundata* were induced to produce microtubers on a revised MS medium supplemented with various concentrations of sucrose, 20 mg/l L-cysteine, 0.5 mg/l kinetin (Table 3). The frequency of tuberization was affected by daylength, which is optimal at 12 and 16 h daylight, depending on the sucrose concentration. The frequency of tuberization and the number of tubers was highest at 5% sucrose and declined at higher and lower concentrations. Shorter daylengths (8 h) and continuous light were not suitable for tuberization. About 80% of the microtubers sprouted after 2 to 3 months of storage under aseptic conditions on moist filter paper. These microtubers were planted in seed beds and could be harvested after 7 months. Half of the harvested tubers weighed ca. 100 g (range 40–400 g) and these could be used as planting material for ware tuber production (Ng 1988).

In vitro tuberization has also been studied in shoot cultures of *D. alata* and *D. bulbifera*. Ammonium ions inhibited microtuber induction in *D. alata* but not in *D. bulbifera*. For *D. alata* best results were obtained with a full strength tuberization medium (specially formulated) supplemented with 2% sucrose, 2.5 μM kinetin, and 0.4% agar at 25 °C under an 8-h photoperiod. For *D. bulbifera*, full-strength MS medium supplemented with 4% sucrose, 2.5 μM kinetin, and 0.4% agar at 25 °C with an 8-h photoperiod gave good results. In the presence of kinetin, *D. bulbifera* also formed microtubers under a 16-h photoperiod. The rate of tuberization was slower in *D. alata* (10 to 12 weeks) than in *D. bulbifera* (6 to 8 weeks). The latter species produced aerial tubers in a continuous fashion from 8 weeks after planting, whilst the former produced an underground tuber following maximum vine growth. Both species consistently produced microtubers with individual weights above 100 mg, which are large enough for direct planting and subsequent growth in unsterilized soils (Mantell and Hugo 1989).

3.4 Axillary Bud or Nodal Culture

The first reports of successful plantlet formation from *Dioscorea* stem nodes were those of Asahira and Nitsch (1968, 1969). In vitro plantlets have subsequently been established from a number of nodal explants (Table 3).

For plantlet formation the following additives at various concentrations have been used: adenine sulfate together with NAA (Chaturvedi 1975); BA followed by NAA (Chaturvedi 1975; Mantell et al. 1978); kinetin alone or with NAA (Lakshmi Sita et al. 1976; Mantell et al. 1978); and BA alone (Heble and Staba 1980b).

Nodal segments grown in vitro will give rise to one to four shoots when supplied with basal medium (BM) and a relatively low level of cytokinins. Propagation becomes more efficient if each node can be induced to form more shoots. Efficiency is further increased if each shoot produced in this way can in turn produce multiple shoots. Multiple shoot induction is usually induced by increasing the cytokinin/auxin ratio, by eliminating auxin, or by changing the type of cytokinin (generally BA is preferred to kinetin) (Forsyth 1982).

Forsyth and Van Staden (1982) described multiple shooting from nodal segments of *D. bulbifera*. The lamina and petioles were removed from 1- to 2-cm nodal segments prior to sterilization. A sterilization technique utilizing 80% ethanol, benzimidazole, and sodium hypochlorite was used. Revised MS medium supplemented with 2% sucrose, 0.7% oxoid agar, and kinetin, BA and/or IBA was used. The cultures were maintained at $25 \pm 2\,^{\circ}C$, 75% relative humidity under a 16-h photoperiod.

Shoot elongation was best on a medium supplemented with 5 mg/l kinetin. On this medium shoots elongated within 3 days (Fig. 1A). Within 8 weeks after inoculation onto this medium, one to two roots developed per explant. When transferred to soil these plantlets readily succumbed to microbial attack, resulting in plantlet death. If, however, subsequent to shoot induction and elongation (promoted by 5 mg/l kinetin), the explant is transferred to MS supplemented with 5 mg/l IBA, many roots formed within 7 days. Whilst the plantlet remained on the auxin-containing medium these roots failed to elongate (Fig. 1B). However, within 48 h of transfer to BM devoid of hormones, the auxin-induced roots had elongated and penetrated the medium (Fig. 1C). Eight-week-old plantlets were successfully transferred directly to sterile soil without any previous hardening off.

When explants were cultured on a medium with 5 mg/l kinetin, an average of five shoots developed per node within 28 days. By doubling the kinetin concentration to 10 mg/l, it was possible to nearly double the number of shoots to eight shoots per node (Table 4). However, a further increase to 20 mg/l kinetin resulted in fewer shoots per node. The tips of these shoots showed signs of "dying back", indicating that the kinetin concentration was supra-optimal.

BA was more effective than kinetin in inducing multiple shoot formation. At 1 mg/l, nine shoots developed per node (Table 4), which is comparable to the number obtained with 10 mg/l kinetin. Doubling the BA to 2 mg/l was detrimental, and malformed shoots developed at higher BA levels. Individual shoots produced were excised aseptically, transferred to fresh medium containing 1 mg/l BA, and induced to form another cluster of five to nine new shoots. Roots were initiated by placing the shoots on basal medium supplemented with 5 mg/l IBA. Thus this method is a rapid and efficient means of propagation (Forsyth 1982; Forsyth and Van Staden 1982).

Mantell et al. (1978) showed that under a 16-h photoperiod nodal segments of *D. alata* and *D. rotundata* consistently produced callus, roots, and shoots; whereas under a 12-h photoperiod growth was variable and plantlet production was poor. In most cases of plantlet formation a 16-h photoperiod was used, although

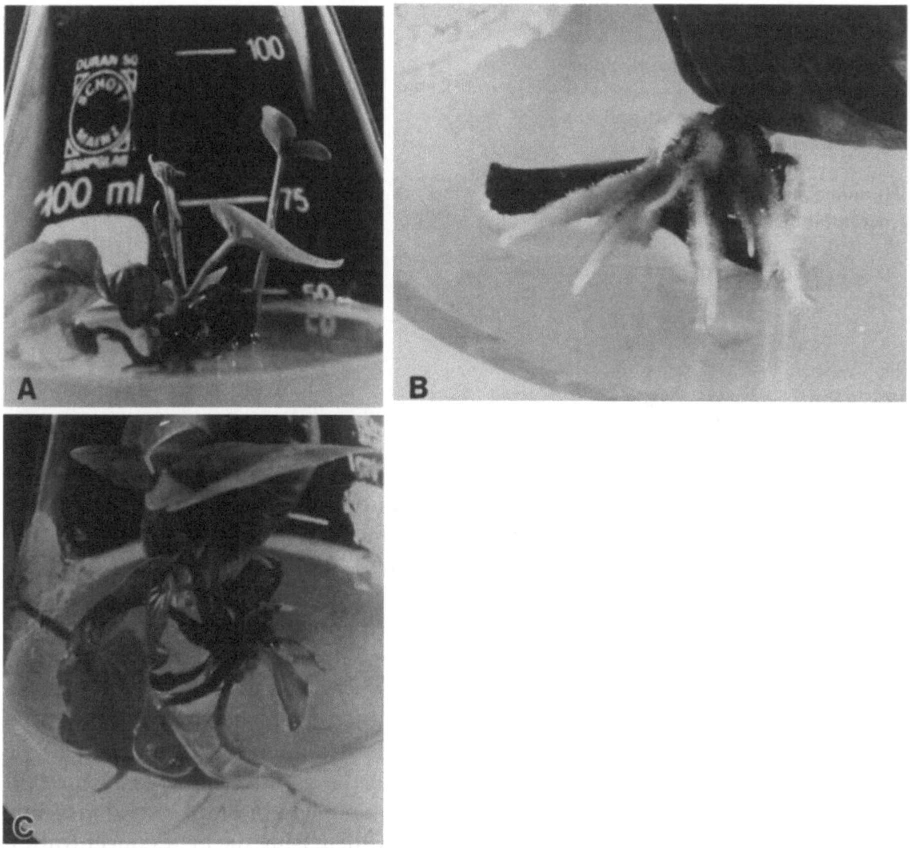

Fig. 1.A. Shoot elongation of *D. bulbifera* with 5 mg/l kinetin. (Forsyth and Van Staden 1982). **B** Short stubby roots formed on *D. bulbifera* on medium containing 5 mg/l IBA. (Forsyth and Van Staden 1982). **C** Plantlet of *D. bulbifera* showing good root and shoot development on hormone-free medium (Forsyth and Van Staden 1982)

Lakshmi Sita et al. (1976) used a 12-h, and Chaturvedi (1975) a 14-h photoperiod.

Although plantlet formation from nodal segments generally required growth regulators, Mantell et al. (1978) were successful with a medium without growth regulators. In most cases, however, shoots were initially formed on media supplemented with cytokinins [0 to 2.5 mg/l BA (Mantell et al. 1978); 2 mg/l BA (Chaturvedi 1975); 5 mg/l kinetin or 1 mg/l BA (Forsyth and Van Staden 1982); 0.1 to 1 mg/l BA (Heble and Staba 1980b)]. In *D. bulbifera* only one growth regulator (kinetin or BA) is required to induce multiple shoot formation, but in *D. sylvatica* preliminary studies indicate that a delicate balance of BA and IBA is needed to optimize shoot production (Forsyth and Van Staden 1982). Shoots are usually rooted by transfer to a second medium supplemented with auxins [0.5 mg/l NAA (Chaturvedi 1975); 5 mg/l IBA (Forsyth 1982; Forsyth and Van Staden 1982)]. In

Table 4. Average number of shoots per node when *D. bulbifera* node sections were grown on basal medium supplemented with BA or kinetin. Five replicates were used. (Forsyth and Van Staden 1982)

Hormone (mg/l) concentration	BA	Kinetin
0	3	3
0.5	6	3
1	9	3
2	3	3
5	3	5
10	3	8
20	2	3

the experiments of Forsyth (1982), root elongation only occurred on a hormone-free medium. In other situations, the nodal segments were rooted and shooted on the same medium [2 mg/l with or without 0.2 mg/l NAA (Lakshmi Sita et al. 1976); 15 mg/l adenine sulfate and 0.1 mg/l NAA (Chaturvedi 1975)].

In order for micropropagation to be viable, the in vitro plantlets formed must be successfully established under field conditions. Plantlets developed from nodal stem segments have been transferred into the soil with success rates of 70 to 100%. Transfer is either directly into sterile soil under controlled temperature and humidity conditions, or after a period of "hardening off", during which the hormones and sucrose levels are reduced so that the plantlets slowly become self-sufficient. "Hardening off" is done either in a liquid medium or on a solid medium. In the latter the available moisture is reduced by increasing the agar concentration to 1% (w/v).

With *D. bulbifera*, 8-week-old plantlets could be transferred directly to a sterile soil mix without the gradual "hardening off" process being carried out. The survival rate was 80% (Forsyth 1982; Forsyth and Van Staden 1982). *D. floribunda* plantlets were transferred to sterile soil either directly or after hardening in Hoagland's solution (Lakshmi Sita et al. 1976). *D. alata* and *D. rotundata* were established directly in a steam-sterilized 4:4:3 peat:sand:soil mixture (Mantell et al. 1978).

Plantlets established from *D. floribunda* callus were "hardened off" in two stages in liquid medium before being transferred to sterile soil under controlled temperature and humidity conditions. The plants were then transferred to the field. Seventy percent of the plantlets survived rooting and the transfer to soil (Sengupta et al. 1984). Chaturvedi had a 100% success rate with *D. floribunda* which were "hardened off" in a liquid inorganic nutrient medium (Chaturvedi 1975).

Plantlets formed from somatic embryos can apparently be established in soil without difficulty but no details were given (Osifo 1988). In vitro microtubers of *D. floribunda* produced roots and shoots and could be detached from the parent plant and transferred to the soil with 100% survival (Sengupta et al. 1984).

The establishment of plants from in vitro-grown plantlets has been successful and will not therefore hinder the micropropagation of *Dioscorea*.

4 Commercial Aspects of Micropropagation

The micropropagation of *Dioscorea* has great potential. Plantlets have been established from numerous explant sources which do not require the sacrifice of the parent plant. The advantages of in vitro plantlet production are that not only a small area is required for the multiplication steps (Mantell et al. 1978) and a greater number of plants can be obtained from each plant compared to conventional propagation methods, as the cultured nodal segments can be repropagated more than once in a single season. Mantell et al. (1978) have estimated that by using their propagation cycle it is possible to obtain 65 000 plantlets from a single node segment within 6 months.

This form of micropropagation has a few disadvantages. Firstly, in vitro plantlets are not easily transported over a distance, as they are vulnerable to unfavorable conditions. They should not be kept for prolonged periods in the dark and transportation should not take more than 1 week (Ng 1988). The problem of poor establishment of in vitro plantlets in the field, particularly in tropical situations, needs further consideration, especially in view of the fact that the specialized handling that is often required for laboratory-to-field transfer of plantlets is absent in rural areas of tropical countries (Mantell and Hugo 1989). Micropropagation via microtubers is an alternative. It has several advantages over in vitro plantlets as their handling is easier and they are not so vulnerable to transportation hazards. Tubers can be produced in large numbers, kept in the dark for prolonged periods without loss of viability, stored until ideal field conditions prevail, and are easily transplanted into the soil, thus ensuring greater plant survival. Thus microtubers are a convenient, efficient method to propagate and distribute *Dioscorea* (Ng 1988; Mantell and Hugo 1989).

5 Protocol

1. Explant:
   Stem nodes or zygotic embryos.
2. Medium
   MS, pH 5.5 to 6, 0.5 to 1% agar, cytokinins, and auxins.
3. Optimal Conditions
   24 to 28 °C, 50 to 75% relative humidity, 16-h photoperiod.

6 Summary and Conclusions

With the increase in demand for steroids, the cultivation of *Dioscorea* spp. with higher diosgenin content is becoming more urgent. The major source of diosgenin, wild yams, is becoming scarce due to over-collection of the tubers by pharmaceutical companies. The supply of wild yams is not unlimited, and an alternative, cheap source of steroids must be found. Conventional propagation, both sexual and asexual, is fraught with problems, including variability and the time required for establishment. In vitro plantlets, tubers, or somatic embryos may be used but will first have to be produced on a large scale. This has several advantages: it overcomes seed dormancy, produces virus-free clonal plantlets, improves the multiplication rate of the in vivo plant, and allows international germplasm exchange. The actual cultivation of the plants to produce extractable tubers in the field is slow and labor-intensive.

An alternative source of diosgenin, which will require considerable manipulation, is from in vitro callus and/or suspension cultures. As indicated, the steroid can be produced in this way. These techniques will eliminate climatic factors and will make the raw material available in producer countries where the plants normally do not grow. With ever-improving biotechnological techniques, viable commercial production must soon become an attainable goal.

Acknowledgments. We thank the FRD, Pretoria, for financial support.

References

Aminuddin, Chowdhury AR (1983) Production of diosgenin in somatic callus tissues of *Dioscorea floribunda*. Planta Med 48: 92–93

Ammirato PV (1982) Growth and morphogenesis in cultures of monocot yam, *Dioscorea*. In: Fujiwara A (ed) Plant tissue culture 1982 Maruzen. Tokyo, pp 169–170

Asahira T, Nitsch JP (1968) Tuberization in vitro: *Ullucus tuberosus* et *Dioscorea*. Bull Soc Bot Fr 115: 345–352

Asahira T, Nitsch JP (1969) Effect of polarity and kinetin on the browning reaction of *Dioscorea batatas* and *Dioscorea japonica*. Planta 84: 292–294

Chaturvedi HC (1975) Propagation of *Dioscorea floribunda* from in vitro culture of single-node stem segments. Curr Sci 23: 839–841

Chaturvedi HC, Srivastava SN (1976) Diosgenin biosynthesis in tuber cultures of *Dioscorea deltoidea*. Lloydia 39: 82–83

Datta SK, Datta K (1984) Chemodifferentiation of diosgenin in *Dioscorea composita*. Phytochemistry 23: 2684–2685

Drapeau D, Blanch HW, Wilke CR (1986a) Growth kinetics of *Dioscorea deltoidea* and *Catharanthus roseus* in batch culture. Biotechnol Bioeng 28: 1555–1563

Drapeau D, Sauvaire Y, Blanch HW, Wilke CR (1986b) Improvement of diosgenin yield from *Dioscorea deltoidea* plant cell cultures by use of a non-traditional hydrolysis method. Planta Med 52: 474–478

Forsyth C (1982) In vitro responses of *Dioscorea bulbifera* L. and *Dioscorea sylvatica* (Kunth) Eckl. MSc Thesis, Univ Natal, Pietermaritzburg

Forsyth C, Van Staden J (1982) An improved method of in vitro propagation of *Dioscorea bulbifera* Plant Cell Tissue Organ Cult 1: 275–281

Forsyth C, Van Staden J (1984) Tuberization of *Dioscorea bulbifera* stem nodes in culture. J Plant Physiol 115: 79–83

Furmanowa M, Guzewska J (1989) *Dioscorea*: in vitro culture and the micropropagation of diosgenin-containing species. In: Bajaj YPS (ed) Biotechnology in agriculture and forestry, vol 7: Medicinal and aromatic plants II. Springer, Berlin Heidelberg New York, pp 162–184

George EF, Sherrington PD (1984) Plant propagation by tissue culture. Handbook and Directory of Commercial Laboratories. Exegetics, England

Grewal S, Atal CK (1976) Plantlet formation in callus cultures of *Dioscorea deltoidea* Wall. Indian J Exp Biol 14: 352–353

Hardman R (1969) Pharmaceutical products from plant steroids. Trop Sci 11: 196–288
Heble MR, Staba EJ (1980a) Steroid metabolism in stationary phase cell suspensions of *Dioscorea deltoidea*. Planta Med (Suppl) 46: 124–128
Heble MR, Staba EJ (1980b) Diosgenin synthesis in shoot cultures of *Dioscorea composita*. Planta Med (Suppl) 46: 120–123
Kalu BA (1989) Seed yam production by minisett technique: evaluation of three *Dioscorea* species in the Guinea and derived Savanna zone of Nigeria. Trop Agric (Trinidad) 66: 83–86
Kaul B, Stohs SJ, Staba EJ (1969) *Dioscorea* tissue cultures III. Influence of various factors on diosgenin production in *Dioscorea deltoidea* callus and suspension cells. Lloydia 32: 347–359
Lakshmi Sita G, Bammi RK, Randhawa GS (1976) Clonal propagation of *Dioscorea floribunda* by tissue culture. J Hortic Sci 55: 551–554
Mantell SH, Hugo SA (1989) Effects of photoperiod, mineral medium strength, inorganic ammonium, sucrose and cytokinin on root, shoot and microtuber development in shoot cultures of *Dioscorea alata* L. and *D. bulbifera* L. yams. Plant Cell Tissue Organ Cult 16: 23–37
Mantell SH, Haque SQ, Whitehall AP (1978) Clonal multiplication of *Dioscorea alata* L. and *Dioscorea rotundata* Poir. yams by tissue culture. J Hortic Sci 53: 95–98
Martin FW, Delpin H (1969) Techniques and problems in the propagation of sapogenin-bearing yams from cuttings. J Agric, Univ Puerto Rico 53: 191–198
Martin FW, Gaskins MH (1968) Cultivation of the sapogenin bearing *Dioscorea* species. US Dept Agric, Prod Res Rep 103: 3–19
Murashige T, Skoog F (1962) A revised medium for rapid growth and bioassays with tobacco tissue cultures. Physiol Plant 15: 473–497
Ng SYC (1988) In vitro tuberization in white yam (*Dioscorea rotundata* Poir). Plant Cell Tissue Organ Cult 14: 121–128
O'Hair SK (1984) Farinaceous crops. In: Martin FW (ed) Handbook of tropical food crops. CRC, Boca Raton, Florida, pp 133–137
Okagami N, Kawai M (1982) Dormancy in *Dioscorea*: differences of temperature responses in seed germination among six Japanese species. Bot Mag Tokyo 95: 155–166
Okoli OO (1978) Stimulating axillary buds in yams (*Dioscorea* spp.). Exp Agric 14: 89–92
Osifo EO (1988) Somatic embryogenesis in *Dioscorea*. J Plant Physiol 133: 378–380
Pal A, Sharma AK (1977) Diosgenin content of old and new world species of *Dioscorea* with special reference to the chromosomal races. Indian J Exp Biol 15: 1109–1112
Purseglove JW (1972) Tropical crops. Monocotyledons I. Longman, London pp 97–117
Rao AN (1969) Tissue culture from bulbils of *Dioscorea sansibarensis*. Can J Bot 47: 565–567
Research Group of Plant Morphology, Kiangsu Botanical Institute (1981) The preliminary experiments on the callus tissue culture of *Dioscorea zingiberensis* Wright. In: Proc Symp Plant tissue culture (Peking 1978), Pitman, London, pp 481–484
Rokem JS, Schwarzberg J, Goldberg I (1984) Autoclaved fungal mycelia increase diosgenin production in cell suspension cultures of *Dioscorea deltoidea*. Plant Cell Rep 3: 159–160
Rokem JS, Tal B, Goldberg I (1985) Methods for increasing diosgenin production by *Dioscorea* cells in suspension Cultures. J Nat Prod 48: 210–222
Ross JH (1972) The flora of Natal. Mem Bot survey of South Africa No 39, Govern Print, Pretoria, RSA
Seibert M, Kadkade P (1980) Environmental factors. A. Light. In: Staba EJ (ed) Plant tissue culture as a source of biochemicals. CRC, Boca Raton, Florida, pp 123–141
Sengupta J, Mitra GC, Sharma AK (1984) Organogenesis and tuberization in cultures of *Dioscorea floribunda*. Plant Cell Tissue Organ Cult 3: 325–331
Sengupta J, Mitra GC, Sharma AK (1985) Excised root culture of *Dioscorea floribunda* for diosgenin synthesis. Indian J Exp Biol 23: 107–109
Singh JP (1978) Effect of nitrogen sources on shoot bud differentiation of *Dioscorea deltoidea* Wall. callus cultures. Biol Plant 20: 436–439
Staba EJ (1977) Tissue culture and pharmacy. In: Reinert J, Bajaj YPS (eds) Applied and fundamental aspects of plant cell, tissue, and organ culture. Springer, Berlin Heidelberg New York, pp 694–702
Staba EJ (1980) Secondary metabolism and biotransformation. In: Staba EJ (ed) Plant tissue culture as a source of biochemicals. CRC, Boca Raton, Florida, pp 59–97
Staba EJ, Kaul B (1971) Production of diosgenin by plant tissue culture technique. US Patent No 3.628.287
Takeda K (1972) The steroidal sapogenins of the Dioscoreaceae. In: Reinhold L, Liwschitz Y (eds) Progress in phytochemistry 3. Wiley London, pp 287–333

Tal B, Gressel J, Goldberg I (1982) The effect of medium constituents on growth and diosgenin production by *Dioscorea deltoidea* cells grown in batch cultures. Planta Med 44: 111–115

Tal B, Rokem JS, Goldberg I (1984a) Timing of diosgenin appearance in suspension cultures of *Dioscorea deltoidea*. Planta Med 50: 239–241

Tal B, Tamir I, Rokem JS, Goldberg I (1984b) Isolation and characterization of an intermediate steroid metabolite in diosgenin biosynthesis in suspension cultures of *Dioscorea deltoidea* cells. Biochem J 219: 619–624

Terui K, Okagami N (1989) Dormancy in *Dioscorea*: Rapid germination of detached embryos from dormant seeds of *Dioscorea tokoro*. Plant Cell Physiol 30: 287–293

Uduebo AE (1971) Effect of external supply of growth substances on axillary proliferation and development in *Dioscorea bulbifera*. Ann Bot 35: 159–163

Van der Zaag P, Fox RL (1981) Field production of yams (*Dioscorea alata*) from stem cuttings. Trop Agric (Trinidad) 58: 143–145

Viana AM, Felippe GM (1986) Effect of light and temperature on the seed germination of *Dioscorea composita*. Rev Bras Bot 9: 109–166

Viana AM, Felippe GM (1987) Endogenous factors on the rooting of leaf cuttings of *Dioscorea composita* Hemsl. Cienc Cult (Sao Paulo) 39: 618–622

Viana AM, Felippe GM (1988) Root formation in cuttings of *Dioscorea composita*. J Agric Sci 110: 451–456

Viana AM, Magalhaes EG, Felippe GM (1987) Diosgenin in leaves of *Dioscorea composita* Hemsl. Rev Bras Bot 10: 33–36

White PR (1954) The cultivation of animal and plant cells. Ronald, New York

II.17 Micropropagation of Glehnia (*Glehnia littoralis* Fr. Schmidt ex Miq.)

N. Hiraoka and M. Oyanagi[1]

1 Introduction

1.1 Botanical Description

The genus *Glehnia* belongs to the family Umbelliferae and comprises only one species, *G. littoralis* Fr. Schmidt ex Miq. (Japanese name: Hamabōfū). *Archangelica officinalis* DC., *G. leiocarpa* Mathias, and *Phellopterus littoralis* Benth. are its synonyms. Some taxonomists regard the North American species, *G. leiocarpa*, as distinct, distinguished by its less pubescent fruit. *G. littoralis* is a perennial herb growing on sandy shores of the northern Pacific countries such as China, Korea, Japan, the USSR, Canada, and the USA (Hiroe 1979). Figure 1 shows a plant at the

Fig. 1. A flowering plant of *Glehnia littoralis* Fr. Schmidt ex Miq.

[1] Niigata College of Pharmacy, 5-13-2 Kamishinei-cho, Niigata 950-21, Japan

Biotechnology in Agriculture and Forestry, Vol. 19
High-Tech and Micropropagation III (ed. by Y.P.S. Bajaj)

flowering stage. The chromosome number in a somatic cell is 2n = 22 (Kitamura and Murata 1978).

1.2 Usage and Chemical Constituents

The root and rhizome of *G. littoralis* are utilized as a crude drug in China (Pharmacopeia of the People's Republic of China 1985), Korea (Lee 1982), and Japan (Namba 1980). The alcoholic extract of the crude drug has antipyretic and analgesic action in rabbits, but the leaf extract and volatile oil from the root have no such effects (Taki and Hatasa 1960; Taki et al. 1960). The polysaccharides prepared from the root showed an immunosuppressive activity (Fang et al. 1987). The sprouting leaves are served as a tasty vegetable in Japan (Kariyone and Koiso 1985).

The root is rich in linear-type furanocoumarins, psoralen, marmesin, xanthotoxol, xanthotoxin, bergapten, imperatorin, isoimperatorin, cnidilin, alloisoimperatorin, 8-(1,1-dimethylallyl)-5-hydroxypsoralen, bergaptin (bergamottin), and 8-geranyloxypsoralen, along with coumarin derivatives, scopoletin, 7-O-(3,3-dimethylallyl)scopoletin, and osthenol-7-O-β-gentiobioside (Sasaki et al. 1980; Seo and Ryu 1976; Yang and Brown 1962). The root cortex has a higher imperatorin and isoimperatorin content than the inner part (Wang 1987). It also contains β-sitosterol (Yang and Brown 1962). The fruit contains bergapten, imperatorin (Yang and Brown 1962), and phellopterin (Noguti and Kawanami 1940) as furanocoumarins, and petroselic, petroselidic, palmitic, and linolic acids as the constituents of the lipids (Kurono and Sakai 1953). The presence of polyacetylenic compounds in the plant was confirmed by ultraviolet and infrared spectra and thin-layer chromatography (Moon and Yoon 1976a, b; Park et al. 1976).

1.3 Conventional Propagation and Need for In Vitro Culture

Wild *G. littoralis* plants grow solely on the sandy seashore. Their habitat is being reduced by construction work along the sea coast and/or plant collection by digging up by the roots for use as a crude drug or vegetable (Hatta et al. 1967a). Cultivation of the plant in the field has now started to meet the demands for the crude drug in China and Japan.

The plant can be propagated vegetatively from rhizome cuttings, but the method is not practical because of its lower proliferation rate and higher labor cost in comparison with sexual propagation (Hatta et al. 1967a). Consequently, seeds are used for practical production of the plants (Hatta et al. 1967a; Institute of Materia Medica, Chinese Academy of Medical Sciences 1979). The storage of seeds by stratification is effective to raise the germination capacity (Matsuoka et al. 1970; Ogawa et al. 1988). The seeds need an after-ripening period of 3–4 months at ca. 5 °C (Ogawa et al. 1988).

Hatta et al. (1967b) collected seeds from wild plants of *G. littoralis* at two different coastal locations and cultivated both plants in the same field. They found significant differences in various characteristics, such as root productivity and yield of aqueous ethanolic extract from the root. The chemical constituents in the roots varied qualitatively and quantitatively, depending on the habitats from which the plant materials had been collected (Oyanagi et al. 1990a, b). Those facts suggest that *G. littoralis* plants include heterogeneous populations. Since no efforts have been made to breed this plant, micropropagation would be the best method to obtain botanically and chemically homogeneous crops that offer the crude drug or vegetable in good and homogeneous quality, provided that an elite mother plant was used as the starting material to raise clonal crops.

2 Micropropagation

2.1 Initiation of Shoot Tip Culture

Axenic shoot culture can be initiated from shoot tips severed from wild adult plants as well as young cultivated seedlings. Both terminal and lateral buds can be used as explants to start shoot cultures. Shoot tips are surface sterilized by the usual method, e.g., treatment with 70% ethanol for 30 s and 2% sodium hypochlorite solution containing 0.1% Tween 20 for 15 min, followed by rinsing.

The shoot multiplication rate increased to more than 4 by the end of the fourth culture. The in vitro shoots were subcultured on LS medium (Linsmaier and Skoog 1965) supplemented with 3% sucrose, 1% agar, 1 μM 1-naphthalene-acetic acid (NAA), and 10 μM 6-benzylaminopurine (BAP).

2.2 Shoot Multiplication

Hormonal requirement for shoot multiplication was tested with various combinations of auxins and cytokinins. As shown in Fig. 2, BAP is the most effective for shoot multiplication of the four cytokinins tested, i.e., BAP, kinetin, N^6-(2-isopentenyl)adenine (2-iP), and zeatin. 2,4-D is inferior in shoot multiplication capacity to the other three auxins, i.e., indole-3-butyric acid (IBA), indole-3-propionic acid (IPA), and NAA. It also caused callus formation at the basal part of the shoots.

Figure 3 shows the effects on shoot multiplication of combinations of the two auxins and BAP in various concentrations. The shoots incubated on the IAA media were spindly, and NAA at 1 μM or higher concentrations stimulated callus formation. Therefore 0.3 μM NAA and 3 μM BAP is the optimal hormonal combination for shoot multiplication. Figure 4A shows a multiple shoot.

Table 1 shows the effects of shoot-inoculation density on multiplication. Double-shoot (not two single shoots) inoculation gave a more stable shoot multiplication rate than single-shoot inoculation. However, the multiplication rate

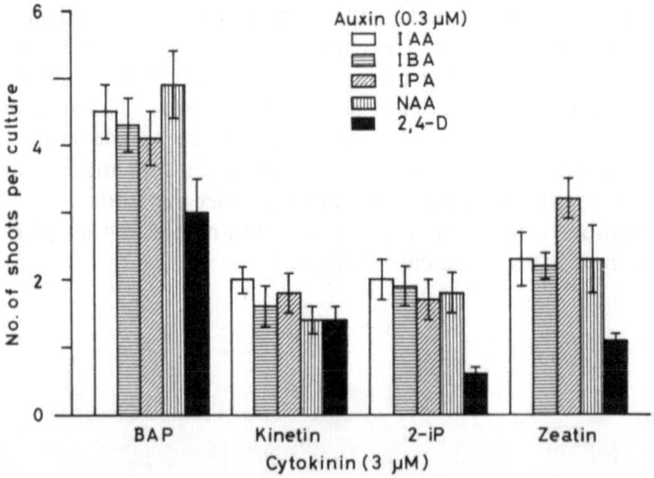

Fig. 2. Effects of auxins and cytokinins on shoot multiplication. The data were recorded after 30 days of incubation and each column represents the mean with standard error from 12 replicate cultures. (Hiraoka and Oyanagi 1988)

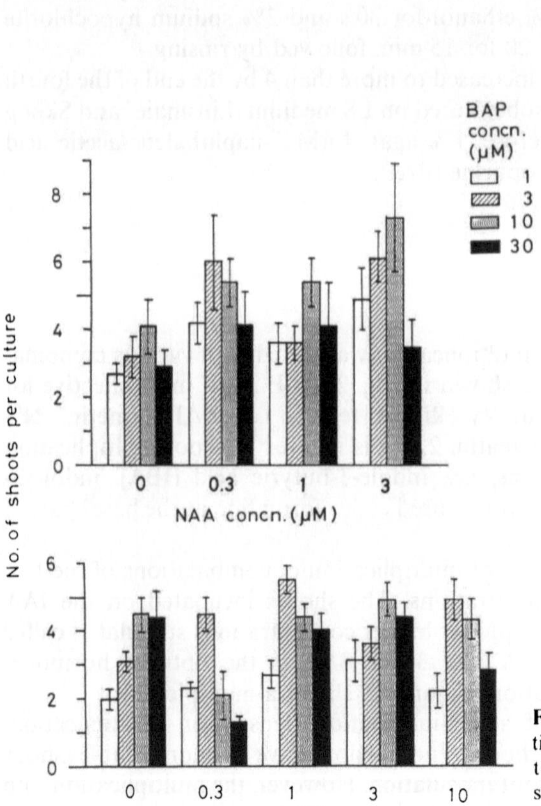

Fig. 3. Effects of auxin-BAP combinations on the multiplication of shoots cultured for 30 days. Each column represents the mean with standard error from 10–15 replicates

Fig. 4A–D. Micropropagation of *G. littoralis* by shoot tip culture. **A** A shoot incubated for 35 days on LS medium supplemented with 1 μM NAA and 10 μM BAP. **B** A rooted shoot 26 days after inoculation of a shoot on LS medium supplemented with 1 μM IBA. **C** A potted plantlet grown for 52 days in a greenhouse. **D** A flowering clonal plant in the 2nd year of cultivation in the field

of the former was lower than of the latter. Single-shoot inoculation is advantageous to obtain more shoots within a given period.

The shoot multiplication rate was more than 4 during the 1st year of subculture, but it decreased to about 3 in the 2nd year and to about 2 after that in the case of Gle 1, 2 and 3 lines (Table 2). Each of these three lines was derived from a single seedling, the seed of which was collected at the same place. The other two lines, Gle H and Gle Z, showed lower multiplication rates even during the first year. This may suggest a shift in the optimal culture conditions or in the multiplication capacity, depending on culture lines.

Table 1. Effects of the inoculum-shoot number on the multiplication rate of *Glehnia littoralis* shoot culture during ten consecutive cultures

Culture no.	Single-shoot inoculation					Double-shoot inoculation				
	N	Mean	SD	V	CV	N	Mean[a]	SD	V	CV
17	13	3.85	2.30	5.31	59.7	15	1.93**	0.90	0.82**	46.6
18	23	3.17	1.11	1.24	35.0	19	2.29**	0.67	0.45*	29.3
19	19	3.26	1.48	2.20	45.4	23	1.96**	0.58	0.34**	29.6
20	16	2.81	1.11	1.23	39.5	14	1.82**	0.54	0.29*	29.7
21	13	3.15	1.28	1.64	40.6	15	1.74**	0.56	0.32**	32.2
22	15	3.20	1.21	1.46	37.8	16	2.03**	0.85	0.72	41.9
23	28	2.25	1.04	1.08	46.2	20	1.98	0.68	0.46*	34.3
24	20	2.90	1.07	1.15	36.9	13	1.73**	0.70	0.48	40.5
25	20	2.15	1.31	1.71	60.9	12	1.75	0.40	0.16**	22.9
26	17	2.35	1.11	1.24	47.2	10	1.70	0.48	0.23*	28.2

N: The number of sample. SD: Standard deviation. V: Variance. CV: Coefficient of variation. *, **: Significantly different from the corresponding single-shoot inoculation by t-test (mean) or F-test (variance), $p < 0.05$ (*), $p < 0.01$ (**).
[a] Number of shoots per culture obtained was divided by 2.

Table 2. Changes in shoot multiplication rate of five shoot culture lines of *Glehnia littoralis* during subculture

Culture no.[b]	Line code[a]				
	Gle 1	Gle 2	Gle 3	Gle H	Gle Z
7–12	> 4	> 4	> 4	1.8 ± 0.2	1.8 ± 0.7
13–18	2.9 ± 0.1[c]	3.1 ± 0.8	3.3 ± 0.1	1.9 ± 0.1	1.5 ± 0.3
19–24	2.9 ± 0.7	3.0 ± 0.4	2.9 ± 0.6	1.6 ± 0.3	1.8 ± 0.2
25–30	2.1 ± 0.3	2.2 ± 0.2	2.1 ± 0.4	1.7 ± 0.3	2.0 ± 0.1
31–36	2.1 ± 0.3	2.2 ± 0.4	2.6 ± 0.6	1.7 ± 0.2	1.5 ± 0.2
37–42	2.0 ± 0.3	1.9 ± 0.6	2.0 ± 0.5	1.6 ± 0.3	1.6 ± 0.1

[a] Each line was derived from a single plant. The mother plants of Gle 1–3, Gle H and Gle Z were collected at Nagasaki, Shizuoka, and Niigata, respectively.
[b] The period of one culture is about 1 month.
[c] Mean ± standard deviation.

2.3 Rooting of Shoots

As is generally observed with other plant species (George and Sherrington 1984), cytokinins were inhibitory to the rooting of shoots. The addition of an auxin promoted both the percentage of rooted shoots and the number of roots per shoot (Fig. 5). The best growth regulators for the rooting medium were 3–10 μM IAA or 1 μM IBA, which induced no callusing.

Table 3 shows the effects of support systems on rooting. The number of roots per shoot was four to six in the 4th week of culture in all systems. The filter paper bridge method was inferior in percentage of rooted shoots to the other three systems, in which more than 90% shoots rooted after the 3rd week of incubation.

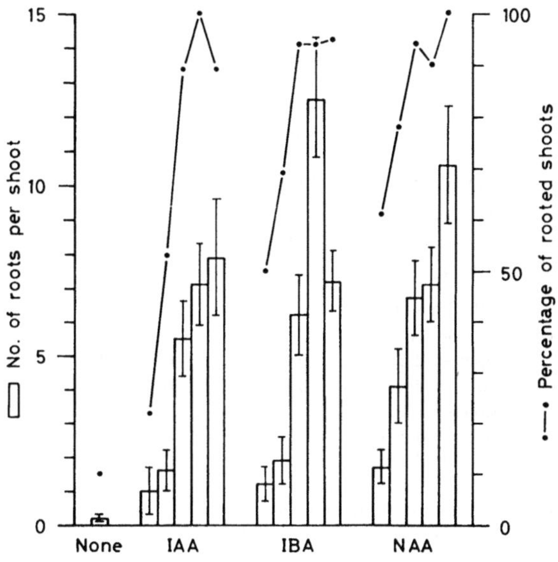

Fig. 5. Effects of auxins on rooting of shoots cultured for 31 days. The concentration of auxin is 0.1, 0.3, 1, 3 and 10 μM *from left to right* within each group of auxin. Each value is the mean of 20 replicate cultures. *Vertical lines* show standard error. (Hiraoka and Oyanagi 1988)

Table 3. Effects of support systems on the percentage of rooted shoots cultured on LS medium containing 1 μM IBA

Support system	Culture period (week)		
	2	3	4
Filter paper	63%	72	72
Gellan gum (0.4%)	79	96	100
Agar (0.7%)	32	92	96
Agar (1.0%)	17	92	92

The shoots on gellan gum medium rooted earlier than those on the other support systems.

2.4 Transfer of Plants

The rooted shoots (Fig. 4B) were transplanted to potting mixture containing sand, e.g., soil-sand (1:2), soil-sand-peat (1:1:1, 3:1:1) or perlite-sand-peat (1:1:1). Covering was not necessary; it even enhanced the deterioration or softening of the plantlets. The percentage of established plantlets was 60–75% under these conditions (Fig. 4C). About half of the leaves developed during in vitro culture remained and were growing 1 month after potting. The potted plants were transferred to the field. In the second year of cultivation, they bloomed and fruited (Fig. 4D).

Table 4. Characteristics of the leaf blade of plants grown in the field and in vitro. (Hiraoka and Oyanagi 1988)

Plant material	Plants in the field	Plantlets in vitro
Surface area per fresh weight (cm^2/g)		
Glehnia littoralis	46 ± 4	91 ± 6
Atractylodes lancea	88 ± 7	131 ± 17
Datura metel	67 ± 8	134 ± 21
Cuticle thickness (μm)		
Glehnia littoralis		
Upper epidermis	0.9 ± 0.1	0.5 ± 0.1
Lower epidermis	0.8 ± 0.1	0.3 ± 0.1
Atractylodes lancea		
Upper epidermis	0.8 ± 0.1	0.3 ± 0.1
Lower epidermis	0.8 ± 0.1	0.2 ± 0.1

Each set of figures represents the mean ± SD from 15 samples (surface area per fresh weight) or 50 observations (cuticle thickness).

2.5 Characteristics of Propagated Plants

No morphological abnormalities were observed in the leaves of plants propagated by this method. The chromosome number in root-tip cells of the clonal plants was exclusively $2n = 22$.

The plantlets of *G. littoralis* had a thicker leaf cuticle and smaller surface area per fresh weight of leaf blade than those of *Atractylodes lancea* and *Datura metel*, both of which needed covering at the acclimatization stage in pots (Table 4). These factors may contribute to the unique characteristic of *G. littoralis* plantlets, that they survive in pots without covering.

3 Conclusions

Sexual propagation is a practical method for *G. littoralis* cultivation. However, micropropagation would offer a complementary tool when more homogeneous quality in the crude drug or the rapid clonal propagation of an elite line is desired, because the plant has heterogeneous characteristics. The micropropagation of the plant can be successfully achieved, as has been illustrated in this chapter.

4 Protocol

1. The shoot tips were dissected from the surface sterilized terminal or lateral buds and incubated at 25 °C in a photoperiod of 16 h/d fluorescent light (about 1000 lx) on LS medium supplemented with 1 μM NAA and 10 μM BAP. Shoot tips from adult plants as well as young seedlings can be used as explants.
2. Multiple shoots were obtained by culturing single shoots on LS medium containing 0.3 μM NAA and 3 μM BAP under the same conditions as above.
3. The shoots are rooted on LS medium with 3–10 μM IAA or 1 μM IBA.
4. The rooted shoots can be potted without covering. After establishment of the plantlets, they are transferred to the field.

References

Fang XD, You M, Ying WB, Sun ZM, Shen PZ, Shi ZY, Ye QW (1987) The immunosuppressive activities of polysaccharides from *Glehnia littoralis* (Bei Sha Shen). Acta Pharm Sin 21: 931–934
George EF, Sherrington PD (1984) Plant propagation by tissue culture. Exegetics, Eversley, England
Hatta R, Goto M, Matsuoka T, Imai S, Watanabe K (1967a) Studies on cultivation of *Glehnia littoralis* Fr. Schmidt. Part I. A fundamental investigation. Takeda Kenkyusho Nempo 26: 84–97
Hatta R, Goto M, Matsuoka T, Nagao Y, Okumura Y (1967b) Studies on cultivation of *Glehnia littoralis* Fr. Schmidt. Part II. Studies on the comparison between two strains. Takeda Kenkyusho Nempo 26: 98–106
Hiraoka N, Oyanagi M (1988) In vitro propagation of *Glehnia littoralis* from shoot tips. Plant Cell Rep 7: 39–42
Hiroe M (1979) Umbelliferae of the world. Ariake, Tokyo
Institute of Materia Medica, Chinese Academy of Medical Sciences (1979) Zhong Yao Zhi. Renmin Weisheng Chupan She, Beijing
Kariyone T, Koiso R (1985) Atlas of medicinal plants. Nippon Rinshosha, Osaka
Kitamura S, Murata G (1978) Colored illustration of herbaceous plants of Japan (Choripetalae). Hoikusha, Osaka
Kurono G, Sakai T (1953) Studies on fatty acids from fruit oil of Umbelliferae plants. IX. Fatty acids from the fruit oil of *Phellopterus littoralis* Benth. Yakugaku Zasshi 73: 608–610
Lee TB (1982) Illustrated flora of Korea. Hyang-moon-sa, Seoul
Linsmaier EM, Skoog F (1965) Organic growth factor requirements of tobacco tissue cultures. Physiol Plant 18: 100–127
Matsuoka T, Okumura Y, Nagao Y (1970) Studies on cultivation of *Glehnia littoralis* Fr. Schmidt. Part III. The storage of seeds. Takeda Kenkyusho Nempo 29: 249–253
Moon CK, Yoon MH (1976a) The screening of polyacetylenic compounds in Korean plants. I. Yakhak Hoe Chi 20: 44–46 (CA 85: 119630)
Moon CK, Yoon MH (1976b) Screening of polyacetylenic compounds in Korean plants. I. Soul Taihakkyo Saengyak Yonguso Opjukjip 15: 83–85 (CA 88: 47502)
Namba T (1980) Colored illustration of Wakan Yaku (the crude drugs in Japan, China and the neighboring countries), vol 1. Hoikusha, Osaka
Noguti T, Kawanami M (1940) Constituents of *Phellopterus littoralis* Benth. (1). Yakugaku Zasshi 60: 57–62
Ogawa T, Mizukami H, Ohashi H (1988) Cultivation and breeding of *Glehnia littoralis* (1). Storage of seeds by stratification and optimum temperature for germination. In: Tomita Y (ed) Proc 35th Annu Meet Jap Soc Pharmacognosy, 16–17 Sep, Niigata
Oyanagi M, Hiraoka N, Tomita Y, Ogawa T, Mizukami H, Ohashi H (1990a) High-performance liquid chromatographic analysis of furanocoumarins in *Glehnia littoralis*. Shoyakugaku Zasshi 44: 219–224
Oyanagi M, Hiraoka N, Tomita Y, Ogawa T, Mizukami H, Ohashi H (1990b) Variability of the

furanocoumarin composition and isozyme pattern in *Glehnia littoralis* of different geographical origin. Shoyakugaku Zasshi 44: 323–327
Park DS, Moon CK, Park NS (1976) The screening of polyacetylenic compounds in Korean plants. Soul Taehakkyo Yakhak Nonmunjip 1: 132–135 (CA 87: 114583)
Pharmacopeia of the People's Republic of China (1985) Pt 1. Renmin Weisheng Chupan She, Beijing
Sasaki H, Taguchi H, Endo T, Yoshioka I (1980) The constituents of *Glehnia littoralis* Fr. Schmidt ex Miq. Chem Pharm Bull 28: 1847–1852
Seo YK, Ryu KS (1976) Components of Glehniae Radix. Kor J Pharmacog 7: 233–235
Taki K, Hatasa M (1960) Pharmacological action of *Phellopterus littoralis* Benth. II. On the analgesic action. Gifu Ika Daigaku Kiyo 8: 471–474
Taki K, Hatasa M, Oguchi S (1960) Pharmacological action of *Phellopterus littoralis* Benth. I. On the antipyretic action. Gifu Ika Daigaku Kiyo 8: 464–470
Wang J (1987) TLC-UV spectrophotometric determination of imperatorin and isoimperatorin in Radix Glehniae. Bull Chin Mater Med 12: 38–40
Yang CH, Brown SA (1962) Furanocoumarins of *Phellopterus littoralis*. Can J Chem 40: 383–385

II.18 Micropropagation of Guayule (*Parthenium argentatum*)

XU HUIZHU and QIAN MINZHI[1]

1 General Account

1.1 Botanical Description

Guayule is a member of the family Compositae, and belongs to the genus *Parthenium* (Fig. 1). It is a rubber-producing shrub, native to a semi-arid area in north-central Mexico and southwestern Texas, USA. The plant, with chromosome numbers of 2n = 36 to 108 or more, has narrow leaves, covered in a drought-protecting white wax, that alternate along the stems, and a canopy of small flowers

Fig. 1. A 4-year old guayule plant growing in Wuhan, China. (Xu Huizhu et al. 1988)

[1] Wuhan Institute of Botany, Academia, Sinica, Wuhan 430074, China

Biotechnology in Agriculture and Forestry, Vol. 19
High-Tech and Micropropagation III (ed. by Y.P.S. Bajaj)

borne on exceptionally long stems. Only about 60 cm high, it develops a taproot that may penetrate the soil more than 6 m, supplemented by extensive fibrous roots to absorb moisture from a considerable area of desert soil and thus withstand periodic drought. It may survive 30 or 40 years under desert conditions where annual rainfall may be less than 250 mm (Rollins et al. 1977).

1.2 Distribution and Importance

In addition to Mexico and the United States, guayule has been successfully grown in Spain, Turkey, Israel, Argentina, south Australia, the USSR, China, and India. Guayule has been one of the continuing sources of commercial rubber. Guayule rubber is not contained in ducts, but in single thin-walled cells, two-thirds of the rubber being in the stems and branches, the remainder in the roots; there is no rubber in the leaves. On native guayule bushes rubber constitutes, on the average, about 10% of the total dry weight of the plant. In the 1940s, strains containing up to 26% rubber were found.

Resin ducts, found throughout the shrub, constitute 10–15% of the total dry weight of the plant, and include terpenes, sesquiterpenes, diterpenes, glycerides, and low molecular weight polyisoprene hydrocarbons. It is conceivable that the resins could prove more valuable than the rubber.

The cuticle wax covering on the leaves of guayule comprises about 2.5% of the weight of the dry leaves. It is a hard wax with one of the highest melting points (76 °C) ever recorded for a natural wax. This is very important today because hard waxes are in universal demand (Rollins et al. 1977).

Our staff, Zhu Xinqiang and his colleagues, have identified the main constituents of the essential oil distilled from fresh leaves of guayule as: 1.8-cineole (22.06%), α-pinene (22.82%), β-pinene (20.78%), borneyl acetate (10.14%), sabinene (6.18%), and camphene (4.56%), which is another valuable by-product of guayule.

Guayule is a hardy plant and will survive drought for several years. Its ability to survive and accumulate rubber without maintenance may make guayule an excellent crop for stopping soil erosion, and even for grazing – both of which are important for increasing the productivity of semi-arid wasteland.

1.3 Conventional Propagation and Breeding

Guayule is usually propagated by nursery-grown seedlings and sometimes by cuttings. Vigorously growing plants bloom and seed continuously throughout summer and fall; a plant can yield several thousand seeds after a single rainfall and, if stored carefully, these can remain viable for several decades.

Guayule has much inherent genetic variability and is amenable to genetic improvement. The guayule types of 2n = 36 are completely sexual and reproduce in the usual way by double fertilization; the guayule plants of higher chromosome numbers reproduce asexually by apomixis. The sexual types of plants can develop hybrids with useful characteristics. These hybrid plants can then be induced into apomictic forms to replicate the characteristics of the hybrid for generation after generation. Guayule can be hybridized with other *Parthenium* species to establish plants with improved characteristics.

1.4 Need for In Vitro Culture

During the first decade of this century, the guayule bush attracted attention as a source of natural rubber. In 1910, about 50% of the US rubber was extracted from wild guayule shrubs.

Concurrent with US efforts, studies on guayule were made in Australia, South Africa, India, Mexico, Morocco, Senegal, and Brazil. In addition, China, Pakistan, Israel, Argentina, Egypt, the USSR, and Saudi Arabia are also interested in guayule; however, there are currently no commercial guayule plantations anywhere in the world. Improving guayule through breeding and selection for increasing the percentage of rubber in the plant, and developing a more rapidly growing shrub are essential for commercial production. Guayule's resin content, disease resistance, ease of defoliation, ability to compete with weeds, and cold and drought tolerance are also amenable to improvement.

In vitro culture could greatly facilitate this improvement. Guayule has considerable genetic variability. In nature it exists in a large number of strains – indeed almost every plant is a separate strain – and plant breeding is the main key to unlocking its potential. With the use of plant tissue culture, pure breeds of guayule with the same genotype could be established, and large-scale propagation of plants for rapid breeding is feasible. Direct selection of new strains through control of culture conditions in vitro, and extraction of rubber and other byproducts from cell cultures will be feasible as the technique of in vitro cell culture develops.

2 Brief Review of Tissue Culture Work

In vitro culture of guayule began with the pioneering work of Arreguin and Bonner in 1950. They studied the effects of leaf extracts and other biochemicals on guayule tissue cultures for the production of rubber, but continued growth of callus past the first culture period was not obtained.

In the past 10 years, other reports on tissue culture work have been published. In 1980, LaBrecque reported that Murashige and Radin were developing somatic embryoids and haploid plants for guayule selection and propagation and cell cultures that might produce rubber. The effects of 2(3,4-dichlorophenoxy)-triethylamine on guayule tissue cultures were studied by Dastoor et al. (1981), and Zavala et al. (1982) reported the callus induction and organogenesis in cultured tissues. Radin et al. (1982) and Pagano and Staba (1983) reported on secondary products from cultures. In 1983 Staba and Nygaard reported that nine different strains of guayule had been established as tissue culture. Shoot cultures developed from germinated seeds, whole plant cultures developed from shoot cultures with regenerated roots, and might possibly be developed from root cultures with nodules that formed shoots. Suspensions of large cell aggregates developed from callus grown in liquid medium. Smith (1983) described also a method for propagation in vitro, whereby transplanting and greenhouse establishment of plantlet cultures from excised buds had been carried out successfully. In 1984, Xu Huizhu and Qian Minzhi reported results of a study on quick propagation. Visser and Trautmann (1984) reported accumulation of unidentified cyclohexane extractable compounds in suspension cultures. In 1986, Norton and Rodriguez summarized measurements of the molecular weight, distribution, amounts of rubber, and regulation of terpenoid rubber in normal and crown gall tissue cultures. In 1989 Trautmann and Visser described a new flow-through system to inhibit browning in callus cultures of guayule. Recently, Dhar et al. (1989) cultured nodal explants and obtained callus and shoots.

3 Micropropagation

3.1 General

In 1981, guayule seeds were introduced into our institute in a small quantity with about 6% germination. The progeny plants produced tiny seeds, often without embryos. The present study began in February 1983 with the in vitro culture of shoot tips and young leaves. By pretreatment of the mother plants in the culture room and by using growth regulators in the medium, whole plants were induced and transplanted successfully. An asexual strain from a single bud was established through experiments over a period of 8 months, and about 2800 plants survived. At the same time, an effective propagational procedure via tissue culture was set up. We breed asexual strains of guayule, selected according to cold tolerance in the field, to adapt to the cold weather conditions in the winter in Wuhan.

3.2 Materials and Methods

Shoots about 1 cm long in tender leaves from juvenile guayule plants about 15 cm in height, obtained by sowing or cutting, were planted in pots and placed in the culture chamber at 26 ± 1 °C and 1000–2000 lx, then soaked shortly in 70% alcohol, and surface sterilized with 0.1% mercurous chloride for 3 min. This treatment was followed by a sterile distilled water rinse three or four times. Finally, shoot tips of about 0.5 cm long and tender leaf segments of about 0.5×0.5 cm^2 in size were cultured on MS medium containing various growth regulators. All media, together with 0.6% agar and 2% sucrose, were adjusted to pH 5.8 prior to being sterilized by autoclaving at 120 °C for 15 min.

3.3 Shoot Induction

Shoot tips were cultured on MS medium containing 0.5 mg/l 6-benzylaminopurine (6-BA), which induced successful growth of the terminal buds. These grew and produced shoot cultures 2–3 cm in height, and then sprouted axillary buds and branched (Fig. 2A) in 20 days.

Shoot cultures were excised aseptically and transferred to MS medium containing both 6-BA and IAA for 10–15 days; they grew vigorously and developed into bushy shrublets without roots as the axillary buds sprouted in class after class (Fig. 2B). Further development of shoot cultures in large quantities was carried out in subcultures of the clumps of excised shoot cultures. It was necessary for development of shoot cultures and decrease in translucent shoot cultures, that the

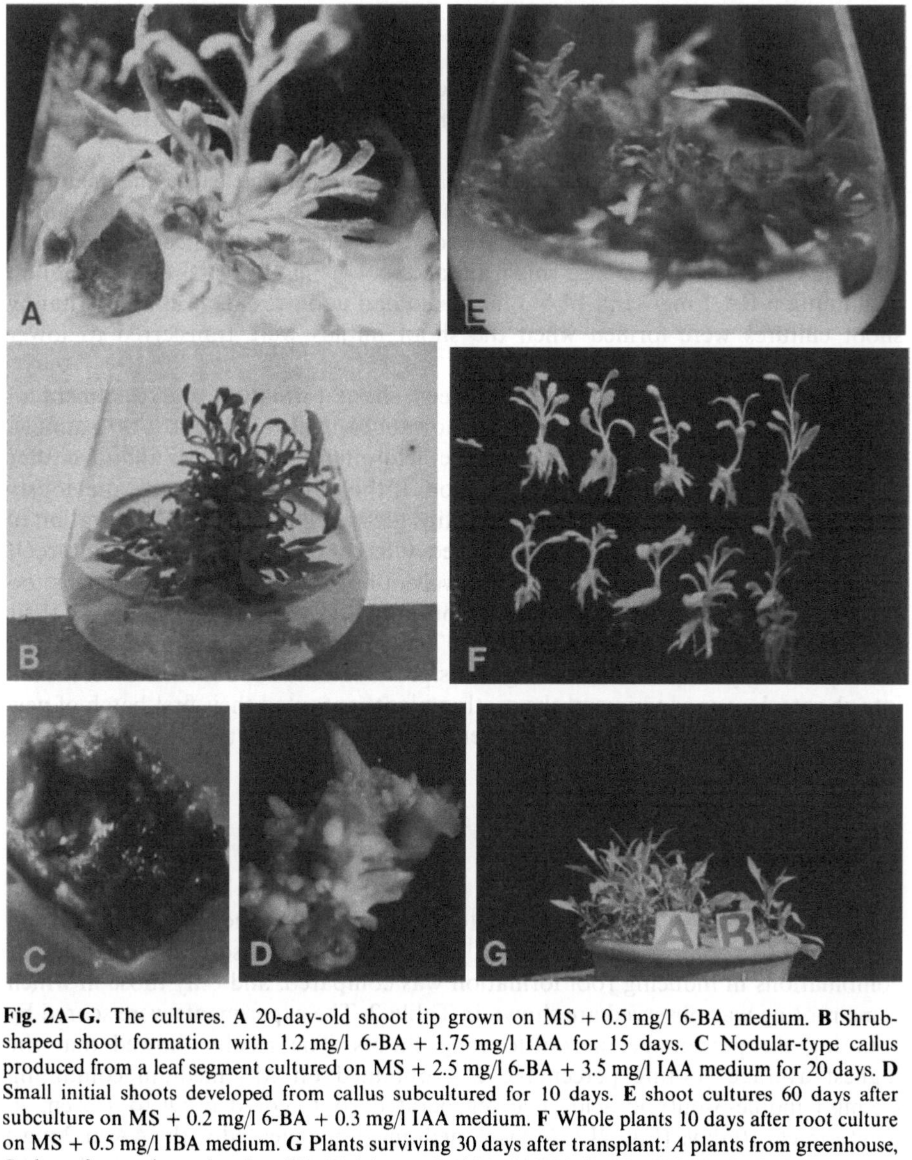

Fig. 2A–G. The cultures. **A** 20-day-old shoot tip grown on MS + 0.5 mg/l 6-BA medium. **B** Shrub-shaped shoot formation with 1.2 mg/l 6-BA + 1.75 mg/l IAA for 15 days. **C** Nodular-type callus produced from a leaf segment cultured on MS + 2.5 mg/l 6-BA + 3.5 mg/l IAA medium for 20 days. **D** Small initial shoots developed from callus subcultured for 10 days. **E** shoot cultures 60 days after subculture on MS + 0.2 mg/l 6-BA + 0.3 mg/l IAA medium. **F** Whole plants 10 days after root culture on MS + 0.5 mg/l IBA medium. **G** Plants surviving 30 days after transplant: *A* plants from greenhouse, *B* plants from culture chamber. [Xu Huizhu et al. 1984a, b, 1986 (**B, G**), 1988 (**A, C, D, E, F**)]

Table 1. Effect of different media on callus induced from leaf segments. (Xu Huizhu 1988)

Media	Hormones (mg/l) 6-BA	IAA	IBA	No. cultured	No. induced	Rate (%)
	1.20	1.75	0	16	10	62.5
MS	2.50	3.50	0	14	12	85.7
	1	0	3	25	11	44
	2.50	3.50	0	14	0	0
N6	1	0	3	20	3	15

concentration of growth regulators in the medium alternated between high and low. In this experiment 6-BA 1.2 mg/l with IAA 1.75 mg/l and 6-BA 0.2 mg/l IAA 0.3 mg/l were used.

Shoot cultures were also induced from leaf segments. The leaf segments on MS medium containing 6-BA 1.2–2.5 mg/l and IAA 1.75–3.5 mg/l for 20–30 days produced some green nodular-type callus on their section or surface (Fig. 2C), and then the callus developed into small initial shoots (Fig. 2D). MS or N6 medium containing 6-BA 1 mg/l and IAA 3 mg/l also had good results (Table 1). Clumpy shoot cultures were formed when the shoot initials were transferred to lower hormone concentrations (Fig. 2E).

A close relation was observed between shoot formation and treatment of mother plants, as well as the levels of hormones in the medium. For example, explant shoot tips taken straight from the field often turned black and died after surface sterilization after the above method. If the mother plants were previously treated in the culture chamber, the shoot tips grew well in culture. The duration of pretreatment of the mother plants produced different effects on the tissue culture. If the pretreatment was too long and the levels of hormones in the medium was too low, the shoot tips could not develop normally and the leaf segments lost their viability, and sometimes turned white. The hormone levels should be raised correspondingly to make the leaf segments produce callus and further differentiate into shoots. In our experiment, the mother plants sprouted their first batch of new shoots after being in the culture chamber for 30–40 days, and these shoots were the best explants.

3.4 Root Induction

When the shoot cultures had grown to 1 cm or longer, they were excised and transferred to root induction medium. The effectiveness of several growth factor combinations in inducing root formation was compared, and only those in which some rooting was observed are shown in Table 2. The experiments were scored by counting the percentage of root-forming shoots. The results showed that the highest average rooting percentage was obtained on MS medium containing 0.5 mg/l indole-3-butyric acid. The regenerated roots could grow to 1–2 cm in length in ca. 10 days (Fig. 2F). A basal medium supplemented with IAA at 2 mg/l also showed good rooting effect. The great majority of shoots rooted in 10 days. In

Table 2. Effect of media on rooting rate of shoot cultures induced from shoot tips. (Xu Huizhu and Qian Minzhi 1984a, b)

Media	H + 2 mg/l IAA			MS + 0.5 mg/l IBA		
Batch	No. cultured	No. rooted	Rate (%)	No. cultured	No. rooted	Rate (%)
1	105	98	93	115	104	90
2	69	69	100	75	74	99
3	25	24	96	50	48	96
4[a]	123	93	76	456	419	92

[a] This batch of shoot culture rooted under varying natural temperature (22–44 °C) and illumination (3000–6000 lx at 10.00 h.) in the greenhouse, others rooted in culture room.

addition, on MS medium containing low concentration of 6-BA, thin root cultures were occasionally induced when the medium was almost depleted.

There was a difference between rooting potential of different shoot cultures induced from shoot tips and from leaf segments. During rooting, the growth of the shoot cultures induced from leaf segments was checked to a certain extent, so that transplantation of the plantlets was not easy.

3.5 Induction of Strong Plant Cultures

Environmental conditions were tested for culture of guayule in vitro. In the high temperature season from July to September in Wuhan, shoot cultures were put into a sunny room for subculture and it was shown that the axillary buds had a great capacity to sprout and a large number of strong shoot cultures were obtained in a short time. This result indicated that diffused sun light was superior to the artificial light provided by fluorescent tubes of 40 W for shoot cultures to grow vigorously.

According to the growth habit of guayule, the shoot cultures were treated under the variance of natural temperature (22–40 °C) and illumination (3000–6000 lx at 10.00 h.) in the greenhouse during the rooting period, and vigorous growth was

Table 3. Survival rate of transplanted whole plants. (Xu Huizhu and Qian Minzhi 1984a, b)

Place of sand bed	Shade screen	Days of shade	Weather	No. transplants	No. survival	Rate (%)
	Tree	30	Sun	300	216	72
Open air[a]	Reed	3	Rain	460	338	73
	Reed	3	Sun	945	882	93
Greenhouse	Reed	30	Rain	987	151	16
	Reed	6	Sun	300	265	88

[a] Covered with plastic film for first 6 days.

obtained. Generally, one shoot culture produced an average of five root cultures. The regenerated roots grew to 1–2 cm long in 10 days on root induction medium, when the plant cultures could be transplanted. The in vitro-raised plants looked tall, strong, and dark green, and a higher percentage of plant cultures survived transplanting than of those not treated in the greenhouse (Table 3 and Fig. 2G).

3.6 Transplantation

After the shoot cultures had established a root system on the agar root induction medium, they were taken out, washed free of all agar, and transplanted into soil mix of sand/soil 2:1 or sand beds of 30 cm in depth (Fig. 3A). The transplanted plants were maintained in the greenhouse or open air out of the wind and exposed to the sun, covered with plastic film at the beginning, until the plant cultures were well established. Their row spacing and spacing in the rows on the sand beds were 8–10 cm and 3–4 cm, respectively.

If the transplanting was done in the high temperature season and sunny weather, the plants were properly shaded. When it was raining, the plastic film cover on the plants was opened to ventilate the sand beds.

It was shown in the experiments that the plant cultures could withstand temperature up to 50 °C and illumination up to 50000 lx in the course of transplantation without harm to their growth. The survival rate became higher as the illumination was intensified; if proper humidity was ensured, strong illumination helped stimulate stem tissue xylogenesis and adaptation. The process of transplantation into open air sand beds is quite simple and thus suitable for propagation on an industrial scale (Table 3).

3.7 Cutting of Tender Branches from Plant Cultures

Cuttings from plant cultures could be used for propagation both to raise propagation numbers and to reduce production costs. When the plants were transplanted, their tender branchlets were excised as cuttings for sand beds beside the transplanted plants. The cuttings began rooting after 8–10 days (Fig. 3B), under varying natural temperature of 12–47 °C, and the best survival rate was as high as 95%. The cuttings taken from plant cultures rooted more easily than those from the mature plants into which the seeds developed.

In cuttings both aeration and moisture were necessary for successful rooting. The air spaces in the sand must not be logged with water, and the air must be kept humid.

3.8 Final Planting

The propagules, consisting of the strong plants derived from transplantation and cutting (Fig. 3C) were dug out, and finally planted in the field on a cloudy day in spring, the best season for final planting. The propagules started to bloom after 3 months and grew to a height of 40–50 cm in 1 year (Fig. 3D).

Fig. 3A–D. Plants from tissue culture. **A** Plant cultures 4 weeks after transplantation into open air sand bed. **B** Rooted plantlets 2 weeks after cutting of tender branches from plant cultures. **C** A vigorous plant 8 weeks after transplantation. **D** 1-year-old guayule plants finally planted in the field. (Xu Huizhu and Qian Minzhi 1984a, b, 1986)

4 Summary and Conclusion

In the present study, by using the cytokinin 6-benzylaminopurine and the auxin indole-3-acetic acid in MS nutrient medium, shoot cultures were induced and recycled to produce greater numbers. Shoot cultures rooted well in indole-3-butyric acid medium. Plant cultures were transplanted successfully, and cuttings from tender branches of plant cultures were also successful.

Whenever vegetative propagation is economically possible, it is generally preferred to sexual propagation, because genetic characteristics are better maintained. The common method of vegetative propagation for guayule is the rooting of cuttings; however, this is far from being universally applicable. Our tests show that the cuttings rooted very slowly and had a low rooting percentage. The success of tissue culture opens a new method for vegetative propagation of guayule; particularly the culture of shoot tips in vitro is applicable for quick propagation.

As a general rule, where tissue culture is used as a means of vegetative propagation, the period of unorganized growth should be kept as short as possible to minimize the chances of genetic change. For this reason, the culture of shoot tips in vitro in guayule is a perfect example, in this study, because the callus cultures are avoided in the course of plant regeneration, and the same genetic characteristics have been obtained practically from the same explant.

Guayule is a shrub native to the semi-arid region where the sun is strong and the temperature varies greatly between day and night. Its cultures are also capable of adapting to varying natural temperatures and strong illumination, and grow vigorously. This characteristic makes the tissue culture of guayule easy. In the summer, there is no need to use a special culture chamber or greenhouse to obtain good results.

Important characteristics of guayule, for example rubber content, water resistance, resistance to cold, etc. are not evident in juvenile plants; sometimes, explants have to be taken from older plants in the field for propagation of high-quality single plants. In this case, the culture of leaf segments is better than that of shoot tips because the leaves can be surface sterilized without harming their development.

5 Protocol

1. Best Explant
   Put the mother plants in the culture chamber until new shoots are growing; excise the shoot tips for culture
2. Best Medium
   Growth medium: MS + 0.5 mg/l 6-BA; subculture medium: MS + 0.2 or 1.2 mg/l 6-BA + 0.3 or 1.75 mg/l IAA; rooting medium: MS + 0.5 mg/l IBA
3. Optimal Condition
   26 °C, 2000 lx to produce shoots; 20–40 °C, 4000 lx to root and become vigorous; open-air sunny sand beds for transplanting

References

Arreguin B, Bonner J (1950) The biochemistry of rubber formation in the guayule. II. Rubber formation in aseptic tissue cultures. Arch Biochem 26: 178–186

Dastoor MN, Schubert WW, Petersen GR (1981) Preliminary results of in vitro propagation of guayule. J Agric Food Chem 29: 686–688

Dhar AC, Kavi Kishor PB, Rao AM (1989) In vitro propagation of guayule (*Parthenium argentatum*) – a rubber-yielding shrub. Plant Cell Rep 8: 489–492

LaBrecque M (1980) Guayule bounces back. Mosaic (NSF) 11: 30–37

Norton RA, Radin DN, Rodriguez E (1986) Rubber synthesis in crown gall and normal tissue cultures of guayule. Guayule – A natural rubber source. Proc 4th Int Conf Guayule Res and Dev, Guayule Rubber Soc, pp 43–50

Norton RA, Rodriguez E (1986) Regulation of terpenoids and crown gall tissue cultures of guayule. Biochemistry and regulation of *cis*-Polyisoprene in Plants. Proc Workshop, Texas A and M Univ, College Stn, Texas Oct 5–6, pp 106–118

Pagano SA, Staba EJ (1983) The effects of growth regulators and environmental conditions on rubber production in guayule callus-suspension and liquid shoot tissue cultures. In: Program and summaries of the Guayule Rubber Society, Inc., 4th Annu Conf, June 20–23, 1983. University of California, Department of Botany and Plant Sciences, Riverside, California, p 29

Radin DN, Behl HM, Proksch P, Rodriguez E (1982) Rubber and other hydrocarbons produced in tissue cultures of guayule (*Parthenium argentatum*). Plant Sci Lett 26: 301–310

Rollins RC, Allen PJ et al. (1977) Guayule: an alternative source of natural rubber. Nat Acad Sci, Washington, DC

Smith MK (1983) In vitro propagation of guayule (*Parthenium argentatum* Gray). Plant Sci Lett 31: 275–282

Staba EJ, Nygaard BG (1983) In vitro culture of guayule. Z Pflanzenphysiol 109: 371–378

Trautmann IA, Visser JH (1989) Development of a liquid flow-through system to inhibit browning in callus cultures of guayule (*Parthenium argentatum* Gray). Plant Cell Tissue Organ Cult 16: 39–46

Visser JH, Trautmann IA (1984) Growth of guayule in suspension culture. Program and summaries, Guayule Rubber Soc, 5th Annu Conf, Washington, DC 84, June 17–21

Xu Huizhu, Qian Minzhi (1984) A study on quick propagation of guayule in vitro. Wuhan Bot Res 2 (2): 306–310

Xu Huizhu, Qian Minzhi (1984a) A study on quick propagation of guayule in vitro. Int Symp Genetic manipulation in crops, Oct 22–26, Beijing, China

Xu Huizhu, Qian Minzhi (1984b) A study on quick propagation of guayule in vitro. Wahan Bot Res 2(2): 306–310

Xu Huizhu, Qian Minzhi (1986) In vitro propagation of guayule. Tissue culture of woody plants and application. Higher education publishing house of China, 501–507.

Zavala ME, Biesboer DD, Mahlberg PG (1982) Callus induction and organogenesis in culture tissues of *Parthenium argentatum*. Phytomorphology 32: 340–344

Zhu Xingiang, Wang Guoliang, Wang Jinfeng, Jia Weijiang (1989) A study on the chemical constituent of essential leaves. Acta Bot Sin 31(11): 895–896

II.19 Micropropagation of *Pinellia ternata*

Y. SHOYAMA[1], I. NISHIOKA[1], and K. HATANO[1,2]

1 Introduction

Pinellia ternata (Araceae) is a perennial herb which grows wild in China, Korea, and Japan. It has a global tuber about 1 cm in diameter below the ground and a bulbil on the underground part of the petiole, and also multiplies in this way. The epidermis of the tuber is removed and dried to give the crude drug. This crude drug, called Pan-hsia in Chinese and Hange in Japanese, is used in approximately 20% of all important prescriptions in Chinese medicine, being prescribed together with other herbal drugs for vomiting, especially for treating morning sickness. The tuber contains homogentisic acid, its glucoside (Hasegawa et al. 1959), 3,4-dihydroxybenzaldehyde, its diglucoside (Suzuki 1969), and ephedrine (Oshio et al. 1978; Fig. 1).

The chromosome variation as observed in Japan is described in Table 1 (Hatano and Shoyama, unpubl.). The main somatic chromosome number is 2n = 104. Differences in the quality and quantity of the crude drug may depend on

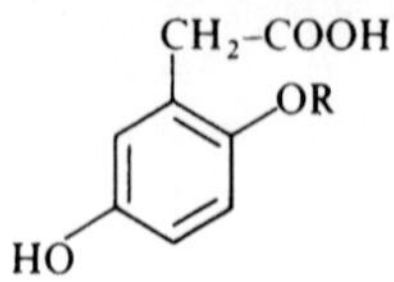

R = H: homogentisic acid
R = glucosyl: homogentisic acid glucoside

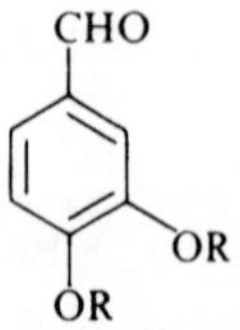

R = H: 3,4-dihydroxybenzaldehyde
R = glucosyl: 3,4-dihydroxybenzaldehyde diglucoside

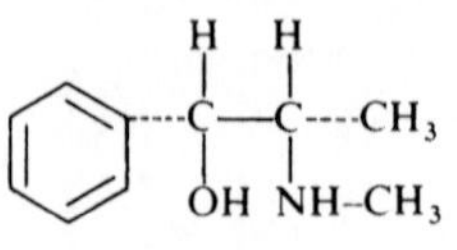

l-ephedrine

Fig. 1. Constituents of the crude drug of *Pinellia ternata*

[1] Faculty of Pharmaceutical Sciences, Kyushu University, Higashiku, Fukuoka 812, Japan
[2] Present address: Cooperate R and D, Plant Technology Institute, Ube Industries Ltd., 1978-5 Ogushi, Ubeshi, Yamaguchi 755, Japan

Biotechnology in Agriculture and Forestry, Vol. 19
High-Tech and Micropropagation III (ed. by Y.P.S. Bajaj)

Table 1. Chromosomal analysis of *Pinellia ternata* grown in Japan

Chromosome number	Growth place (prefecture)
2n = 96	Tokushima
104	Hokkaido, Aichi Hiroshima, Fukuoka
108	Yamaguchi
118	Yamanashi
128	Okayama

the strain, although there is still some confusion about the relation between these constituents and the pharmacological activities of the drug. The large-scale cultivation of this plant has not been established as yet because of the difficulty in collecting seedlings or bulbils, which are mainly used for the traditional propagation.

In this chapter three procedures of micropropagation are presented: (1) direct micropropagation using tuber segments, (2) propagation in liquid medium through a protocorm-like body induced from in vitro tuber segments, and (3) the culture of callus induced from in vitro leaf segments.

2 Direct Propagation from Tuber Segments – Procedure I (Shoyama et al. 1983a)

2.1 Materials and Methods

Tuber segments of *Pinellia ternata* (2n = 104), measuring 1 cm in diameter, were taken from a greenhouse-grown collection. The explants were washed with tap water, surface sterilized with 3% NaOCl containing 0.1% Tween 80, and washed twice thoroughly with sterilized water. The explants were cut aseptically (2-mm cubes) and the epidermis peeled off. Twenty segments were cultured individually and the average was maintained in all cases. The basal medium consisted of MS salts (Murashige and Skoog 1962) with (in mg/l): myoinositol, 100; glycine, 2; sucrose, 30000; agar, 8000 (solid medium) or without agar (liquid medium) supplemented with auxins such as 2,4-D, NAA, IAA, and kinetin as a cytokinin in concentrations and combination as shown in Table 2 and Figs. 3, 4, 5. All culture tubes contained 30 ml of medium adjusted to pH 5.5 before autoclaving. The cultures were exposed to continuous light from cool white fluorescent tubes at a temperature of 25 ± 1 °C for 3 months.

2.2 Results

As a preliminary survey, the effects of 2,4-D concentration on regeneration of the tuber segments were investigated (Table 2). With 2,4-D (1 mg/l), a tuber segment

Table 2. Effect of 2,4-D on the regeneration of *Pinellia ternata*. (Shoyama et al. 1983a)

2,4-D concentration (mg/l)	Callusing per segment (%)	Regenerated plantlets per segment	fresh weight per segment (mg)
0	7.5	0.7	27
0.1	62.5	0.3	32
1	54.2	1.4	1129
3	100.0	0	132
5	90.0	0	75
10	44.4	0	32

produced 1.4 regenerated plantlets per segment. The supplements with less than 1 mg/l showed a small amount of regeneration. Moreover, compared with the weight of the segment, the results clearly show that the correct concentration of 2,4-D is 1 mg/l for regeneration. If the tuber segment was cultured with a concentration greater than 1 mg/l, plantlets were not formed, but callusing occurred in all media with 3 mg/l, or a higher concentration, as shown in Table 2. Thus, if callus multiplication is required a higher concentration of 2,4-D, ca. 3 mg/l, may be useful.

The relation of auxin and kinetin to the regeneration of the tuber segment was investigated. Within 3 weeks the tuber tissue gave rise to large tissue masses, which produced many buds like bulbils or protocorm bodies (Fig. 2), on which first many roots and then shoots appeared. As shown in Fig. 3, the 2,4-D and kinetin

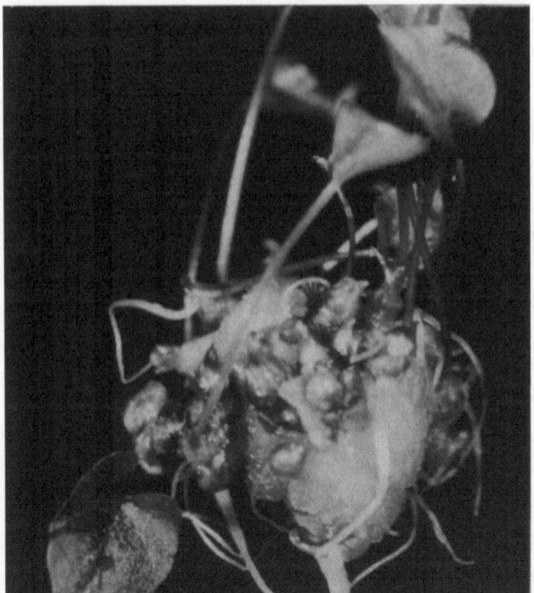

Fig. 2. Protocorm-like body induced from tuber segment

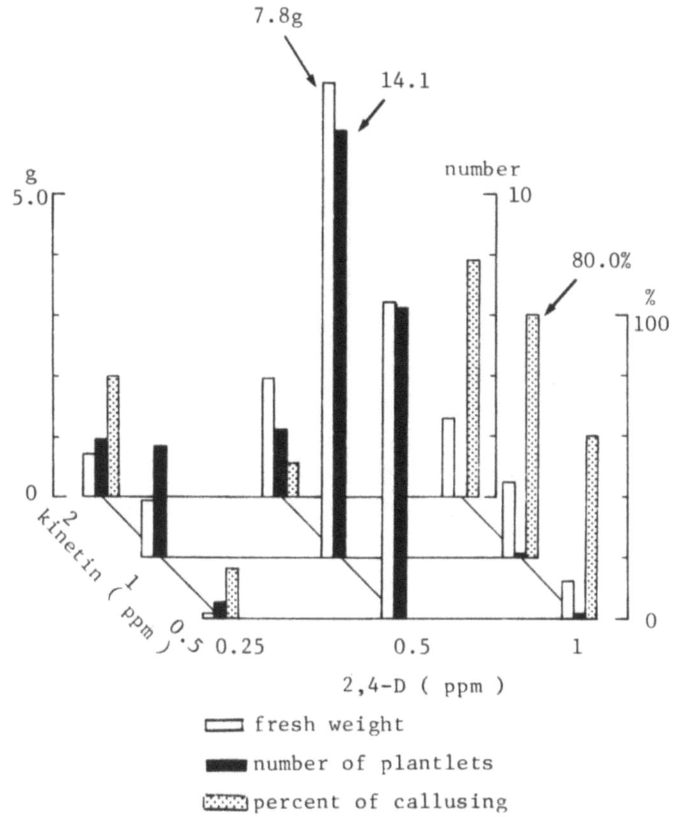

Fig. 3. Effect of varying concentrations of 2,4-D and kinetin on regeneration, tissue weight, and callus formation in tuber culture

concentration clearly influenced the number of plantlets regenerated. The experiments with supplements of 2,4-D (0.5 mg/l) in the presence of kinetin (1 mg/l) were the most effective for both numbers of regenerated plantlets and weight of the tissue. Under these conditions, 14 regenerated plantlets per segment were obtained. With increasing or decreasing concentrations of kinetin, however, the number of regenerated plantlets decreased. From these results we conclude that the propagation of plantlets is inversely proportional to prolific callus induction, and that tissue weight per segment is directly proportional to the regeneration of tubers. Supplementing 2,4-D (0.25 mg/l) with kinetin (1 mg/l) also favored regeneration of plantlets (Fig. 3), but the numbers were smaller than when using 0.5 mg/l. On the other hand, the addition of 2,4-D (1 mg/l) resulted only in callusing. Although preliminary experiments have shown that 1 mg/l of 2,4-D was suitable for regeneration when used alone, in this case the interrelation of kinetin with 2,4-D may occur to promote regeneration.

Concerning NAA, the number of plantlets regenerated increased with NAA concentration (Fig. 4). NAA (2 mg/l) with kinetin (1 mg/l) favored regeneration and 7.4 plantlets per segment were obtained (Fig. 4). However, the number of plantlets

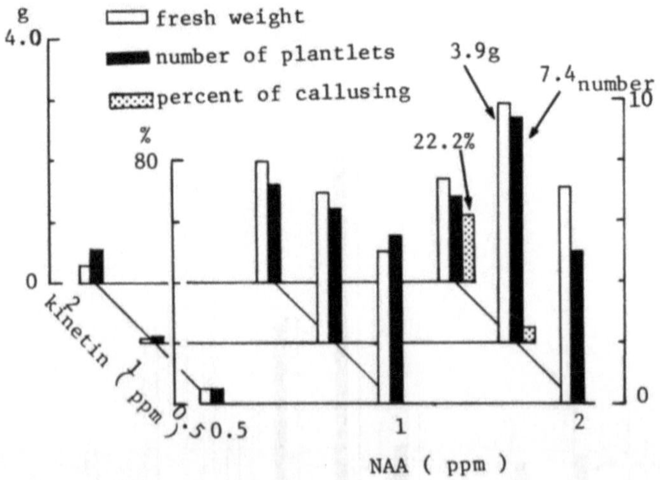

Fig. 4. Effect of varying concentrations of NAA and kinetin on regeneration, tissue weight, and callus formation in tuber culture

regenerated per segment is approximately half that obtained when using both 2,4-D and kinetin (0.5–1 mg/l), as mentioned above. Since callusing (5%) appears in this case, it can be expected that an addition of more than 2 mg/l may promote callusing. In addition, an increase in kinetin concentration results in callus formation.

Growth with IAA and kinetin generally favored callus formation rather than regeneration (Fig. 5). The speed of callus growth was, however, extremely slow compared with induction with 2,4-D. This shows that IAA is not appropriate for propagating this plant.

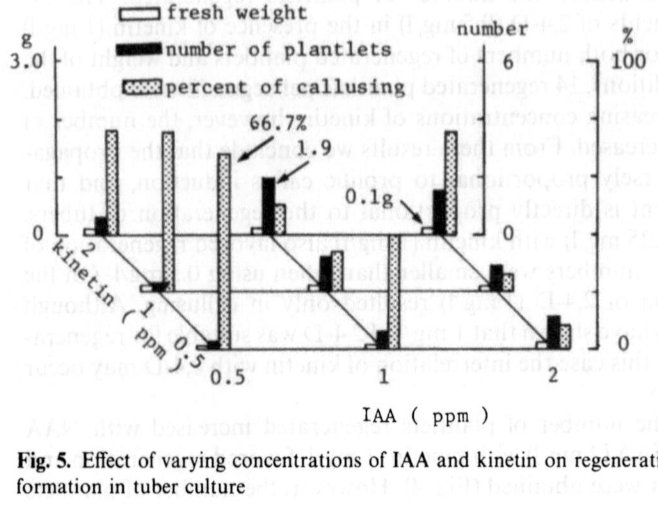

Fig. 5. Effect of varying concentrations of IAA and kinetin on regeneration, tissue weight, and callus formation in tuber culture

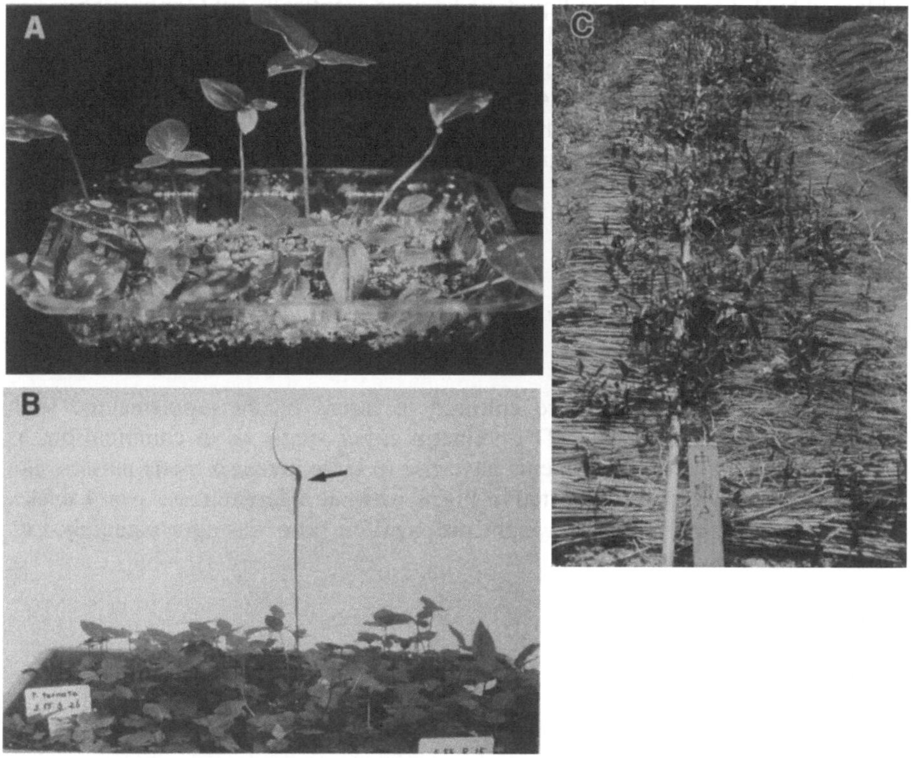

Fig. 6. A Transplantation of regenerated plants to vermiculite in the greenhouse. **B** Cultivation of regenerated plants in the greenhouse for 5 months. *Arrow* shows flowering. **C** Cultivation of regenerated plants in the field

After 3 months of culture, the buds, already rooted and shooted, were separated individually and transplanted to pots filled with vermiculite in the greenhouse. The transplantation was perfectly successful in all cases. All plants matured and some flowered within 5 months (Fig. 6).

3 Propagation in Liquid Medium Through Protocorm-Like Bodies – Procedure II (Shoyama et al. 1983b)

3.1 Materials and Methods

Tuber segments were aseptically taken from a medium supplemented with 0.5 mg/l 2,4-D and 1 mg/l kinetin (procedure I). The explants were aseptically cut (2-mm cubes) and the epidermis peeled off. For multiplication of tissue masses ten segments (or in subcultures 22 segments) were individually cultured in liquid medium for 4 weeks and the average was maintained in all cases. Tissue masses obtained by 4-week culture in liquid media were aseptically cut (2-mm cubes). Ten segments were cultured on solid media for 3 months.

Root tips obtained from regenerated plants were pretreated with 0.002 M 8-hydroxyquinoline at 19 °C for 4 h and then with 0.1% colchicine at 19 °C for 3 h, fixed in Newcomere solution, hydrolyzed for 5 min in 1 N HCl, and Feulgen stained. Chromosomes in root tips were counted in seven cells.

3.2 Results

3.2.1 Induction and Propagation of Protocorm-Like Bodies from Tuber Segments in Liquid Medium

When the tuber segments were cultured in liquid media supplemented with different concentrations of 2,4-D or kinetin either singly or in combination, as shown in Fig. 7, the tuber segments gave rise to large greenish tissue masses, like protocorm-like bodies, as indicated in Fig. 8, without differentiation over 4 weeks. With 0.25 mg/l 2,4-D the fresh weight propagation ratio was approximately 190.

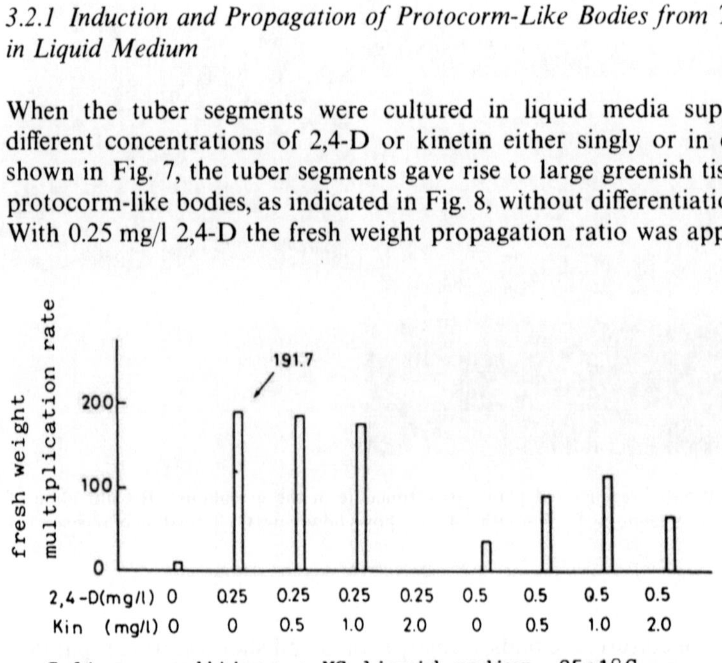

Fig. 7. Effects of 2,4-D and kinetin on multiplication of tissue mass from tubers

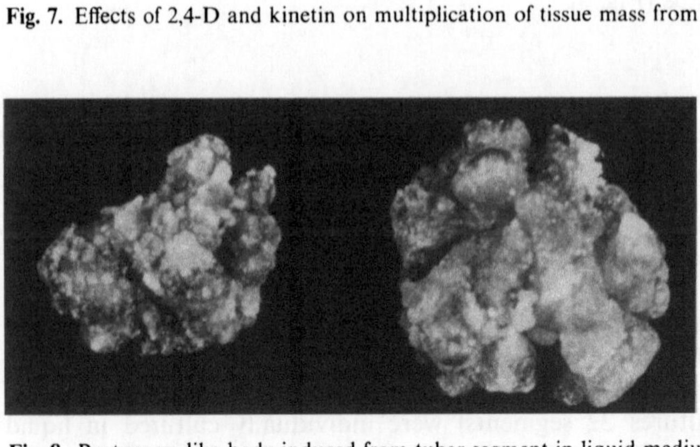

Fig. 8. Protocorm-like body induced from tuber segment in liquid medium

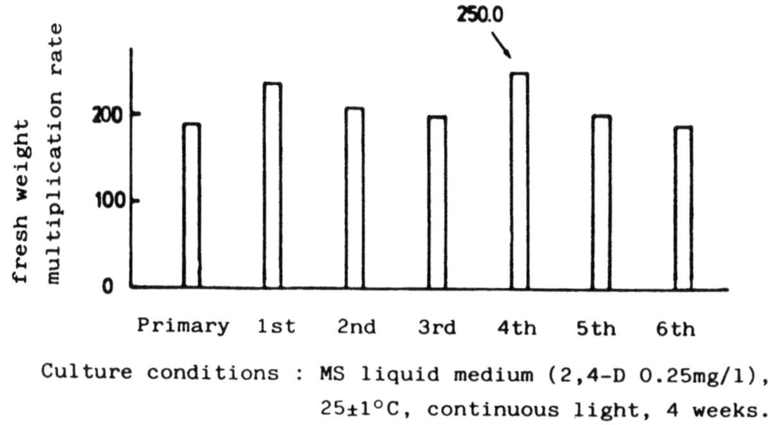

Fig. 9. Multiplication of subcultured tissue mass

Addition of kinetin inhibited growth slightly, and of 2 mg/l kinetin strongly. The addition of 0.5 mg/l 2,4-D either singly or in combination with kinetin had slightly inhibitory effects on the fresh weight propagation ratio when compared to the combination of 0.25 mg/l 2,4-D and kinetin. From these results it became evident that the suitable concentration of 2,4-D in liquid media was 0.25 mg/l.

Since in callus cultures the growth rate and regeneration stability of the callus generally diminish and finally disappear (Shono 1965), the fresh weight propagation rate of tissue masses in individual generation was investigated (Fig. 9). Individual subcultures were carried out under the same conditions as for the primary cultures. The fresh weight propagation ratio in individual generations was approximately 200 times, and about 250 times in the fourth generation. It was clearly shown that the fresh weight propagation rate was continuous, at least for 5 months.

3.2.2 Differentiation from Tissue Mass on Solid Medium

In order to regenerate plantlets, tissue masses cultured in liquid media were sectioned into approximately 200 segments and then transferred to solid media supplemented with a combination of 2,4-D and kinetin, as shown in Fig. 10. After 2 weeks the tissue segments gave rise to large greenish tissue masses; these yielded many buds, on which first many roots, then shoots appeared. 0.25 mg/l 2,4-D and 1 mg/l kinetin gave the maximum number of plantlets regenerated, about 15 plantlets per segment (Fig. 11). The addition of 1 mg/l 2,4-D with kinetin favored callusing rather than regeneration, as with the results of procedure I. From these results it is estimated that medium supplemented with 0.5 mg/l 2,4-D and 1 mg/l kinetin is favorable for plantlet regeneration from tissue masses.

Next, the regeneration of tissue masses in individual generation was investigated. Although primary culture showed that the combination of 0.5 mg/l 2,4-D and 1 mg/l kinetin was most effective for regeneration, it was more favorable in the later

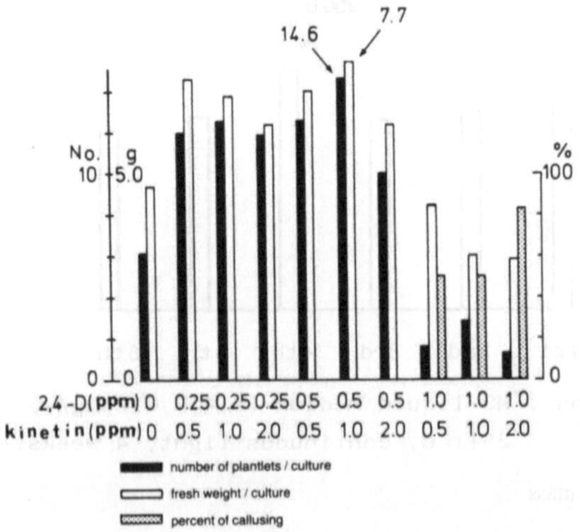

Fig. 10. Effect of 2,4-D and kinetin on regeneration from protocorm-like body segment

Fig. 11. Regeneration from protocorm-like body segment

subcultures to use the combination 2,4-D and kinetin (0.25–2 mg/l). Approximately 12 plantlets per segment were obtained in individual generation (Fig. 12).

After 3 months of culturing, the regenerated buds were separated from one another and transferred to vermiculite in the greenhouse. The transplantation was perfectly successful in all cases, as shown in procedure I. Plants obtained from tuber culture had a normal chromosome number of 2n = 104, and showed no phenotypic abnormalities (Fig. 13).

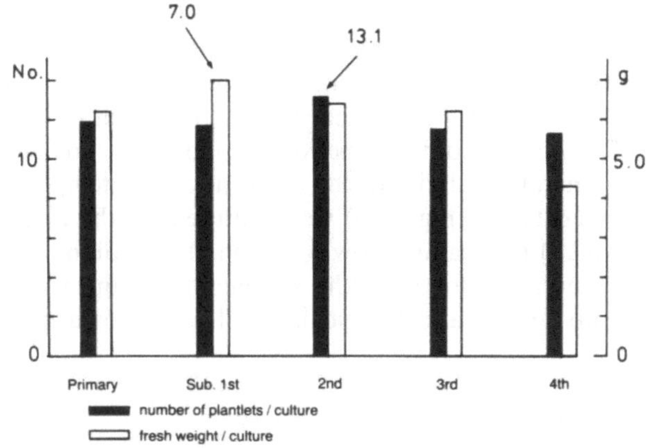

Fig. 12. Number of regenerated plantlets in individual generations

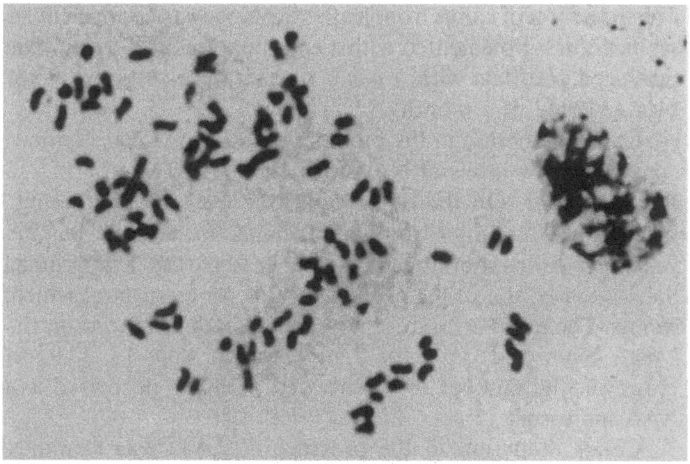

Fig. 13. Chromosomes of regenerated plantlets of *Pinellia ternata*

4 Propagation by Callus Culture of Leaf Segment – Procedure III (Hatano et al. 1986)

4.1 Materials and Methods

Leaves obtained from the in vitro culture of tuber segments as indicated in the previous procedure was aseptically cut into square pieces (4 × 4 mm) and the veins removed. Ten segments were individually cultured as indicated in Figs. 14 and 15 for 3 months and the averages of fresh weight of culture and the number of shoots regenerated were determined in all cases. For the maintenance of callus culture, the callus was transferred to fresh medium every 3 months and subcultured. The callus obtained in the medium supplemented with 2,4-D (2 mg/l) plus kinetin (1 mg/l) was divided into 200-mg blocks. Ten segments were cultured in NAA and kinetin, and five segments in 2,4-D and kinetin, as indicated in Figs. 17 and 18. Low temperature storage was carried out in cold room (4 °C), refrigerator (0 °C), or freezer (− 17 °C), for different periods, as indicated. To determine the growth rate and the regeneration ability of the callus after low temperature storage, ten stored callus segments (200 mg each) were cultured on the medium supplemented with 2,4-D (2 mg/l) (for callus growth) or on the medium with NAA (1 mg/l) (for plantlet regeneration) for 3 months. Regenerated plants were transferred to pots containing a mixture of sand:vermiculite (1:1) and grown in the greenhouse.

4.2 Results

4.2.1 Callus Induction

The induction of callus from leaflet explants was confirmed. Leaf segments cultured on medium supplemented with a combination of IAA and kinetin (0.5–2 mg/l each) produced plantlets with a weak regeneration response without callus formation over a period of 3 months.

Callus induction in the presence of NAA was investigated (Fig. 14). In medium containing low levels of NAA (0.5–1 mg/l) and kinetin (0.5–2 mg/l), callus formation was poor. On the other hand, the medium containing NAA at 2 mg/l and kinetin at 0.5–2 mg/l stimulated callus formation. In this instance, a friable yellowish callus developed at the end of the leaf segments within 2 weeks. Subsequently, the callus produced buds on its surface, which gradually grew into shoots. The numbers of regenerated explants produced on the medium containing 2 mg/l NAA and 0.5, 1, or 2 mg/l kinetin were 13, 13.9, and 10.8, respectively. However, the number of regenerated plantlets decreased as the concentration of NAA increased.

Callus induction in the presence of 2,4-D was examined (Fig. 15). Medium supplemented with 0.25 mg/l 2,4-D and kinetin (0.5, 2 mg/l) gave neither good

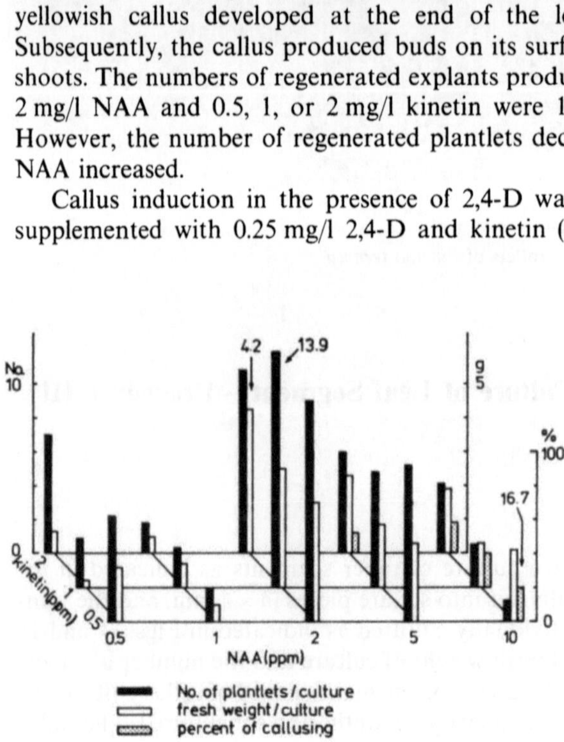

Fig. 14. Callus induction from leaf segment on medium supplemented with NAA and kinetin

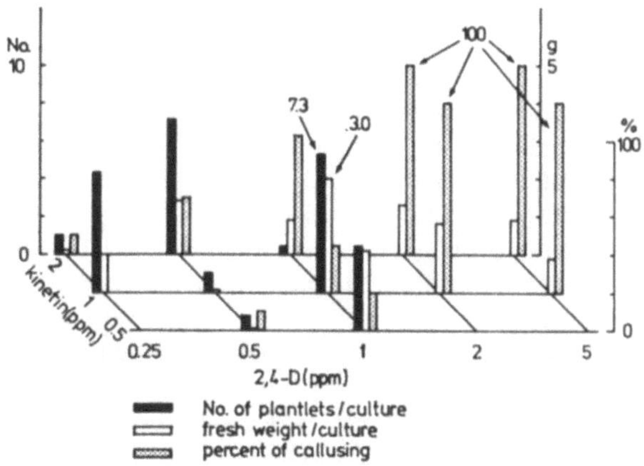

Fig. 15. Callus induction from leaf segment on medium supplemented with 2,4-D and kinetin

callusing nor effective differentiation. In media containing the following combinations: 0.5 mg/l 2,4-D and 2 mg/l kinetin, or 1 mg/l 2,4-D and 1 mg/l kinetin, callus formation occurred to form shoot buds. In these media the growth of callus was rapid. The leaf segments also callused in the medium supplemented with 2,4-D (2 mg/l) and kinetin (1–2 mg/l). Callusing was initiated along the edges of the leaf and finally involved the whole leaf within 3 months. The callus was pale white and friable, as indicated in Fig. 16, and provided the source of callus for further culture. This callus was subcultured every 3 months, without the occurrence of regeneration. Addition of 2,4-D (4 mg/l) produced a good callusing rate, but it slightly inhibited the multiplication of callus in comparison with the callus growth in the medium supplemented with 2 mg/l of 2,4-D and 1 mg/l kinetin, as indicated in Fig. 15.

Fig. 16. Callus induced from leaf segment on the medium supplemented with 2,4-D

4.2.2 Regeneration of Plantlet from Callus

In medium without hormones 47.8 plantlets per culture were regenerated from the callus segment within 3 months, as shown in Fig. 17. In the medium supplemented with kinetin (1 mg/l) alone the callus segment produced 50.7 plantlets per callus segment. If the NAA concentration in the medium was above 1 mg/l, the number of regenerated plantlets decreased. The regeneration of plantlets was observed without callus propagation even when 10 mg/l of NAA was added to the medium. Although cytokinins stimulated both the regeneration of plantlets from callus and multiple shoot formation, in general, we found that the addition of kinetin to the medium containing NAA inhibited the regeneration of plantlets. This phenomenon was found over the whole range of NAA concentrations investigated.

Subsequently, the effects of 2,4-D on the regeneration of plantlets from callus were determined (Fig. 18). The medium containing 2,4-D at 0.5 mg/l was shown to be more inhibitory for plantlet regeneration than hormone-free medium. The number of regenerated plantlets was inversely proportional to 2,4-D concentration. As was the case with NAA, and addition of kinetin decreased the number of regenerated plantlet per segment. If the callus segment was inoculated to a medium supplemented with kinetin at more than 2 mg/l, plantlets were not formed; only callus propagation occurred (Fig. 18). Thus, for callus proliferation, a higher concentration of 2,4-D at or above 2 mg/l must be used.

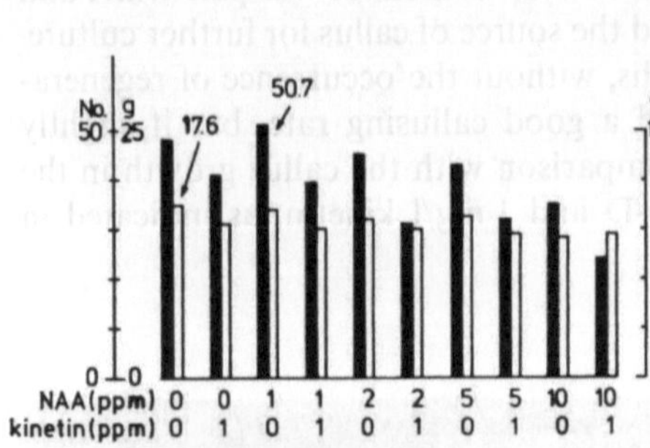

Fig. 17. Effect of NAA and kinetin on regeneration from callus segment

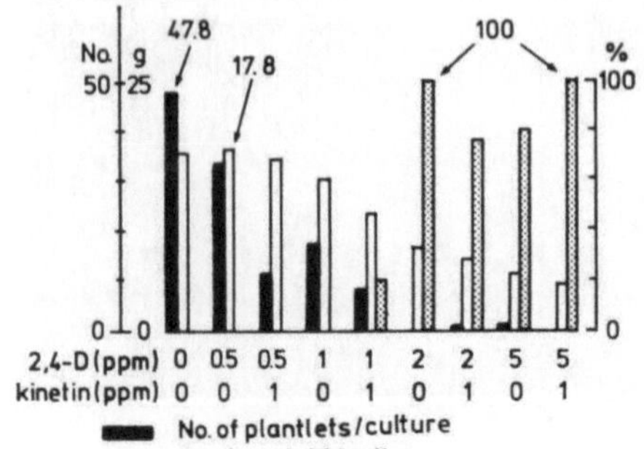

Fig. 18. Effect of 2,4-D and kinetin on regeneration from callus segment

The callus segments could be subcultured to multiply the callus, retaining its regeneration potential. The number of plantlets obtainable for callus did not diminish even after several subcultures.

The stability of chromosomes of *Anthurium scherzerianum*, belonging to the same family as *Pinellia ternata*, has been confirmed in callus culture (Geier 1982). As a result of chromosomal analysis as in procedure II, plantlets obtained from callus segments showed the normal chromosome number of 2n = 104. Therefore, the callus culture of this plant may be a useful technique for micropropagation without chromosomal variation as well as for tissue mass culture as in procedure II.

4.2.3 Low temperature Storage of Callus

The effects of cold storage of callus on the production of secondary products and organogenesis have been studied. According to Hiraoka and Kodama (1982), the resistance of callus to low temperature storage depended on its strain. Therefore, the callus was stored under three different conditions and then its growth and regeneration capabilities were assayed (Fig. 19).

When the callus segment was stored at − 17 °C without any cryoprotectant, within a short period all callus segments died. If the callus was stored at 0 °C, the capacity for regeneration and callus propagation rapidly decreased and finally disappeared after 20 weeks of storage. On the other hand, the storage of callus segments at 4 °C was found to provide the callus with a higher capacity for callus multiplication than the control culture (approximately 70 times, as shown in Fig. 18). However, the regeneration capacity was somewhat lowered. These results

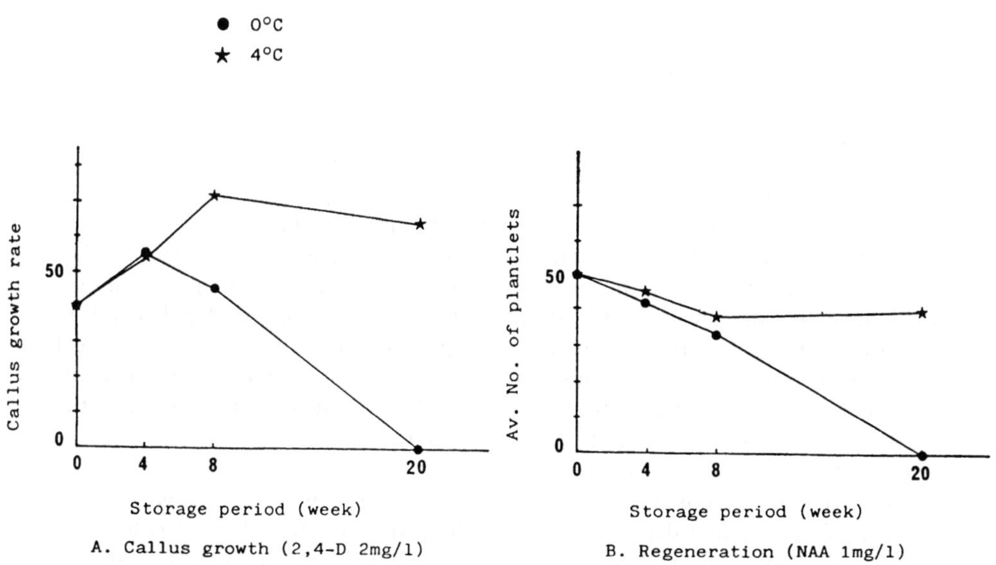

Fig. 19. Effect of low temperature storage on callus multiplication and regeneration capacity of callus culture conditions: MS medium, 25 ± 1 °C, continuous light, 12 weeks, callus segment weight ca. 200 mg

suggested that the long-term storage of callus at 4 °C might be possible without loss of organogenetic capacity.

5 Analysis of Secondary Metabolites in Callus and Protocorm-Like Body

The tuber of *Pinellia ternata* contains homogentisic acid and its glycoside, 3,4-dihydroxybenzaldehyde, which have been isolated as irritating substances (Hasegawa et al. 1959; Suzuki 1969), and ephedrine (Oshio et al. 1978). The production of secondary metabolites by cell culture of various plants has been investigated by many groups (see Bajaj 1991). In these studies, it is well known that differentiation and production of secondary metabolites are in general parallel.

In this study the protocorm-like body seems to be an intermediate stage between callus and regenerated plantlet. Therefore, it is expected that the protocorm-like body may biosynthesize the secondary metabolites contained in the tuber. After the callus and the protocorm-like body were extracted with methanol, individual extracts were partitioned with organic solvent and surveyed by TLC and HPLC. The results show that the callus and the protocorm-like body do not contain secondary metabolites. Therefore, the individual stage between callus and regenerated plants, which do contain these compounds, should be analyzed.

6 Conclusion

Figure 20 shows schematically the micropropagation system of *Pinellia ternata.* The tuber can be used for direct propagation and about 2 500 000 plantlets are obtained from a single tuber in 1 year. When liquid culture is used, however, it is theoretically possible to obtain more than 4×10^{23} clonal plants in 1 year from a single tuber. It has been confirmed that the fresh weight propagation rate of tissue masses and the regeneration rate from these are almost continuous for at least 7 years. On the other hand, callus can be easily induced from leaf segments rather than tuber segments. Since the propagated callus differentiates to give plantlets which have the same chromosome number ($2n = 104$), it is clear that callus culture can also be routinely used for clonal multiplication without any phenotypical abnormality. Furthermore, in callus stored at 4 °C both propagation and regeneration abilities do not decrease. It may be possible to cultivate this plant with the advantage of having an alternative method for the maintenance of the tuber or bulbil without seasonal factors.

Many Chinese medicines are still based upon the wild resources, but the qualities of these crude drugs are uneven, depending on the place of collection, seasonal changes and/or the strain. This is one reason why the breeding of medicinal plants is needed. It is obvious that in order to achieve a consistent quality in the drug, and to overcome the shortage of natural resources, clonal propagation is necessary.

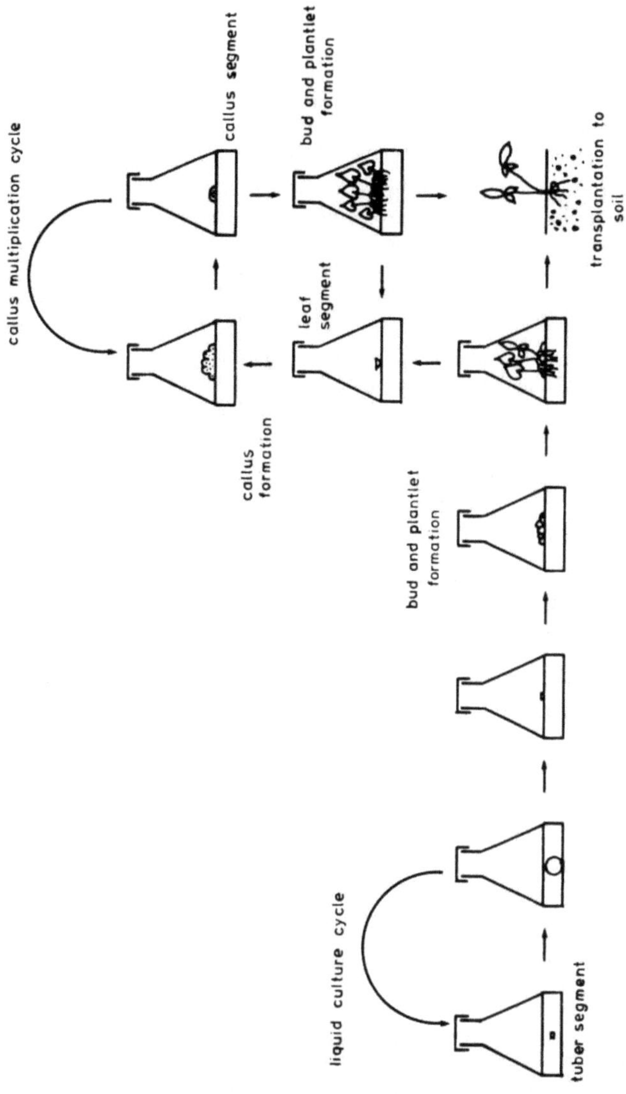

Fig. 20. Multiplication system of *Pinellia ternata*

7 Protocols

1. Direct Propagation from Tuber Segments
   Culture the tuber segments (2-mm cube) on MS agar medium supplemented with 2,4-D (0.5 mg/l) and kinetin (1 mg/l) under continuous light at 25 ± 1 °C for 3 months to regenerate 14 plantlets.

2. Propagation in Liquid Medium Through Protocorm-Like Body
   a) Induction and propagation of protocorm-like body: culture the tuber segment (2-mm cube) on MS liquid medium supplemented with 2,4-D (0.5 mg/l) and culture under continuous light at 25 ± 1 °C for 4 weeks to form protocorm-like body. Culture the protocorm-like body segments (2 mm cube) on MS agar medium supplemented with 2,4-D (0.5 mg/l) and kinetin (1 mg/l) under continuous light at 25 ± 1 °C for 3 months to regenerate 15 plantlets.

3. Propagation by Callus Culture of Leaf Segment
   a) Induction and maintenance of callus: culture the in vitro leaf segment (4 × 4 mm square) on MS agar medium containing 2,4-D (2 mg/l) and kinetin (1–2 mg/l) under continuous light at 25 ± 1 °C for 3 months to induce callus.
   b) Regeneration of plantlet from callus: transfer callus (200 mg block) to MS agar medium and culture under continuous light at 25 ± 1 °C for 3 months to regenerate approximately 50 plantlets.

References

Bajaj YPS (ed) (1991) Biotechnology in agriculture and forestry, vol 15, Medicinal and aromatic plants III. Springer, Berlin Heidelberg New York Tokyo

Geier T (1982) Morphogenesis and plant regeneration from spadix fragments of *Anthurium scherzerianum* cultivated in vitro. In: Fujiwara A (ed) Plant tissue culture 1982. Jpn Assoc Plant Tissue Cult, Maruzen Tokyo, pp 137–138

Hasegawa C, Sakamoto Y, Ichihara K (1959) On the relationship between homogentisic acid and the egumi-taste of Bamboo shoots and Hange. Proc Jpn Acad 35: 246–251

Hatano K, Shoyama Y, Nishioka I (1986) Multiplication of *Pinellia ternata* by callus culture of leaf segment. Shoyakugaku Zasshi 40: 188–192

Hiraoka N, Kodama T (1982) Effect of non-frozen cold storage on the growth, organogenesis, and secondary metabolite production of callus cultures. In: Fujiwara A (ed) Plant tissue culture 1982. Jpn Assoc Plant Tissue Cult, Maruzen Tokyo, pp 359–360

Murashige T, Skoog F (1962) A revised medium for rapid growth and bioassays with tobacco tissue culture. Physiol Plant 15: 473–497

Oshio H, Tsukui M, Matsuoka T (1978) Isolation of l-ephedrine from *Pinellia ternata*. Chem Pharm Bull 26: 2096–2097

Shono K (1965) Changes in organ forming capacity of carrot root calluses during subcultures. Plant Cell Physiol 6: 403

Shoyama Y, Hatano K, Nishioka I (1983a) Rapid and simple multiplication of *Pinellia ternata* by tissue culture. Planta Med 47: 103–105

Shoyama Y, Hatano K, Nishioka I (1983b) Clonal multiplication of *Pinellia ternata* by tissue culture. Planta Med 49: 14–16

Suzuki M (1969) Studies on the irritating substance of *Pinellia ternata* Breitenbach (Araceae). Arzneim Forsch 19: 1307–1309

Section III Fibre and Oilseed Crops, and Grasses

III.1 Micropropagation of Cotton (*Gossypium* Species)

Y.P.S. Bajaj[1] and M.S. Gill[2]

1 General Account

1.1 Botany

Cotton belongs to the genus *Gossypium*, family Malvaceae. This genus comprises 39 species (Fryxell 1984), which have been categorized into seven genome groups based on the cytology of interspecific hybrids (Beasley 1942; Phillips and Strickland 1966; Edwards and Mirza 1979). The A genome contains only two diploid ($2n = 2x = 26$) cultivated species (*G. arboreum* L. and *G. herbaceum* L.), each of which comprises several races. In the B, C, D, and E genomes there are 4, 9, 12, and 4 species, respectively. *G. longicalyx* Hutchinson and Lee has been included in the F genome and *G. bickii* Prokhanon in the G genome. There are six tetraploid ($2n = 4x = 52$) species, including the cultivated types *G. hirsutum* L. and *G. barbadense* L., all of which have AD genomic constitution. The cotton plant is a typical woody perennial shrub, its growth habit is indeterminate and it develops simultaneously both vegetatively and reproductively (Mauney 1984).

1.2 Distribution and Importance

Cotton is the most important natural source of fiber currently used in the textile industry. It is also a valuable source of oil, used primarily for shortening, cooking oil, salad dressing, mayonnaise, and margarine. Different cultivated species of cotton contain 14.6–25.6% oil on dry seed weight basis (Pandey 1976). Cotton seed meal contains a high percentage of protein and is rich in important essential amino acids like lysine, methionine, and tryptophan. (El-Nockrashy et al. 1969), and is used as a fertilizer and for cattle feed. The hull, which is the outer covering of the seed from which the cotton fibers arise, is used as roughage in cattle feed. Linters are the short fibers clinging to the seed after ginning. These are removed at the oil mill to become a valuable source of cellulose for plastics, synthetic fibers, explosives, and padding for cushions.

The cotton plant is native to the tropics and subtropics, but the annual types selected during domestication have allowed it to be grown in temperate regions

[1] Former Professor of Tissue Culture at the Punjab Agricultural University, Ludhiana, India. (Present address: A-135 New Friends Colony, New Delhi 110065, India)
[2] Punjab Agricultural University, Ludhiana, India

Biotechnology in Agriculture and Forestry, Vol. 19
High-Tech and Micropropagation III (ed. by Y.P.S. Bajaj)

Table 1. Important cotton-growing countries in the world. (FAO 1987)

Country	Total area (1000 ha)	Yield (kg/ha)	Production (1000 mt)
China	4910	2560	12570
USA	4067	2082	8469
USSR	3460	2341	8100
India	8200	462	3790
Pakistan	2513	1526	3836
Brazil	1973	848	1672
Nigeria	450	244	110
Egypt	445	2472	1100
Argentina	300	1077	323
Australia	148	3716	550
Israel	43	4000	172

with long warm summers. Today cotton is grown virtually all over the world (Table 1) in tropical latitudes, and as far north as 43°N lat. in the USSR and 45°N in the People's Republic of China. In India, cotton has been an important product for more than 3000 years.

G. herbaceum var. *africanum* is the only wild-growing cotton with spinnable lint, and is probably the ancestor of all linted cottons in both the Old and New Worlds. Presently, there are two major Old World species of cultivated cotton, i.e. *G. arboreum* L. and *G. herbaceum* L.

The New World cottons (tetraploids) are natural amphidiploids that contain a set of chromosomes from one of the two A genome species and a chromosome set from a species of the D genome group (Beasley 1940; Harland 1940; Skovsted 1934, 1937). Gerstel (1953) has demonstrated that the A genome of the natural amphiploids differs from *G. herbaceum* by two reciprocal translocations and from *G. arboreum* by three reciprocal translocations, thus indicating that *G. herbaceum* contributed the A genome to the New World cottons. Cytogenetic studies (Phillips 1962, 1963) and seed electrophoresis of the known D genome species (Cherry et al. 1970; Johnson and Thein 1970) have shown that *G. raimondii* Ulbrich is most closely related to the contributor of the New World D genome. *G. raimondii* grows wild in the Pacific foothills of the Andean Cordillera in Peru.

Presently, about 90% of commercial cotton is obtained from *G. hirsutum* L., known most widely as upland cotton, which is grown in almost all the cotton-growing areas of the world (Lee 1984). *G. barbadense* L. cultivars contribute about 8% of the total world production. *G. barbadense* L. cultivars are grown in the Nile river drainage area (Egyptian cotton), in Peru (Tangvis cotton), and in restricted areas of the southwestern United States (Pima cotton). Cultivation of *G. arboreum* and *G. herbaceum* is restricted to Asia and some parts of Africa.

1.3 Conventional Propagation and Breeding

The conventional method of propagation in cotton is mostly through seed. To maintain the purity of the varieties, precautions have to be taken to prevent cross-pollination. Sometimes, propagation through cuttings (Balasubramanian and Kanniyan 1952; Parmar et al. 1978) is also done to propagate sterile or rare plant types.

Due to the often cross-pollinated nature of cotton, the breeding procedures are, with slight variations, those generally employed with self-pollinated crops. Conventional breeding approaches largely involve hybridization between two or more varieties followed by selection in segregating generations. Various forms of population improvement are also used for the development of new varieties, as well as populations with improved segregational potentials. Selection is also required to maintain varietal purity. Genes for bacterial blight resistance and fiber length and strength have been transferred from *G. barbadense*. The wild species of *Gossypium* also possess a number of genes for fiber quality, disease and insect resistance, stress resistance, and desirable morphological traits. However, genetic manipulations to transfer genes between species are not always easy due to various incompatibility barriers, cytogenetic differentiation, and ploidy level differences. During the past two decades, considerable interest has been focused on the commercial exploitation of hybrid vigor. In India, a large number of hybrids are under cultivation, but their spread is limited due to lack of suitable pollination techniques. The hybrid seed is produced by manual pollination and is very costly. Genetic improvement in cotton is sought for upgrading both quantitative yield and fiber quality. Development of insect-resistant/tolerant varieties is also considered important. More than 100 species of insects (Sohi 1964) such as jassids, white fly, bollworms, etc. attack cotton and adversely affect the yield as well as fiber quality. Varieties with increased resistance or escape mechanism are desirable, since chemical control is both expensive and deleterious to the fiber. Several bacterial and fungal parasites infect cotton and reduce its yield and quality. Common diseases are *Fusarium* wilt, *Verticillium* wilt, bacterial blight, *Ascochyta* blight, root rot, and boll rot. The improved varieties are required to have genes for resistance to these diseases.

The protein contained in cotton seed is unfit for consumption by nonruminants because of the presence of gossypol (Sood and Singh 1974) and related perpenoids in the glands of the cotton seed. The chemical removal of gossypol is costly. Genes for glandless cotton are available, but these varieties are more susceptible to insects. A wild species, *G. bickii* found in Australia (Fryxell 1965) contains genes for "glanded plant and glandless seed" and can be used for developing gossypol-free lines.

2 Need for Biotechnology in Cotton Improvement

In the past, the varieties and hybrids developed by conventional breeding methods have contributed significantly to the increase of cotton production. Excellent varieties incorporating high yield, earliness, compact plant type, resistance to some of the insects and diseases, storm-proof bolls, and better fiber quality have been released for cultivation in different cotton-growing areas in the world; but now it appears that a plateau has been reached and the routine breeding methods are not able to break the impasse. The development of new materials with resistance to insects, especially bollworms, has not met with the requisite level of success. The lack of genetic variation for resistance and practical problems faced in the utilization of wild species are the main causes for the relatively small success through conventional breeding. For this reason, in vitro methods promise a better potential in cotton-breeding programs. The authors have compiled literature, reviewed the subject, and emphasized the need for the use of biotechnology (Bajaj and Gill 1987). Embryo and ovule culture have been used to obtain hybrids between different species, thus paving the way for transfer of desirable genes into the cultivated types. The fusion of somatic cells via protoplast fusion can further widen the scope and limits of interspecific hybridization. Anther culture-derived haploid plants are known to produce homozygous diploids within a shorter period. Genetic variability can be induced in callus and cell suspension cultures, and screening for resistance to antibiotics and herbicides can be done at cell level.

Micropropagation techniques (Bajaj 1991) can be employed for collection and maintenance of germplasm, rapid multiplication of seed of new varieties, production of clones for evaluation against diseases, etc., and for propagating male sterile lines. For literature see Table 2.

3 Micropropagation

The cotton plant is generally cultivated as an annual crop and is propagated through seed. The varieties suffer rapid deterioration due to natural crossing (0.2–80%) and through mechanical mixtures during ginning. Therefore, the need for suitable means of vegetative propagation has always been felt. The plants obtained through cloning are true to type and can be used for producing nucleus seed for maintaining the purity of a variety. Moreover, clonal propagation can also be employed for the collection and maintenance of germplasm, maintenance of male sterile lines, propagation of wild species, and for pathological studies on clones. The conventional method of vegetative propagation in cotton by cuttings (Balasubramanyan and Kanniyan 1952; Parmar et al. 1978) shows a very slow rate of propagation. In vitro methods of micropropagation have been used in a number of crop plants, trees, and ornamentals to obtain higher rates of multiplication, and these techniques can also be utilized in cotton. Since cotton callus cultures are known to undergo genetic erosions and chromosomal variation (Bajaj and Gill 1985; Bajaj et al. 1986), the following methods of micropropagation for clonal propagation are discussed.

3.1 Plants from Shoot Tips and Meristem Cultures (Fig. 1, Tables 3, 4)

Shoot tip/meristem culture is one of the simplest methods of in vitro propagation and the plants obtained are true to type. In cotton, the in vitro culture of excised meristems of *G. hirsutum* was first reported by Chappel and Mauney (1967). The meristematic tissues cultured on Pfeffer's medium supplemented with 1 ppm GA_3

Table 2. Micropropagation studies on cotton (for more references on various aspects of in vitro culture see Bajaj and Gill 1987)

Species	Explant	Medium	Response	Reference
Shoot tip/meristem culture				
G. arboreum *G. hirsutum* and three interspecific hybrids	Shoot tips (1 cm) Meristem (1 mm)	MS + IAA + kinetin	Complete plants, in some cases multiple shoots	Bajaj and Gill (1986)
Embryo culture				
Species	Explant	Medium	Response	Reference
G. hirsutum L.	Hybrid embryos	White's (1939)	Seedlings with rudimentary cotyledons	Beasley (1940)
G. hirsutum L.	27-day-old embryos	White's (1943)	Normal seedlings	Lofland (1950)
G. hirsutum L.	12–14-day-old embryos	White's (5 times) + adenine sulfate (40 mg/l) + CH (150 mg/l) + NaCl (7 mM/l) + sucrose (20 g/l)	Seedlings in some cases	Mauney (1961)
G. hirsutum L.	12–14-day-old embryos	Same as above (Mauney 1961) with calcium malate or ammonium malate	Vigor of seedlings increased	Mauney et al. (1967)
G. arboreum L. × *G. stocksii* *G. arboreum* × *G. anomalum* *G. harbaceum* × *G. stocksii*	15-day-old embryos	MS + CH (250 mg/l) + IAA (1.5 mg/l) + K (0.5 mg/l) + sucrose 3%	Hybrid plantlets that matured upon transfer to field, generally self-sterile	Gill and Bajaj (1984a, b)
G. arboreum × *G. anomalum*	15-day-old embryos	MS + (CH 250 mg/l) + IAA (1.5 mg/l) + K (0.5 mg/l) + sucrose 3%	Hybrid plants backcrossed to *G. arboreum*, large variation seen in backcross progenies	Gill and Bajaj (1986)

Table 2. (*Contd.*)

Species	Explant	Medium	Response	Reference
		Embryo culture		
G. herbaceum × *G. hirsutum*	Immature hybrid embryos	MS and Beasley and Ting (1973 medium + Vit B_1, B_6, nicotinic acid	Normal growth of embryos was observed in hormone-free medium	Azizkhodozhaev and Umarov (1988)
G. arboreum × *G. anomalum*	Immature hybrid embryos	MS + CH (250 mg/l) + IAA (1.5 mg/l) + kin (0.5 mg/l) + sucrose 3%	Hybrids resembled the wild parent more closely	Gill and Bajaj (1991)
		Ovule culture		
Species	Explant	Medium	Response	Reference
G. hirsutum L.	2-day-old ovules	Not given	Seedlings	Beasley et al. (1971)
G. hirsutum L.	6-day-old ovules	White's (1943) + CH (1000 mg/l) + GA (5 mg/l)	Ovules surpassed size of seeds developed in vivo, no germination	Joshi and Johri (1972)
G. hirsutum L.	5-day-old ovules	Murashige and Skoog's (1962)	Embryos with cotyledons 1/3 of normal size	Eid et al. (1973)
G. hirsutum L.	2-day-old ovules	Beasley and Ting's (1973) (BT)	Good ovule and fiber development	Beasley and Ting (1973)
G. hirsutum L.	2-day-old ovules	BT + NH_4NO_3 (15 mM) + IAA (5 μm) + GA (5 μm) + K (0.05 μm) + sucrose (40 g/l)	Mature embryos and seedlings	Stewart and Hsu (1977a)
G. hirsutum × *G. arboreum* *G. hirsutum* × *G. herbaceum*	2-day-old ovules	BT + NH_4Cl (15 mM) + 4% sucrose	Mature viable embryos	Stewart and Hsu (1978)
Hybrid ovules from crosses involving tetraploid and diploid cultivated and wild species	3-day-old ovules	MS + CH (250 mg/l) + IAA (1.5 mg/l) + K (0.5 mg/l) + sucrose 3%	Hybrid plants that matured upon transfer to field	Gill and Bajaj (1984a)

G. hirsutum L. × *G. arboreum* L.	3-day-old ovules	MS + LH (250 mg/l) + IAA (1 mg/l) + K (0.2 mg/l) + sucrose 3%	Weak hybrid seedlings obtained	Gill and Bajaj (1987)
G. arboreum L. × *G. hirsutum*	3-day-old ovules	Same as Gill and Bajaj (1984a)	Weak and abnormal hybrid seedlings	Gill and Bajaj (1987)
G. hirsutum × *G. sturtianum*	2-day-old ovules	Basal salts + vitamin + myoinositol + sucrose 4%	Hybrid and backcross generations were obtained	Altman et al. (1987)
G. hirsutum × *G. arboreum*	8–12-day-old ovules	Modified Beasley and Ting (1973) medium	Hybrid plants were obtained	Thengane et al. (1986)

Somatic embryogenesis

Species	Explant	Medium for callus induction	Medium for somatic embryogenesis	Response	Reference
G. hirsutum L.	Stem segments	Basal medium + various combinations and concentrations of auxins, cytokinins and sugars	Basal medium and zeatin	Nodulation occurred	Williams (1978)
G. klotzschianum Andress	Hypocotyl segments	As devised by Smith et al. (1977)	Preculture in a liquid medium containing 2-iP (10 mg/l) then cultured in presence of 10–15 mM glutamine	Somatic embryoids differentiated	Price and Smith (1979)
G. hirsutum L.	Hypocotyl segments	LS + NAA (2 mg/l) + K (1 mg/l) + glucose 3%	LS medium lacking hormones and NH_4NO_3, but double conc. of KNO_3 and GA	Somatic embryos differentiated	Davidonis and Hamilton (1983)
G. hirsutum L.	Hypocotyl sections	MS + NAA (2 mg/l) + K (1 mg/l) + glucose 3%	MS + NAA (2 mg/l) + kin (1 mg/l) + sucrose 3%	Somatic embryos and then complete plants were obtained	Shoemaker et al. (1986)

Table 2. (*Contd.*)

Somatic embryogenesis					
Species	Explant	Medium for callus induction	Medium for somatic embryogenesis	Response	Reference
G. hirsutum L.	Hypocotyl segments	MS + myoinositol (100 mg/l) + B_5 vitamins + glucose 3% + 2,4-D and K	MS + myoinositol (100 mg/l) + B_5 vitamins + 2,4-D (0.1 mg/l) + K (0.5 mg/l) + sucrose 3%	Somatic embryogenesis was detected in cultivars Coker 201 and Coker 315	Trolinder and Goodin (1987)
G. hirsutum	Stem and leaf explants from seedlings and mature plants	MS + vitamins + myoinositol + glucose 3% + different concentrations and combinations of 2,4-D, NAA, K, BA, and 2-iP	Immature tissues – MS + 2,4-D (0.1 mg/l) + K (0.5 mg/l) Mature tissues – MS + NAA (2 mg/l) + K (0.1 mg/l) Leaf tissues MS + 2,4-D (0.1 mg/l) + 2-iP (1 mg/l)	Somatic embryos differentiated in mature and immature tissues	Trolinder and Goodin (1988a, b)
G. gossypioides (Ulbrich) Standley	Stem segments	MS + IAA (2 mg/l) + K (1 mg/l)	MS + NAA (0.1 mg/l) + 2-iP (3 mg/l)	Somatic embryos and plants were regenerated	Tan et al. (1988)
G. hirsutum L.	Hypocotyls inoculated with *Agrobacterium tumefaciens* strain LBA 4404 carrying pCMC 1204	Medium containing kanamycin	Regeneration medium	Genetically transformed plants resistant to kanamycin	Umback et al. (1987)

maintained normal growth and the tissues enlarged and expanded for 8 weeks. Although new leaf primordia were initiated, the root system failed to develop. The GA_3-stimulated leaf expansion was further enhanced by the addition of guanine or RNA as a source of purines and pyrimidines. Sucrose content higher than 0.06 molar concentration caused distortion of foliar primordia. Auxins, other growth substances, and vitamins had no effect on meristematic development. Coconut milk, when used as supplement, helped to maintain a green condition in the meristem.

Our work (Bajaj and Gill 1986) on shoot tip (1 cm) and meristem (1 mm) cultures of *G. arboreum*, *G. hirsutum*, and three interspecific hybrids, *G. arboreum* × *G. anomalum*, *G. arboreum* × *G. stocksii*, and *G. herbacium* × *G. stocksii*, is briefed here. The shoot tips (approx 1.5 cm) taken from the field-grown plants were surface sterilized in 0.1% mercuric chloride for 5 min, and then washed repeatedly in sterilized distilled water. The shoot tips (1 cm) and meristems (1 mm) were excised aseptically and cultured on MS medium (Murashige and Skoog 1962) supplemented with various combinations of IAA, NAA, kinetin, or BA. The meristems cultured on MS + IAA (0.5 mg/l) + kinetin (1–2 mg/l) grew slowly and formed callus (Fig. 1A). In the case of shoot tips, the unfolding of the leaves occurred after 4–5 days in both the species (Fig. 1C, D). However, after 10–14 days of culture, these leaves either turned brown or remained green, but all withered. New axillary outgrowths were formed after 20–25 days in *G. arboreum* and 26–30 days in *G. hirsutum*. The response of shoot tips cultured on different media is shown in Tables 3 and 4. When IAA was used in combination with kinetin, both the species responded best at 2 mg/l kinetin level. With an increase in the concentration of kinetin, growth of shoots slowed down and multiple shoots were formed (Fig. 1B). However, if the shoots were allowed to remain too long on higher concentrations of kinetin (4–6 mg/l) the leaves showed swelling and distorted shape, eventually leading to necrosis and death of the shoot. The in vitro-grown shoots (Fig. 1C, D) formed roots in 4 weeks (Fig. 1E) when transferred to the medium containing 2 mg/l IAA. The plants thus obtained continued further growth upon transfer to the soil (Fig. 1F). The shoot tips of interspecific hybrids responded best on MS + kinetin (1 mg/l) and complete plants were regenerated (Table 4), the success rate being 41.6% to 72.7% in different hybrids. *G. arboreum* × *G. anomalum* hybrids gave the best response, and this hybrid also showed the maximum vigor in the field.

Altman et al. (1989) devised an in vitro method for collecting the germplasm of various wild species of *Gossypium*, especially those where viable seeds could not be obtained. They used 0.5 × Linsmaier and Skoog (1965) inorganic salts supplemented with 10 g/l glucose, 0.5 g/l casein hydrolysate, 15 mg/l rifamycin, 15 mg/l trimethoprim, 9 g/l agar, 1 g/l of the fungicide Tilt MBC, and 1 mg/l L-naphthalene acetic acid. The cuttings obtained from various species were surface sterilized in the field itself and cultured on the above medium without any further precautions to assure asepsis. The cuttings were transported in the field in the culture tubes for up to 3 weeks. The cultured shoots grew vigorously, and adventitious roots were initiated at the basal ends. Upon return to the laboratory, the cuttings were rinsed with a 4% bleach solution, treated with Rootone F on their basal ends, and placed in a moistened, sterile soil mix of peat: sand: vermiculite (1:2:1) supplemented with dolomite lime and Osmocote slow release nutrients. The overall response was 10%

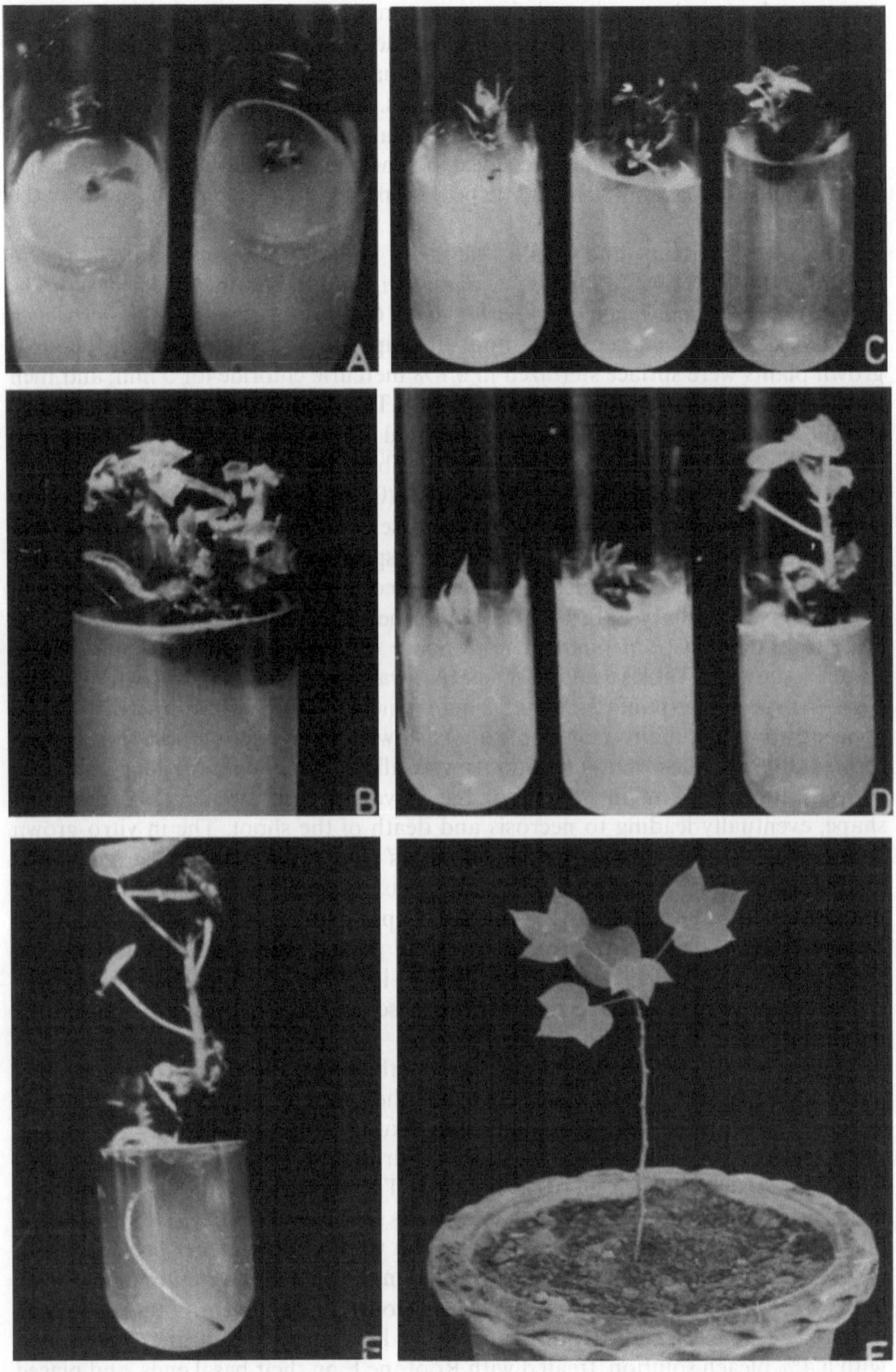

Fig. 1A–F. In vitro culture of shoot tips and meristems of cotton. **A** Two- and 4-week-old cultures of meristems (1 mm) of *Gossypium arboreum* on MS + IAA (0.5 mg/l) + kinetin (1 mg/l). **B** Multiple shoot

shoot formation from six *Gossypium* species. The rooting of cuttings was also possible.

3.2 Plants from Embryos and Ovules (Figs. 2, 3, Tables 5, 6)

The technique of embryo/ovule culture can be employed for propagation of species/genotypes from which viable seeds cannot be obtained. Seedlings from cultured ovules of *G. hirsutum* were first obtained by Stewart and Hsu (1977b, 1978), whereas successful embryo culture involving 12–14-day-old embryos was first reported by Mauney et al. (1967).

In our studies (Gill and Bajaj 1984a, b; 1987) plants of *G. hirsutum*, *G. arboreum*, *G. herbaceum*, and various interspecific hybrids were obtained from 3-day-old ovules and 15-day-old embryos cultured on MS supplemented with different concentrations and combinations of IAA, kinetin, coconut milk (CM), casein hydrolysate (CH), sucrose, and agar (0.7%).

The response of 3 days after pollination (DAP) ovules of the three species in culture is given in Table 5. Some of the self-pollinated ovules germinated to produce seedlings even on the medium devoid of hormones, CH and CM. However, germination was considerably enhanced when the medium was supplemented with either CH or CW alone or in combination with IAA and kinetin. On almost all the media, the ovules started to proliferate within 1 week of culture. The ovules from the diploid species produced profuse, friable callus (Fig. 2A). The *G. hirsutum* callus was slow-growing and compact. Maximum callus was produced on media containing IAA (1 and 1.5 mg/l) + kinetin (0.5 mg/l) and CH (250 mg/l). Along with the mass of callus, the ovules kept growing, and these could be seen either as lying on or embedded in the callus. The ovules started germinating after about 50 days in the diploid species, and 60 days in *G. hirsutum*. At the time of germination, the size of the *G. arboreum* or *G. herbaceum* ovules was smaller than those developed in situ, whereas *G. hirsutum* ovules attained a size almost equal to the in vivo-developed ovules. Seedlings of *G. hirsutum* were also more vigorous. The best response was observed on MS + IAA (1.5 mg/l) + kinetin (0.5 mg/l) + CH (250 mg/l) for *G. arboreum* and *G. herbaceum*, whereas for *G. hirsutum* the best medium was MS + IAA (1 mg/l) + kinetin (0.2 mg/l) + CH (250 mg/l).

The 15 DAP embryos responded better than 12 DAP embryos. In the latter case, the embryos generally turned brown and died, only occasionally forming weak seedlings. In *G. hirsutum* 6% sucrose proved to be better than 3% sucrose for younger embryos. The 15 DAP embryos produced callus or plants on most of the media. The best response with respect to plant formation was observed on MS + IAA (1.5 mg/l) + kinetin (0.5 mg/l) + CH (250 mg/l) + sucrose 3% in *G. arboreum* and *G. herbaceum*. The 12 and 15 DAP embryos of *G. hirsutum* responded best on MS + IAA (1.5 mg/l) + CW (7%) and sucrose 6%. *G. arboreum* embryos produced multiple shoots (Fig. 3A, B) on MS + IAA (1.0 mg/l) + BA

formation on MS + NAA (0.5 mg/l) + BA (2 mg/l). **C, D** Shoot tip (1 cm) culture of *G. arboreum* and *G. hirsutum* on MS + IAA (0.5 mg/l) + kinetin (2 mg/l), *left* at the time of culture; *right* after 4 weeks. **E** An in vitro-obtained plantlet of *G. hirsutum* in which rooting has been induced on a medium supplemented with IAA (2 mg/l). **F** Same, 3 months after transfer to a pot. (Bajaj and Gill 1986)

Table 3. Shoot tip culture of *Gossypium arboreum* and *G. hirsutum* on various media. (Bajaj and Gill 1986)

	G. arboreum				*G. hirsutum*			
Medium (mg/l)	No. of shoots cultured	No. of growing shoots	% Growth	Remarks	No. of shoots cultured	No. of shoots growing	% Growth	Remarks
1. MS + IAA (0.5) + K (1)	11	3	27.2	Good growth and callus	16	7	43.7	Good growth
2. MS + IAA (0.5) + K (2)	13	5	38.4	Very good growth	13	6	46.2	Very good growth
3. MS + IAA (0.5) + K (4)	11	4	36.4	Good growth	11	4	36.4	Poor growth
4. MS + IAA (0.5) + K (6)	13	2	15.4	Multiple shoots	11	3	27.2	Poor growth

Table 4. Shoot tip culture of various interspecific hybrids of diploid cotton on MS + kinetin (1 mg/l). (Bajaj and Gill 1986)

Cross	No. of explants cultured	No. of shoots growing	% response	Remarks
G. arboreum × *G. anomalum*	11	8	72.7	Very good growth, vigorous shoot formation
G. arboreum × *G. stocksii*	12	5	41.6	Good growth
G. herbaceum × *G. stocksii*	11	6	54.5	Good growth

(2 mg/l). These shoots could be separated and subcultured on medium to induce rooting. It was also observed that when 15 DAP embryos of *G. arboreum* and *G. herbaceum* were kept in complete dark for the first 15 days of culture, plant formation was considerably increased (Table 6). When light was provided during the first 15 days, the embryos turned green and produced weak and abnormal seedlings.

3.3 Plants from Hybrid Embryos/Ovules (Tables 7, 8)

The response of hybrid ovules (3 DAP) from various interspecific crosses (Table 7) cultured on different media was similar to their respective maternal parent, though the development was somewhat slower than the self-pollinated ovules (Gill and Bajaj 1987). In the diploid × diploid crosses, plants were obtained even on medium devoid of growth regulators; however, in tetraploid × diploid crosses, no plants were obtained without the addition of auxins and cytokinins.

The hybrid embryos from the crosses between various diploid species (Table 8) also produced plants when cultured on MS + IAA (1.5 mg/l) + kinetin (0.5 mg/l) + CH 250 mg/l + sucrose 3%. The response of hybrid embryos in culture was almost similar to their respective maternal parents (Gill and Bajaj 1984a, b).

3.4 Plants Through Somatic Embryogenesis

Somatic embryogenesis is one of the fastest methods for large-scale propagation. In cotton, Price and Smith (1979) first observed somatic embryos in suspension cultures of *G. klotzschianum*, a wild species. Embryogenesis occurred after a preculture of the callus on a medium containing 2-iP (10 mg/l). The embryoids formed roots and vestigial leaves. Glutamine (10–15 mM) was necessary for embryoid formation. Presence of 2,4-D in the medium resulted in increased vigor of the suspension cultures and subsequently in the formation of many embryoids, but was not necessary for somatic embryogenesis. In *G. hirsutum* the percentage of

Table 5. Percentage of ovules giving seedlings and the callus growth response of excised ovules (3 days after pollination) of *Gossypium arboreum* and *G. hirsutum* cultured on various media (figures in parenthesis indicate the number of ovules cultured). (Gill and Bajaj 1987)

Medium (mg/l)	*G. arboreum* Germination (%)	*G. arboreum* Callus[a] development	*G. hirsutum* Germination (%)	*G. hirsutum* Callus[a] development	*G. herbaceum*[a] Germination (%)	*G. herbaceum*[a] Callus[a] development
MS basal	17.6 (125)	+	8.3 (96)	–	7.9	+
MS + CH 250	29.2 (96)	+	22.2 (99)	–	18.8	+
MS + CW 7%	21.2 (80)	+	25.4 (110)	–	18.0	+
MS + CH 250 + IAA 0.5 + K 0.2	22.2 (108)	+	23.8 (122)	–	16.6	+
MS + CH 250 + IAA 0.5 + K 0.5	23.7 (93)	+	25.0 (120)	–	12.5	+ +
MS + CH 250 + IAA 1 + K 0.2	28.1 (121)	+	44.4 (133)	+ +	22.5	+ + +
MS + CH 250 + IAA 1 + K 0.5	20.3 (103)	+ + +	24.1 (108)	+ +	19.4	+ + +
MS + CH 250 + IAA 1.5 + K 0.5	37.5 (112)	+ + +	18.8 (117)	+ +	25.0	+ + +

[a] – no callus; + little callus; + + good callus; + + + profuse callus.

Fig. 2A–D. In vitro culture of excised ovules of *G. arboreum*, and the transfer of plants to field. **A** A mass of callus obtained from an ovule cultured 3 DAP on MS + IAA (2 mg/l) + kinetin (1 mg/l), 4-week-old culture. **B, C** Plantlets obtained from excised ovules initially cultured 3 DAP on MS + IAA (1.5 mg/l) + kinetin (0.5 mg/l) + CH (250 mg/l), and then transferred to half-strength MS + 1% sucrose. **D** Embryo-derived plants of *G. arboreum* 5 months after transfer to the field; note the normal growth and flowering. (Gill and Bajaj unpublished)

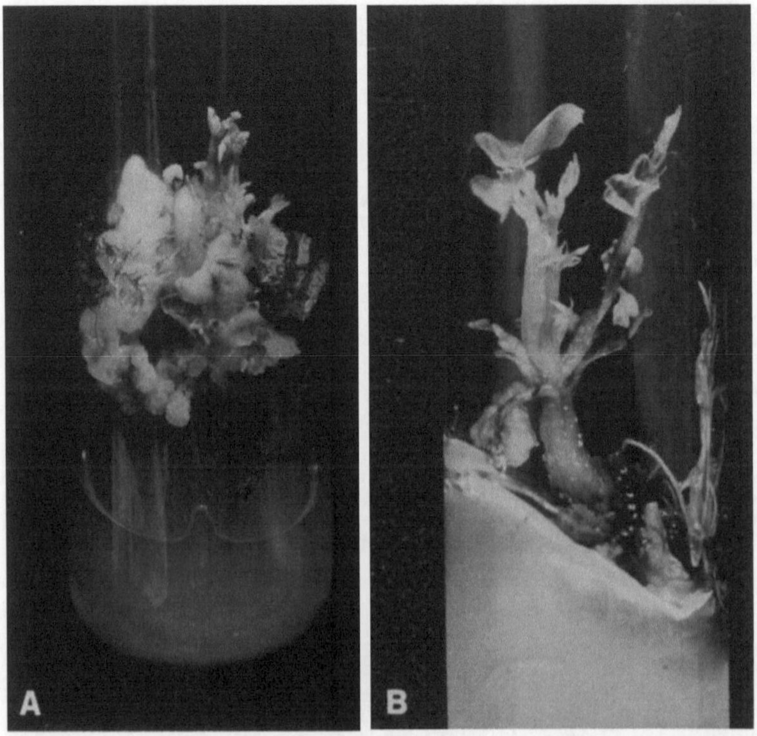

Fig. 3A, B. Induction of multiple shoots. **A** *Gossypium arboreum* embryo cultured on MS + IAA (1 mg/l) + BAP (2 mg/l) showing the initiation of buds and multiple shoots. **B** A hybrid embryo of *G. arboreum* × *G. anomalum* 15 days after culture on MS + CH (500 mg/l) + IAA (1.5 mg/l) + kinetin (1 mg/l) showing multiple shoots (Gill and Bajaj unpublished)

Table 6. Effect of light and dark treatments on the in vitro growth response of cotton embryos cultured (15 days after pollination) on MS + IAA (1.5 mg/l) + CH (250 mg/l) + kinetin (0.5 mg/l) + sucrose 3%. (Gill and Bajaj 1984b)

Species	Treatment	No. of embryos cultured	No. of plants formed	% Plant formation
G. arboreum	Dark (15 days)	36	27	75.0
	Light	32	14	43.7
G. herbaceum	Dark (15 days)	33	20	60.6
	Light	34	13	38.2

callus forming proembryoids increased to about 30% by prolonged culture without NAA and kinetin (Davidonis and Hamilton 1983). Proembryoid development was further enhanced by transfer to media lacking NH_4NO_3 but containing double the standard KNO_3 and GA. Shoemaker et al. (1986) evaluated 17 cultivars of *G. hirsutum* for callus initiation and maintenance and identified two cultivars, Coker 201 and Coker 315, as embryogenic. Embryogenic callus was routinely

Table 7. Percentage germination and callus growth response of hybrid ovules cultured on various media (data based on 100 ovules cultured in each experiment). (Gill and Bajaj 1987)

Media (mg/l)	*G. hirsutum* cv. F414 × *G. arboreum* cv. G27		*G. arboreum* cv. G27 × *G. hirsutum* cv. F414	
	Germination (%)	Callus[a] development	Germination (%)	Callus development
MS + CH 250 + IAA 1 + K 0.2	27.5	+	10.7	+ +
MS + CH 250 + IAA 1.5 + K 0.5	14.3	+	34.7	+ + +
SH (basal)	0	−	0	−
SH + IAA 0.5	0	+ + +[b]	0	+ +
SH + IAA 1	0	+ + +[b]	0	+ +

[a] − no callus; + little callus; + + good callus; + + + profuse callus.
[b] Fibers were also formed in these cultures.

Table 8. In vitro growth response of the hybrid embryos cultured (15 days after pollination) on MS + IAA (1.5 mg/l) + CH (250 mg/l) + kinetin (0.5 mg/l) + sucrose 3%. (Gill and Bajaj 1984)

Cross	No. of embryos cultured	No. of plants formed	% Plant formation
1. *G. arboreum* × *G. stocksii*	31	22	71
2. *G. arboreum* × *G. anomalum*	74	51	69
3. *G. herbaceum* × *G. stocksii*	39	19	48

obtained within 6 weeks by initiating callus on glucose media for 3–4 weeks followed by transfer to the same media containing 3% sucrose. The mature embryos germinated upon transfer to auxin-free medium with reduced sucrose levels. Trolinder and Goodin (1987) obtained somatic emryogenesis in all the eight strains of *G. hirsutum* studied, the callus being initiated from hypocotyl tissues of 3-day-old seedlings. Globular embryos were detected after 6 weeks in culture. The calli were subcultured in growth regulator-free liquid medium and after 3–4 weeks embryos were isolated by sieving, and plated onto semi-solid medium. The mature embryos germinated on sterile vermiculite saturated with medium. Later, Trolinder and Goodin (1988a) reported that explants of 3-day-old seedlings formed somatic embryos in 100% of the cultures when treated with 2,4-D (0.1 mg/l) and kinetin (0.5 mg/l). Media for the induction of somatic embryos from mature stem and leaf tissues were also devised (Trolinder and Goodin 1988b). Addition of 1.9 g/l KNO_3 to MS resulted in best embryo proliferation and maturation. Regeneration of plants from a wild species, *G. gossypioides* (Ulbrich) Standley, has been reported by Tan et al. (1988). The callus was induced from cut stem segments on MS medium containing IAA (2 mg/l) and kinetin (1 mg/l), and plant regeneration was achieved when IAA was replaced by NAA (0.1 mg/l) and 2-iP (3 mg/l).

3.5 Culture of Anthers and Protoplasts

The advances made in the field of anther and pollen culture (Bajaj 1983, 1990) have made haploid production easy in many crop plants; however, in cotton more work needs to be done. Production of callus from anthers of *G. hirsutum*, *G. barbadense* (Barrow et al. 1978; Hsi and Wu 1981; Bajaj 1982), and *G. arboreum* (Bajaj 1982, Bajaj and Gill 1989) has been reported. In the callus cultures obtained by Hsi and Wu (1981), root and shoot-like structures were observed, but complete plants could not be obtained. In our work (Bajaj and Gill 1989), excised anthers of a diploid Asiatic cotton (*G. arboreum*) in culture underwent pollen embryogenesis and profuse proliferation. Cold treatment (4 °C) promoted nuclear division in pollen. The callus showed a wide range of chromosome numbers varying from haploid (n = 13) to hexaploid and aneuploid, but plants were not obtained. In protoplast culture, as in anther culture, technology still needs to be refined to regenerate complete cotton plants.

4 Summary and Conclusions

Cotton is generally propagated through seed. The cultivated varieties suffer rapid deterioration due to occasional cross-pollination caused by honeybees or other insects, and also due to mechanical mixing of seed at the ginneries. Therefore, the maintenance of the genetic identity of the varieties, and particularly a large number of genetic collections, is very difficult. Clonal propagation can be helpful for preserving the identity of various genotypes, but the conventional methods of propagation through vegetative propagation in cotton are slow and unreliable. However, in vitro methods of propagation can be applied for micropropagation.

In cotton, plants have been produced through in vitro culture of shoot tips, meristems, young ovules, embryos, and from callus cultures. Shoot tip/meristem culture is one of the most widely used methods of in vitro propagation, and the plants obtained are true to type. Such plants can be employed for various types of pathological and genotype environment interaction studies. They have also been used for collection of germplasm of wild species where seeds could not be obtained. Shoot tip/meristem culture can also be employed for the multiplication of interspecific hybrids to grow large populations of such hybrids, which will increase the chances of recovering desirable recombinants.

The culture of immature ovules and embryos seems to be the most promising technique for the production of interspecific hybrids which are otherwise difficult or impossible to obtain. Embryo culture can also be employed for the propagation of species which produce inviable embryos or with seeds having long dormancy.

Micropropagation through somatic embryogenesis in callus/suspension cultures is one of the fastest methods of propagation, but the plants thus derived may be genetically variable. Somatic embryogenesis, although reported in cotton, is restricted to a few genotypes of *G. hirsutum*. Such studies need to be extended to other species and varieties. Anther culture and protoplast technology still need to be developed to be employed for micropropagation. It is concluded that the progress made in the biotechnology of cotton, especially in vitro culture, has far-reaching implications in the improvement and micropropagation of cotton.

5 Protocol

The established protocols for micropropagation in cotton through shoot tip culture, somatic embryogenesis, embryo, and ovule culture are presented below. Anther culture, protoplast culture, and other aspects of cotton tissue culture are not included, since more experimentation is necessary before they can be employed for micropropagation.

1. Shoot Tip Culture (Bajaj and Gill 1986).
   a) Explants (0.5–1.5 cm) are obtained from field- or greenhouse-grown healthy and normal plants
   b) Surface sterilize the shoot tips by dipping in 0.1% mercuric chloride for 5 min and then washing repeatedly in sterilized water
   c) Culture the shoot tips in MS medium supplemented with IAA (0.5 mg/l) and kinetin (2 mg/l), sucrose 3%, and agar 0.7%
   d) Keep the cultures in an illuminated room (approx. 2000 lx) at 25 ± 1 °C
   e) To induce rooting, transfer the growing shoots to medium containing MS + 2 mg/l IAA

2. Ovule Culture (Gill and Bajaj 1987).
   a) Harvest bolls 3 days after pollination
   b) Surface sterilize by dipping in ethanol and flaming
   c) Excise the ovules and culture seven or eight ovules in each culture tube containing agar-solidified MS medium with 1.5 mg/l IAA, 0.5 mg/l kinetin and 250 mg/l CH for diploid species and MS + IAA (1 mg/l) + kinetin (0.2 mg/l) + CH (250 mg/l) for *G. hirsutum*. Culture at 29 ± 18 °C for 8–10 weeks

3. Embryo Culture (Gill and Bajaj 1984b).
   a) Harvest bolls 15 days after pollination
   b) Surface sterilize by dipping in ethanol and flaming
   c) Excise the embryos in a sterilized Petri dish and culture on agar-solidified MS medium (1962) fortified with kinetin (0.5 mg/l) + IAA (1.5 mg/l) + CH (250 mg/l) + sucrose 3%
   d) Incubate the cultures at 29 ± 1 °C in the dark for first 15–25 days and then transfer to a well-illuminated room. Transfer the plantlets to pots containing a mixture of sterilized sand and soil when the first pair of leaves appear. Cover the pots with beakers or polythene bags for the first few days and water regularly with a nutrient solution.

4. Somatic Embryogenesis (Trolinder and Goodin 1987).
   a) Initiate callus cultures from hypocotyl segments of aseptically germinated seeds on MS medium with 100 mg/l inositol, B5 vitamins, 30 g/l glucose, 0.1 mg/l kinetin, and 0.1 mg/l 2,4-D
   b) Subculture the callus after 1 month in liquid medium devoid of growth regulators using 100 mg callus per 10 ml of liquid medium. Rotate the liquid cultures at 120 rpm under 30 $\mu E\, m^2/s$ light and 28 ± 2 °C temperature
   c) Sieve the embryogenic suspensions after 3–4 weeks using a mesh size of 20 and 30 and resuspend the cell clumps/embryos of 40 mg/l and pipet these onto the surface of 25 × 100-mm Petri dishes containing 40 ml of basal medium with 1.9 g/l KNO_3, 1.6 g/l Gelrite and 0.75 g/l Mg_2Cl_2. Incubate the plates at 28 ± 2 °C
   d) Germinate the mature embryos (10 mm in size) in 25 × 150 mm culture tubes containing one culture tube cap (Kaput) of 2 vermiculite saturated with 10–12 ml of Stewart and Hsu (1977b) medium containing 0.1 mg/l IAA

References

Altman DW, Stelly DM, Kohel RJ (1987) Introgression of the glanded plant and glandless seed trait from *Gossypium sturtranum* willis into cultivated Upland cotton using ovule culture. Crop Sci 27: 888–889

Altman DW, Fryxell PA, Koch SD, Howell CR (1989) *Gossypium* germplasm conservation augmented by tissue culture techniques for field collection. Econ Bot 43 (4)

Azizkhodozhaev A, Umarov ZM (1988) Effect of growth regulators and vitamins on growth of immature hybrid embryos of cotton in vitro. Uzb Biol Zh 2: 17–19 (in Russian)

Bajaj YPS (1982) Survival of anther and ovule-derived cotton callus frozen in liquid nitrogen. Curr Sci 51: 139–140

Bajaj YPS (1983) In vitro production of haploids. In: Evans DA, Sharp WR, Ammirato PV, Yamada Y (eds) Handbook of plant cell culture, Vol I. Macmillan, New York, pp 228–287

Bajaj YPS (ed) (1990) Biotechnology in agriculture and forestry, vol 12: Haploids in crop improvement I. Springer, Berlin Heidelberg New York Tokyo

Bajaj YPS (ed) (1991) Biotechnology in agriculture and forestry, vol 17: High-tech and micropropagation I. Springer, Berlin Heidelberg New York Tokyo

Bajaj YPS, Gill MS (1985) In vitro induction of genetic variability in cotton (*Gossypium* spp.) Theor Appl Genet 70: 363–368

Bajaj YPS, Gill MS (1986) Micropropagation and germplasm preservation of cotton (*Gossypium* spp.) through shoot tip and meristem culture. Indian J Exp Biol 24: 561–562

Bajaj YPS, Gill MS (1987) Biotechnology of cotton improvement. In: Crocomo J et al. (eds) Biotechnology of plants and microorganisms. Ohio State Univ Press, Columbus, pp 118–151

Bajaj YPS, Gill MS (1989) Pollen-embryogenesis and chromosomal variation in anther culture of a diploid cotton (*Gossypium arboreum* L.). SABRAO J 21: 57–63

Bajaj YPS, Gill MS, Mohapatra D (1986) Somaclonal and gametoclonal variation in wheat, cotton and brassica. In: Semal J (ed) Somaclonal variations and crop improvement. Nijhoff, Dordrecht, pp 160–169

Balasubrahmanyan R, Kanniyan K (1952) Vegetative propagation of cotton. Indian Cott Gr Rev 4: 184–186

Barrow JF, Katterman FRH, William D (1978) Haploid and diploid callus from cotton anthers. Crop Sci 18: 619–622

Beasley CA, Ting IP (1973) The effects of plant growth substances on in vitro fibre development from fertilized cotton ovules. Am J Bot 60: 130–139

Beasley CA, Ting IP, Feigen LA (1971) Improvements in fibre yield may come from test tube cotton. Calif Agric 2: 6–8

Beasley JO (1940) Hybridization of American 26-chromosome and Asiatic 13-chromosome species of *Gossypium.* J Agric Res 60: 175–187

Beasley JO (1942) Meiotic chromosome behaviour in species, species hybrids, haploids and induced polyploids of *Gossypium.* Genetics 27: 25–54

Chappel EJ, Mauney JR (1967) Culture of apical meristem of *Gossypium hirsutum* in vitro. Phyton 24: 93–100

Cherry JP, Katterman FRH, Endrizzi JE (1970) Comparative studies of seed proteins of *Gossypium* by gel-electrophoresis. Evolution 24: 431–77

Davidonis GH, Hamilton RH (1983) Plant regeneration from callus tissues of *Gossypium hirsutum* L. Plant Sci Lett 32: 89–93

Edwards GA, Mirza MA (1979) Genomes of the Australian wild species of cotton. II. The designation of a new G-genome for *Gossypium bickii.* Can J Genet Cytol 21: 367–372

Eid AAH, Delange E, Waterkeyn L (1973) In vitro culture of fertilized cotton ovules. I. The growth of cotton embryos. Cellule 69: 361–371

El-Nockrashy AS, Simmons JG, Frampton VL (1969) A chemical survey of seeds of the genus *Gossypium* Phytochemistry 8: 1949–1958

FAO (1987) Production yearbook, FAO, Rome

Fryxell PA (1965) A revision of the Australian species of *Gossypium* with observations on the occurrence of *Thespia* in Australia (Malvaceae). Aust J Bot 13: 71–102

Fryxell PA (1984) Taxonomy and germplasm resources. In: Kohel RJ, Lewis CF (eds) Cotton. Am Soc Agron, Madison, Wisconsin, USA

Gerstel DU (1953) Chromosomal translocations in interspecific hybrids of the genus *Gossypium.* Evolution 7: 234–244

Gill MS, Bajaj YPS (1984a) In vitro production of interspecific hybrids in cotton. Curr Sci 53: 102–104

Gill MS, Bajaj YPS (1984b) Interspecific hybridization in the genus *Gossypium* through embryo culture. Euphytica 33: 305–311

Gill MS, Bajaj YPS (1986) In vitro production of interspecific hybrids in *Gossypium* and their utilization in backcrossing. Z Pflanzenzücht 96: 174–176

Gill MS, Bajaj YPS (1987) Hybridization between diploid (*Gossypium arboreum*) and tetraploid (*Gossypium hirsutum*) cotton through ovule culture. Euphytica 36: 625–630

Gill MS, Bajaj YPS (1991) In vitro hybridization between *Gossypium arboreum* L. and *G. anomalum* Wawra and Peyr, and their backcross progeny. Plant Tissue Culture (in press)

Harland SC (1940) New polyploids in cotton by the use of colchicine. Trop Agric (Trinidad) 17: 53–54

Hsi YL, Wu IK (1981) Callus formation induced from the anther culture of upland cotton (*Gossypium hirsutum* L.). Abstr Proc Symp Plant tissue culture. Pitman, London, pp 239–240

Johnson BL, Thein MM (1970) Assessment of evolutionary affinities in *Gossypium* by protein electrophoresis. Am J Bot 57: 1081–1092

Joshi PC, Johri BM (1972) In vitro growth of ovules of *Gossypium hirsutum* L. Phytomorphology 22: 195–207

Lee JA (1984) Cotton as a world crop. In: Kohel RJ, Lewis CF (eds) Cotton. Am Soc Agron, Madison, Wisconsin, USA

Linsmaier EM, Skoog F (1965) Organic growth factor requirements of tobacco tissue cultures. Physiol Plant 18: 100–127

Lofland HB (1950) In vitro culture of cotton embryos. Bot Gaz 3: 301–307

Mauney JR (1961) The culture of immature cotton embryos. Bot Gaz 122: 205–209

Mauney JR, Chappel J, Ward BJ (1967) Effects of malic acid-salts on growth of young cotton embryos in vitro. Bot Gaz 122: 205–209

Mauney JR (1984) Anatomy and morphology of cultivated cottons. In: Kohel RJ, Lewis CF (eds) Cotton. Am Soc Agron, Madison, Wisconsin, USA

Murashige T, Skoog F (1962) A revised medium for rapid growth and bio-assays with tobacco tissue cultures. Physiol Plant 15: 473–497

Pandey SN (1976) Utilization of cotton seed and its products. Cotton Devel 6: 4–9

Parmar AL, Shroff VN, Deshpande NV (1978) Studies on vegetative propagation in cotton. Crop Improv 5: 73–76

Phillips LL (1962) Segregation in new allopolyploids of *Gossypium*. IV Segregation in New World × Asiatic and New World ×,wild American hexaploids. Am J Bot 49: 51–67

Phillips LL (1963) The cytogenetics of *Gossypium* and the origin of New World cottons. Evolution 17: 460–469

Phillips LL, Strickland MA (1966) The cytology of a hybrid between *Gossypium hirsutum* and *G. longicalyx*. Can J Genet Cytol 8: 91–95

Price HJ, Smith RH (1979) Somatic embryogenesis in suspension cultures of *Gossypium klotzschianum* Andress. Planta 145: 305–307

Shoemaker RC, Couche LJ, Galbraith DW (1986) Characterization of somatic embryogenesis and plant regeneration in cotton (*Gossypium hirsutum* L.). Plant Cell Rep 3: 178–181

Skovsted A (1934) Cytological studies in cotton. II Two interspecific hybrids between Asiatic and New World cottons. J Genet 28: 407–24

Skovsted A (1937) Cytological studies in cotton. IV Chromosome conjugation in interspecific hybrids. J Genet 34: 97–134

Smith RH, Price HJ, Thaxton JB (1977) Defined conditions for the initiation and growth of cotton callus in vitro. I. *Gossypium arboreum*. In Vitro 13: 329–334

Sohi GS (1964) Pests of cotton. In: Entomology in India. Entomol Soc of India, New Delhi, pp 111–148

Sood DR, Avtar Singh (1974) About gossypol in cotton. Cotton Dev 3(4) 15–17

Stewart JM, Hsu CL (1977a) In ovulo embryo culture and seedling development of cotton (*Gossypium hirsutum* L.). Planta 137: 113–118

Stewart JM, Hsu CL (1977b) Influence of phytohormones on the response of cultured cotton ovules to (2-chloroethyl) phosphonic acid. Physiol Plant 39: 79–85

Stewart JM, Hsu CL (1978) Hybridisation of diploid and tetraploid cottons through in-ovulo embryo culture. J Hered 69: 404–408

Tan XL, Qian YO, Chein Y (1988) Effect of explant source and cultural conditions on plant regeneration in *Gossypium gossypioides* (Ulbrich) Standley. Acta Genet Sin 15: 81–85

Thengane S, Paranjpe SV, Khaspe SS, Mascarenhas AF (1986) Hybridisation of *Gossypium* species through in ovulo embryo culture. Plant Cell Tissue Organ Cult 6: 209–219

Trolinder NL, Goodin JR (1987) Somatic embryogenesis and plant regeneration in cotton. (*Gossypium hirsutum* L.) Plant Cell Rep 6: 231–234

Trolinder NL, Goodin JR (1988a) Somatic embryogenesis in cotton (*Gossypium*). I. Effects of source of explant and hormone regime. Plant Cell Tissue Organ Cult 12: 31–42

Trolinder NL, Goodin JR (1988b) Somatic embryogenesis in cotton (*Gossypium*). II. Requirements for embryo development and plant regeneration. Plant Cell Tissue and Organ Cult 12: 43–53
Umbeck P, Johnson G, Barton K, Swain W (1987) Genetically transformed cotton (*Gossypium hirsutum* L.) plants. Bio-Technol 5: 263–266
White PR (1939) Potentially unlimited growth of excised plant callus in an artificial nutrient. Am J Bot 26: 59–64
White PR (1943) A handbook of plant tissue culture. J Cattell, Lancaster
Williams MD (1978) Conditions for optimal growth and differentiation of *Gossypium hirsutum* L. tissue cultures. Diss Abstr Int B 39(5): 2041 B

III.2 Tissue Culture and Micropropagation of Jute (*Corchorus* spp.)

A.S. Islam[1,2], M.M. Haque[1], M. I. Hoque[1], and Z.I. Seraj[1]

1 General Account

Commercial jute fibers are derived from the cultivars of the two species of *Corchorus*, namely *C. olitorius* and *C. capsularis* (family: Tiliaceae). The fibers originate from the secondary phloem. The fibers extracted from *C. olitorius* are golden in color and superior in tensile strength to those of *C. capsularis*, which yields white fibers. Commercially, the former is known as Tossa and the latter as White jute.

The cultivars of both species grow to a height of 9–10 feet and are sparsely branched. The leaves of *C. capsularis* are bitter to the taste, while those of *C. olitorius* are sweet. The leaves and young stems contain abundant mucilage. The flowers of both species are yellow but those of *C. olitorius* are two to three times larger than in *C. capsularis*. The fruits of *C. olitorius* are elongated, with conspicuous ridges and furrows, while those of *C. capsularis* are spherical. Both species differ also in their seed color and size; while the seeds of *C. capsularis* are chocolate brown and large, those of *C. olitorius* are greenish black and smaller compared to the other species. The number of seeds in *C. olitorius* ranges from 200–250, while in *C. capsularis* the number varies from 40–50.

One more characteristic feature which distinguishes the two species is that while *C. capsularis* can grow in low-lying innundated areas, the other species is susceptible to flood water.

Both species are short-day plants, requiring a period of 11 h or less for flowering. When sowing is completed in the right season, i.e., for White and Tossa varieties in mid-March and mid-April, respectively, they attain their full height and mature in between 110 and 130 days. The chromosome number of all known cultivars of both species is $2n = 2x = 14$.

Wild forms of *C. olitorius* are distributed all over the tropical and subtropical regions of the world, particularly in Asia and Africa. On the other hand, wild races of *C. capsularis* are confined to the Indo-Pak subcontinent. According to Kundu (1951), cultivars of *C. olitorius* originated in Africa and those of *C. capsularis* in the eastern part of the Indo-Pak subcontinent, i.e., in the region now known as Bangladesh. Jute cultivation is being extended to more and more countries of the world, such as Nepal, Thailand, Indonesia, Brazil, and China (Table 1).

[1] Department of Botany, University of Dhaka, Dhaka-1000, Bangladesh
[2] Department of Botany, The University of Texas at Austin, Austin, Texas 78713-7640, USA

Biotechnology in Agriculture and Forestry, Vol. 19
High-Tech and Micropropagation III (ed. by Y.P.S. Bajaj)

Table 1. Acreage, yield and production of jute and jute-like fibres in the principal jute growing countries in 1985 according to FAO (1985)

Country	Area harvested (1000 ha)	Yield (kg/ha)	Production (mt)
Bangladesh	932	1660	1547
Burma	51	1029	52
China	850	2424	2060
India	1260	1719	2200
Indonesia	23	1036	24
Nepal	3	2333	7
Thailand	209	1311	274

Both species are multiplied by means of seeds. There is no dormancy, and seeds from freshly harvested fruits and those collected in the previous year germinate readily, almost to the extent of 98–99%. However, both species can be propagated by means of cuttings from the near-mature plant. This is done particularly for breeding work because fiber qualities are at their best in plants showing 50% small immature fruits and 50% various flowering stages. Breeders therefore harvest their selected plants from the field at the above stage.

In order to obtain seeds from the selected individuals, their tops are cut before harvesting, and planted in the soil. These cuttings grow further to maturity for a period of another 2 weeks or more. The seeds are collected from the mature fruits of these cuttings. Genetic variability is limited in both species, more so in *C. capsularis*, which is completely self-pollinated. The cultivars of *C. olitorius* show 10% cross-pollination. Since selection has limited scope in bringing about improvement in both species, ionizing radiation and chemical mutagens have been used in the past to produce desirable varieties. However, the use of ionizing radiation did not prove very fruitful. For instance, during the last 30 years or so, only one variety of *C. capsularis* (Atom Pat 38) has been released by the Bangladesh Institute of Nuclear Agriculture through the use of ionizing radiation. In India also, among a number of varieties released out of irradiated material, only one, namely JRC 7447, has proved acceptable to farmers. Its cultivation is still limited because of the high input of fertilizers required to produce its maximum yield.

The need to hybridize the two species was felt for a long time (Finlow 1921) because cultivars of each species have some desirable characteristics not present in the other. For instance, Tossa is resistant to mosaic symptoms, anthracnose (caused by *Colletotricum corchori*), and partially resistant to nematodes, whereas White is resistant to dieback disease (caused by *Gleosporium singulata*) and partially tolerant to both kinds of mite. White is tolerant to flood but Tossa is susceptible. The color of the fibers of the latter is creamy white and those of Tossa golden. The quality of fibers is better in Tossa compared to White. In this context breeders wanted to breed jute varieties suitable for flood-prone low-lying land but with qualities of fiber like Tossa.

With the aim of combining the good qualities of both species, crosses were made between the two by a number of workers (Islam and Rashid 1960; Swaminathan et al. 1961; Iyer et al. 1961; Islam 1964; Das 1968; Haque 1970; Raut and

Naik 1983). Hybridization between cultivars of the two species succeeded, but no recombinants survived in F_3s to give rise to plant types combining the best qualities of the two species. Only maternal type plants, i.e., skewed type of inheritance, was observed in F_3s and onwards.

The characters required to be combined in the hybrid derivatives of the two jute-yielding species are as follows:

Corchorus olitorius (Tossa)	*C. capsularis* (White)
1. Immunity to mosaic virus. *C. capsularis* is susceptible to the extent that in certain years more than 50% plants are affected in some areas	Immunity to dieback disease caused by *Glomerlla singulata*; *olitorius* is susceptible. (The new name of the causative organism is *Gleosporium*)
2. Immunity to anthracnose caused by *Colletotricum corchori*. *C. capsularis* is susceptible	Partial immunity to root rot caused by *Rhizoctonia solani*. *C. olitorius* is severely affected by root rot
3. Partial immunity to nematode particularly in the high land	Partial immunity to mite. *C. olitorius* is severely affected by mite
4. Fibers of much better tensile strength than *C. capsularis*. Fetches at least 1–1½ times better price than *C. capsularis* fiber of weaker strength	Immunity to waterlogging. *C. olitorius* is susceptible to waterlogging so much so that it dies when submerged under floodwater. Nonshattering habit of fruit. The fruits of *C. olitorius* dehisce upon maturity, losing a sizable portion of seeds before harvest

C. capsularis is self-pollinated while the other species, *C. olitorius*, is cross-pollinated to the extent of 10%. Probably for this reason only a few cultivars have been obtained by jute breeders through selection during the past 70 years or so.

Since tissue culture-derived plants of a number of crops are known to produce a wide range of variation (Evans and Sharp 1986; see also Bajaj 1990), it was considered worthwhile to use this technique for production of promising lines in both White and Tossa jute through somaclones. Such lines with valuable agronomic characters can be easily utilized in the breeding of desirable characters.

It was further thought that if attempts to obtain regenerants from suitable explants of jute cultivars proved successful, the technique could be tried to produce somaclonal variants from F_1 hybrids of *C. olitorius* × *C. capsularis*, which invariably revert to maternal types in advanced generations (Hoque et al. 1988). Attention was therefore focused on tissue culture. Iyer and Raina in 1972 succeeded for the first time in producing callus from the anthers of the two jute species, but the callus did not differentiate. Islam et al. (1978) reported success in isolating protoplasts from two jute varieties of *C. capsularis*, namely D154 and Lalnaris, and the O4 variety of *C. olitorius*. They did not observe any dividing protoplast. Das et al. (1981a) claimed success in obtaining shoot differentiation from the protoplast-derived callus of *C. olitorius* and *C. capsularis*, but so far they have not succeeded in establishing any protocol for obtaining regenerants except very infrequently (pers. commun.) Haque and Takeuchi (1983) tried the method recommended by the above authors, i.e., use of seawater to alleviate mucilage from their protoplast preparation, but without success. When cotyledons of germinating seedlings were grown aseptically in simple solidified agar containing 3% sucrose, there was good recovery of protoplasts ($10^6/mm^2$), as against a small number ($10^3/mm^2$) in the control in which MS medium was added to agar.

Islam et al. (1981) tried to obtain polyhaploid through anther culture of a spontaneous amphiploid, *C. olitorius* × *C. depressus* in an attempt to restore the fiber qualities which deteriorated following chromosome doubling. They obtained profuse callusing from the anthers but in the differentiating medium only roots developed. In the same year, Sen et al. (1982, pers. commun.) reported obtaining differentiation of jute plants through shoot-tip culture. In the next year, Das (1983) and Sultana (1983) obtained shoot differentiation when they cultured shoot tip of the amphiploid *C. olitorius* × *C. depressus* and the natural tetraploid *C. siliquosus*, respectively, in MS (Murashige and Skoog 1962) medium containing equal proportions of cytokinin and auxin. They observed root development from the base of the same callus which supported shoots. Since vascular connection between the shoot and the root was absent in their material, none of their plants survived after they were transplanted from the test tube to the soil. The above workers also failed to ascertain whether the shoot developed from the pre-existing primordia in the dissected shoot tip.

2 In Vitro Culture and Micropropagation

2.1 Methods of Sterilization

Explants were sterilized in 0.1% $HgCl_2$ for 20 min and then washed three or four times in sterile distilled water under a laminar flow. The seeds were first sterilized in absolute alcohol for 30 s to 1 min and then washed once in sterile distilled water. The seeds were then immersed in 0.1% $HgCl_2$ for 15 min and washed in sterile distilled water.

2.2 Media Composition

MS (Murashige and Skoog 1962), modified MS (MMS), UM (Uchimiya and Murashige 1974), SH (Schenk and Hildebrandt 1972), and Potato Extract (PE) (Jiajun and Junwen 1977) were used with or without supplements of specific auxins, cytokinins, vitamins, and amino acids, as required for specific experiments.

The medium was prepared from the stock solutions which were maintained separately for (1) macronutrients, (2) micronutrients, and (3) organic constituents. Auxins, cytokinin, gibberellin, antiauxin, and amino acid stock solutions were prepared as needed; 0.8% Difco or Sigma bacto agar was found to be the best for supporting solid medium. Except for hormone, gibberellin, and cytokinin, all other stock solutions were stored at 4 °C in a refrigerator. The stock solution of the first-mentioned substances was preserved in a deep freeze. The solvents used for this group were the standard reagents such as 0.1N NaOH for NAA and BAP, 70% ethanol for IAA, IBA, GA_3, and 2-iP, and 0.1N HCl for kinetin and adenine sulfate. Since the composition of all media is well documented except PE, the composition of only the latter is given below in mg/l.

KNO_3 1000; $(NH_4)_2SO_4$ 100; $Ca(NO_3)_2$, $4H_2O$ 100; $MgSO_4$, $7H_2O$ 125; KH_2PO_4 200; KCl 35; thiamine 1; sucrose 30000. To the above mixture 5, 10, and 20% (v/v) potato extract was added and designated PIII, PII, and PI. The effect of double, half, and one-fourth strength MS was seen on the nature and rate of callus growth. pH, unless stated specially, was adjusted to 5.8 before autoclaving.

2.3 Embryo Culture and Wide Hybridization

Islam (1964) cultured hybrid embryos (from the cross *C. capsularis* × *C. olitorius*, using the former as female) on solidified agar medium containing Knop's ingredients, 0.5 mg/l IAA and 0.5 mg/l kinetin. Only one F_1 hybrid was obtained, but the plant did not grow to maturity and died before flowering. In reciprocal combination, Haque (1970) obtained only one F_1 hybrid plant through embryo culture. Islam and Abbasi (1965) obtained only one hybrid plant through embryo culture between colchicine-induced tetraploid *C. olitorius* and a natural New World tetraploid *C. siliquosus*. Using *C. aestuans* as mother in crosses with *C. olitorius*, Haque (pers. commun.) tried to grow incompletely developed hybrid embryos in simple culture medium, but their growth was very slow and they all died before they were transplanted to the soil. Arbani (pers. commun.) attempted to raise F_1 hybrids between the two species at *tetraploid* level using colchicine-induced tetraploids of *C. olitorius* and *C. capsularis* through embryo culture, but unfortunately none which included a few healthy seedlings survived following their transplantation to the pot.

Except in the cross *C. aestuans* × *C. olitorius*, there was extremely low or no survival of the F_1 hybrids outside the culture medium. In other words, failure to obtain any F_1 hybrid resulted not from genetical incompatibility between the species involved but due to lack of knowledge in handling the transplanted seedlings from the test tubes to the soil.

Techniques have now been developed in our laboratory to save a large number of the tissue culture-derived plants. This was done by covering them with a slightly perforated plastic bag after their transplantation to the pot. The cover was removed in subsequent days for a greater and greater length of time until they were sufficiently hardened to remain without cover. Since most of the seedlings can be transplanted to the pot successfully, embryo culture can now be used with much confidence to produce a large number of hybrids between cultivars of the two jute species in the search for desirable recombinants, as well as hybrids in combinations in which only a few F_1 hybrids were hitherto obtained with great difficulty. Furthermore, with the formulation of many rich media such as KM now routinely employed in tissue culture, it will be worthwhile to try these media (unknown to scientists in the 1960s) to rescue underdeveloped embryos in difficult or hitherto unattainable cross combinations such as *C. olitorius* × *C. aestuans*, *C. olitorius* × *C. trilocularis*, etc.

2.4 Anther Culture

Following the abortive attempts of Iyer and Raina (1972) to obtain differentiation in anther-derived callus of *C. capsularis* and *C. olitorius*, Jahan and Chowdhury (1980) embarked upon a project of bringing down the chromosome number of a spontaneous amphiploid of the cross, *C. olitorius* × *C. depressus* to the diploid level through anther culture. The fibers of the amphidiploid suffered in quality, perhaps on account of chromosome duplication, and there was thus an urgent need to produce polyhaploids in an attempt to improve the fiber qualities. This cross was originally designed to transmit genes for photoinsensitiveness from a diploid wild species, *C. depressus*, to cultivars of *C. olitorius*. Some segregants proved to be photoinsensitive like the diploid wild parent, but with poor fiber qualities, making them unacceptable commercially. All our efforts to produce polyhaploid through anther culture ended in failure. Flower buds, 2 days before anthesis, were selected for anther culture. Anther from such buds contained uninucleate pollen grains. Anther culture with other jute species has shown that the uninucleate stage of pollen gives the best results in androgenesis. Pollen grains from flower buds of different stages were examined in 1% acetocarmine and only those expected to open in the following 2 days were found to contain the right stage.

In MS medium, callus production was poor. Eighty percent of the anthers of O/D (*C. olitorius* × *C. depressus*) callused profusely in R_2 medium supplemented with 0.5 mg/l NAA/IAA and 2 mg/l kinetin. Anthers of three more varieties in addition to O/D were tried in MS with IAA/NAA and kinetin as supplements in different proportions. With 1 mg/l each of NAA and kinetin all four materials callused well, the highest production (75%) being observed in Chinchura Green (CG), followed by O4 (70%), O/D (65%), and Assam red (60%). Variety D154 of *C. capsularis* formed good callus. In IAA- and kinetin-supplemented MS medium, roots developed from the callus before they were subcultured.

2.5 Leaf Explants

Leaf discs from 1-month-old seedlings grown in axenic culture of 1 cm in diameter were cut with a sterilized cork borer and placed with their upper surface on the agar medium. In MS supplemented medium callus developed readily along the cut surface in all the material (Table 2) in 2 weeks. The supplements that proved best for callusing in *C. olitorius* × *C. aestuans* hybrids and c-4 × *C. olitorius* were 1 mg/l IAA and 0.3 mg/l BAP. Callus turned hard and green when the above materials were grown in MS medium with 1 mg/l NAA + 2 mg/l BAP. The callus of O/D gave two shoots when grown in 0.1 mg/l IAA and 2 mg/l zeatin. Three out of eight callus cultures of leaf explants of c-4 × *C. olitorius* raised in 0.2 mg/l IAA and 2 mg/l zeatin gave rise to small shoots when transferred to MS medium containing 0.1 mg/l IAA. None of these shoots developed to a size suitable for transplantation.

Table 2. Shoot development from the callus of leaf explants in MS medium. Callus developed from all leaf discs

Plant material	Supplements (mg/l)	No. of leaf discs callused	No. of shoots formed
C. olitorius var. O4	IAA 1 + BAP 0.3	10–12	—
	IAA 0.1 + NAA 0.01	10–12	—
C. olitorius × *C. aestuans*	IAA 1 + BAP 0.3	10–12	—
	IAA 0.1 + NAA 0.01	10–12	—
C. olitorius × *C. depressus*	IAA 0.1 + NAA 0.01	10–12	—
	IAA 0.2 + Zeatin 2	10–12	2
c-4 × *C. olitorius*	IAA 0.2 + Zeatin 2	10–12	3
	NAA 1 + BAP 2	10–12	—

c-4 × = Colchicine-induced tetraploid.

2.6 Root Explants

Root segments, 1.5 cm in length, of 3-day-old seedlings were cultured in MS medium containing either 1 mg/l IAA + 0.3 mg/l BAP or 0.5 mg/l NAA + 0.5 mg/l BAP. Explants of *C. capsularis* var. D154 and var. Lalnaris did not callus, while those of *C. olitorius* vars. O4, Assam red, c-4 × *C. olitorius* and O/D callused and produced roots profusely, some of which were negatively geotropic. None produced shoots in differentiating medium containing 2 mg/l zeatin. *C. trilocularis* – a wild species – callused poorly in the beginning but after subculturing in the same medium, its growth improved.

2.7 Use of Anti-auxin to Promote Shoot Formation

In an attempt to inhibit abundant root formation and simultaneously induce shoot differentiation in O/D, Islam (1981) used several combinations of abscisic acid (ABA) in MS with 1 mg/l zeatin. Both 10^{-5} and 10^{-6} lm ABA inhibited root growth but no shoots were seen to develop from the callus. Triiodobenzoic acid (TIBA) at the same concentrations arrested the formation of root but had no effect on shoot development. In higher doses, both ABA and TIBA were found to be lethal to callus growth. Furthermore, without supplements of zeatin, ABA and TIBA adversely affected the development of callus. No root was seen in callus raised in total darkness in medium containing 0.1 mg/l NAA and 2 mg/l BAP.

2.8 Shoot Tip Culture

Das and Sen (1981) reported successful virus-free plantlet formation from the meristem-derived calli of *C. capsularis.* They observed that suitable combinations of IBA and kinetin (K) stimulated growth of shoot explants without callus formation in certain *Corchorus* species, while substitution of IBA with NAA produced only callus. These calli, on transfer to IBA and K, gave rise to a number of shoots. When none of our explants, i.e., leaf, root, and anther, produced shoots with any degree of reliability, shoot tips of a large number of varieties and hybrids were cultured in various media with various supplements in different combinations and concentrations. A 0.5–1-mm-diameter shoot tip of a number of cultivars and hybrids of jute species was cultured in various media, such as MS, modified MS (MMS), and SH. Among the various hormone combinations added to the basic medium, that of IBA and kinetin in the ratio of 1:1 was found to induce shoot formation. The highest frequency of shoots in the three species and interspecific hybrids (Table 3) was obtained in MS medium supplemented with 0.4 mg/l each of kinetin and IBA. It has been observed in the past that a high or equal proportion of auxin:cytokinin induces callus and that of cytokinin:auxin promotes shoot formation. In the above material, Islam et al. (1982) observed shoot formation only when the ratio of cytokinin and auxin was equal. No shoot formation was observed when the amount of cytokinin was increased.

In all instances, shoot development was preceded by callus. Since no shoot was found to develop from the second or subsequent subculture, it could not be ascertained whether the shoots regenerated from the callus or developed from pre-existing shoot primordia. Root was found to develop from the base of the callus. It was realized only during the transfer of the plantlet from the test tube to the soil that there was no vascular connection from the shoot to the root via callus. The plantlets died soon after transfer to the pot, primarily because they did not have roots to establish themselves on the soil.

Since doubt existed as to whether shoots differentiated from the callus or developed from the primordia, we decided to settle this point by inducing shoot formation in the second and subsequent subcultures. Callus was raised from shoot tips of D154, a variety of *C. capsularis*, in the following two media: MS + 0.5 mg/l BAP + 0.5 mg/l IBA + 0.1 mg/l 2,4-D and MS + 0.5 mg/l isopentanyl adenine (iP) + 0.5 mg/l IBA + 200 mg/l casein hydrolysate (CH). In both media callus developed profusely within 2 weeks. Callus clumps, regardless of their origin, were transferred to either of the two differentiating media, namely 0.5 mg/l BAP

Table 3. The effect of various supplements in MS medium on root and shoot differentiation from the callus (Fig. 1)

Supplements	Material	No. of shoot tips inoculated	% Callus induction	% Shoot regeneration	% Root regeneration
0.4 mg/l K	*C. siliquosus*	56	42.85	58.33	16.00
+ 0.4 mg/l	*C. olitorius* C.G.	35	39.28	—	50.00
IBA	*C. capsularis* D154	25	36.00	—	32.00
	A/O	25	48.00	—	68.00
	O/D	98	76.53	3.06	48.98
2 mg/l K	*C. siliquosus*	28	42.86	32.14	32.14
2 mg/l IBA	*C. olitorius* C.G.	25	36.00	—	32.00
	C. capsularis D154	70	50.00	2.86	44.29
	A/O	35	42.86	5.71	60.00
	O/D	78	62.82	6.41	55.31

A/O = *C. aestuans* × *C. olitorius* hybrid (2n = 14).
O/D = Amphiploid of *C. olitorius* × *C. depressus* hybrid (2n = 28).

Table 4. Effect of different supplements in MS on shoot regeneration from shoot tip-derived calli. The numbers shown in the first row in the second, third, and fourth columns refer to calli raised in supplements: 0.5 mg/l IBA + 0.5 mg/l BAP + 0.1 mg/l 2,4-D, while those shown in the second row refer to calli raised in 0.5 mg/l-iP + 0.5 mg/l IBA + 200 mg/l CH

Supplements (mg/l)	Materials	No. of subcultures producing callus	No. of callus recallusing	No. of shoots regenerated per callus (average)	Time taken for regeneration from the initiation of callus (in days)
0.5 BAP +	O4	24 + 24	15 + 15	—	—
25 tyrosine	C.G.	24 + 24	15 + 15	—	—
	D154	24 + 24	18 + 17	6 + 4	49–55
	CVL-1	24 + 24	17 + 16	—	—
	C. siliquosus	24 + 24	15 + 16		
0.1 IBA +	O4	20 + 22	11 + 9	—	—
50 glutamine	C.G.	20 + 22	10 + 8	—	—
	D154	20 + 22	13 + 12	4 + 3	60–67
	CVL-1	20 + 22	11 + 10	—	—
	C. siliquosus	20 + 22	11 + 11	—	—

+ 25 mg/l tyrosine or 0.1 mg/l IBA + 50 mg/l glutamine. After subculture, multiple shoots were found to develop from clumps of the callus tissue (Table 4) of D154 within 3 weeks (Rahman 1984).

2.9 Regeneration from Cotyledon-Derived Callus

Callus from the cotyledons of D154 and O4 of variety *C. capsularis* and *C. olitorius*, respectively, was obtained in MS supplemented by 0.5 mg/l BAP and 25 mg/l tyrosine. When the cotyledon-derived callus of D154 was subcultured, multiple shoots regenerated from it. From the initiation of callus, it took about 30–35 days for multiple shoots to develop. No differentiation took place in O4.

2.10 Anatomical Studies Through Differentiating Callus

At the next stage of investigation, 12-µm-thick serial sections of the shoot-producing callus were made and stained in saffranin and fast green. This was carried out to study the mode of development of shoots from the callus. Before the emergence of shoots, these bodies of differentiating callus showed organized masses of tissue with distinct cambial layers embedded in them. The cambial layers, after repeated divisions, developed into torpedo like-bodies from which shoots eventually regenerated. The origin of shoots from the second and third subculture, together with anatomical evidence, left no room for doubt that differentiation took place from the organized callus and not from any pre-existing primordia (Rahman et al. 1985). It is difficult to interpret without further investigation why almost similar results were obtained in the same variety of jute, i.e., D154 using one of the two amino acids: tyrosine or glutamine, the former in combination with BAP and the latter with IBA. Probably in the same biosynthetic pathway connected with differentiation, tyrosine and glutamine are produced at two different stages and therefore addition of any one of them triggers the process.

2.11 Multiple Shoot Formation and Origin of Somaclonal Variation

In an attempt to enhance the frequency of shoot formation, we have obtained interesting results (Das et al. 1986). Seeds of varieties D154, CVL-1 of *C. capsularis*, and varieties O4, CG of *C. olitorius*, and *C. olitorius* × *C. aestuans* hybrids were placed in a Petri dish for germination at 17.00 h after sterilizing the seeds in the usual way described earlier in the chapter and at 10.00 h. On the following morning, i.e., after 17 h, the testa of the seeds, along with their endosperm, was removed. The endosperm-detached seed was then placed on a sterile slide and a portion of the root tip was removed. After this operation, cotyledons were removed from some seeds and in others were left intact. The total length of seeds with or without cotyledons was about 1 mm. They were placed in MS medium supplemented with a 0.5 mg/l BAP and 25 mg/l tyrosine. The medium was solidified with 0.8% agar. Throughout the experiment, the temperature regime was 28 ± 2 °C day and 23 ± 2 °C night. The light regime was in the neighborhood of 8000 to 10000 lx. All test tubes were placed under a 12-h photoperiod regime. Callus developed from these inocula in 10 days, and in another 10 days multiple shoots originated from the callus in the same medium if IBA (0.5 mg/l) was added to it (Table 5). Without IBA the callus had to be transferred to the same medium for shoot formation. The number of shoots was greater in the subculture compared with shoots developed directly from the explant. The number of shoots was 30–35 in D154 (Fig. 1), 10–15 in CVL-1 (Fig. 3), and in O4 it was of the order 4–6 shoots. Only CG produced no shoots (Fig. 2). The clumps of multiple shoots were devoid of roots (Fig. 3).

At the next stage, the multiple shoots were removed from the test tube to a conical flask containing the same medium. These organically connected shoots were separated and transferred to the medium found best for rooting in another series of experiments. This medium consisted of half-strength MS and 0.2 mg/l IBA (Fig. 4). Within 3 weeks the separated shoots developed good roots and were ready for transfer to the pot (Fig. 5, Table 6). The same procedure was followed for transplantation as has been described earlier, viz. use of sterilized soil and covering of the transplanted seedlings with perforated polythene bags. The seedlings were gradually exposed to natural environment by removing the polythene bag for a greater and greater length of time until it was not necessary to cover them. Young tissue culture-derived seedlings are shown in Fig. 5.

Das et al. (1987) raised R_2 population consisting of 731 plants. Some of the R_2 plants were different in respect of the following characters, (1) smooth-surfaced fruits, (2) larger number of fruits/cluster compared to the mother plant (Fig. 6), (3) less tapering stem than the parental material, and (4) early flowering.

Table 5. Response of different additives in MS for regeneration of shoots from different parts of germinating seeds in different varieties of *C. capsularis* and *C. olitorius*

Media	Material	Explant	No. of explants	No. of callus clumps formed	% Callus	Multiple shoot initiation		No. of shoot callus (approx.)
						No.	%	
MS + 0.5 mg BAP + 25 mg/l tyrosine	D154	Cotyledon	80	15	14.75	4	26.67	3
		Plumule	75	20	26.67	17	85.00	12
		Cotyledon + plumule	100	90	90.00	85	94.44	40
	A/O	Cotyledon	45	3	6.67	00	—	—
		Plumule	41	4	9.76	1	25.00	9
		Cotyledon + plumule	72	35	48.61	25	71.43	30
	O4	Plumule	35	1	2.85	00	—	—
		Cotyledon + plumule	55	10	18.18	6	60.00	27
	CVL1	Cotyledon + plumule	45	20	44.44	16	80.00	5

Table 6. Response of root formation and establishment of plantlets

Media	Material	No. of regenerated shoot inoculated	No. of shoots showing root formation	% Root formation	No. of plants transplanted to soil	No. of plants established	% Plants established
MS + 0.1 mg/l IAA + 0.1 mg/l IBA	D154	60	48	80.00	12	8	66.67
	CVL1	16	3	18.75	—	—	—
MS + 0.2 mg/l IAA	D154	50	26	52.00	6	—	—
	CVL1	11	2	18.18	—	—	—
MS + 0.2 mg/l IBA	D154	63	51	80.95	35	23	67.64

Fig. 1. Multiple shoot (30–35) regeneration from the callus of *C. capsularis* var. D154 after 3 weeks on MS + 0.5 mg/l BAP + 25 mg/l tyrosine
Fig. 2. No shoot regeneration in C.G. variety of *C. olitorius* in the same medium within the same period
Fig. 3. Multiple shoots (15–20) regenerated from the callus of *C. capsularis* var. CVL-1 in MS medium containing 0.5 mg/l BAP + 25 mg/l tyrosine within 3 weeks
Fig. 4. Root induction on half-strength MS + 0.2 mg/l IBA
Fig. 5. Two-month-old plantlet of D154 established in soil
Fig. 6. Portion of a branch of a tissue culture-derived plant of D154 showing four fruits per cluster

2.12 Shoot Formation in *C. olitorius*

Mention has already been made that a comparatively small number of multiple shoots originated from the callus of O4 in the same medium found suitable for cultivars of *C. capsularis* (Table 6). However, these shoots were weakly developed and degenerated before their transfer to the soil. In an attempt to prevent senescence, Islam and Ahmed (unpubl.) added an antioxidant NDGA (nordihydroguaiaretic acid), 3 mg/l, to the rooting medium containing 0.3 mg/l IBA. By adding this supplement they were able to obtain eight well-developed, well-rooted plantlets. When transferred to the soil, two of them survived and grew to maturity. Thus the addition of an antioxidant such as NDGA has opened up the possibilities of saving the shoots from degeneration and helping them to establish as fully mature plants. Thus, using tissue culture technique, it would be possible to produce somaclonal variants among cultivars of *C. olitorius*, which yields better-quality jute fibers.

2.13 Inheritance of the Multiple Shoot Formation Trait

In the preceding section, it has been shown that while from cultivars of *C. capsularis* such as D154 and CVL-1 30–35 multiple shoots differentiated from their callus, the number of such shoots regenerating from the callus of cultivars of the other species, *C. olitorius*, was only a few. Moreover, in the latter the shoots were weak and they died before transfer to the soil. Since it has been questioned for 25 years whether the progeny of the above cross were true hybrids because of their close resemblance with the female parent, it was decided to study the segregation pattern of the shoot-forming character of the above two parents. It was argued that if F_1s showed a few weakly developed shoots they would be considered to have arisen as a result of self-pollination; if, on the other hand, they showed an intermediate number or nearly the same number of shoots as the male parent, it would indicate that the character in question is polygenic/dominant and that the progeny were truly hybrid in nature.

When cotyledon-attached plumules of F_1 seeds were cultured in MS medium supplemented with tyrosine (25 mg/l) and BAP (0.5 mg/l), they produced an intermediate number, i.e., 15–20 shoots (Islam and Hoque 1987). When the calli raised from the seeds of F_2 were cultured in the above medium, they produced a varying number of shoots. Their number ranged from a few, as in the female parent, to nearly as many as in the male, while in the majority the number fell somewhere in the middle group between 10 and 25.

From the above observation, two conclusions can be drawn: (1) the inheritance of multiple shoot-forming character is polygenic in nature, and (2) the F_1 progeny were genuine hybrids because the multiple shoot-forming character could be transmitted only from the male parent *C. capsularis*.

In the light of the above results, it is suggested that in a situation in which an interspecific hybrid closely resembles the female parent, the shoot-regenerating ability of both the species should be studied. In the case where one species is found to produce multiple shoots and the other a few or none at all, the hybrid nature of the progeny could be easily established by using the former as the male parent and studying the pattern of inheritance in F_1 and F_2.

2.14 Use of Nordihydroguaiaretic Acid (NDGA) in Preventing Senescence of Regenerants of *C. olitorius*

Earlier it was mentioned that the success of Das et al. (1986) in obtaining regenerants in BAP-tyrosine supplemented MS medium from different parts of embryonal segments of 20-h-old germinating seeds of jute was confined only to cultivars of *C. capsularis*. Their regenerants from cultivars of *C. olitorius* did not grow well and died before they were large enough for transfer to the rooting medium. In an attempt to save the *C. olitorius* regenerants, Ahmed and coworkers (1989) of the Bangladesh Atomic Energy Commission, repeated the regeneration experiment using NDGA, an antioxidant, following the same method and using the medium as suggested by Das et al. (1986). When the above workers (Ahmed et al. 1989) transferred the poorly developed regenerated shoots of a cultivar of *C. olitorius*, namely, O 4 into the rooting medium (MS + 0.2 mg/l IBA) containing 3 mg/l NDGA, the plantlets grew well and large enough for transplantation into the soil. They obtained a number of mature O 4 plants which fruited and gave them viable seeds.

2.15 Regenerants from Callus Tissue and Their Biochemical Study

Working with two cultivars of *C. capsularis* D-154 and CVL-1 and one cultivar of *C. olitorius*, namely O 4, Seraj and coworkers (pers. comm.) obtained only callus when they supplemented MS medium with tyrosine and 2,4-D, using 5–7-mm-long hypocotyl segments of 7–10-day-old seedlings. None of these callus pieces differentiated into shoots in subsequent subcultures with or without the above supplements. Their explants also callused profusely when they added one more supplement, namely, 3 mg/l NDGA to MS medium, but this time the callus pieces of two varieties (D-154 and CVL-1) of *C. capsularis* differentiated into single or multiple shoots when they were subcultured on just MS medium without any supplements (Fig. 7D). In some instances, a second subculture in supplement-free MS was found necessary for regeneration. The variety O 4 of the second species, *C. olitorius* did not differentiate when subjected to the same treatment.

Furthermore, they observed that the embryogenic callus cells in the medium containing tyrosine alone or with added NDGA were green, thin-walled, and rich in starch grains. Starch grains were larger in the cells initiated in MS supported by a combination of BAP, tyrosine, and NDGA compared to when NDGA was excluded from the medium (Fig. 7A, B). Nonembryogenic callus cells, on the other hand, in tyrosone and 2,4-D-supplemented medium were thick-walled and showed hardly any starch grains (Fig. 7C). That embryogenic callus cells are rich in starch grains was recently shown by Nayak and Sen (1989) in *Paspalum scorbiculatum*.

2.16 Biochemical Aspects of Differentiation

In an attempt to get an idea about the factor(s) contributing to differentiation, Seraj and coworkers (pers. comm.) determined the average specific activity of glyoxalase-1 in both embryogenic and nonembryogenic callus. This enzyme has been correlated not only with cell division in *Datura* and pea (Ramaswamy et al. 1983, 1984) but has been shown to decrease in amount in tissues undergoing differentiation, such as in *Brassica* (Sethi et al. 1988).

The results reported by the above workers (Seraj, pers. comm.) agree with those reported by Sethi et al. (1988) with respect to the two varieties (D-154 and CVL-1) of *C. capsularis* in that the average specific activity of glyoxalase-1 in nonembryogenic calli of the two cultivars of *C. capsularis* was many times higher than that

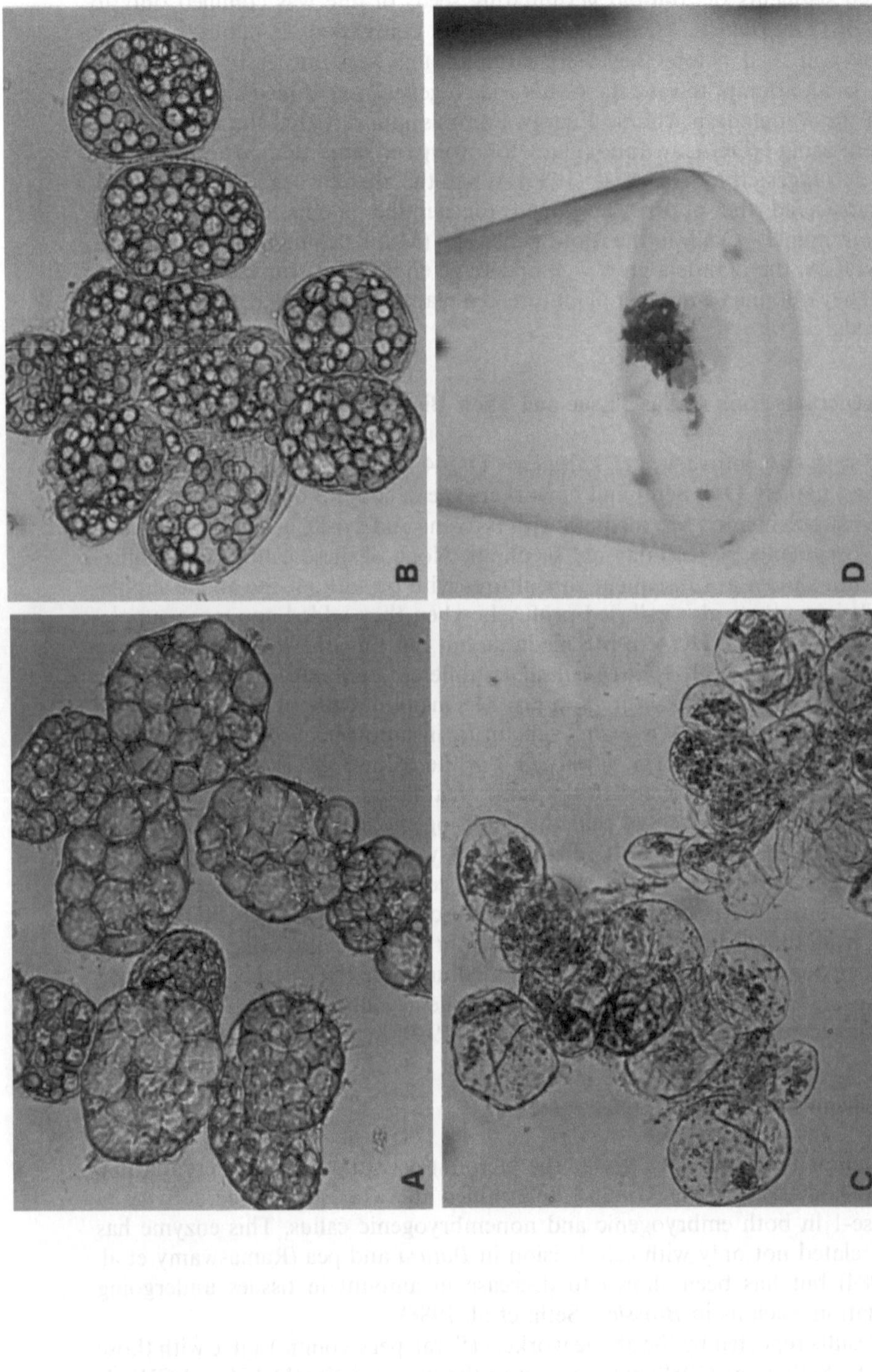

Fig. 7. **A** A few thin-walled cells from the embryogenic callus of jute *C. capsularis* var. (D154) initiated in BAP-supplemented medium containing 25 mg/l tyrosine and 3 mg/l NDGA; note that the thin-walled cells are rich in large starch grains. **B** A few cells from the embryogenic callus of the same variety initiated in the above medium without NDGA; as in **A** the cells are thin-walled and rich in starch grains but the latter are much smaller in size. **C** A few thick-walled cells from the nonembryogenic callus of the same variety of jute. Note that they are conspicuous by the absence of starch grains. **D** Multiple shoots in the first subculture only in MS from the callus of the same variety initiated in MS + 0.5 mg/l BAP + 25 mg/l tyrosine and 3 mg/l NDGA. (A.B. Sarker, unpubl.)

found in the embryogenic callus (Fig. 8). In the tyrosine and BAP supplemented MS medium, the enzyme showed a slightly higher level of activity compared to that observed in the medium containing NDGA in addition. Interestingly, the level of the enzyme activity did not differ significantly between the differentiating calli of the two cultivars of *C. capsularis* (D-154 and CVL-1) and nonembryogenic callus of *C. olitorius* cultivar (O 4), although calli of all three varieties were initiated on the same media supplements, namely, on BAP, tyrosine, and NDGA supplemented MS medium (Fig. 8). The results reported by earlier workers (Das et al. 1986;

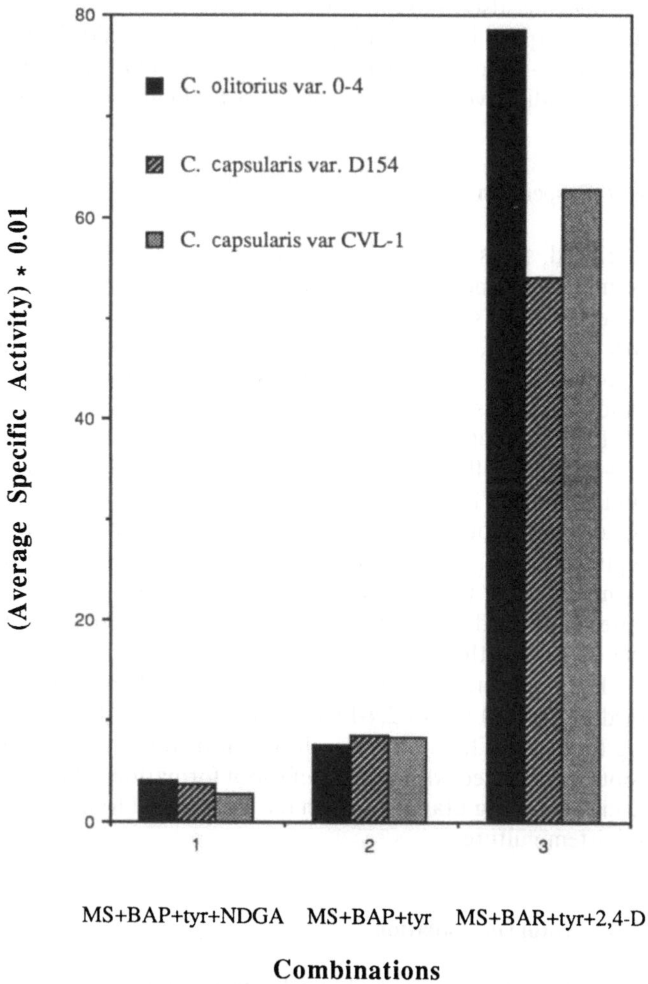

Fig. 8. Histograms (*1–3*) showing average specific activity (enzyme unit/mg protein/min) of glyoxalase-1 in calli derived from hypocotyl segments of two cultivars of *C. capsularis* (D154 and CVL-1) and one cultivar of *C. olitorius* (O-4) initiated in MS using three different combinations of supplements: 2,4-D + tyrosine; BAP + tyrosine; and BAP + tyrosine + NDGA. Note that in the embryogenic calli represented in *1* and *2*, the level of enzyme activity is greatly reduced compared to that observed in the nonembryogenic callus supported by 2,4-D and tyrosine (*3*). (A.B. Sarkar, unpubl.)

Ahmed et al. 1989) may appear contradictory but the age of the explants may easily account for this difference. The explants of Seraj and coworkers consisted of 5–7-mm-long hypocotyledonary segments of 7–10-day-old seedlings, whereas those of the earlier workers (Das et al. 1986; Ahmed et al. 1989) were embryonal parts including hypocotyls of 20-h-old germinating seeds. This could be one of the main reasons for the discrepancy in the results.

In this context it will be worthwhile to determine the glyoxalase-1 level in the embryogenic callus of the material, such as *Carica papaya*, in which embryogenic callus is reported to be initiated in a medium containing a high concentration of 2,4-D and glutamine (Fitch and Manshardt 1990). If the embryogenic callus of *C. papaya* is also found to contain a lower level of the above enzyme compared to its counterpart, i.e., nonembryogenic callus, it would then indicate that the enzyme glyoxalase-1 is one of the key factors in controlling differentiation; and that its level will be indicative of whether or not a particular callus tissue is embryogenic.

2.17 Suspension Culture

Sterilized seeds of D154 were inoculated in plain agar with 2% sucrose for germination under sterile conditions. Hypocotyl, cotyledons, and roots were dissected from 4-day-old seedlings and placed in callusing medium consisting of MS + 2 mg/l 2,4-D, 0.25 mg/l K, 9.9 mg/l thiamine, 9.5 mg/l pyridoxine, 4.5 mg/l nicotinic acid, 20.0 mg/l casein hydrolysate, with 0.8% agar. Only hypocotyl produced greyish white soft calli profusely.

Pieces of soft callus tissue weighing about 1 g each were then transferred to 250-ml conical flasks, each containing about 80 ml liquid medium of the composition as above. The flasks were shaken in a rotatory shaker at a speed of 80 rpm under continuous illumination at 2000 lx. Within 10 days of inoculation in the liquid medium, the calli produced good suspension culture with profuse small clumps of cells. In another 4 weeks these clumps gave rise to aggregated cells. The latter developed into numerous tiny balls of white tissue, which ultimately settled at the bottom of the flask.

For regeneration, the white balls were then transferred to solidified Kao medium with 0.1 mg/l 2,4-D + 0.2 mg/l zeatin + 1 mg/l NAA + 2 g/l sucrose and 1 g/l glucose. These balls developed into tough and green calli. In 4 weeks only roots regenerated with no sign of shoot formation. The experiment will be repeated using tyrosine/glutamine, which has been found to promote shoot formation in the meristem culture.

2.18 Protoplast Isolation

As the first step to somatic hybridization, attempts were made by Islam et al. (1978) to isolate protoplasts from two varieties of *C. capsularis*, namely D154 and Lalnaris – a semi-wild species – and one variety of *C. olitorius*, i.e., O4. All three plant species presented difficulties in the isolation of protoplasts by exuding mucilage from the cut surface of the leaf. In addition to mucilage, the isolated protoplasts became contaminated by bacteria. Contamination was eliminated in

Table 7. Protoplast yield per square millimeter in different combinations of enzyme solution

Species	MS				Sucrose 3% + agar 0.75%			
	CI	CII	CIII	CIV	CI	CII	CIII	CIV
Corchorus olitorius var. CG	10^2–10^3	10^3	10^3	10^3	10^5–10^6	10^4	10^4	10^4
C. capsularis var. D154	$10^{1.5}$	$10^{1.5}$	$10^{1.5}$	$10^{1.5}$	10^5–10^6	10^4	10^4	10^4

CI = Macerozyme R10 (1%) + Cellulase Onozuka R10 (2%).
CII = Macerozyme R10 (1%) + Pectinase (2%).
CIII = Cellulase Onozuka R10 (2%) + Pectinase (1%).
CIV = Cellulase Onozuka R10 (2%) + Driselase (1%).

two ways (1) by adding to the culture 40 μg gentamycine or (2) by germinating jute seeds in axenic culture. Due to mucilage, the density of isolated protoplasts was always poor, being in the neighborhood of only 10^2 to $10^3/mm^2$. When cotyledons from the test tube-grown seedlings were used for protoplast isolation, the mucilage exudation was greatly reduced, but it was not at a level to promote division of protoplasts with newly generated cell wall. In the meantime, Das et al. (1981) successfully removed mucilage from protoplast mixture by washing it with 80% seawater. Using the same substance, Haque and Takeuchi (1983) tried to remove mucilage, but without success. They used a one-step method, i.e., a mixture of Cellulase Onozuka R10 (2%) and Macerozyme R10 (1%). Because of the accumulation of abundant mucilage, it was very difficult to filter the material even through a 140-μm sieve filter. In one of their experiments, Haque and Takeuchi (1983) grew the seeds in a 0.75% agar medium with just 3% sucrose, adding no other nutrients to the medium. The cotyledons in such test tube-grown plants in which there was only sucrose gave a good yield of protoplasts without any mucilage. The yield per mm^2 was approximately 10^6, as against 10^2 to 10^3 (Table 7).

In KM medium the protoplasts were seen to divide and form a small group of cells. Das et al. (1981a) obtained differentiation in protoplast-derived callus of the two cultivated jute species. No further publications of these authors were available to confirm their earlier report.

Now that an improved technique to isolate jute protoplasts in good density has been developed and the protoplasts have been seen to divide in some media, we are continuing our experiments to induce shoot formation via callus, with the aim of producing somatic hybrids between the cultivars of the same, and ultimately between the two, jute species.

3 Summary and Conclusion

The objectives set for the tissue culture of jute in 1977 have been only partially accomplished. This modest success has nevertheless created hope that this technique can be profitably employed in the improvement of jute cultivars.

By culturing 1-mm-long germinating seeds with decapitated root tips, it is now possible to obtain multiple shoots numbering 30–40 in the cultivars of *C. capsularis* and 6–12 in some cultivars of *C. olitorius*. The clump of multiple shoots can be separated and rooted easily in the rooting medium. By applying this method, it may not be difficult to raise a large somaclonal population of some of the cultivars of both species and screen them for desirable traits since it is urgently required to find diversified uses of jute. The technique will also be very helpful to multiply hybrid seeds of some interspecific crosses which have been reported successful. Somaclonal variants arising from hybrid seeds of the cross, *C. olitorius* × *C. capsularis* may prove to preserve certain promising combinations of the two species. In straight breeding, all recombinants are lost in the advanced generations.

Using plumules of germinating seeds and adding tyrosine, BAP, and IBA in MS medium, rapid multiplication of breeders' seed of new and promising lines of jute may now be possible by inducing multiple shoot formation in them. This will cut down the number of years ordinarily required for the release of a new variety.

Now that the role of tyrosine in combination with BAP has been definitely proved in the differentiation of shoots, it will be worthwhile to try the above chemicals to induce shoot formation in anther-derived callus for the production of haploids.

From the foregoing discussion it is evident that tissue culture in jute has already proved to be an important and powerful tool for attempting to produce desirable varieties. Research in jute tissue culture has reached the stage that progress from this point towards accomplishment of ultimate objectives, namely producing jute varieties with desirable characteristics, does not seem unattainable.

References

Ahmed G, Hossain ABMM, Islam MS (1989) Regeneration of multiple shoots in jute *Corchorus olitorius* (var. O-4) from cotyledon and hypocotyl explants of germinating seeds. Indian J Exp Biol 27: 334–337

Bajaj YPS (ed) (1990) Biotechnology in agriculture and forestry, vol 11. Somaclonal variation in crop improvement I. Springer, Berlin Heidelberg New York Tokyo

Das B (1983) Development and differentiation of jute plants through shoot tip culture. M.Sc. Project, Dep Bot, Dhaka University

Das B, Haque MM, Islam AS, Rahman MH, Hoque MI (1986) In vitro plantlet development in jute. In: Rao AN, Mohan Ram HY (eds) Proc Nayudamma Mem Symp on Agricultural applications in biotechnology, Madras, India, pp 106–114

Das B, Hoque MI, Haque MM, Islam AS (1987) Somaclonal variation in *Corchorus capsularis* jute. In: Islam AS, Haque MM (eds) Proc Reg Workshop on Tissue culture of tropical crop plants, Dhaka, pp 63–67

Das K, Sen SK (1981) In: Symp on Crop Improvement through Tissue Culture (poster session). Bose Inst, Calcutta

Das K, Kumar PM, Sen SK (1981a) Progress in application of protoplast technology in tackling genetical problems of cultivated jute species. In: Islam AS (ed) Proc Int Workshop on Improvement of Tropical Crops Through Tissue Culture. Dhaka, pp 74–80

Das K, Sinha RR, Sen SK (1981b) Regeneration of plantlets from leaf callus of *Corchorus* species through shoot and bud formation. Genet Res Group, Bot Dep, Bose Inst, Calcutta

Das PK (1968) Development studies in the jute cross with special reference to the reciprocal combination: *Corchrous capsularis L.* × *C. olitorius* L. Nucleus 11 (1): 34–41

Evans DA, Sharp WR (1983) Single gene mutations in tomato plants regenerated from tissue culture. Science 221: 949–951

Evans DA, Sharp WR (1986) Application of somaclonal variation. Biotechnology 4: 528–532
Fao (1985) Production yearbook, FAO, Rome
Finlow RS (1921) Historical notes on experiment with jute. Beng. Agric. J India 16: 265–279
Fitch M, Manshardt RM (1990) Somatic embryogenesis and plant regeneration from immature zygotic embryos of papaya (*Carica papaya* L.). Plant Cell Rep 9: 320–324
Haque MM (1970) Attempts to breed better and disease resistant jute (*Corchorus*) strains through inter- and intraspecific crosses. Ph.D. Thesis, Bot Dep, Sind University, Pakistan
Haque MM, Takeuchi M (1983) Alleviation of mucilage problem in protoplast isolation of jute species. Bangladesh J Bot 12: 178–181
Hoque M, Imdadul, Haque MM, Islam AS (1988) Confirmation of *Corchorus olitorius* × *C. capsularis* hybrid through tissue culture and biochemical tests. Bangladesh J Bot 17(1): 71–79
Islam AS (1964) A rare hybrid combination through application of hormone and embryo culture. Nature (London) 201: 320
Islam AS (1981) Production of desirable jute plants through tissue culture. Proc. Int. Symp on Tissue culture of economically important plants. (Ed. A.N. Rao). COSTED Singapore, pp 159–161
Islam AS, Abbasi J (1965) A tetraploid jute hybrid c-4x jute, *C. olitorius* New World wild jute, *C. siliquosus*. Sind Univ Sci Res J 2: 21–29
Islam AS, Rashid A (1960) First successful hybrid between the two jute-yielding species, *C. olitorius* × *C. capsularis*. Nature (London) 185(4708): 258–260
Islam AS, Chowdhury MKU, Mustafa MG, Haque MM (1978) Isolation of protoplasts in two species of jute. Bangladesh J Bot 7(2): 71–77
Islam AS, Haque MM, Jahan R, Chowdhury MKU, Ahmed G, Rahman H (1981) Attempt to produce polyhaploid from a spontaneous amphidiploid of the jute hybrid *C. olitorius* × *C. depressus*. Bangladesh J Bot 10(2): 63–68
Islam AS, Rahman MH, Sultana C, Das B, Islam ABMS (1982) Successful plantlet differentiation from shoot tip-derived callus of *Corchorus*. Bangladesh J Bot 11(2): 185–187
Islam AS, Hoque M, Imdadul (1987) Study of the inheritance of differential regenerating ability of the two jute species. In: Proc SCAMAP Regional Workshop on Plant tissue culture and biotechnology of medicinal and aromatic plants, Lucknow, India, 14–16 Dec (Abstr)
Iyer RD, Raina SK (1972) The early ontogeny of embryoids and callus from pollen and subsequent organogenesis of anther cultures of *Datura metel* and rice. Planta (Berlin) 104: 146–156
Iyer RD, Sulbha K, Swaminathan MS (1961) Fertilization and seed development in crosses between *C. olitorius* and *C. capsularis*. Indian J Genet Plant Breed 21(3): 192–200
Jahan R, Chowdhury MKU (1980) Preliminary investigation into jute tissue culture, Annu Rep submitted to Univ Grants Commission, Dhaka University
Jiajun Z, Junwen O (1977) Studies on simplified potato medium for pollen plant induction in wheat and rice. Inst Genet, Acad Sin, Beijing
Kundu BC (1951) Origin of jute. Indian J Genet 11: 95–99
Murashige T, Skoog F (1962) A revised medium for rapid growth and bioassays with tobacco tissue culture. Physiol Plant 15: 473–497
Nayak P, Sen SK (1989) Plant regeneration through somatic embryogenesis from suspension cultures of a minor millet, *Paspalum scorbiculatum*. Plant Cell Rep 8: 296–299
Rahman H (1981) Tissue culture studies in the genus *Corchorus*. M.Sc. Thesis, Bot Dep, Dhaka University
Rahman H, Islam ABMS (1983) Production of genetic variability in cultivated jute through tissue culture. Proc 15th Int Genet Congr, New Delhi
Rahman SMZ (1984) Studies of the initiation and maintenance of explant callus and organogenesis in the genus *Corchorus* L. M.Sc. Thesis, Bot Dep, Dhaka University
Rahman SMZ, Hadiuzzaman S, Haque MM, Islam AS (1985) Shoot formation in *Corchorus capsularis* var. D154, from unorganised callus. Bangladesh J Bot 14(2): 141–145
Ramaswamy O, Pal S, Guha-Mukherjee S, Sopory SK (1983) Presence of glyoxalase-1 in pea. Biochem Int 7: 307–318
Ramaswamy O, Pal S, Guha-Mukherjee S, Sopory SK (1984) Correlation of glyoxalase-1 activity with cell proliferation in *Datura* callus culture. Plant Cell Rep 3: 121–124
Raut RN, Naik G (1983) Interspecific hybridization in cultivated jute. Proc FAO Expert Consultation on Jute and Kenaf Improvement, Calcutta, 5–9 Sep; Dhaka 12–15 Sep
Schaeffer GW, Sharpe FT, Cregan PB (1984) Variation for improved protein and yield from rice anther culture. Theor Appl Genet 67(5): 383–390
Schenk RU, Hildebrandt AC (1972) Medium and techniques for induction and growth of monocotyledonous and dicotyledonous cell cultures. Can J Bot 50: 199–204

Scowcroft WR, Larkin PJ (1982) Somaclonal variation: A new option for plant improvement. In: Vasil IK, Scowcroft WR, Frey KJ (eds) Plant improvement and somatic cell genetics. Academic Press, New York, pp 159–178
Sen SK, Das K, Sinha RR (1981) Recovery of virus-free plantlets of cultivated jute species. Genet Res Group, Bot Dep, Bose Inst, Calcutta
Sethi U, Basu A, Guha-Mukherjee S (1988) Control of cell proliferation and differentiation by regulating polyamine biosynthesis in cultures of *Brassica* and its correlation with glyoxalase-1 activity. Plant Sci 56: 167–175
Sultana C (1983) In vitro plantlet formation of *Corchorus* species through shoot tip and leaf explants. M.Sc. Project, Dep Bot, Univ Dhaka
Swaminathan MS, Iyer RD, Sulbha K (1961) Morphology, cytology and breeding behaviour of hybrids between *C. olitorius* and *C. capsularis*. Curr Sci 30: 67–68
Uchimiya H, Murashige T (1974) Evaluation of parameters in the isolation of viable protoplasts from cultured tobacco cells. Plant Physiol 54: 936–944

III.3 Micropropagation of Sunflower (*Helianthus annuus* L.)

A. Cavallini and M.C. Lupi[1]

1 General Account

The common sunflower (*Helianthus annuus* L.) is a member of the Compositae (Asteraceae) family. Its name (from the Greek helios, sun, and anthos, flower) is derived from the heliotropic movement of the sunflower's head (nutation). Grown for its seed, this species is one of the most important suppliers of vegetable oil, and is now fairly widely cultivated throughout the world.

1.1 Botany

The genus *Helianthus* includes about 100 species (Watson 1929), the majority of which are native to North America. Heiser (1957) and Heiser et al. (1969) described 67 species of *Helianthus*, 50 native to North America, and 17 to South America. Probably the genus *Helianthus* developed from its closest genus, *Viguiera*, or from *Viguiera*-like ancestors. Other close relatives of *Helianthus* are *Tithonia* and *Phoebanthus* (Heiser 1978). The genus provides two food plants, *H. annuus*, the sunflower, and *H. tuberosus*, the topinambour or Jerusalem artichoke. Several varieties of *H. annuus*, as well as other species of the genus, are sometimes cultivated as ornamentals.

The domestic sunflower is annual, 1 to 3 m tall, has a single stem, and usually a single head (the capitulum) – which varies in diameter from 10 to more than 30 cm. Some types have additional heads on short branches near the top of the main stem. Branching is a characteristic of wild sunflowers, which can be from less than 1 to 5 m tall, and whose branches originate from the bottom to the top of the primary stem.

Many small flowers are aggregated together in the head (Fig. 1). The outside of the head is circled by the ray flowers, petal-like structures, usually yellow. Ray flowers are always sterile and play no role in reproduction, except for attracting pollinating insects. Within the head lie smaller flowers, the disk-flowers; they have five fused petals, which surround the reproductive organs. When the ray flowers have opened, a band of disk flowers adjacent to ray flowers begins to mature; maturation goes on from the outside to the center of the head. Each disk flower may develop an achene, the fruit, which contains the seed. A mature domestic plant

[1] Department of Agricultural Plant Biology, Genetics Section, Via Matteotti 1/B, 56100 Pisa, Italy

Biotechnology in Agriculture and Forestry, Vol. 19
High-Tech and Micropropagation III (ed. by Y.P.S. Bajaj)

Fig. 1. Mature inflorescence of sunflower (*Helianthus annuus* L.)

has usually from 250 to 1500 seeds, arrayed in a spiral pattern. The seeds are 7–15 mm long, of various colors and color patterns; they are larger than those of the wild species (3 to 6 mm in length).

1.2 Cytology

The genus *Helianthus* has a basic chromosome number of x = 17 (Whelan 1978), and contains diploid, tetraploid, and hexaploid species. *Helianthus annuus* L. is a diploid species (2n = 2x = 34); chromosomes are small and not easily distinguishable from one another, except for two chromosome pairs, showing satellites. An idiogram of *H. annuus* karyotype was proposed by Vranceanu et al. (1974), based on the position of the centromere and the presence or absence of chromosomes with satellites.

Cavallini and Cremonini (1985) and Cavallini et al. (1986) showed the occurrence of chromosomal mosaicism (aneusomaty) in both shoot and root tip of different cultivars and lines of sunflower: somatic chromosome number ranged from 17 to 34, and cells with 28–30 chromosomes were the most frequent. The origin of aneuploid cells was found in mitotic aberrations, occurring during embryo development; in the course of plant development, the percentage of aneuploid cells decreased (diplontic selection) until, at the pre-meiotic stage, only diploid cells could be found, and normal meiosis occurred, showing 17 bivalents (Cremonini and Cavallini 1986).

Chromosomal mosaicism was also reported by Gundaev (1971; cited in Whelan 1978) in sunflower twin embryos. Such chromosomal variability may be related to the supposed polyploid origin of this species: an ancestral polyploid condition would justify the loss of chromosomes, genes in the euploid cells being in multiple copies; in this sense, the experiments by Jackson and Murray (1983), showing the appearance of quadrivalents in colchicine-treated microsporocytes of this species, indicated that some, if not all, of the chromosomes were quadruplicated in the "diploid" sunflower. The allopolyploid origin of *H. annuus* was presumed also by Heiser and Smith (1955), who proposed *Viguiera* species with n = 8 and n = 9 as progenitors of cultivated sunflowers.

As to nuclear DNA content, this species shows a strict control of diploid condition, i.e., polyploid cells occur only in such ephemeral tissues as the anther tapetum (Cavallini and Cionini 1986). By contrast, a great intraspecific variation (20–25%) was found in 4C nuclear DNA (Cavallini et al. 1986). No apparent correlation was found between nuclear DNA changes and plant phenotype.

1.3 Distribution and Importance

Sunflower is one of the four major annual crops grown for edible oil. The others are soybean (*Glycine max* L.), rapeseed (*Brassica campestris* L. and *B. napus* L.), and groundnut (*Arachis hypogaea* L.). Cotton (*Gossypium hirsutum* L.) is a fiber crop, but its seed by-product contributes significantly to the world's edible oil supply.

Sunflower has been the main source of edible vegetable oil in Russia and other eastern European countries for decades; in the last few years, production has increased in other countries, such as the USA and Canada. In Table 1, annual production of the four major oil crops is reported for the course of the century: in comparison to rapeseed and groundnut, soybean and sunflower seed production has increased at a faster rate.

Significant amounts of sunflower seed are used for other purposes than oil. Some large seed is used as whole, roasted seed or dehulled and sold as confectionery "nuts". Smaller seeds are used in food for pet birds and small animals.

Archelogical evidence reveals the use of sunflower among American Indians (Heiser 1955). The sunflower was introduced to Europe by early Spanish explorers (Putt 1978), and gradually moved eastward and northward in the European continent. Peter the Great introduced the sunflower into Russia in the 18th century (Zukovsky 1950). Following the initial identification and use, its cultivation

Table 1. Mean annual world production (× 1000 t) of four major oilseed crops. (Compiled from Putt 1978 and Mielke 1987)

	1934–38	1967–71	1983–87	1983–87 / 1934–38 ratio
Soybean	11 900	44 974	92 952	7.8
Groundnut	5300	17 711	13 112	2.5
Rapeseed	3910	6128	16 932	4.3
Sunflower	2448	9886	17 751	7.3

Table 2. Production and country-wise distribution of sunflower in 1986/87. (Compiled from Mielke 1987)

	Production (× 1000 t)	Area (× 1000 ha)
France	1904	829
Greece	164	79
Italy	380	160
Spain	890	906
Hungary	857	391
Romania	900	550
Yugoslavia	450	107
USSR	5258	3848
USA	1214	791
Argentina	2250	1882
Australia	145	179
South Africa	265	323
Turkey	940	683
China	1440	1100
India	400	800
Other countries	1170	1083
World total	18627	13711

expanded rapidly and, by the beginning of the 20th century, sunflower was becoming a major crop in Russia. Other European countries which early showed interest in sunflower oil production were Hungary, France, Romania, Bulgaria, Spain, and Yugoslavia (Putt 1978). In recent years high sunflower seed production has been achieved also in Italy (Benvenuti and Vannozzi 1988).

In North America, it is generally accepted that the present cultivated sunflower stems from material reintroduced from Russia after the crop became widely grown there in the latter part of the 19th century (Semelczi-Kovacs 1975). Until 1940, sunflower was considered only as a silage crop; it evolved as an oilseed crop after the breeding work of the Canada Department of Agriculture (Putt 1978). In the 1960's, the introduction of USSR cultivars that had very high oil content made sunflower seed economically attractive for processing and marketing by the US oilseed processing industry, which promoted sunflower breeding and cultivation in the USA.

Although less important than in Europe and North America, a significant interest exists in this crop on all continents. A very high production of sunflower seed has been obtained in Argentina, which began exporting sunflower seed as early as the beginning of this century. Sunflower oil is also of great commercial importance in Australia, India, China, Turkey, and South Africa (see Table 2).

1.4 Conventional Propagation and Breeding

Sunflower is a highly cross-pollinated crop, pollination occurring primarily by insects (Putt 1940). Most breeding methods for maize and other cross-pollinated crops are applicable to sunflower with certain modifications because of its flowering process and morphological characteristics.

Major objectives in sunflower breeding include improved seed yield, earlier maturity, shorter plant height, uniformity of plant type, disease and insect resistance, and, in oilseed types, high oil percentage (Fick 1978). The emphasis on specific objectives differs for various production areas and for different breeding programs. In the last few years, improved oil quality, protein percentage, and protein quality have become additional and increasingly important breeding objectives.

Although practised by the earliest growers, breeding and selection work to improve sunflower at experimental stations was initiated in 1910 in the USSR (Pustovoit 1964), where mass selection was commonly used during the early stages of cultivar improvement (Gundaev 1971). During the 1920's, a widely used and highly successful method for improving sunflower cultivars was developed by Pustovoit (Pustovoit 1964): the method is a form of recurrent selection that includes progeny evaluation and subsequent cross-pollination among progenies with superior characteristics. The method was especially effective in increasing the oil percentage in the seed from about 30% in the early 1920's to over 50% in the 1970's (Fick 1978).

Inbreeding as a method for improving sunflower was used to develop lines with improved oil percentage, strong single stems, and disease and insect resistance (Voskoboinic and Soldatov 1974). As with other cross-pollinated crops, breeders soon realized that the value of inbreeding was to develop lines with certain desirable characteristics for subsequent crossing to produce synthetic cultivars or interline hybrids. Hybrids were first obtained through mechanical or chemical emasculation; the discovery of genetic male sterility (Gundaev 1966) and the linkage of male sterility to genetic markers (Leclercq 1966) allowed the production of the first commercial hybrids in 1970 in Romania (Vranceanu et al. 1988). Cytoplasmic male sterility was then discovered by Leclercq (1968) from an interspecific cross involving *H. petiolaris* Nutt. and *H. annuus* L. This source of male sterility was shown to be very stable and is now the source used almost exclusively in breeding programs around the world (though other sources of sterility have been found to overcome cytoplasmic uniformity; Benvenuti and Vannozzi 1988). After the identification of genes for fertility restoration of the male sterile cytoplasm (Kinman 1970), the first hybrids were made available for commercial production in the USA in 1972 (Fick 1978). Presently, these hybrids account for some 99% of the sunflower production in the USA and other western countries, while in the USSR and China, sunflower seed production obtained from open-pollinated cultivars is 80 and 90%, respectively (Benvenuti and Vannozzi 1988).

1.5 Need for In Vitro Culture

Cell and tissue culture may complement the existing breeding programs. Regeneration and/or propagation of interspecific *Helianthus* hybrids (Bohorova et al. 1985) may allow the more rapid introduction of new genes or cytoplasmic factors in cultivated varieties. Moreover, micropropagation (Lupi et al. 1987), if normal plants are regenerated in large number, may be used to propagate particular genotypes and/or inbred lines, and homozygous plants may also be produced through culture of anthers or unpollinated ovaries and ovules (see Sect. 2). Plants regenerated via callus formation (Greco et al. 1984; Paterson and Everett 1985; Lupi et al. 1987) may show new genetic variability. Finally, tissue culture techniques might be used for foreign gene introduction via *Agrobacterium* in order to select new genotypes (see Moyne et al. 1989; Schrammeijer et al. 1990).

2 Sunflower Tissue Culture

Shoot organogenesis from callus was first described by Sadhu (1974) from a single callus obtained by culturing stem pith on White's medium containing 1 mg/l indoleacetic acid (IAA) for 10 weeks. Plant regeneration was reported but the cultivar utilized was not named. Greco et al. (1984) grew leaf, cotyledon, and hypocotyl segments on Murashige and Skoog (1962) medium (MS) supplemented with different concentrations of 2,4-dichlorophenoxyacetic acid (2,4-D) and 6-benzyl-aminopurine (BAP); whereas 2,4-D alone produced poorly developed nodular callus, BAP alone was shown to induce abundant growth of compact, green callus, which then regenerated many shoots, some of which flowered precociously in vitro. Similar results were reported by Lupi et al. (1987), who obtained callus formation from hypocotyl and cotyledon explants cultured on medium containing 2.0 mg/l naphthaleneacetic acid (NAA) plus 0.5 mg/l BAP; these calli regenerated shoots when transferred on medium containing 0.5–2 mg/l BAP. Power (1987) observed shoot regeneration from dormant mature and immature zygotic embryos of sunflower inbreds and hybrids cultured on BAP-plus NAA-containing media; the developmental stage of the zygotic embryo was shown to be critical and regeneration was achieved from both callused and noncallused tissues. Recently, Witrzens et al. (1988) and Espinasse and Lay (1989) also showed that immature embryos as source of explants allow the highest frequencies of shoot regeneration through culture on BAP-containing media. Finally, Bohorova et al. (1985) obtained shoot regeneration from different explants of *Helianthus* interspecific hybrids, using MS medium supplemented with 0.1 mg/l NAA, 0.1 mg/l BAP, 0.01 mg/l gibberellic acid (GA_3) and 40 mg/l adenine.

Plant regeneration via somatic embryogenesis was reported by Chandler and Jan (1983), who cultured immature zygotic embryos of interspecific and intergeneric hybrids of *Helianthus*. The embryogenetic pattern of regeneration was then reported by Paterson and Everett (1985): 100 experimental inbred lines were tested for morphogenetic ability and the use of one of these genotypes (SS415B) allowed rapid regeneration of plants from hypocotyl-derived calli, via an embryogenic route. Embryoids developing from vascular tissues were observed also by Knopp and Mix (1986) in cultures of cotyledons and hypocotyl fragments of different sunflower cultivars. The largest effects on regeneration were related to the nitrate concentration in the media and to the age of the seedlings when the explants were taken (Paterson and Everett 1985). Ethylene also plays an important role: inhibiting ethylene production in young seedlings promoted regeneration from the hypocotyl-derived calli (Paterson and Adams 1987). Finally, immature embryo culture on high sucrose (12%) media led Finer (1987) to obtain direct somatic embryogenesis, without callus formation. The somatic embryos proliferated from the explant surface and germinated after transfer to a low sucrose medium. Somatic embryogenesis was also reported from thin cell layers (Pelissier et al. 1990) and callus liquid cultures (Prado and Berville 1990) of sunflower hypocotyls.

To obtain haploids and/or homozygous lines, cultures of both anthers and unfertilized ovules were undertaken by several authors. Bohorova et al. (1980) and Alissa et al. (1985) cultured anthers from interspecific hybrids and obtained plantlets by direct androgenesis or by regeneration from calli. Callus formation was stimulated by supplying the medium with 2,4-D and kinetin (K) (Bohorova et al. 1980, 1985), while direct androgenesis was observed when using media containing zeatin (Bohorova et al. 1980). A high number of regenerants were haploids (Bohorova et al. 1980) but also aneuploid and polyploid plants occurred (Alissa et al. 1985).

Mix (1985), using anthers from normal cultivated sunflowers, observed that the percentage of anthers forming callus ranged from 15 to 80%, depending on the genotype, and obtained seven plantlets, three of which were haploid. Mezzarobba and Jonard (1986) reported that callus organogenesis and anther embryogenesis were greatest when using anthers between the diad and tetrad stage of PMC as explants, and increasing temperature increased anther embryogenesis under conditions of 8–16 days darkness; a total of seven diploid and two haploid plants were produced by these authors.

Mix (1985) also reported callus formation from unpollinated ovaries, ranging from 10 to 85% depending on the genotype, and subsequently obtained 19 diploid plantlets from calli of various cultivars. Gynogenetic plants were also obtained through culture of unfertilized ovules, with an average rate of five plants per 1000 explants cultured on MS medium supplemented with 10% sucrose and NAA 2 mg/l (Gelebart and San 1987). Chromosome counts on the root meristems of the gynogenetic plants showed that the majority were haploid but others were diploid or mixoploid; during plant growth, the number of haploid cells in one and the same plant decreased because of spontaneous doubling, until, at the floral stage, the plants had become diploid (Gelebart and San 1987).

Concerning protoplast culture, Binding et al. (1981) reported regeneration of sunflower from protoplasts isolated from cultured shoot tips, although no details were presented. More recently, Lenee and Chupeau (1986), Bohorova et al. (1986), Bergounioux et al. (1988), Guilley and Hahne (1989), and Dupuis et al. (1990) reported protoplast isolation and culture of cultivated and wild sunflowers. In these reports, protoplasts were obtained from many sources (hypocotyls, cotyledons, seedling roots, mesophyll, stems), and callus was obtained in different frequencies from protoplast, though plantlet regeneration is not reported. Moyne et al. (1988) reported both callus and embryoid formation from hypocotyl protoplasts; however, these embryoids failed to develop into mature embryos.

Finally, in vitro culture has been successfully used to allow the embryo development of interspecific crosses in the *Helianthus* genus, in the case of pronounced embryonal incompatibility (Bohorova et al. 1982; Alissa et al. 1986). This technique might be useful to facilitate the introduction of wild characteristics in cultivated sunflowers, and cytogenetic studies of the *Helianthus* genus (see also Robinson and Everett 1990; Bohorova and Atanassov 1990; Yang et al. 1990; Jonard and Mezzarobba 1990).

3 Micropropagation

Shoot tip culture allows rapid multiplication of sunflower plants. Shoots may develop directly from meristem tips or via callus formation and subsequent organogenesis. Different micropropagation techniques will be described in this section.

3.1 Materials and Sterilization

Open-pollinated cultivars (Trifi et al. 1981; Greco et al. 1984; Lupi et al. 1987), commercial hybrids (Trifi et al. 1981), and selfed lines (Paterson 1984; Cavallini and Lupi 1987) have been used to study sunflower micropropagation. Bohorova et al. (1985) used micropropagation techniques to multiply experimental hybrids such as *H. annuus* × *H. decapetalus*, *H. annuus* × *H. hirsutus* and *H. annuus* × *H. tomentosus*.

Seeds were sterilized in calcium or sodium hypochlorite at various concentrations (5 to 15%) for different periods. In some cases, supplementary washes were made in dishwashing detergent (Paterson 1984) or in 70% ethanol (Greco et al. 1984; Lupi et al. 1987) before rinsing seeds in distilled sterile water. Seeds were aseptically cultivated to achieve primary explants.

Trifi et al. (1981) germinated seeds on Knop's solid medium (Gautheret 1959) at 23 °C and with a 12-h photoperiod (10 000 lx); both shoot tips (after 10 days) and nodes (after 15 days) were used as explants. Paterson (1984) germinated seeds aseptically in the dark in tubes containing 0.5% sucrose plus 0.8% agar aqueous solution; after 3–4 days, plantlets were exposed to light (1500–3000 lx) for 2 days and then 5–8-mm-long shoot tips were dissected and cultured. Greco et al. (1984) excised shoot apices (2–3 mm in length) from seedlings with the first pair of noncotyledonary leaflets, previously germinated on solid MS medium under continuous light (1500 lx) at 25 °C. Lupi et al. (1987) cultivated seeds on solid MS medium at 24 °C under continuous light (2500 lx); 0.5–1-mm-long shoot tips containing the apical dome plus five to six leaflets were excised from 7-day-old plantlets and cultured.

Field-grown plants were used as a source of explants by Bohorova et al. (1985): shoot apices were surface sterilized by immersion in 3% calcium hypochlorite for 20 min, followed by three washes in sterile distilled water.

3.2 Culture Media and Conditions

Different media were used to initiate the cultures. Various authors used MS medium as base culture medium. Trifi et al. (1981) cultured explants in tubes containing 20 ml MS medium supplemented with 3% sucrose, 0.6% agar, 0.5 mg/l NAA, and 0.5 mg/l BAP; pH was adjusted at 5.8; the culture was performed under the same photoperiod as seed germination. Paterson (1984) used Petri dishes containing solid MS medium supplemented with 0.5% sucrose and different phytohormones including BAP, K, zeatin (Z), IAA, NAA, 2,4-D, indole-butyric

Table 3. Callus formation and/or shoot production from vegetative apices of *Helianthus annuus* cultured on different media (growth factor concentrations in mg/l). (Lupi et al. 1987)

Culture media	No. of apices	No. of apices with callus	Relative callus growth	No. of shoots[a]	No. of shoots per apex (average) ± S.E.
MS BAP 2 NAA 0.02	20	20	(+++)	0	0.0
MS BAP 2 NAA 0.02 $GA_3$10	22	22	(++++)	0	0.0
MS BAP 2 $GA_3$10	18	10	(+)	163	9.0 ± 2.0
MS NAA 0.02 $GA_3$10	24	12	(+)	28	1.1 ± 0.1
MS $GA_3$10	24	0	(−)	38	1.6 ± 0.2
MS $GA_3$5	18	0	(−)	27	1.5 ± 0.2
MS BAP 1	20	20	(++)	70	3.5 ± 0.3
MS K 2 NAA 0.02	32	32	(+++)	131	4.1 ± 0.5
MS K 2 $GA_3$5	25	21	(+)	245	9.8 ± 1.0
MS K 2	28	28	(+++++)	0	0.0
MS hormone-free	22	0	(−)	22	1.0 ± 0.0

[a] Shoots counted after 40 days of culture.
(+): relative callus growth rated on a scale of (+) to (+++++); (−): no callus.

acid (IBA), GA_3, and others; explants were placed under fluorescent light (1500–3000 lx; 16 h light/8 h dark) at 26 °C.

Greco et al. (1984) established their cultures in Petri dishes containing agarized MS medium supplemented with BAP and/or 2,4-D at different concentrations, under continuous light (1500 lx).

Bohorova et al. (1985) used modified MS medium containing different combinations of BAP, NAA, K, 2,4-D, IBA, GA_3, IAA, sucrose, and various vitamins and amino acids; hybrid shoot tips were cultured in tubes at 26 °C with 16-h photoperiod (2500 lx).

Lupi et al. (1987) cultured shoot tips on MS solid medium with different combinations of growth regulators (see Table 3) at 24 ± 1 °C under continuous light of 2500 lx, in Petri dishes; the explants were transferred at 20-day intervals to fresh media.

3.3 Culture Establishment and Media

In vitro culture of shoot tips usually determines both callus formation and plant regeneration. In most cases callus is induced at the cut edge of the explant and precedes adventitious bud formation.

For direct multiplication, Trifi et al. (1981) described rapid propagation through in vitro culture of nodes isolated from both normal plants and plantlets obtained through shoot tip culture; these plantlets showed two kinds of morphogenetic pathway: slowly developing rosette plants and rapidly growing plantlets with elongated internodes.

The method used by Trifi et al. (1981) was tested by Paterson (1984), who reported strong differences in both callus formation and regeneration responses among the line tested and concluded that this method is not suitable for a large number of inbred lines. In his experiments, BAP and K were shown to promote the most active shoot multiplication, but high BAP concentrations promoted callogenesis and the formation of very small rosette plants. The highest number of shoots resulted from media containing 0.1 to 1 mg/l BAP. When auxins were added to the media, shoot growth was completely suppressed and callus formation was promoted (Paterson 1984).

Very intense callus proliferation is reported by Greco et al. (1984) when shoot tips were cultured on MS medium supplemented with BAP 1 and 5 mg/l. Some 76% of calli, subcultured on MS medium supplemented with BAP 1 mg/l, regenerated several shoots.

In our experiments (Lupi et al. 1987; Cavallini and Lupi 1987), the effect of many combinations of BAP, NAA, GA_3, and K at different concentrations was tested on shoot tips with two or three leaf primordia. After 40 days of culture, the number of shoots were counted (Table 3). The lack of hormones in the culture

Fig. 2A–D. Callus and plantlet regeneration from vegetative apices of *Helianthus annuus*. **A** Callus formation after 20-day culture on MS + BAP 2 + NAA 0.02. **B, C** Plantlets regenerated after 30-day culture on MS + K 2 + NAA 0.02 (*arrow* indicates a precocious flower bud). **D** Rooted plantlet after 20-day culture on liquid half-strength MS + GA_3 1 (hormone concentrations in mg/l)

medium promoted growth of the shoot apex to form one plantlet. BAP plus GA_3 resulted in shoot multiplication but the most effective promoter of multiplication appeared to be K (2 mg/l) plus GA_3 (5 mg/l). BAP plus NAA treatment completely suppressed shoot growth and promoted copious callus formation (Fig. 2A). Nevertheless, this callus, after transfer to medium containing GA_3 (5 or 10 mg/l), showed adventitious shoot bud regeneration. On the contrary, the callus formed in presence of K plus NAA led to bud development and no transfer to GA_3 medium was needed (Fig. 2B, C). When K alone was present in the culture medium at the concentration of 2 mg/l, only callus growth was observed; this callus, after transfer to medium containing 5 mg/l GA_3, formed roots but no shoots. When other media were tested, the number of shoots produced was on average 1.4–4.1 per apex (Table 3).

Similar results were reported by Bohorova et al. (1985) in their experiments on multiplication of interspecific sunflower hybrids: when cultured on MS medium supplemented with BAP 0.1 mg/l plus NAA 0.1 mg/l, shoot tips formed semi-organized calli with meristematic structures. When calli were transferred on MS medium containing BAP plus NAA plus GA_3 a few shoots were produced (one to three shoots per explant). Morphogenetic ability of these cultures is reported to decrease after the third subculture, but shoot regeneration ability was restored by transferring callus onto MS medium supplemented with L-glutamine 800 mg/l plus L-asparagine 800 mg/l.

3.4 Rooting of the Shoots

One major problem with adventitious shoots is the limited amount of rooting (Paterson 1984; Bohorova et al. 1985; Lupi et al. 1987). A number of techniques and media were tried to induce root formation, with little response.

Micropropagated shoots were generally excised at the base and transferred either on solid media with 0.8% agar (Paterson 1984; Lupi et al. 1987) or on filter paper bridges placed in tubes containing liquid medium (Lupi et al. 1987). Different media were tested: hormone-free MS, hormone-free White (1943) and hormone-free Nitsch (1951) (N) medium, supplemented or not with different concentrations of auxins (IAA, IBA, NAA), 2,4-D, and GA_3, and different sucrose levels (Paterson 1984; Lupi et al. 1987).

According to Paterson (1984), none of these treatments had any significant effect on rooting; auxins promoted callus formation only at the base of the shoots; as BAP was shown to promote shoot differentiation but inhibit the subsequent root formation, a shortened cytokinin treatment was used to induce multiplication but no positive effect was achieved in the subsequent rooting of shoots (Paterson 1984).

Also in our experiments, especially on solid rooting media, the most frequent response was callus induction at the shoot base. A relatively high percentage of rooting was obtained when using half-strength liquid MS medium plus 1.0 mg/l GA_3 or liquid N medium plus 0.1 mg/l IBA (Table 4; Fig. 2D). Multiple shoots induced by K plus NAA and K plus GA_3 were placed on these media and maintained either in the dark or under continuous light. As shown in Table 4, the percentage of rooted plantlets was higher in the dark than in the light.

Table 4. Rooting and flower bud development in shoots of *Helianthus annuus* after 20 days of culture on liquid half-strength MS medium plus 1 mg/l GA_3 (R_1), or on liquid N medium plus 0.1 mg/l IBA (R_2), in the dark or under continuous light treatment. Growth factor concentrations in mg/l. (Lupi et al. 1987)

Micropropagation media	Rooting media	Light			Dark		
		No. of shoots	Shoots with roots %	Shoots with flower buds %	No. of shoots	Shoots with roots %	Shoots with flower buds %
MS K 2 $GA_3$5	R_1	20[a]	30 ± 2[b]	100	32	59 ± 5	50 ± 5
	R_2	34	29 ± 2	70 ± 7	22	73 ± 8	23 ± 4
MS K 2 NAA 0.02	R_1	20	10 ± 5	100	21	62 ± 8	43 ± 8
	R_2	26	38 ± 1	58 ± 4	18	44 ± 7	22 ± 6

[a] Total number of shoots tested.
[b] Mean and S.E. of three experiments (6–12 shoots per experiment).

In most papers (Henrickson 1954; Trifi et al. 1981; Paal et al. 1981; Paterson 1984; Greco et al. 1984; Lupi et al. 1987; Witrzens et al. 1988), precocious flowering is reported in sunflower plants cultivated in vitro (see Fig. 2B). This is a general problem of sunflower tissue culture. In our experiments (Lupi et al. 1987), in vitro flowering is reduced depending on the rooting medium and the environmental conditions at which rooting is induced (see Table 4). The best rooting and the lowest percentage of flowering was obtained with the shoots derived from medium containing GA_3 plus K, using the latter in the dark.

3.5 Transfer to Greenhouse

Rooted sunflower plants were transferred to pots containing standard soil compost for sunflower and covered with transparent plastic tubes for 1–2 weeks to prevent

Fig. 3A, B. Formation of well-developed roots in micropropagated plants of *H. annuus* cultivated in a hydroponic system. For details see text. (Lupi et al. 1987)

desiccation while they become established (Lupi et al. 1987) in a greenhouse with a 16-h photoperiod at 25 °C.

In order to improve the frequency of survival and, at the same time, to shorten the time of the plantlets in vitro, Lupi et al. (1987) changed the methodology as follows: the plantlets rooted in N liquid medium containing IBA 0.1 mg/l were moved to half-strength Hoagland's salt solution and placed in a culture room with 16-h photoperiod (20 000 lx) at 25 °C. After 1 week, the plantlets were transferred and maintained in an appropriate hydroponic system (Fig. 3) using Hoagland's solution at the same environmental conditions and covered with transparent plastic boxes. This plastic covering was removed after 10 days but the plantlets were maintained in culture until numerous and well-developed roots were formed (Lupi et al. 1987; Fig. 2B). A mature micropropagated plant is shown in Fig. 4.

This method allowed a higher survival percentage of micropropagated plantlets; nevertheless, many shoots showed an abnormal morphology (stunted and etiolated shoots) and, above all, for many plants precocious flowering resulted in seed set failure. When seeds from micropropagated plants were obtained, they were successfully germinated; this progeny did not flower precociously and showed normal morphology (Lupi et al. 1987).

Fig. 4. Micropropagated plant of *H. annuus* after transfer to soil

Table 5. Chromosome counts in primary calli and regenerated vegetative apices of *Helianthus annuus* at different stages of in vitro culture on MS medium plus K 2 mg/l and NAA 0.02 mg/l. Flowering shoots were obtained after 20 days of culture on MS medium plus BAP 2 mg/l and NAA 0.02 mg/l and 10 days of culture on MS medium plus GA_3 5 mg/l. (Compiled from Cavallini and Lupi 1987)

Growth stage	No. of individuals	No. of cells examined	Mean percentage of cells in each chromosome number class										
			17–18	19–20	21–22	23–24	25–26	27–28	29–30	31–32	33–34	> 34	tetraploid
Primary explant	4	65	1.5	1.5	3.1	4.6	13.8	18.5	23.0	12.3	21.5	—	—
Callus (2 d)	5	80	6.2	5.0	5.0	10.0	28.7	12.5	16.2	5.0	11.2	—	—
Callus (4 d)	5	72	6.9	4.2	12.5	4.2	19.4	15.3	12.5	8.3	15.3	—	1.4
Callus (12 d)	5	53	1.9	5.7	5.7	7.5	15.1	20.8	15.1	7.5	20.8	—	—
Callus (20 d)	5	65	7.7	—	7.7	7.7	13.8	13.8	9.2	12.3	20.0	3.1	4.6
1-mm-long shoots	10	206	0.5	0.5	—	3.0	8.2	20.7	25.2	20.7	21.2	—	—
3–4-cm-long plantlets	10	251	—	—	—	—	0.5	11.3	20.9	27.1	40.2	—	—
Flowering plantlets[a]	10	239	—	—	—	—	—	1.8	10.4	28.6	59.3	—	—
Flowering shoots	10	222	—	—	0.4	0.5	—	10.7	8.6	23.8	56.0	—	—

[a] After 50 days of culture.

3.6 Cytological Status of Meristem-Derived Callus and Plantlets

There are only a few reports in the literature about the nuclear conditions of callus in sunflower. Butcher et al. (1975) reported increasing frequency of polyploid nuclei with aging of the callus, but the occurrence of aneuploid nuclei was not considered.

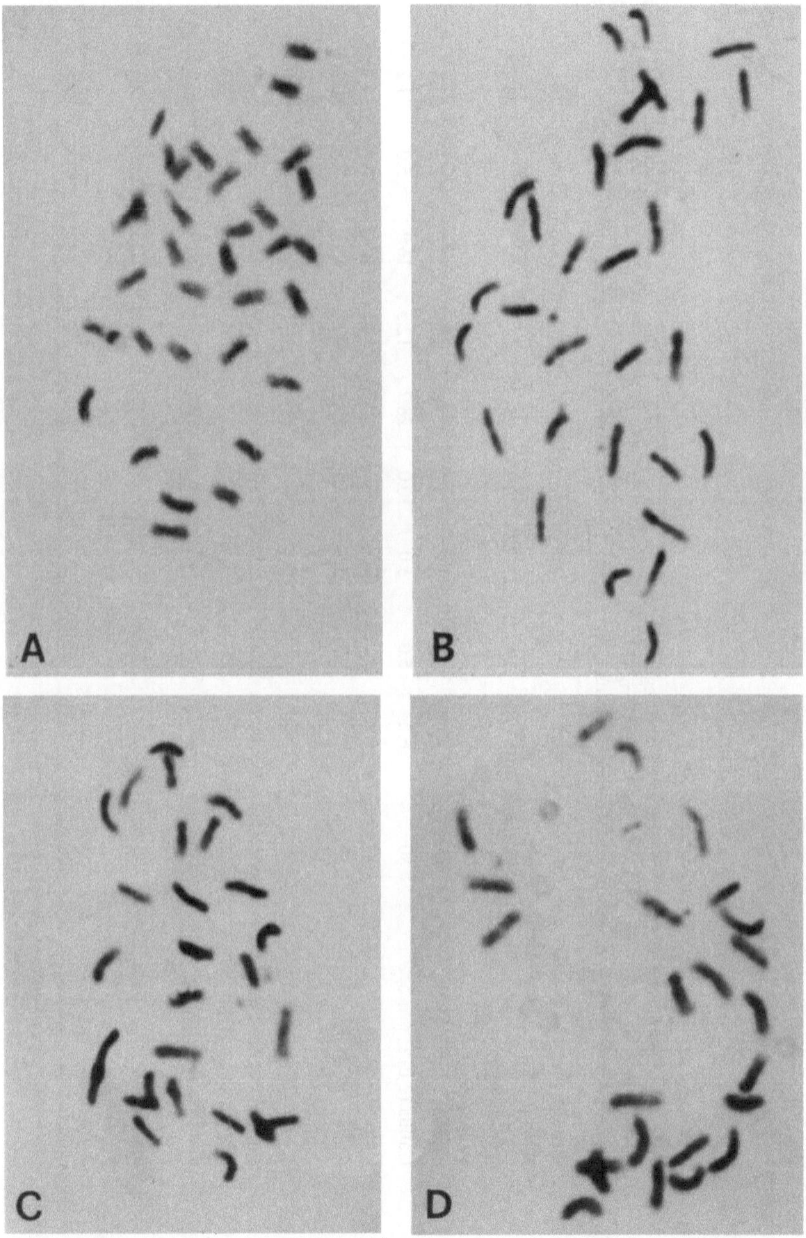

Fig. 5A–D. Chromosome numbers in apices of 1-mm-long micropropagated shoots of *H. annuus*. **A** Diploid metaphase (2n = 34). **B–D** Hypodiploid prometaphases (30, 27, and 24 chromosomes, respectively)

Table 6. Percentages (± S.E.) of diploid cells in different in vivo and in vitro developmental stages of a selfed sunflower line; growth factor concentrations in mg/l. (Cavallini and Lupi 1987)

In vivo			In vitro					
			MS K2 NAA 0.02			MS BAP 2 NAA 0.02 (20 d) + MS $GA_3$5 (10 d)		
Plant developmental stage	Days of growth	% Diploid cells (± S.E.)	Stage	Days of culture	% Diploid cells (± S.E.)	Stage	Days of culture	% Diploid cells (± S.E.)
Seedling shoot apex	6	23.2 ± 4.8	1 mm long shoot	21	21.2 ± 1.9			
Plant at the 6th leaf pair (shoot apex)	26	42.9 ± 7.2	3–4 cm long plantlet (shoot apex)	40	40.2 ± 2.8			
Flowering plant (shoot apex)	40	72.4 ± 6.6	Initial floral bud	50	59.3 ± 3.0	Precocious floral bud	30	56.0 ± 2.2

In our experiments, a small increase in the frequency of aneuploid nuclei, compared to that in the shoot apex in vivo (Cavallini and Cremonini 1985; see Sect. 1.2), was found in the first days of culture (Table 5). Since mechanisms of aneuploid cell production (metaphase noncongression, anaphasic chromosome bridges, lagging chromosomes) were rare in the callus, most of the aneuploid cells present therein resulted from mitotic propagation of aneuploid cells pre-existing in the explant (Cavallini and Lupi 1987).

The chromosomal status of regenerants at the first developmental stage studied reflects the chimeric nature of the primary callus (Table 5, Fig. 5). Chromosomal number mosaicism in regenerated plants has been reported for many species, of both diploid and polyploid origin (D'Amato 1985). In the sunflower line studied, all regenerants were aneusomatics, with a high frequency of hypodiploid cells, at least at the developmental stages analyzed, and the time course of aneusomaty during in vitro plant development fully matches that of in vivo development: a selective advantage of diploid over aneuploid cells (diplontic selection) takes place during plant growth (Table 5; Cavallini and Lupi 1987).

It is worth noting that when floral induction is very precocious, the chromosome number mosaicism is immediately comparable to that of normal noninduced floral buds, both in vivo and in vitro (see Table 6). The processes underlying this very rapid diplontic selection are not known.

4 Summary and Conclusions

Large differences in response, for regeneration and/or somatic embryogenesis, were observed, strictly depending on the genotype used, and only a few genotypes (ca. 1% of lines, according to Paterson and Everett 1985) were able to regenerate plants. Even when shoot regeneration was achieved, most regenerated plants were stunted and abnormal, rooted with difficulty, and showed precocious flowering in vitro (Paterson 1984; Greco et al. 1984; Lupi et al. 1987).

In micropropagation, plantlets regenerated from shoot apices usually also showed these developmental problems. The method described by Lupi et al. (1987) resulted in a high frequency of shoot formation from vegetative apices cultured on MS medium supplemented with 2 mg/l K plus 5 mg/l GA_3. Rooting at a relatively high percentage was achieved onto N medium containing 0.1 mg/l IBA, and a technique determining higher survival of the rooted plants was set up (see Sect. 5. Protocol). This method was shown to be suitable for numerous genotypes.

Unfortunately, most propagated plants were abnormal and weak, and gave small quantities of seed. This strongly reduces the possibilities for utilizing micropropagation techniques in breeding programs.

Probably, in vivo, the perfect organization of apical meristems ensures a correct developmental pattern, in spite of chromosomal mosaicism (Cavallini and Cremonini 1985); in contrast, in vitro organogenetic processes in presence of insufficient control over the endogenous levels of growth substances together with wide chromosomal mosaicism (Cavallini and Lupi 1987) might determine developmental abnormalities in regenerated plants. Further studies are, therefore, necessary in order to establish culture conditions more suitable to achieve normal plants in higher frequencies.

5 Protocol

1. Seeds are scratched to remove seed coats, surface sterilized with 70% ethanol solution for 2 min, followed by immersion in a 5% aqueous solution of sodium hypochlorite for 20 min, and rinsed in sterile distilled water.
2. Individual seeds are placed in glass tubes onto MS medium solidified with 0.8% agar, pH 5.6. Cultures are maintained at 24 °C under continuous fluorescent light (2500 lx).
3. After 1 week, shoot tips are aseptically excised under a dissection microscope and placed onto MS solid medium supplemented with 2 mg/l K plus 5 mg/l GA_3 in Petri dishes under conditions as in step 2.
4. After 20 days of growth, explants are transferred to 100-ml Erlenmeyer flasks (20 ml per flask) containing fresh media of the same or different composition (for hormones) to increase or induce bud regeneration.
5. Micropropagated shoots are excised at the base and transferred onto filter paper bridges in tubes containing rooting liquid medium (N medium plus 0.1 mg/l IBA) in the dark for 20 days.
6. Rooted shoots are moved to glass tubes containing half-strength Hoagland's salt solution, placed in a culture room with 16-h photoperiod (20000 lx) at 25 °C for a week and then transferred in an appropriate hydroponic system, under colorless transparent plastic to prevent desiccation, until numerous roots are formed.
7. Plantlets are potted in standard soil compost for sunflower in a greenhouse with 16-h photoperiod at 25 °C.

Acknowledgment. We wish to thank Prof. F. D'Amato for critical reading of the manuscript. This study was supported by MURST, Italy.

References

Alissa A, Serieys H, Jonard R (1985) Sur les possibilités de régéneration d'espèces sauvages et d'hybrides interspecifiques du genre *Helianthus* par androgenèse in vitro. CR Acad Sci Paris 300: 25–30

Alissa A, Jonard R, Serieys H, Vincourt P (1986) La culture d'embryons isolés in vitro dans un programme d'amelioration du tournesol. CR Acad Sci Paris 302: 161–164

Benvenuti A, Vannozzi GP (1988) Obiettivi e risultati del miglioramento genetico del girasole. In: Benvenuti A (ed) Stato attuale e prospettive delle colture oleaginose erbacee in Italia, Pacini Editore, Pisa, pp 49–68

Bergounioux C, Perennes C, Brown SC, Sarda C, Gadal P (1988) Relation between protoplast division, cell-cycle stage and nuclear chromatin structure. Protoplasma 142: 127–136

Binding H, Nehls R, Kock R, Finger J, Mordhorst G (1981) Comparative studies on protoplast regeneration in herbaceous species of the Dicotyledoneae class. Z Pflanzenphysiol 101: 119–130

Bohorova NE (1982) In vitro plant development of seeds of *Helianthus* interspecific hybrids. CR Acad Bulg Sci 35: 105–107

Bohorova NE, Atanassov AI (1990) Sunflower (*Helianthus annuus* L.): In vitro interspecific hybridization. In: Bajaj YPS (ed) Biotechnology in agriculture and forestry, vol 10. Legumes and oilseed crops I. Springer, Berlin Heidelberg New York, pp 453–471

Bohorova NE, Atanassov AI, Antonova GP (1980) In vitro isolation of anthers from interspecific hybrids in the *Helianthus* genus. CR Acad Bulg Sci 33: 1545–1548
Bohorova NE, Atanassov AI, Georgieva-Todorova J (1985) In vitro organogenesis, androgenesis and embryo culture, in the genus *Helianthus*. Z Pflanzenzücht 95: 35–44
Bohorova NE, Cocking EC, Power JB (1986) Isolation, culture and callus regeneration of protoplast of wild and cultivated *Helianthus* species. Plant Cell Rep 5: 256–258
Butcher DN, Sogeke AK, Tommerup IC (1975) Factors influencing changes in ploidy and nuclear DNA measurements in sunflower seedlings, crown gall tumours, and habituated callus tissues. Photochem Photobiol 27: 532–540
Cavallini A, Cremonini R (1985) Aneusomaty in sunflower (*Helianthus annuus* L.). Z Pflanzenzücht 95: 118–124
Cavallini A, Cionini PG (1986) Nuclear DNA content in differentiated tissues of sunflower (*Helianthus annuus* L.). Protoplasma 130: 91–97
Cavallini A, Lupi MC (1987) Cytological study of callus and regenerated plants of sunflower (*Helianthus annuus* L.). Plant Breed 99: 203–208
Cavallini A, Zolfino C, Cionini G, Cremonini R, Natali L, Sassoli O, Cionini PG (1986) Nuclear DNA changes within *Helianthus annuus* L.: cytophotometric, karyological and biochemical analyses. Theor Appl Genet 73: 20–26
Chandler JM, Jan CC (1983) Production of interspecific sunflower hybrids through tissue culture. Agronomy (Abstr) 1983: 58
Cremonini R, Cavallini A (1986) Origin and fate of aneusomaty in sunflower (*Helianthus annuus* L.). Plant Breed 97: 89–92
D'Amato F (1985) Cytogenetics of plant cell and tissue cultures and their regenerates. CRC Crit Rev Plant Sci 3: 73–112
Dupuis JM, Pean M, Chagvardieff P (1990) Plant donor tissue and isolation procedure effect on early formation of embryoids from protoplasts of *Helianthus annuus* L. Plant Cell Tissue Organ Cult 22: 183–189
Espinasse A, Lay C (1989) Shoot regeneration of callus derived from globular to torpedo embryos from 59 sunflower genotypes. Crop Sci 29: 201–205
Fick GN (1978) Breeding and genetics. In: Carter J (ed) Sunflower science and technology. Am Soc Agron, Madison, Winsconsin, USA, pp 279–338
Finer JJ (1987) Direct somatic embryogenesis and plant regeneration from immature embryos of hybrid sunflower (*Helianthus annuus* L.) on a high sucrose-containing medium. Plant Cell Rep 6: 372–374
Gautheret R (1959) La culture de tissus. Techniques et réalisations. Page 863. Masson, Paris
Gelebart P, San LH (1987) Obtention des plantes haploïdes par culture in vitro d'ovaires et d'ovules non fecondés de tournesol (*Helianthus annuus* L.). Agronomie 7: 81–86
Greco B, Tanzarella OA, Carrozzo G, Blanco A (1984) Callus induction and shoot regeneration in sunflower (*Helianthus annuus*). Plant Sci Lett 36: 73–77
Guilley E, Hahne G (1989) Callus formation from isolated sunflower (*Helianthus annuus*) mesophyll protoplasts. Plant Cell Rep 8: 226–229
Gundaev AI (1966) Prospects of selection in sunflower for heterosis. Sb Rab Maslichn Kult 3: 15–21
Heiser CB (1955) The origin and development of cultivated sunflower. Am Biol Teacher 17: 161–167
Heiser CB (1957) A revision of the South American species of *Helianthus*. Brittonia 8: 283–295
Heiser CB (1978) Taxonomy of *Helianthus* and origin of domesticated sunflower. In: Carter J (ed) Sunflower science and technology. Am Soc Agron, Madison, Winsconsin, USA, pp 31–54
Heiser CB, Smith DM (1955) New chromosome numbers in *Helianthus* and related genera (Compositae). Proc Indiana Acad Sci 64: 250–253
Heiser CB, Smith DM, Clevenger S, Martin WC (1969) The North American sunflowers. Mem Torrey Bot Club 22: 1–218
Henrickson CE (1954) The flowering of sunflower explants in aseptic culture. Plant Physiol 29: 536–538
Jackson RC, Murray BG (1983) Colchicine induced quadrivalent formation in *Helianthus*: evidence of ancient polyploidy. Theor Appl Genet 64: 219–222
Jonard R, Mezzarobba A (1990) Sunflower (*Helianthus* spp.): Anther culture and field studies on haploids. In: Bajaj YPS (ed) Biotechnology in agriculture and forestry, vol 10. Legumes and Oilseed Corps I. Springer, Berlin Heidelberg New York, pp 485–501
Kinman ML (1970) New developments in the USDA and state experiment breeding programs. In: Proc 5th Int Sunflower Conf, Memphis, Tennessee, USA, pp 181–183

Knopp E, Mix G (1986) In vitro Spross- und Wurzelregeneration aus Gewebeteilen der Sonnenblume (*Helianthus annuus* L.). Landbauforsch Völkenrode 36: 127–129

Leclercq P (1966) Une stérilité male utilisable pour la production d'hybrides simples de tournesol. Ann Amelior Plantes 16: 135–144

Leclercq P (1968) Une stérilité male cytoplasmique chez le tournesol. Ann Amelior Plant 19: 99–106

Lenée P, Chupeau Y (1986) Isolation and culture of sunflower protoplasts (*Helianthus annuus* L.): factors influencing the viability of cell colonies derived from protoplasts. Plant Sci 43: 69–75

Lupi MC, Bennici A, Locci F, Gennai D (1987) Plant formation from callus and shoot tip culture of *Helianthus annuus* L. Plant Cell Tissue Organ Cult 11: 47–55

Mezzarobba A, Jonard R (1986) Effects du stade de prélèvement et des prétraitements sur le developpement in vitro d'anthères prélevées sur le tournesol cultivé (*Helianthus annuus* L.). CR Acad Sci Paris 303: 181–186

Mielke T (1987) Prospect for oilseed, fats and oils in 87/88. In: Proc 8th Europ Outlook Conf, London, pp H(1)-H(18)

Mix G (1985) Antheren- und Ovarienkultur von Sonnenblumen (*Helianthus annuus* L.). Landbauforsch Völkenrode 35: 153–156

Moyne AL, Thor V, Pelissier R, Bergounioux C, Freyssinet G, Gadal P (1988) Callus and embryoid formation from protoplasts of *Helianthus annuus*. Plant Cell Rep 7: 437–440

Moyne AL, Tagu D, Thor V, Bergounioux C, Freyssinet G, Gadal P (1989) Transformed calli obtained by direct gene transfer into sunflower protoplasts. Plant Cell Rep 8: 97–100

Murashige T, Skoog F (1962) A revised medium for rapid growth and bioassays with tobacco tissue culture. Physiol Plant 15: 473–497

Nitsch JP (1951) Growth and development in vitro of excised ovaries. Am J Bot 38: 566–576

Paal HA, Kurnic E, Szobo L (1981) Plantlet regeneration from in vitro shoot tip culture of sunflower. Novenytermeles 30: 201–209

Paterson KE (1984) Shoot tip culture of *Helianthus annuus*. Flowering and development of adventitious and multiple shoots. Am J Bot 71: 925–931

Paterson KE, Adams DO (1987) The role of ethylene in the regeneration of *Helianthus annuus* (sunflower) plants from callus. Physiol Plant 71: 151–156

Paterson KE, Everett NP (1985) Regeneration of *Helianthus annuus* inbred plants from callus. Plant Sci 42: 125–131

Pelissier B, Bouchefra O, Pepin R, Freyssinet G (1990) Production of isolated somatic embryos from sunflower thin cell layers. Plant Cell Rep 9: 47–50

Power CJ (1987) Organogenesis from *Helianthus annuus* inbreds and hybrids from the cotyledons of zygotic embryos. Am J Bot 74: 497–503

Prado E, Berville A (1990) Induction of somatic embryo development by liquid culture in sunflower (*Helianthus annuus* L.). Plant Sci 67: 73–82

Pustovoit VS (1964) Conclusions of work on the selection and seed production of sunflowers. Agrobiology 5: 672–697

Putt ED (1940) Observations on morphological characters and flowering processes in the sunflower (*Helianthus annuus* L.). Sci Agric 21: 167–179

Putt ED (1978) History and present world status. In: Carter J (ed) Sunflower science and technology. Am Soc Agron, Madison, Winsconsin, USA, pp 1–30

Robinson KEP, Everett NP (1990) Sunflower (*Helianthus annuus* L.): Establishment of cultures, transformation, and the regeneration of plants. In: Bajaj YPS (ed) Biotechnology in agriculture and forestry, vol 10. Legumes and oilseed crops I. Springer, Berlin Heidelberg New York, pp 434–452

Sadhu MK (1974) Effect of different auxins on growth and differentiation in callus tissue from sunflower stem pith. Indian J Expt Biol 12: 110–111

Schrammeijer B, Sijmons PC, Van Den Elzen PJM, Hoekema A (1990) Meristem transformation of sunflower via *Agrobacterium*. Plant Cell Rep 9: 55–60

Semelczi-Kovacs A (1975) Acclimatization and dissemination of the sunflower in Europe. Acta Ethnogr Acad Sci Hung 24: 47–88

Trifi M, Mezghani S, Marrakchi M (1981) Multiplication vegetative du tournesol (*Helianthus annuus*) par culture in vitro. Physiol Veg 71: 99–102

Voskoboinik LK, Soldatov KI (1974) The research trends in the field of sunflower breeding for heterosis at the VNIIMK. In: Proc 6th Int Sunflower Conf, Bucharest, Romania, pp 383–389

Vranceanu AV, Stoenescu F, Ulinici A, Iliescu H, Paulian F (1974) Sunflower. Edit Acad Rep Soc, Bucharest, Romania, p 322

Watson EE (1929) Contributions to a monograph of the genus *Helianthus*. Pap Mich Acad Sci 9: 305–475

Whelan EDP (1978) Cytology and interspecific hybridization. In: Carter J (ed) Sunflower science and technology. Am Soc Agron, Madison, Winsconsin, USA, pp 339–370

White PR (1943) Nutrient deficiency studies and an improved inorganic nutrient medium for cultivation of excised tomato roots. Growth 7: 56–65

Witrzens B, Scowcroft WR, Downes RW, Larkin PJ (1988) Tissue culture and plant regeneration from sunflower (*Helianthus annuus*) and interspecific hybrids (*H. tuberosus* × *H. annuus*). Plant Cell Tissue Organ Cult 13: 61–76

Yang HY, Zhou C, Yan H (1990) In vitro production of haploids in *Helianthus*. In: Bajaj YPS (ed) Biotechnology in agriculture and forestry, vol 10. Legumes and oilseed crops I. Springer, Berlin Heidelberg New York, pp 472–484

Zukovsky PM (1950) Cultivated plants and their wild relatives. Commonw Agric Bur, Farnham Royal, England

III.4 In Vitro Culture and Micropropagation of Ryegrass (*Lolium* spp.)

J. CREEMERS-MOLENAAR and L.J. BEEREPOOT[1]

1 Introduction

1.1 Ryegrasses and Their Importance

The ryegrasses (*Lolium* spp.) can be classified into eight different species. However, since these species are partly or fully interfertile, this classification is not always unequivocal (Breese and Tyler 1986). Two of these species, perennial ryegrass (*Lolium perenne* L.) and Italian ryegrass (*Lolium multiflorum* Lam.) are agriculturally very important. Strictly annual forms of Italian ryegrass are often referred to as Westerwolds ryegrass (*Lolium multiflorum* Lam. ssp. *westerwoldicum*). Hybrid ryegrass (*Lolium* × *hybridum* Housskn.) is derived from artificial or naturally occurring crosses between Italian and perennial ryegrass. Perennial and Italian ryegrass are diploid (2n = 14), outbreeding and self-incompatible species. No naturally occurring tetraploids have ever been found. However, by means of colchicine treatment, tetraploid varieties (4n = 28) have been made artificially (Wit 1959).

In temperate areas, perennial ryegrass and Italian ryegrass, together with their synthetic hybrid, make up the major part of the newly sown grasses. Breese and Tyler (1986) state that they account for more than 85% of seed sales of agricultural grasses in Great Britain. According to estimates for the grass seed use in the EEC in 1987–1988, 73% of the total use of more than 1 million kg seed per year is ryegrass seed. This includes amenity grasses.

However, grass seed use, although very important for motivating trade and research, does not give a complete view of the importance of ryegrasses and the area under cultivation, since a major part of the grasslands are only resown once in every 4–10 years. Even pastures which are never resown can contain up to 100% ryegrass under favorable climatic conditions and with proper management.

1.2 Distribution and Use of Ryegrasses

Perennial ryegrass and Italian ryegrass are widely used as forage species in countries with temperate climates and sufficient rainfall in the growing season, e.g.,

[1] Barenbrug Holland BV, Stationsstraat 40, 6678 AC Oosterhout, The Netherlands. (Correspondence address: Dr. J. Creemers-Molenaar, Centre for Plant Breeding and Reproduction Research (CPRO-OLO), P.O. Box 117, 6700 AC Wageningen, The Netherlands)

Biotechnology in Agriculture and Forestry, Vol. 19
High-Tech and Micropropagation III (ed. by Y.P.S. Bajaj)

western Europe, New Zealand, and parts of Australia. In the USA, it is grown west of the Sierra Nevada and the Cascade Range, in the southern humid areas, and northwards along the Atlantic coast (Frakes 1973). In Canada some ryegrass is used in the Atlantic Provinces and the south coastal part of British Columbia.

Ryegrasses are well suited for intensive farming practices and respond well to high fertilizer applications. Compared to other forage grasses, they produce forage with a high palatability and digestibility. Because of their quick establishment, they are more suitable than other grasses for overseeding and sod-seeding in an existing sward. Ryegrasses are also used as a green manure crop and for soil conservation purposes.

In those areas where severe frost, summer drought, or heat prevents the survival of perennial ryegrass, Italian or Westerwolds ryegrass can be used as an annual forage crop or as a catch crop. For this reason, the use of Italian ryegrass is geographically more widespread than the use of perennial ryegrass.

Perennial ryegrass is used pure or with other species in seed mixtures for permanent grassland used for pasture, silage, or hay. It can withstand heavy grazing quite well. The persistence varies with the climate, farming practices, and occurrence of pests. In favorable areas it seems to persist indefinitely, in other areas it is said to be only "quite satisfactory for 3 or 4 years" (Frakes 1973).

Perennial ryegrass is used all over the world as a valuable species for home lawns, sports fields, and recreational grounds. Its main advantages over other amenity grass species are the excellent wear tolerance and the fast germination and establishment. Disadvantages are the limited drought and heat tolerance and its demand for frequent cutting and fertilization. Modern cultivars of amenity ryegrass are finer-leaved and grow more slowly than forage ryegrasses.

Italian ryegrass is used to give a high production of good quality silage or hay for 1 or 2 years. Persistent varieties of Italian ryegrass and hybrid ryegrass can last up to 3 years under favorable conditions. Westerwolds ryegrass is a short-living type, and will persist from only one cut up to a full growing season.

1.3 Diseases and Pests

Not much effort is spent on the control of pests and diseases in grasses compared to many other crops. Chemical control is not often profitable and toxic residuals can cause problems for the grazing animal. Therefore, resistance breeding is the obvious solution to most disease problems. The outbreeding grasses, especially when they are sown in mixtures of species and varieties, are less vulnerable to epidemic diseases than genotypically uniform crops like cereals. Frequent cutting or grazing can stop spreading of diseases because infected plant parts are removed and less favorable conditions for infection are left (Plumb 1988).

The most important disease in ryegrasses in most areas, crown rust (*Puccinia coronata*), is less damaging under intensive management with frequent cuts than under conditions giving a relatively slow growth rate of ryegrass. Crown rust occurs mainly in late summer and autumn and is promoted by high temperatures and high air humidity. A minor attack reduces the palatability of ryegrass, heavier attacks cause important yield losses and even plant death. Considerable genetic variation exists, and selection for resistance is relatively easy, both in the field and in greenhouse tests.

Winter damage in perennial or biennial ryegrasses can be caused by frost but also by different snow molds (*Gerlachia nivalis* and *Typhula* spp.), especially in areas with prolonged snow cover in winter. There are differences in susceptibility between varieties. A selection method for increasing resistance to snow mold has been described by Jonsson and Nilsson (1986).

In a perennial grass sward, individual tillers do not live for years. New tillers grow from axillary buds on older tillers. Because of this continuous vegetative propagation, viruses are quite common, especially in old grass swards. Of the many different viruses affecting ryegrasses, barley yellow dwarf virus and ryegrass mosaic virus are the most important. They cause great damage and sometimes death to individual plants. In a sward, however, effects are less catastrophic because uninfected neighbors will fill the gaps left by slower-growing virus-infected plants. Differences in susceptibility between varieties do exist; however, breeding for virus resistance or tolerance is difficult. Virus resistance might be among the first important goals for genetic manipulation techniques.

Insects can be especially harmful during the establishment of newly sown grass into an old grassland after plowing or sod-seeding. Frit fly larvae (*Oscinella frit*) bore into grass tillers and feed especially on the young tissue around the meristem. Leatherjackets, larvae of craneflies (*Tipula* spp.), are common in northern Europe. They feed on underground parts of ryegrass plants and on seedlings. Resistance to these insects is unknown in ryegrasses. Genetic engineering might offer possibilities for the introduction of insect resistance.

Only some of the most common pests and diseases are mentioned here. More comprehensive information is given by Plumb (1988).

1.4 Breeding and Propagation

Commercial ryegrass varieties are panmictic populations of heterozygous plants, expressing a certain level of uniformity for morphological characteristics such as date of ear emergence, length of culms and growth habit. Most modern varieties are based on a limited number of selected plants or clones. They are often referred to as synthetic varieties.

Seed production after random mating between the originally selected clones is often called syn-1. This seed can be sown to raise the next seed generation, syn-2, etc. Commercial seed is usually syn-5 or syn-6 seed. During successive generations, varieties should be stable in performance and morphological characteristics, which is, however, not always the case (Hayward and Abdullah 1985). Producing seed free of weeds requires care and the application of different herbicides. If resistance to glyphosate could be introduced by means of genetic manipulation techniques, production of good quality seed would be cheaper, easier, and with less environmental problems.

Selection of parent plants for new varieties usually starts in heterogeneous populations. These populations can be F_2 or F_3 generations derived from crosses, ecotypes collected from very old pastures, or existing varieties. Wild ecotypes provide a seemingly limitless gene-pool of variation on which the breeder can draw (Tyler 1987).

Selection methods can be based on mass selection between plants or clones (Table 1). Mass selection is a cheap and easy method, enabling a breeder to cope with large numbers of genotypes. A disadvantage is that performance in spaced-plant conditions, even with narrow spacing, can be quite different from performance in a sown sward. Also, it is not easy to estimate yield differences between plants or clones. Because of the heterozygous nature of the tested plants, phenotypically good plants can reveal bad characteristics in the next generation. Nevertheless, mass selection for phenotypic performance has proven to be successful for developing varieties in ryegrasses at our breeding station.

Several selection methods have been published for use in ryegrasses involving tests of sown progeny of plants instead of testing the plants themselves. Raising enough seed for use in replicated field trials is possible using a topcross or a polycross. Utz and Oettler (1978) used a topcross in which an excess pollencloud from a sown variety pollinates the genotypes to be tested. Frandsen and Frandsen (1948) and Schaepman (1952) used a polycross method in which progeny seed is harvested from clones, each plant of which is positioned randomly in such a way that every genotype is pollinated by most of the other genotypes of the polycross group.

Both the topcross and the polycross require that mother clones can be kept alive until the progeny tests have been finished. Because of severe frost in some years, virus diseases, or lack of persistence, this can be very difficult in trial fields. Meristem culture, as described in Section 2.5, is very useful for this purpose.

Table 1. One of the breeding methods used by Barenbrug Holland B.V. for forage perennial ryegrass, based on mass selection

Year	
1	Cross between two varieties
2	Production of F_2 seed
3	±400 F_2 plants per cross are vegetatively multiplied to clones Clones are planted in a selection trial with a between-plant distance of 12.5 cm. Cutting and fertilizing are according to farming practice
4, 5	Observation of persistence, winter hardiness, disease resistance, and early spring growth
6	Transplanting of 10% superior clones to disease nursery About half of them are rejected, mainly on crown rust susceptibility In autumn, the remaining 5% are planted in an observation trial, ten plants per clone, distance between clones 50 cm
7	Observation of morphological characteristics such as heading date Formation of groups of three to six similar clones Transplanting to irrigated plots in rye just before flowering Syn-1 seed production in isolation
8	Sowing of yield trials at several locations
9, 10	Yield assessment and observation of trials Candidate varieties looking better than the control varieties are sown for seed multiplication in the autumn of year 10
11	Harvest of syn-2 seed, which is used to start National List tests

One of the ways to increase the selection response in breeding programs is by maximizing the additive genetic variation between selection units (Hallauer and Miranda 1981). Selection between homozygous plants would be much more effective than between heterozygous ones. However, inbreeding in the self-incompatible ryegrasses is difficult. Utz and Oettler (1978) managed to obtain only 284 S6-lines by single seed descent, starting with 2600 initial selfings.

1.5 The Application of Biotechnology in Ryegrass Breeding

Plant biotechnology covers the area of research that combines tissue culture and molecular techniques with the aim of obtaining plants with improved agronomical traits. In this respect the availability of these techniques should be considered as an additional tool in current breeding programs. The possible benefits of biotechnology for grass breeding will be discussed below, a detailed description of the methods and the achievements in *Lolium* will be given in Section 2 of this review.

Micropropagation. Plants can be regenerated "in vitro" from cultured meristems. In ryegrass breeding this technique is used to preserve plant genotypes and to eliminate viruses.

Somatic Hybridization and Cybridization. Combining techniques for protoplast culture and regeneration with methods for protoplast fusion offers additional possibilities for plant breeding. Protoplast fusion can be accomplished between protoplasts from one species (intraspecific), between protoplasts from related species (interspecific), or between distantly related species (intergeneric). After fusion the fused protoplasts (hybrid cells) need to be regenerated to mature (hybrid) plants. Initially, it was expected that this technique could be used to overcome sexual crossing barriers to produce novel species. However, it was shown that although hybrid plants could be obtained, e.g., between crop plants and distantly related wild species, the hybrids often express undesirable traits like low seed fertility (Glimelius et al. 1991). A more promising way to exploit hybrid plants from such fusions is to use them as bridges to transfer valuable traits (e.g., disease resistance) from the alien species to the crop plant.

By asymmetric fusion hybrids are obtained which contain the cytoplasm genomes (chloroplasts, mitochondria) of both parental lines and the nuclear genome of only one of the two lines. To discriminate, such hybrids are referred to as cybrids. Asymmetric cybridization involves fragmentation or elimination of the nuclear DNA of one of the parental lines prior to fusion. This can be accomplished by irradiation (gamma, X-ray) or enucleation by centrifugation, respectively. The mixed cytoplasms in the cybrid cells will segregate during subsequent mitotic divisions, allowing the proliferation of cells with novel combinations of nuclear and cytoplasm genomes. This direct method for cytoplasm transfer has proven valuable for the transfer of mitochondrion- and chloroplast-encoded traits, e.g. within *Brassica* species (Glimelius et al. 1991), circumventing the need of repeated backcrosses.

For the commercial production of hybrid seeds, male sterility in one of the parental lines is indispensible. In perennial ryegrass a type of male sterility occurs, which can be stably maintained and is maternally inherited (cytoplasmic male sterility, CMS). This sterile line could be used as cytoplasm donor in asymmetric fusion experiments to transfer male sterility to other perennial ryegrass breeding lines.

Haploids. Haploid plants can be regenerated from cultured anthers, microspores, and ovaries. The production of homozygous lines by anther culture could enhance selection efficiency in perennial ryegrass (see Sect. 1.4). If hybrid varieties of perennial ryegrass become available (Utz and Oettler 1978), anther culture will be useful for producing inbred lines. In breeding programs forced self-fertilization is used to make inbred lines. However, this method selects strongly against the naturally occurring self-incompatibility in *Lolium*. When such inbred lines are polycrossed or topcrossed, a certain unpredictable proportion of the progeny seed can come from self-fertilization, leading to wrong estimations of breeding values (Utz and Oettler 1978). Producing homozygous plants by anther culture could possibly overcome this problem, since no selection against self-incompatibility takes place.

Transformation. Successful transformation is achieved when genes of interest are transferred, stably integrated, and expressed in the transformed plant and its offspring (Potrykus 1990). A limited number of agronomically important genes are now available and have been transmitted successfully to an even more limited number of mainly dicotyledoneous crop species (Vasil 1990). These genes confer resistance to several herbicides, insects, and viruses, and as discussed in Section 1.3 transformation of ryegrass with these traits could lead to improved varieties.

2 In Vitro Approaches

2.1 Introduction to Tissue Culture Studies in *Lolium*

As discussed in Section 1.5, the application of tissue culture and genetic manipulation techniques in plant breeding programs of *Lolium* offers promising perspectives. A well-established method for clonal propagation and long-term storage of genotypes by meristem tip culture has been available for more than 10 years. Recently, research has concentrated on the culture of anthers, and haploid plants have been obtained from several *L. perenne* varieties. Callus with a capacity to regenerate plants can be initiated readily from all (somatic) explants of *Lolium* that have been tested.

However, techniques like somatic hybridization and direct gene transfer require the availability of single cells and protoplasts, as well as a reliable system for plant regeneration from single cells and protoplasts (see reviews, Vasil 1988; Bajaj 1989).

In *Lolium* successful establishment of morphogenic cell suspension cultures has been reported only recently and plants have also now been regenerated from cell-suspension-derived protoplasts of *L. perenne* and *L. multiflorum*.

The following sections describe the culture conditions, techniques, culture media, and results of recent advances concerning tissue culture studies on *Lolium*.

2.2 Sterilization of the Explants

2.2.1 Seeds

If young leaves or roots are to be used as the explant material, plantlets can best be grown under aseptic conditions. The seeds are surface sterilized with a 50% (v/v) commercial sodium hypochlorite solution (5–7% final available chlorine) containing 0.01% (v/v) Tween 80 for 20 min. After rinsing several times in sterile tap water, the seeds are germinated under sterile conditions. For callus initiation from seeds or embryos the same sterilization procedure should be followed. Heavily contaminated seeds can be treated two times with a 100% (v/v) sodium hypochlorite solution, interrupted by soaking in sterile water for 2–7 days (Dalton 1988a). Dehusking of the seeds in 50% (v/v) sulfuric acid prior to sterilization has been shown to enhance callus initiation from mature seeds in *L. perenne* (Torello et al. 1983).

2.2.2 Plants

Tillers from field- or greenhouse-grown plants are surface sterilized by dipping the tillers into 70% (v/v) ethanol for 2–10 s, and then into a 50% (v/v) sodium hypochlorite solution containing 0.01% (v/v) Tween 80 for 20 min. After rinsing five or six times in sterile tap water, the immature inflorescences or meristem tips are excised aseptically. For the culture of anthers, spikes are sterilized as described for immature inflorescences or with 0.1% (w/v) mercuric chloride for 8–20 min, followed by three washes in sterile water.

2.3 Composition of Culture Media

2.3.1 Basic Constituents

Table 2 shows that MS medium (Murashige and Skoog 1962) is the basal medium that has been used most often for micropropagation (meristem tips) and callus induction from explants of *Lolium*. For callus induction, the MS medium is usually supplemented with 0.8% agar, 3% sucrose, varying concentrations of growth regulators (2,4-D), IAA, kinetin, zeatin, and/or BAP) and complex nutrients (coconut milk and casein hydrolysate). For the direct regeneration of plantlets from meristem tips, a simple MS-based medium supplemented with 0.2 mg/l kinetin can be used. The pH of the medium is adjusted to 5.8 by the addition of 0.1 M HCl or NaOH. The complete medium can be sterilized for 15 min at 121 °C, but to prevent pH and other medium changes it is recommended to filter sterilize the nutrients as a concentrated solution, which can be added to the autoclaved agar solution at a suitable temperature.

Table 2. Summary of the tissue culture studies in *Lolium* spp.

Plant species	Inoculum	Medium[a] (mg/l)
Lolium multiflorum	Meristem tip	MS + kin (0.2)
L. multiflorum	Meristem tip	MS + 2,4-D (0.1) + kin (0.2)
L. perenne	Meristem tip	MS + 2,4-D (0.1) + kin (0.2)
L. multiflorum	Meristem tip	MS + 2,4-D (2) + BAP (0.2) + CH (100)
L. multiflorum	Root	LS + 2,4-D (5)
L. multiflorum	Root	MS + 2,4-D (2) + BAP (0.2) + CH (100)
L. temulentum	Root	LS + 2,4-D (5)
L. perenne	Root	LS + 2,4-D (5)
L. rigidum	Immature embryo	MS + 2,4-D (10) + CH (0.1%)
L. perenne	Mature embryo	MS + 2,4-D or Dicamba + BAP (0,5)
L. multiflorum	Immature embryo	MS + 2,4-D (2) + BAP (0.2)
L. perenne + *L. multiflorum*	Immature seed	NO + 2,4-D (1.5) + IAA (6.5) + zeatin (0.25)
L. perenne	Mature seed	MS + 2,4-D (10)
L. perenne	Mature seed	MS + 2,4-D (5)
L. multiflorum	Immature inflorescence	MS + 2,4-D (2)
L. multiflorum	Immature inflorescence	MS + 2,4-D (2) + BAP (0.2) + CH (100)
L. perenne	Immature inflorescence	MS + 2,4-D (2) + BAP (0.2) + CH (100)
L. multiflorum	Immature inflorescence	MS + 2,4-D (5)
L. perenne	Immature inflorescence	MS + 2,4-D (5)
L. multiflorum	Node	MS + 2,4-D (2)
L. multiflorum × *Festuca arundinacea*	Internode	MS + 2,4-D (2/4)
hybrids	Leaf	MS + 2,4-D (2/4)
L. multiflorum	Leaf	MS + 2,4-D (2) + BAP (0.2) + CH (100)
L. perenne × *L. multiflorum*	Suspension culture	Mod MS + IAA (6.5) + 2,4-D (1.5) + kin (2.15)
L. multiflorum	Suspension culture	MS + 2,4-D (2)
L. multiflorum	Suspension culture	MS + 2,4-D (5/6)

L. multiflorum	Suspension culture	MS + 2,4-D (5/2)
L. perenne	Suspension culture	MS + 2,4-D (5/6)
L. multiflorum	Protoplasts	MS + 2,4-D (0–0.25)
L. multiflorum	Protoplasts	Mod MS + 2,4-D (0.1) + BAP (0.1) + glucose + mannitol
L. perenne	Protoplasts	Mod MS + 2,4-D (0.1) + BAP (0.1) + glucose + mannitol
L. perenne	Protoplasts	RY-2[b] + 2,4-D (4.4) + glucose + FC (0.8%) + CM (50%)
L. multiflorum	Anther	Complex medium[b]
L. perenne	Anther	Complex medium[b]
L. temulentum	Anther	Complex medium[b]

[a] The simple medium for initial culture is given. Different media have been used for maintenance and plant regeneration. Numbers in parenthesis are mg/l.
[b] The media used are composed of parts of formerly described (MS, LS) media with many additives. For detailed description, see references.
[c] For detailed references, see section References.

Growth response	Reference[c]
Plants	Dale (1975, 1980a)
Plants	Dale (1977a, b)
Plants	Dale (1977a)
Callus and plants	Jackson and Dale (1988)
Callus and roots	Atkin and Barton (1973)
Callus and plants	Jackson et al. (1986), Jackson and Dale (1988)
Callus	Atkin and Barton (1973)
Callus	Atkin and Barton (1973)
Callus and plants	Skene and Barlass (1983)
Callus, embryos, and plants	Dale (1980b)
Callus and plants	Schmidt and Posselt (1990), Schmidt (1991)
Callus and plants	Ahloowalia (1975)
Callus and roots	Torello et al. (1983)
Callus and plants	Torello and Symington (1984)
Callus, embryos, and plants	Dale et al. (1981)

Table 2. (*Contd.*)

Growth response	Reference[c]
Callus and plants	Dale and Dalton (1983)
Callus and plant	Dale and Dalton (1983)
Callus, embryos, and plants	Creemers-Molenaar et al. (1988b)
Callus, embryos, and plants	Creemers-Molenaar et al. (1988b)
Callus, embryos, and plants	Dale et al. (1981)
Callus and plants	Kasperbauer et al. (1979)
Callus and plants	Kasperbauer et al. (1979)
Callus and plants	Jackson and Dale (1988)
Callus and embryos	Ahloowalia (1975)
Callus and plants	Jones and Dale (1982)
Callus, embryos, and plants	Dalton (1988b), Creemers-Molenaar et al. (1989)
Callus, embryos, and plants	Rajoelina et al. (1990)
Callus, embryos, and plants	Dalton (1988b), Creemers-Molenaar et al. (1989), Zaghmout and Torello (1990)
Callus	Jones and Dale (1982)
Callus and plants	Dalton (1988a, b)
Callus and green plantlets	Dalton (1988a, b)
Callus and plants	Creemers-Molenaar et al. (1989)
Callus and plants	Niizeki (1977), Nitzsche and Wenzel (1977), Pagniez and Demarly (1979), Bante et al. (1990, 1991)
Callus and plants	Stanis and Butenko (1984), Olesen et al. (1988), Bante et al. (1990, 1991), Boppenmeyer et al. (1989), Halberg et al. (1990), Hayward et al. (1990)
Callus and albino plantlets	Rose et al. (1987)

2.3.2 Growth Regulators

In tissue culture studies on cereals and grasses, auxins have proven to be indispensable for triggering callus initiation and maintaining established callus cultures. In *Lolium*, 2,4-D (2,4-dichlorophenoxyacetic acid) is effective for callus initiation and callus maintenance. The optimal concentration depends on the explant type and ranges from 2–10 mg/l. The effect of the 2,4-D concentration on callus induction and regeneration has been determined for immature inflorescence cultures in *L. perenne* and *L. multiflorum* (Creemers-Molenaar et al. 1988b; Fig. 1). It was concluded that for the concentration range tested (2.5–15 mg/l) no optimal concentration could be determined for callus induction. However, in *L. perenne* higher concentrations of 2,4-D during the callus initiation period resulted in the regeneration of more albino shoots and less green shoots (Fig. 1). Similar results were obtained with another synthetic auxin, Dicamba (3,6-dichloro-2-methoxy benzoic acid), in suspension-derived callus of rice (Zimny and Lörz 1986).

Several studies compared the effect of different auxins on callus induction, maintenance, and plant regeneration from explants in cereals and grasses. In sugarcane the synthetic auxin picloram (4-amino-3,5,6-trichloropicolinic acid) was superior to 2,4-D for the induction of morphogenic callus and the maintenance of long-term regeneration-competent callus (Fitch and Moore 1990). In rice the phenoxyacetic acid derivative 4FPA (4-fluorophenoxyacetic acid) was superior to 2,4-D with respect to callus induction and long-term maintenance of the regeneration competence (Yasuda et al. 1990).

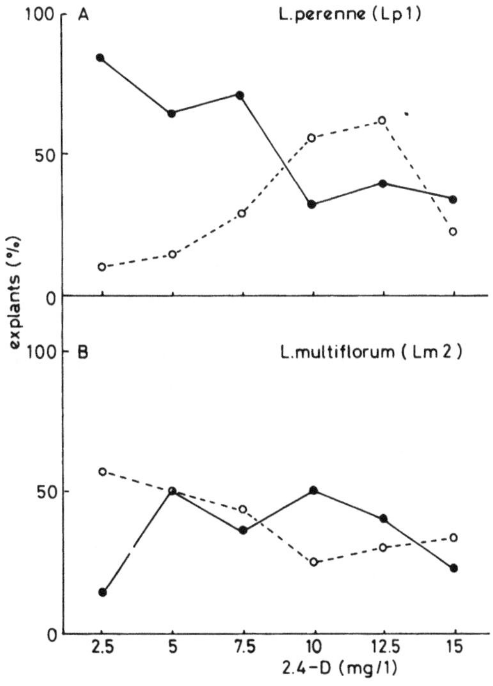

Fig. 1. The effect of different 2,4-D concentrations on the regeneration frequency of immature inflorescence-derived explants in one *L. perenne* and one *L. multiflorum* variety. The percentages are the means of 8 to 14 explants with compact callus; up to nine pieces of compact callus were tested per explant. ●——● green shoots, ○— — —○ albino shoots. (Creemers-Molenaar et al. 1988b)

Very few studies compared the effect of different auxins on callus induction from explants in *Lolium*. Schmidt (1991) showed that the proportion of embryo-derived calli that formed regenerants increased when dicamba was used instead of 2,4-D in the callus induction medium. However, the effect of different auxins was dependent on the genotype. For callus induction from leaf tip and embryo explants from *L. multiflorum* dicamba was more effective than 2,4-D (Conger et al. 1982). NAA and picloram were also more effective, but these auxins had to be applied at higher concentrations and the calli showed a high tendency to form roots. In the sole presence of the natural auxin IAA, no callus was formed on leaf tip or embryo explants of *L. multiflorum*. Callus induction frequency and callus growth from embryos of *L. multiflorum* on media with 2,4-D or another synthetic auxin, 2,4,5-T, were comparable, but again more roots were formed in the presence of 2,4,5-T (Conger et al. 1978).

However, once a suitable explant is chosen, it is not callus induction but the loss of regeneration competence in long-term callus and suspension cultures which is a problem in *Lolium*. Considering the above-discussed positive effects of picloram and 4FPA in sugarcane and rice, respectively, the application of these auxins in callus and suspension cultures in *Lolium* needs further research.

In addition to 2,4-D, the callus induction medium for explants of *Lolium* has also been supplemented with low concentrations of cytokinins (Table 2). Schmidt and Posselt (1990) showed that the frequency of plant regeneration from mature embryos increased with the concentration of BAP (0–0.5 mg/l) in the callus induction medium. The number of regenerants per callus was not influenced by BAP. Furthermore, the effect of BAP was dependent on the composition of the basal MS medium (type of auxin and concentration of thiamine). The effects of cytokinins in callus cultures of other gramineous species are variable. In wheat callus cultures cytokinins inhibited growth (Dudits et al. 1975). In protoplast-derived callus of *Panicum maximum*, zeatin favored the formation of organized embryoids (Lu et al. 1981). Low concentrations of kinetin and zeatin induced shoot formation from protoplast-derived callus in rice (Kyozuka et al. 1987).

2.4 Physical Culture Conditions

Most tissue culture studies of *Lolium* are performed at a temperature of 25 °C; however, with reference to the light regime no consistency in conditions is evident.

During the initial culture period callus cultures from explants of *Lolium* may be incubated either in the dark (Creemers-Molenaar et al. 1988b) or in diffuse light (Kasperbauer et al. 1979; Dale et al. 1981). Suspension cultures of *L. multiflorum* and *L. perenne* were kept in diffuse light continuously (Dalton 1988a; Creemers-Molenaar et al. 1989). Such light-grown suspension cultures often contained cell aggregates with green areas, but other growth characteristics were the same for dark- and light-grown cultures (Creemers-Molenaar et al., unpubl.). For the regeneration of shoots, callus may be cultured in continuous light (Dale 1980b; Dale and Dalton 1983; Jackson et al. 1986; Creemers-Molenaar et al. 1988b) or in a light/dark regime (Skene and Barlass 1983; Torello and Symington 1984).

The effects of different light regimes on tissue cultures of *Lolium* have not been reported, but obviously need further attention.

2.5 Meristem Culture

For the preservation of grass genotypes, meristem culture has been used (Dale 1975, 1977a, 1980a). Meristem culture can also be used to eliminate ryegrass mosaic virus and other viruses in valuable ryegrass plants (Dale 1977b).

At our breeding station, meristem culture is applied to plants which have produced seed for progeny tests during the summer. In autumn, plants are put inside an unheated, but frost-free greenhouse. Insecticides are applied against frit fly larvae (*Oscinella frit*), because they feed on the inner parts of tillers, where the meristems are situated. Two weeks before meristem culture work starts, the plants are transferred to a greenhouse at 20 °C to stimulate growth and development of the shoot tips.

From about ten healthy, vegetative tillers of every genotype, the basal 8 cm is cut. Roots are cut back to ± 1 cm. These tillers are surface sterilized as described in Section 2.2. Shoot tips are excised aseptically out of the tillers under a stereomicroscope. For virus elimination purposes, the dissected part should be as small as possible, containing only the meristem tip. For preservation purposes, larger parts of shoot tips, 1–2 mm long, may be excised (Fig. 2). Per genotype, eight shoot tips are placed each in a glass tube containing about 12.5 ml solid culture medium, as described in Section 2.3.1. Glass tubes containing meristems are placed in a growth chamber at 20 °C with ± 2000 lx continuous light from fluorescent Philips TL 83 lamps.

Not all shoot tips develop into plants, and some may still be infected. Therefore, eight shoot tips per genotype are cultured, to obtain at least three successful ones. Normally, within a few days the first leaf starts to develop. After about 8 weeks, most shoot tips have developed into complete shoots with roots. At this stage, three tubes with successfully established, uninfected shoots per genotype are put into cold storage at 4 °C and continuous light, ± 1000 lx (Philips TL 57 RS). They can remain in cold storage for 18 months before subculturing is needed.

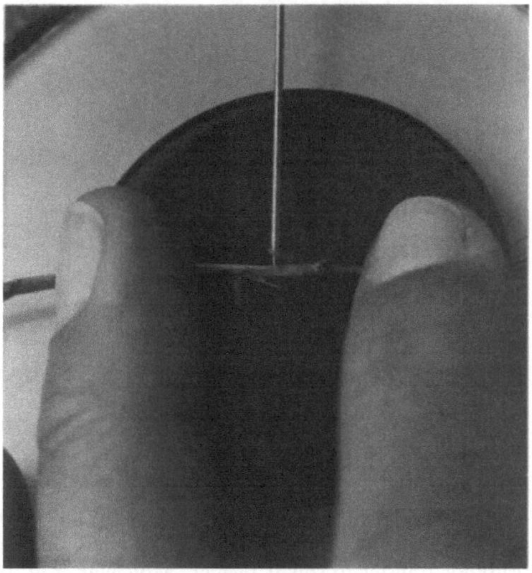

Fig. 2. A shoot tip, excised from a ryegrass tiller, to be used in meristem culture

During cold storage, the plants grow slowly and form several new tillers, which enables some vegetative multiplication. A 2-cm piece of a tiller, containing the shoot tip, is placed with the lower half in the fresh culture medium. Usually, growth starts immediately and after 4 weeks plants are put back in cold storage again.

Subculturing can be repeated as often as desired. At our breeding station, we have kept Italian ryegrass clones alive for 7 years, without any loss of vigor. This would have been very difficult, if not impossible, using trial fields or greenhouses.

Reestablishment of plants in soil is relatively easy. Glass tubes with the desired plants are collected from the cold storage and plants are taken out of their tubes. Culture medium is removed under running tap water. Plants are put in normal, unsterilized potting compost. They remain covered for about a week to keep the air humidity high. Several hundreds of selected clones so far have been reestablished and no genotype has ever failed to establish itself in soil again.

2.6 Callus Initiation and Plant Regeneration

Cultured explants of *Lolium* initially form a watery type of callus, which is considered to be nonmorphogenic. However, part of these watery calli have been

Fig. 3A–C. Callus initiation and plant regeneration from immature inflorescence explants of *L. perenne*. **A** Watery (*w*) and compact callus (*c*), 4 weeks after culture initiation. **B** Regeneration of a green shoot from compact callus. **C** Plantlets obtained from immature inflorescence-derived callus, ready for transfer to soil. (Creemers-Molenaar et al. 1988b)

observed to form centers of compact morphogenic callus (Fig. 3A). Plant regeneration has been shown to originate mainly from compact callus via the germination of somatic embryos (embryogenesis). However, less frequently, shoots have also been observed to regenerate directly from compact callus (organogenesis).

Table 2 shows that callus cultures have been established from almost all plant parts of *L. multiflorum* and *L. perenne*. A representative time course of the procedure for callus initiation and plant regeneration is given in Table 3. Depending on the explant type, culture periods will differ from this example, but the developmental process is the same.

Vasil (1987) reviewed factors which primarily influence the callus response and regeneration from explants of gramineous species. It was concluded that the physiological and the developmental state of the explant at the time of exision control the morphogenic response to a larger extent than its genotype. In cultured tissues of gramineous species only the meristematic cells are able to proliferate and form regeneration-competent cells. The presence of too many differentiated cells in the explant might suppress the morphogenic response of competent cells (Vasil 1987). The importance of the developmental stage and the type of explant was demonstrated in perennial and Italian ryegrass (Dale 1980b; Joarder et al. 1986; Creemers-Molenaar et al. 1988b). In *L. perenne*, explants from immature inflorescences at a defined developmental stage showed a high response for callus induction (80–100%) and compact callus formation (56–100%) (Creemers-Molenaar et al. 1988b). In comparison, less than 10% of cultured mature embryos of *L. perenne* formed compact callus (Creemers-Molenaar, unpubl.).

Table 3. Time course of the procedure for callus initiation and plant regeneration from immature inflorescence explants of *L. multiflorum* and *L. perenne*

Week	
1	After sterilization of the tillers, the immature inflorescences are excised and 2-mm-long segments are cultured on solid MS medium, supplemented with 5 mg/l 2,4-D. The cultures are incubated in the dark at 25 °C
4	Explants which have formed watery callus are transferred to fresh medium
6–8	At the end of the second culture period, a number of explants have formed areas of white, compact callus (Fig. 3A). To enhance the proliferation of compact callus, this is excised and subcultured two times onto fresh medium, at 2-weekly intervals
10	At the end of the fourth subculture period, most of the compact calli have become nodular, showing the onset of somatic embryogenesis. To stimulate further development of pro-embryogenic structures, compact and nodular calli are subcultured onto MS medium with 0.2 mg/l 2,4-D
13	At this stage well-developed somatic embryos can be observed. Compact and embryogenic calli are subcultured onto MS medium without 2,4-D in the light
16–19	During this period many shoots regenerate asynchronously from the callus. The majority of the shoots are green, but albino shoots also are often observed. Shoots regenerate by the germination of somatic embryos or directly from compact callus (Fig. 3B). After 3–6 weeks green shoots are separated from the callus and transferred to culture vessels with the same medium. The cultures are kept at 20 °C under continuous light from fluorescent Philips TL 83 lamps (Fig. 3C)
19–22	Rooted plantlets are transferred to pots with soil in the greenhouse, under conditions that prevent evaporation
23–26	Plants are grown in the greenhouse or in the field, without special care for maintenance

Plant regeneration has been achieved from all explant types in *Lolium* which have been tested. Both green and albino shoots have been observed to regenerate from most explant types. The occurrence of albinism will be further discussed in Section 2.9.

2.7 Cell Suspension Cultures

2.7.1 Establishment of Cell Suspension Cultures

In gramineous species, suspension cultures have proven to be the most suitable starting material for the isolation, culture, and regeneration of protoplasts. Such suspension cultures should be finely dispersed, fast-growing, and morphogenic. Suspension cultures with the capacity to regenerate green plants have been

Fig. 4A–C. Suspension culture of *L. perenne*. **A** Finely dispersed suspension culture in liquid MS medium. **B** Somatic proembryos attached to watery callus. **C** Plant regeneration from a suspension culture 2 weeks after plating on solid MS medium without 2,4-D. (Creemers-Molenaar et al. 1989)

obtained from several gramineous species, including: *Oryza sativa, japonica* genotypes (Abdullah et al. 1986; Toriyama et al. 1986; Yamada et al. 1986; Kyozuka et al. 1987, 1988) and *indica* genotypes (Lee et al. 1989; Wang et al. 1989; Datta et al. 1990), *Zea mays* (Kamo and Hodges 1986; Vasil and Vasil 1986; Rhodes et al. 1988, Prioli and Sondahl 1989), *Saccharum officinarum* L. (Ho and Vasil 1983), *Triticum aestivum* L. (Ahuja et al. 1982; Harris et al. 1988; Vasil et al. 1990), *Dactylis glomerata* (Gray et al. 1984), *Festuca arundinacea* Schreb. (Dalton 1988a, b; Rajoelina et al. 1990; Takamizo et al. 1990), *Festuca rubra* L. (Zaghmout and Torello 1989), *Lolium perenne* (Dalton 1988a, b; Creemers-Molenaar et al. 1989; Zaghmout and Torello 1990), and *Lolium multiflorum* (Dalton 1988a, b; Creemers-Molenaar et al. 1989).

In *Lolium*, suspension cultures are initiated using immature inflorescence-derived compact callus or directly with 30–40 seed-derived embryos in liquid MS medium, supplemented with 5–6 mg/l 2,4-D. The cultures are incubated on a rotary shaker (100–140 rpm, 15 mm amplitude) in the dark or in continuous indirect light (200–400 lx) at 25 °C. Established suspension cultures are obtained 10–20 weeks after initiation and mainly consist of small pieces of compact callus with densely cytoplasmic cells (Fig. 4A). Young suspension cultures (< 20 weeks) are often embryogenic, with somatic proembryos attached to the compact cell aggregates or freely suspended in the medium (Fig. 4B). These suspension cultures are maintained by the weekly subculture of small cell aggregates to fresh medium.

Within a given plant variety, different cell suspension cultures, each initiated with a number of genotypes, exhibit much variation with respect to growth characteristics and the amenability to become established. The results reported by Creemers-Molenaar et al. (1989) give some indication that immature inflorescence-derived callus is more suitable for suspension culture initiation than seed-derived embryos. In *L. perenne* the percentage of suspension cultures initiated from embryos and callus that became established was 0–38% and 20–100%, respectively.

2.7.2 Plant Regeneration from Cell Suspension Cultures

Young suspension cultures of *L. perenne* and *L. multiflorum* (< 20 weeks after initiation) have a high potential for plant regeneration. After plating on solid MS medium without 2,4-D, up to 200 green shoots/g fresh weight suspension cell aggregates have been obtained in *L. perenne* (Creemers-Molenaar et al. 1989). Figure 4C shows the regeneration of shoots from suspension cell aggregates. However, most suspension cultures lose the ability to regenerate green shoots within 20–30 weeks after initiation (Fig. 5). Most suspension cultures also regenerate albino shoots, and the frequency of albino shoots that regenerate tends to increase with suspension culture age, preceding complete loss of the regeneration potential (Fig. 5). Loss of the regeneration capacity in long-term suspension cultures has been observed in most cereals and grasses and has stimulated research to maintain or restore the regeneration potential:

1. In a 2-year-old suspension culture of rice, a 3-month pretreatment with 1.5% NaCl resulted in the enrichment of embryogenic cells, thus restoring the high

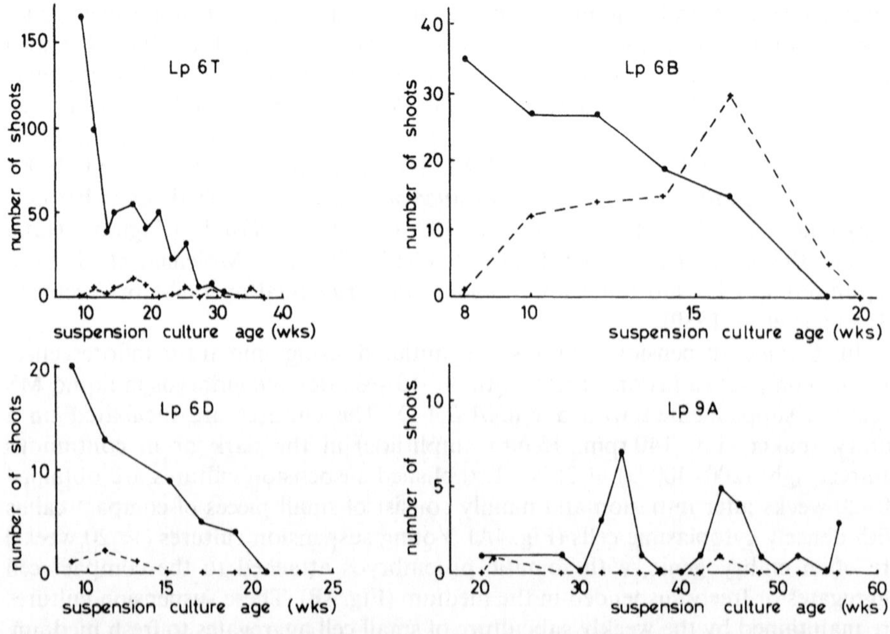

Fig. 5. Plant regeneration from suspension cultures of two varieties of *L. perenne* (Lp6 and Lp9), in relation to suspension culture age. Lp6 suspension cultures were initiated from mature embryos, Lp9 from immature inflorescence-derived callus. Three weeks after plating on MS medium without 2,4-D, 12 pieces of compact callus (5 × 5 mm) were subcultured on the same medium. After 3–4 weeks the number of green and albino shoots were determined from two replicate plates. + --- + albino shoots, ●——● green shoots. (Creemers-Molenaar et al. 1989)

regeneration competence of the culture (Binh and Heszky 1990). Preliminary results with suspension cultures of perennial ryegrass did not confirm these results (Creemers-Molenaar, unpubl.). NaCl treatment resulted in either browning and death of the culture or the slow proliferation of compact friable callus. This callus showed no characteristics of plant regeneration upon plating on solid medium.

2. In wheat suspension cultures lowering of the 2,4-D concentration induced the formation of roots. Removal of the root-forming callus and culture of the remaining callus at reduced 2,4-D levels resulted in the establishment of a long-term (2.5 years) regeneration-competent suspension culture (Wang and Nguyen 1990).

3. Vasil et al. (1990) stated that in wheat only the use of carefully selected, aged embryogenic callus leads to the establishment of long-term (1-year) regeneration-competent suspension cultures.

4. Creemers-Molenaar and van Oort (1990) showed that fast-growing suspension cultures of perennial ryegrass can be stored for at least 6 weeks at 4 °C without loss of the regeneration potential. When the same cultures were kept under standard conditions, the regeneration capacity decreased as usual. Thus, cold storage of perennial ryegrass suspension cultures permits the prolonged use of the cultures and is time-saving with respect to subculture regimes.

5. In callus cultures of alfalfa the replacement of part or all of the sucrose by maltose increased the formation of embryos and the regeneration of green plants

(Strickland et al. 1987). The culture of wheat anthers in liquid medium with maltose instead of sucrose resulted in increased callus induction from microspores and increased the regeneration of green shoots from these calli (Orshinsky et al. 1990). In perennial ryegrass anther culture the replacement of sucrose by maltose improved the percentage of anthers that formed callus (Bante 1990, 1991). Replacement of part or all of the sucrose by maltose in the suspension culture medium did not affect the regeneration potential of perennial ryegrass suspension cultures (Creemers-Molenaar and van Oort 1990).

6. In perennial ryegrass, the frequency of immature inflorescence-derived calli forming albino shoots increased with increasing concentrations of 2,4-D in the callus induction medium (Creemers-Molenaar et al. 1988b). Lowering of the 2,4-D concentration to 2.5 mg/l or partial replacement of 2,4-D by the weaker auxin NAA in the suspension culture medium of perennial ryegrass did not improve the competence for plant regeneration (Creemers-Molenaar, unpubl.).

7. Promising results have been obtained by Shillito et al. (1989), who demonstrated that the morphogenic potential of a maize suspension culture was retained after cryopreservation. After cryopreservation the suspension cultures could be re-established, and protoplasts isolated from such suspensions were able to regenerate into plants again. Regarding the short regeneration period of perennial ryegrass suspension cultures, and the more or less fruitless efforts to prolong this period, cryopreservation might be an essential step to maintain regeneration-competent suspension cultures in perennial ryegrass.

In conclusion, there is as yet no general procedure available for the establishment of long-term regeneration-competent suspension cultures in cereals and grasses. Methods, as reported in the literature, often are only applicable for certain species, or for certain genotypes. Consequently, optimal tissue culture conditions will have to be determined for every monocot species.

2.8 Protoplasts

2.8.1 Isolation of Protoplasts

Lolium protoplasts are isolated from actively growing suspension cultures by treating the cells with an enzyme mixture containing 1% (w/v) Cellulase Onozuka RS and 0.1% (w/v) Pectolyase Y-23 in CPW-13M medium (Frearson et al. 1973) for 4–6 h with gentle agitation. In addition, Meicelase, Macerozyme R10, and Driselase have been used. For overnight incubation the concentration of the enzymes is reduced by half. Pure protoplast suspensions are obtained by sieving and several washings of the protoplast-enzyme mixture (Fig. 6A). Protoplast yield from *L. perenne* suspension cultures varies from 0.1–14 million/g fresh weight, and is dependent on the growth characteristics of the suspension culture.

2.8.2 Protoplast Culture and Regeneration

Protoplasts of *Lolium* are cultured in liquid medium (Dalton 1988a, b; Creemers-Molenaar et al. 1989) or in agarose-solidified medium (Creemers-Molenaar

Fig. 6A–C. Callus formation and plant regeneration from protoplasts of *Lolium perenne*. **A** Freshly isolated suspension culture protoplasts. **B** Microcalli in agarose-solidified RY-2 medium after subculture of agarose segments onto fresh medium. **C** A green plant regenerated from a protoplast. (Creemers-Molenaar et al. 1989)

et al. 1989) in the dark (Fig. 6B). As described for rice (Yamada et al. 1986), *Lolium* protoplasts also require a culture medium with a high osmolality (890–920 mOs/kg), mainly adjusted with glucose. Although in *Lolium* the plating efficiency (expressed as the percentage of plated protoplasts that form microcalli) has been shown to be as high as 2.2% in liquid RY-2 medium, such values can only be achieved with protoplasts isolated from fast-growing suspension cultures that have lost the potential for plant regeneration. Regeneration-competent suspension cultures yield few protoplasts ($< 10^6$/g FW) with low plating efficiencies (0–0.1%). Furthermore, additional treatments like electrofusion or PEG (polyethyleneglycol) fusion have been shown to reduce the plating efficiency of protoplasts. In perennial ryegrass several modifications of the RY-2 medium improve the plating efficiency (Creemers-Molenaar 1991; Creemers-Molenaar et al. 1990, 1992A):

1. The addition of conditioned medium (50% V/V) to RY-2 medium improved the plating efficiency of protoplasts from aged suspension cultures and was indispensable for the proliferation of microcalli from protoplasts from young suspension cultures.

2. The addition of 10–30 mg/l ASA (O-acetylsalicylic acid) to conditioned RY-2 medium (RY-C) further enhanced the plating efficiency 1.5–3 times.

3. The addition of different combinations of antioxidants to the CPW-isolation and the RY-C culture medium improved the plating efficiency up to five times compared with protoplasts isolated and cultured without antioxidants (Creemers-Molenaar and van Oort 1990).

Plant regeneration from *Lolium* protoplasts is obtained after several subcultures of compact callus to regeneration medium in the light (Creemers-Molenaar et al. 1989). Using this procedure, soil-established plants have been regenerated from protoplasts of several suspension culture lines of *L. perenne*, representing four varieties. A gradual decrease in the 2,4-D concentration (RY-C4.4 = conditioned RY-2 medium with 4.4 mg/l 2,4-D) in the successive regeneration media and a gradual adaptation to light conditions (RY-C4.4-dark, MS2.5-dark, MS0-indirect light, MS0-direct light) improved the proliferation of microcalli and plant regeneration frequency (Creemers-Molenaar, unpubl.). The improved subculture regime, together with the use of conditioned RY-2 medium, resulted in the regeneration of 334 green plants per 3.5×10^5 plated protoplasts.

2.9 Production of Haploid Plants Through Anther Culture

Haploid green plants have been regenerated from anther culture in several gramineous species, including *L. perenne* and *L. multiflorum*, but the majority of the regenerated plants are albino. Important progress was made by Olesen et al. (1988), who tested 65 745 anthers from 30 clones of ten *L. perenne* varieties for callus induction and plant regeneration. The anthers containing microspores in the late uninucleate stage were cultured onto PII-substrate (Wang and Hu Han 1984) solidified with 0.3% Gelrite and supplemented with 2.5% potato extract. For plant regeneration, calli were subcultured to regeneration medium 190-2 (Wang and Hu Han 1984) in the light. Twelve of the 30 clones, representing five varieties, produced anther-derived green plants. It was shown that in *L. perenne* the genotype of the donor plants largely determined the response of cultured anthers with respect to callus induction and green plant regeneration (Olesen et al. 1988; Halberg et al. 1990; Bante et al. 1991). Recently, Halberg et al. (1990) showed that in *L. perenne*, clones with a superior anther culture response could be obtained through the sexual crossing of primary responsive clones. Next, these superior clones might be used as inducers of anther culture response through crossing with non-responding breeding material. Of course, a disadvantage of this approach is the need for several cycles of backcrossing. In addition, attempts have been made to improve the anther culture response by modifying the culture conditions (Bante et al. 1990, 1991). The percentage of anthers forming callus or embryos could be increased by: (1) substitution of sucrose by maltose, trehalose, or maltotriose (2–6x); (2) reduction of the ammonium nitrate concentration in the induction medium (2x); (3) cold pretreatment of the spikes (4–5x); and (4) incubation of the anthers in 2% CO2 (4–5x). When the former modifications were used in combination, up to 35% of the anthers formed callus.

The application of anther culture in ryegrass breeding will only be successful if the method works well for a wide range of genotypes. The genotype-dependent response and the observation that the majority of the regenerated shoots are albinos make clear that further research is necessary.

2.10 Albinism

In tissue cultures of the Gramineae, the occurrence of albino shoots is often encountered and can hinder progress in the development of genetic manipulation techniques. In *Lolium* albino shoots are observed quite early after culture initiation and the frequency increases with culture age. Up to 42% of immature inflorescence-derived compact calli of *L. perenne* produced albino shoots (Creemers-Molenaar et al. 1988b). In *L. multiflorum* 50% of the shoots that regenerated from embryo-derived callus cultures were albino (Dale 1980b). In addition to green plants, albino plantlets were regenerated from a callus culture, which originated from one triploid embryo of hybrid origin [2x (*Lolium multiflorum* Lam.-*perenne* L.) × 4x (*L. perenne* L.)] (Ahloowalia 1983). The effect of aging on the occurrence of albino shoots is observed in callus cultures maintained on solid medium over prolonged periods (Dale et al. 1981), but is most pronounced in suspension cultures of *L. perenne* and *L. multiflorum* (Dalton 1988a, b; Creemers-Molenaar et al. 1989). In addition to aging, other factors have been reported to enhance the occurrence of albino shoots in *Lolium*: (1) higher culture temperature (Jackson and Dale 1988), (2) growth conditions of the donor plants, i.e., field conditions versus greenhouse conditions (Fig. 7; Creemers-Molenaar et al. 1988b), (3) genotype (Olesen et al. 1988; Creemers-Molenaar et al. 1988b), and (4) 2,4-D concentration of the culture medium (Fig. 1) (Creemers-Molenaar et al. 1988b).

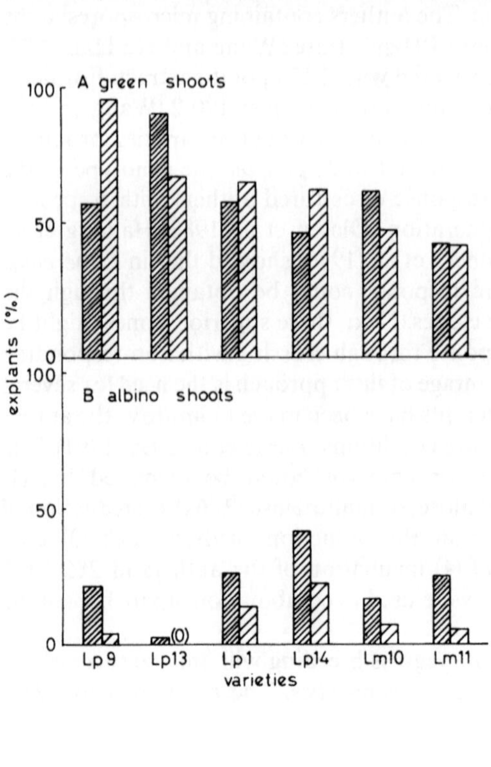

Fig. 7. Regeneration frequency of immature inflorescence explants in dependence on the growth conditions of the donor plants in *L. perenne* (Lp) and *L. multiflorum* (Lm). The percentages are the means of 12 to 38 explants with compact callus. Observations were made after subculture of three pieces of compact callus per explant. *Light shading* greenhouse-grown plants; *heavy shading* field-grown plants. (Creemers-Molenaar et al. 1988b)

Several of the above-mentioned factors also have been observed to influence the emergence of albino shoots in other gramineous species (Van der Valk et al. 1988, 1989; Wang et al. 1978; Lo et al. 1980; Chandler and Vasil 1984).

A molecular characteristic of albinism, as demonstrated for wheat and barley albino plantlets, is the presence of large deletions in the chloroplast DNA (Day and Ellis 1984, 1985). In wheat anther culture it was shown that the inheritance of the response, with respect to the percentage of green and albino shoots that regenerate, was dependent on the action of different classes of nuclear genes (Tuvesson et al. 1989). However, the cause of albinism and the possible involvement of nuclear and cytoplasmic genomes have not yet been elucidated.

2.11 Progress in Somatic Hybridization, Cybridization, and Transformation

In perennial ryegrass electrofusion between suspension-derived protoplasts and mesophyll-derived protoplasts has been described (Creemers-Molenaar et al. 1988a). However, fused protoplasts did not divide and, subsequently, no microcalli have been obtained. Further experiments revealed that, compared with electrofusion, suspension culture-derived protoplasts of perennial ryegrass showed better survival after polyethylene glycol (PEG) treatment. Using polyethylene glycol, gamma-irradiated protoplasts of an *L. perenne* CMS line were fused with iodoacetamide-inactivated protoplasts of a fertile *L. perenne* variety (Creemers-Molenaar et al., 1992b). From these fusion experiments cybrid calli were obtained that contained the nucleus of the recipient line and the mitochondrial genome of the donor line. Up until now, plant regeneration has not been achieved from these cybrid calli.

In *Lolium* transformed callus was obtained after PEG treatment of suspension-derived protoplasts (Portrykus et al. 1985). Until now, no transformed plants have been regenerated in any *Lolium* species.

3 Conclusion

Important progress has recently been made in the development of tissue culture techniques in *Lolium*. Individual plants, representing valuable genotypes, can routinely be cloned and stored in vitro by meristem tip culture. Successful plant regeneration from callus, suspension cultures, and protoplasts has been achieved in *L. perenne* and *L. multiflorum*. Furthermore, anther culture-derived plants have been obtained from several varieties of *L. perenne* and *L. multiflorum*. However, the occurrence of albinism and the limited regeneration frequency of green shoots from protoplasts and anther culture still are a serious limitation and need further investigation. In conclusion, although much work still has to be done, the achievements so far in the tissue culture of *Lolium* are most promising with respect to the future application of genetic manipulation techniques in plant breeding programs.

References

Abdullah R, Cocking EC, Thompson JA (1986) Efficient plant regeneration from rice protoplasts through somatic embryogenesis. Bio/Technol 4: 1087–1090

Ahloowalia BS (1975) Regeneration of ryegrass plants in tissue culture. Crop Sci 15: 449–452

Ahloowalia BS (1983) Spectrum of variation in somaclones of triploid ryegrass. Crop Sci 23: 1141–1147

Ahuja PS, Pental D, Cocking EC (1982) Plant regeneration from leaf base callus and cell suspensions of *Triticum aestivum*. Z Pflanzenzücht 89: 13–144

Atkin RK, Barton GE (1973) The establishment of tissue cultures of temperate grasses. J Exp Bot 24: 689–699

Bajaj YPS (1989) Genetic engineering and in vitro manipulation of plant cells – technical advances. In: Bajaj YPS (ed) Biotechnology in agriculture and forestry, vol 9. Plant protoplasts and genetic engineering II. Springer, Berlin Heidelberg New York, pp 1–25

Bante I, Sonke T, Tandler RF, van den Bruel AM, Meijer EM (1990) Anther culture of *Lolium perenne* L. and *Lolium multiflorum* L. Acta Bot Neerl 39(1): 103

Bante I, Sonke T, Tandler RF, van den Bruel AM, Meijer EM (1991) Anther culture of *Lolium perenne* and *Lolium multiflorum*. In: Sangwan BS, Sangwan RS (eds) The impact of biotechnology in agriculture. Int Conf, Amiens, France 10–12 July 1989 (in press)

Binh DQ, Heszky LE (1990) Restoration of the regeneration potential of long-term culture in rice (*Oryza sativa* L.) by salt pretreatment. J Plant Physiol 136: 336–340

Boppenmeier J, Zuchner S, Foroughi-Wehr B (1989) Haploid production from barley yellow dwarf virus resistant clones of *Lolium*. Plant Breeding 103: 216–220

Breese EL, Tyler BF (1986) Patterns of variation and the underlying genetic and cytological architecture in grasses with particular reference to *Lolium*. In: Styles BT (ed) Intraspecific classification of wild and cultivated plants. Clarendon, Oxford, pp 53–69

Chandler SF, Vasil IK (1984) Optimization of plant regeneration from long term embryogenic callus cultures of *Pennisetum purpureum* grass. J Plant Physiol 117: 147–156

Conger BV, Carabia JV, Lowe KW (1978) Comparison of 2,4-D and 2,4,5-T on callus induction and growth in three Gramineae species. Environ Exp Bot 18: 163–168

Conger BV, Hilenski LL, Lowe KW, Carabia JV (1982) Influence of different auxins at varying concentrations on callus induction and growth from embryo and leaf-tip explants in Gramineae. Environ Exp Bot 22: 39–48

Creemers-Molenaar J (1991) Regeneration from protoplasts of perennial ryegrass; progress and applications. In: den Nijs APM, Elgersma A (eds) Fodder crops breeding: achievements, novel strategies and biotechnology. Proc 16th Meet Fodder Crops Section of Eucarpia, Wageningen, The Netherlands, 16–22 Nov 1990, pp 123–128

Creemers-Molenaar J, Loeffen JPM, Zaal MACM (1988a) Protoplasts from *Lolium perenne* L.: procedures for plant regeneration and electro-fusion. Proc 2nd Int Congr Plant molecular biology, Jerusalem, 13–18 Nov, p 430

Creemers-Molenaar J, Loeffen JPM, Van der Valk P (1988b) The effect of 2,4-dichlorophenoxyacetic acid and donor plant environment on plant regeneration from immature inflorescence-derived callus of *Lolium perenne* L. and *Lolium multiflorum* L. Plant Sci 57: 165–172

Creemers-Molenaar J, Van der Valk P, Loeffen JPM, Zaal MACM (1989) Plant regeneration from suspension cultures and protoplasts of *Lolium perenne* L. Plant Sci 63: 167–176

Creemers-Molenaar J, Van Oort Y (1990) Antioxidants influence plating efficiency and microcallus growth of protoplasts in *Lolium perenne* L. In: Nijkamp HJJ, van der Plas LHW, van Aartrijk J (eds) Progress in plant cellular and molecular biology. Proc 7th Int Congr on Plant tissue and cell culture. Kluwer, Amsterdam, pp 44–49

Creemers-Molenaar J, van Eeuwyk FA, Krens FA (1992A) Culture optimization for perennial ryegrass protoplasts. J Plant Physiol 139: 303–308

Creemers-Molenaar J, Hall RD, Krens FA (1992B) Asymmetric protoplast fusion aimed at intraspecific transfer of cytoplasmic male sterility (CMS) in *Lolium perenne* L. Theor. Appl. Genet. (In Press)

Dale PJ (1975) Meristem tip culture in *Lolium multiflorum*. J Exp Bot 26: 731–736

Dale PJ (1977a) Meristem tip culture in *Lolium, Festuca, Phleum* and *Dactylis*. Plant Sci Lett 9: 333–338

Dale PJ (1977b) The elimination of ryegrass mosaic virus from *Lolium multiflorum* by meristem tip culture. Ann Appl Biol 85: 93–96

Dale PJ (1980a) A method for in vitro storage of *Lolium multiflorum* Lam. Ann Bot 45: 497–502

Dale PJ (1980b) Embryoids from cultured immature embryos of *Lolium multiflorum*. Z Pflanzenphysiol 100: 73–77

Dale PJ, Dalton SJ (1983) Immature inflorescence culture in *Lolium, Festuca, Phleum* and *Dactylis*. Z Pflanzenphysiol 111: 39–45

Dale PJ, Thomas E, Brettell RIS, Wernicke W (1981) Embryogenesis from cultured immature inflorescences and nodes of *Lolium multiflorum*. Plant Cell Tissue Organ Cult 1: 47–55

Dalton SJ (1988a) Plant regeneration from cell suspension protoplasts of *Festuca arundinacea* Schreb. (tall fescue) and *Lolium perenne* L. (perennial ryegrass). J Plant Physiol 132: 170–175

Dalton SJ (1988b) Plant regeneration from cell suspension protoplasts of *Festuca arundinacea* Schreb., *Lolium perenne L.* and *L. multiflorum* Lam. Plant Cell Tissue Organ Cult 12: 137–140

Datta SK, Peterhans A, Datta K, Potrykus I (1990) Genetically engineered indica-rice recovered from protoplasts. Bio/Technology 8: 736–740

Day A, Ellis THN (1984) Chloroplast DNA deletions associated with wheat plants regenerated from pollen: possible basis for maternal inheritance of chloroplasts. Cell 39: 359–368

Day A, Ellis THN (1985) Deleted forms of plastid DNA in albino plants from cereal anther culture. Curr Genet 9: 671–678

Dudits D, Nemet G, Haydu Z (1975) Study of callus growth and organ formation in wheat (*Triticum aestivum*) tissue cultures. Can J Bot 53: 957–963

Fitch MMM, Moore PH (1990) Comparison of 2,4-D and picloram for selection of long-term totipotent green callus cultures of sugarcane. Plant Cell Tissue Organ Cult 20: 157–163

Frakes RV (1973) The ryegrasses. In: Heath ME, Metcalfe DS, Barnes RF (eds) Forages, 3rd edn. Iowa State Univ Press, Ames, Iowa, pp 307–313

Frandsen HN, Frandsen KJ (1948) Polycross-metoden. Nord Jordbruksforsk 7–8: 239–261

Frearson EM, Power JB, Cocking EC (1973) The isolation, culture and regeneration of *Petunia* leaf protoplasts. Dev Biol 33: 130–137

Glimelius K, Fahlesson J, Landgren M, Sjodin C, Sundberg E (1991) Gene transfer via somatic hybridization in plants. TibTech 9: 24–30

Gray DJ, Conger BV, Hanning GE (1984) Somatic embryogenesis in suspension and suspension-derived callus cultures of *Dactylis glomerata*. Protoplasma 122: 196–202

Halberg N, Olesen A, Tuvesson IKD, Andersen SB (1990) Genotypes of perennial ryegrass (*Lolium perenne* L.) with high anther-culture response through hybridization. Plant Breeding 105: 89–94

Hallauer AR, Miranda JB (1981) Selection: theory. In: Hallauer AR, Miranda JB (eds) Quantitative genetics in maize breeding. Iowa State Univ Press, Ames, Iowa, pp 159–204

Harris R, Wright M, Byrne M, Varnum J, Brightwell B, Schubert K (1988) Callus formation and plantlet regeneration from protoplasts derived from suspension cultures of wheat (*Triticum aestivum* L.). Plant Cell Rep 7: 337–340

Hayward MD, Abdullah IB (1985) Selection and stability of synthetic varieties of *Lolium perenne* L. The selected character and its expression over generations of multiplication. Theor Appl Genet 70: 48–51

Hayward MD, Olesen A, Due IK, Jenkins R, Morris P (1990) Segregation of isozyme marker loci amongst androgenetic plants of *Lolium perenne* L. Plant Breeding 104: 68–71

Ho W, Vasil IK (1983) Somatic embryogenesis in sugarcane (*Saccharum officinarum* L.): growth and plant regeneration from embryogenic cell suspension cultures. Ann Bot 51: 719–726

Jackson JA, Dale PJ (1988) Callus induction, plant regeneration and an assessment of cytological variation in regenerated plants of *Lolium multiflorum* L. J Plant Physiol 132: 351–355

Jackson JA, Dalton SJ, Dale PJ (1986) Plant regeneration from root callus in the forage grass *Lolium multiflorum*. In: Withers LA, Alderson PJ (eds) Plant tissue culture and its agricultural applications. Butterworths, London, pp 85–89

Joarder OI, Joarder NH, Dale PJ (1986) In vitro response of leaf tissues from *Lolium multiflorum* – a comparison with leaf segment position, leaf age and in vivo mitotic activity. Theor Appl Genet 73: 286–291

Jones MGK, Dale PJ (1982) Reproducible regeneration of callus from suspension culture protoplasts of the grass *Lolium multiflorum*. Z Pflanzenphysiol 105: 267–274

Jonsson HA, Nilsson C (1986) Plant breeding for resistance to snow moulds in grasses. Nord Jordbruksforsk 68: 596

Kamo KK, Hodges TK (1986) Establishment and characterization of long-term embryogenic maize callus and cell suspension cultures. Plant Sci 45: 111–117

Kasperbauer MJ, Buckner RC, Bush LP (1979) Tissue culture of annual ryegrass × tall fescue F_1 hybrids: callus establishment and plant regeneration. Crop Sci 19: 457–460

Kyozuka J, Hayashi Y, Shimamoto K (1987) High frequency plant regeneration from rice protoplasts by novel nurse culture methods. Mol Gen Genet 206: 408–413

Kyozuka J, Otoo E, Shimamoto K (1988) Plant regeneration from protoplasts of indica rice: genotypic differences in culture response. Theor Appl Genet 76: 887–890

Lee L, Schroll RE, Grimes HD, Hodges TK (1989) Plant regeneration from indica rice (*Oryza sativa* L.) protoplasts. Planta 178: 325–333

Linsmaier EM, Skoog F (1965) Organic growth factor requirements of tobacco tissue cultures. Physiol Plant 18: 100–127

Lo PF, Chen CH, Ross JG (1980) Vegetative propagation of temperate forage grasses through callus culture. Crop Sci 20: 363–367

Lu C, Vasil V, Vasil IK (1981) Isolation and culture of protoplasts of *Panicum maximum* Jacq. (guinea grass): somatic embryogenesis and plantlet formation. Z Pflanzenphysiol 104: 311–318

Murashige T, Skoog F (1962) A revised medium for rapid growth and bioassays with tobacco tissue cultures. Physiol Plant 15: 473–497

Niizeki M (1977) Haploid, polyploid and aneuploid plants from cultured anthers and calluses in species of *Nicotiana* and forage crops. J Fac Agric 58: 343–466

Nitzsche W, Wenzel G (1977) Haploids in plant breeding. In: Fortschritte der Pflanzenzüchtung 8. Advances in plant breeding. Parey, Berlin, pp 46–48

Olesen A, Andersen SB, Due IK (1988) Anther culture in perennial ryegrass (*Lolium perenne* L.). Plant Breeding 101: 60–65

Orshinsky BR, Mcgregor LJ, Johnson GIE, Hucl P, Kartha K (1990) Improved embryoid induction and green shoot regeneration from wheat anthers cultured in medium with maltose. Plant Cell Rep 9: 365–369

Pagniez M, Demarly Y (1979) Obtention d'individus androgenetiques par culture in vitro d'antheres de Ray-grass d'Italie (*Lolium multiflorum* Lam.). Ann Amelior Plantes 29: 631–637

Plumb RT (1988) The effects of pests and diseases on grasses. In: Jones MB, Lazenby A (eds) The grass crop. Chapman and Hall, London, pp 277–309

Potrykus I, Saul MW, Petruska W, Paszkowski J, Shillito RD (1985) Direct gene transfer to cells of a gramineous monocot. Molec Gen Genet 199: 183–188

Potrykus I (1990) Gene transfer to cereals: an assessment. Bio/Technology 9: 535–542

Prioli LM, Sondahl MR (1989) Plant regeneration and recovery of fertile plants from protoplasts of maize (*Zea mays* L.). Bio/Technology 7: 589–594

Rajoelina SR, Alibert G, Planchon C (1990) Continuous plant regeneration from established embryogenic cell suspension cultures of Italian ryegrass and tall fescue. Plant Breeding 104: 265–271

Rhodes CA, Lowe KS, Ruby KL (1988) Plant regeneration from protoplasts isolated from embryogenic maize cell cultures. Bio/Technology 6: 56–60

Rose JB, Dunwell JM, Sunderland N (1987) Anther culture of *Lolium temulentum, Festuca pratensis* and *Lolium* × *Festuca* hybrids. I. Influence of pretreatment, culture medium and culture incubation conditions on callus production and differentiation. Ann Bot 60: 191–201

Schaepman H (1952) Application of the polycross test to grass breeding. Euphytica 1: 105–111

Schmidt F (1991) Entwicklung eines Gewebekultursystems bei *Lolium perenne* (L.) als Voraussetzung zur in vitro- Resistenzselektion. PhD Thesis, Universität Hohenheim

Schmidt F, Posselt UK (1990) Reproducible regeneration of callus cultures derived from mature embryos in *Lolium perenne*. In: den Nijs APM, Elgersma A (eds) Fodder crops breeding: achievements, novel strategies and biotechnology. Proc 16th Meet Fodder Crops Section of Eucarpia, Wageningen, The Netherlands, 16–22 Nov 1990, pp 205–206

Shillito RD, Carswell GK, Johnson CM, DiMaio JJ, Harms CT (1989) Regeneration of fertile plants from protoplasts of elite inbred maize. Bio/Technology 7: 581–587

Skene KGM, Barlass M (1983) Regeneration of plants from callus cultures of *Lolium rigidum*. Z Pflanzenzücht 90: 130–135

Stanis VA, Butenko RG (1984) Developing viable haploid plants in anther culture of ryegrass. Dokly Biol Sci 275: 249–251

Strickland SG, Nichol JW, McCall CM, Stuart DA (1987) Effect of carbohydrate source on alfalfa somatic embryogenesis. Plant Science 48: 113–121

Takamizo T, Suginobo K, Ohsugi R (1990) Plant regeneration from suspension culture derived protoplasts of tall fescue (*Festuca arundinacea* Schreb.) of a single genotype. Plant Science 72: 125–131

Torello WA, Symington AG (1984) Regeneration from perennial ryegrass callus tissue. Hortic Sci 19: 56–57

Torello WA, Mancino L, Troll J (1983) Initiation, growth, and maintenance of callus tissue derived from mature caryopses of perennial ryegrass (*Lolium perenne* L.). Rasen Turf Gazon 1: 4–7

Toriyama K, Hinata K, Sasaki T (1986) Haploid and diploid plant regeneration from protoplasts of anther callus in rice. Theor Appl Genet 73: 16–19

Tuvesson IKD, Pedersen S, Andersen SB (1989) Nuclear genes affecting albinism in wheat (*Triticum aestivum* L.) anther culture. Theor Appl Genet 78: 879–883

Tyler BF (1987) Description and distribution of natural variation in forage grasses. In: INRA (eds) Proc Eucarpia fodder crops section meeting, Lusignan, pp 13–22

Utz HF, Oettler G (1978) Performance of inbred lines and their top crosses in perennial ryegrass (*Lolium perenne* L). Z Pflanzenzücht 80: 223–229

Van der Valk P, Zaal MACM, Creemers-Molenaar J (1988) Regeneration of albino plantlets from suspension culture derived protoplasts of *Poa pratensis* L. (Kentucky bluegrass). Euphytica Suppl: 169–176

Van der Valk P, Zaal MACM, Creemers-Molenaar J (1989) Callus induction and plant regeneration from immature inflorescences and mature seeds of *Poa pratensis* L. (Kentucky bluegrass). Plant Cell Rep 7: 644–647

Vasil IK (1987) Developing cell and tissue culture systems for the improvement of cereal and grass crops. J Plant Physiol 128: 193–218

Vasil IK (1988) Progress in the regeneration and the genetic manipulation of cereal crops. Bio/Technology 6: 397–402

Vasil IK (1990) The realities and challenges of plant biotechnology. Bio/Technology 8: 296–301

Vasil V, Vasil IK (1986) Plant regeneration from friable embryogenic callus and cell suspension cultures of *Zea mays* L. J Plant Physiol 124: 399–408

Vasil V, Redway F, Vasil IK (1990) Regeneration of plants from embryogenic suspension culture protoplasts of wheat (*Triticum aestivum* L.). Bio/Technology 8: 429–434

Wang C, Sun C, Chu C, Wu S (1978) Studies on the albino pollen plantlets in rice. In: Proc Symp on Plant tissue culture, Pittman, Peking, pp 149–160

Wang D, Miller PD, Sondahl MR (1989) Plant regeneration from protoplasts of indica type rice and CMS rice. Plant Cell Rep 8: 329–332

Wang WC, Nguyen HT (1990) A novel approach for efficient plant regeneration from long-term suspension culture of wheat. Plant Cell Rep 8: 639–642

Wang X, Hu Han (1984) The effect of potato II medium for *Triticale* anther culture. Plant Sci Lett 36: 237–239

Wit F (1959) Chromosome doubling and the improvement of grasses. Agraria 11: 97–115

Yamada Y, Zhi-Qi Y, Ding-Tai T (1986) Plant regeneration from protoplast-derived callus of rice (*Oryza sativa* L.). Plant Cell Rep 5: 85–88

Yasuda T, Miyano S, Yamamoto Y, Uchida N, Yamaguchi T (1990) Analogues of phenoxyacetic acid and the regeneration of calluses from seeds of indica rice. Plant Cell Physiol 31(6): 763–766

Zaghmout OMF, Torello WA (1989) Somatic embryogenesis and plant regeneration from suspension cultures of red fescue. Crop Sci 29: 815–817

Zaghmout OMF, Torello WA (1990b) Somatic embryogenesis and plant regeneration from embryogenic suspension cultures of perennial ryegrass. In Vitro Cell Dev Biol 26: 419–424

Zimny J, Lorz H (1986) Plant regeneration and initiation of cell suspensions from root-tip derived callus of *Oryza sativa* L. Plant Cell Rep 5: 89–92

III.5 Micropropagation of Zoysiagrass (*Zoysia japonica* Steud.)

J.M. Al-Khayri[1], F.H. Huang[1], L.F. Thompson[2], and J.W. King[2]

1 Introduction

1.1 Crop Distribution and Usage

Zoysiagrass (*Zoysia* Willd.), chromosome number 20, is a member of the Gramineae family (Forbes 1952) and belongs to the subfamily Eragrostoideae. It originated in tropical eastern Asia and has now been introduced and distributed throughout the warm humid, warm semi-arid, and transitional regions of the world (Hanson et al. 1969; Beard 1973). In the USA, three species of *Zoysia* are used for turfgrass: *Z. japonica* Steud., *Z. matrella* (L.) Merr., and *Z. tenuifolia* Willd., ex Tring.; respectively, they are known by the common names Japanese lawngrass or Korean lawngrass, Manilagrass, and Mascarenegrass or Korean velvetgrass (Beard 1973).

Z. japonica Steud. was introduced to the USA prior to 1895 and has become adapted throughout a wide area. It is grown in the United States from coastal areas of New England, south to Florida, westward to eastern portions of the Great Plains, and under irrigation in the southwestern states. Zoysiagrass is best adapted to areas that have a long, warm growing season; it is a sod-forming perennial that develops a uniform, dense, high-quality turf (Hanson et al. 1969).

Zoysiagrass is used primarily for lawns and can be used on such intensively utilized areas as playgrounds, athletic fields, fairways, and airfields (Beard 1973). To some extent, it is used for erosion control (Hanson et al. 1969).

1.2 Morphology and Growth Habit

Zoysiagrass leaves and stems are distinctly upright; the species has a fibrous and moderately deep root system (Beard 1973). The leaves are rolled in a bud shoot with flat blades 2 to 4 mm wide that are pubescent just above the base and glabrous below; ligules are composed of fringes of 0.2-mm-long hairs; auricles are absent; sheaths are round to slightly flattened, glabrous with fascicled hairs at the throat, and split with overlapping margins. The stems are round, ranging from erect to ascending from the base, strongly stoloniferous, and somewhat rhizomatous. The inflorescences are short, terminal, and spike-like (Beard 1973; Hanson et al. 1969).

[1] Department of Horticulture and Forestry/ [2] Department of Agronomy, University of Arkansas, Fayetteville, AR 72701, USA

Biotechnology in Agriculture and Forestry, Vol. 19
High-Tech and Micropropagation III (ed. by Y.P.S. Bajaj)

In relation to warm-season turfgrass species, *Z. japonica* surpasses most others in color retention at low temperature (Forbes and Ferguson 1947) and in winter hardiness (Forbes and Ferguson 1947; Hanson et al. 1969). During winter dormancy, a distinct brownish white tan color is displayed. Of the commonly used turfgrasses, it exhibits excellent drought and heat hardiness, good shade tolerance, and the most wear tolerance during active growth. Its dense growth habit results in superior resistance to weed invasions (Beard 1973).

The species grows best on well-drained soils having a pH of 6 to 7, but tolerates a broad range of soil fertility and salinity (Juska 1959). Zoysiagrass persists on infertile sites but responds to fertilizers (Hanson et al. 1969). It is free of major disease problems as compared to other turfgrasses, but Rhizoctonia brown patch, Sclerotinia dollar spot, Puccinia rust, *Helminthosporium* leaf spot, and *Fusarium* blight have been observed on zoysiagrasses (Beard 1973). The species is more resistant to attack by insects (Childers 1947) and more tolerant to most turfgrass herbicides than other turfgrasses (Burt 1965).

1.3 Conventional Propagation

In the past, zoysiagrass establishment was limited to vegetative means of sprigging or spot sodding since seeds were not available. Although some seeds of *Z. japonica* have been imported from Korea, vegetative establishment has persisted because seed germination is poor, less than 60%, and slow, requiring up to several weeks (Hanson et al. 1969). Recently, a new seed scarification technique involving potassium or sodium hydroxide solution and light treatments has resulted in germination as high as 95.3% after only 7 days. The development of this successful seed treatment permits the establishment of zoysiagrass from seeds. Seedling progeny, however, show considerable variation and diverse growth habits (Portz 1981).

1.4 Crop Improvement

Only a few improved varieties of *Z. japonica* have been developed. Meyer zoysia (Z-52), selected for leaf width and winter hardiness, was released by the Crops Research Division of the United States Department of Agricultural Research Service (USARS) and the US Golf Associations (USGA) Green Section in 1951 (Beard 1973; Hanson et al. 1969). In 1955 the Crops Research Division and Georgia Agricultural Experiment Station (AES) released Emerald, a hybrid between *Z. japonica* × *Z. tenuifolia*, combining the greater winter hardiness and faster rate of spread of *Z. japonica* with finer leaves, denser turf, and dark green color of *Z. tenuifolia* (Hanson et al. 1969). The Midwest variety was released in 1963 by the Indiana AES (Beard 1973). The most recently released cultivars of *Z. japonica* have been El Toro in 1984 by the California AES and Belair in 1985 by the USDA-ARS at Maryland. El Toro is characterized by medium-green color, medium-coarse leaf texture and medium-low density. It has a superior adaptation to drought. Belair possesses medium-green color, medium leaf texture, and medium density. Its drought resistance is excellent (Beard 1987).

2 In Vitro Culture Methodology

A tissue culture system for the regeneration of *Z. japonica* was developed by Al-Khayri et al. (1987a, b, 1989a, b), the technicalities of which are described in this chapter.

2.1 Explant Disinfection and Incubation Conditions

Mature caryopses of *Z. japonica* Steud. (common zoysiagrass) were surface sterilized in 70% ethanol for 2 min, then in 2.6% sodium hypochlorite containing 0.1% (w/v) Tween 20 for 20 min, and then were rinsed with sterile distilled water three times. With the aid of a stereomicroscope (× 10), seed coverings were removed and embryos were separated from the endosperm by a transverse cut above the scutellum.

Under aseptic conditions, excised embryos were plated onto agar medium. The culture tubes were incubated in a culture room which was maintained at 25 ± 5 °C. Callus was initiated under a 16-h photoperiod of cool-white fluorescent light (65 $\mu E/s/m^2$) and under complete darkness. The dark condition was simulated by wrapping individual culture tubes with aluminum foil. To induce plant regeneration, the calli were subjected to light only upon transfer onto regeneration medium.

2.2 Medium Preparation

Cytokinins have been reported to cause no significant improvement in callus culture of grasses (Chen et al. 1979; Lo et al. 1980); therefore, auxin was emphasized in our study, namely 2,4-dichlorophenoxyacetic acid (2,4-D). The content of 2,4-D in the medium was determined by the culture purpose. For callus initiation and maintenance, 2,4-D was necessary in the medium, but elimination or reduction of 2,4-D was required for plant regeneration. The medium used for callus initiation and growth was supplemented with several 2,4-D concentrations to determine the optimum level appropriate for zoysiagrass culture. Effectiveness of MS (Murashige and Skoog 1962) and N6 (Chu et al. 1975) basal media was tested. The MS salt was supplemented with 100 mg/l inositol and 0.4 mg/l thiamine HCl and the N6 salt with 2 mg/l glycine, 1 mg/l thiamine HCl, 0.5 mg/l pyridoxine HCl, and 0.5 mg/l nicotinic acid. In addition, media were supplemented with 30 g/l sucrose and 8 g/l agar. Media used for callus induction also contained 1, 2.5, 5, 7.5, or 10 mg/l of 2,4-D. To induce callus differentiation and subsequent plant regeneration, the 2,4-D was either omitted or reduced to 0.5 mg/l. The media were adjusted to pH 5.8 with 1N and 0.1N NaOH, dispensed into culture tubes, and autoclaved for 20 min at 10.4×10^4 Pa and 120 °C.

2.3 Experimental Design

For callus induction, a three-factor experimental design was employed with two levels of medium (MS or N6), five levels of 2,4-D concentration (1, 2.5, 5, 7.5, or

10 mg/l), and two levels of light (dark and light). For plant regeneration, the culture medium factor was replaced by the 2,4-D elimination approach (direct and gradual) in the MS medium, since only MS medium was effective in producing plantlets. Each treatment was applied to 30 caryopses (replications). The treatments were evaluated based upon callus fresh weight, frequency of regeneration, and number of plants regenerated per callus. The data were subjected to an analysis of variance (ANOVA) and the means separated by LSD at the 5% level of significance.

2.4 Callus Induction

After 7 to 14 days of culturing, calli proliferating from the embryo regions of the caryopses became apparent. To determine the most appropriate cultural requirements, calli were induced under variable conditions. The effectiveness of the induction factors tested was based upon callus growth expressed in callus fresh weights obtained 12 weeks after culturing. Analysis of variance (Table 1) showed significant influence due to medium and 2,4-D concentration, as well as a significant two-way interaction for medium and 2,4-D concentration, and a significant three-way interaction for medium, 2,4-D concentration, and light.

The two media, MS and N6, were selected to be tested because of their efficacy in regeneration of other monocots. A successful tissue culture system employing MS medium has been reported for a number of monocotyledons including fescue (Torello et al. 1984), Guinea grass (Lu and Vasil 1982), tall fescue (Lowe and Conger 1979), and maize (Lu et al. 1983). Regeneration on N6 medium has been reported for common bermudagrass (Ahn et al. 1985) and rice (Chu et al. 1975). Zoysiagrass mean fresh weight of the calli grown on MS medium significantly exceeded that induced on N6 medium (Table 2).

The concentration of 2,4-D greatly affected calli fresh weight, resulting in maximum callus weight with 1 mg/l of 2,4-D incubated under light (Table 2). A general tendency for an inverse relationship between callus weight and 2,4-D level was noticed, particularly among calli induced on MS medium.

The effect of light on callus growth varied depending on medium and 2,4-D concentration. Light-induced callus was larger than dark-induced callus on MS medium, but on N6 medium dark induction produced larger calli weight than that

Table 1. Analysis of variance of callus fresh weight

Source	DF	Mean square	P-value
Medium	1	1.628	0.0001 S[a]
Light	1	0.001	0.7048 NS
2,4-D level	4	0.599	0.0001 S
Medium × light	1	0.024	0.0572 NS
Light × 2,4-D	4	0.012	0.1255 NS
Medium × 2,4-D	4	0.417	0.0001 S
Medium × 2,4-D × light	4	0.021	0.0155 S
Error	380	0.007	

[a] S = significant; NS = not significant at 5% level.

Table 2. Effects of medium, 2,4-D concentration, and light on callus fresh weight (Al-Khayri et al. 1989b)

2,4-D level (mg/l)	Callus fresh weight (g)			
	MS medium		N6 medium	
	Light	Dark	Light	Dark
1.0	0.486a[a]	0.373b	0.029f–j	0.061ef
2.5	0.155c	0.140cd	0.054e–j	0.056e–i
5.0	0.063ef	0.094de	0.010g–j	0.036f–j
7.5	0.059e–g	0.064ef	0.005h–j	0.008g–j
10.0	0.057e–g	0.056e–i	0.005h–j	0.004j
{Av, light}	{0.164a}	{0.145a}	{0.021b}	{0.033b}
[Av, medium]	[0.154a]		[0.027b]	

[a] Values followed by a common letter are not significantly different; type of parenthesis, or no parenthesis, denotes values being compared. LSD = 0.051 for comparing individual treatments, LSD = 0.023 for comparing averages of light { }, and LSD = 0.016 for comparing averages of medium [ ].

induced in the light. The weight difference due to light using a common medium, however, was not significant. Significant differences exist wherein light, 2,4-D level, and medium were considered. On MS medium containing 1 mg/l of 2,4-D the callus cultured under light grew significantly larger than that grown in the dark; the light condition, however, resulted in similar callus weights on MS medium with 2,4-D concentrations higher than 2.5 mg/l (Table 2). The most favorable set of conditions for callus induction was MS medium supplemented with 1 mg/l of 2,4-D under light incubation.

2.5 Plant Regeneration

Complete elimination of 2,4-D from the medium, or reducing its concentration to a lower level, has been observed to induce morphogenesis and plant regeneration in a number of grasses (Nishi et al. 1968; Chen et al. 1979; Ahn et al. 1985; Artunduaga et al. 1988). Both methods were examined for zoysiagrass regeneration; calli were either cultured directly on MS medium free of 2,4-D or placed on MS medium containing 0.5 mg/l of 2,4-D for 2 weeks and then transferred onto a 2,4-D-free medium. For convenience, the first method will be referred to as a direct elimination and the latter as a gradual elimination. Three to five weeks later, both methods of hormonal elimination resulted in plantlet regeneration on the MS medium, but no plant formation occurred on the N6 medium. The regeneration efficiency of the system was based upon two parameters, percentage of calli-regenerated plantlets (regeneration frequency), and the number of plantlets regenerated per callus (regeneration rate).

The ability of callus to exhibit plant formation averaged 22% of total number of calli cultured on regeneration medium (Table 3). Zoysiagrass is a cross-pollinating species and, thus, has highly heterozygotic embryos. The genotypic differences

Table 3. Effects of 2,4-D level and light used for callus induction, and method of 2,4-D elimination, on subsequent plant regeneration frequency. (Al-Khayri et al. 1989b)

2,4-D[a] (mg/l)	Direct elimination				Gradual elimination				
	Light		Dark		Light		Dark		Average
	N[b]	%[c]	N	%	N	%	N	%	%
1.0	26	42	22	59	25	20	21	43	41
2.5	23	13	27	26	20	35	26	27	25
5.0	24	17	25	48	23	22	28	32	30
7.5	21	5	24	8	27	7	22	5	6
10.0	23	4	25	4	22	5	23	9	6
{Average}	{16%}		{29%}		{18%}		{23%}		
[Av, elimination]			[23%]				[21%]		
Overall average					22%				

[a] Concentration of 2,4-D used for callus induction.
[b] N is number of calli treated for regeneration.
[c] Percentage of calli regenerated plantlets.

exhibited by the caryopses may be reflected, to a certain extent, by the regenerability of calli they produce. In tissue culture of tall fescue, which is also cross-pollinating, an average of only 18% of the calli produced exhibited plant regeneration (Lowe and Conger 1979).

The regeneration frequency of zoysiagrass calli resulting from each induction treatment and the 2,4-D elimination method are shown in Table 3. Regeneration frequency was affected most by the level of 2,4-D in the primary culture. It was hindered by concentrations above 5 mg/l and favored by 1 mg/l of 2,4-D.

Another factor affecting regeneration frequency was the light condition during callus induction. The percentage of dark-induced callus demonstrating competency was higher than that of light-induced callus.

Direct and gradual elimination of 2,4-D from the culture medium produced similar regeneration frequencies (Table 3). Lowering 2,4-D in the culture medium from 5 to 1 mg/l without subsequent culturing on the 2,4-D-free medium also has been reported to induce plant formation in orchardgrass (Conger and Carabia 1978). Calli of tall fescue induced on a medium with 9 mg/l 2,4-D and maintained on 5 mg/l were induced into morphogenesis by transferring them onto a medium containing 0.5 mg/l. Shoot formation, however, was enhanced by transferring the differentiating calli to a medium without 2,4-D (Lowe and Conger 1979). Similarly, little bluestem calli also produced higher plant regeneration on a 2,4-D-free medium than on a medium with 0.1 mg/l (Songstad et al. 1986).

The number of plantlets regenerated per callus was tested to determine efficacy of various treatments used to induce callus as well as methods of 2,4-D elimination on subsequent plant regeneration. Analysis of variance indicates that the main factors – light, 2,4-D concentration, and 2,4-D elimination method – significantly influenced regeneration rate (Table 4). The two-way interaction of 2,4-D concentration × its elimination method suggests that effectiveness of the 2,4-D elimination method is dependent upon the auxin concentration in primary cultures.

Table 4. Analysis of variance of plant regeneration rate (number of plants per callus) on MS medium

Source	DF	Mean square	P-value
Light	1	153.13	0.0023 S[a]
2,4-D concentration	4	977.17	0.0001 S
Elimination method	1	78.13	0.0282 S
Light × 2,4-D	4	39.62	0.0455 S
Light × elimination	1	16.25	0.3144 NS
2,4-D × elimination	4	92.19	0.0002 S
Elimin. × 2,4-D × light	4	4.78	0.8779 NS
Error	180	15.96	

[a] S = significant; NS = not significant at 5% level.

The number of regenerated plants per callus are recorded in Table 5. In general, direct removal of 2,4-D was more effective than its gradual elimination in producing a higher number of plants per callus, but when callus was induced with 2.5 mg/l of 2,4-D, the reverse was true. The number of plants per callus decreased steadily as concentration of 2,4-D in primary cultures increased, with 1 mg/l producing the highest number and with 7.5 and 10 mg/l of 2,4-D producing negligible regeneration. Dark callus induction proved to be significantly superior in the subsequent regeneration rate when direct 2,4-D elimination was used. With gradual elimination of 2,4-D, dark induction also resulted in a higher number of regenerants per callus than light induction, but the difference was not significant.

The regeneration process commenced 3 to 5 weeks following culturing on regeneration medium and continued for another 8 weeks. Most of the regeneration, however, occurred between the 5th and 8th week. Albinos, which seem to be a common event in the regeneration of monocots (Lo et al. 1980; Torello et al. 1984;

Table 5. Effects of 2,4-D level and light used for callus induction and method of 2,4-D elimination on plant regeneration rate. (Al-Khayri et al. 1989b)

2,4-D level (mg/l)	Mean no. of plants per callus				
	Direct elimination		Gradual elimination		
	Light	Dark	Light	Dark	Average
1.0	9.6b–d[a]	16.0a	7.9c–e	10.8bc	11.1a
2.5	5.6ef	7.4c–f	8.5b–e	9.5b–d	7.5b
5.0	8.7b–e	12.0b	4.1f	6.3d–f	7.8b
7.5	0.3g	0.3g	0.3g	0.1g	0.3c
10.0	0.1g	0.2g	0.1g	0.1g	0.1c
{Av, light}	{4.9b}	{7.2a}	{4.2b}	{5.4b}	
[Av, elimination]		[6.0a]		[4.8b]	

[a] Values followed by a common letter are not significantly different; type of parenthesis, or no parenthesis, denotes values being compared. LSD = 3.5 for comparing individual treatment, LSD = 1.6 for comparing averages of light { }, LSD = 1.8 for comparing averages of 2,4-D concentrations ( ), and LSD = 1.1 for comparing averages of regeneration method [ ].

Yeh and Chang 1986), comprised 9% of total regenerated plants. Normal regenerated plants possessing shoot and root systems grew actively when transferred to soil. To test the viability of regenerated plants, they were transferred to the field, where they survived over winter and continued to grow.

2.6 Establishment of Mature Plants

Young regenerated plantlet clusters were transferred to separate tubes for further growth (Fig. 1). The plantlets in a cluster were carefully separated once they reached 3 to 6 cm in height. They were washed with water to remove residues of agar medium from root regions and transplanted into potting soil. In the transferring process, the fragile plantlets sometimes folded their leaves, reducing evapotranspiration, but they overcame this initial wilt soon after the pots were placed in plastic bags which maintained humidity. The plants were gradually acclimated by removing the plastic bag and applying tap water as needed. In 2 to 3 weeks the plants became well established and were transplanted into larger pots or flats and placed in a greenhouse (Fig. 2). Stolon development was observed after 6 to 8 weeks of growth in the soil. The plants grew normally and survived overwintering under field conditions.

Fig. 1. Regenerated plantlets 8 weeks old with roots and shoots growing on MS agar medium prior to transfer to soil

Fig. 2. Regenerated zoysiagrass plants growing in potting soil during acclimation and later in the greenhouse

2.7 Histology

Investigation of the callus formation and the developmental process of callus into embryo and ultimately into a plant was the purpose of the histological studies. Light microscopy and scanning electron microscope (SEM) were utilized in this study. Microscopic slides for light microscopy were prepared according to the standard paraffin embedding techniques (Constabel 1982). The samples were prepared for SEM by fixing in Karnovsky's fixative for 2 h, washing twice (20 min each) in 0.05 M cacodylate buffer (pH 7.2), and post-fixing in 1% osmiumtetroxide (OsO_4) for 1 h. The samples were washed for 1 to 2 min in distilled water and then dehydrated with a gradual series of ethanol concentrations of 30, 50, 70, 80, and 95%, for 20 min each change. After the dehydration, the specimens were critical point dried in a Sorvall critical point drying system, using liquid carbon dioxide as a transitional fluid. They were then removed from the unit, mounted on aluminum stubs and gold coated (175 Å) by sputtering with a Polaron E6100 SEM coating unit. The specimens were viewed at 30 kV with an ISI-60 scanning electron microscope. A Polaroid film (type 55 positive/negative) was used to obtain the micrographs.

Embryogenic and Nonembryogenic Calli. Visual observation revealed that two types of calli proliferated from zoysiagrass caryopsis; these calli, embryogenic and nonembryogenic, are similar to those described in the tissue cultures of other grasses (Haydu and Vasil 1981; Heyser et al. 1985; Lu et al. 1983; Torello et al. 1984; Vasil et al. 1984). One type, designated as nonembryogenic callus (NC), was soft, friable, and whitish yellow. The other type, embryogenic callus (EC), was white, opaque, smooth, glossy, and compacted. The softness and friability of the whitish yellow callus provide an easy and reliable means of distinguishing the NC from the EC, which tends to be white and compacted. The EC has been described as consisting of small, generally rounded cells rich in cytoplasm with prominent nuclei and starch grains (Yeh and Chang 1986). Our SEM study showed that the EC was composed of small, but not necessarily round, cells (Fig. 3). The smaller cell size of the EC may contribute to the surface smoothness and the compactness of the callus. The NC was composed of larger cells of various shapes (Fig. 4).

Formation of Embryogenic Complex. Upon the elimination of 2,4-D from the culture medium, callus organization and morphogenesis commenced. In agreement with investigations of other monocotyledonous plants, the nonembryogenic callus responded by producing roots and root hairs (Peggy and Vasil 1983; Yeh and Chang 1986) (Fig. 5), while somatic embryos developed in the embryogenic callus (Vasil et al. 1984; Cobb et al. 1985; Yeh and Chang 1986). The embryogenic callus surface became increasingly convoluted following the elimination of 2,4-D and eventually gave rise to clusters of globular structures referred to as embryogenic complexes (Fig. 6).

Development of Somatic Embryos. The somatic embryos (embryoids) appeared as shiny globular structures clustered together on the surface of the embryogenic complex (Fig. 7). In *Zea mays* L. smooth and shiny globular structures on the surface of the callus formed either singly or in clusters (Vasil et al. 1984). We have observed formation of embryoids mostly in clusters, but in one case a single embryo was noticed. In most cases each embryoid had a distinct suspensor giving it a club-shaped appearance similar to that described for embryoids produced from the *Pennisetum americanum* callus (Vasil and Vasil 1982). At this stage the general appearance of the embryoids indicated nonsynchronical development, which may explain the long period of continuous regeneration lasting 9 weeks. Following the globular stage, a well-defined notch appeared along the median section of each embryoid, delineating the outer scutellar region from the inner margins of the shoot meristem (Fig. 8) and later developing into a coleoptile. Both radial and bilateral symmetry were observed at this stage. In some cases, secondary embryoids formed in the scutellum of a primary embryoid (Fig. 9).

Maturation of Somatic Embryos. The formation of coleoptiles marked the end of the early developmental stages of the embryoids. This stage was followed by a well-differentiated embryoid formation with a longitudinal slit along the periphery of the coleoptile and a well-defined coleorhiza (Fig. 10). In the final stage of the embryoid development, all parts of the embryoids were well defined, with the coleoptile surrounded by the scutellum and the radical on the underside (Fig. 11).

Regeneration of Plants from Somatic Embryos. Roots extended and emerged from somatic embryos grown on the 2,4-D-free MS medium. Root emergence was

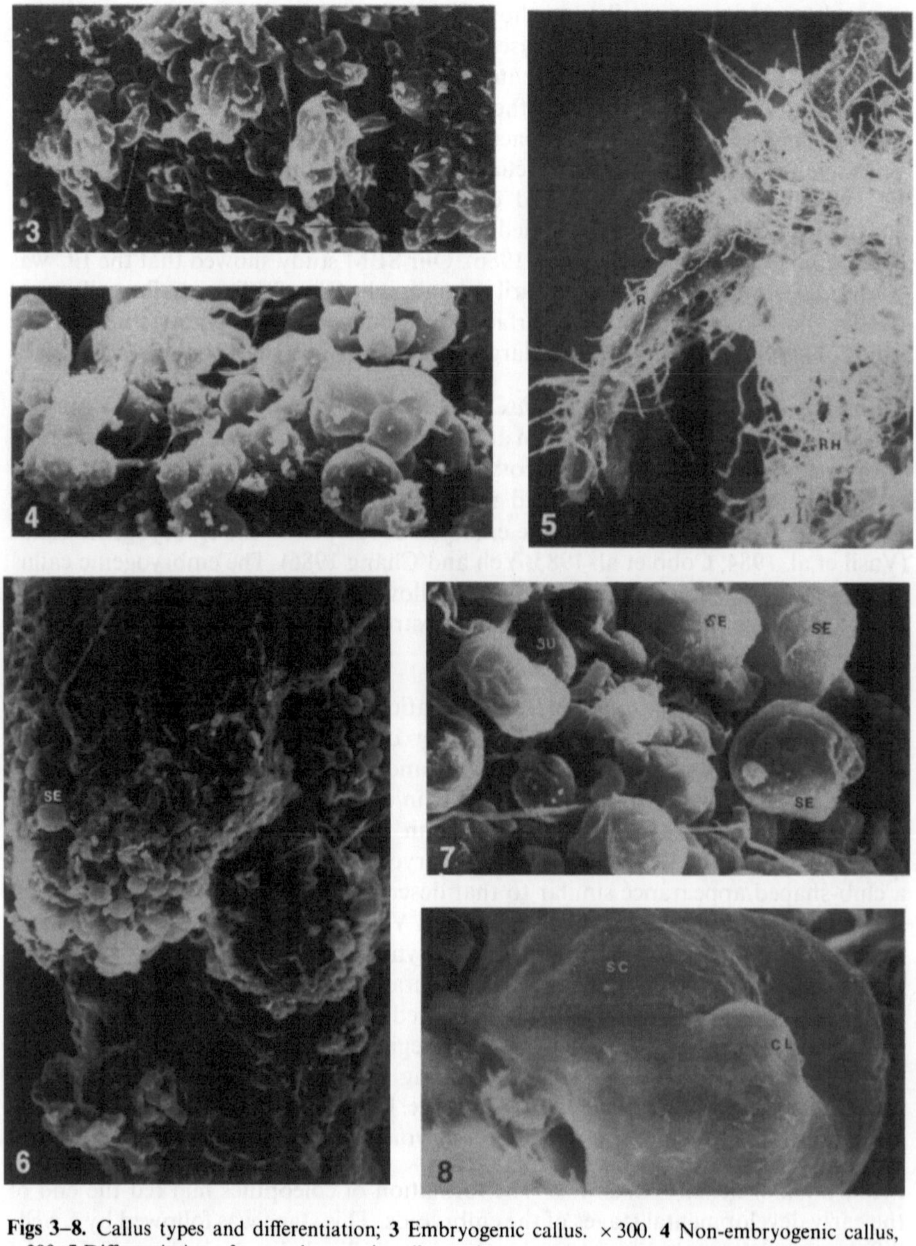

Figs 3–8. Callus types and differentiation; **3** Embryogenic callus. × 300. **4** Non-embryogenic callus, × 300. **5** Differentiation of nonembryogenic callus: only roots (*R*) and root hairs (*RH*) were produced. × 55. **6** Group of globular somatic embryos (*SE*) forming embryogenic complexes. × 75. **7** Early stages of somatic embryo (*SE*) formation showing suspensors (*SU*). × 200. **8** Later stage of SE development showing coleoptile (*CL*) and scutellum (*SC*). × 750

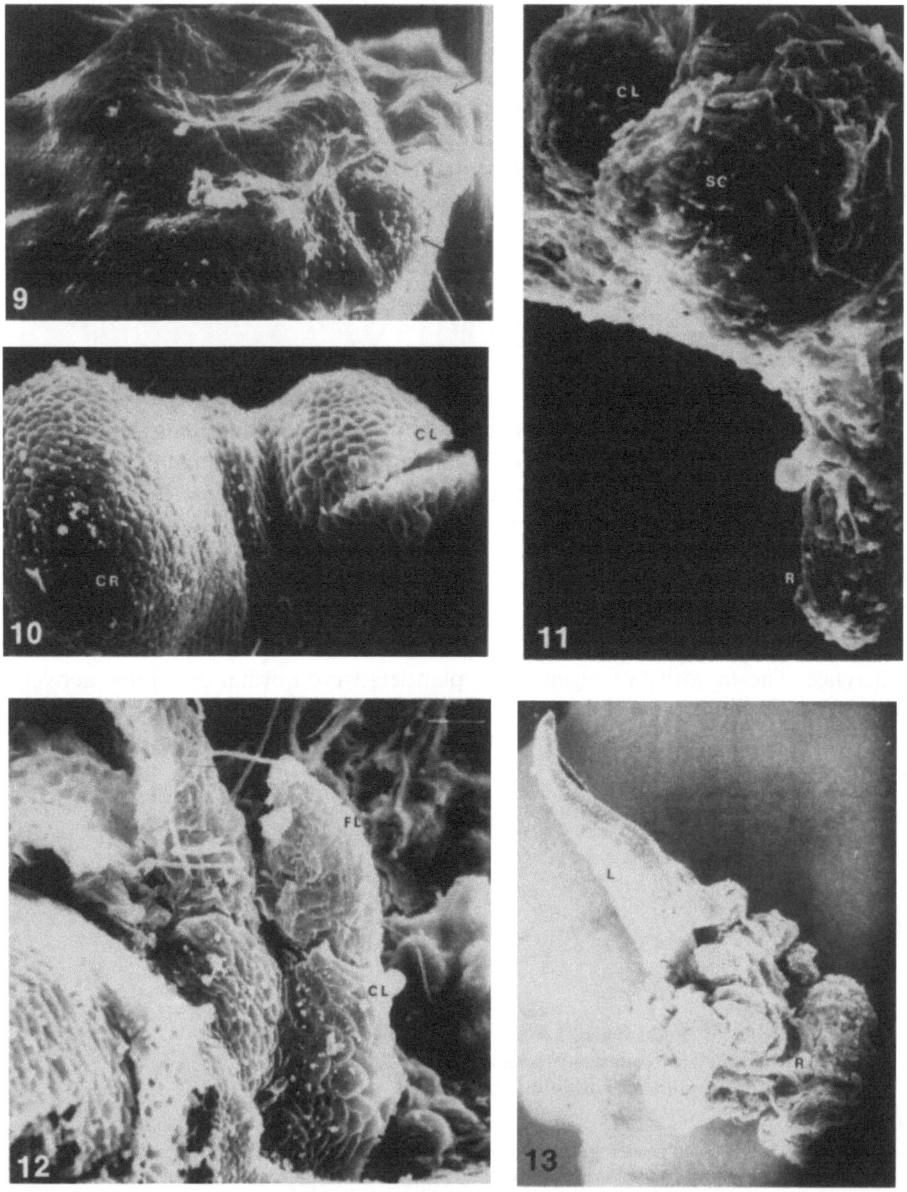

Figs 9–13. Somatic embryogenesis and plant regeneration; **9** Somatic embryo (*SE*) showing secondary SE (*arrow*). × 900. **10** Mature SE showing a coleoptile (*CL*) with a lateral slit and coleorhiza (*CR*). × 130. **11** Germination of SE showing emerging root (*R*) and elongating coleoptile (*CL*), which is surrounded by scutellum (*SC*). × 70. **12** Germinating SE showing elongating coleoptile (*CL*) and emergence of first leaf (*FL*). × 110. **13** Regeneration of plantlets showing roots (*R*) and leaves (*L*). × 12

accompanied by the elongation of the coleoptile (Fig. 11). Germination of the somatic embryos continued with the same process by which zygotic embryos germinate. The first leaf emerged from the coleoptile, thus initiating photosynthesis (Fig. 12). Young plantlets then formed with shoots and roots (Fig. 13). In most cases, each coleoptile contained one shoot meristem, but the presence of multiple shoot meristems was occasionally observed. The somatic embryos that germinated on the 2,4-D-free medium produced normal plants, except for an occasional albino.

3 Summary and Conclusions

An in vitro culture system for zoysiagrass is described for inducing calli on MS medium supplemented with 2,4-D (1, 2.5, 5 mg/l). In our experiment callus growth was greatest with 1 mg/l of 2,4-D; its fresh weight declined as the auxin level increased. Mature zoysiagrass embryos provide an excellent source of explants for callus induction and subsequent plant regeneration. Embryogenesis was enhanced by initiation of callus in the dark, but calli grew larger under the light condition. Whether the 2,4-D concentration in the regeneration medium was gradually decreased or completely eliminated in one transfer step made no significant difference. The majority of regenerated plantlets were normal and grew actively when transplanted into soil, but a small number of albinos evolved. Histological study provided evidence for the occurrence of somatic embryogenesis leading to zoysiagrass regeneration.

4 Protocol

1. *Explant Disinfection*
   a) Place seeds in 70% ethanol for 2 min, then decant
   b) Immerse seeds in 2.6% sodium hypochlorite (50% Clorox)
   c) Rinse three times with sterile distilled water
2. *Dissection*
   a) With the aid of a stereomicroscope and under aseptic conditions, remove the seed covering
   b) Cut caryopses into two sections by an incision above the scutellum region (about the middle of the caryopses)
   c) Discard sections containing the endosperm; culture only sections containing embryos
3. *Callus Induction*
   a) Culture embryo-containing sections on MS medium supplemented with 1 mg/l of 2,4-D, 30 g/l sucrose; solidify with 8 g/l agar and adjust pH to 5.8
   b) Incubate cultures at 25 °C preferably under dark for 4 to 6 weeks
4. *Plant Regeneration*
   a) Separate callus from the original explant
   b) Transfer calli onto a 2,4-D-free medium (the same composition as the induction medium)
   c) Incubate cultures at 25 °C under a 16-h photoperiod of cool fluorescent light (65 $\mu E/m^2/s$)

5. *Acclimation*
   a) Transplant regenerated plantlets to moist potting soil and maintain at high humidity for 1 week by placing the pots in plastic bags
   b) After the removal of the plastic bags, transfer to a green house for few weeks, then transplant to the field when conditions are favorable for zoysiagrass establishment

References

Ahn BJ, Huang FH, King JW (1985) Plant regeneration through somatic embryogenesis in common bermudagrass tissue culture. Crop Sci 25: 1107–1109

Al-Khayri JM, Huang FH, Thompson LF, King JW (1987a) *In vitro* plant regeneration of zoysiagrass (*Zoysia japonica* Steud.). In Vitro 23 (3): 28A

Al-Khayri JM, Huang FH, Thompson LF, King JW (1987b) Somatic embryogenesis in plant regeneration of zoysiagrass. Abst Am Soc Agron 147

Al-Khayri JM, Huang FH, Thompson LF, King JW (1989a) In vitro plant regeneration of zoysiagrass. Ark Farm Res 38 (2): 11

Al-Khayri JM, Huang FH, Thompson LF, King JW (1989b) In vitro plant regeneration of zoysiagrass. Crop Sci 29: 1324–1325

Artunduaga IR, Taliaferro CM, Johnson BL (1988) Effect of auxin concentration on induction and growth of embryogenic callus from young inflorescence explants of Old World bluestem (*Bothriochloa* spp.) and bermuda (*Cynodon* spp.) grasses. Plant Cell Tissue Organ Cult 12: 13–19

Beard JB (1973) Turfgrass: science and culture. Prentice -Hall, Engelwood Cliffs, NJ, pp 143–146

Beard JB (1987) New warm season turfgrass cultivars. Grounds Maintenance, April, pp 10–11

Burt EO (1965) Warm-season turfgrass and preemergence herbicides. Golf Course Rep 33: 54–56

Chen CH, Lo PF, Ross JG (1979) Regeneration of plantlets from callus cultures of Indiangrass. Crop Sci 19: 117–118

Childers NF (1947) Manila grass for lawns. Federal Exp Stn Puerto Rico, Circular 26: 1–16

Chu CC, Wang CC, Sun CS, Hsu C, Yin KC, Chu CY, Bi FY (1975) Establishment of an efficient medium for anther culture of rice through comparative experiments on nitrogen sources. Sci Sin 18: 659–668

Cobb BG, Vanderzee D, Loescher WH, Kennedy RA (1985) Evidence for plant regeneration via somatic embryogenesis in the grasses *Echinochloa muricata* and *E. crusgalli* var. oryzicola. Plant Sci 40: 121–127

Conger BV, Carabia JV (1978) Callus induction and plantlet regeneration in orchardgrass. Crop Sci 18: 157–159

Constabel F (1982) Histological methods. In: Wetter LR, Costabel F (ed) Plant tissue culture methods. NRC, Ontario, Canada, pp 34–37

Forbes I (1952) Chromosome numbers and hybrids in *Zoysia*. Agron J 44: 194–199

Forbes I, Ferguson MH (1947) Observations on the zoysiagrasses. Greenkeepers Rep 15: 7–9

Hanson AA, Juska FV, Burton GW (1969) Species and varieties. In: Hanson AA, Juska FV (eds) Turfgrass science. Am Soc Agron, Madison, WI, pp 407–409

Haydu Z, Vasil IK (1981) Somatic embryogenesis and plant regeneration from leaf tissues and anthers of *Pennisetum purpureum* Schum. Theor Appl Genet 59: 269–273

Heyser JW, Nabors MW, Makinnon C, Dykes JK, Demontt JK, Kautzmann DC, Mujeeb-Razi A (1985) Long-term high-frequency plant regeneration and the induction of somatic embryogenesis in callus culture of wheat (*Triticum aestivum* L.). Z Pflanzenzücht 94: 218–233

Juska FV (1959) The response of Meyer zoysia to lime fertilizer treatment. Agron J 51: 81–83

Lo PF, Chen CH, Ross JG (1980) Vegetative propagation of temperate forage grasses through callus culture. Crop Sci 20: 363–367

Lowe KW, Conger BV (1979) Root and shoot formation from callus cultures of tall fescue. Crop Sci 19: 397–400

Lu CY, Vasil IK (1982) Somatic embryogenesis and plant regeneration in tissue cultures of *Panicum maximum* Jacq. Am J Bot 69: 77–81

Lu CY, Vasil V, Vasil IK (1983) Improved efficiency of somatic embryogenesis and plant regeneration in tissue cultures of maize (*Zea mays* L.). Theor Appl Genet 66: 285–289

Murashige T, Skoog F (1962) A revised medium for rapid growth and bioassays with tobacco tissue cultures. Physiol Plant 15: 473–497

Nishi T, Yamada Y, Takabashi E (1968) Organ differentiation and plant restoration in rice callus. Nature (London) 219: 508–509

Peggy OA, Vasil IK (1983) Improved efficiency and normalization of somatic embryogenesis in *Triticum aestivum* (wheat). Protoplasma 117: 40–44

Portz HL (1981) Zoysiagrass establishment from seed. USGA Green Section Record, Sep/Oct, pp 10–13

Songstad DD, Chen CH, Boe AA (1986) Plant regeneration in callus cultures derived from young inflorescences of little bluestem. Crop Sci 26: 827–829

Torello WA, Symington AG, Rufner R (1984) Callus initiation, plant regeneration, and evidence of somatic embryogenesis in red fescue. Crop Sci 24: 1037–1040

Vasil V, Vasil IK (1982) Characterization of an embryogenic cell suspension culture derived from cultured inflorescences of *Pennisetum americanum* (pearl millet, Gramineae). Am J Bot 69: 1441–1449

Vasil V, Vasil IK, Lu CY (1984) Somatic embryogenesis in long-term callus cultures of *Zea mays* L. (Gramineae). Am J Bot 71: 158–161

Yeh ML, Chang WC (1986) Plant regeneration through somatic embryogenesis in callus culture of green bamboo (*Bambusa oldhamii* Munro.). Theor Appl Genet 73: 161–163

Subject Index

MIX
Papier aus verantwortungsvollen Quellen
Paper from responsible sources
FSC® C105338

If you have any concerns about our products,
you can contact us on
ProductSafety@springernature.com

In case Publisher is established outside the EU,
the EU authorized representative is:
Springer Nature Customer Service Center GmbH
Europaplatz 3, 69115 Heidelberg, Germany

Printed by Libri Plureos GmbH
in Hamburg, Germany